2nd International Conference on Physics and Engineering Mathematics (ICPEM 2021)

Journal of Physics: Conference Series Volume 2148

Wuhan, China
13 – 14 November 2021

ISBN: 978-1-7138-5119-6
ISSN: 1742-6588

Printed from e-media with permission by:

Curran Associates, Inc.
57 Morehouse Lane
Red Hook, NY 12571

Some format issues inherent in the e-media version may also appear in this print version.

Institute of Physics
Dirac House, Temple Back
Bristol BS1 6BE UK

Phone: 44 1 17 929 7481
Fax: 44 1 17 920 0979

techtracking@iop.org

Additional copies of this publication are available from:

Curran Associates, Inc.
57 Morehouse Lane
Red Hook, NY 12571 USA
Phone: 845-758-0400
Fax: 845-758-2633
Email: curran@proceedings.com
Web: www.proceedings.com

TABLE OF CONTENTS

Preface

The Program Committee warmly welcomes all our distinguished delegates and guests to attend the 2021 2nd International Conference on Physics and Engineering Mathematics (ICPEM 2021), the Online International Conference to be held on November 13th-14th, 2021. ICPEM is an annual event that focuses on the various state-of-the-art advances and innovations in Applied Physics and Engineering Mathematics. It aims to provide a forum for researchers, engineers and academicians to exchange new ideas and application experiences, establish business or research relations and find global partners for future collaboration in the fields of Physics and Engineering Mathematics.

The ICPEM 2021 conference received around 141 submissions from all over the world; all submissions went through a rigorous peer review process where each submitted paper was reviewed by at least two experts from the area. Based on the review reports, 69 papers were selected for oral presentation in the conference and included in the Proceedings of 2021 2nd International Conference on Physics and Engineering Mathematics.

On behalf of the Program Committee of the 2021 2nd International Conference on Physics and Engineering Mathematics, we would like to thank all the people who have contributed to the conference this year. In particular, we would like to thank all referees for their review work. We would like to thank the keynote speakers and session chairs for spending their valuable time to support the conference. We also would like to thank all the authors for contributing their papers to the conference. Finally, we are particularly grateful to all the organizers and sponsoring institutions for the generous support they have offered to this conference.

We hope and believe that every participant will enjoy the academic atmosphere brought by the contributors of the conference. We look forward to a more successful conference next year.

The Organizing Committee of ICPEM2021

Conference Chairs
Prof. H. Eugene Stanley, Boston University, USA
Prof. Grado, Aniello, INAF- Osservatorio Astronomico di Capodimonte, Italy
Dr. Fan Jiangdi, JD Duz Academy for Superconductivity, China
Dr. Roberto Zivieri, University of Messina, Italy

Editors
Dr. M. Ajmal Khan, RIKEN Cluster for Pioneering Research (CPR), Japan
Prof. Dr. Osman ADIGUZEL, Department of Physics, Firat University, Turkey
Prof. João Rodrigues, Universidade da Madeira, Portugal
Prof. CHEE-KEONG TAN, Clarkson University, USA

Program Chairs
Dr. Grigoras Marian, National Institute of Research and Development for Technical Physics (NIRDTP), Romania
Prof. Chumin Wang National Autonomous University, Mexico
Prof. Beizhang, Beihang University, China

Technical Program Committee
Dr. Mirela Petruța Șuchea, CEMATEP, Hellenic Mediterranean University, Greece and IMT Bucharest, Romania
Prof. GIUSEPPE CIABURRO, Università degli Studidella Campania, Italy
Prof. Giuseppe Failla, University of Reggio Calabria, Italy.
Prof. Adil El Baroudi, Arts et Métiers Sciences et Technologies, France,
Prof. Farabi TEMEL, Konya Technical University, TURKEY
Prof. Fabrizio Ambrosino, University of Campania Luigi Vanvitelli, Caserta-Italy
Prof. Emin Taner ELMAS, IGDIR UNIVERSITY, Turkey
Prof. Ekta Goel, NIT Warangal, Telangana, India
Prof. Anita Grozdanov, University Ss Cyril and Methodius, Skopje
Dr. Ouertani Beya, University of Carthage & CRTEn Tunisia
Prof. Abdolah Amirany, Shahid Beheshti University
Dr. Dolat khan, City University of Science & Information Technology Peshawar, Pakistan
Dr. Quah Hock Jin, Institute of Nano Optoelectronics Research and Technology (INOR), Universiti Sains, Malaysia
Prof. Hany Salah Abdel-Azeem Mesbah, National Research Institute of Astronomy and Geophysics, Egypt
Prof. Forat Yasir AlJaberi, Al-Muthanna University, AlMuthanna, Iraq,
Prof. Ambreen Afsar Khan, International Islamic University, Islamabad, Pakistan
Dr. Mostafa, Ural Federal University, Russia
Prof. Mohamed Reda Ali Mohamed, Benha University, Egypt
Dr. Najiyah Safwa binti Khashi'ie, Universiti Teknikal Malaysia Melaka, Malaysia
Prof. Ioan-Lucian Popa, "1Decembrie 1918" University of Alba Iulia, Romania
Prof. Nouredine Zekri, Université des Sciences et Technologie d'Oran, Algeria
Prof. Elias C. Aifantis, Michigan Technological University, USA

Dr. Prem Kumar Singh, Gandhi Institute of Technology and Management, India
Dr. Mohd Amiruddin Abd Rahman, Universiti Putra Malaysia, Malaysia

Peer review declaration

All papers published in this volume of Journal of Physics: Conference Series have been peer reviewed through processes administered by the Editors. Reviews were conducted by expert referees to the professional and scientific standards expected of a proceedings journal published by IOP Publishing.

- **Type of peer review: Single-blind / Double-blind / Triple-blind / Open / Other (please describe)**
  1. ICPEM Editors perform an initial check of the manuscript's suitability upon receipt, and use a software tool to finish the plagiarism analysis, manuscripts are out of conference topics will be rejected directly, generally, authors will receive the result within 3-5 working days in this round.
  2. Only the manuscripts passed the initial checking can be submitted to reviewers, ICPEM Editorial Office will then organize the peer-review process performed by independent experts. Papers will be strictly and thoroughly peer-reviewed by experts and reviewers.
  3. All regular papers are reviewed by at least two reviewers, but usually by three or more, and rated considering: Relevance, Originality, Technical Quality, Significance and Presentation of the submissions; There are four results: 1, Accept; 2, Accept after Minor Revisions; 3, Reconsider after Major Revisions; 4, Reject.
  4. Authors have 2-3 weeks to make minor or major revisions after received the comments from reviewers. Usually, one round of major revisions is allowed.
  5. Only the submission passed the peer review and accepted by reviewers will be included in the conference proceeding finally.

- **Conference submission management system:**
  Online Email System
- **Number of submissions received:**
  141
- **Number of submissions sent for review:**
  116 (25 papers out of the conference scope are rejected directly)
- **Number of submissions accepted:**
  69
- **Acceptance Rate (Number of Submissions Accepted / Number of Submissions Received X 100):**
  49%
- **Average number of reviews per paper:**
  2-3
- **Total number of reviewers involved:**
  164
- **Any additional info on review process:**
- **Contact person for queries:**
  Name : Josh Sheng
  Affiliation: Hubei Zhongke Research Institute of Nature Science, China
  Email : icpem2020@163.com

Sound Field Control in Surround Screen Speaker Array by WFS and CBT Algorithms

Rui Huan and Xi Zhao

Communication University of China, No.1 Dingfuzhuang East Street, Beijing, China

Abstract. This paper proposes a method to solve the problem that the sound reproduction system cannot work when the movie screen is made by a sound-proof material such as LED. It is demonstrated in an array of 192 speakers to surround a screen for sound reproduction, called surround screen speaker array. The speaker array is built in an actual cinema. The sound field control algorithms are implemented by mixers. In order to improve the uneven sound field distribution and sound field aliasing caused by the speaker array, two algorithms WFS and CBT are used in this paper. A new control algorithm is proposed and demonstrated to improve the uniformity of the sound field distribution and reduce the sound field interference.

1. Introduction

With the development of display technology, LED has satisfied the requirements of cinema screen and applied to many large screens for cinemas [1, 2]. As a movie screen, the display effect of LED is better than traditional screen. The material of screen used in the traditional cinemas is generally a kind of sound-permeable material. However, when the movie screen is a sound proof material such as LED screen, the traditional main channel of sound reproduction system cannot work. Array of speakers to surround the screen for sound reproduction, called surround screen speaker array, is proposed in this paper to solve this problem. The surround screen speaker array is mainly used to replace the main sound channel of a traditional cinema, so the traditional 5.1 sound system can be combined with this new system [3]. This new combination can solve the LED movie screen's sound-proof problem, but there are two problems need to be solved. Firstly, the sound field distribution of the speaker array is non-uniform. Secondly, the interference of sound fields from the speaker array usually occurs [4].

Corresponding sound field control algorithms can solve these problems. There were many studies on the sound field control of speaker arrays in the past. The HOA algorithm proposed by Gerzon [5] used a spherical harmonic function to represent the reproduced sound field. The wave field synthesis algorithm proposed by Berkhout [6] could make the speaker array form a virtual sound source during playback and form a correct sound image in the entire sound field to control the sound field. Kirkeby and Nelson [7] proposed a plane wave reproduction method based on sound pressure matching, which used the least squares method to minimize the error between the sound field generated by the speaker array and the desired sound field. The CBT algorithm proposed by Keele [8] used each speaker unit in the speaker array to perform delay processing to make the delay of each speaker unit of the line array equal to the delay of the circular array itself, thereby changing the sound field distribution. Zhang W [9] utilized the mode domain method to control the sound field of multiple sound zones, and independently and simultaneously reproduce multiple sound fields in different spatial regions in the same space. Gauthier P A [10] used the frequency domain optimization algorithm for sound field control and performs sound field reconstruction in free field or closed space.

In this paper, new algorithms are proposed and demonstrated in an array of 192 speakers to surround the screen for sound reproduction, called surround screen speaker array. The speaker arrays of the center channel are arranged in horizontal rows. According to the sound wave superposition theorem, the sound field is more complicated, so this article mainly analyzes the sound field of the center channel. Two different control algorithms, WFS [11] and CBT [8] are used for the center channel. Finally, new control algorithms are proposed and demonstrated to solve the problem of uneven sound field distribution and sound field interference.

2. Experiment Setup

Figure 1. Speaker array configuration surround the screen.

As shown in figure 1, 192 speakers are used to form a speaker array surrounding the screen to replace the main channel in the cinema sound reproduction system. 64 speaker units are arranged in the upper and lower rows, respectively. 32 speaker units are arranged in left and right columns, respectively. All speakers are the Soundking KM5B. Each speaker unit has a width of 0.21 m, which leads to the screen width of 13.44 m and the height of 6.72 m. The combination of the upper speaker array and the lower speaker array replace the center channel of the main speaker, the left speaker array on the left column replaces the left channel, and the right speaker array on the right column replaces the right channel.

The measurement is conducted in an actual cinema. The measurement area is in the audience seat area. There are 14 rows of audience seats, and each row has 32 seats. The spacing distance among seats in the same row is 0.6 meters. The interval distance between rows of seats is 1 meter. The measurements are conducted in every row. 17 points per row are carried out. So the measuring point interval is $1.2\text{m}\times 1\text{m}$. The sound level meter used for the measurement is NTI, and the sound level meter uses a type I measuring microphone. The speaker array play 10 seconds pink noise and record the octave equivalent sound pressure level Lzeq. The measurements are in compliance with the ISO 2696-2015 [12] standard for the measurement of sound field in cinemas.

3. Sound Field Control Algorithms For Surround Screen Speaker Array

3.1. WFS Algorithm

The upper speaker array and the lower speaker array are controlled by the same algorithm. The acoustic energy ratio of the upper and the lower speaker arrays is 1:1. The WFS algorithm uses the 2.5D [9] algorithm and the driver function formula is

$$D(x_0,t) = w\delta\left(t - \frac{|x_0 - x_s|}{v}\right) * \left(f(t) * \hat{s}(t)\right)$$

Where x0 denotes the position of each speaker unit, xs denotes the virtual sound source position, w denotes the weighting value, $\hat{s}(t)$ denotes the pre-filter function, and f(t) is the original signal. The virtual sound source is set at a position 5 meters behind the speaker array.

3.2. CBT Algorithm
The arc shape is set to 10 degrees in the CBT algorithm.

4. Experimental Results and Analyses
The simulated sound field map and the actual sound field map measured of different algorithms at different frequency bands are given below. The simulated map is marked with S mark in the figure.

4.1. WFS Algorithm

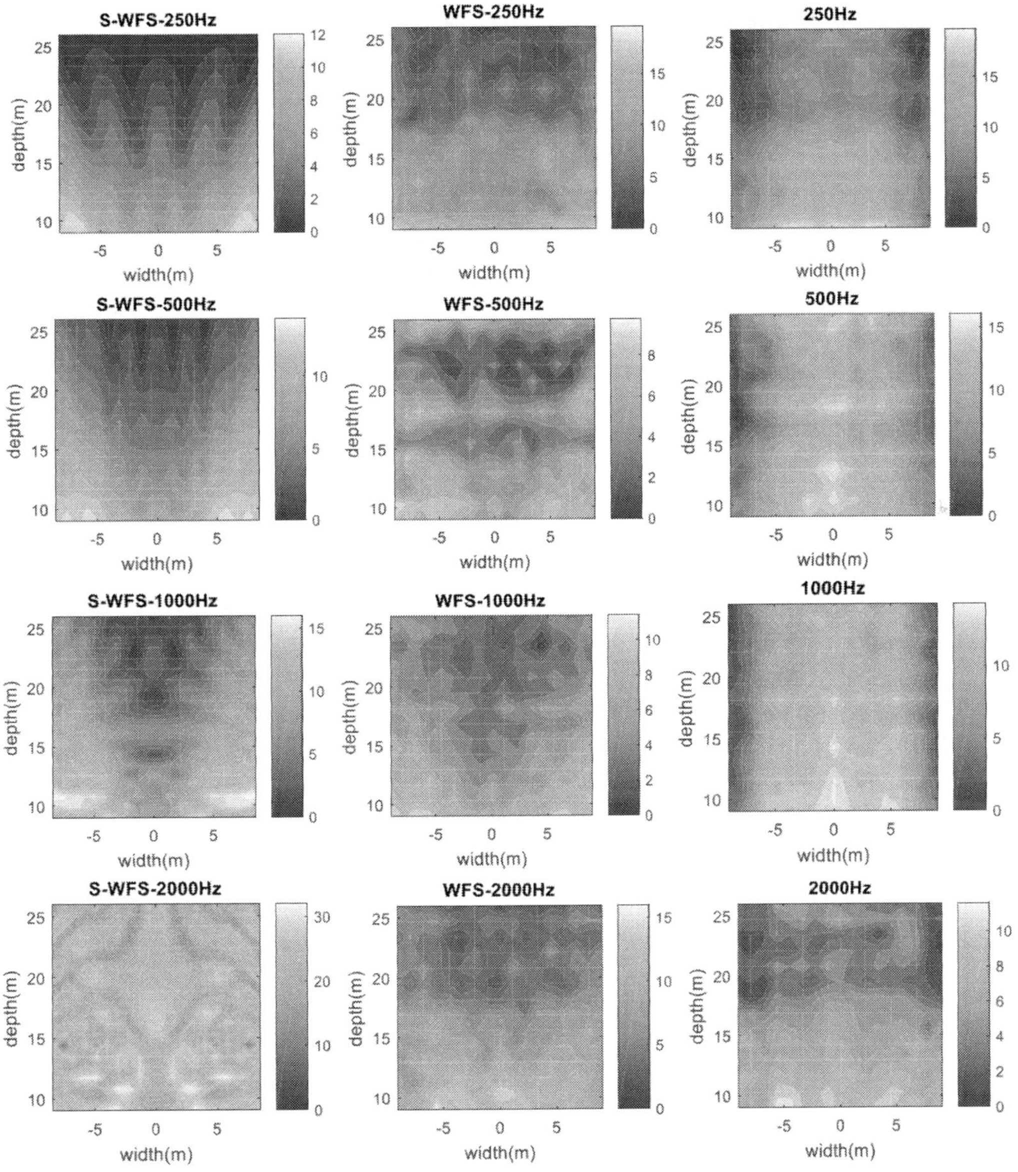

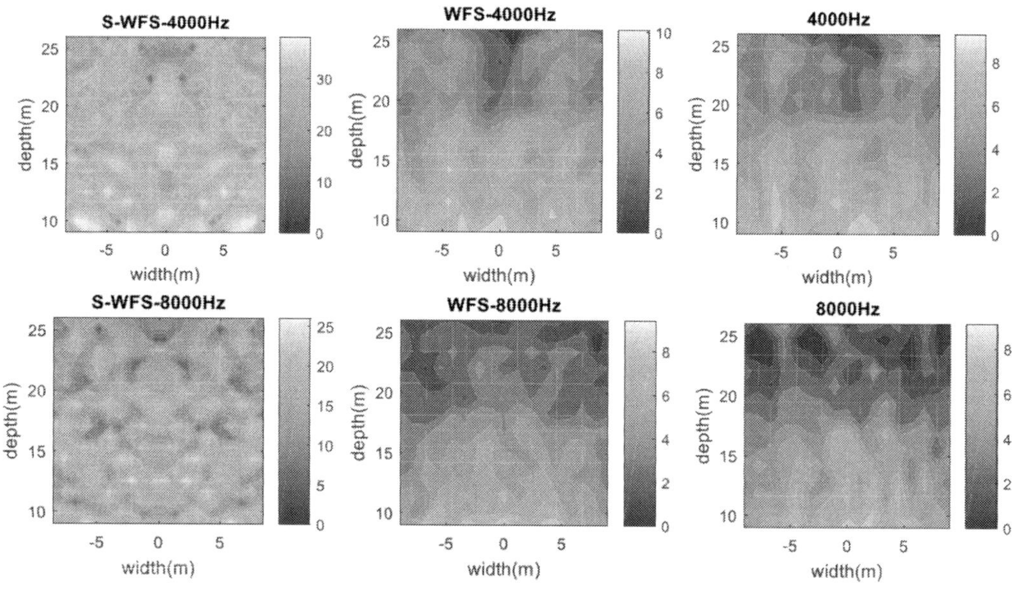

Figure 2. Comparisons of sound fields in WFS algorithm.

As shown in figure 2, below 1000Hz, the measured sound field processed by the WFS algorithm is closer to the sound field of the theoretical simulation, and the energy distribution is relatively average. However, at 500Hz, there is a horizontal energy band in the middle of the audience area, which is similar to the original sound field map without the algorithm. Above 2000Hz, the measured sound field map processed by the WFS algorithm is more similar to the original sound field map processed without the algorithm. It is indicated that in the frequency range below 1000 Hz, the WFS algorithm can control the sound field of the speaker array, while in the frequency range above 2000 Hz, the WFS algorithm has no effect on the sound field control of the actual speaker array.

4.2. CBT Algorithm

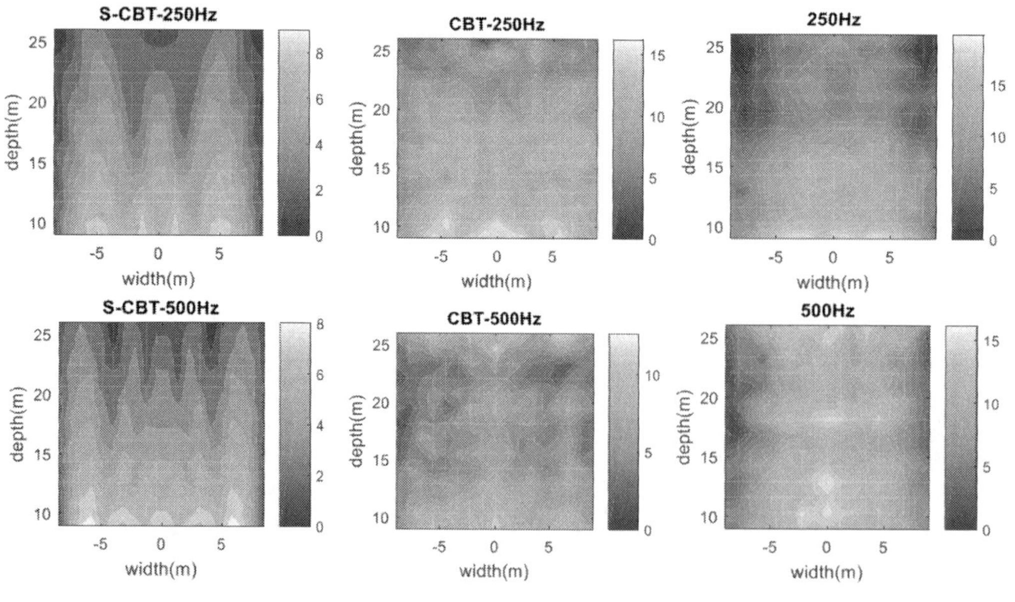

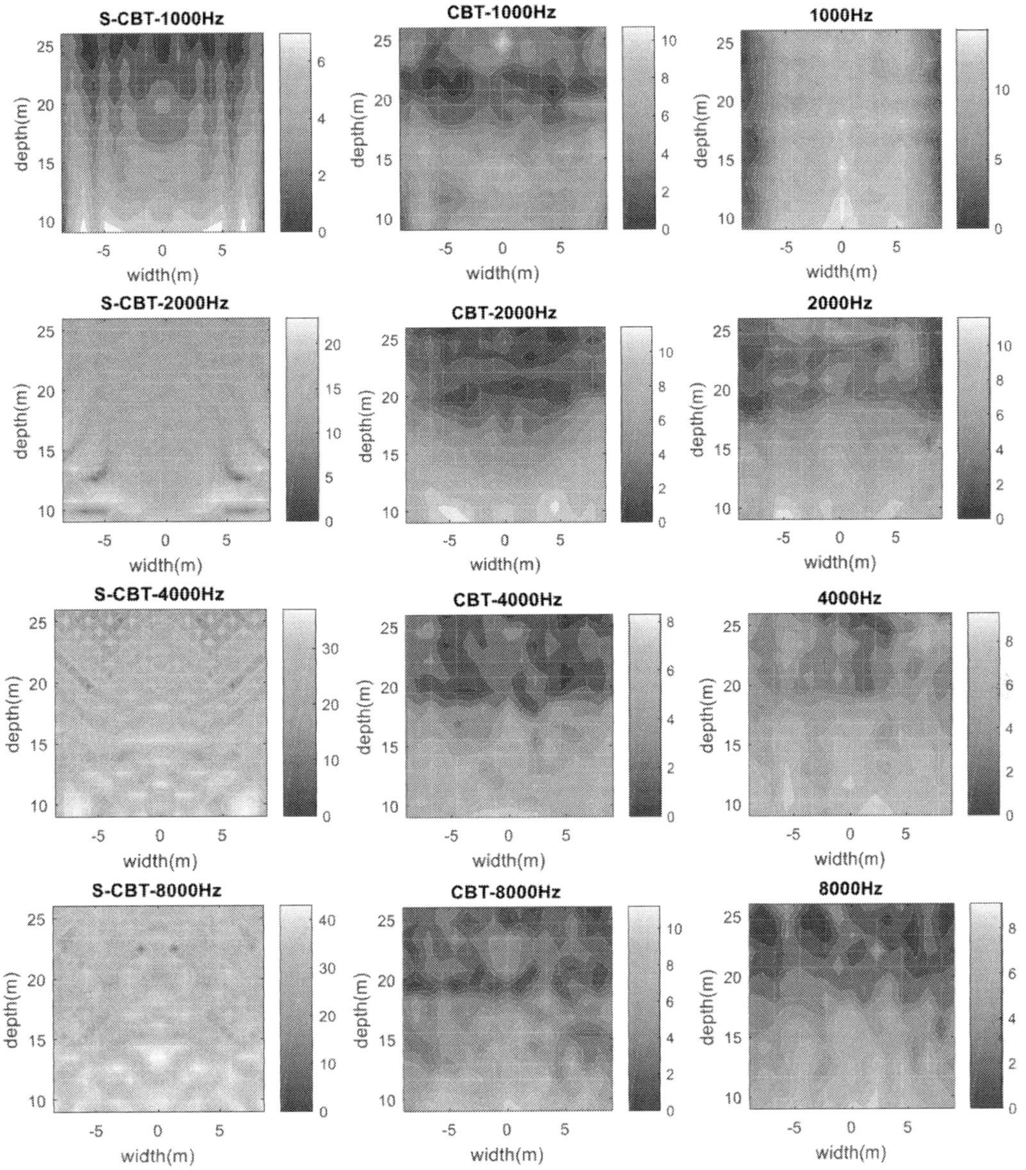

Figure 3. Comparisons of sound fields in CBT algorithm.

As shown in the figure 3, below 1000Hz, the energy of the original sound field is concentrated and the simulated sound field processed by the CBT algorithm has vertical light and dark stripes, while the energy distribution of the measured sound field processed by the CBT algorithm is relatively uniform, which is not similar to the two cases mentioned above. It is indicated that the CBT algorithm changes the original sound field distribution below 1000 Hz, but it doesn't achieve the effect of algorithm control. Above 2000 Hz, the measured sound field processed by the CBT algorithm is similar to the original sound field, there is no obvious interference fringe and the energy distribution is more dispersed than the original sound field.

5. Conclusions

In order to solve the soundproof problem of LED screen in the cinema, this paper uses 192 speaker units to form a surround screen speaker array to reproduce the main sound channel. The combination of the upper and lower side speaker array replaces the center channel, the left side speaker array replaces the left channel and the right side speaker array replaces the right channel. The sound field of the center channel is controlled by two different algorithms and the actual cinema sound field is measured. After comparative analysis, it is found that the WFS algorithm has better control effect on the sound field in the low frequency range, and CBT algorithm has better control effect on the sound field in the high frequency range. So an overall control algorithm is proposed.

The low frequency adopts the WFS algorithm, the high frequency adopts the CBT algorithm, and the high frequency and low frequency crossover points are 2000 Hz.The formula is as follows:

$$\mathrm{Y(w)} = \mathrm{AS(w)F(w)}e^{-jw\frac{|x_0-x_s|}{v}} \qquad \frac{w}{2\pi} < 2000$$

$$\mathrm{Y(w)} = \mathrm{F(w)}e^{-jw(1-cos\theta)r} \qquad \frac{w}{2\pi} \geq 2000$$

X0 denotes the position of each speaker unit, Xs denotes the virtual sound source position, A denotes the weight value, and r and Θ are the parameter related to x0. S (w) is a pre-filter function, F (w) is the original signal, and Y (w) is the signal processed by the algorithm.

This algorithm can distribute the sound field as a whole evenly. The WFS algorithm can obtain the correct sound image at low frequency, and the CBT algorithm makes the sound field distribution more uniform at high frequency.

References

[1] Abramov V S, Puisha A É, Polyakova I P, et al. 2003 LED modules for large screens [J] *Journal of Optical Technology* **70** (7) 492-494.

[2] Matthews. 2008 Jermey N.A. Are organic LEDs ready for the big screen? [J] *Physics Today* **61** (11) 26-27.

[3] ITU. 1994 ITU-R BS 775-1 Multi-channel stereo-phonic sound system without accompanying picture [S]

[4] Olson H.F. 1957 *Acoustical Engineering* [M].

[5] Gerzon M.A. 1985 Ambisonics in multichannel broadcasting and video [J] *Journal of the Audio Engineering Society* **33** (11) 859-871.

[6] Berkhout A.J, Vries D.D, Vogel P. 1993 Acoustic control by wave field synthesis [J] *The Journal of the Acoustical Society of America* **93** (5) 2764-2778.

[7] Kirkeby O, Nelson P.A. 1993 Reproduction of plane wave sound fields [J] *The Journal of the Acoustical Society of America* **94** (5) 2992-3000.

[8] Keele D.B. 2000 The application of broadband constant beamwidth transducer (CBT) theory to loudspeaker arrays [C] *Audio Engineering society convention 109.*

[9] Zhang W, Abhayapala T.D, Betlehem T, et al. 2016 Analysis and control of multi-zone sound field reproduction using modal-domain approach [J] *The Journal of the Acoustical Society of America* **140** (3) 2134-2144.

[10] Gauthier P.A, Berry A, Woszczyk W. 2005 Sound-field reproduction in-room using optimal control techniques: Simulations in the frequency domain [J] *The Journal of the Acoustical Society of America* **117** (2) 662-678.

[11] Spors S, Rabenstein R, Ahrens J. 2008 The Theory of Wave Field Synthesis Revisited [C] *The 124th Convention of the Audio Engineering Society convention paper 7358.*

[12] ISO. 2015 ISO 2696-2015 Cinematography--B-chain electro-acoustic repose of motion-picture Control rooms and indoor theatres--Specifications and measurements [S].

A Modified SIR Model for the COVID-19 Epidemic in China

Jiale Wang, Yang Liu, Xusheng Liu and Keming Shen

School of Science, East China University of Technology, Nanchang 330013, China.
Email: shen_keming.ecut@hotmail.com

Abstract. The analysis of the coronavirus disease 2019 (COVID-19) is of great importance to deeply understand the dynamics of this coronavirus spread. Based on the complexity of it, a modified susceptible-infected-removed (SIR) model is applied to analyse the time dependence of active and hospitalized cases in China. The time evolution of the virus spread in different provinces was adequately modelled. Changeable parameters among them have been obtained and turned to be not naively independent with each other. The non-extensive parameter was found to be strongly connected with the freedom of systems. Taken into the prevention and treatment of disease, more measures by the government lead to higher values of it.

1. Introduction

These The coronavirus disease 2019 (COVID-19) has been a kind of global public health emergency [1] since it occurred in December 2019 in Wuhan, the capital of Hubei Province in China. The outbreak subsequently spread rapidly throughout China and elsewhere. More and more intervention strategies were implemented in China in order to contain this outbreak. It has caused significant disruption to the social and economic structure. Nevertheless, it is still unknown whether these policies have had an impact and how long they were able to remain there. On the other hand, the case abroad developed worse, which has reversely affected the epidemic progression in China then. It is thus critical to assess the effects of these control measures on the development of COVID-19.

Nowadays in the market there have already existed plenty of models, for example, susceptible-infected-removed (SIR), susceptible-infected-susceptible (SIS), and susceptible-exposed-infected-removed (SEIR) models, which have been proposed to predict the epidemic progression. However, all of these mathematical models of infectious disease shared some of ideal hypotheses: a) the reproduction number R is often proposed to be exponential which is coming from the classical Boltzmann-Gibbs (BG) statistics; b) the internal interactions are always neglected when considering the propagation of disease, such as the normal human contacts in daily life; c) the total number is recognized as a constant, namely the system is a closed one; d) *etc.*.

Though a large number of works [2-5] have made it possible to research on the development of COVID-19 within the previous models quantitatively, there still remains uncertainty in the theoretical framework, especially for the influences of so many control measures by our government and else. Consequently, this work would like to modify the simple SIR model in order to study the COVID-19 epidemic in China with respect to the non-extensive statistics. The non-extensive statistics, firstly proposed in [6] has attracted lots attention and discussions [7]. Instead of the normal exponential function, a generalized q-exponential function, defined as [6, 7],

$$\exp_q(x) := [1 + (1 - q)x]^{\frac{1}{1-q}} \tag{1}$$

was introduced to deal with the problems for the appearance of critical fluctuations due to interactions or non-ideal cases. One non-extensive parameter q is to account for all possible factors violating assumptions of the usual BG cases. Its inverse function is also listed [6, 7],

$$\ln_q(x) := \frac{x^{1-q}-1}{1-q} \tag{2}$$

Both of them return to the usual exponential and logarithm function with $q \to 1$.

The purpose of this paper is to clarify the non-extensive effects on the fitting parameters of the modified SIR model for the COVID-19 epidemic. We focus on the situations in some provinces in China, where both similarities and differences appear. For instance, we have analysed the situations in some coastal provinces, tier-one-provinces, inland provinces and so on. In order to research on the effects of import cases from abroad, the situations in Taiwan and Hong Kong are also well studied within our model.

This paper is organized as follows. In Section 2 we introduce the theoretical framework, where the non-extensive q-SIR model is stated as well as the normal SIR one. Their comparisons are also explored afterwards. The consequences for various disease spreading situations in different provinces are exhibited in Section 3; more detailed discussions on the results are also contained. Section 4 is our brief summary and outlook.

2. Theoretical Framework

As a classical compartmental model in epidemiology, the SIR model was created in 1927 [8] in which a fixed population was considered with only three compartments:

- $S(t)$, susceptible, is used to represent the individuals not yet infected with the disease at time t, or those susceptible to the disease of the population.
- $I(t)$, infected, denotes the individuals of the population who have been infected with the disease and are capable of spreading the disease to those in the susceptible category.
- $R(t)$, recovered, is the compartment used for the individuals of the population who have been infected and then removed from the disease, either due to immunization or due to death. Those in this category are not able to be infected again or to transmit the infection to others.

Using them one can establish the SIR epidemic model without so-called vital dynamics (birth and death, sometimes called demography) described above as follows [9-12]:

$$\frac{dS(t)}{dt} = -\beta\, I(t)S(t) \tag{3}$$

$$\frac{dI(t)}{dt} = \beta\, I(t)S(t) - \gamma\, I(t) \tag{4}$$

$$\frac{dR(t)}{dt} = -\gamma\, I(t) \tag{5}$$

where β is the average number of contacts per person per time, multiplied by the probability of disease transmission in a contact between a susceptible and an infectious subject, and $\gamma\, I(t)$ is assumed to describe the transition rate between $I(t)$ and $R(t)$. The first differential equation is the fraction of those contacts between an infectious and susceptible individual which result in the susceptible person becoming infected. Worth to mention that in which there is no removal from the infectious compartment ($\gamma = 0$), the SIR model reduces to a very simple SI model, which has a logistic solution, in which every individual eventually becomes infected.

Considering the SIR system is closed:

$$S(t) + I(t) + R(t) = N \tag{6}$$

it is possible to derive its analytic solution in implicit form [13]:

$$I(t) = S_0 + I_0 - S(t) + \frac{1}{\sigma}\ln\frac{S(t)}{S_0} \tag{7}$$

where $\sigma = \beta/\gamma$ is the basic reproduction number (also called basic reproduction ratio). S_0 and I_0 denote the initial numbers of, respectively, susceptible and infectious subjects. It is derived as the expected number of new infectious (sometimes called secondary infections) from a single infection in a population where all subjects are susceptible. [14, 15]

Note that the function, $F = \beta I$, in the above model, is recognized as the force of infection, which models the transition rate from the compartment of susceptible individuals to the compartment of infectious individuals. However, for large classes of communicable diseases it is more realistic to consider a force of infection that does not linearly depend on the absolute number of infectious subjects. As for the stock of susceptible population, it has shown us that $S \approx N$ for most of cases due to the large population in SIR system. Therefore, these evolution equations need to be reconsidered.

Similar to the non-extensive statistics, the generalized evolution equations of COVID-19 are modified as follows:

$$\frac{dS(t)}{dt} = 0 \tag{8}$$

$$\frac{dI(t)}{dt} = b\, I^q(t) - \gamma\, I(t) \tag{9}$$

$$\frac{dR(t)}{dt} = -\gamma\, I(t) \tag{10}$$

whose solutions are listed as:

$$S(t) = N \tag{11}$$

$$I(t) = A \cdot \{1 - B\exp[-\gamma(1-q)t]\}^{\frac{1}{1-q}} \tag{12}$$

$$R(t) = \frac{A}{\gamma(2-q)}\{1 - B\exp[-\gamma(1-q)t]\}^{1+\frac{1}{1-q}} \tag{13}$$

With such a q-power modified term of infectious individuals in Eq.(12), we then explore the time dependence of active and hospitalized cases in various kinds of provinces in China, typically for the number of infectious individuals. More precisely, the exact solution of $I(t)$ should be:

$$I(t) = \left\{\frac{b}{\gamma} - \exp[\gamma(1-q)(t_0 - t) + \frac{1}{\gamma(q-1)}\ln(b - \gamma I_0^{1-q})]\right\}^{\frac{1}{1-q}}.$$

Similar for the distribution of $R(t)$. Without the loss of generality, we re-write them as above for simplicity. Worthy to mention that for such a distribution, there are already some other similar works appearing to describe satisfactorily the available data for COVID-19. For instance, C. Tsallis and U. Tirnakli [16] applied one q-statistical functional form in order to predict the peaks around the world. The numerical results will be shown in the next section.

3. Results and Discussions

We start our discussions with presenting in Figure1 the epidemic progression in several provinces in China from January 2020 to January 2021. Data are taken from [17]. According to the report of China CDC, January 1 2020 was considered as the 1st day of a contagion period, so we focused on these data sets. It indicates that all the year of 2020 there appeared almost ZERO new death individuals in China (seen as the green dots and lines in figure 1), which nicely shows the effects caused by control strategies by the government. While both the confirmed and recovered infections vary heavily among all the provincial data.

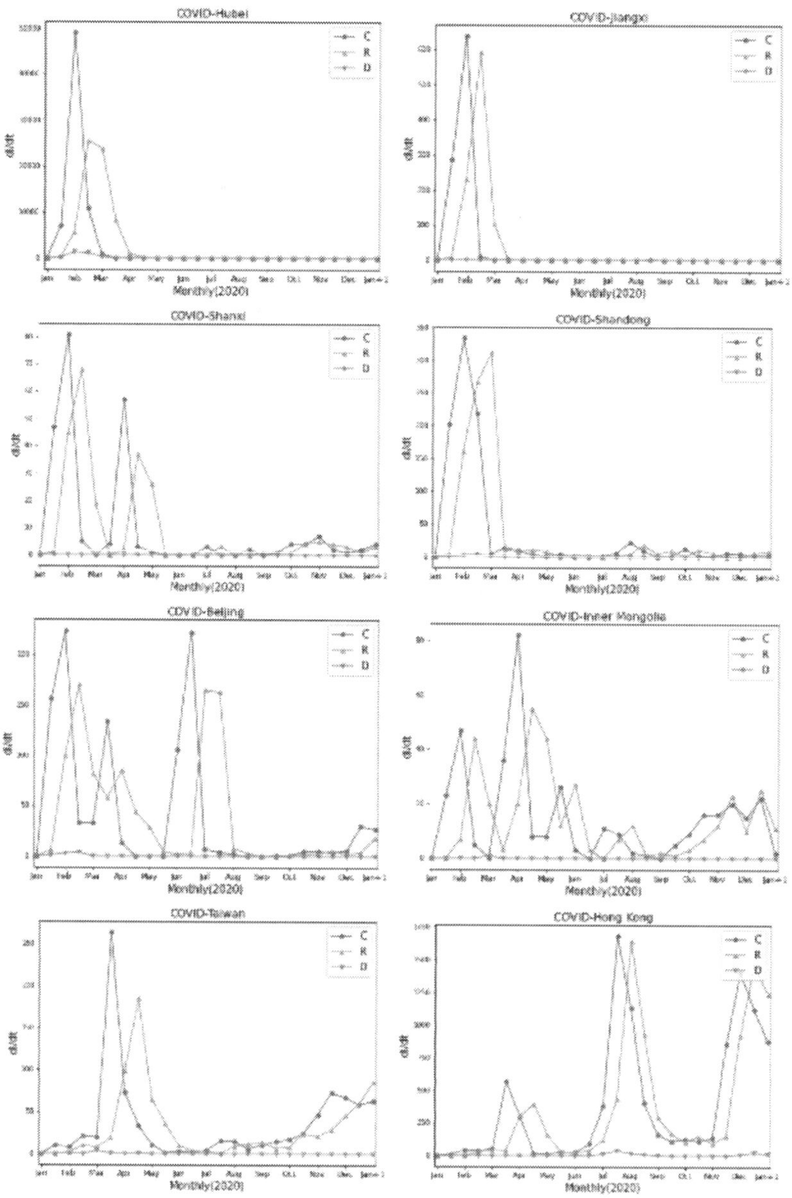

Figure 1. Actual monthly data of Confirmed Infections (C), Recovered Cases (R) and Dead Individuals (D) for several provinces (Hubei, Jiangxi, Shanxi, Shandong, Beijing, Inner Mongolia, Taiwan and Hong Kong, respectively).

For the upper two plots, they show the epidemic progression for provinces like Hubei and Jiangxi. From them we could see that all data of infections (both C and R) vanish from around May of 2020. This is largely due to the effectiveness of quarantine and control measures in China, especially the lock-down of Wuhan city from January 23, 2020 to April 8, 2020.

The cases in Shanxi and Shandong are distributing similarly as above but some exceptions where a slight outbreak happened there. This could be obviously seen as in the next two panels. However, there are still many provinces or cities in which the COVID-19 still exists and even arises a secondary infection. Because of the frequent contacts and interchanges with outside, the size of system in cities like Beijing and Inner Mongolia is very huge as well as its complexity. That is to say, they are more like open systems.

On the other hand, it is instructive to plot the distributions in Hong Kong and Taiwan, where the effects from abroad cannot be neglected indeed. The lower two panels show how the imported cases work on the spread of this contagious disease.

In virtue of it the validity of approximation, $S_0 \approx N$, is nicely supported by real data in China [12], and the development of recovered individuals in SIR model strongly relies on the case of infectious ones, in this work we shall focus on the possible interpretations of spread of infections within the non-extensive effect. Furthermore, all the evolutionary tendency of infectious individuals are to be fitted by the distribution:

$$\begin{aligned} I(t) &= A \cdot \{1 - B \exp[-\gamma(1-q)t]\}^{\frac{1}{1-q}} \\ &= a \cdot \{1 - b \exp[-ct]\}^{d} \end{aligned} \tag{14}$$

Worthy to note that the large values of time t could recover an exponential function whereas it leads to a Tsallis q-exponential distribution, cf. Eq. (1) when$t \to 0$. On the other hand, γ still gives out the information on the average time for which an individual is infectious. The non-extensive parameter q tells us the departure from the ideal situation, in which the distribution behaves as an exponential function.

In our programs the fitting function has been simplified as shown in Eq.(14) with the new parameters, c and d being q-dependent. By using it we analysed the monthly spread of infections from January of 2020 to January of 2021. In order to better understand the mechanism of the infectious numbers, we summed up all the previous new confirmed cases as the number of infections by then. All the annual situation was analysed since we would like to see the connections or differences among different provinces.

In table I all the fitting parameters in Eq.(14) are tabulated for various kinds of provincial data. Almost the same values of normalization constant, b, indeed tell us that there is no big difference among these provinces except for Taiwan and Hong Kong, which were largely affected by the imported cases. The weak fluctuation appeared in the fittings for Beijing's infectious distributions. This could be nicely explained by its speciality of tier-one-city.

On the other hand, as shown in Eq.(14), the production of parameters c and d explores the information of the average time, γ. It varies between the normal and high-risk areas for COVID-19 in China. As for the non-extensive parameter, q=1-1/d, there was no big fluctuation among most of provincial data as well, especially for the relatively close provinces such as Beijing and Inner Mongolia. This indicates the common spread mechanism among most parts in China. While we cannot neglect the big divergence for the cases in Taiwan and Hong Kong.

Table 1. The values of fitting parameters in Eq.(14) in the modified SIR model for different provinces:

Provinces	a	b	c	d
Hubei	100000	1.010	1.023	0.20
Jiangxi	1500	1.009	0.886	0.21
Shanxi	290	1.009	0.851	0.16
Shandong	1150	1.009	0.892	0.16
Beijing	1050	1.101	12.394	0.92
Inner Mongolia	550	1.015	4.040	0.92
Taiwan	50	0.456	-1.708	-2.4
Hong Kong	80	0.653	-1.141	-2.4

At the same time, one can easily observe an obvious connection between the values of fitting parameter a and the infectious peaks in all provinces. For example, the situation in Hubei was nearly the most serious in China at the beginning of the year 2020 and even the government took the control measures by locking down some cities in Hubei for some time. This could be well seen by the maximum number of parameter, $a = 100000$, of all.

In order to illustrate the fittings of this modified SIR model for the COVID-19 pandemics in China more clearly, fitting results are plotted in figures 2-5. Our model is proved to be an alternative tool to investigate the spread of infectious individuals for COVID-19 in China.

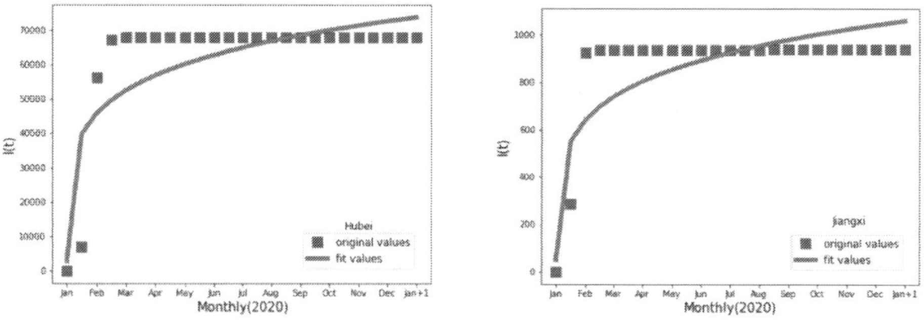

Figure 2. Fittings on the monthly evolution of infectious individuals for Hubei and Jiangxi.

Specifically, Figure 2 lists the fittings on the monthly evolution of infectious individuals for the provinces of Hubei and Jiangxi. We could see that the number of infections reached an almost constant from March of 2020, which is shown in our fit values (seen as the red lines). No doubt that the shape looks similar since the Chinese government took all the control measures for the whole country. The numerical difference at the peak lights upon their own outbreaks at the initial period of COVID-19 in China, for example the case was much better in Jiangxi at that time.

A little difference happened to the cases for Shanxi and Shandong, as shown in figure 3. As reported by the local governments, there existed some few infectious cases at the end of autumn in 2020, which has been proved to be the consequence of some people who were previously infected abroad and then returned home. This leads to a better fitting consistency where there were more non-extensive interactions in the system.

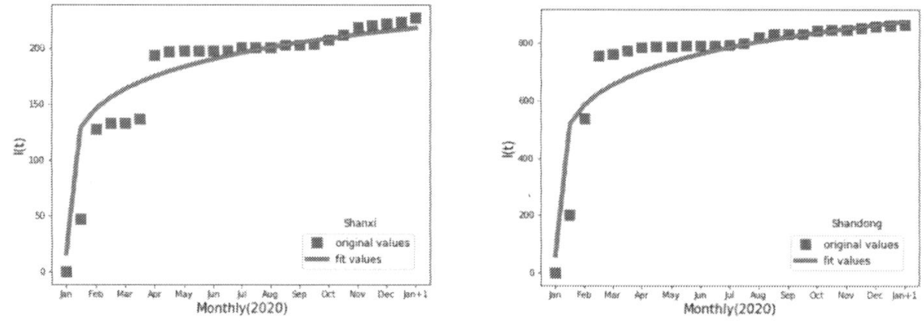

Figure 3. Fittings on the monthly evolution of infectious individuals for Shanxi and Shandong.

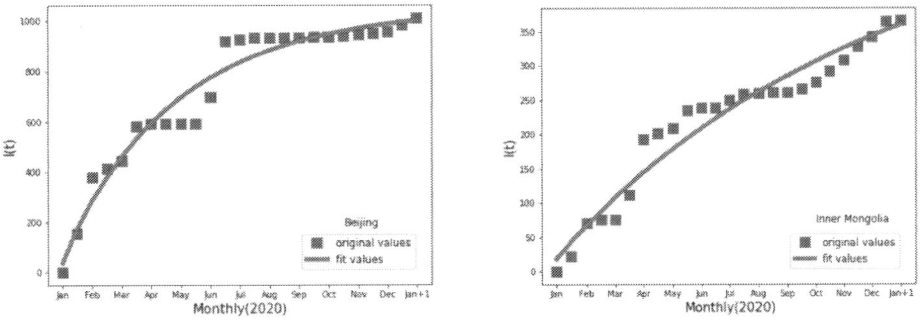

Figure 4. Fittings on the monthly evolution of infectious individuals for Beijing and Inner Mongolia.

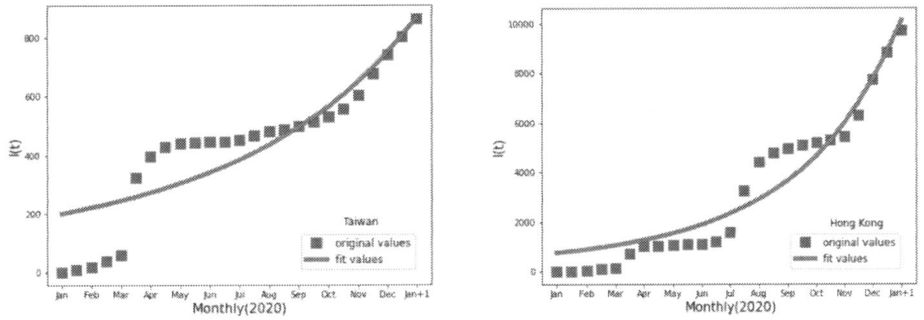

Figure 5. Fittings on the monthly evolution of infectious individuals for Taiwan and Hong Kong.

As expected, the spread in the tier-one-city, Beijing, is more complicated. Because of the frequent connections with outside, it is much more easily to see that the non-extensive effects cannot be neglected. As a result, our model is more proper to describe their disease development. In Figure4 the distribution cf. Eq.(14) nicely coincides with the original values of infectious individuals. Similar cases happened to the provinces like Inner Mongolia and Xinjiang because of their special geographical locations.

Finally it is also important to explore the results for the cases in Taiwan and Hong Kong, where the imported factors are more inevitable. Moreover, most controlling measures taken in mainland of China were NOT well applied in these parts. Even when a secondary outbreak happened, the local government somehow neglected it. Thus the situation there became much worse at the end of 2020. Figure5 agrees with the discussions above, that our modified SIR model is typically for the system which meets with more interferences from outside.

4. Summary and Outlook

To summarize, we have calculated the non-extensive SIR model to account for the complexity of the epidemic spreads of the COVID-19 in China. Especially the Chinese government has taken many emergency actions in order to help reduce the contagion probability and cure the infectious individuals. On the other hand, the imported disease cases unavoidably affects the exploration of COVID-19 in China. All lead to it that the system for this epidemic transmission should be considered within the non-extensive application rather than the classical ideal one. By the q-version we have modified the original SIR model to investigate the time evolution of infectious individuals in different provinces in China.

It is of importance to research on the fittings within our modified SIR model for the various provincial data sets. All the corresponding fitting parameters illuminate that the non-extensive application works well on the spread of infectious individuals in mainland of China. Compared with the cases in several typical provinces, such as Hubei, Beijing and others, we indeed found out the common properties among different places in mainland of China. This is due to the fact that all the effective actions were taken in all corresponding cities. While it is worthy to note the cases in Taiwan and Hong Kong, for which the system prefers to an ideal one since nearly no effective measures were taken or no enough notations were caused by the local government at the initial time.

Last not the least, a complete understanding of the physics of non-extensive parameters in the study of COVID-19 should be solved in general context. In this paper we have tried to illuminate it from the comparisons among various fittings of provincial data sets in China. Our model, on the other hand, successfully provides another method to investigate on problems in the scientific researches on the epidemic disease such as COVID-19. Meanwhile, our researches could deeply help understand the physical explanation of the Tsallis non-extensive parameter q, which is also what we will pay attention to in the future.

Acknowledgments

This research is partly supported by the funding for the Doctoral Research of ECUT Nos. DHBK2019211 and DHBK2019200 and by NSFC under Grant No. 11947020.

References

[1] WHO: Coronavirus disease 2019 (COVID-19): situation report 2020 (51).
[2] Yang Z F, et al. 2020, Modified SEIR and AI prediction of the epidemics trend of COVID-19 in China under public health interventions *J. Thorac. Dis* **12** (3) 165-174.
[3] Teles P, 2020, A time-dependent SEIR model to analyse the evolution of the SARS-covid-2 epidemic outbreak in Portugal, *Bull World Health Organ.* E-pub: 7 April 2020.
[4] Ahmad Z, et al. 2020 A report on COVID-19 epidemic in Pakistan using SEIR fractional model, *Scientific. Reports* **10** (1) 22268.
[5] Radha M, Balamuralitharan S. 2020 A study on COVID-19 transmission dynamics: stability analysis of SEIR model with Hopf bifurcation for effect of time delay, *Adv. in Diff. Equa.* 523.
[6] Tsallis C. 1988 Possible generalization of Boltzmann-Gibbs statistics, *J. Stat. Phys.* 52 479-487.
[7] Lists of many applications of non-extensive statistics are available at http://tsallis.cat.cbpf.br/biblio.htm.
[8] Kermack W O, McKendrick A G. 1991 Contributions to the mathematical theory of epidemics—I, *Bulletin of Mathematical Biology* **53** (1-2) 33–55.
[9] Kenah E and Robins J M. 2007 Network-based analysis of stochastic SIR epidemic models with random and proportionate mixing, *J. Theor. Biol* **249** (4) 706-22.
[10] Kenah E and Robins J M. 2007 Second look at the spread of epidemics on networks, *Phys. Rev. E* 76 036113.
[11] Cauchemez S, Ferguson N M. 2008 Likelihood-based estimation of continuous-time epidemic models from time-series data: application to measles transmission in London, *J. R. Soc., Interface* **5** (25) 885-97.
[12] Tao Y. 2020 Maximum entropy method for estimating the reproduction number: An investigation for COVID-19 in China and the United States, *Phys. Rev. E* 102 032136.
[13] Harko T, Lobo F S N, Mak M K. 2014 Exact analytical solutions of the Susceptible-Infected-Recovered (SIR) epidemic model and of the SIR model with equal death and birth rates, *Applied Mathematics and Computation* 236 184–194.
[14] Bailey, Norman T J. 1975 *The mathematical theory of infectious diseases and its applications* (2nd ed.), London: Griffin. ISBN 0-85264-231-8.

[15] Sonia A, Nunn, Charles. 2006 *Infectious diseases in primates: behavior, ecology and evolution. Oxford Series in Ecology and Evolution,* Oxford [Oxfordshire]: Oxford University Press.
[16] Tsallis C, Tirnakli U. 2020 Predicting COVID-19 Peaks around the World *Fron. Phys.* 8 217.
[17] National Health Commission of the People's Republic of China, The latest information on the infections of COVID-19 [EB/OL], http://www.nhc.gov.cn/xcs/yqtb/list_gzbd.shtml.

Research on Isolated Cloud Point Fast-Filtering Algorithm for Digital Twin

Hongjun Li[1, 2], Xiaorui Guo[1,2], Kang Yang[1,2], Chi Zhang[1] and Zhihan Zhang[1,2,*]

[1] School of Mechanical Engineering and Automation, Wuhan Textile University, room N107, Nanhu campus, Wuhan, China
[2] Industrial Detonator Intelligent Assembly Hubei Engineering Research Center, room N107, Nanhu campus, Wuhan, China
Email: 1051093701@qq.com

Abstract. With the establishment of the digital twin stereoscopic warehouse concept, the importance of twin data has become increasingly prominent for digital twin system. Aiming at the problem of low accuracy in obtaining twin data from warehouse stocks, an isolated point cloud filtering algorithm combine with digital signal processing is proposed. The algorithm can retrieve the coordinate value of isolated point in original twin point cloud by constructing a twin point cloud fitting model, thereby filter out the point information of isolated region and obtain the warehousing twin data. The experiment results show that the algorithm can filter all isolated points while keeping the characteristics of original twin point cloud. The method provides accurate twin data support for digital twin stereoscopic warehouse.

1. Introduction

Digital twin uses information, such as physical models, data acquisition, historical data and so on, to achieve high-fidelity virtual mapping from physical space to digital virtual space[1].It was proposed by Grieves[2] in 2003.With the information wave represented by big data, internet of things, virtual reality, cloud computing and other technologies sweeping the world[3], the concept of digital twins has been introduced into intelligent warehousing, which has profoundly affected the planning, construction and governance of warehousing. In 2019, Tao [4] proposed ten major applications of digital twin, the theoretical model of digital twin stereoscopic warehouse was built for the problems of low warehouse utilization and throughput to be improved. Obtaining complete and accurate twin data from physical entities is a significant cornerstone for constructing the five-dimensional digital twin warehouse stereoscopic model.

As a new equipment of active three-dimensional spatial information acquisition sensor, LiDAR determines the distance information between the laser source and the target by detecting light pulses or modulated pulses[5], which has the advantages of high efficiency and accuracy. And it has been widely applied in the surveying and mapping fields such as stockpiling measurement, hydrological modeling, and forest investigation [6].Due to the complexity of spatial environment, noise is inevitably introduced into the twin point cloud. Isolated points, without any connection, is the typical noise in twin point cloud, which were caused by the occlusion of non-target objects.

In order to solve the isolated points in the original twin point cloud, we propose a twin point cloud filtering algorithms which is combined with digital signal processing technology. Under the premise of maintaining the main characteristics of twin point cloud, the algorithm filters out all isolated points and provides accurate twin data for the system of digital twin stereoscopic warehouses.

2. Principle of Isolated Data Filtering Algorithm

2.1. Algorithm Framework

The algorithm base on the feature that each twin points belong to same twin point cloud frame collected by LiDAR have the same X coordinate value in the Cartesian coordinate system. The adjacent twin points height difference series ,as a discrete non-periodic time domain signal, use discrete Fourier transform from time domain to frequency domain to obtain the twin point cloud fitting model. When we solve the twin point cloud fitting model, we use it to retrieve the position information of isolated points from original twin point cloud. Isolated points in the twin point cloud can be filtered out quickly and accurately. The algorithm is mainly divided into three steps: twin point cloud preprocessing, twin point cloud fitting model establishing and isolated point retrieving. The overall process as follows:

1) A twin point cloud frame(n twin points) are extracted from the original twin point cloud, and the height difference of adjacent twin points (difference of Z coordinate value) is calculated according to the order of Y coordinates . Compose a data set $\Delta z_{(n-1)}$.

2) $\Delta z_{(n-1)}$ is used discrete Fourier transform to deal with the height difference data set, and the characteristic frequency signal caused by isolated points are filtered out after converting the time domain signal to the frequency domain signal. The inverse Fourier transform is performed on the low-pass filtered frequency domain signal to obtain the filtered time domain signal, which is restored height information to construct a twin point cloud fitting model.

3) Retrieve the position coordinates of the isolated point in the original twin point cloud through the twin point cloud fitting model, and filter out the point information of the isolated area to obtain the twin data that does not contain the isolated points.

The details of the specific steps in the processing flow are shown in figure 1:

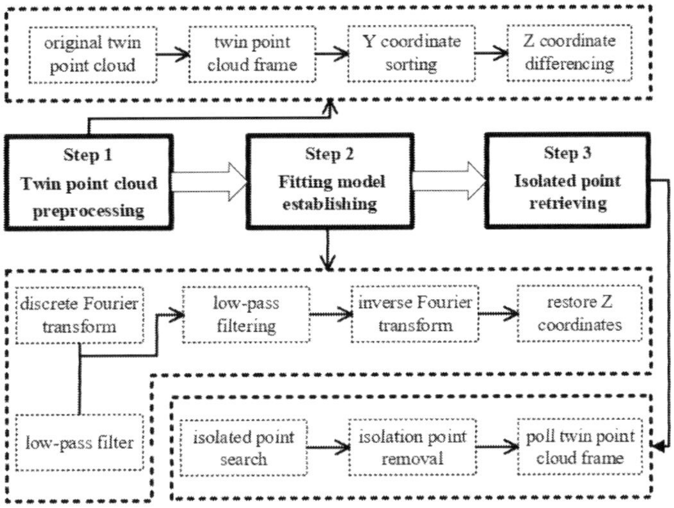

Figure 1. Flow chart of isolated data filtering processing.

Twin point cloud preprocessing (Step 1): According X coordinate value, algorithm extract a twin point cloud frame from the twin point cloud data file. The bubble sort method is used to reorder the Y coordinate value from small to large, and Z coordinate value is corresponded with original Y coordinate value. So we obtain the height data set $z_{(n)}$ .Set index number for height difference between two adjacent point clouds.

Twin point cloud fitting model establishing (step 2): Convert of height difference data set $\Delta z_{(n-1)}$

from time domain signal to frequency domain signal by discrete Fourier transform. A low-pass filter is constructed to filter out high frequency signals in frequency domain. The time domain signal $\Delta\widetilde{z}_{(n-1)}$ is restored by inverse Fourier transform. All Z coordinate value of the fitting model point are derived by previous basis point, and the first basis point is $z_{(1)}$.

Isolated point retrieving (Step 3): Difference between the twin fitting model in step 2 and the original twin point cloud. If the difference value exceeds the threshold, the point is treated as an isolated point and filtered from the original twin data. Finally, we obtain the target twin data.

2.2. Twin Point Cloud Preprocessing

At the scanning process of vehicle LiDAR, Twin points in the twin point frame have the same X coordinate value .Therefore, the single-frame twin point cloud can be extracted by retrieving twin points with same X coordinate value from original twin point. Due to the existence of occlusions in the actual measurement process, the data storage order cannot be carried out accordance with the metabolic direction of the Y coordinate value. If the subsequent algorithmic steps is directly carried out, isolated data information in the fitting model cannot be filtered. Therefore, we can obtain a twin point cloud frame $P_{(i)} = (x, y_{(n)}, z_{(n)})$ consist of n twin points which had been sorted according to the Y coordinate value. Based on $z_{(1)}$ (figure 2(a)), the height difference series is obtained as follows:

$$\Delta z_{(m)} = z_{(m+1)} - z_{(m)}, 1 \le m \le n-1 \quad (1)$$

2.3. Twin Point Cloud Fitting Model Establishing

The purpose of this processing step is obtaining twin point cloud fitting model without isolated points. Because the isolated points is far away from the subject of twin point cloud in the spatial domain, the data fluctuates violently in the height difference series $\Delta z_{(m)}$ .According to the digital signal filtering method, $\Delta z_{(m)}$ is regarded as discrete sampling signal, which is converted into frequency domain signal by discrete Fourier transform[7], and the frequency domain filtering of height difference series is realized by low-pass filter. The filtering model can be formulated as follows:

$$F'_{(k)} = H_{(k)} F_{(k)} \quad (2)$$

$F_{(k)}$ is Fourier transform of height difference series. $H_{(k)}$ is the transform function of low-pass filter[8]. $F'_{(k)}$ is generated by filtering high frequency information of $F_{(k)}$ .Than, inverse Fourier transform[7] is performed on $F'_{(k)}$ to obtain $\Delta\widetilde{z}_{(m)}$, a new height difference series with uniform variation of data values(figure 2(b)).Substitute to equation(1) to obtain single frame twin point cloud fitting model as follow:

$$\widetilde{z}_{(n)} = \begin{cases} z_{(1)} & n = 1 \\ z_{(n-1)} + \Delta\widetilde{z}_{(n-1)} & n > 1 \end{cases} \quad (3)$$

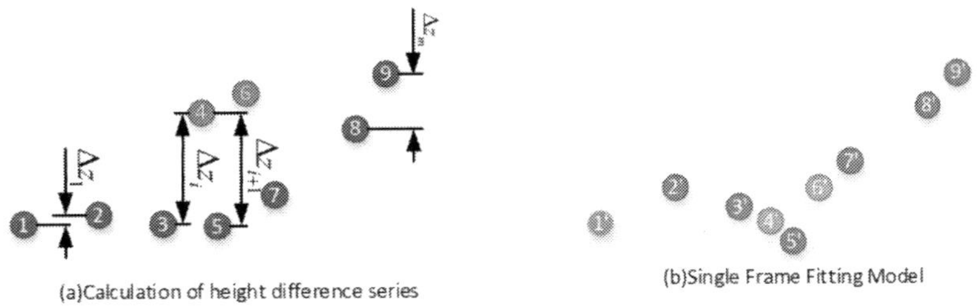

Figure 2. Algorithm Diagram 1.

2.4. Isolated Point Retrieving
The twin point cloud fitting model is obtained by frequency domain filtering contains two messages. Firstly, when the X and Y coordinates of the original isolated points have not changed, the Z coordinate value has been greatly corrected, but it is still unable to determine the specific spatial position of the isolated point. Secondly, frequency domain filtering cause Z coordinate value of non-isolation points distortion in a certain range. Based on this information, the twin point cloud fitting model is compared with the original twin point cloud (figure 3(a)).Threshold comparison using the difference of Z coordinates of points with the same index number in two data sets. Comparative methods can be formulated as follows:

$$K \geq \left|\widetilde{z}_{(n)} - z_{(n)}\right| \tag{4}$$

If the Z coordinate difference between two points, have common index umber i, is less than the threshold K, the point i in the original twin point cloud frame is considered to be a non-isolation point. Otherwise, it is determined as an isolated point. We can obtain the twin data height set $z_{(p)}$ when all isolated points height information in $z_{(n)}$ are filtered out. Twin data frame $\widetilde{P}_{(i)} = (x, y_{(p)}, z_{(p)})$ is constructed (figure 3(b)). $y_{(p)}$ is a set of Y coordinate value corresponding to $z_{(p)}$ in $P_{(i)}$,and the X coordinate value is fixed in the single-frame twin point cloud. The target twin data is obtained when original twin point cloud filter all isolated points.

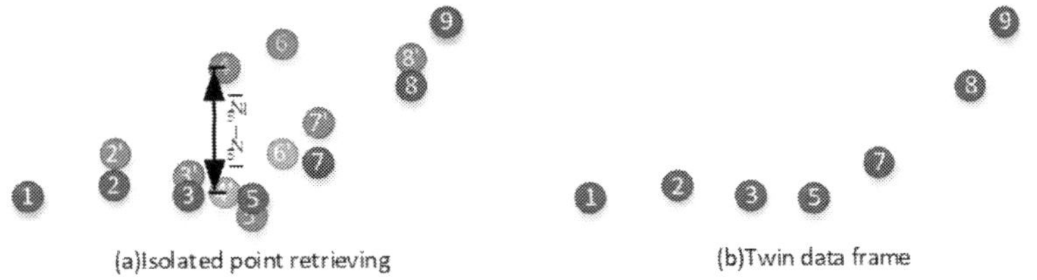

Figure 3. Algorithm Diagram 2.

3. Algorithm Example and Analysis

3.1. Experimental Results
In order to validate that the algorithm can satisfy the isolated data filtering of the original twin point cloud obtained in the complex storage space environment, the measured original twin point cloud of a cement factory in Zhejiang Province are selected for experimentation. Figure 4(a) shows the original twin point cloud of the sand mound slope obscured by the beam of the factory building. Figure 4(b) is a single-frame twin point cloud fitting model. It is obvious that under the premise of small distortion of non-isolation points in the fitting model obtained by algorithm steps 1 and 2, the Z coordinate value

of isolated point is greatly corrected.

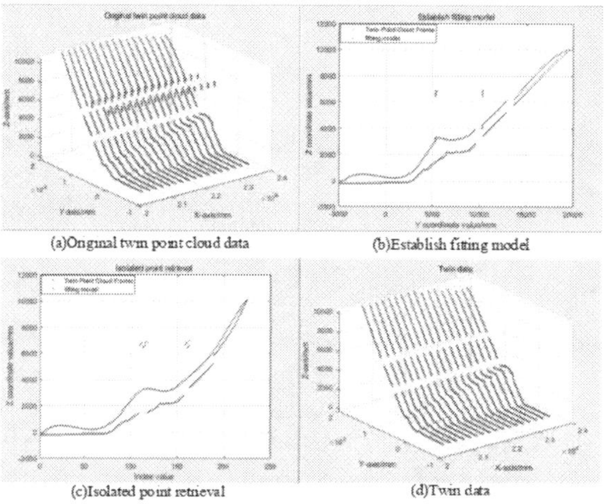

(a)Original twin point cloud data (b)Establish fitting model
(c)Isolated point retrieval (d)Twin data

Figure 4. Twin point cloud 1.

figure4(c) shows the process of isolated point position retrieving in the original twin point cloud according to the twin point cloud fitting model. By setting an appropriate threshold, the isolated points in the original twin point cloud can be completely filtered, while the characteristics of non-isolation points are still maintained. Figure 4(d) shows the target twin data.

We choose another set of measured original point cloud for the experiment, as shown in Figure 5(a).Twin point cloud fitting model (figure 5(b)) is obtained when the algorithm completes process from step1 to step2 .Finally, the intact twin data are obtained as shown in figure 5(c).

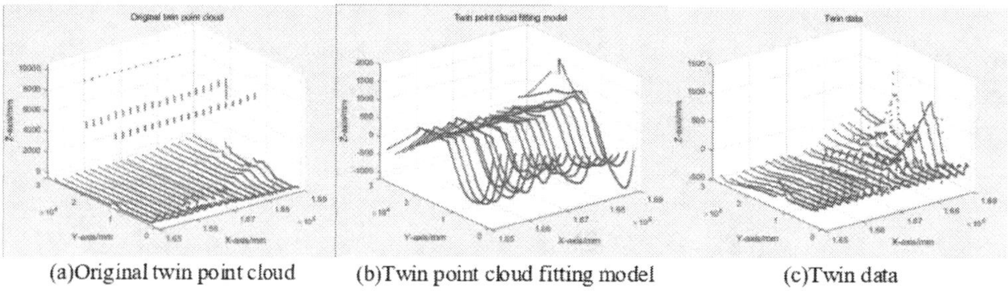

(a)Original twin point cloud (b)Twin point cloud fitting model (c)Twin data

Figure 5. Twin point cloud 2.

3.2. Algorithm Efficiency Evaluation

To validate the efficiency of the proposed algorithm, according to the filtering accuracy evaluation formula proposed in literature [9]:

$$P_1 = \frac{N_{TP}}{N_{TP} + N_{FP}} \tag{5}$$

N_{TP},the number of non-noise points are correctly judged as non-noise points. N_{FP} is the number of noise points mistaken for non-noise points. According to the filtering total error formula proposed in literature [10]:

$$P_2 = \frac{N_{TP}' + N_{FP}}{N} \tag{6}$$

N_{TP}' is the number of non-noise points mistaken for noise points, which had been filtered in the Step3 of algorithm, N is the total of points in the original twin cloud. Statistics of experimental results are shown in table 1:

Table 1. Algorithm filtering statistical table.

Parameter	Data set 1	Data set 2	Data set 3
N	3615	3176	8197
N_{TP}	3363	2937	7891
N_{TP}'	12	35	260
N_{FP}	0	0	0
P_1	1	1	1
P_2	0.0033	0.0110	0.0317

The results, three experiment of isolated points filtering, show the algorithm can remove all isolated points in the original twin point cloud, and the filtering accuracy P_1 reaches 100%.Under this premise, the algorithm keep characteristics of original twin point cloud as far as possible, the rate of filtering total error is less than 0.05.Therefore the veracity of twin data is greatly guaranteed.

There are two reasons why non-isolation twin points are filtered. Firstly, if the amount of twin points in the single twin point cloud frame is even(amount of height difference data is odd),the algorithm will remove the last point when discrete Fourier transform is performed(figure 6(a) and (b) show the No.10 point is removed from twin point frame).Secondly, when the threshold K is setting small, few non-isolation twin points could be mistakenly removed in the process of Isolated point retrieving(figure 6(c)).The filtering result with small error as show in figure 6(d).

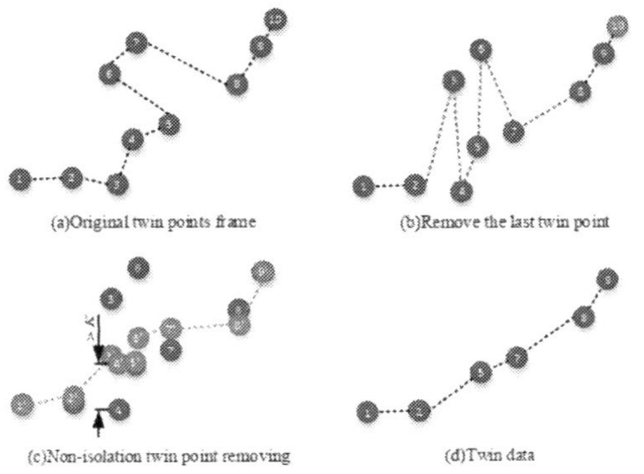

Figure 6. Algorithm Diagram 3.

4. Conclusion

In order to solve the problem of twin data acquisition during the establishment of digital twin stereoscopic warehouse, a high-precision twin point cloud filtering algorithm is proposed in this paper. All the isolated points in the original twin point cloud can be filtered by constructing twin point cloud fitting model. It provides a reliable support for the virtual models and information expansion in the digital twin stereoscopic warehouse system. According to the favorable experiment results, the fast-filtering algorithm has the characteristics of simple implementation and accurate isolated points filtering. It also provides new research ideas for point cloud filtering.

References

[1] Song Y, Shi Y W, Yu J S, Tang D Y, Tao F. 2019 *J. Computer Integrated Manufacturing Systems* **25** 1559-67.
[2] Grieves M and Vickers J. 2017 Digital twin: mitigating unpredictable, undesirable emergent behavior in complex system [J] *Springer International Publishing*.
[3] Gu J X, Yang B S, Dong Z and Yang C H. 2020 *J Bulletin of Surveying and Mapping* **6** 134-140.
[4] Tao F et al. 2019 *J* Computer *integrated manufacturing systems* **25** 1-18.
[5] Huang Z W, Liu F, Hu G W 2017 *J Acta Optica Sinica* **37** 353-362.
[6] Liu B et al 2015 *J Laser & Infrared* **45** 117-122.
[7] Zhou L Q and Su F 2005 *M Fundamentals of Digital Signal Processing*.
[8] Wang D W, Jia R C and Wang H Y 2012 *J Modern Electronics Technique* **35** 71-72.
[9] Zhang J M, Chen F J and Long J L 2021 *J Laser & Optoelectronics Progress* **58** 0610015.
[10] Huang S Y, Liu L M, Dong J and Fu X J 2020 *J Opto-Electronic Engineering* **47** 190688.

Two Entangled Electrons How to Link to Each Other?——What's the Magnetic Force State on Both Sides?

Zhong Guan

Jiangxi Provincial Water Conservancy Planning Design and Research Institute, 330029, Nanchang, China
Email: 908844275@qq.com

Abstract. There discovered the maximum possible magnetic induction in nature, equal to the magnetic induction at the poles of an electron's spin, When the spin magnetic moments of two electrons are close to each other, they act on each other with the maximum possible magnetic induction , and finally entered the maximally entangled state after the energy drops. By this time, the spin magnetic moments on both sides situated in anti-parallel, between them there existed four invisible magnetic circuit, and each magnetic circuit just contain a fluxon. No matter how far the distance between the spins, owing to the inalienability of fluxon, no magnetic flux leakage (coupling degree 100%), so these four magnetic circuit will always existed, maintaining the maximally entangled state system immutably. This is the material basis for the entangled state to be existed, nothing to do with "spooky action at a distance". In this paper, a visual schematic diagram has drawn to describe these, and the magnetic force state, force relationship and "light barrier" problem are analyzed.
Keywords. Qubit experiment, Maximally Entangled State, Maximum Possible Magnetic Induction, Fluxon, Anti-parallel Magnetic Moment, Quantization of Magnetic Moment Direction.

1. Introduction

About the two entangled electrons operated in what states, and what is happening between them exactly (?), quantum mechanics has its own logic. But people cannot help but ask: Are there existed invisible "cables" between a pair of entangled electrons? Can you use visualization graphics to depict the links between them? Is there force relationship between two entangled electrons? Why they separated long distances (many light years, for example) can remain the entangled state? In this paper, we walk on the edge of classical electromagnetism and quantum mechanics, and try to make further exploration on these problems.

2. What's The Entangled State between Two Electrons

The Mr. Bao carries the electron spin σ[1], and the Miss Ai carries the electron spin τ, while they use their own apparatus for their self-carried electron spins σ and τ to do qubit experiment[2] respectively, Whether the two quantum systems are related to each other or not, they can be regarded as being in a same composite system,

Mr.Bao discovers in his experiment that the spin components $\sigma_x, \sigma_y, \sigma_z$[1] he measured, each of which is a random variable, that is: if the first observation is +1, the next one may be -1 (i.e. the opposite), but it may also be +1; The spin components τ_x, τ_y, τ_z from the observatons of Miss Ai, are also the random variables; This is where quantum mechanics is completely different from classical mechanics.

2.1. The Maximally Entangled State[2]

In the initialization phase, the two spins σ and τ are uniformly prepared to be "up" (along the axis $\boldsymbol{Z}$). And then, bring them close to each other, causing the two spin magnetic moments of both σ and τ to interact. Waiting for enough time, they will emit a photon, and then the direction of their spin axes will be transformed into anti-parallel state. At this time, the energy level of the composite system drops to the lowest level, making the composite system into the maximally entangled state, called "singlet state". The state equation of "singlet state" is expressed by the following formula [2]:

The state vector:

$$|\psi>=(|\boldsymbol{ud}>-|\boldsymbol{du}>)/\sqrt{2} \tag{1}$$

Where $\boldsymbol{u}=$ the up normalized basis vector, $\boldsymbol{d}=$ the down normalized basis vector.

The important thing is that once the two spins σ and τ are in the maximally entangled state, the above singlet state equation always holds no matter how far they are distributed (say, many light years). From this equation, the following important relationships can soon be derived [2]:

(1) The mathematical expected value of the components of spin σ : $<\sigma_x>,<\sigma_y>,<\sigma_z>$ are all equal to zero;

(2) The mathematical expected value of the components of spin τ : $<\tau_x>$,$<\tau_y>,<\tau_z>$, are all equal to zero;

(3) The expected value of the product of operators:

$$<\sigma_x\tau_x>=-1,<\sigma_y\tau_y>=-1,\ <\sigma_z\tau_z>=-1.$$

The predictions of quantum mechanics agree perfectly with the experiments. Although the average value of these components is zero across multiple measurements, but there is a fixed correlation between their product each time: $\sigma_x\tau_x=-1,\ \sigma_y\tau_y=-1,\ \sigma_z\tau_z=-1$; That is, If the spin component τ_x measured by Miss Ai is equal to +1, she instantly knows that the far away spin component σ_x of Mr.Bao must be -1, and If she has measured the τ_x is equal to -1, then she immediately knows that σ_x far away is +1; and so on. Between two distant sides, such correlativity is established instantaneously, there no time delay, the correlation coefficient being equal to 1; the correlation is entanglement, and that is the maximally entangled state (i.e., singlet state).

2.2. Different Degrees of Entanglement[2]

The maximally entangled state is only one of the four entangled states in the composite system. The other three states are different degrees of entangled states (called triplet states), and the energy states are relatively high. One of the two subsystems in the composite system, can calculate the density matrix [2] of his (or her) side of the system solely, by all of his (or her) own measurements, so as to judge the state of the whole composite system and the degree of entanglement. See the reference [2] for details. More importantly: When Mr. Bao manipulates the magnetic field to evolve the state of his spin σ, there is no change [2] in the density matrix of Miss Ai (i.e., the entangled state of the composite system remains unchanged), that is called "locality".

2.3. Measure Are Also Entanglement

The measuring apparatus itself is a quantum system[2], which contains special electron spin for measurements; each time a measure started, it creates an entanglement with the measured spin, spitting out a record; like this, the required set of qubits are obtained by successive measurements.

The entire composite system not only includes the measured electron spin on both sides, but also the measuring apparatus (and, deeply, the observer's eye) [2].

So that's an overview, but the main thing that people are interested in is the maximally entangled state. Even though the logic of quantum mechanics is impeccable and its conclusions unchangeable, there have been questions about whether the maximally entangled states involve the action at a distance?

3. How Much Magnetic Flux and Magnetic Induction-- Produced by an Electron Spin

Classical mechanics and quantum mechanics each have their own logic, but there existed universal principles at the margins.

3.1. How Much the Magnetic Flux Produced by Magnetic Moment of an Electron Spin?

According to the view of classical mechanics, matter is infinitely divisible, and can use calculus mathematics to define physical quantities such as moment of momentum and magnetic moment, regardless of its size. According to this view, the spin magnetic moment of an electron should be caused by many (or countless) more subtle loops of current on its surface (or body) around its central axis, and the vector sum of the magnetic moments of these microring currents is just the electron spin magnetic moment $\boldsymbol{m}_e$:

$$m_e = \sum_{i}^{n} S_i \Delta I_i \bullet e_i \tag{2}$$

Where S_i = the area of the $\boldsymbol{i}$ -th ring current in an electron (m^2), ΔI_i = the quantity of the $\boldsymbol{i}$ -th ring current in an electron(A), e_i = The unit vector in the normal direction of S_i , n = The total number of microcurrent loops in an electron (which can tend to be infinite). However, according to the definition of formula (2) to calculate, there will appear a big problem: Substituting in the known value m_e of the magnetic moment of the electron's spin, it is possible to calculate the linear speed at which the charge on the electron's equator rotates about its central axis, much faster than c the speed of light. A lot of people said, "It doesn't make sense.". This will be discussed in the next section.

By reference the calculation method of additional magnetic field, generated by magnetic medium, under the action of external magnetic field [3], as shown In figure 1, in order to obtain the average magnetic field intensity $\boldsymbol{H}_B$ in the space near the electron poles, a group of equivalent circular current with the same magnetic moment ($= \mu_e$) is used to replace all the microcurrent loops in the electron (surface and body). The equivalent circular current is distributed on a cylindrical surface, whose central axis is recombined with the central axis of the electron spin, and the cylindrical surface and the top and bottom surfaces are tangent to the electron surface, just to catch the electron. Next, a rectangle ampere-loop $\overline{\boldsymbol{ABCD}}$ is selected. . One side of the rectangle, $\overline{\boldsymbol{AB}}$, is parallel to(or coincide with) the center line of the cylinder, the other side $\overline{\boldsymbol{CD}}$ is close to the outer surface of the cylinder. The sides $\overline{BC}$ and $\overline{\boldsymbol{DA}}$ of the rectangle, are close to the outer surface of the cylinder's top and bottom. Thus, according to the ampere loop theorem [3]:

$$\Sigma \Delta I_i = \oint_L \boldsymbol{H}_B \bullet d\boldsymbol{L} = \boldsymbol{H}_B \bullet \overline{\boldsymbol{AB}} + 0 = 2r_e \boldsymbol{H}_B ,$$

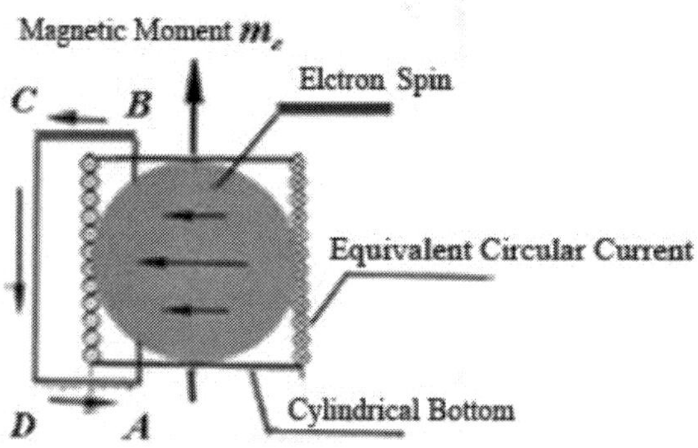

Figure 1. Schematic Diagram of Equivalent Circular Current of Electron Spin.

Multiply both sides by the area of the bottom of the cylinder, πr_e^2, can be obtained the average magnetic field intensity H_B on the top and bottom surfaces of the cylinder, and used to represent the average magnetic field near the poles, caused by the electron spin, as follows:

$$H_B = \frac{\mu_e}{V_B}, (A/m) \tag{3}$$

Where $V_B = 2r_e \bullet \pi r_e^2 =$ volume (m^3) of cylinder surrounded by the equivalent circular current, $r_e =$ classical electron radius (m), $\mu_e =$ magnetic moment of the electron spin.

The above formula(3) is the same form as one formula [3]of classical electromagnetics, which is used to calculate the additional magnetic field intensity generated by the magnetic medium, under the action of the external magnetic field.

Using formula (3) to compute the magnetic flux generated by the magnetic moment of the electron spin, are as follows:

$$\Phi_e = BS_e = \mu_0 H_B S_e = \mu_0 \frac{\mu_e}{V_B} S_e = \frac{1}{2}\mu_0 \frac{\mu_e}{r_e}, (Wb) \tag{4}$$

In the formula, $S_e =$ the area (m^2) of bottom of the cylinder surrounded by equivalent circular currents, $B =$ magnetic induction(Wb/m^2) on area S_e, $\mu_0 =$ vacuum permeability ($4\pi \times 10^{-7}\ H/m$), $\mu_e = 9.2847700 \times 10^{-24} A \bullet m^2$, $r_e = 2.81794092 \times 10^{-15} m$, and after substituting the original data, the magnetic flux generated by magnetic moment of an electron spin can be obtained: $\Phi_e = 2.070232558 \times 10^{-15}\ Wb$.

In the experiment of discovering the second type of superconductor [3], it has been found that the magnetic flux is quantized, and the increase or decrease of the magnetic flux can only be carried out in one part and one part. Each part is called a flux quantum, which is the smallest unit of the magnetic flux and cannot be divided into smaller parts with the magnitude of $\Phi_0 = 2.06783461 \times 10^{-15} Wb$.

People can immediately find that the magnetic flux generated by the electron spin magnetic moment obtained above from classical electromagnetism is very close to a flux quantum $\Phi_e \approx \Phi_0 (= 2.07 \times 10^{-15} Wb)$, and accords well. Therefore, it can be said:

"The magnetic flux produced by the magnetic moment of an electron spin is a fluxon, which cannot be divided into smaller parts."

3.2. Rationality Test of Classical Electromagnetism Electron Spin Model

3.2.1. Do the Charge and Electron Body, Spin Synchronously? A classical model [3] of an electron is a uniformly charged spherical shell with a radium of $\boldsymbol{R}$, and a charge of e, when it rotates about an axis through the center with an angular velocity of ω_e, how can we deduce the expression of its magnetic moment?

According to the idea of formula (2), the spherical coordinates are established on the surface of the spherical shell, assuming that $\varphi =$ the longitude on the sphere, and $\theta =$ the latitude on the sphere; It is very easy to obtain the total magnetic moment of the electron:

$$m_e = 2\int_0^{\pi/2} S_\theta di_\theta = 2\int_0^{\pi/2} \frac{e}{4} R^2 \omega_e \cos^3\theta d\theta = -\frac{1}{3} e\omega_e R^2 \tag{5}$$

Where $\boldsymbol{S}_e =$ the latitudinal circle area (circular current area) at the latitude to be θ, and $\boldsymbol{di}_e =$ the circular current on the microring domain of $\boldsymbol{d}\theta$ at the latitude to be θ. Substituting the original data: $m_e = \mu_e$, $R = r_e$, $e = 1.60217733\times10^{-19}(C)$, the angular velocity of the charge rotating around the central axis of the elctron can be found:

$\omega_e = 3\,\mu_e / e r_e^{\,2} = 2.189364416\times10^{25}(1/s)$ (amazing!)

At the equater of the electron, the linear speed at which the charge rotates:

$\boldsymbol{v}_e = \omega_e r_e = 6.169499577\times10^{10}(m/s) = \boldsymbol{205.9c}$ ($\boldsymbol{c} =$ speed of light).

Some people think that the above "faster-than-light" calculation results, contrary to the special relativity, is unacceptable, and judge the classical electromagnetic electron spin model is incorrect and undesirable.

However, there is another way to think about it. The "Law of conservation of charge" is one of the most basic conservation laws of nature that man has discovered. According to this law: "The value of the charge does not change with the speed of motion, independent of the reference system.".

That is, the electric charge can move at any speed (Such as faster than light) without changing its value; but the charge carrier (the electron body), as it moves near the speed of light, its inertial mass approaches infinity! It can be seen that the above ω_e and $\boldsymbol{v}_e (= 205.9c)$, being the rotated angular velocity and the linear speed on the equater, which should be generated by the charge of electron, and cannot be produced by the carrier(electron body), this two things should be distinguished. Then, inside the electron the motion of the charge and the carrier should out of sync, the speed of the charge is greater than the carrier. The intrinsic spin magnetic moment of the electron should be caused by the high speed spin motion of the charge inside the electron rather than the contribution of the charge carrier. In this way, the above contradiction no longer exists.

3.2.2. Electric Charge May Be A Separate Subject. A hypothesis should be tenable if it leads to results consistent with reality. In the preceding section, based on the electron spin model of classical electromagnetism, it is concluded that the "the spin magnetic flux of a electron is a magnetic flxon", $\Phi_e \approx \Phi_0$, which is in good agreement and should not be an accident coinsidence. It can be seen that the electron spin model of classical electromagnetism is still desirable.

In this case, we should adopt the analysis in the previous section and regard the charge inside the electron and the carrier as two different things. Charge should be existed as a separate subject with independent basic properties distinct from its carrier, rather than just a property of the elementary particle. The basic properties of charge, besides coulumb force and electromagnetic field source, include: charge conservation, no inertial mass, can be divided into smaller parts, can do asynchronous spin motion in the carrier, etc.

3.3. The Maximum Possible Magnetic Induction in Nature Bmp

The magnetic induction at the surface of an atomic nucleus is about $10^{12}\boldsymbol{T(Wb/m^2)}$ [3], and that at the surface of a neutron star is about $10^8\boldsymbol{T}$ [3], both of which are far beyond what can be achieved in artificial laboratories [3] (such as the Tokamak fusion machine or the European LHC Hadron Collider), but they are still not the upper limits of magnetic induction in nature. Please look at the magnetic induction near the two poles of electron spin, which can be calculated according the previous (3) and (4):

$$\boldsymbol{B} = \mu_0 H_B = \mu_0\, \mu_e / 2r_e \pi r_e^2 = \boldsymbol{8.2986126\times 10^{13}\ Wb/m^2}$$

And according to "the magnetic flux of an electron spin is exactly one fluxon", it can also be calculated according to the magnetic flux on unit area, $\Phi_0/\boldsymbol{S_e}$:

$$\boldsymbol{B} = \Phi_{.0}/\boldsymbol{S_e} = \Phi_0/\pi r_e^2 = 8.289004\times 10^{13}\ \boldsymbol{Wb/m^2}$$

The above two results are basically consistent, indicating that the magnetic induction near the two poles of the electron spin is about one order of magnitude greater than the magnetic induction on the nuclear surface.

In the experiment of discovering the second kind of superconductor [3], it has been proved that magnetic flux quantized, and each fluxon has its definite value [3] $\Phi_0(=h/2e,\ h = Planck's\ Constant)$ contained in a small single magnetic circuit; when the magnetic field is enhanced, it can only increase the number of these small single magnetic circuits, and cannot increase the magnetic flux Φ_0 of each magnetic circuit. This should be suitable for every type of magnetic field. Therefore, the upper limit of magnetic induction in nature, should be the magnetic induction on a small cross-section only containing one flux quantum, which is equivalent to the magnetic induction near the two poles of electron spin, which can be called the maximum possible magnetic induction B_{mp} :

$$B_{mp} = \Phi_0/S_e = h/2eS_e \qquad (6)$$

Where e = electron charge, $\boldsymbol{Se}$ = equatorial croos-section area of the electron, same as before. The recommend value for B_{mp} is $8.289\times 10^{13}\boldsymbol{Wb/m^2}$.

4. The Magnetic Linking Relationship between Two Entangled Electrons

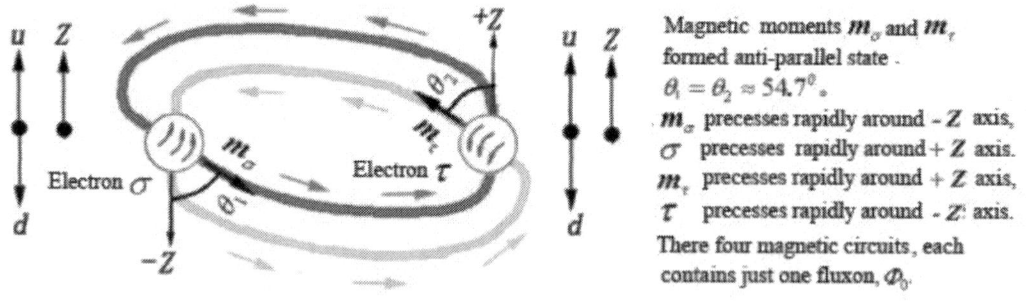

Figure 2. Schematic Diagram of Magnetic Linking---Maximally Entangled State (Singlet State).

It is generally accepted in academia that quantum mechanical processes cannot be represented or described by visual graphics, and all such attempts ended in failure. However, it seems that not everything is as it should be.

As shown in the diagram, the composite system has fallen to the lowest energy state, i.e.,the maximally entangled state, singlet state. The spin magnetic moment $\boldsymbol{m}_{\sigma}$ of electron σ and the spin magnetic moment $\boldsymbol{m}_{\tau}$ of electron τ are in anti-parallel state. Each other acts on the other side with its generated B_{mp} of "possible maximum magnetic induction in nature", as the external magnetic field for the other side; and there created two ring magnetic circuits (or four branch magnetic circuits) respectively, represented by red and green lines in the figure; and the magnetic flux in each magnetic circuit is exactly equal to one flux quantum Φ_0.

The spin magnetic moment $\boldsymbol{m}_{\sigma}$ processes at a high speed around the - Z axis (external magnetic field) direction, with its spatial angle to be the θ_1, and its spatial average direction to be - Z axis. Since the electron has a negative charge, so its spin axis points in the oppsite direction, with gyroscopic precession around the +Z axis,and its spatial average direction pointing in the +Z axis, or **u** direction(up).

Similary, the spin magnetic moment $\boldsymbol{m}_{\tau}$ precesses at a high speed around $+\boldsymbol{Z}$ axis (external magnetic field) direction, with a spatial angle of θ_2, and its spatial average pointing in the $+\boldsymbol{Z}$ axis direction; therefore the τ spin axis rotates around the $-\boldsymbol{Z}$ axis in a gyroscopic precession, and its spatial mean orientation recombines with the $-\boldsymbol{Z}$ axis, that is, the **d** direction(down).

According to the Stern-Gerlach experiment[4][5], under the action of the external magnetic field, the spatial orientation of the magnetic moment of the electron is quantized, and there are only two possible values of the included Angle $\theta_1\ (or\theta_2)$ to be 54.7^0 **or** 125.3^0; if $\theta_1 and \theta_2$ are taken as arrays, there should be 4 possible permutations, where $\theta_1 = \theta_2 = 54.7^0$ corresponding to the lowest sum of the magnetic potential energy of the system, as shown in the figure above. The composite system is in the lowest energy state, being the maximally entangled state, or the singlet entangled state, the state vector $|\psi\rangle = (|\boldsymbol{ud}\rangle - |\boldsymbol{du}\rangle)/\sqrt{2}$; the other three arrangement have higher energies and being partially entangled states.

5. Magnetic Force State And Magnetic Circuit Properties

5.1. Magnetic Induction of External Magnetic Field

Electron spin σ and τ provide the"external magnetic field" to each other, passing through the other's body. Its magnetic induction $B = B_{mp}$, to be the maximum possible induction, according to formula (6), the calculated value used is: $8.289004\times10^{-13}\boldsymbol{Wb/m}^2$.

5.2. Resultant Force of Magnetic Forces

According to classical electromagnetism, in a uniform external magnetic field, the resultant force of the magnetic forces on the closed current is equal to zero [3]. Therefore, acting on the electron spin σ or τ, the resultant force of the magnetic forces under the external magnetic field, is equal to zero.

5.3. Magnetic Torque

(1)The magnetic torque on the electron spin σ (or τ) in the external magnetic field, can be calculated by using the formula of the magnetic torque on the current-carrying coil in the uniform magnetic field [3]:

$$M = m\times B \tag{7}$$

In the formula $\boldsymbol{m}$ = magnetic moment current-carrying coil. By substituting $\boldsymbol{m}=\boldsymbol{m}_\sigma$ (or $\boldsymbol{m}_\tau$), $B=B_{mp}$, the magnetic torque acting on the electron spin σ (or τ) can be obtained.

As shown in figure 2, When the Angles between the magnetic moment $\boldsymbol{m}_\sigma$, m_τ, and their respective external magnetic field directions, $\theta_1=\theta_2=54.7^0$, the composite system is in the maximlly entangled state, at this time: $\boldsymbol{M}=\mu_e BSin54.7^0=6{,}281\times10^{-10}\,N\bullet m$; it forces the two electrons to rotate around their respective external magnetic fields B_{mp} in a gyroscopic precession.

(2)The magnetic torques above produce two super-strong magnetic couples consisted of two pairs of magnetic forces, whose directions respectively parallel to the axes of the two spin's magnetic moments $\boldsymbol{m}_\sigma$, m_τ. Since the two spin's magnetic moments are in the antiparallel state, the two super-strong magnetic couples are also in an antiparallel state. No matter how far apart they are, the corresponding forces between them are always "equal in size and opposite in direction", and the magnitude of forces remain unchanged, but the directions are generally not in the same straight line. Moreover, the existence of each of the two magnetic couples depends on the external magnetic field Bmp provided by the other side.

5.4. Magnetic Potential Energy
The calculation method of magnetic potential energy of current-carrying coil in uniform magnetic field can be adopted [3]:

$$W_m=-\boldsymbol{m}\bullet\boldsymbol{B} \tag{8}$$

As shown in Figure2, when the Angles between the spin magnetic moments of two electrons and their external magnetic fields, $\theta_1=\theta_2=54.7^0$, the magnetic potential energy of electron σ (or τ) is $\boldsymbol{W_m}=\mu_e\boldsymbol{BCos54.7^0}=-\boldsymbol{4.447\times10^{-10}J}$; the sum of the magnetic potential energy of the two electron spins in the composite system is $\boldsymbol{W}_{m\Sigma}=-8.894\times10^{-10}\boldsymbol{J}$, corresponding to the lowest energy state of the system, being the maximally entangled state, singlet state.

5.5. Magnetic Circuit Properties
(1) In each magnetic circuit of the single-fluxon above , the Kirchhoff's first theorem[6], namely the continuity principle of magnetic circuit, still holds: $divB=0$; but Kirchhoff's second theorem, or magnetic Ohm's Law[6], based on linear model, no longer holds.

(2) Macroscopic Properties. Since the scale of the electron is extremely small, if a small area perpendicular to the direction of the magnetic circuit is made at any point along the path of the above magnetic circuits, from the macroscopic point of view, it can be considered that all four magnetic pass through here, and the algebraic sum of the four round-trip magnetic fluxons at the point is zero. Therefore, the macroscopic magnetic field intensity H = 0. In the other words: on the magnetic path between two entangled electrons, the macroscopic magnetic field intensity is zero everywhere, and the magnetic potential difference[6] between any two points is zero.

6. Brief Summaries
The basic experiments in this paper are qubit experiment [2], Stern-Gerlach experiment [4] [5] and type II superconductor experiment [3], and the classical electromagnetic model [3] of electron spin is not abandoned. The main results are as follows:

(1)There found the "maximum possible magnetic induction " in nature, B_{mp}, equal to the magnetic induction on the cross-section of a magnetic circuit containing single fluxon, and equivalent to the magnetic induction at the two poles of an electron's spin, with about one order of magnitude greater than the magnetic induction on the nuclear surface.

When two electron's spins are close to each other, each acts on the other with the " maximum possible magnetic induction" generated by itself, and after energy drops(by emitting a photon or other means), formed the entangled state between them.

(2) Between the two electron spins which are already entered the maximally entangled state, there existed four invisible magnetic circuits, and the magnetic flux of each circuit is just one fluxon Φ_0; and no matter how far the distance between them is elongated, based on the inseparability of magnetic fluxon, there is no magnetic leakage (magnetic loss is zero, coupling degree of both sides is 100%), the four magnetic circuits will always exist and maintain the maximally entanglement state of the system unchanged.

The spin magnetic moments (anti-parallel) of two sides, make gyroscopic precession around the magnetic field B_{mp} provided by the other side, driving the four magnetic circuits to follow and created a crosslink network diagram for the maximally entangled state (dynamic). This is the material basis for the existence of singlet entanglement.

(3)The statistical results of the qubit experiments above, just represent the quantum properties of the system, where these key expressions" $\sigma_x\tau_x = -1$, $\sigma_y\tau_y = -1$, $\sigma_z\tau_z = -1$ " representing the correlativity of the maximum entangled state, also belongs to the characteristics of the quantum system(that is, the statistic information of product between two spin's components), and never mean there the "signal energy" transfered from one side to the other side of the singlet magnetic chain network. Since it does not involve the transfer of energy, it should not be restricted by the [7] "light barrier". As a result, these two distant electron spins interact instantaneously, with no time interval, according to their inherent correlation.

(4) In the composite system motioned above, the resultant force of the magnetic forces acting on each electron's spin is equal to zero; as the distance lengthens, the Coulomb force to be ignored. Under the action of external magnetic field $\boldsymbol{B}_{pm}$, the magnetic torque on each electron spin reaches $6.281\times10^{-10}\,\boldsymbol{N}\bullet\boldsymbol{m}$, forming two super-strong magnetic couples; and their directions of the magnetic forces are parallel to the axes of the respective spin magnetic moments. The force directions of the two magnetic couples also formed the anti-parallel state. No matter how far apart they are, the corresponding forces between them are always "equal in size and opposite in direction" with the same magnitude value, but the directions are generally not on a same straight line. The existence of each force pair is inseparable from the external magnetic field $\boldsymbol{B}_{pm}$ provided by the other side.

References

[1] Dirac P A M. 2021 The Principles of Mechanics (in Chinese Translated by Ling D B), First Edition Beijing: *China Machine Press* 146-147.

[2] Leonard Susskind 2021 Stanford University Open Class (Video), Quantum Mechanics (All 10 Lectures): Lecture 6 Entangled State, Lecture 7 Density Matrix. https://www.bilibili.com/video/av53480408/.

[3] Zhang S H. 1999 Tsinghua University Text books (in Chinese): University Physics (Third Book), Electromagnetism, Second Edition Beijing: *Tsinghua University Press* 206-212, 226, 255-258,293,359-361.

[4] Friedrich B and Herschbach D. 2003 Stern and Gerlach: How a Bad Cigar Helped Reorient Atomic Physics [J] *Physics Today* **56** (12) 53-59.

[5] Cheng S Z and Jiang Z Y. 1979 Reference Books for Higher Education (in Chinese): General Physics (Third Book), Second Edition Beijing: *People's Education Press* 108-111.

[6] Shen H, Hu Z G etc. 1987 Electrical Engineer Hand Book (in Chinese), part 2: Fundamental Electric Engineering, First Edition Beijing: *China Machine Press* 5 23 88.

[7] Einstein A. 1979 The Meaning of Relativity (in Chinese Translated by Lee H) *Beijing: Science Press* 16-22.

Molecular Dynamics Simulations of Structural Changes for a Molten $Ag_{54}Cu_1$ Cluster during Cooling

Zhijing Zhang

York Mills Collegiate Institute, ON, M3B 1W6, Canada
Email: sally.zhijing@gmail.com

Abstract. Structural changes of an $Ag_{54}Cu_1$ cluster had been computationally studied by molecular dynamics approaches. Packing transition was demonstrated by analytical tools including potential energy, atomic density profiles, and shape factor as well as visually packing images. During the process of temperature decreasing, this cluster preferentially assumes icosahedral geometry. Copper atom usually has an atomic position inside a cluster. As temperature decreases, its position will change. Potential energy shows different temperature regimes in the structural transformation. Atomic density profile gives packing pattern in different region. Shape factor presents the morphology changes of this cluster.

1. Introduction

Nanometallic clusters can be used in many different fields such as optics, magnetism and catalysis [1]. This is because they have a lot of varieties in their morphologies, and this can directly leads to the result that they have a high degree of tunability. These are so interesting which made it attracts many researchers study on it. On the research of these nanometallic clusters, researchers had used different techniques in both theories and experiments for the structures of binary alloy clusters. Those researches show that as the size of those clusters changes, their structures and properties have significant differences [2, 3]. Especially when the atomic number included in one cluster is less than 100, because the ratio of surface/volume is large, their structures show greater diversities. In those small sized clusters, the clusters that included magic number are most attractive, for example, clusters made of 55 atoms. About those magic clusters, when they are doped with an element from other group, they will show some new properties. The problem of how to get small sized alloy clusters with specific structure is becoming a significant issue on producing new nanomaterials. Copper has the same $d^{10}s^1$ valence electron structure with silver, and because of that, copper is usually used as a doping element for silver cluster. Copper and silver alloy clusters can be used widely to improve the effectiveness of reaction, and the selectivity of catalyst [4-10]. The structural transformation in binary cluster of copper and silver containing magic number of 55 atoms is hard to be observed in laboratory. Computational approaches become the most efficient way to do the research on structural transformation during the process of cooling, where molecular dynamics (MD) is a way that suit the packing structure and atomic movements in those small sized clusters. The MD simulations have been used widely in phase transformation in metal and alloy [11-14].

In this work, MD simulations are used for studying structural transformation of a molten $Ag_{54}Cu_1$ cluster during cooling process. Potential energy, atomic density profile, shape factor and visualized parking structures describes the structure and morphology of this cluster at different temperature.

2. Model and Methods

The interaction among Ag and Cu atoms is described by the EAM form, which was proposed by Williams [15]. Potential energy E_{tot} is being determined by the following equation

$$E_{tot} = \sum_{i} F_i(\rho_e) + \frac{1}{2}\sum_{\substack{i,j \\ i\neq j}} \phi_{ij}(r_{ij}) \tag{1}$$

where $\phi_{ij}(r_{ij})$ is a two-body central potential between atoms i and j with a separation of r_{ij} in between, and $F_i(\rho_e)$ is the embedding energy of atom i with the electron density ρ_e. The density value comes from the superposition and also the sum of the electron density from the nearest neighbor atoms of the atom i. ρ_e is the electron density of the neighbor atom j of the atom i. The simulations were done with the NVT ensemble using Andersen thermostat. Through the process of solving Newton's motion equations, we were able to get the positions and velocities of each atom. Also, we used a predictor-corrector algorithm to integrate the equations of motion. Throughout the simulations, a time step of 1.6×10^{-15} s is used. To accumulate statistics, before running, the system was first equilibrated in 485 000 time steps. Also the atomic trajectories and energy recorded in the subsequent 5 000 time steps were used to get the thermodynamic equilibrium values. Firstly, we constructed a $32.7\times 32.7\times 32.7$ nm^3 bulk faced center cubic (FCC) Ag crystal (the lattice constant a_0 is 4.09 Å), 55 atoms are extracted from this constructed Ag crystal. To make the box size of the simulated central cell large enough to avoid the interaction of the atoms in the central cell with the other atoms in its 26 neighbor imaging cells under periodic boundary conditions, the size of the MD simulation cell is $32.7\times 32.7\times 32.7$ nm^3. This cluster was subjected to structural relaxation at 1235K to obtain a molten state, and one silver atom was replaced with one copper atom. Then it decreased to 1000K, and decreased 50K each time until it reaches 300K.

The following values were determined in the simulations.

$$\rho(r_i) = \langle N_i \rangle / N, \tag{2}$$

where < > denotes the average over the entire trajectory, the atomic density profile $\rho(r_i)$ is calculated by dividing the system into some layers along the r direction from the cluster center and accumulating a histogram of the number, N_i of atoms in each layer, where $\langle N_i \rangle$ is the average number of atoms in layer i.

The moment of inertia can reflect atomic positions and mass distribution. As a tensor, it can be determined by:

$$I_{ij} = \sum_{iatom=1}^{N} m_{Cu/Ag}^{i}\left(x_{iatom}^{i} - x_c\right)\left(x_{iatom}^{i} - x_c\right) \quad (i \text{ or } j = 1,2,3) \tag{3}$$

$$x_c = \sum_{iatom=1}^{N} m_{Cu/Ag}^{i} \times x_{iatom}^{i} / \sum_{iatom=1}^{N} m_{\frac{Cu}{Ag}}^{i} \quad (i = 1,2,3) \tag{4}$$

where $m_{Cu/Ag}$ is the mass of Cu or Ag atom, x_c is the mass center, x^i or x^j is the coordinate of the *iatom*. i (or j) is 1, 2, 3 respectively corresponds to the x, y and z axes. I^1, I^2 and I^3 are three values of principle axes that are obtained by diagonalization of tensor component, where I^1 and I^3 stands for the maximum and minimum valued. The shape factor can be defined as:

$$F_{shape} = I^1 / I^3 \tag{5}$$

3. Results and Discussions

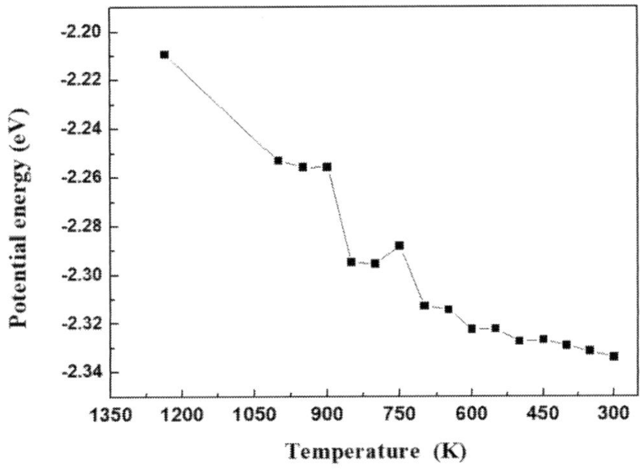

Figure 1. The potential energy change as temperature decreasing.

From the figure 1, it is clear that as temperature is decreased from 1235K to 300K, the potential energy has shown a pattern of decreasing. From 1235K to 1000K, although the potential energy decreased greatly, the atoms are packed disorderly at the two temperatures. The change of potential energy comes from the obvious change of cluster shape, where the ellipsoid shaped cluster becomes more round. After that, the decrease speed is slowed down until it reaches about 950K. From there, the energy is increased with a very small amplitude. Then there comes the sharp decline in the graph. If we look at the picture of this cluster, we can clearly see that at 850K, it becomes an almost perfect icosahedron. After that, the decreasing in potential energy begins to slow down, and there comes a short increase. In meanwhile, this cluster has changed a bit. Then, the potential energy begins to decrease slowly with only two short increases until the temperature finally reaches 300K.

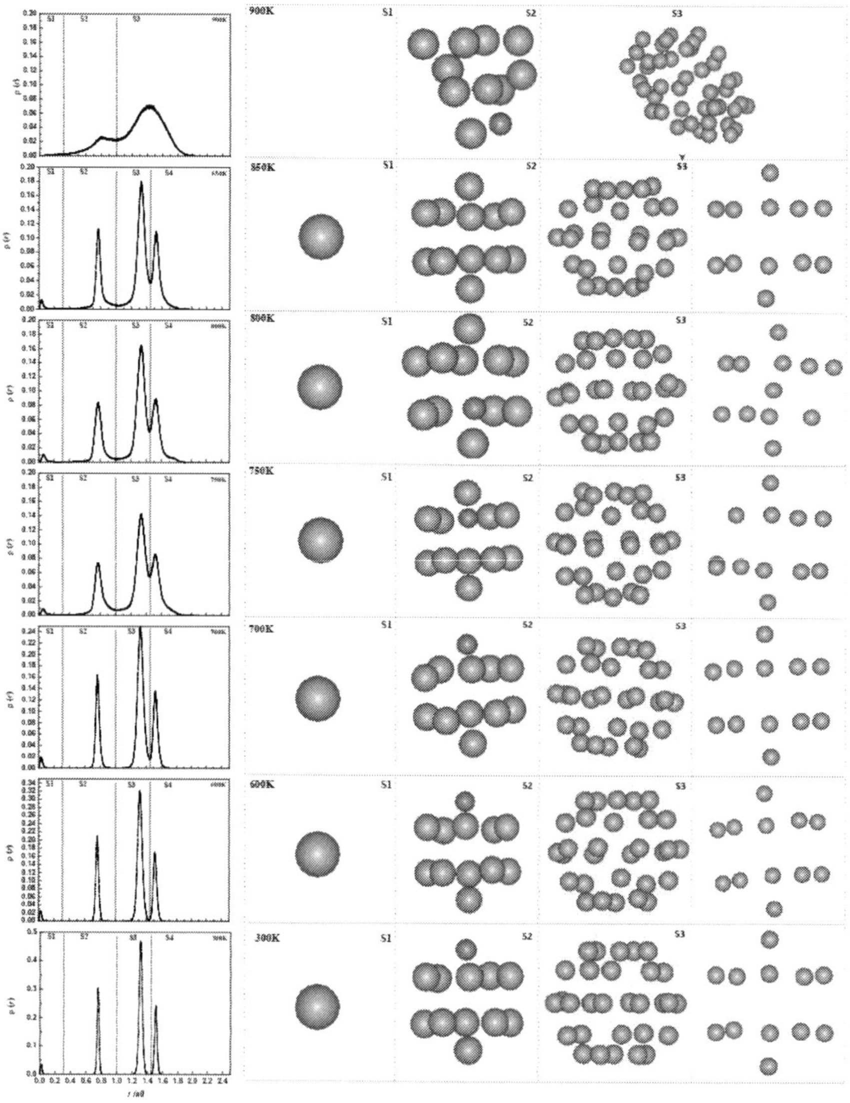

Figure 2. Graph on the left shows the atomic densities of the cluster, to make things more clear, we divide the graph into four shells. The images on the right describe the amount of atoms and the structure in each shell.

As shown in figure 2, when the temperature is 900K, the graph has two broaden peaks, where the first shell in 900K is empty, with no atom there. In the second shell, there are 11 atoms there, which is included a copper atom in it, and the other atoms in the outer part of this cluster. At 850K, there are four distinct peaks, and the peaks become narrower. It can be noted there is a small peck in the center of this cluster, implying the first shell is not empty anymore, instead, one silver atom shift to the first shell. There's also a change happens in the second shell compared with that at 900K, the copper left the second shell, and it shift to the third shell. Now, there are 12 atoms in the second shell and 30 atoms in the third shell. In the fourth shell, there are 12 atoms. As the temperature is decreased to 800K, although there are still 12 atoms in the four shell, one silver atom leaves the lattice position of icosahedron at 850K. In addition, the copper atom shifts back to the second shell. At 750K, compare to the graph above, the third and fourth peak seems to be closer together. The first shell still has 1 atom, 12 atoms in the second shell, 31 in the third shell, and 11 in the fourth shell. Correspondingly, there is a jump of potential energy as illustrated in figure 1. As the temperature decreases further, the peaks of the density profiles become higher and thinner, where the third and fourth peak seems more separate from one another. The atomic packing in the different shells shows that the first shell still has 1 atom, 12 atoms in the second shell, 30 in the third shell, and 12 in the fourth shell. Here, there atoms are at their icosahedral positions.

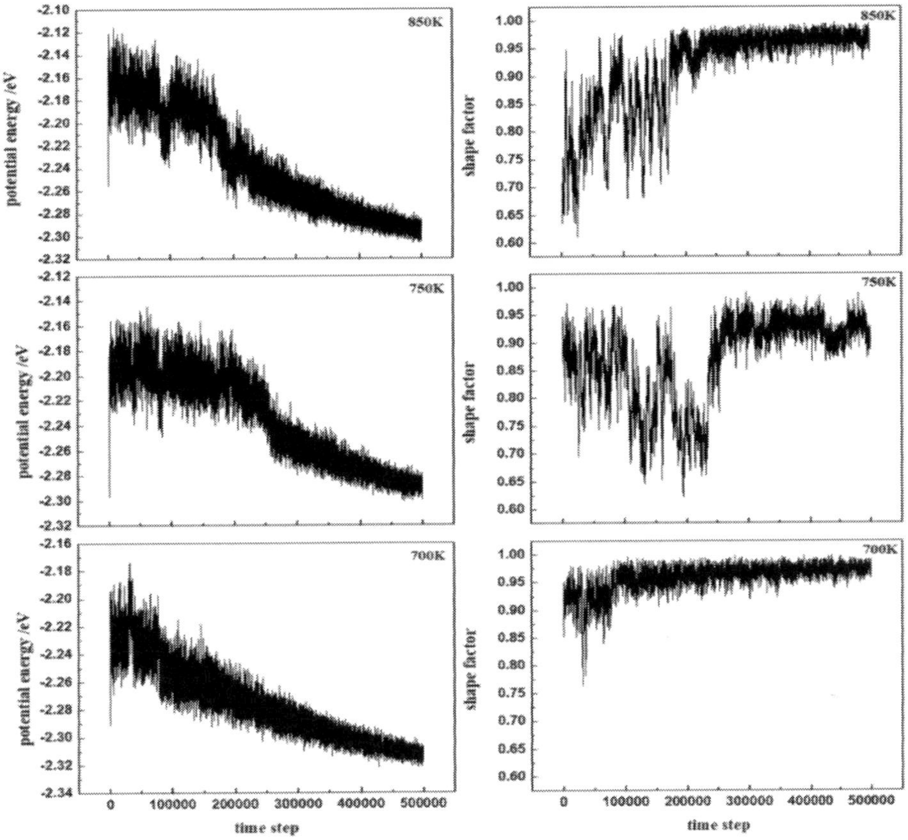

Figure 3. Graphs on the left show how potential energy changes as time step increases. Graphs on the left show shape factor changes as time step increases.

Figure 3 illustrates variations of the potential energy and shape factor with the time steps at 800K, 750K, and 700K. As illustrated in this graph, there is a significant change about 200 000 steps at 800K, suggesting that the structural transition occurs from molten state to icosahedral packing. In the meantime, the cluster turns into a shape that looks likely to a ball. Then the shape of this cluster stops to shift in a wide range. At 750K, the shape of this cluster shows significant changes. At first, this cluster presents ball shape, and then its shape is really different to a ball. After 300 000 time steps, its shape restores to the ball shape. At 700K, it is clear that the shape of the cluster is becoming more stable. Only at the beginning, the shape changes apparently. Correspondingly, potential energy fluctuates before 100 000 time steps.

4. Conclusion

In this paper, MD simulations were performed to study the structural changes during cooling the molten $Ag_{54}Cu_1$ cluster. Structural changes came from interchanging movements among the atoms, which significantly affected potential energy to change accordingly. The present simulations visually showed the changes with the crystallization from the packing patterns in different shells obtained from atomic density profiles. The transformation of structure of this cluster happened within a temperature range. After transformed from molten state to icosahedron, there were still changes in positions of the atoms inside the cluster. At the same time, the shape of this cluster also had a significant change. This type of clusters' structures usually significantly changed under high temperature. When the molten cluster was at high temperatures, the copper atom located inside the cluster. As the temperature decreased, the copper atom would begin to shift out the cluster. Keeping on the decrease of the temperature, the copper atom moved to the inside of this cluster.

Acknowledgments
The author thanks Dr. Lin Zhang provides instruction and useful discussions.

References
[1] Muñozflores B M, Kharisov B I, Jiménezpérez V M , et al. 2011 Recent Advances in the Synthesis and Main Applications of Metallic Nanoalloys[J] *Industrial & Engineering Chemistry Research* **50** (13) 7705-7721.
[2] Chen H J, Li Z W, Zhao Y B, 2004 Progress on the preparation of nanosized alloy materials[J] *Progress in Chemistry* **16** (5) 682-686.
[3] Ferrando R, Jellinek J, Johnston R L, 2008 Nanoalloys:From theory to applications of alloy clusters and nanoparticles[J] *Chemical Review* **108** (3) 845-910.
[4] Delogu F, Arca E, Mulas G, 2008 Numerical investigation of the stability of Ag-Cu nanorods and nanowires[J] *Physical Review B* **78** (2) 024103.
[5] Oezaslan M, Hasch´e F Strasser P, 2011 In situ observation of bimetallic alloy nanoparticle formation and growth using high-temperature XRD[J] *Chemistry of Materials* **23** (8) 2159-2165.
[6] Sopousek J, Pinkas J, Broz P, 2014 Ag-Cu colloid synthesis:Bimetallic nanoparticle characterisation and thermal treatment[J] *Journal of Nanomaterials 2014* 1-13.
[7] Aguado A, López J M, 2011 Identifying structural and energetic trends in isovalent core-shell nanoalloys as a function of composition and size mismatch[J] *Journal of Chemical Physics* **135** (13) 134305.
[8] Laasonen K, Panizon E, Bochicchio D, 2013 Competition between icosahedral motifs in AgCu, AgNi, and AgCo nanoalloys: A combined atomistic DFT study[J] *Journal of Physical Chemistry. C* **117** (49) 26405-26413.
[9] Langlois C, Wang Z W, Pearmain D, 2010 HAAD-STEM imaging of CuAg core-shell nanoparticles[J] *Journal of Physics Conference Series* **241** 012043.
[10] Liu J H, Zhang L, 2020 Strain-induced packing transition of Ih Cun@Ag55-n (n=0,1,13,43) clusters from atomic simulations[J] *Mathematical Biosciences and Engineering* **17** (6) 6390-6400.
[11] Zhang L, Xu S N, Zhang C B, Qi Y, 2009 Molecular dynamics investigation of structural changes accompanying with freezing a molten Cu135 cluster on cooling[J] *Computational Materials Science* **47** (1) 162-167.
[12] Zhang L, Zhang C B, Qi Y, 2008 Local structure changes of 54-, 55-, 56-atom copper clusters on heating[J] *Physics Letter A* **372** (16) 2874-2880.
[13] Zhang L, Zhang C B, Qi Y. 2009 Molecular-dynamics investigation of structural transformation of a Cu201 cluster in its melting process[J] *Physica B-Condensed Matter* **404** (2) 205-209.
[14] Zhang L, Fan Q N. 2011 Molecular dynamics study showing the effect of quenching temperature on structural changes of a molten Cu411 cluster[J] *Physica Scripta* **84** (4) 045303.
[15] Williams P L, Mishin Y, Hamilton J C. 2006 An embedded-atom potential for the Cu-Ag system[J] *Modelling and Simulation in Materials Science and Engineering* **14** (5) 817-833.

Journal of Physics: Conference Series **2148** (2022) 012006 doi:10.1088/1742-6596/2148/1/012006

Research on Multi-scene and Multi-station Fusion Mode Based on Layered Theory of Energy Internet

Xiaoliang Yang, Honggang Zhao and Hongyuan Ren

Beijing Jion Bright Digital Power Technology Co., LTD., Beijing, 100085, China
Email: yxlallok@126.com

Abstract. Multi-station integration is a new form of business in the development of energy Internet and a brand new practice of power iot. It is an innovative form that strengthens the interaction between charge and storage of source network on both sides of energy supply and demand, and enhances the flexibility of power grid by taking substation, energy storage station, distributed energy station and other resources as the core. However, there is not a set of construction mode that can guide implementation, copy and popularize. Therefore, guided by the existing construction practice of multi-station fusion in China and combined with multi-user scenarios, this paper studies the multi-station fusion mode, proposes a multi-station fusion planning system based on EIST theory, gives the fusion mode under different business scenarios, and synchronously constructs a new ecological business chain with multi-station fusion as the core. It aims to make full use of the innate advantages of substation in energy flow convergence, realize the "integration of energy flow, business flow and data flow", comprehensively support the transformation of digital power grid, and practice the national development strategy of "digital economy" and "digital China".

1. Introduction

Substation is the power transmission hub, is an important place in the power system for voltage conversion, energy concentration and distribution, energy flow control and voltage adjustment, plays a role of link between power plants and users. At the same time, in the context of the development of energy Internet, substation can also serve as an important information node and relay station for energy conversion between users and power grid, which transforms the traditional single source-load mode into a friendly interaction mode of "source-load mode and load-load mode".

"Multi-station integration" is an important content of implementing the national "new infrastructure" development strategy. The so-called multi-station integration means to explore and utilize densely distributed substation resources to construct and operate data center station, energy storage station, distributed energy station, 5G base station, Beidou ground enhancement station, etc., which can support smart grid business internally, assist new infrastructure externally, and promote economic and social development [1-3].

"Multi-station" refers to the infrastructure and system platforms related to information communication and energy environment, including 5G communication base station, Beidou Foundation Reinforcement Station, distributed new energy power station [4-7] and environmental monitoring station, in addition to the traditional three stations [5-8]. "Convergence" means to realize the integration of energy, information communication, government affairs and other fields through the construction of "multiple stations", and realize the coordination of power grid enterprises, communication operators, and relevant government commissions and bureaus as service subjects.

These guidelines, written in the style of a submission to *Material Science and Engineering: Conf. Ser.*, show the best layout for your paper using Microsoft Word. If you don't wish to use the Word template provided, please use the following page setup measurements.

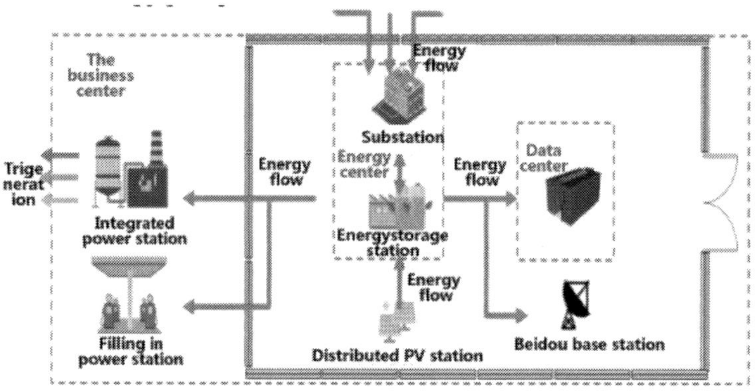

Figure 1. Schematic diagram of multi-site fusion energy flow.

In this paper, 110kV substation will be taken as the main research object. According to the construction scale of substation, user energy characteristics, urban planning land nature and other factors, different fusion scenarios will be set up to carry out multi-station fusion research in multi-scenes, and the "N+N" multi-mode multi-station fusion configuration scheme will be given. It will lay a foundation for the whole network to realize the construction and implementation of the source-network-load interactive system, and help the grid transform to an intelligent, digital, clean and efficient new power system.

2. Analysis of the Current Situation of Multi-Station Integration Construction in China

Since early 2019, the state grid company should explore the use of substation resources construction operating charge for electricity (storage) station and data center mode, namely "multistation fusion" concept, after the mail tunnels industry group was established "multistation fusion" working group, responsible for coordinating the internal unit from technology, construction, operation and business model, investment and financing to carry out in-depth study, With the goal of completing the top-level scheme design of "multi-station integration", selecting and constructing pilot projects and putting them into operation, pooling resources and cooperating. At present, the construction situation and functions of "multi-station integration" demonstration projects carried out by power grid companies of some provinces and cities in China are as follows:

2.1. State Grid Gansu Power Supply Company

On April 23, 2020, its power supply company and China telecom company, gansu province, gansu province lanzhou 110 kv sand flat variable data center formally completed delivery, through to the substation reusable site resources utilization, the redevelopment of the old warehouse 632 square meters, the production of a total of 172 face 7 kw rack, is within its system first outdoor type of large and medium-sized multistation data fusion center.

Built the project put into operation, it is its power in gansu "5 g + grid innovation application" to carry out actively explore an important step, for local new infrastructure development in gansu province has provided a reference, promotion of ideas, for new infrastructure build cross-industry enterprise cooperation pattern model, realize the complementary advantages and win-win cooperation between enterprises, It is of great significance to support local economic and social development.

2.2. State Grid Hefei Power Supply Company

On April 28, 2020, the two data centers of Zhonghai Open and close Office and Yulong Road Substation of State Grid Hefei Power Supply Company in Binhu New District were officially put into operation. By not changing the original overall layout, solar photovoltaic panels were installed in the roof space of the station to build an 88 kW PHOTOVOLTAIC power station. 16 sets of DC quick charging piles and 8 sets of AC charging piles are equipped to meet the needs of vehicles in the economic and economic zone of Hefei. At the same time, the original xinlu ev charging station and newly allocated batteries will be comprehensively utilized to build an energy storage station with a capacity of 5.14 MW.

Station is not only the anhui electric power first "multistation fusion" project, but also domestic advanced green renewable "micro grid" demonstration project, photovoltaic power plant green electricity supply of electric vehicle charging station, data center and communication station, excess electricity deposited in the on-site energy storage station, the formation of "light ChuChong micro network system integration. At the same time, it is a response to the implementation of the national "new infrastructure" strategy, which will play an important role in promoting the construction of smart city in Hefei.

2.3. State Grid Shanghai Power Supply Company

On April 15, 2019, the 35-kilovolt Yinjiabang Substation "multi-station integration" demonstration project was officially started in Zhangjiang Science City, Pudong, Shanghai. It is based on the traditional substation, equipped with 3.78 MW / 7.65 MW/energy storage system and a data center with 24 server cabinets. At the same time, a 5G tower and six communication screens are reserved to form the multi-station integration mode of "substation + energy storage station + data center".

The energy storage station can effectively cut peak load and fill valley of the power grid, so that the maximum load rate of the main transformer of Yin Jia Bang substation can be reduced to about 70% from the current heavy load state, which greatly improves the power supply capacity of the regional power grid. Data center can store and utilize a large amount of equipment status information to support the development of comprehensive energy service. In the later stage, it will be equipped with super foundations such as 5G base stations to provide effective resources and means for the construction of digital power grids and the expansion of comprehensive energy services.

To sum up, local construction features are the transformation and upgrading of traditional substation or switch station and other old station site resources, using their idle space, based on the construction and operation mode of "unified planning, unified design, characteristic construction", and their business types and ways are different. However, a multi-station integration construction scheme has not been constructed to facilitate the replication and promotion of construction across the province and the country.

To sum up, local construction features are the transformation and upgrading of traditional substation or switch station and other old station site resources, using their idle space, based on the construction and operation mode of "unified planning, unified design, characteristic construction", and their business types and ways are different. However, a multi-station integration construction scheme has not been constructed to facilitate the replication and promotion of construction across the province and the country.

3. Multi-Station Integrated Energy Interconnection Planning System Based on the Layered Theory of Energy Internet

3.1. The Concept of Complex Adaptation Theory

Complex adaptive system (CAS), the core of which is adaptation produces complexity, complexity is the condition of intelligence. Its basic characteristics include agent, aggregation, nonlinearity, factor flow, diversity, feature identification, internal model and building blocks. The first five are common characteristics of complex adaptive systems, which will play a role in adaptation and evolution. The last three are the mechanisms by which individuals interact with their environment.

Complexity is generated, not given; The internal cause of complexity is that the system or things adapt to the environment for survival and development, and complexity emerges in the adaptation. So adaptability creates complexity. In order to adapt to the continuous development of the urban order, the energy system must constantly adapt to the new form, and study the new ideas to solve the energy problem under the new situation.

3.2. Layered Theory of Energy Internet

Energy Internet Stratification Theory (EIST) is based on the theory of complex adaptation, combining with each project to gradually summarize experience and expand the types of projects, forming the core idea of paying attention to terminal Energy users and progressive evolution layer by layer. Combined with system theory, information theory and cybernetics, the multi-station integrated energy interconnection planning system is built, giving full consideration to the increasing proportion of distributed energy and the inevitable trend of the development of the Internet of things. The core part of the theoretical system is to build a planning framework suitable for energy Internet, which is divided into three layers of energy cell, function unit, operation unit and an energy supply and demand center, referred to as the "Trinity Principle". Based on the characteristics and development trend of multi-station integration construction, the specific hierarchical definition is proposed:

Energy cell layer: the area formed by the boundary of 10kV line/power supply area on/off, which contains all power equipment, sensing equipment, network equipment, intelligent terminal and corresponding energy users, etc.

Functional unit layer: it is composed of multiple energy cells (10kV line/switch) and formed by the boundary of substation power supply area, that is, a multi-station fusion substation. Including distributed energy station, energy storage station, big data center and other types of energy facilities, can be divided according to the power supply unit area in grid.

Operation unit layer: it is composed of multiple functional units (substations) and formed by the boundary of the management scope of municipal power company. Intelligent energy dispatching, intelligent operation and maintenance service cloud platform of municipal power grid is built to realize intelligent energy dispatching, intelligent operation and maintenance and intelligent early warning within the region. It can be divided according to the grid area of power supply in grid. At the same time, it can flexibly split and merge functional units according to the changes of different energy mechanisms, so as to meet the demand of optimal power supply in the region.

Energy supply (Demand) center (provincial power grid Company, Internet of Things as "cloud" level): it is an area bounded by the management scope of provincial power company. It is the comprehensive management of supply and demand from lower-level energy to higher-level energy. By building a smart energy management platform, it can complete the intelligent scheduling and energy transaction of energy supply and demand balance.

3.3. Multi-Station Fusion Planning System Based On EIST Theory

Based on the theory of complex adaptation and stratification, according to the trinity principle, a multi-station fusion planning system based on EIST theory is proposed. Urban multi-station fusion planning will highlight the hub and fusion characteristics of substation, define a functional unit by a single substation, and define different subordinate energy facilities as an energy cell. Considering the underlying energy demand and supply capacity, and considering the reliability and economy of energy transmission, the energy cells are divided into self-supporting functional units, externally supplied functional units and mixed functional units. Specific definitions are as follows:

Self-contained functional unit: refers to an independent building or area, which uses its own energy production to meet energy use needs and does not participate in energy trading and energy dispatching. Usually, photovoltaic, wind power and energy storage systems can be configured in the region to jointly solve the problems of cold, heat and electricity supply in the form of electric energy substitution.

External supply functional unit: refers to the unit with small energy load demand and poor economic efficiency of independent construction of energy equipment, relying on the nearby large energy station to meet the regional cold, heat and electricity load demand in the form of external energy supply, and participating in energy trading and energy dispatching.

Hybrid functional unit: refers to the functional unit with large area and load. Its energy supply capacity can not only meet its own energy supply demand, but also supply energy for surrounding functional units. It can also purchase energy from other cells for regional supply based on energy price, which is a comprehensive functional unit. At the same time, participate in energy dispatching, adjust the overall peak load of the region and optimize the load curve. Generally, a variety of energy sources such as cold and heat combined power supply, new energy, energy storage, heat storage and cold storage can be considered in the region to meet the energy needs of multiple functional units in the form of integrated energy allocation and supply.

Multiple substation function units can be formed into energy networks with different sizes and functions according to different operation units and modes. A large amount of operation and transaction data can be quickly calculated by using low-delay characteristics of 5G network through underlying data storage and edge computing. Finally, the event results will be returned to the municipal energy management platform, unified operation and scheduling, multi-business operation, forming a multi-station integrated business system for digital power grid. The system can effectively realize the flexible interaction between the side source, the network and the load of the substation, and the diversity of energy sources also makes it possible for the power grid company to provide better energy services to the user side. The schematic diagram of multi-station integration service planning system for digital power grid is as follows:

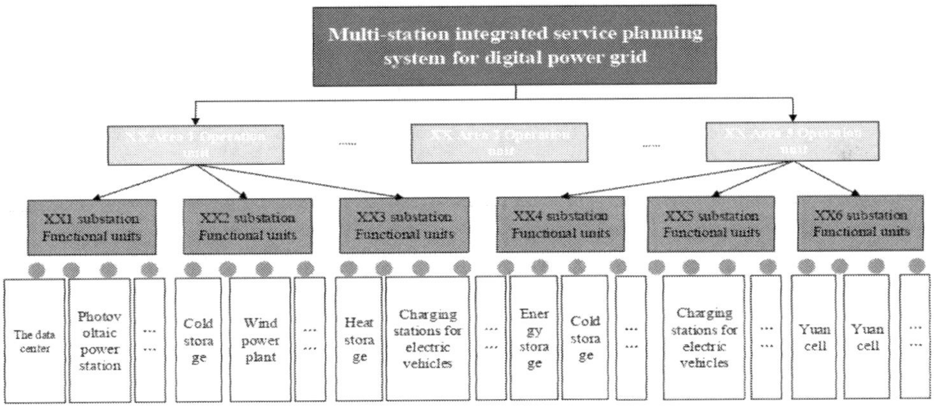

Figure 2. Schematic diagram of multi station integrated service planning system for digital power grid.

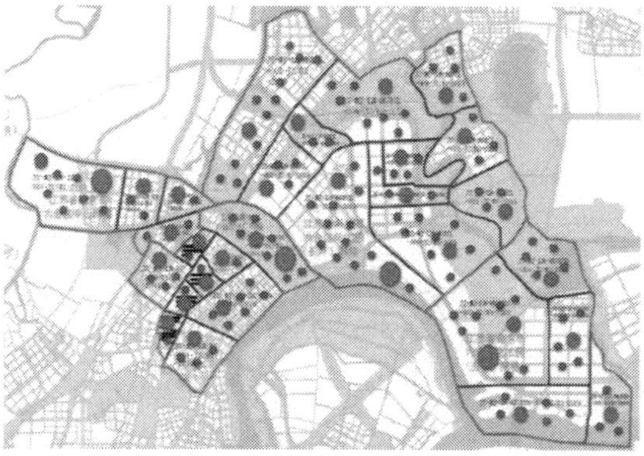

Figure 3. Schematic diagram of hierarchical division of regional multi station fusion.

4. Research on Fusion Mode of Multi-Station Fusion in Multi-Scene

Facing the future development of digital power grid, intelligent substation has become an irreversible trend. Therefore, substation network and information construction can provide information support for power grid scheduling optimization, operation management, energy transaction, comprehensive energy innovation business, etc., and provide guarantee for information interaction between power network and multi-energy system. The digital construction of substation needs data center station and 5G base station as important support.

Therefore, when considering the integration scheme design of multiple energy stations in multiple scenarios, the basic configuration mode of "substation + data center station +5G base station" will be adopted. On this basis, different business application scenarios will be set according to different user types and load characteristics in the region. The "3+N" substation fusion mode is proposed to meet the requirements of cold, heat, electricity and user-side services.

Schematic diagram of "3" station

Diagram of "N" station

Figure 4. 3+N station schematic diagram.

According to GB50059-2011 code for Design of 35kV ~ 110kV Substations, 110kV substations generally determine the capacity and number of main transformers according to the power supply conditions, load properties and power capacity of surrounding areas. In addition, in order to facilitate fault or maintenance, they can flexibly switch operation and reduce the scope of power failure and power failure loss. Generally, two or more main transformers are configured to achieve the principle of reliable power supply, convenient maintenance and economic operation. According to the actual construction of 110kV substation, considering the characteristics of substation and the diverse business conditions on the user side within the power supply range, a multi-scene mode and multi-station fusion configuration scheme for digital power grid is proposed. The detailed design is shown in the following table:

Table 1. Design of multi scene and multi station fusion mode.

Type	Scene classification	Multi-station fusion mode design
Design of multi-station fusion mode based on user side service	Service scenario 1: The main power supply area is industrial	The fusion mode of "3+ energy storage station + distributed photovoltaic power station" is adopted
	Service scenario 2: The main power supply areas are commercial and office areas	The integration mode of "3+ energy storage station + distributed photovoltaic power station + charging pile charging station + cold storage station/combined cold, heat and electricity power supply station + Beidou base station" is adopted
	Service scenario 3: The main power supply area is residential	The integration mode of "3+ energy storage station + distributed photovoltaic power station + charging pile charging station + heat storage station + Beidou base station" is adopted
	Business Scenario 4: The main power supply area is a mixture of industrial, commercial, residential, and public facilities	The integration mode of "3+ energy storage station + distributed PHOTOVOLTAIC power station + charging pile charging station + heat storage station/cold storage station/combined cold, heat and electricity power supply station + Beidou base station" is adopted
Fusion mode design based on different types of substation	New substation: maximum utilization of resources, maximum economic benefit, maximum scale effect, maximum interaction of source network load	New energy storage station, photovoltaic, charging pile charging station, data center, cold storage station, heat storage station, 5G base station, Beidou base station, DC microgrid, etc

	Transform substation: improve power supply reliability and equipment utilization rate	New energy storage, photovoltaic, data center, DC microgrid, etc
	Conventional substations: incremental resource utilization and optimization	New energy storage, photovoltaic, charging pile charging station, data center, cold storage station, heat storage station, 5G base station, Beidou base station, DC microgrid, etc

In addition, in order to reflect the social attributes of power enterprises, some substations will have the function of popular science, and the appearance of the substation will be designed to facilitate the public to understand the operation of the power system. Therefore, this kind of multi-station integration mode design will be on top of the basic design, and should also add science station, landscape station, etc.

5. Conclusion

As an emerging business in the field of energy, multi-station fusion has been applied in Shanghai, Nanjing, Tianjin, Chongqing, Xiamen and other places. However, the current substation fusion model, business application scenarios, construction and operation modes are not mature enough to form large-scale popularization and promotion. Therefore, this paper analyzes the construction characteristics and differences of multi-station fusion in various provinces and cities in China, summarizes the commonalities, studies the fusion combination mode of multi-station fusion in multiple scenarios, and innovatively proposes the "3+N" combination mode based on different user-side business types and the combination design mode based on different types of substation. The research results will serve as the practical support criteria for the substation centered regional energy Internet innovation business model, and the theoretical planning system and multi-scene fusion mode will be applied to guide the construction of multi-station fusion system in different scenarios. This model can be built into a new model of multi-station fusion construction that can be replicated and promoted in the whole country, and help the industrialization development of multi-station fusion business. At the same time, it strengthens the digital infrastructure construction of power grid through multi-station fusion technology, strengthens the basic computing and operation capacity of digital power grid, and pushes forward the construction target of digital power grid, source network load interaction system and new power system.

Reference

[1] Zhang Y, Wang B Y, Li R, Mi X R, et al. 2019 Research on Business Mode and Development Path of Multi-station Integration[J] *Distribution & Utilization* **6** (11) 62-65.
[2] Xu W B, Cheng H F, Bai Z H, Miao C H, Sun F C, et al. 2019 Optimal Design and Operation of Energy Storage Power Station under Multi-station Fusion Mode[J] *Distribution & Utilization* **11** (1) 84-91.
[3] LU L. 2018 Research on edge computing technology for 5G network[J] *Information and Communications Technology and Policy* **8** (11) 1-6.
[4] Xiang T C, Li G D. 2017 A two-phase integral optimization strategy for network restoration with photovoltaic generation [J] *Electric Power Engineering Technology* **36** (5) 21-28.
[5] Zhou Y, Wang D H. 2016 Coordinated optimization on electric vehicles and renewable energy[J] *Guangdong Electric Power* **29** (10) 42-48.
[6] Zhao X F, Lin W J, Liu Y B. 2013 Application and Analysis of Distributed Power Supply System in Data Center [C] *Proceedings of 2013 Information and Communication Network Technology Business Development Conference* **6** (8) 593-599.
[7] Chen F, Zhang H, Chen Y Q, et al. 2013 Application research of new technology in data center power supply system[J] *Journal of Power Supply* **11** (2)93-100.

[8] Sun C B, Song Y, Cai C, et al. 2019 Multi-objective optimal configuration scheme for distributed energy storage in distribution network considering multi-scenario requirements[J] *Electric Power Construction* **40** (7) 48-56.

Research on Automatic Profile Generation Based on 3D Model

Mingwen Chi

Department of Information Engineering,Baotou Light Industry Vocational Technical College, 19 Jianhua road, Qingshan district, Baotou.
Email: 154878935@qq.com

Abstract. In this paper, the technology of profile generation based on 3D model is studied. The main steps are as follows: (1) the location where the profile needs to be generated in 3D model design; (2) Using 3D data cutting technology to realize the generation of geological lines in profile; (3) Read the basic exploration data related to profile position in the database; (4) According to the data generated in the first three steps, the cross-section is automatically drawn after data coordinate transformation. The above method can quickly generate the geological profile of any location according to the 3D geological model, which is helpful for geological analysis and provides reference data for engineering design.
Keywords. 3D model; Section; automatically generate.

1. Introduction

Geological maps are a scientific means of graphical representation of geological phenomena [1]. Complicated geological maps are used in basic geological survey, resource exploration, engineering investigation, mine design and exploitation, urban planning and other aspects [2, 3]. There are many kinds of geological maps, among which the geological profile is the most basic expression of geological content, the basis and forerunner of the whole geological work results, and the map frequently drawn by geological engineers [4, 5].

Geological profile is the traditional way of drawing the manual way, late mature GIS theory development of the last century, people started to interactive graphics vector method is adopted to improve the compilation of geological maps, automation of semi-automatic maps are drawn began to grow, and most USES the manual intervention, semi-automatic rendering with the aid of expert knowledge in developing professional geological map [6, 7]. In this process, most of the geological maps compiled are independent of each other, and even if there is any connection, it is still fuzzy inference, unable to achieve accurate positioning, and unable to eliminate the incompatibility of data between several sections [8]. At present, some profile makers directly connect boreholes with different exploration lines, which ensures the compatibility of data between sections, but leads to another new problem. For example, the pinch-out processing of some lens bodies presents serious irrationality, that is, it is possible to connect pinch-out points directly to the corresponding horizons of adjacent boreholes. Although the incompatibility of data is avoided in this way, the principle of violating geological inference appears. Therefore, it is very necessary to develop a set of semi-automatic mapping software of geological section directly based on 3D digital geological body, which can not only guarantee the compatibility of data between multiple sections, but also make accurate inference according to expert knowledge [9, 10].

2. Data Source and Generation Process of Profile

At present, database is basically used to manage basic exploration data in the drawing of cross-section at home and abroad, and engineering geological cross-section is established automatically or manually according to the exploration data (mainly drilling data), and its data storage is also carried out in a two-dimensional environment. This paper proposes that the profile data includes not only the basic exploration data from the database, but also the cut data from the 3D model, and can also receive external data. Therefore, there are three main sources of data for automatic profile drawing in this paper: (1) basic exploration data; (2) Cutting data of 3D model; (3) Other external data. According to the characteristics of the above data sources, the automatic generation process of the profile is designed :(1) read the basic exploration data in the database according to the position of the profile exploration line; (2) Using the exploration line and the 3D model to cut, to obtain the 3D profile geological lines; (3) coordinate transformation from three-dimensional data to two-dimensional plane data; (4) Read other profile data that need to be supplemented; (5) Automatic formation of profile vector data and profile file according to the above data.

3. Database Data Reading

According to the traditional way of data source, first of all, the basic exploration data will be stored in the database. The data mainly involved in the drawing of the profile include drilling data and all kinds of experimental data. This part of data is the supplementary content of the profile.

In this paper, the data of borehole and adit are taken as examples to illustrate the method of data acquisition. Firstly, the spatial coordinates of borehole and adit data are read from the database, and then converted into planar coordinates through coordinate transformation. Here, borehole is taken as an example to describe. There are two kinds of drilling holes: straight hole and inclined hole. The drawing of straight hole is determined according to the coordinate of drilling orifice and hole depth. But for inclined hole, it can be determined by conversion according to survey parameters recorded by borehole. The specific conversion method is as follows: Figure 1 shows the vertical projection schematic diagram of the borehole, and figure 2 shows the horizontal projection schematic diagram of the borehole. Among them, the thick line is the projection line of the borehole, the other lines are auxiliary lines, n is the number of survey segments, d is the deviation distance of the borehole, L is the projection length of the borehole elevation, φ is the azimuth Angle of the borehole, θ is the inclination Angle of the borehole, Li is the length of each survey segment, and θ I is the inclination Angle of each survey segment.

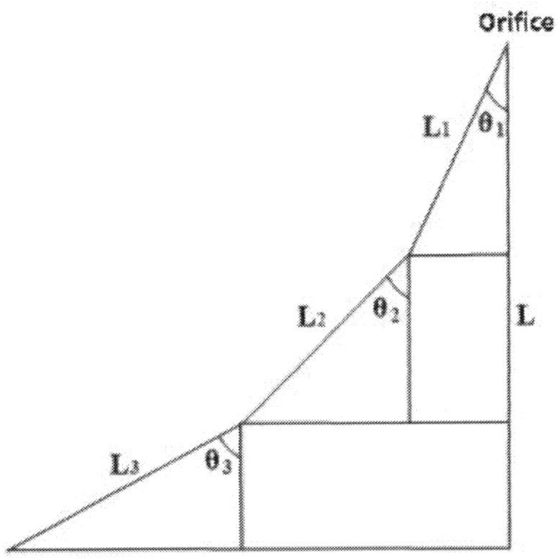

Figure 1. Vertical projection of borehole.

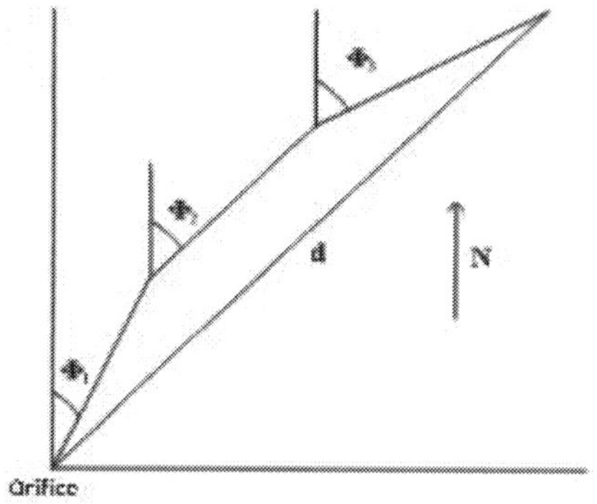

Figure 2. Horizontal projection of borehole.

The specific length and location marking of the borehole drawn in the profile can be calculated by the following formula:

$$d = \sqrt{\left(\sum_{i=1}^{n} L_i \sin\theta_i \sin\Phi_i\right)^2 + \left(\sum_{i=1}^{n} L_i \sin\theta_i \cos\Phi_i\right)^2}$$

$$L = \sum_{i=1}^{n} L_i \cos\theta_i$$

$$\Phi = \arcsin \frac{\sum L_i \sin\theta_i \sin\Phi_i}{\sum L_i \sin\theta_i \cos\Phi_i}$$

$$\theta = \arcsin \frac{\left(\sum L_i \sin\theta_i \sin\Phi_i\right)^2 + \left(\sum L_i \sin\theta_i \cos\Phi_i\right)^2}{\sum L_i \cos\theta_i}$$

4. Cut and Obtain 3D Model Data

On the basis of the established 3D model, the relevant stratum or structural information can be obtained by cutting the profile position lines to provide the basic geological data for the profile. The main technology is the intersecting operation of the object. The basic process is as follows: firstly, the position of the exploration profile is determined according to the exploration line; generally, lines are used to store the sequence of position points; surfaces are used to represent strata or structural units; and then, the intersecting line between lines and surfaces is used to determine the relevant stratigraphic or structural information of the profile position. For example, as shown in figure 3, LINE1 is used to represent the position of the exploration line, while Polygons represents the sequence of surfaces that constitute geological bodies, in which the sequence points of LINE1 determine the position of the broken section. In the 3D model, the elevation of sequence points has no practical significance and is only set for visualization. Secondly, Line1 and Polygons are used to obtain the intersecting line LINECs, so that LINECs can be used as geological line data in the exploration section. According to this principle, this paper needs to solve the main problem is the algorithm of line and surface intersection, line and surface intersection, can be regarded as the line projection to the surface of the operation. In addition, in order to more intuitively show the corresponding relationship between the exploration profile and other exploration profile information, the intersection of the exploration line and other geological lines in the profile is also required, so the problem of intersection between lines is also needed to be solved.

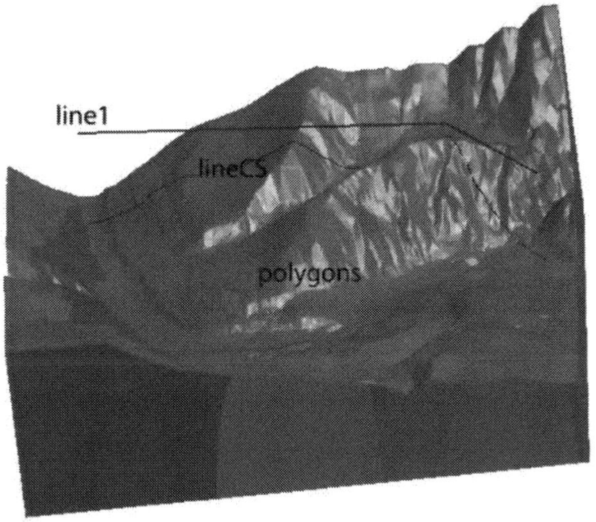

Figure 3. 3D model to obtain exploration data.

The intersection algorithm of spatial line and surface and the algorithm of line and line intersection are used to form spatial profile data, as shown in figure 4.

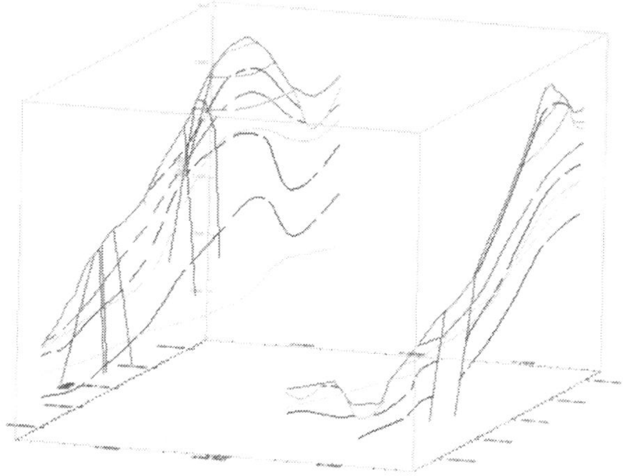

Figure 4. 3D profile line data obtained by 3D cutting.

5. Extract Other Data

In general, the object management of 3D modeling software system adopts layer management, and the data is extracted directly through layers. According to the needs of different functions, it can be accurately obtained according to attributes. According to the different properties of the object, it can be extracted in the following ways:

(1) According to the color extraction, objects of different types and purposes can be marked with different colors according to the user's operation habits, which can be obtained by RGB color value or by the color number defined in AutoCAD. The former method can distinguish more types of objects, up to 255*255*255. The latter method can only distinguish 256 kinds of color numbers;

(2) According to the type of the object to extract, this method is generally only used in the basic editing process, can be divided into: text, dot symbol, line, polygon, regular surface and volume;

(3) According to the attribute acquisition, this method is the most critical extraction method of the system. The system adopts the B-Rep model and uses the coding method to distinguish the category of the object. Each object has its own attribute value, which can uniquely identify its category. This can be obtained by the property name when the object is extracted. In addition, an object can have several attributes, with double or multiple "identities". In order to distinguish the two attributes, the system stipulates that the two attributes are separated by commas (,), so as to avoid repeated storage of data.

(4) According to the batch extraction of layers, the system can define one or several layers as professional layers, which are specially used to store one or several types of data. When lifting, the system can directly batch extract according to the layers. This method is efficient, easy for users to operate, and is also a common method of data extraction in the system.

6. Application Examples

Based on the 3D model and basic data of a hydropower station, the automatic drawing and verification of the cross-section are carried out in this paper. The geological line data in the 3D model obtained by cutting the 3D model is shown in figure 5. After converting the 3D data, the profile formed by reading the basic exploration data in the database is shown in figure 6.

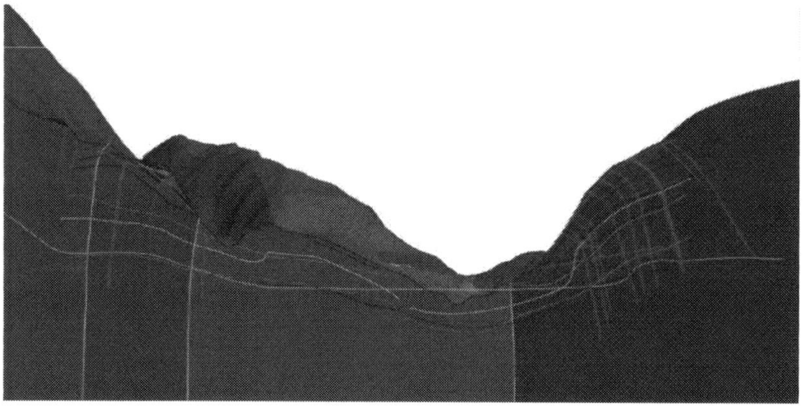

Figure 5. Geological profile model with cut geological lines.

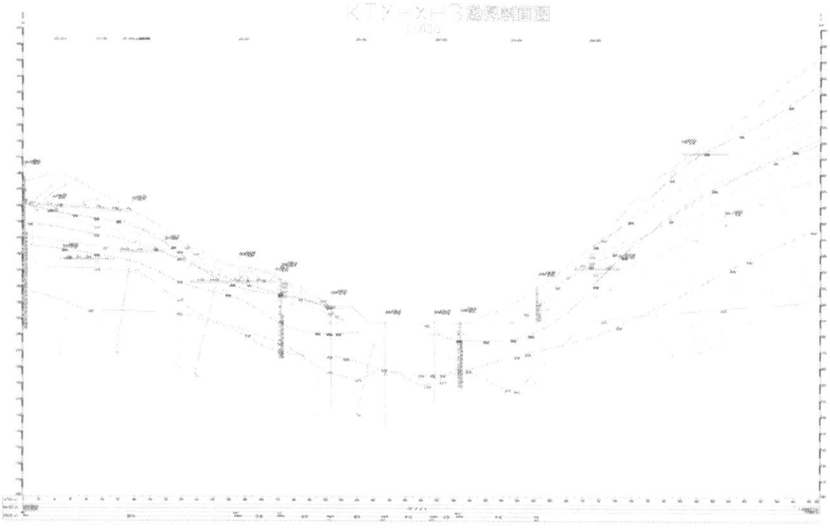

Figure 6. Automatically generated geological data profile.

Acknowledgments
The research of this paper was funded by the Science and Technology Research Project of Higher Education in Inner Mongolia Autonomous Region (Project No. : NJZY20162).

References

[1] Li G J, Liu L, Niu Z P. 2021 Research and Development of 3D Geological Model Technology for Waterway Engineering Based on Civil 3D [J/OL] *Port & waterway engineering* 1-6 https://doi.org/10.16233/j.cnki.issn1002-4972.20210630.036.
[2] Li C. 2004 Computer Automatic Generation Technology of Geological Section and Histogram [D] *Xi 'an* University *of Science and Technology*.
[3] Wu C L, Liu G, Tian Y P, et al. 2005 The theory and method of informationization of geological and mineral exploration [J] *Earth Science* **30** (3) 359-365.
[4] Wu C L, Wang X Q, Liu G, et al. 1996 Design principle and application of geological and mineral point source information system [M] *Wuhan: China University of Geosciences Press*.
[5] Di W M. 2002 computer aided drawing of geological profile and plan [J] *Gold science and technology* 36-39.
[6] Zhang Y H. 2020 A new method of Drawing Topographic line of geological section by programming with LISP language [J] *China Science and technology information* (16) 92-94.
[7] Wang D Z. 2000 Development and application of engineering geological profile software [J] *Engineering geological computer application* 2 6-9.
[8] Farin G. 1993 Curves and Surface for Computer Aided Geometric Design A Practical Guide [M] 3rd Edition. Acedemic Press.
[9] Li X J, He Y F, Wang G C, Xu L P, Chen Y. 2017 stratigraphic distribution model self matching rofile dynamic generation algorithm [J] *Geophysical and geochemical exploration* **41** (05) 939-945.
[10] Cai Q, Wang Y N. 2017 study on automatic generation method of hydrogeological profile based on Component GIS [J] *Management and science and technology of small and medium-sized enterprises* (last Xunyuan) (05) 141-142.

Bayesian Model Updating Based on Kriging Surrogate Model and Simulated Annealing Algorithm

Zenghui Wang, Hong Yin and Zhenrui Peng*

School of Mechanical Engineering, Lanzhou Jiaotong University, Lanzhou 730070, Gansu Province, China

*Email: pzrui@163.com

Abstract. Aiming at the problem of difficulty in selecting the proposal distribution and low computational efficiency in the traditional Markov chain Monte Carlo algorithm, a Bayesian model updating method using surrogate model technology and simulated annealing algorithm is proposed. Firstly, the Kriging surrogate model is used to mine the implicit relationship between the structural parameters to be updated and the corresponding dynamic responses, and the Kriging model that meets the accuracy requirement is used to replace the complex finite element model to participate in the iterative calculation to improve the model updating efficiency. Then, the simulated annealing algorithm is introduced to reorganize the Markov chains from different proposal distributions to obtain high-quality posterior samples, which are used to estimate the parameters posterior distributions. Finally, a space truss structure is used to verify the effectiveness of the proposed method.

Keywords. Bayesian inference; model updating; Markov chain Monte Carlo algorithm; Kriging model; simulated annealing algorithm

1. Introduction

In recent years, finite element model updating (FEMU) technology has gradually become a research hotspot at the structural dynamics domain, and is widely used in structural damage identification [1]. Most of the current model updating methods are deterministic methods, however, the application of deterministic methods is limited due to uncertainty factors in the structural materials, boundary conditions and the test process [2].

The uncertainty model updating methods based on Bayesian statistical theory comprehensively considers historical data and experts' experience to suppose the prior distributions, and combines the statistical information of the measured responses to continuously update the prior distribution to make it close to the real posterior probability distribution, which is widely used in uncertainty model updating [3-4]. Beck et al. [5] introduced Bayesian theory into model updating for the first time. In order to estimate the parameters posterior distributions, he also proposed a Markov chain Mento Carlo (MCMC) algorithm [6-7], and verified the effectiveness of the MCMC algorithm through a two-degree-of-freedom steel frame structure, and then the Bayesian method has been developed rapidly. Ching et al. [8] proposed the Transition MCMC (TMCMC) method, which samples from a string of intermediate probability density functions (pdf) to estimate the posterior distributions, which can avoid sampling directly from the complex posterior pdf, which improved sampling efficiency. Cheung et al. [9] proposed a Hybrid MCMC (HMCMC) method, which shows great potential in the Bayesian model updating with high-dimensional uncertainty parameters. However, most of the above methods are based on a single Markov chain, which means that these methods will rely too much on the selection of the proposal distribution variance, the sampling efficiency of single-chain MCMC method is low when meets high parameter dimension, the quality of the obtained posterior samples is poor.

To address these problems, a Bayesian FEMU method using Kriging surrogate model technology and simulated annealing algorithm is proposed in this paper. Firstly, select the parameters to be updated; use the uniform experimental design method to construct the training set samples; use the training set samples as the inputs of the Kriging model, and the corresponding dynamic responses as the outputs to construct the accurate Kriging model; then, use simulated annealing mechanism and Metropolis criteria to reorganize Markov chains from different proposal distributions to obtain high-quality posterior samples. The parameter posterior pdf is obtained by statistical analysis of the Markov chain's stationary phase. Lastly, the viability of the method is demonstrated through a space truss structure.

2. Basic Theory and Enhancements

2.1. FEMU theory based on Bayesian Inference

Bayesian FEMU methods combine prior information and test data, the posterior pdf of the parameters is inferred by the MCMC algorithm. This process can be explained as [10-11]:

$$\begin{aligned} p(\boldsymbol{\theta} \mid \boldsymbol{x}) &= \frac{p(\boldsymbol{x} \mid \boldsymbol{\theta})\pi(\boldsymbol{\theta})}{\int_{\theta} p(\boldsymbol{x} \mid \boldsymbol{\theta})\pi(\boldsymbol{\theta})d\boldsymbol{\theta}} \\ &= \mathrm{c} \cdot p(\boldsymbol{x} \mid \boldsymbol{\theta})\pi(\boldsymbol{\theta}) \end{aligned} \tag{1}$$

where $\boldsymbol{x}$ is test data; $\boldsymbol{\theta}$ is the vector composed of all the parameters to be updated; $\pi(\boldsymbol{\theta})$ is the prior distribution, usually taken as the generalized unbiased uniform distribution $\pi(\boldsymbol{\theta})=1$; $p(\boldsymbol{x} \mid \boldsymbol{\theta})$ is the given conditional distribution, usually called the likelihood function; c is a constant factor, usually called a regularization constant. Assuming N independent experiments, the likelihood function in the Bayesian equation can be expressed as:

$$\begin{cases} p(\boldsymbol{x} \mid \boldsymbol{\theta}) = \dfrac{1}{(\sqrt{2\pi \mathbf{cov}_y})^N} \exp[-L(\boldsymbol{\theta})] \\ L(\boldsymbol{\theta}) = \sum_{i=1}^{N} (y_i - y(\boldsymbol{\theta})_i)^{\mathrm{T}} \mathbf{cov}_y^{-1} (y_i - y(\boldsymbol{\theta})_i) \end{cases} \tag{2}$$

where y_i is test response; $y(\boldsymbol{\theta})_i$ is model calculated response; $\mathbf{cov}_y$ is test information covariance matrix. Substituting the above equations into equation (1), the pdf can be written as:

$$p(\boldsymbol{\theta} \mid \boldsymbol{x}) = \mathrm{c}' \cdot \exp[\sum_{i=1}^{N} (y_i - y(\boldsymbol{\theta})_i)^{\mathrm{T}} \mathbf{cov}_y^{-1} (y_i - y(\boldsymbol{\theta})_i)] \tag{3}$$

where c' is a constant. In actual problems, there is usually no explicit response expression $y(\boldsymbol{\theta})$ and the integration computation is more difficult. Therefore, the pdf is calculated using the MCMC to overcome complex integration computation problem.

2.2. Kriging Surrogate Model

Kriging model is a surrogate model, which is on the basis of stochastic theory. It has good fitting and prediction ability to nonlinear problems and error estimation function. Kriging model regards unknown function as the concrete realization of Gaussian process, which includes linear regression and non-parametric part [12]:

$$y(\boldsymbol{x}) = \sum_{i=1}^{q} \beta_i f_i(\boldsymbol{x}) + z(\boldsymbol{x}) = \boldsymbol{f}^{\mathrm{T}}(\boldsymbol{x})\boldsymbol{\beta} + z(\boldsymbol{x}) \tag{4}$$

where $f(x)$ is a polynomial function; β is a regression model coefficient matrix, and $z(x)$ is a static stochastic process obeys normal distribution. The correlation between responses $z(x)$ is expressed by a function with spatial distance between two sample points, and the Gaussian function which fits the actual engineering practice is selected as the correlation function, which is in the following form:

$$R(x_i, x_j) = \exp(-\sum_{k=1}^{m} \theta_k \left| x_i^k - x_j^k \right|^2) \tag{5}$$

where x_i^k and x_j^k are the k component of any two sample points; m is the design variables; θ_k is correlation coefficient.

Maximum likelihood is introduced to estimate model parameters:

$$\hat{\beta} = (F^{\mathrm{T}} R^{-1} F)^{-1} F^{\mathrm{T}} R^{-1} Y \tag{6}$$

$$\hat{\sigma}^2 = \frac{1}{n}(Y - F\beta)^{\mathrm{T}} R^{-1} (Y - F\beta) \tag{7}$$

where F is sample point matrix; Y is response column vector; R is the correlation matrix; the elements are $R_{ij} = R(x_i, x_j)$ $(i, j = 1,2,\mathrm{L}\ ,n)$; n is test points number.

2.3. Simulated Annealing Algorithm

The goal of simulated annealing (SA) algorithm, a probabilistic algorithm, is to obtain an approximate solution of a energy function. The Metropolis sampling can be seen as the core of SA, which yields stochastic samples by a proposal pdf, the mathematical description of which is unknown, but whose value can be calculated [13]. A rapid annealing plan is adopted in SA, which temperature is set at $\alpha \in (0,1)$, build on this and decline at an exponential rate. The relationship between the temperature of the next anneal cycle c+1 and the temperature of the current anneal cycle c is as: $T_{c+1} = \alpha T_c$.

When the initial and final temperatures are given, the cooling rate can be obtained by the relation $\alpha = (T_N / T_0)^{1/(N_c - 1)}$, with N_c the number of annealing cycles. The new random candidate k^{c+1} are generated by the following equation:

$$k^{c+1} = k^c + \omega * (U^{k^c} - L^{k^c}) \tag{8}$$

$$\omega = \mathrm{sgn}(u - \frac{1}{2}) T_c [(1 + \frac{1}{T_c})^{|2u-1|} - 1] \tag{9}$$

where U^{k^c} and L^{k^c} are the upper bound and the lower bound of parameters at the current anneal cycle c; $\omega \in [-1,1]$ is generated by the uniform distribution $u \in [0,1]$.

2.4. Multi-Chains Simulated Annealing Reorganization Algorithm

Combining delay rejection and adaptive Metropolis (DRAM) strategies, Kriging surrogate model and simulated annealing algorithm, the specific steps of Bayesian model updating method are as follows.

(1) Select the initial sample points θ_i^0 of M Markov chains randomly by the prior information, and set the random sampling times is T;

(2) Set the initial variance C_{0s} (s=1,2,⋯,M) of M proposal distributions randomly, and arrange the variance values of the proposal distributions sequentially, which are used to sample to produce Markov chains of different sources;

(3) Obtain M Markov chains by parallel sampling according to the DRAM algorithm;

(4) Take the initial sample point of the first Markov chain as the initial sample of simulated annealing algorithm, generate two sample points in the sample points of the remaining (M-1) Markov

chains at the same iterative step randomly, which are used as the upper and lower boundary values U and L of the disturbance, respectively, and a new candidate sample point is obtained by equation (8);

(5) Determine whether to accept the new candidate sample by Metropolis criterion, and terminate the reorganization when the set sampling times T is reached;

(6) Analyse the stationary phase of the reorganized Markov chain statistically to obtain the updated parameter values.

3. Example Analysis

The structural model shown in figure 1 is a common space truss structure in engineering, which consists of 28 nodes, 66 rod units and 48 degrees of freedom. The constraint condition is that 4 supports (node number 1, 8, 9, 16) are fixed. When analyzing the structure, only the Y and Z directional DOF of each node are considered, which are numbered in turn. The truss is divided into three parts: the upper chords, the middle chords and the lower chords. In most cases, the damage will lead to significant changes in structural stiffness, mainly manifested in the uncertainty of elastic modulus. The elastic modulus of three groups are selected as the parameters to be updated, which are $\theta_1=E_1/E_0$, $\theta_2=E_2/E_0$ and $\theta_3=E_3/E_0$, respectively. The initial elastic modulus E_0 is 190 GPa, and it is assumed that the measured mean value of each parameter obeys the normal distribution matrix [1.5, 2, 1.2], and the standard deviation is all 0.1. Set the ranges of parameters are [1, 2], [1.5, 2.5] and [0.7, 1.7], respectively.

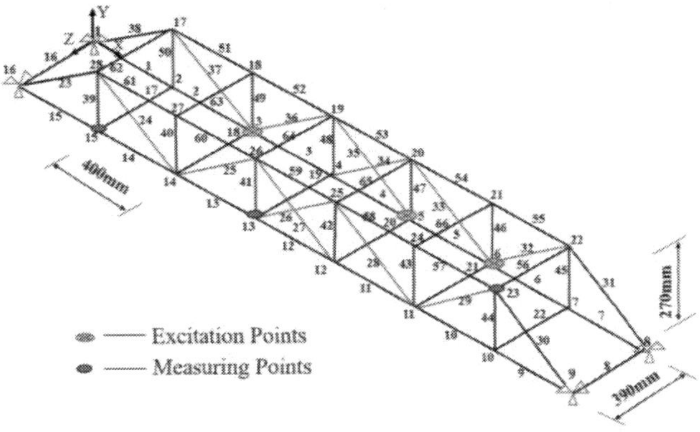

Figure 1. Space truss model.

Kriging model is constructed using the method in Section 2.2, which meets the precision requirements, and the RMSE value of the first order frequency is selected to test the Kriging model accuracy, which are listed in table 1. It can be seen that Kriging model has high prediction accuracy and can replace the initial FE model for iterative calculation.

Table 1. Error of the constructed Kriging model.

Order	Minimum error	Maximum error	Average error
1	1.32×10^{-12}	4.30×10^{-8}	1.70×10^{-9}
2	7.16×10^{-12}	2.34×10^{-7}	9.25×10^{-9}
3	2.21×10^{-11}	7.23×10^{-7}	2.86×10^{-8}

The Markov chains are obtained by standard MH algorithm and the proposed algorithm, respectively, set the sampling time is 20000. The Markov chains obtained by the two methods are shown in figure 2 and figure 3. It can be seen that the phenomenon of "sampling stagnation" is always occurs in Markov chains. However, when using the proposed algorithm to sample, the Markov chains can always maintain a swing state, and the mixing performance of samples is significantly better than the standard MH algorithm.

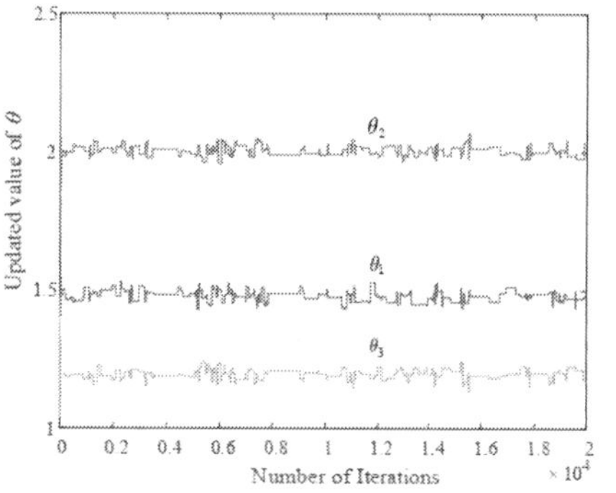

Figure 2. Markov chains obtained by standard MH.

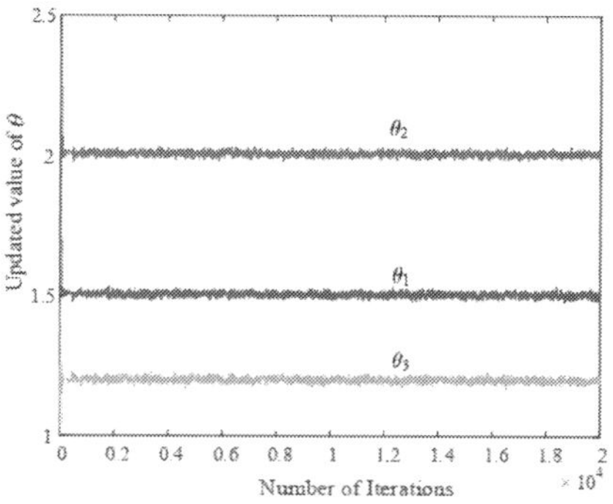

Figure 3. Markov chains obtained by proposed method.

The posterior samples in figure 3 are analysed using the normal probability test knowledge in probability statistics: because the samples in non-stationary period will seriously disrupt statistical analysis and affect estimation accuracy of parameter means, the first 10% samples of Markov chains in figure 3 are eliminated, and then the rest samples are tested for normal probability.

It can be seen from figure 4 that most posterior samples of the three parameters obtained by the proposed algorithm can fall on the assumed normal distribution line after removing inferior samples such as nonstationary samples. Therefore, the quality of samples is higher and the assumption of normal distribution holds.

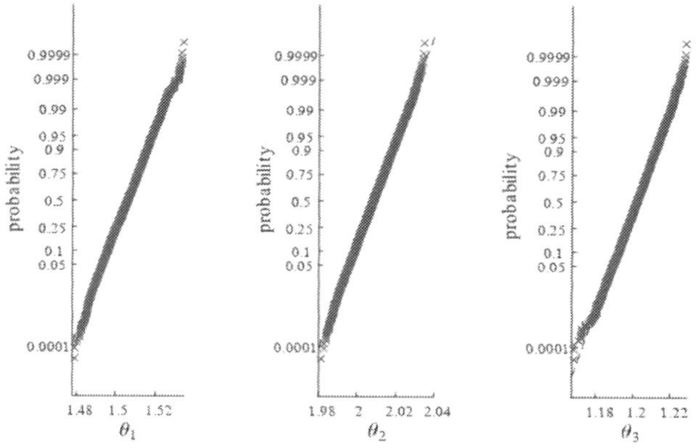

Figure 4. Normal probability test of four parameters.

Table 2. Comparison of the updating results for the four parameters of the space truss model.

Parameter	True value	Initial value	Pre-updating error/%	Updated value	Updated error/%
θ_1	1.500	1.953	30.187	1.506	0.421
θ_2	2.000	2.399	19.930	2.009	0.443
θ_3	1.200	1.469	22.417	1.202	0.141

The mean value of parameters is estimated by the stable phases of corresponding Markov chains, and the updated results of the mean values are listed in table 2, where the updated errors of mean values are all less than 0.5%, and the updated errors of the three parameters are all better than the standard MH algorithm, which indicates that the proposed algorithm achieves good results in the uncertainty model updating.

4. Conclusions

Compared with the traditional standard MH algorithm, "sampling stagnation" stage of Markov chains generated by the proposed algorithm is significantly reduced, which can better reflect the distribution characteristics of parameter space to be updated and get better updating accuracy, which is of great significance for the FEMU study of engineering structures.

Acknowledgments

This work is supported by the National Natural Science Foundation of China (51768035); Collaborative Innovation Team Project of Gansu Province (NO.2018C-12); Lanzhou Talent Innovation and Entrepreneurship Project (NO.2017-RC-66). Outstanding Graduate Student "Innovation Star" Project of Gansu Province (2021CXZX-580).

References

[1] Soize C. 2013 Stochastic modeling of uncertainties in computational structural dynamics-recent theoretical advances [J] *Journal of Sound Vibration* **332** (10) 2379-2395.

[2] Mosavi A A, Sedarat H, O'connor S M, et al. 2014 Calibrating a high-fidelity finite element model of a highway bridge using a multi-variable sensitivity-based optimization approach [J] *Structure and Infrastructure Engineering* **10** (5) 627-642.

[3] Mustafa S, Debnath N, Dutta A. 2015 Bayesian probabilistic approach for model updating and damage detection for a large truss bridge [J] *International Journal of Steel Structures* **15** 473-485.

[4] Yan W J, Chronopoulos D. 2020 Bayesian inference for damage identification based on analytical probabilistic model of scattering coefficient estimators and ultrafast wave scattering simulation scheme [J] *Journal of Sound and Vibration* **468** 115083.

[5] Beck J L, Katafygiotis L S. 1998 Updating models and their uncertainties I: Bayesian statistical framework [J] *Journal of Engineering Mechanics* **124** (4) 455-461.

[6] Beck J L, Au S K. 2002 Bayesian updating of structural models and reliability using Markov chain Monte Carlo simulation [J] *Journal of Engineering Mechanics* **128** 380-391.

[7] Peng Z R, Zheng J, Bai Y, et al. 2020 A model updating method based on an improved MCMC algorithm [J] *Journal of Vibration and Shock* **39** (4) 236-245.

[8] Ching J Y, Chen Y C. 2007 Transitional Markov chain Monte Carlo method for Bayesian model updating, model class selection, and model averaging [J] *Journal of Engineering Mechanics* **133** (7) 816-832.

[9] Cheung S H, Beck J L. 2009 Bayesian model updating using hybrid Monte Carlo simulation with application to structural dynamic models with many uncertain parameters [J] *Journal of Engineering Mechanics* **135** (4) 243-255.

[10] Das N, Debnath A. 2018 A Bayesian finite element model updating with combined normal and lognormal probability distributions using modal measurements [J] *Applied Mathematical Modelling* **61** 457-483.

[11] Yin H, Wang Z H, Cao M M, et al. 2021 An improved finite element model updating method based on the singular values of frequency response functions [J] *Mathematical Problems in Engineering* https://doi.org/10.1155/2021/5543549.

[12] Peng Z R, Cao M M, Liu M D. 2020 Model updating method based on wavelet decomposition of acceleration frequency response function [J] *Acta Aeronautica et Astronautica Sinica* **41** (7) 181-191.

[13] Astroza R, Nguyen L T, Nestorovic T. 2016 Finite element model updating using simulated annealing hybridized with unscented Kalman filter [J] *Computers and Structures* **177** 176-191.

Spatial High Precision Scanning Control Method Based on Feedforward Closed-Loop Model Reference Adaptive

Lanjie Guo, Hao Wang and Li Song

Beijing Institute of Space Mechanics & Electricitt,
No. 104 Yard, Youyi Road, Haidian District, Beijing 100094, China.
Email: lanjieguoedu@163.com

Abstract. The time-modulated Fourier transform spectrometer realizes spectrum detection by scanning the optical path of the corner mirror. During the scanning process, the servo system is required to have high-precision and low-speed characteristics. Aiming at the fluctuation of scanning speed caused by spatial micro-vibration during scanning, a closed-loop model reference adaptive control algorithm based on feedforward is studied. The permanent magnet synchronous linear motor is used to drive the angle mirror to move back and forth along the guide rail to achieve large optical path and high-precision scanning with the maximum optical path difference of ± 34cm, the speed stability ≥ 99%.

1. Introduction

Fourier transform infrared spectrometer is widely used to explore the composition of planetary atmosphere and the characteristics of solid surface due to its wide spectral range, high spectral resolution, accurate wavenumber and radiance calibration and high detection reliability[1-2]. The high-resolution spectrometers "Fy- 3", "Fy- 4" and "GF-5" launched in China in recent years are all time-modulated Fourier transform spectrometers. Because of their different scanning methods and maximum scanning optical path difference, different spectral resolutions are obtained [3-4]. Compared with the above loads, the direct driving method of permanent magnet synchronous linear motor and linear guide rail is adopted in this paper, which overcomes the disadvantages of limited swing angle of swing arm scanning and insufficient to provide large optical path difference, so that the spectral resolution and moving mirror travel are greatly improved [5-6]. In view of the interferometer working in a space environment, the micro-vibration in space causes velocity fluctuations, and it is difficult to achieve 99% speed stability. This paper studies the adaptive control method to achieve low-speed smooth motion of the corner mirror, and proposes a closed-loop model based on feedforward.

2. Controlled Object

The voltage balance equation of permanent magnet synchronous linear motor is

$$L_a \frac{di_a(t)}{dt} + R_a i_a(t) = u(t) - K_e \frac{dl_1(t)}{dt} \tag{1}$$

and the motor equation of motion is

$$\frac{d^2 l_1(t)}{dt^2} = \frac{K_t i_a(t)}{m} - B_m \frac{1}{m}\frac{dl_1(t)}{dt} - \frac{T_L(t)}{m} \tag{2}$$

where $i_a(t)$ is current, R_a is resistance, L_a is inductance, and $u(t)$ is motor voltage; K_t and K_e are thrust constant and back EMF constant respectively; m , $l_1(t)$, B_m are load mass, load moving displacement and viscous friction coefficient respectively; $T_m(t)=K_t i_a(t)$ is the motor thrust, $T_L(t)$ is the load resistance. Let $v=\frac{dl_1(t)}{dt}$ be the speed of load movement, $a=\frac{d^2 l_1(t)}{dt^2}$ be the acceleration of load movement, Therefore, equation (2) can be rewritten as,

$$ma=K_t i_m(t)-T_L(t)-B_m v \tag{3}$$

The state equation of the motor is established according to equations (1) ~ (3).
Define the position variable and speed variable are system state variables, let $x_1(t)=l_1(t)$ and $x_2(t)=v(t)$ the expression of the state is

$$\dot{x}(t)=Ax(t)+B(g(t)+u(t)) \tag{4}$$

and the expression of the output is

$$y(t)=Cx(t) \tag{5}$$

where $A=\begin{bmatrix} 0 & 1 \\ 0 & -\frac{K_t K_e}{R_a m} \end{bmatrix} \in R^{2\times 2}$, the state vector is $x(t)=\begin{bmatrix} x_1(t) \\ x_2(t) \end{bmatrix} \in R^2$, $B=\begin{bmatrix} 0 \\ \frac{K_t}{R_a m} \end{bmatrix} \in R^2$, $C=[1\ 0] \in R^2$;
$g(t)=-\frac{R_a m}{K_t}T_L-\frac{B_m R_a m}{K_t}x_2(t)$.
Next, the control strategy will be designed based on the above controlled object.

3. Closed-Loop Model Reference Adaptive Control Algorithm Bsed on Feedforward

According to the closed-loop model reference adaptive control principle, the control quantity is adjusted according to the deviation between the reference model output and the system output. There is a certain time delay, which reduces the robustness of the system. Based on the known information of interferometer control system - reference velocity and acceleration, a closed-loop model reference adaptive control strategy based on feedforward is proposed quantity with a certain gain. The whole control strategy block diagram is shown in figure 1.

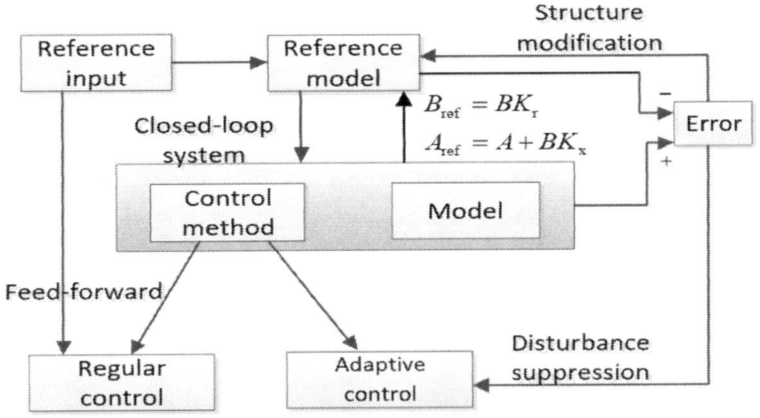

Figure 1. Block diagram of feedforward based closed-loop model reference adaptive control.

For equation (4), the state feedback adaptive rate is designed so that system $x(t)$ globally uniformly asymptotically tracks $x_{\text{ref}}(t) \in R^2$ of the reference model, and the reference model is written as

$$\dot{x}_{\text{ref}}(t) = A_{\text{ref}} x_{\text{ref}}(t) + B_{\text{ref}} r(t) + \gamma e(t) \tag{6}$$

Where $A_{\text{ref}} \in R^{2\times2}$ is Hurwitz matrix, $r(t) \in R$ is bounded reference information, $\gamma \in R^{2\times2}$ is positive feedback coefficient to be designed, and $\gamma^{\text{T}} = \gamma$. It is assumed that there are ideal known control gains $K_{\text{r}} \in R$ and $K_{\text{x}} \in R^2$, which satisfy $A_{\text{ref}} = A + BK_{\text{x}}$ and $B_{\text{ref}} = BK_{\text{r}}$.
The system perturbation is

$$g(t) = \theta^{\text{T}}(t)\varphi(x(t)), \quad g(t) \in R \tag{7}$$

where $\theta(t) \in R^m$ is unknown constant matrix, $\varphi(x(t)) \in R^m$ is an uncertain regression vector.
Adaptive control uses the adaptive law to make $x(t)$ globally consistent and asymptotically track the reference state $x_{\text{ref}}(t)$. It is necessary to ensure that the output of system remains globally consistent and bounded during servo tracking. Therefore, given any bounded instruction $r(t)$, the choice of control input $u(t)$ must satisfy $e(t) = x(t) - x_{\text{ref}}(t)$ asymptotically approaching 0 globally, that is, $\lim_{t\to\infty}\|x(t) - x_{\text{ref}}(t)\| = 0$ is established.
The conventional control law is

$$u_{\text{F0}}(t) = K_{\text{x}} x(t) + K_{\text{r}} r(t) + K_1 r_{\text{v}} + K_2 r_{\text{a}} \tag{8}$$

The adaptive control law is $u_{\text{a}}(t) = -\hat{\theta}^{\text{T}}(t)\varphi(x(t))$, and the total control law of the system is $u_{\text{f}}(t) = u_{\text{F0}}(t) + u_{\text{a}}(t)$, where, r_{v} and r_{a} represent the reference speed and reference acceleration respectively, K_1 and K_2 represent the feedforward gains. $\hat{\theta}(t)$ is the estimation of the ideal weight vector $\theta(t)$, defined $\dot{\hat{\theta}}(t)$ as the first derivative of the weight vector $\hat{\theta}(t)$, the update rate of the weight vector is $\dot{\hat{\theta}}(t) = \Gamma\varphi(x)e^{\text{T}}PB$, $\Gamma = \Gamma^{\text{T}} > 0$ is the adaptive rate, $P = P^{\text{T}} > 0$ and $Q = Q^{\text{T}} > 0$ satisfy the algebraic Lyapunov equation satisfying equation (8),

$$PA_{\text{ref}} + A_{\text{ref}}^{\text{T}} P = -Q \tag{9}$$

Bring equations (7) ~ (8) into equation (4-5), let $\Delta\theta(t) = \hat{\theta}(t) - \theta(t)$.
Define Lyapunov energy function:

$$V(e(t), \Delta\theta(t)) = e^{\text{T}}(t)Pe(t) + \Delta\theta^{\text{T}}(t)\Gamma^{-1}\Delta\theta(t) \tag{10}$$

Where (10) is satisfied $P = P^T > 0$, and the first derivative of Lyapunov energy function is obtained

$$\dot{V}(e(t), \Delta\theta(t)) = -e^{\text{T}}(t)Qe(t) - 2e^{\text{T}}(t)\gamma Pe(t) \le 0 \tag{11}$$

so the $x(t)$ is uniformly bounded , $\dot{x}(t)$ and $\dot{e}(t)$ are bounded
As

$$V(e(0), \Delta\theta(0)) \le \lambda_{\max}(P)e^{\text{T}}(0)e(0) + \frac{1}{\lambda_{\min}(\Gamma)}\Delta\theta^{\text{T}}(0)\Delta\theta(0) \tag{12}$$

in $[0, t]$, take integral at both ends of $\dot{V}(e(t), \Delta\theta(t)) = 2e^{\text{T}}(t)P(A_{\text{ref}} - \gamma)e(t)$, we have

$$\int_0^t \dot{V}(e(\tau_1),\Delta\theta(\tau_1))\mathrm{d}\tau_1 = \int_0^t (2e^{\mathrm{T}}(\tau)P(A_{\mathrm{ref}}-\gamma)e(\tau))\mathrm{d}\tau_1 \tag{13}$$

Define $\dot{V}(e(t),\Delta\theta(t)) = -e^T Qe - 2\gamma P \le 0$, the left side of equation (13) is less than zero, so

$$V(e(0),\Delta\theta(0)) - V(e(t),\Delta\theta(t)) \ge 2\lambda_{\min}(P)\lambda_{\min}(A_{\mathrm{ref}}-\gamma)\int_0^t \|e(\tau)\|^2 \mathrm{d}\tau_1 \tag{14}$$

Simultaneous (13) ~ (14), we have

$$\lambda_{\max}(P)e^{\mathrm{T}}(0)e(0) + \frac{1}{\lambda_{\min}(\Gamma)}\Delta\theta^{\mathrm{T}}(0)\Delta\theta(0) \ge 2\lambda_{\min}(P)\lambda_{\min}(A_{\mathrm{ref}}-\gamma) \tag{15}$$

$$\int_0^t \|e(\tau)\|^2 \mathrm{d}t\tau_1 \le \frac{\lambda_{\max}(P)e^{\mathrm{T}}(0)e(0) + \frac{1}{\lambda_{\min}(\Gamma)}\Delta\theta^{\mathrm{T}}(0)\Delta\theta(0)}{2\lambda_{\min}(P)\lambda_{\min}(A_{\mathrm{ref}}-\gamma)} \tag{16}$$

Therefore, the error L_2 is expressed as

$$\|e(t)\|_{L_2} \le \sqrt{\frac{\lambda_{\max}(P)e^{\mathrm{T}}(0)e(0) + \frac{1}{\lambda_{\min}(\Gamma)}\Delta\theta^{\mathrm{T}}(0)\Delta\theta(0)}{2\lambda_{\min}(P)\lambda_{\min}(A_{\mathrm{ref}}-\gamma)}} \tag{17}$$

From equation (17), it can be seen that $\|e(t)\|_{L_2}$ depends on parameters Γ and γ, the influence of initial value $\Delta\theta(0)$ of parameter estimation error on tracking error is reduced by increasing Γ, and by adjusting γ to reduce system tracking error, which is also the aspect that the closed-loop model reference adaptive control is better than that of the open-loop.

Next, the steady-state characteristics of the feedforward based closed-loop model reference adaptive control algorithm will be obtained by analyzing L_∞.

The control quantity of equation (17) is brought into the state space model of the controlled object

$$\begin{aligned}\dot{x}(t) &= Ax(t) + B\left(K_{\mathrm{x}}x(t) + K_{\mathrm{r}}r(t) + K_1 r_{\mathrm{v}} + K_2 r_{\mathrm{a}} - \hat{\theta}^{\mathrm{T}}(t)\varphi(x(t))\right) \\ &= (A + BK_{\mathrm{x}})x(t) + BK_{\mathrm{r}}r(t) + B\left(K_1 r_{\mathrm{v}} + K_2 r_{\mathrm{a}}\right) - B\hat{\theta}^{\mathrm{T}}(t)\varphi(x(t))\end{aligned} \tag{18}$$

The solution of the above formula is

$$\begin{aligned}x(s) &= \left(sI\text{-}A_{\mathrm{ref}}\right)^{-1} B_{\mathrm{ref}}r(s) + \left(sI\text{-}A_{\mathrm{ref}}\right)^{-1} B\left(K_1 r_{\mathrm{v}} + K_2 r_{\mathrm{a}} - \hat{\theta}^{\mathrm{T}}(s)\varphi(x(s))\right) \\ &\quad + \left(sI\text{-}A_{\mathrm{ref}}\right)^{-1} x(0)\end{aligned} \tag{19}$$

The solution of the reference model $\dot{x}_{\mathrm{ref}}(t) = A_{\mathrm{ref}}x_{\mathrm{ref}}(t) + B_{\mathrm{ref}}r(t) + \gamma e(t)$ is

$$x_{\mathrm{ref}}(s) = \left(sI\text{-}A_{\mathrm{ref}}\right)^{-1} B_{\mathrm{ref}}r(s) + \left(sI\text{-}A_{\mathrm{ref}}\right)^{-1}\gamma e(s) + \left(sI\text{-}A_{\mathrm{ref}}\right)^{-1} x_{\mathrm{ref}}(0) \tag{20}$$

Suppose that $x_{\mathrm{ref}}(0) = x(0)$, subtract equations (19) and (20) to obtain

$$e(s) = \frac{B\left(K_1 r_{\mathrm{v}} + K_2 r_{\mathrm{a}} - \hat{\theta}^{\mathrm{T}}(s)\varphi(x(s))\right)}{\left(sI\text{-}A_{\mathrm{ref}} + \gamma\right)} \tag{21}$$

For a uniformly asymptotically stable MIMO system, when the input and output are bounded, it is finite

2148 (2022) 012009 doi:10.1088/1742-6596/2148/1/012009

$$\|e(s)\|_{L_\infty} \le \left\|(sI\text{-}A_{\text{ref}}+\gamma)^{-1}B\right\|_{L_1} \left\|K_1 r_v + K_2 r_a - \hat{\theta}^{\text{T}}(s)\varphi(x(s))\right\|_{L_\infty} \tag{22}$$

The upper limit of tracking error $e(t)$ can be reduced by feedforward control parameters K_1 and K_2.

4. Simulation Results

$$\underbrace{\begin{bmatrix} 0 & 1 \\ 0 & -6.1615 \end{bmatrix}}_{A} + \underbrace{\begin{bmatrix} 0 \\ 0.5187 \end{bmatrix}}_{B} K_x = \underbrace{\begin{bmatrix} 0 & 1 \\ -\omega_n^2 & -2\xi\omega_n \end{bmatrix}}_{A_{\text{ref}}}, \underbrace{\begin{bmatrix} 0 \\ 0.5187 \end{bmatrix}}_{B} K_r = \underbrace{\begin{bmatrix} 0 \\ \omega_n^2 \end{bmatrix}}_{B_{\text{ref}}} \tag{23}$$

Let the feedback gains are $K_x = [-30.8482\ -1.2311]$, $K_r = [30.8482]$, and the adaptive law gains are $\Gamma = 10$, $Q = \begin{bmatrix} 10 & 0 \\ 0 & 10 \end{bmatrix}$.

The planning curves of the angle mirror movement are shown in the follwing figure, which includes the displacement curve, the velocity curve, and the acceleration curve. Next, curves will be used as the input of the control system to verify the control algorithm.

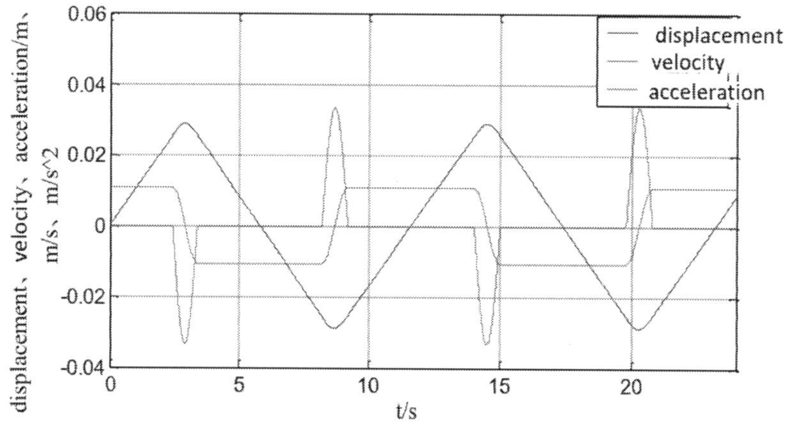

Figure 2. Displacement, velocitt and acceleration planning curves.

The thermal flutter disturbance is selected as the micro vibration model, and the micro vibration is composed of micro vibration sources with different frequencies and amplitudes. Set 20Hz as the dominant frequency of micro-vibration, superimpose two micro-vibration components of 10Hz and 30Hz, the amplitudes are 0.25 times and 0.5 times of the dominant frequency, and the initial phases are shifted by 30° and 45°, respectively. The expression is

$$\begin{aligned} d_f(t) &= \sum_{i=1}^{k} A_{fi} \sin(\Omega_i t + \Phi_i) \\ &= A_{f1}\sin(40\pi t) + \frac{3}{4}A_{f1}\sin(20\pi t + \frac{\pi}{6}) + \frac{1}{2}A_{f1}\sin(60\pi t + \frac{\pi}{4}) \end{aligned} \tag{24}$$

where A_{f1} is the undetermined constant.

In the simulation of suppressing micro-vibration disturbance, different A_{f1} values are taken to simulate the two algorithms. This paper analyzes and compares the two algorithms through the displacement output and velocity output, and estimate the maximum space micro-vibration disturbance torque that can be suppressed by the control algorithm.

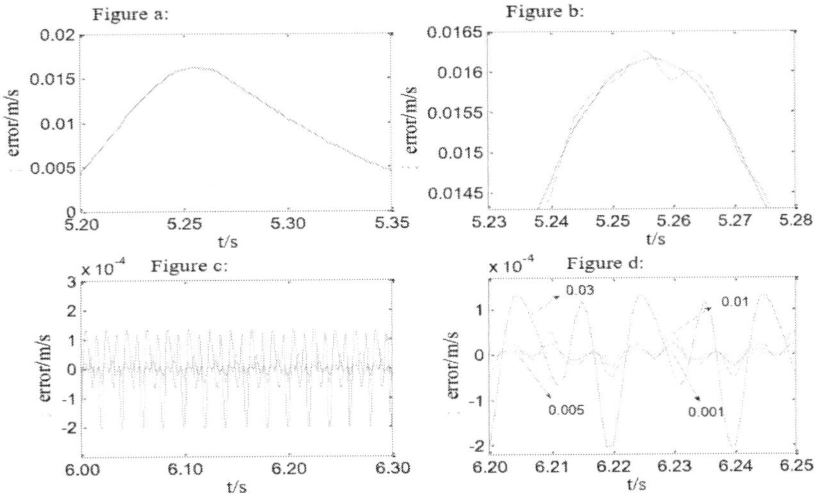

Figure 3. Velocity error curve of the closed-loop model reference adaptive control under different micro vibration amplitudes.

Figure 3 shows the control effects of the closed-loop model reference adaptive control algorithm on these disturbances when A_{f1} is 0.001N, 0.005N, 0.01N and 0.03N are taken in the micro vibration model (24). Figure a shows the velocity error curve of the corner mirror movement in the u-turn section, and the curve is locally enlarged, as shown in Figure b; Figure c is the velocity error of intercepting the uniform velocity section, and the curve is locally enlarged, as shown in Figure d. Because the disturbance is always in a fluctuating state, when the system is stable, the angle mirror can not achieve complete uniform scanning, and the velocity error is the fluctuating value. The velocity fluctuations in the uniform section caused by different micro vibration disturbance torque amplitudes are 99.63%, 99.58%, 97.57% and 96.24%.

Figure 4 shows the velocity errors of the control algorithm with feedforward to suppress different micro vibrations under the same simulation environment as the above conditions. The velocity fluctuations in the uniform section caused by different micro vibration disturbance torque amplitudes are 99.73%, 99.62%, 98.14% and 97.02%.

Comparing figure 2 and figure 3, the two control algorithms in the uniform speed section are not much different in suppressing chattering. The advantage of the adaptive algorithm with feedforward control in the u-turn section is more obvious, and the speed error is one order of magnitude lower than that of the non-feedforward control algorithm. The u-turn angle mirror adopts sine function motion, the u-turn section velocity fluctuation with feedforward control strategy is small, the response is fast, the time to adjust to the uniform speed is short, the optical path scanning efficiency is improved, and the system has good robustness.

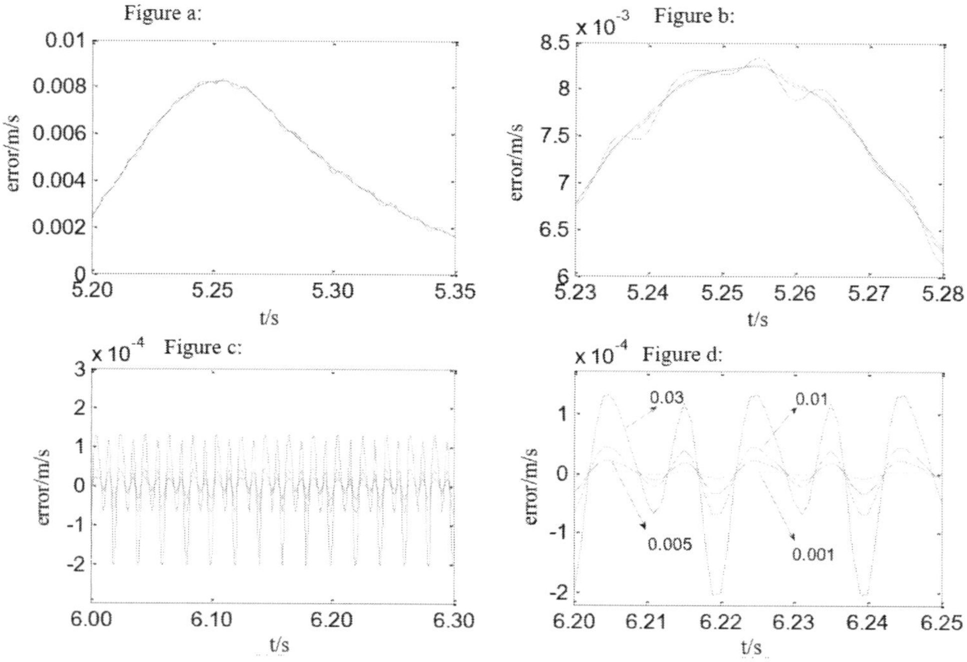

Figure 4. Velocity error curve with feedforward control under under different micro vibration amplitudes.

5. Conclusion

The translational optical path reciprocating scanning system based on adaptive control strategy studied in this paper has long scanning stroke, small motion quality and short optical path, which is more suitable for large field of view space-borne detection instruments. It has laid the foundation for the development of time-modulated Fourier spectrometers with high-precision and high-resolution spectral detection capabilities.

References

[1] Lin L, Tan Y, Lin K. 2018 The influence of flywheel micro vibration on space camera and vibration suppression[J] Mechanical *Systems and Signal Processing* **100** (1) 360-370.

[2] Wu P F, Xu Z L, Wang X Y, et al. 2018 Micro-vibration environment parameters measurement and analysis of a novel satellite in orbit[J]. *International Symposium of Space Optical Instrument and Application* 305-314.

[3] Yun H, Liu L, Li Q, et al. 2020 Investigation on two-stage vibration suppression and precision pointing for space optical payloads[J]. *Aerospace Science and Technology* 96 105543.

[4] Guo L J, Wang H, Wang C, Ma W P, Lin Z. 2019 Optical Path Scanning Control System Based on Modified Repetitive Control[J] *Spacecraft Recovery & Remote Sensing* **40** (1) 32-40.

[5] Hou L Z, Xu P M, Wang C Q. 2018 Design and Implementation of Scan-arm Corner Cube Interferometer with Large OPD and High Robustness[J] *Spacecraft Recovery & Remote Sensing* **39** (3) 51-59.

[6] Cui Y L, Feng Z W, Liu J Y. 2019 Micro-vibration influencing characteristic on image qualitt and solutions for high-resolution satellites[J] *Advances in Astronautics Science and Technology* 2 7–13.

Mathematical Relation between Concurrence and Intensity of a Photon in the Quantum Delayed-Choice Experiment

Firat Diker

Faculty of Engineering and Natural Sciences, Sabanci University, Orta Mahalle, Tuzla, Istanbul, Turkey.
Email: firatdiker@sabanciuniv.edu

Abstract. I provide the explicit mathematical expression of the correlation between concurrence and the intensity of a photon in the quantum delayed-choice experiment. This shows us that one may observe the wave-particle duality and entanglement simultaneously.

1. Introduction

Since the pioneering work by Einstein, Podolsky, and Rosen [1], quantum nonlocality has been widely discussed [2-4]. It is directly in connection with entanglement, which is the main ingredient for many quantum information processing tasks and experiments such as the quantum delayed-choice experiment (QDCE) [5] adapted from Wheeler's experiment on the notion of wave-particle duality [6,7]. This fundamental characteristic has been discussed broadly by the scientific community [6,8]; however, nobody had found a clear answer to the question as to whether a photon knows in advance how to act under certain conditions. To answer this question, Wheeler proposed a gedanken experiment [6,7] testing the hidden variable model. The result of this experiment is that a photon may act as a wave or particle, depending on the arrangement of the experimental apparatus. The photon cannot know what it will be confronted with because the changes are made while it is already inside the interferometer. The delayed-choice method has been realized experimentally [9,10]. Wheeler's proposal has been adapted into the mod-based optical scheme where one benefits from entanglement. This adaptation allows us to observe the particle and wave nature of a photon through its intensity by accessing mode-based quantum states [5,11]. The same experiment is also realizable by using polarization-based photons [12].

2. Method

The QDCE allowed us to observe both the particle and wave aspects of a photon. This experiment transforms a two-photon state by using single- and two-qubit quantum gates (figure 1). The essential part of the setup is the control-Hadamard (c-Had) gate, which is a two-qubit quantum gate transforming the target qubit if the control qubit is in state $|1\rangle$. States $|0\rangle$ and $|1\rangle$ correspond to spatial modes. As mentioned earlier, one makes use of entangled two-qubit states to observe the wave-particle duality. The c-Had gate entangles two photons when applied, resulting in the following state:

$$\Psi(\alpha,\phi) = \cos\alpha\,(\tfrac{1}{\sqrt{2}}(|0\rangle_\mathrm{s} + \mathrm{e}^{i\phi}|1\rangle_\mathrm{s}))|0\rangle_\mathrm{a} + \sin\alpha\,(\cos\tfrac{\phi}{2}|0\rangle_\mathrm{s} - \mathrm{i}\sin\tfrac{\phi}{2}|1\rangle_\mathrm{s})|1\rangle_\mathrm{a} \quad (1)$$

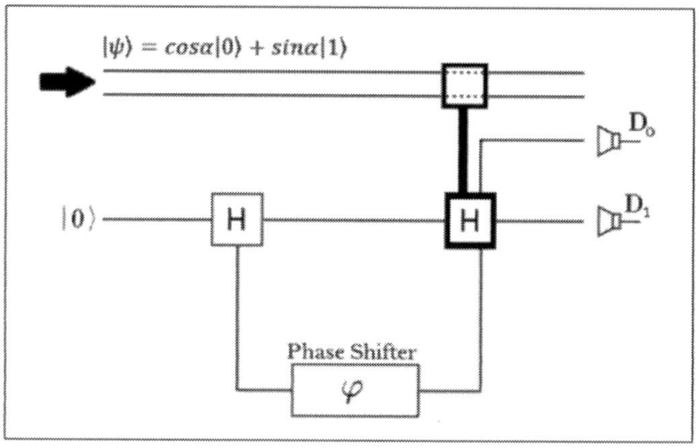

Figure 1. The ancillary photon on the upper side of the figure and the system photon on the lower side of the figure are sent into the optical circuit. The beam splitter acts as a Hadamard gate, and the system photon is split into two modes. Then, one of the modes is phase-shifted. This wave-like property can be directly seen by measuring the intensity of the system photon. The c-Had gate acts on both photons. As a result, the biphotonic state is obtained. Intensities measured at the end depend on the phase (wave-like property) and the relative amplitude (particle-like property).

Where a and s represent the ancillary photon and the system photon, respectively. Due to superposition, the global state of the system in equation (1) includes both cases (whether the second beam splitter acts or not). One can easily calculate the intensity I_0 at detector D_0,

$$I_0(\alpha,\phi) = \frac{1}{2}\cos^2\alpha + \cos^2\frac{\phi}{2}\sin^2\alpha \quad (2)$$

which is a function of the relative phase between two modes, and relative amplitude factors. When the system photon is detected, its intensity informs us of both the wave and particle nature of a photon.

The general expression of the global state on which we perform measurements is given in equation (1). To the best of our knowledge, nobody has investigated this experiment from the point of concurrence implicitly embedded in the intensity equation. We use the general concurrence expression for two-qubit states, $C(\Phi) = \langle\Phi|\tilde{\Phi}\rangle$ [13]. We apply this to equation (1) and obtain

$$C(\alpha,\phi) = \frac{1}{\sqrt{2}}\left(|\sin 2\alpha|\sqrt{1+\sin^2\phi}\right). \quad (3)$$

Concurrence is a function of relative phase and amplitude. This implies a correlation between the degree of entanglement and the intensity of the system photon. Both equation (2) and equation (3) are trigonometric functions, each of which can be defined in terms of other (together with α and ϕ). Following this, we derive the equation,

$$I_0(\alpha,\phi) = \frac{1}{2}\left(\cos^2\alpha + \cos^2\frac{\phi}{2}\,\frac{C(\alpha,\phi)^2}{\cos^2\alpha\,(1+\sin^2\phi)}\right). \quad (4)$$

The intensity equation is correlated with concurrence. Notice that concurrence is not independent of other trigonometric terms since equation (2) and equation (3) are the functions of the same parameters. One observes not only wave- and particle-like properties but also the entanglement between the ancillary photon and the system photon. The expression we have found in equation (4) is the direct observation of entanglement through intensity. The c-Had gate is the crucial component entangling photons.

3. Conclusion

One may observe two unique features of quantumness: the wave-particle duality and entanglement because they are strongly correlated in the QDCE. The deterministic generation of entanglement between the wave and particle states of two photons has been realized [15]; however, entanglement (together with wave and particle properties) is already possible to observe in the QDCE. The QDCE has been generalized in a multi-path scheme, including generalized control-Hadamard gates [16]. It is an open problem whether simultaneous observation of entanglement and the wave-particle duality in a generalized scenario is possible. Our result paves the way for a better understanding of the connection between entanglement and the wave-particle duality.

References

[1] Einstein A, Podolsky B and Rosen N. 1935 *Phys. Rev.* **47** 777-80.
[2] Herbert N and Karush J. 1978 Found. Phys. 8 313-17.
[3] Clauser J F and Shimony A. 1978 Rep. Prog. Phys. 41 1881-927.
[4] Stapp H P 1982 Phys. Rev. Lett 49 1470-74.
[5] Ionicioiu R and Terno D R. 2011 Phys. Rev. Lett 107 230406.
[6] Wheeler J A and Zurek W H. 1984 Quantum Theory and Measurement (Princeton University).
[7] Wheeler J A and Marlow A R. 1978 Mathematical Foundations of Quantum Mechanics (Academic).
[8] Feynman R P, Leighton R B and Sands M L. 1965 Lecture notes on Physics (Addison-Wesley).
[9] Jacques V, Wu E, Grosshans F, Treussart F, Grangier P, Aspect A and Roch J F. 2007 Science 315 966-68.
[10] Manning A G, Khakimov R I, Dall R G and Truscott A G. 2015 Nat. Phys 11 539-42.
[11] Peruzzo A, Shadbolt P, Brunner N, Popescu S and O'Brien J L. 2012 Science 338 634-37.
[12] Diker F. 2018 J. Opt. Soc. Am. B 35, 284-86.
[13] Wootters W K. 2001 Quantum Inf. Comput **1** (1) 27-44.
[14] Clauser J F, Horne M, Shimony A and Holt R A. 1969 Phys. Rev. Lett 23 880-84.
[15] Rab A S, Polino E, Man Z X, An N B, Xia Y J, Spagnolo N, Franco R L and Sciarrino F. 2017 Nat Commun **8** (1) 1-7.
[16] Chen X et al. 2021 Nat. Commun **12** (1) 1-10.

Simulation Study of Single-Event Burnout for the 4H-SiC VDMOSFET with N+ Split Source

Jiancheng Zhou, Chenghao Yu* and Ying Wang

Key Laboratory of RF Circuits and Systems, Ministry of Education, Hangzhou Dian zi University, Hangzhou, China.

*Email: yuchenghao@hdu.edu.cn

Abstract. Silicon Carbide (SiC) power MOSFET is the next generation device in the supply system of spacecraft. However, the current degradation or catastrophic failure of the power device could be induced when a drain voltage exceeds critical condition. In this article, an improved VDMOSFET structure for the Single-Event Burnout (SEB) is demonstrated. The improved power VDMOSFET includes a P+ shielding region at the JFET region. Meanwhile, forming a CSL layer by ion-implantation at the JFET to reduce the specific on-resistance. The device is etched in both sides to form trench and then implanting N-type impurities at the side walls of the trench to form the N+ split source (SDS-VDMOSFET). The 2-D numerical simulator Silvaco Atlas was used to study the SEB performance for the 1.2 kV-rated SiC SDS-VDMOSFET in a high linear energy transfer (LET) value of 0.5 pC/μm. The simulation results show that the improved structure can effectively reduce the peak lattice temperature induced by heavy-ion and increase the SEB threshold voltage compared with the standard VDMOSFET. Furthermore, the improved structure also presents a lower specific on-resistance. As a result, the maximum temperature of the standard VDMOSFET has exceeded 3000 K at a drain voltage of 400 V. However, the maximum temperature of the improved VDMOSFET is only 2090 K at a drain voltage of 800 V.

1. Introduction

Silicon carbide (SiC) power MOSFETs have presented several compelling characteristics, such as, high thermal conductivity, high-power, and high critical breakdown voltage [1-3]. With the continuous development of the aerospace technology, especially for the deep space probe and other extreme conditions, the SiC MOSFETs are significantly suitable. However, many researchers have found that SiC power MOSFETs are susceptible to heavy-ion radiation in the space environment [4-6]. The high-energy heavy-ion can cause serious radiation damage in the SiC MOSFET even at the drain voltages below the breakdown voltage (BV). For example, the catastrophic Single-Event-Burnout (SEB) and the Single-Event Gate Rupture (SEGR) can be induced by the high-energy heavy-ion.

Catastrophic failure has been found to occur at about half the rated BV for the commercial VDMOSFET devices by the heavy-ion irradiation experiments [7-9]. So far, many researches have been carried out for the SiC VDMOSFET to investigate the mechanism of SEB event [5-9]. The most effective and common way to improve the anti-irradiation ability of the SiC MOSFET is adding a buffer layer between the epilayer and substrate to decrease the strong electric field and high impact ionization rate after an ion's strike. The lattice temperature of the device after a heavy-ion's impact could also be decreased further by optimizing the buff layer parameters. It has been proved by Lu et al.

[10] that the SEB threshold voltage could be effectively improved by adding an optimized buff layer. But it is inevitable that the on-resistance of the device would also increase.

In this article, a way to improve the SEB performance at high linear energy transfer (LET) value and decrease the specific on-resistance ($R_{on,sp}$) of the device is proposed. In the structure, a N+ split source is proposed to introduce a horizontal electric field to make the distribution of hole current induced by heavy-ion become more uniform that the SEB performance can be improved. In addition, the multi-buffer layers are also added to modulate the strong electric field as mentioned above. Furthermore, the Current spreading Layer (CSL) is added to suppress the $R_{on,sp}$, and the P+ shielding region at the JFET region is used to reduce the electric field of the gate oxide are both considered. As a result, the maximum lattice temperature is simulated at 2090 K with a LET value of 0.5 pC/μm when the drain voltage is 800V.

2. Description of Device Structure and SEB Mechanism

Figure 1(a) shows a cross-sectional view of the 4H-SiC Standard-VDMOSFET, and figure 1(b) shows a schematic cross section of the improved VDMOSFET. The total thickness of the optimized multi-buffer layers (four buffer layers) is 5.2 μm, and the thickness of each buffer layer is equal. The depth of P-region and N+ split source is 1.6 and 1.0 μm, respectively. The trench will be filled with the metal to form the source electrode. The doping concentrations of the four buffer layers increase sequentially with the same multiplication factor. The CSL, drift, and substrate regions are all uniformly doped with the concentration of 1.0×10^{17}, 8.0×10^{15}, and 5×10^{18} cm^{-3}. In the P- base region, the gauss doping profiles is used and the peak doping concentrations is 3.1×10^{18} cm^{-3} with a depth of 0.5 μm. The main parameters are demonstrated in the table 1.

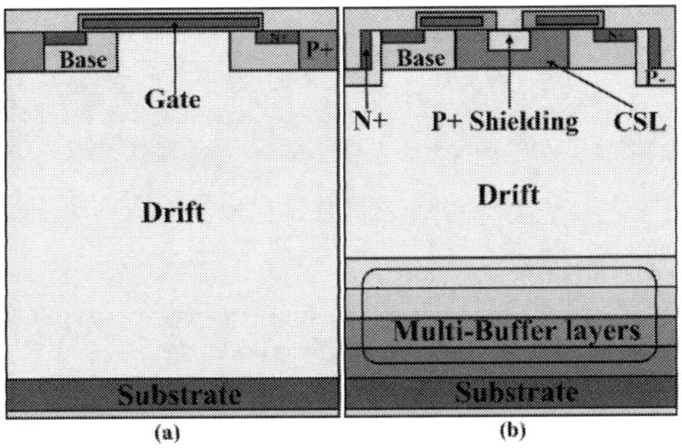

Figure 1. Cross-sectional view of 1.2-kV SiC power VDMOSFET.

(a) Standard-VDMOSFET (b) SDS-VDMOSFET.

Table 1. The main parameters.

Structure parameters	Standard-VDMOSFET	SDS-VDMOSFET
Drift thickness(μm)	10.0	8.8
CSL doping concentration(cm^{-3})	--	1.0×10^{17}
P+ shielding doping concentration(cm^{-3})	--	1.0×10^{18}
N+ split source Thickness(μm)	--	0.3

N+ split source concentration(cm^{-3})	--	1.0×10^{18}
P- base concentration(cm^{-3})	--	5.0×10^{17}

The fabrication process of the SDS-VDMOSFET is shown in figure 2 (a~h). Firstly, the desired epitaxial wafer is prepared by forming the multiple-buffer layers and drift layer with the chemical vapor deposition. Secondly, the ion implantation is used to form the P- base and the P+ shielding at the JFET region. Afterwards, the N+ Source is formed by the ion implantation. Next, the trench is formed by the etching process. Then, at the bottom and side wall of the trench, the P+ shielding region and N+ split source is obtained by the multiple-ion implantation. Finally, the trenches are filled with metal, and it is also deposited on the surface of the device to form the electrodes.

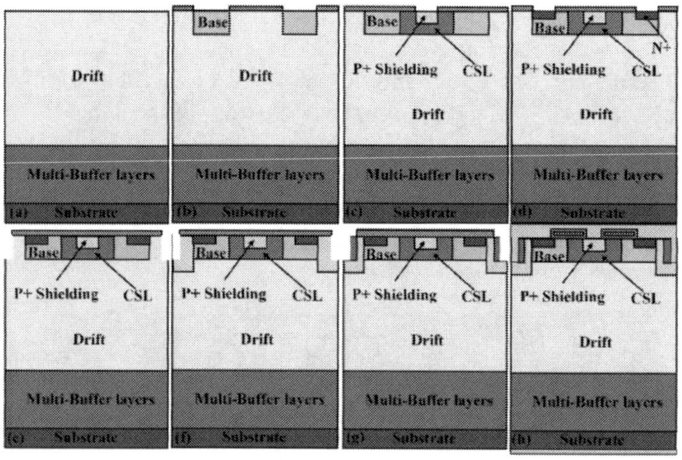

Figure 2. The fabrication process of the SDS-VDMOSFET.

3. Numerical Simulation Results and Discussion

In this section, the simulation results of the Standard-VDMOSFET and SDS-VDMOSFET are comparatively investigated with a series of photographs and analysis.

3.1. Basic Electric Characteristic

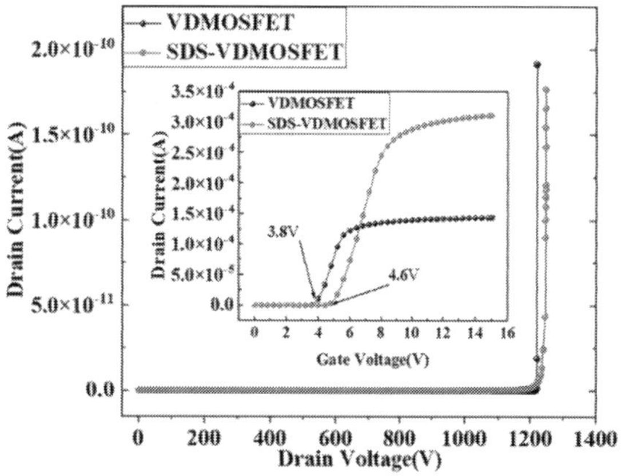

Figure 3. The breakdown voltage characteristics of structures.

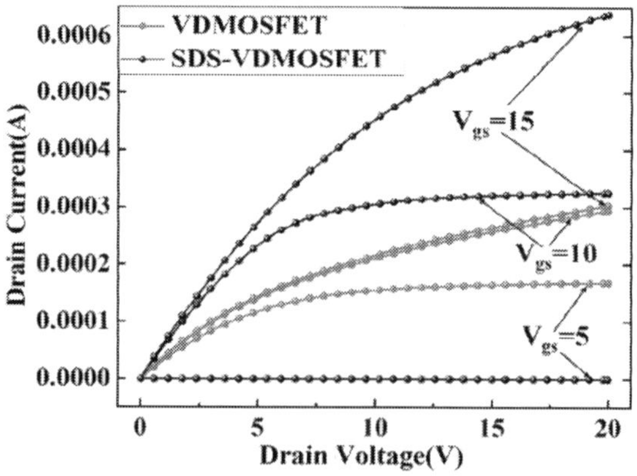

Figure 4. The ON-state characteristic curves of structures.

Figure 3 shows the BV characteristics of the two structures. The BVs of the Standard-VDMOSFET and SDS-VDMOSFET are 1220 and 1230 V, respectively. The threshold voltages of the two devices show different values of 3.8 and 4.6 V because of the different doping concentration in the P-base regions. However, there is a big gap in their $R_{on,sp}$, although the doping profiles in the drift regions are the same. Figure 4 shows the ON-state performance of the two devices. As a result, the $R_{on,sp}$ of the Standard-VDMOSFET and SDS-VDMOSFET are 3.7 and 2.1 m$\Omega\cdot$cm^2 at V_{gs}=15 V.

3.2. SEB Characteristic

The SEB characteristics of the two structures were simulated by using the 2-D simulator Silvaco ATLAS. The simulation is based on some physical models which include the ANALYTIC model, the concentration dependent mobility (CONMOB) model, the parallel electric field dependent mobility (FLDMOB) model. A bandgap narrowing (BGN) model was needed due to the presence of highly doped regions. The recombination models include the SRH model with doping and temperature dependency and the AUGER model. The incomplete ionization model and the Selberherr model were used to calculate the impact ionization rate, as well as the lattice self-heating model (LAT.TEMP) to compute the device temperature.

The heavy ion model is described by (1), the track charge generation rate for a vertical track is the function of the spatial and temporal Gaussian. The parameters used in simulation are list in table 2.

$$rate(x,y) = \frac{LET}{q\pi\omega_0 T_C} exp\left(-\frac{(x-x_0)^2}{\omega_0^2}\right)\cdot exp(-\frac{(t-T_0)^2}{T_C^2}) \quad (1)$$

Table 2. The parameters used in simulation.

LET (pC/μm)	Track Radius ω_0 (μm)	Initial Charge Generation Time T_0 (s)	Temporal Gaussian Function Width T_C (s)	Lateral Ion Striking Position x_0 (μm)
0.5	0.05	4×10^{-12}	2×10^{-12}	x_0

4. SEB Simulation Results and Discussion

4.1. Simulation Results for the Standard-VDMOSFET

In SiC power MOSFETs, the SiC material will directly sublimate rather than melt when the temperature is higher than 2100~2200 K. In order to find the most sensitive position of the Standard-VDMOSFET, as shown in figure 5, we have simulated the SEB characteristics of the device at different positions. At last, we found that the maximum temperature of the device has exceeded 4000 K which locates at the drift/substrate junction. The electric field in the oxide layer also far exceeds the critical value when the heavy-ion strikes in the center of the JFET region, as shown in this figure. This demonstrates that the SEB and SEGR events both occur in the Standard-VDMOSFET at V_{ds}=300 V. So, the most sensitive position is in the P- base area of the Standard-VDMOSFET.

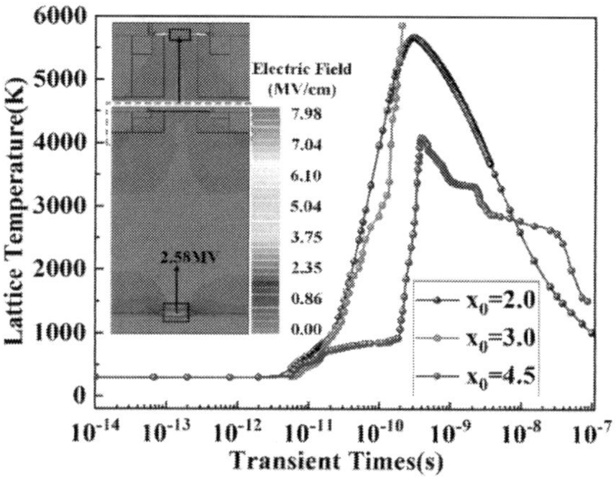

Figure 5. The maximum temperature of Standard-VDMOSFET at different position when V_{ds}=300 V LET=0.5 pC/μm.

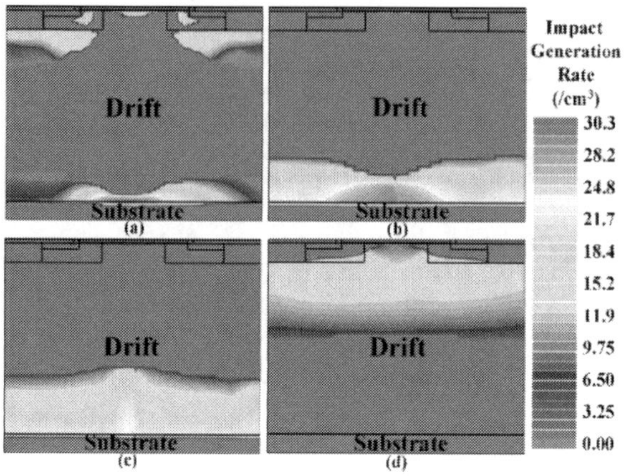

Figure 6. The impact generation rate distribution at different times. (a) t=0.1ns (b) t=0.5ns (c) t=1ns (d) t=10ns.

The impact generation rate of the Standard-MOSFET after the heavy-ion strikes in the JFET region at V_{ds}=300 V is also illustrated in figure 6. From the figure, we can see the highest impact generation rate is located in the drift/substrate homogeneous junction due to the strong electric field which could also lead to the self-heating effect to make the device burnout.

4.2. SEB Performance for the SDS-VDMOSFET

The characteristics of a SEB has been studied with the different drain voltage while keeping the gate voltage constant at 0 V. The electric field at the drift/substrate homogeneous junction can be decreased by adding the multi-buffer layers. And the electric field of the oxide can also be reduced by forming the P+ shielding region in the JFET region. In the figure 7, the electric field distribution of the device is transformed to the upper region and gradually decrease. Eventually, the largest electric field is located in the P+ shielding which prevents the gate oxide from failure.

As shown in the figure 8, a part of hole current flows out of the device from the N+ split source region, which can make the distribution of the hole current induced by heavy-ion more uniform to suppress the operation of the parasitic BJT. The presence of the N+ split source region reduces the hole current flowing through the N+ source of the SDS-VDMOSFET and also suppresses the temperature in the N+ source region, as illustrated in the figure 8.

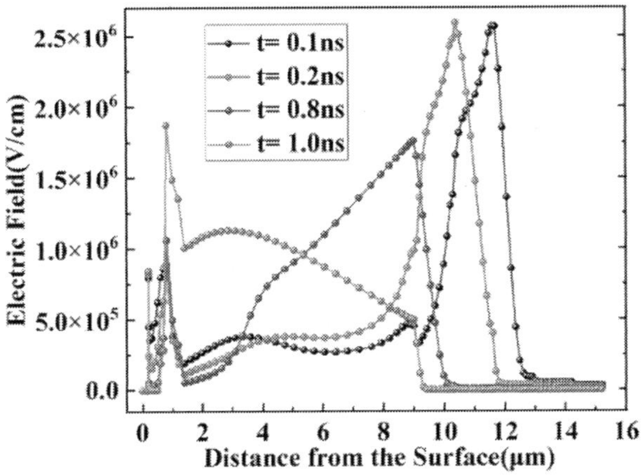

Figure 7. The electric field distribution along the ion track when the V_{ds}=800 V and LET=0.5 pC/μm at x_0=4.5 μm.

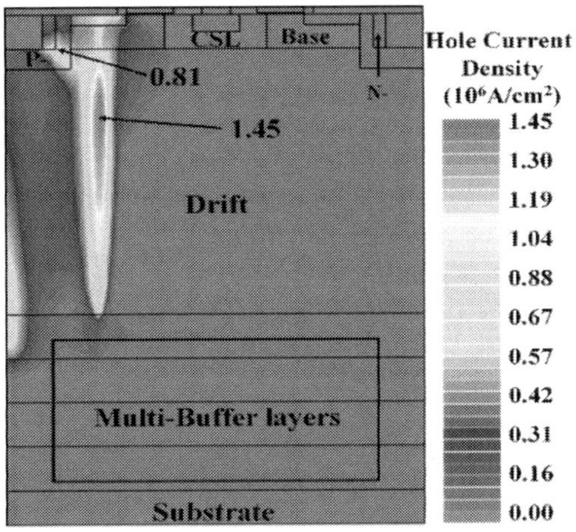

Figure 8. The hole current distribution of the SDS-VDMOSFET.

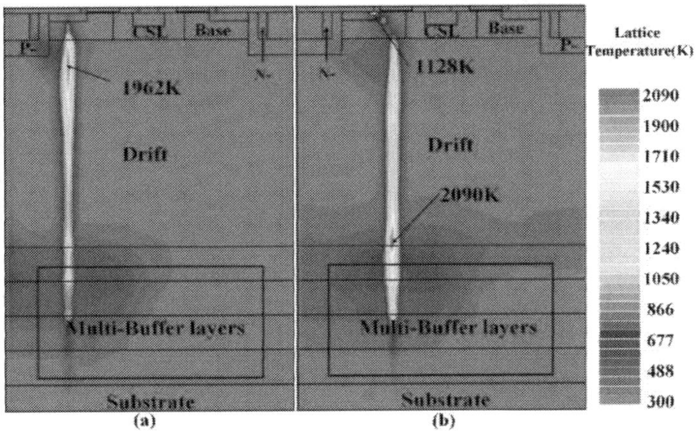

Figure 9. Lattice temperature distribution of the SDS-VDMOSFET at V_{ds}=800 V LET=0.5 pC/μm. (a) ion strikes position x_0=2.0 μm (b) ion strikes position x_0=3.0 μm.

In figure 9 (a), the maximum temperature of the device is 1962 K when the ion incidences in the source region. In the figure 9 (b), the maximum temperature is 2090 K which locates in the drift/buffer homojunction. The temperature of the electrode contact is only 1128 K at this condition, so it does not have a significant effect on the source contact metal.

5. Conclusion

In this article, the anti-irradiation performance of the 1.2-kV VDMOSFETs with the N+ split source and multi-buffer layers is studied. Numerical simulations are used to illustrate the SEB effect of the device. The final results demonstrate that the lattice temperature of the device could be reduced to 2090 K at V_{ds}=800 V with LET=0.5 pC/μm. This method not only increases the SEB threshold voltage, but also suppress the occurrence of the SEGR effect. At the same time, the $R_{sp,on}$ is reduced to 2.1 mΩ·cm^2 by adding a CSL region. So, the proposed device could effectively reduce the energy loss of the system.

References

[1] Elasser A, Chow T P. 2002 Silicon carbide benefits and advantages for power electronics circuits and systems [J] *Proceedings of the IEEE* **90** (6) 969-986 DOI: 10.1109/jproc.2002.1021562.

[2] Bhatnagar M, and Baliga B J. 1993 Comparison of 6H-SiC, 3C-SiC, and Si for power devices [J] *IEEE Transactions on Electron Devices* **40** (3) 645-655 DOI: 10.1109/16.199372.

[3] Parashar S, Kumar A and Bhattacharya S. 2018 High Power Medium Voltage Converters Enabled by High Voltage SiC Power Devices [J] *International Power Electronics Conference (IPEC-Niigata 2018 -ECCE Asia)* 3993-4000, DOI: 10.23919/IPEC.2018.8506674.

[4] Imaizumi M and Miura N. 2015 Characteristics of 600, 1200, and 3300 V Planar SiC-MOSFETs for Energy Conversion Applications [J] *IEEE Transactions on Electron Devices* **62** (2) 390-395 DOI: 10.1109/TED.2014.2358581.

[5] Zhou X, Jia Y, Hu D and Wu Y. 2019 A Simulation-Based Comparison Between Si and SiC MOSFETs on Single-Event Burnout Susceptibility [J] *IEEE Transactions on Electron Devices* **66** (6) 2551-2556 DOI: 10.1109/TED.2019.2908970.

[6] Krishnamurthy S, Kannan R, Hussin F A and Yahya E A. 2019 Enhanced Trench Shielded Power UMOSFET for Single Event Burnout Hardening [J] *IEEE Regional Symposium on Micro and Nanoelectronics (RSM)* 91-94 DOI: 10.1109/RSM46715.2019.8943495.

[7] Zerarka M, Austin P, Toulon G, Morancho F, Aebess H, Tasselli J. 2012 Behavioral Study of Single-Event Burnout in Power Devices for Natural Radiation Environment Applications [J] *IEEE Transactions on Electron Devices* **59** (12) 3482-3488 DOI: 10.1109/TED.2012.2222889.

[8] Shoji T, Nishida S, Hamada K, Tadano H. 2015 Analysis of neutron-induced single-event burnout in sic power mosfets [J] *Microelectronics Reliability* **55** (s9–10) S0026271415001730 DOI: 10.1016/j.microrel.2015.06.081.

[9] Martinella C et al. 2021 Impact of Terrestrial Neutrons on the Reliability of SiC VD-MOSFET Technologies [J] *IEEE Transactions on Nuclear Science* **68** (5) 634-641 DOI: 10.1109/TNS.2021.3065122.

[10] Lu J et al. 2020 Impact of Varied Buffer Layer Designs on Single-Event Response of 1.2-kV SiC Power MOSFETs [J] *IEEE Transactions on Electron Devices* **67** (9) 3698-3704 DOI: 10.1109/TED.2020.3008398.

Thermodynamics of Generalized Fermi Systems in a Harmonic Trap

Yushan Li

School of Physics and Electronic Engineering, Heze University, Heze, Shandong, 274015, China
Email: lysh507@163.com

Abstract. Thermodynamics of the generalized ideal Fermi systems in the two- and three-dimensional harmonic traps are respectively calculated by the Tsallis entropy in this paper. The influences of the trap and q-number on the thermodynamic parameters (epically the heat capacity) are analysed in detail. The results yield a well agreement with the classical cases.

1. Introduction

There are two subsystems named fermions and bosons in the nature. Fermions obey Fermi-Dirac distribution. Bosons obey Bose-Einstein distribution. This results that they appear completely different physical properties, especially at the region of low temperature. Under the region of low temperature, Bose system in a harmonic trap shows the unique Bose-Einstein condensation (BEC) phenomenon. Fermi system also can produce the similar BCS (Bardeen, Cooper and Schrieffer)-BEC crossover in certain special cases due to their mutual attraction of the nearest particles. Therefore, by studying the characteristics of Fermi system in an external potential, the researchers can better understand the BEC phenomenon.

Recently, many researchers develop and improve q-generalized entropy theory proposed by Tsallis through introducing the atomic interactions. The new q-generalized entropy theory yields a better case with the actual system. It is necessary to study the thermodynamic properties of Fermi system by using the extensive statistical theory [1-8]. In this paper, we firstly discuss the thermodynamic properties of the generalized ideal Fermi system in the free space based on the Fermi-Dirac distribution function in section 2. Furtherly, we obtain the corresponding cases of the generalized ideal Fermi system in the two- (2D) and three-dimensional (3D) harmonic traps in section 3 and section 4, respectively. Lastly, we calculate the related particle number, internal energy and heat capacity and discuss the influence of changing q-number.

2. Theory and Method

According to the non-generalized statistical theory, Tsallis entropy is expressed by

$$S_q = \left\langle \ln_q \frac{1}{p_i} \right\rangle = \sum_{i=1}^{w} p_i \ln_q \frac{1}{p_i} = \frac{1-\sum_{i=1}^{w} p_i^q}{q-1} . \tag{1}$$

After approximate factorization Eq. (1), one can obtain the following generalized Fermi-Dirac distribution function

$$n_q = \frac{1}{[1+(q-1)\beta(\varepsilon-\mu)]^{1/(q-1)}+1}. \tag{2}$$

In Eq. (2), n_q is the average particle number of the micro states with energy ε. μ is the chemical potential. q is the extensive parameter. When q is fixed to 1, Eq. (2) is transitioned into the classical Fermi Dirac distribution, which yields to

$$a_l = \frac{\omega_l}{e^{\alpha+\beta\varepsilon_l}+1}. \tag{3}$$

Because of the limitation of Pauli exclusion principle, the Fermi system should have the limitation $0\leq n_q \leq 1$. Therefore, according to Eq. (2), one can obtained

$$1+(q-1)\beta(\varepsilon-\mu) \geq 0 \tag{4}$$

Eq. (4) is rewritten as the following expression for arbitrary q

$$\varepsilon \begin{cases} \leq \mu+\dfrac{1}{(1-q)}\beta & (q\leq \ ) \\ \geq \mu-\dfrac{1}{(q-1)\beta} & (q\geq \ ) \end{cases}. \tag{5}$$

When it satisfies the condition $\varepsilon-\mu>0$, the equation will become $[1+(q-1)\beta(\varepsilon-\mu)]^{1/(q-1)}>1$. When it satisfies the condition $\varepsilon-\mu<0$, the equation becomes $[1+(q-1)\beta(\varepsilon-\mu)]^{1/(q-1)}<1$. One can introduce a generalized fugacity z_q, which is denoted by

$$z_q = \left[1+(1-q)\beta\mu\right]^{1/(1-q)}. \tag{6}$$

Thus, we obtain

$$[1+(1-q)\beta(\varepsilon-\mu)]^{1/(q-1)} = z_q^{-1}[1+(q-1)\beta z_q^{q-1}\varepsilon]^{1/(q-1)}. \tag{7}$$

By Eq. (2), Eq. (4) and the Taylor series $\frac{1}{1+z}=\sum_{i=0}^{\infty}(-1)^i z^i$, which reads

$$n_q = \begin{cases} \sum_{i=0}^{\infty}(-1)^{i-1} z_q^{i}[1+(q-1)\beta z_q^{q-1}\varepsilon]^{-i/(q-1)} & (\varepsilon>\mu) \\ 1/2 & (\varepsilon=\mu) \\ \sum_{i=0}^{\infty}(-1)^i z_q^{-i}[1+(q-1)\beta z_q^{q-1}\varepsilon]^{i/(q-1)} & (\varepsilon<\mu) \end{cases}. \tag{8}$$

3. Generalized Fermi Systems in a 2D Harmonic Trap

For the generalized ideal Fermi system in the 2D isotropic harmonic trap, the single particle Hamiltonian is written as

$$H = \frac{p_x^2+p_y^2}{2m}+\frac{1}{2}m\omega^2\left(x^2+y^2\right). \tag{9}$$

For the case of $q \leq 1$ and $\mu < 0$, one can obtain the analytical expressions of particle number, internal energy and heat capacity as follows

$$N = \frac{4g\pi^2 (k_B T)^2}{(h\omega)^2} f_{q,2}(z_q), \tag{10}$$

$$E = 4k_B TN \frac{f_{q,3}(z_q)}{f_{q,2}(z_q)}, \tag{11}$$

$$C_{V,q} = 4Nk_B \left[3\frac{f_{q,3}(z_q)}{f_{q,2}(z_q)} - 2\frac{f_{q,2}(z_q)}{f_{q,1}(z_q)} \right]. \tag{12}$$

The above-mentioned generalized Fermi integral $f_{q,D}(z_q)$ satisfies

$$f_{q,D}(z_q) = \sum_{i=0}^{\infty} \frac{(-1)^{i-1} z_q^{i+(1-q)D} \Gamma\left(\frac{i}{1-q}+1\right)}{(1-q)^D \Gamma\left(\frac{i}{1-q}+D+1\right)}. \tag{13}$$

For the case of $\mu \geq 0$, one can obtain the analytical expressions of particle number, internal energy and heat capacity as follows

$$N = \frac{4g\pi^2 (k_B T)^2}{(h\omega)^2} f_{q,2}(z_q), \tag{14}$$

$$E = 2k_B TN \frac{f_{q,3}(z_q)}{f_{q,2}(z_q)}, \tag{15}$$

$$C_{V,q} = k_B N \left[3\frac{f_{q,3}(z_q)}{f_{q,2}(z_q)} - 2\frac{\partial f_{q,3}(z_q)/\partial z_q}{\partial f_{q,2}(z_q)/\partial z_q} \right]. \tag{16}$$

Generalized Fermi integral $f_{q,D}(z_q)$ satisfies

$$\begin{aligned} f_{q,D}(z_q) = & \sum_{i=0}^{\infty} \frac{(-1)^i z_q^{-i+(1-q)D} \left(1-z_q^{q-1}\right)^D H\left[\frac{i}{1-q}, D, D+1, 1-z_q^{q-1}\right]}{(1-q)^D D} \\ & + \sum_{i=1}^{\infty} \frac{(-1)^{i-1} z_q^{(1-q)(D-1)} H\left[1-D, \frac{i}{1-q}+1, \frac{i}{1-q}+2, z_q^{q-1}\right]}{(1-q)^D \left(\frac{i}{1-q}+1\right)} \end{aligned} \tag{17}$$

Here, $H[a,b,b+1,C] = b\int_0^1 (1-Ct)^{-a} t^{b-1} dt$ is the hypergeometric function. The parameters a, b, C are all independent on t.

4. Generalized Fermi Systems in a 3D Harmonic Trap

For the generalized ideal Fermi system in the 3D isotropic harmonic trap, the single particle Hamiltonian is written as

$$H=\frac{p_x^2+p_y^2+p_z^2}{2m}+\frac{1}{2}m\omega^2\left(x^2+y^2+z^2\right). \tag{18}$$

For the case of $q\leq 1$ and $\mu<0$, one can obtain the analytical expressions of particle number, internal energy and heat capacity as follows

$$N=\frac{8g\pi^3(k_BT)^3}{(h\omega)^3}f_{q,3}(z_q), \tag{19}$$

$$E=9Nk_BT\frac{f_{q,4}(z_q)}{f_{q,3}(z_q)}, \tag{20}$$

$$C_{V,q}=9Nk_B\left[4\frac{f_{q,4}(z_q)}{f_{q,3}(z_q)}-3\frac{f_{q,3}(z_q)}{f_{q,2}(z_q)}\right]. \tag{21}$$

Generalized Fermi integral $f_{q,D}(z_q)$ satisfies Eq. (13).

For the case of $\mu\geq 0$, one can obtain the analytical expressions of particle number, internal energy and heat capacity as follows

$$N=\frac{4g\pi^3(k_BT)^3}{(h\omega)^3}f_{q,3}(z_q), \tag{22}$$

$$E=3Nk_BT\frac{f_{q,4}(z_q)}{f_{q,3}(z_q)}, \tag{23}$$

$$C_{V,q}=Nk_B\left[4\frac{f_{q,4}(z_q)}{f_{q,3}(z_q)}-3\frac{\partial f_{q,4}(z_q)/\partial z_q}{\partial f_{q,3}(z_q)/\partial z_q}\right]. \tag{24}$$

Generalized Fermi integral $f_{q,D}(z_q)$ satisfies Eq. (17).

Absolutely, thermodynamics of generalized ideal Fermi system in harmonic trap are related to the generalized parameter q. Although they have the similar expression form, the generalized Fermi integrals are completely different, which lead to the major differences. The numerical results of heat capacity versus temperature are shown in figure 1 by setting particle number as a fixed parameter. Fig. 1(a), (b) and (c) depict the cases of free space, 2D and 3D harmonic traps, respectively. It can be seen that heat capacity is the approximate monotone increasing function of temperature, which increases with the rising of temperature. When the generalized parameter tends to one, that is $q\rightarrow 1$, the system is transitioned to the classical Fermi-Dirac distribution. The results calculated by generalized Tsallis statistics are consistent with the classical equipartition theorem [9-11]. Due to the limitation of Pauli principle, the value of the heat capacity converges to zero near the absolute temperature. The numerical difference of heat capacity becomes bigger as the temperature increases, which needs more energy to stimulate the same number of particles.

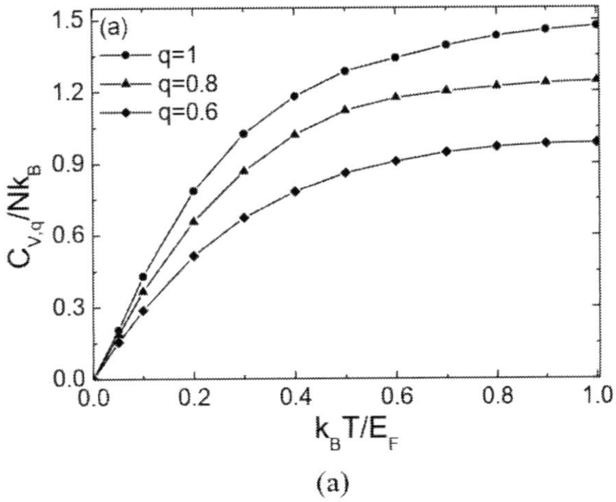

(a)

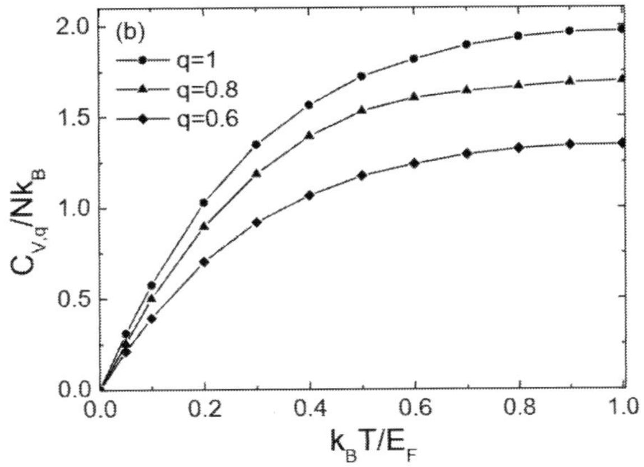

(b)

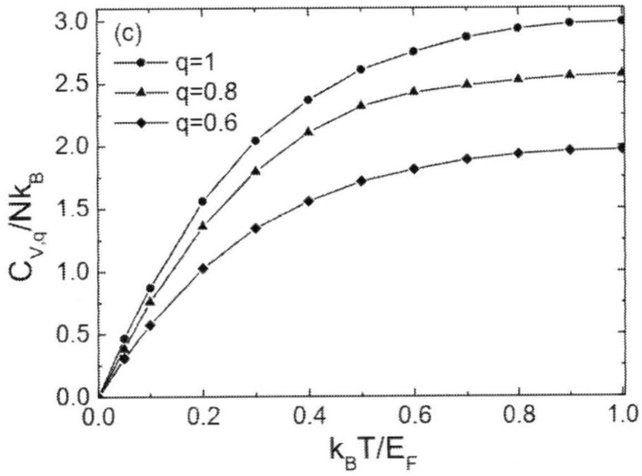

(c)

Figure 1. Heat capacity of generalized Fermi system versus temperature in (a) free space, (b) 2D and (c) 3D traps, respectively.

5. Conclusions

Starting from the generalized Fermi-Dirac distribution function, the thermodynamic properties of generalized ideal Fermi system in free space, 2D and 3D harmonic traps are discussed based on the non-extensive Tsallis theory. Through the analytical and numerical results of heat capacity, the relevant characteristics of the system are reported in detail.

Acknowledgment

This work was supported by the National Natural Science Foundation of China (Grant No. 11804085).

References

[1] Ou C, Chen J. 2003 Thermostatistic properties of a q-generalized Bose system trapped in an n-dimensional harmonic oscillator potential [J] *Phys. Rev. E* **68** 026123.

[2] Nobre F, D Tsallis C. 2004 Metastable states of the classical inertial infinite-range interaction Heisenberg ferromagner: Role of initial conditions [J] *Physica A* **344** 587.

[3] Huang Z, Ou C, Chen J. 2009 Nonextensive and extensive thermostatistic properties of Fermi systems trapped in different external potentials [J] *Chin. Phys. B* **18** 1380.

[4] Sadeghi P, Khademi S, Darooneh A H. 2012 Tsallis entropy in phase-space quantum mechanics [J] *Phys. Rev. A* **86** 012119.

[5] Bagci G, B Oikonomou T. 2013 Tsallis power laws and finite baths with negative heat capacity, *Phys. Rev. E* **88** 042126.

[6] Bagci G, B Oikonomou T. 2016 Validity of the third law of thermodynamics for the Tsallis entropy, *Phys. Rev. E* **93** 022112.

[7] Deppman A 2016. Thermodynamics with fractal structure, Tsallis statistics, and hadrons, *Phys. Rev. D* **93** 054001.

[8] Ford I, J Eyre, R W. 2015 Work relations for a system governed by Tsallis statistics, *Phys. Rev. E* **92** 022143.

[9] Li Y. 2014 Thermodynamic properties of ideal gases in power-law potential with quasi-classical approximation, *Chin. J. Low Temp. Phys.* **36** 310.

[10] Li Y. 2016 Rotating ideal Fermi gases under a harmonic trap, *Physica B* **481** 38.

[11] Li Y, Gu Q. 2016 The particle flow oscillations of rotating non-interacting gases in a two-dimensional harmonic trap, *Phys. Lett. A* **380** 353.

Periodic Pattern Detection of Printed Fabric Based on Deep Learning Algorithm

Zhong Xiang*, **Yujia Shen, Zhitao Cheng, Miao Ma and Feng Lin**

Faculty of Mechanical Engineering and Automation, Zhejiang Sci-Tech University, Hangzhou, China

*Corresponding author, e-mail: xz@zstu.edu.cn

Abstract. Printed fabric patterns contain multiple repeat pattern primitives, which have a significant impact on fabric pattern design in the textile industry. The pattern primitive is often composed of multiple elements, such as color, form, and texture structure. Therefore, the more pattern elements it contains, the more complex the primitive is. In order to segment fabric primitives, this paper proposes a novel convolutional neural network (CNN) method with spatial pyramid pooling module as a feature extractor, which enables to learn the pattern feature information and determine whether the printed fabric has periodic pattern primitives. Furthermore, by choosing pair of activation peaks in a filter, a set of displacement vectors can be calculated. The activation peaks that are most accordant with the optimum displacement vector contribute to pick out the final size of primitives. The results show that the method with the powerful feature extraction capabilities of the CNN can segment the periodic pattern primitives of complex printed fabrics. Compared with the traditional algorithm, the proposed method has higher segmentation accuracy and adaptability.

1. Introduction

Printed fabric is one of the most important textiles in the field of textile industry, it is generally constructed from periodic primitive pattern (PPP), which is commonly consist of the same elements. At present, the PPPs are generally extracted by experienced workers, which is extremely costly and time-consuming. In addition, human eyes are unable to recognize the unit pixels accurately since the unit pixels that make up a digital image are tiny. To meet the needs of digital manufacturing in textile industry, it's essential to find a way to accurately extract the PPP from the printed fabric automatically, so as to accelerate the pattern design process or to improve the accuracy of defects detection [1].

There have been a large amount of work in the field of periodic pattern analysis. Generally, they can be mainly divided into as following 3 categories: (1) spectral-based methods, (2) statistical-based methods, (3) traditional learning-based methods. For example, early in 1980, Terzopoulus et al. [2] tried to analyze periodic textures structure by using the gray-level co-occurrence matrix (GLCM). However, it was computationally intensive and had poor robustness [3]. The approach using autocorrelation function to calculate peaks was proposed by Lin et al. [4]. This method had a certain tolerance to local distortion, but it required to smooth the autocorrelation function, which led to long calculation time [4]. Oh et al. [5] used distance matching function to analyze periodic texture's structure, which has a small amount of calculation consumption. However, the DMF method is very sensitive to noise such as geometric distortion. A method of using sum and difference histogram (SDH) for texture analysis was proposed by Unser et al. [6]. The advantage of this method was that it greatly reduced the calculation time and had a stable effect. Matsuyama et al. [7] used Fourier transform to analyze the textures. The period of repeat pattern primitives could be determined by the pulse distribution of Fourier spectrum. However, the peak of Fourier spectrum was not prominent when the pattern contained fewer periodic primitives, which made the method based on Fourier transform can only segment the periodic pattern primitives of the woven fabric. Kuo et al. [8] converted the printed fabric into a full-color image, and employed the fuzzy mean clustering algorithm to segment the pattern to obtain the pattern of the printed fabric. Finally, the proposed approach using the Hough transform completed the segmentation of periodic pattern primitives. However, this method was only suitable for patterns with relatively clear color boundaries, but got trouble in effectively identifying patterns with rich colors and complex textures.

The above methods often require a manual design of feature extractors, which have poor adaptability and are only effective for some simple periodic pattern primitives. Therefore, it is hard to adapt to complex printed fabrics, which is unable to meet the needs of textile industry. In recent years, the application of deep learning algorithms has become increasingly widespread, which has promoted the rapid development of computer vision, especially in image classification [9], object detection [10], and semantic segmentation [11]. However, few researchers apply it to the pattern primitive's extraction of printed fabrics. Sun Y. et al. [12] found that the use of convolutional neural network can replace traditional hand-designed feature extractors through visualization, alignment and key point prediction. Lettry et al. [13] proposed a method of using the position information in the feature layer of the CNN to find the spatial periodic rules, which has a excellent impact on the extraction of periodic pattern primitives in natural scenes. However, the primitives of periodic patterns in natural scenes are relatively simple [14], which are usually composed of regular geometric objects with clear boundaries. The methods in natural scenes cannot be used directly due to the abundant texture and shape features. Li J. et al. [15] found that CNNs can implicitly learn spatial location information. The CNN learns the characteristic information of the primitives, and uses this characteristic information to segment the periodic pattern primitives of the printed fabric. In order to segment fabric primitives, this paper proposes a novel CNN method as a feature extractor, which enables to learn the pattern primitive feature information and determine whether the printed fabric has periodic pattern primitives. Furthermore, the network is used to calculate the activation peak and segment the periodic pattern primitives of the printed fabric.

The main contributions of this work are summarized as follows: in the first part, we proposed a classification network for printed fabrics and used the transfer learning algorithm for training, so that the convolutional neural network can learn more abundant features. In the second part, we used the convolution layer of the trained classification network as the feature extractor. Through calculating the activation peak, the network effectively segmented the pattern primitives. The results show that our algorithm has better segmentation effect than the existing methods.

2. Primitive Feature Extraction

2.1. Pre-processing

The input belongs to the printed fabric patterns, which is captured by industrial cameras. However, an obvious problem here is that the printed fabrics are placed manually during the collection process, and the surface of the printed fabrics is prone to noise such as dust and wrinkles. We therefore utilize a pre-processing step to align the patterns.

According to the previous research [16] that the range of fabric inclination angle α does not exceed 20°, we used the Radon Transform with a suitable inclination angle. In order to improve the segmentation effect of periodic primitives, we made use of the Non-Local Variations algorithm from Dekel et al. [17], which was based on the structural differences of the periodic regions in the whole image. The input image I reduced the small changes between the periodic structures after T transformation processing, which can be expressed as:

$$T\{I\} = J + N \tag{1}$$

where the J is the corrected image, N is the noise signal. This paper set the parameter iteration=20 according to human experience.

2.2. Feature Extraction

The goal of this step is to utilize the activations of filters in the convolutional layers, rather than traditionally hand-crafted. In this section, we proposed a CNN model based on spatial pyramid pooling. The printed fabrics were divided into two categories according to whether the patterns have periodic patterns. The two types of fabric patterns in the fabric data set are shown in figure 1. This paper puts to use a spatial pyramid pooling module, which allowed the AlexNet to learn multi-scale features and extract abundant features information.

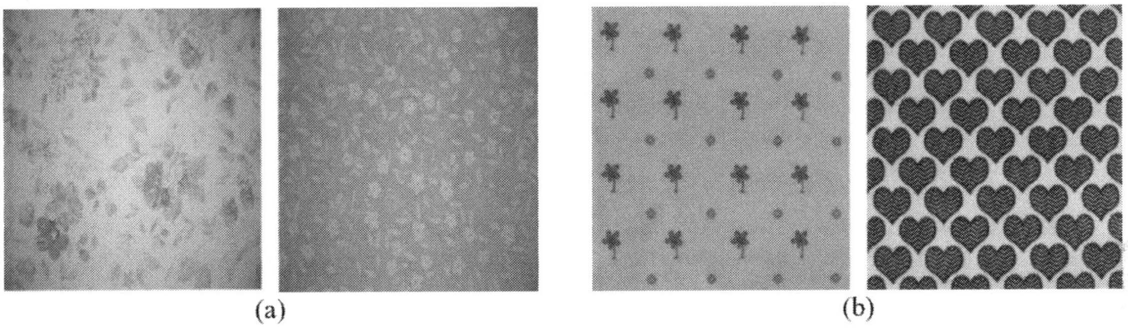

Figure 1. (a) Printed fabric without periodic primitives. (b) Printed fabric with periodic primitives.

2.2.1. Network Structure. Networks with more zero padding can implicitly learn the spatial location information, which makes use of the location information of patches [12]. We did not choose novel deep convolutional neural networks such as VGG-16 or ResNet34, which stacked more layers and deepened the model depth. The network to deepen is prone to overfitting because of the simple samples learning. In addition, the above networks will increase the amount of computation of periodic primitives' segmentation. Therefore, we used AlexNet as the base feature extract network.

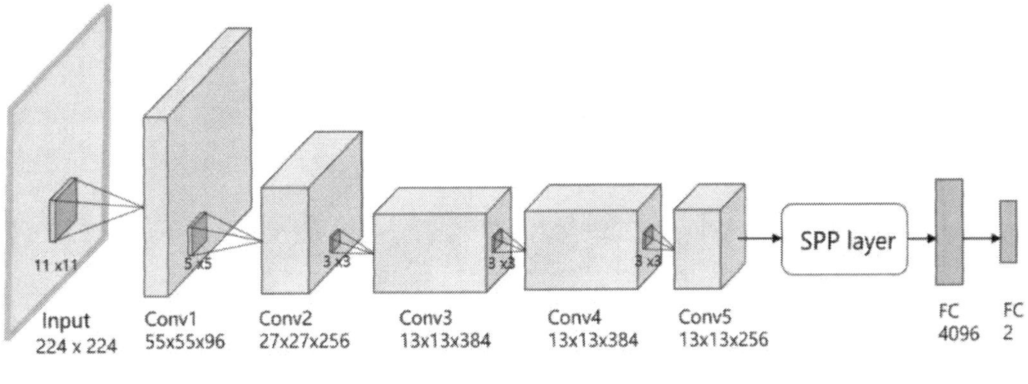

Figure 2. Improved AlexNet structure.

In order to make the network better adapt to segmentation tasks and avoid overfitting, fc1 was removed. Additionally, the last pooling layer was replaced as the Spatial Pyramid Pooling (SPP) [18] layers, which helped the network learn more spatial features. The improved network structure is shown in figure 2. There are a total of eight layers, specifically composed of five convolutional layers, a spatial pyramid pooling module, and two fully connected layers. Between Conv1 and Conv5, all convolution layers are followed by the Relu activation function. The Dropout layers are also used to avoid overfitting. The configuration of each layer can be seen in table 1.

Table 1. The configuration of the improved AlexNet network.

Layer	Output size	Number of filters	Size of kernel	Stride	Pad
Input	224×224×3	-	-	-	-
Conv1	55×55×96	96	11×11	4	2
Max pool1	27×27×96	-	3×3	2	-
Conv2	27×27×256	256	5×5	1	2
Max pool2	13×13×256	-	3×3	2	-
Conv3	13×13×384	384	3×3	1	1
Conv4	13×13×384	384	3×3	1	1
Conv5	13×13×256	256	3×3	1	1
SPP Layer	5376	-	-	-	-
FC1	4096	4096	1×1	1	-
FC2	2	2	1×1	1	-
Output	2	-	-	-	-

2.2.2. Spatial Pyramid Pooling. The module structure used in this paper is shown in figure 3. The input image passed through the last convolutional layer to obtain the feature map, where the size of the feature map was q×q. Then the feature map was used as the input of the spatial pyramid pooling layer, which would be divided into different patches with the size k×k. In our experiments, we set the parameter k to 4, 2, 1, and the sliding window size to ceil [q/k]×ceil[q/k] for pooling, and the stride to floor[q/k]×floor[q/k]. Through the step of pooling at different scales, we got feature maps with the three different shape size of 16×256, 4×256, 1×256 (256 is the number of Conv5 filters). Subsequently, the fixed-dimensional vectors are the input to the FC1, which made the network adapt to the input image of any scale. And because of the use of different down-sampling steps, the network has obtained different receptive fields, which was help to capture information at different scales and develop powerful primitive feature learning capability.

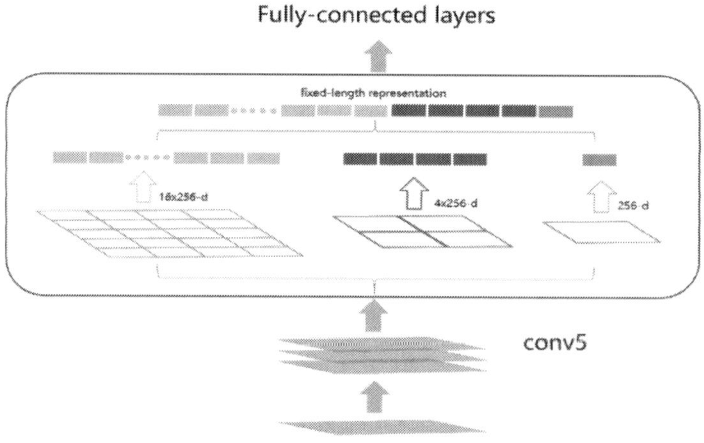

Figure 3. Spatial pyramid pooling module.

2.3. Periodic Pattern Primitive Extraction

The shallow layers of a convolutional neural network usually learn low-level features, such as edges.

As the network deepens, it learns higher-level features representation, such as objects and complex forms. Therefore, CNN can be considered as a set of filters with different levels of abstraction, and these filters will be activated when specific features appear in the image area. In this section, the convolutional layer of the previously trained classification network was used as a feature extractor to segment the periodic pattern primitives of the fabric. Its advantage was that it does not require any prior knowledge of human beings and can learn independently, which helped to adaptively generate a hierarchy of features of the input image.

2.3.1. Acquisition of Displacement Vector. The input image would generate regular local peaks in each feature layer of the network, and each set of peaks corresponded to a set of displacement vectors. Therefore, the displacement vector that appeared most frequently in different feature layers was most likely to be the size of the periodic pattern primitive. The Hough voting algorithm was used in this paper to vote for each pair of displacement vectors and determine the appropriate size of primitives.

When the pre-processed image was input to the trained network, regular activation peaks would be generated after the convolution operation, which were obtained by the non-maximum suppression algorithm. In this step, the 5 convolutional layers of the improved AlexNet in section 2.2.1 were used for calculation. Specifically, l騆 denotes all convolutional layers, the number l represents the selected layer, f_lòF_l denotes a filter of the selected layer. The symbol P means the position of the activation peak of a filter, and pǒP_{f_l}, p^m, p^nǒP_{f_l} each represents the location of the local activation peak, P_{f_l} is the activation peak set of f_l. The activation map is shown in figure 4.

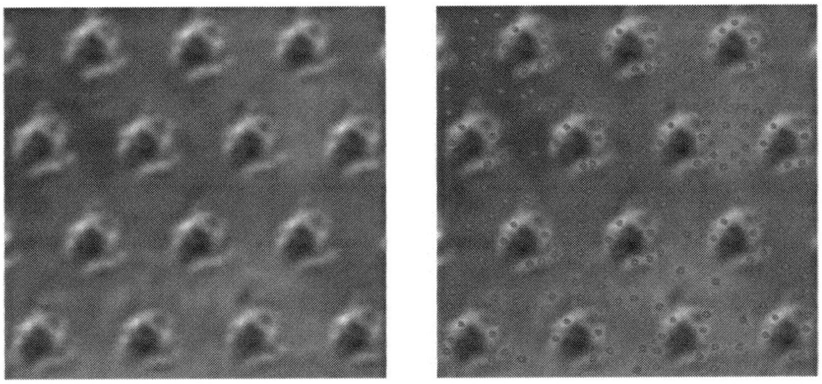

Figure 4. The local maxima of the activation map.

From each pair of activation peaks in each filter, a set of displacement vectors can be calculated, which is indicated by $s^{m,n}$. At this time, each set of displacement vectors is the potential size of the periodic pattern primitives. The formula of the displacement vector is as follows:

$$S_{f_l} = \left\{ s^{m,n} : \left| p^m - p^n \right|, m \neq n \right\} \quad (2)$$

where $|.|$ represents the element-wise absolute value of vector and S_{f_l} is the set of displacement vectors of a filter f_l.

The activation peaks that are most accordant with the optimum displacement vector [19] contributes to pick out the final size of primitives. Therefore, we fused vectors linking pairs of activation peaks by a Hough-like voting space $\mathrm{H}: \mathrm{R}^2 \rightarrow \mathrm{R}$. The filters in each layer have different sizes. Therefore, it is incapable of directly voting for the displacement of different filters. Therefore, it is assumed that each vote obeys a two-dimensional normal distribution centered on $s^{m,n}$, and σ_l is the covariance matrix of the l-th convolutional layer. The formula for the entire voting process is as follows:

$$H_{(x,y)} = \sum_x \sum_{\substack{f_l \in F_l \\ l \in L}} \frac{1}{|S_{f_l}|} \sum_{s^{m,n} \in S_{f_l}} H_{f_l,m,n} \tag{3}$$

$$H_{f_l,m,n} = \frac{1}{2\pi\sigma_l^2} \exp\left(-\frac{1}{2} \| x - s^{m,n} \|_2^2 \right) \tag{4}$$

where (x, y) corresponds to the pixel coordinates of the input image.

In this step, the optimal spatial displacement vector s^* was obtained, which contributed to the size of the periodic pattern primitive. The formula is as follows:

$$s^* = \left(\operatorname*{argmax}_x H_{(x,0)}, \operatorname*{argmax}_y H_{(0,y)} \right) \tag{5}$$

2.3.2. Periodic Primitive Segmentation. After finding the optimal displacement vector s^*, the next step was to find the regions of the input image that were more accordant. Therefore, filter selection aimed to pick the filters with activation peaks most accordant with the selected displacement vector. The consistent votes are then attributed with weights w_{f_l}:

$$w_{f_l} = \sum_{s^{m,n} \in S_{f_l}} \left(\frac{1}{|S_{f_l}| + \gamma} \right) \exp\left(-\frac{\| s^{m,n} - s^* \|^2}{2\beta^2} \right) \tag{6}$$

where the parameter γ is a flat prior and β means the radius of the neighborhood considered at l-th layer as the close surrounding of the selected displacement vector.

The filters with weights larger than $\varphi \times w_{f_l}^*$ would be kept, where $w_{f_l}^*$ was the highest weight among the filters. And the parameter φ was set to 0.65 according to experience [12].

In order to obtain the relative position of the displacement vector, the two-dimensional rectangular coordinate (x, y) of the filter needs to be simplified first. The formula is as follows:

$$\mathrm{M}\ (x, y) \square \left(x \bmod s_x^*, y \bmod s_y^* \right) \tag{7}$$

However, 煹 (x, y) would produce a certain offset, which is marked as $o^* = (o_x, o_y)$.The formula is as follows:

$$o^* = \arg\min_o \sum_{m,n} \left(w_{m,n} \| M\left(s^{m,n} - o\right) - \frac{s^*}{2} \| \right) \tag{8}$$

Finally, the IPM algorithm [20] was used to obtain the centroid coordinates of the periodic pattern primitives. Therefore, the final periodic pattern primitive was segmented by the centroid coordinates and the primitive size which was determined previously.

3. Experiments

In this section, the experimental details are discussed, and the classification results are analyzed later. Then three different types of images are tested for segmentation. In order to better analyze the performance of the algorithm in this paper, a requisite analysis is carried out with the other algorithm. All experiments were performed on a workstation configured with Intel Xeon Gold 5115 CPU(2.50GHz),

64GB RAM and two NVIDIA GTX 1080Ti GPUs.

3.1. Image Classification Results

A total of 600 images of printed fabrics have been collected in this paper. There are 400 images in the training set, including 200 fabrics patterns with periodic primitives and 200 patterns without periodic primitives. The experiment adopted the transfer learning method, and used the pre-trained AlexNet network in the Pytorch deep learning framework to initialize the parameters. For the reason of small samples, operations such as translation and zoom were used to augment the image data. Finally, the number of training images was 4000, and the number of testing images was 50 of each kind. Adam was the optimized algorithm in this experiment, and the overall learning rate for the training was 0.001.

The results are shown in table 2. After transfer learning, the classification accuracy has been significantly improved. The vital basis for segment the periodic pattern primitives of printed fabric is whether the primitives in the image recursively appear after a certain distance. Therefore, Spatial Pyramid Pooling (SPP) module contributed the network to learn multi-scale features and improve the accuracy of the network classification of whether it contained periodic primitives. Then the convolutional layers would be feature extractors for subsequent primitive segmentation.

Table 2. Classification results.

Method	AlexNet	improved AlexNet
without pretrained	52%	62%
pretrained	86%	94%

3.2. Segmentation of Three Kinds of Periodic Primitives

If the primitive contains plenty of elements and complex features, segmentation will be very arduous. According to the number of elements contained in the periodic pattern primitives, the printed fabrics were divided into three categories. As more elements the periodic primitive contained, the more difficult it was to segment.

The HCDH method [17] is one of the researches on the periodic primitives of fabric. This algorithm converts the color image into a gray image, then it uses the sum-difference histogram (SDH) algorithm to calculate the homogenous features, which contributes to determine the size of the periodic primitive according to the distance between two adjacent local maximum homogenous values. However, it is only limited to patterns with simple periodic primitives, and does not take into account the plentiful texture characteristics of the fabric and the colors.

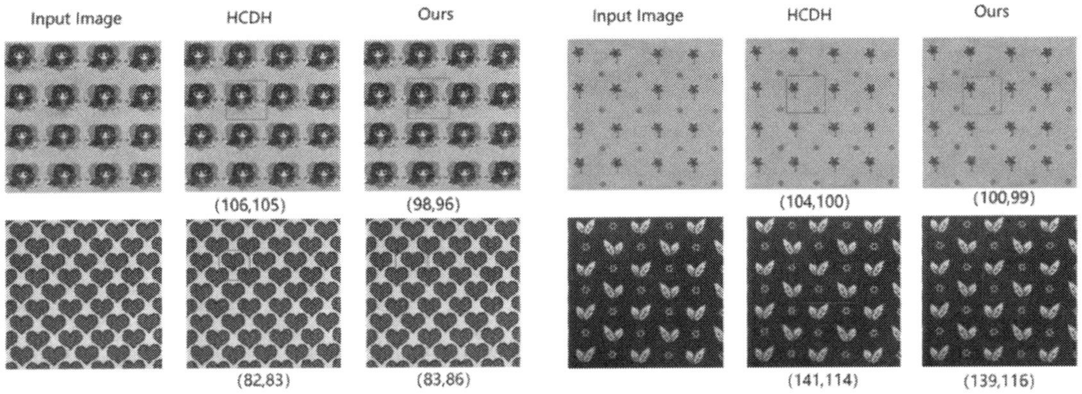

Figure 5. Simple periodic primitives.

The input image size was 400 × 400. Figure 5 shows the result of simple images. The method in this paper and the comparison algorithm could both accurately segment the complete periodic primitives.

However, the HCDH algorithm is invalid for the complex periodic primitives, which contains plentiful elements and features. Obviously, the homogeneity function reaction of the HCDH method is almost a straight line on the curve, and the primitives of the printed fabric pattern cannot be obtained according to the interval between the peaks. In deep learning, the shallow layer can usually obtain low-level features such as edges, while the deep layer can learn more complex feature information, such as complex shapes. The result of our method is shown in figure 6.

It is of great significance to be able to identify periodic primitives with color features. The elements in these primitives have the same shape but different colors. However, the input image of the HCDH algorithm is a gray image, which is impossible to learn the color features. The image we input to the network contains three channels, and the method in this paper can extract the color features of the image. The result of our method is shown in figure 7.

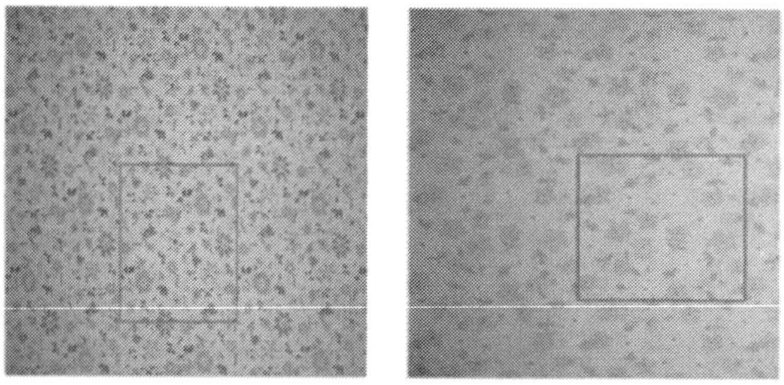

Figure 6. Complex periodic primitives.

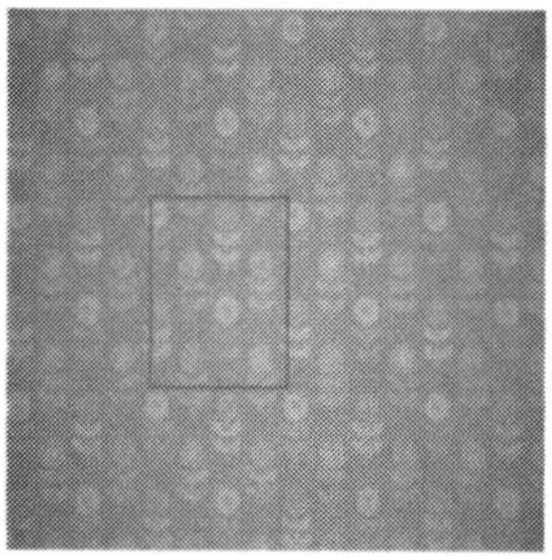

Figure 7. Fabric patterns with different color characteristics.

3.3. Quantitative Evaluation

The effects of the two algorithms are quantitatively analyzed in this section. Cosine similarity is an important index to evaluate whether images are similar. Therefore, if the segmentation effect is precise, the similarity of adjacent periodic pattern primitives will be very high. We judge that whether the segmentation effect is good according to the cosine similarity value of adjacent periodic pattern primitives. Based on experience, the segmentation effect is regarded as accurate if the cosine similarity value greater than 0.95.

The segmentation accuracy of our algorithm and the HCDH algorithm in the data set is shown in table 3. There is no comparison here on fabrics with color features because the HCDH algorithm cannot identify the primitives. Additionally, the simple samples are further divided into 2 categories. Simple periodic primitives A means the single primitive has just one element while primitives B means the single primitive has 2 or 3 elements. On the first type of fabrics, the segmentation accuracy of our algorithm is 10% higher than that of the HCDH algorithm, and it is 14.3% higher on the second type of fabrics. As the fabric becomes more complex, the more difficulty it is to identify. On the third type of fabric, the segmentation accuracy of the HCDH algorithm is only 10.5%, which is almost impossible to identify, while the segmentation accuracy of our algorithm is 89.5%. The proposed method in this paper has high accuracy and adaptability, which benefits from the powerful feature extraction ability of CNNs.

Table 3. Segmentation accuracy.

References	HCDH (%)	Ours (%)
simple periodic primitives A	87.5	97.5
simple periodic primitives B	78.6	92.9
complex periodic primitives	10.5	89.5

4. Conclusion

In this paper, a CNN model for classification of periodic printed fabrics is proposed, and the SPP module is added to AlexNet to improve the feature extraction ability of fabrics consists of PPPs. The results show that the classification algorithm has high classification accuracy. Then, from each pair of activation peaks in each filter, a set of displacement vectors can be calculated. The activation peaks that are most accordant with the optimum displacement vector contribute to pick out the final size of primitives. Finally, the Hough voting and the IPM algorithm are used for complete primitive segmentation. Compared with the traditional algorithm, the results show that the algorithm in this paper has higher accuracy and adaptability. Although the performance of it has reached a high level, it takes a longer time compared to the traditional method and has obvious limitations in practical textile industry. Therefore, in the future, we will further optimize the algorithm to reduce computational consumption, which will be suitable for practical applications.

References

[1] Chang X, Gu C, Liang J, et al. 2018 Fabric Defect Detection Based on Pattern Template Correction [J] *Mathematical Problems in Engineering* (2018-3-22) 2018(PT.3) 1-17.
[2] Zucker S W, Terzopoulos D. 1980 Finding structure in co-occurrence matrices for texture analysis [J] *Computer graphics and image processing* **12** (3) 286-308.
[3] Parkkinen J, Selkäinaho K, Oja E. 1990 Detecting texture periodicity from the cooccurrence matrix [J] *North-Holland* **11** (1) 43-50.
[4] Lin H C, Wang L L, Yang S N. 1997 Extracting periodicity of a regular texture based on autocorrelation functions [J] *North-Holland* **18** (5) 433-443.
[5] Oh G, Lee S, Shin S Y. 1999 Fast determination of textural periodicity using distance matching function [J] *Pattern Recognition Letters* **20** (2) 191-197.
[6] Unser M. 1986 Sum and Difference Histograms for Texture Classification [J] *IEEE Transactions on Pattern Analysis & Machine Intelligence* **8** (1) 118.
[7] Matsuyama T, Miura S I, Nagao M. 1983 Structural analysis of natural textures by Fourier transformation [J] *Computer Vision Graphics & Image Processing* **24** (3) 347-362.
[8] Kuo C-FJ, Shih C-Y, Lee J-Y. 2005 Repeat Pattern Segmentation of Printed Fabrics by Hough Transform Method [J] *Textile Research Journal* **75** (11) 779-783.
[9] Wang J, Zheng Y, Wang M, et al. 2020 Object-scale adaptive Convolutional Neural Networks for high-spatial resolution remote sensing image classification [J] *IEEE Journal of Selected Topics in Applied Earth Observations and Remote Sensing* **14** 283-299.
[10] Han Y Y, Han Y D. 2021 A Deep Lightweight Convolutional Neural Network Method for Real-Time Small Object Detection in Optical Remote Sensing Images [J] *Sensing and Imaging* **22** (1) 1-15.
[11] Ji J, Lu X, Luo M, et al. 2020 Parallel Fully Convolutional Network for Semantic Segmentation [J] *IEEE Access* **9** 673-682.
[12] Sun Y, Wang X, Tang X. 2013 Deep Convolutional Network Cascade for Facial Point Detection [C] *Proceedings of the IEEE Conference on Computer Vision and Pattern Recognition (CVPR)* 3476-3483.
[13] Lettry L, Perdoch M, Vanhoey K, et al. 2017 Repeated Pattern Detection Using CNN Activations [C] *2017 IEEE Winter Conference on Applications of Computer Vision (WACV)* IEEE 47-55.
[14] Wu C, Frahm J M, Pollefeys M. 2010 Detecting Large Repetitive Structures with Salient Boundaries [J] *Lecture Notes in Computer Science* **6312** 142-155.

[15] Li J, Zhao J, Jing L, et al. 2018 Using Convolutional Layer Features for Indoor Human Activity Recognition based on Spatial Location Information [J] *DEStech Transactions on Computer Science and Engineering* (csae).

[16] Pan R R, Gao W D, Liu J H, et al. 2010 Automatic Inspection of Woven Fabric Density of Solid Colour Fabric Density by the Hough Transform [J] *Fibres & Textiles in Eastern Europe* **81** (4) 46-51.

[17] Dekel T, Michaeli T, Irani M, et al. 2015 Revealing and modifying non-local variations in a single image [J] *ACM Transactions on Graphics (TOG)* **34** (6) 1-11.

[18] He K, Zhang X, Ren S, et al. 2015 Spatial Pyramid Pooling in Deep Convolutional Networks for Visual Recognition [J] *IEEE Transactions on Pattern Analysis & Machine Intelligence* **37** (9) 1904-1916.

[19] Rodriguez-Pardo C, Suja S, Pascual D, et al. 2019 Automatic extraction and synthesis of regular repeatable patterns [J] *Computers & Graphics* **83** 33-41.

[20] Lizarraga-Morales R A, Sanchez-Yanez R E, Ayala-Ramirez V. 2013 Fast texel size estimation in visual texture using homogeneity cues [J] *Pattern Recognition Letters* **34** (4) 414-422.

A Triangular Spectral Element Method for the 2-D Viscous Burgers Equation

Mengyu Zhang and Hua Wu*

Department of Mathematics, Shanghai University, Shanghai, 200444, China
*Email: hwu@staff.shu.edu.cn

Abstract. A triangular spectral element method is established for the two-dimensional viscous Burgers equation. In the spatial direction, a new type of mapping is applied. We splice the local basis functions on each triangle into a global basis function. The second-order Crank-Nicolson/ leap-frog (CNLF) method is used for discretization in the time direction. Due to the use of a quasi-interpolation operator, the nonlinear term can be handled conveniently. We give the fully discrete scheme of the method and the implementation process of the algorithm. Numerical examples verify the effectiveness of this method.
Keywords. Burgers equation, triangular spectral element method, quasi-interpolation.

1. Introduction

The spectral method is one of the important numerical methods for solving partial differential equations. Its main advantage is high accuracy, especially for infinitely smooth problems, the convergence rate can reach infinite order. Because of its advantages, the spectral method is widely used in scientific and engineering computations. In recent years, the triangular element spectral method has been developed [1,2,3], that is, the spectral method is used in the triangular element while maintaining the high accuracy. Through the realization of the triangular element spectral method, due to the flexibility of the triangle itself, the application of triangulation to divide the region boundary into triangles can make the spectral method better applied to complex regions.

Recently, a new type of mapping is proposed in [4]. This mapping is a one-to-one mapping, which maps a vertex of a quadrilateral to the midpoint of the hypotenuse of the triangle, making the distribution of nodes more uniform than frequently used Duffy mapping [5]. This new type of mapping is applied in elliptic equation and a class of irrational basis functions is constructed on a triangular element in [6]. In [7], comparing Duffy mapping with this new type of mapping. By introducing auxiliary variables, a mixed triangular spectral element method is constructed in complex regions.

The Burgers equation is a classic nonlinear second-order partial differential equation which describes dissipation on the energy balance. It originates from the turbulence model [8]. It is often used in engineering computations [9] and as a model equation for testing numerical methods.

Let $\Omega = [-1,1]^2$ .Consider the following two-dimensional viscous Burgers equation.

$$\begin{cases} \dfrac{\partial u}{\partial t}+\nabla\cdot F(u)-\nu\nabla^2 u=f(x,y,t), & (x,y)\in\Omega, t\in(0,\mathrm{T}], \\ u(x,y,t)=0, & (x,y)\in\partial\Omega, t\in(0,\mathrm{T}], \\ u(x,y,0)=u_0(x,y), & (x,y)\in\Omega\cup\partial\Omega. \end{cases} \quad (1)$$

where T and ν are the positive constants. F(u) is a function of a certain smoothness class. Under the above conditions, the solution of (1) is unique.If the Dirichlet boundary condition is nonhomogeneous, we can convert the boundary condition to homogeneous, and then consider the above situation.

In this paper, we present a triangular spectral element method for (1) and give its semi-discrete scheme and fully discrete scheme. We use polynomials of different degrees and triangulations of differrent numbers to approximate the unknown functions. Numerical results show the effectiveness of this method.

2. Triangular Spectral Element Method

In this section, we discrete spatial direction through triangular spectral element method and discrete time through the second-order Crank-Nicolson/leap-frog (CNLF) scheme to obtain the corresponding fully discrete scheme.

2.1. Preliminaries

Let $(\cdot,\cdot)_\Omega$ be the inner product of $L^2(\Omega)$. For an integer $m>0$, let $H^m(\Omega)$ be the standard Sobolev space. Define

$$H_0^1(\Omega)=\left\{v\middle|H^1(\Omega):v|_{\partial\Omega}=0\right\}.$$

For an integer $N>0$, let $\square=\{(\xi,\eta):-1\le\xi,\eta\le1\}:=\Lambda^2$ and $\mathrm{P}_N(\Lambda)$ be the polynomial of degree less than N in domain Λ .Considering the new type of mapping in [4], let $\Gamma=\{\triangle_k\}_{k=1}^K$ be triangulations of domain Ω, and we usually assume the triangle $\triangle_k$ to be shaped regular. For the convenience of calculation, we denote M as the number of quadrilateral divisions in x and y direction and get triangles through connecting with the left diagonal of these quadrilaterals.

The new type of mapping $T_k:\square\to\triangle_k$ is as follow

$$\begin{cases} x=x_{Ak}\dfrac{(1+\xi)(3-\eta)}{8}+x_{Bk}\dfrac{(1-\xi)(1-\eta)}{4}+x_{Ck}\dfrac{(3-\xi)(1+\eta)}{8}, \\ y=y_{Ak}\dfrac{(1+\xi)(3-\eta)}{8}+y_{Bk}\dfrac{(1-\xi)(1-\eta)}{4}+y_{Ck}\dfrac{(3-\xi)(1+\eta)}{8}, \end{cases} \quad \forall(\xi,\eta)\in\square. \quad (2)$$

We respectively chose $A(x_{Ak},y_{Ak}),B(x_{Bk},y_{Bk}),C(x_{Ck},y_{Ck})$ as the three vertices of $\triangle_k$. How-ever, this kind of mapping will bring singularity [4] in calculation. We divide the domain in this way and arrange the singular point at the midpoint of the shared side of the two triangles, making it easier to calculate.

Let J_k and J_k^{-1} be the Jacobian matrix and Jacobian inverse matrix of the mapping T_k. Denote J_k is the corresponding Jacobian determinant. In this article, by mapping T_k the function $u(x,y)$ in $\triangle_k$ and the function $\hat{u}^k(\xi,\eta)$ in $\square$ can be matched, which is

$$u(x,y)\big|_{\square_k} \circ T_k = \hat{u}^k(\xi,\eta), \quad \forall (x,y) \in \Omega, \forall (\xi,\eta) \in \square.$$

2.2. Approximation Space and Quasi-Interpolation Operator

As defined in [10], we chose the spectral approximation space with the consistency condition as follow

$$V_{N,\Gamma} := \left\{ \omega \in H^1(\Omega), \hat{\omega}^k \in \left[\mathrm{P}_N(\Lambda)\right]^2, \left(\partial_\xi \hat{\omega}^k + \partial_\eta \hat{\omega}^k\right)\Big|_{(1,1)} = 0 \right\}, \tag{3}$$

where $1 \# k \quad K$. Next, a set of global basis functions of $V_{N,\Gamma}$ are constructed by splicing local basis functions.

Let $\{h_j, j = 0,1,\ldots,N\}$ be the Lagrange interpolation basis functions based on Legendre-Gauss-Lobatto (LGL) points corresponding to the interval Λ. And we denote $\phi_{ij}(\xi,\eta) = h_i(\xi)h_j(\eta)$ and $\mu_j = -\dfrac{h_j'(1)}{2h_N'(1)}, j = 0,1,\ldots,N-1$. Let $\breve{h}_j(z) := h_j(z) + \mu_j h_N(z), 0 \le j \le N-1$, we chose local basis functions of the triangle $\square_k$ as

$$\hat{\psi}_{ij}(\xi,\eta) := \begin{cases} h_i(\xi)h_j(\eta), & 0 \le i,j \le N-1, \\ \breve{h}_i(\xi)h_N(\eta), & 0 \le i \le N-1, \\ h_N(\xi)\breve{h}_j(\eta), & 0 \le j \le N-1. \end{cases} \tag{4}$$

We define $\psi_{ij}^k(x,y) := \hat{\psi}_{ij}(\xi,\eta) \circ T_k^{-1}$. The C^0 global basis functions are chosen as following, let $\{(\xi_i,\eta_j)\}_{i,j=0}^N$ and $\{\omega_{ij}\}_{i,j=0}^N$ are the LGL points and the corresponding weights in Λ^2. Define

$$\Omega_N := \left\{(x,y) : (x,y) = T_k\left(\xi_i,\eta_j\right), 0 \le i,j \le N, 1 \le k \le K\right\},$$

We can know that Ω_N is a set of all the LGL points in $\overline{\Omega}$. Let N_Ω is the number of elements in Ω_N, and Ω_N can also be expressed in order as

$$\Omega_N = \{(x,y) : (x,y) = (x_\kappa, y_\kappa), \kappa = 1,2,\cdots,N_\Omega\},$$

where $\kappa = \kappa(i,j,k), 0 \le i,j \le N, 1 \le k \le K$. And we define Ω_χ is the set of all singular points in $\overline{\Omega}$, K_Ω is the number of points contained in Ω_χ.

We chose basis functions corresponding to the internal points of domain $\square_k$ as

$$\varphi_\kappa(x,y) = \begin{cases} \psi_{ij}^k(x,y), & (x,y) \in \square_k, \\ 0, & \text{other.} \end{cases}$$

For the LGL points on the edge of the triangles which are jointed, such as the edge $\gamma_{k'k''} := \partial\square_{k'} \cap \partial\square_{k''}, k', k'' \in \{1,\ldots,K\}$. The corresponding basis function is

$$\varphi_{\kappa}(x,y)=\begin{cases}\psi_{ij}^{k'}(x,y), & (x,y)\in\square_{k'},\\ \psi_{i'j'}^{k''}(x,y), & (x,y)\in\square_{k''},\\ 0, & \text{other.}\end{cases}$$

where $\psi_{ij}^{k'}(x,y)$ and $\psi_{i'j'}^{k''}(x,y)$ are the local basis functions corresponding to the same LGL point of jointed triangles $\square_{k'}$ and $\square_{k''}$,respectively.

Similarly, the local basis functions on each triangle can be spliced into a global basis function. In this way, non-nodal basis functions can be represented by nodal basis functions. So they still have the property of tensor product, making them easy to calculate. We next introduce a quasi-interpolation operator from space $C(\Omega)$ to the spectral approximation space $V_{N,\Gamma}$.

For any $u(x,y)\in C(\Omega)$, let $u_{\kappa}=\hat{u}^{k}(\xi_i,\eta_j), 0\le i,j\le N, 1\le k\le K$,where $\kappa=\kappa(i,j,k)$. Define the quasi-interpolation operator by using the global basis functions of $V_{N,\Gamma}$, which leads to $\mathrm{I}_{N,\Gamma}: C(\Omega)\rightarrow V_{N,\Gamma}$

$$\mathrm{I}_{N,\Gamma}u(x,y):=\sum_{\kappa=1}^{N_{\Omega}-K_{\Omega}}u_{\kappa}\varphi_{\kappa}(x,y), \qquad \forall(x,y)\in\Omega. \tag{5}$$

2.3. Scheme

The weak form to (1) is: Find $u\in H_0^1(\Omega)$, for any $v\in H_0^1(\Omega)$ such that

$$\begin{cases}\left(\partial_t u, v\right)_{\Omega}-(F,\nabla\cdot v)_{\Omega}+\nu(\nabla u,\nabla v)_{\Omega}=(f,v)_{\Omega},\\ (u(x,y,0),v)_{\Omega}=\left(u_0(x,y),v\right)_{\Omega},\end{cases} \quad \forall t\in(0,\mathrm{T}]. \tag{6}$$

Define $V_{N,\Gamma}^{0}:=V_{N,\Gamma}\bigcap H_0^1(\Omega)$. The semi-discrete scheme to (1) is: Find $u_N\in V_{N,\Gamma}^0$, for any $v_N\in V_{N,\Gamma}^0$ such that

$$\begin{cases}\left(\partial_t u_N, v_N\right)_{\Omega}-\left(F,\nabla\cdot v_N\right)_{\Omega}+\nu\left(\nabla u_N,\nabla v_N\right)_{\Omega}=\left(f,v_N\right)_{\Omega},\\ \left(u_N(x,y,0),v_N\right)_{\Omega}=\left(u_0(x,y),v_N\right)_{\Omega},\end{cases} \quad \forall t\in(0,\mathrm{T}]. \tag{7}$$

The second-order Crank-Nicolson/leap-frog (CNLF) scheme is applied to discretize in the time direction. Let $S_t=\{k\tau: k=1,2,\ldots n_t, t=n_t\tau\}$, we can give the fully discrete scheme to (1), which is: Find $u_N\in V_{N,\Gamma}^0$, for any $v_N\in V_{N,\Gamma}^0$ such that

$$\begin{cases}\left(\left(u_N\right)_{\hat{t}}, v_N\right)_{\Omega}-\left(F,\nabla\cdot v_N\right)_{\Omega}+\nu\left(\nabla\bar{u}_N,\nabla v_N\right)_{\Omega}=\left(f,v_N\right)_{\Omega},\\ \left(u_N(x,y,0),v_N\right)_{\Omega}=\left(u_0(x,y),v_N\right)_{\Omega},\end{cases} \quad \forall t\in S_{T-\tau}. \tag{8}$$

where τ is the step length in the time direction. The definitions of $u_{\hat{t}}(t)$ and $\bar{u}(t)$ are as follow

$$u_{\hat{t}}(t)=\frac{u(t+\tau)-u(t-\tau)}{2\tau}, \bar{u}(t)=\frac{u(t+\tau)+u(t-\tau)}{2}.$$

The specific algorithm of (8) is given below, let $u_N(x,y,t)=\sum_{\kappa=1}^{N_\Omega-K_\Omega}u_\kappa(t)\varphi_\kappa(x,y)$, $U=\left(u_1(t),u_2(t),\ldots,u_{N_\Omega-K_\Omega}(t)\right)^{\mathrm{T}}$ .Substitute v_N for the global basis functions of $V_{N,\Gamma}$ in (8), which leads to

$$\begin{cases} M\dfrac{\mathrm{dU}}{\mathrm{dt}}+\nu KU=L(t)+\alpha(t), \\ U(0)=\mathrm{I}_{N,\Gamma}u_0. \end{cases} \tag{9}$$

where $M=\left(\varphi_\kappa,\varphi_{\kappa'}\right)_\Omega$ and $K=\left(\nabla\varphi_\kappa,\nabla\varphi_{\kappa'}\right)_\Omega,\kappa,\kappa'\in\{1,\ldots,N_\Omega-K_\Omega\}$ .They are the mass matrix and the stiffness matrix, respectively. $L(t)=\left(F,\nabla\cdot\varphi_{\kappa'}\right)_\Omega$, $\alpha(t)=\left(f,\varphi_{\kappa'}\right)_\Omega$ .Then the second-order CNLF scheme is used to discretize (9) in the time direction. Let U^t is the value of U at time t , we can get

$$\begin{cases} M\dfrac{U^{t+\tau}-U^{t-\tau}}{2\tau}+\nu K\dfrac{U^{t+\tau}+U^{t-\tau}}{2}=L(t)+\alpha(t), \\ U^0=\mathrm{I}_{N,\Gamma}u_0. \end{cases} \tag{10}$$

By simplifying the first equation of (10), we get

$$(M+\tau\nu K)U^{t+\tau}=\eta^{t+\tau}, \tag{11}$$

where $t=\tau,2\tau,\ldots,T-\tau$. On the right-hand side of the equation (11), $\eta^{t+\tau}$ is a known value based on calculation from the value at time t and time $t-\tau$. We calculate initial value by

$$U^\tau=\mathrm{I}_{N,\Gamma}\left(u_0+\tau\left.\frac{\partial u}{\partial t}\right|_{t=0}\right).$$

In this way, $U^{2\tau},U^{3\tau},\ldots,U^T$ can be computed successively by (11).

Remark **2.1** All the integrations in this section are calculated using the high-precision Gaussian quadrature formula as

$$(u,v)_\Omega\approx\sum_{k=1}^{K}\sum_{i,j=0}^{N}\left.\left((uv)\big|_{\Omega_k}\circ T_k\cdot J_k\right)\right|_{(\xi_i,\eta_j)}\omega_{ij}.$$

3. Numerical Examples

In this section, we give the results of some numerical examples. And we report L^2 -norm which is defined by

$$\mathrm{E}_{L^2}=\left(\sum_{i,j=0}^{N}\sum_{k=1}^{K}\left[\left(\hat{u}^k-(\hat{u}_N)^k\right)^2J_k\right]\bigg|_{(\xi_i,\eta_j)}\omega_{ij}\right)^{\frac{1}{2}},$$

Example 3.1 Consider (1) with $f(x,y,t)=0,F(u)=\dfrac{u^2}{2}$, which leads to

$$\begin{cases} u_t + u\left(u_x + u_y\right) = \nu\left(u_{xx} + u_{yy}\right), & (x,y)\in\Omega, t\in(0,\mathrm{T}], \\ u(x,y,t) = 0, & (x,y)\in\partial\Omega, t\in(0,\mathrm{T}], \\ u(x,y,0) = u_0(x,y), & (x,y)\in\Omega\cup\partial\Omega. \end{cases} \quad (12)$$

The initial condition $u_0(x,y)$ is taken from the exact solution:

$$u(x,y,t) = \frac{1}{2} - \tanh\left(\frac{x+y-t}{2\nu}\right). \quad (13)$$

The example is calculated by (8). From the table 1 and the table 2, we can see that for different viscosity coefficients n, the method maintains the high precision. When taking a smaller viscosity coefficient n, we can increase N and M to keep the algorithm highly accurate.

Table 1. E_{L^2} at $n = 1, t = 10^{-4}, \mathrm{T} = 1$ of the method (8).

N	M=1	M=2	M=4	M=6
2	2.32e-4	3.25e-5	3.93e-6	1.11e-6
4	8.87e-7	2.68e-8	8.30e-10	1.10e-10
6	3.02e-9	3.83e-11	2.20e-11	2.21e-11
8	2.88e-11	2.67e-11	2.24e-11	2.24e-11

Table 2. E_{L^2} at $n = 0.5, t = 10^{-4}, \mathrm{T} = 1$ of the method (8).

N	M=1	M=2	M=4	M=6
4	1.64e-5	8.12e-7	2.63e-8	3.19e-9
6	1.67e-7	3.44e-9	5.04e-10	5.13e-10
8	5.68e-9	5.44e-10	5.10e-10	5.16e-10
10	5.77e-10	5.43e-10	5.13e-10	5.18e-10

Example 3.2 Consider (12), the initial condition $u_0(x,y)$ is taken from the exact solution [11, 12]:

$$u(x,y,t) = \left(1 + \exp\left(\frac{x+y-t}{2\nu}\right)\right)^{-1}. \quad (14)$$

As in [11], we denote norm infinity and norm relative of errors as follow

$$\mathrm{E}_{L^\infty} = \max_{1\le i\le N_\Omega - K_\Omega}\left|u^{(i)} - u_N^{(i)}\right|, \mathrm{E}_{rel} = \left(\frac{\sum_{i=1}^{N_\Omega - K_\Omega}\left(u^{(i)} - u_N^{(i)}\right)^2}{\sum_{i=1}^{N_\Omega - K_\Omega}\left(u^{(i)}\right)^2}\right)^{\frac{1}{2}},$$

where $u^{(i)}$ and $u_N^{(i)}$ are the exact value and the computed value of the solution u at LGL point i.

From the table 3 and the table 4, we list norm infinity and norm relative of errors about solving (12) by the ChSC method [11] and the method (8), respectively. In the time direction, Runge–Kutta method of order four is used in [11]. We can find our method has better spatial accuracy at the same situation and can reach high precision when N is large enough and t is small enough.

Table 3. Norm infinity and norm relative of errors in [11].

N	t	n	E_{L^∞}		E_{rel}	
			T=0.05	T=0.25	T=0.05	T=0.25
5	0.0050	1.00	8.94e-8	1.19e-7	1.50e-7	1.70e-7
10	0.0005	1.00	7.45e-7	8.05e-7	8.50e-7	9.82e-7
10	0.0010	0.10	1.37e-6	2.09e-6	2.15e-6	2.63e-6
10	0.0050	0.10	1.28e-6	5.84e-6	2.49e-6	4.02e-6
30	0.0005	0.01	4.14e-5	4.32e-3	4.65e-5	2.26e-3

Table 4. Norm infinity and norm relative of errors of the method (8) $(M = 1)$.

N	t	n	E_{L^∞}		E_{rel}	
			T=0.05	T=0.25	T=0.05	T=0.25
5	0.0050	1.00	8.70e-9	6.42e-9	7.19e-9	7.69e-9
10	0.0005	1.00	8.29e-11	6.67e-11	7.35e-11	7.86e-11
10	0.0010	0.10	4.98e-7	1.41e-6	5.01e-7	1.34e-6
16	0.0005	0.10	6.38e-8	1.62e-7	8.45e-8	1.60e-7
64	0.0001	0.01	9.28e-7	3.15e-5	5.50e-7	1.39e-5

4. Conclusion

In this paper, a triangular spectral element method for the two-dimensional viscous Burgers equation is developed. Some numerical examples show the effectiveness of this method. We compare it with other method to show the superiority.We will continue the above research, verify the stability and convergence of this method. And we are about to apply the method to discontinuous problem.

References

[1] Shen J, Wang L L and Li H Y. 2009 A triangular spectral element method using fully tensorial rational basis functions *SIAM J. Numer. Anal* **47** (3) 1619-1650.

[2] Chen L Z, Shen J and Xu C J. 2011 A triangular spectral method for the Stokes equations *Numer. Math-theory Me* **4** (2) 158-179.

[3] Pasquetti R and Rapetti F 2010 Spectral element methods on unstructured meshes: which interpolation points? *Numer Algorithms* **55** (2-3) 349-366.

[4] Li Y Y, Wang L L, Li H Y and Ma H P. 2011 A new spectral method on triangles *J. Comput. Methods Sci. Eng* **76** 237-246.

[5] Duffy M G 1982 Quadrature over a pyramid or cube of integrands with a singularity at a vertex *SIAM J. Numer. Anal* **19** (6) 1260-1262.

[6] Samson M, Li H Y and Wang L L 2013 A new triangular spectral element method I: implementation and analysis on a triangle *Numer. Algorithms* **64** (3) 519-547.

[7] Zhou B Z, Wang B, Wang L L and Xie Z Q 2019 A new triangular spectral element method II: mixed formulation and hp-error estimates *Numer. Math-theory Me* **12** (1) 72-97.

[8] Burger J M 1949 A mathematical model illustrating the theory of turbulence *Adv. Appl. Mech* **1** (2) 171–175.
[9] Lees M, Forsythe G E and Wasow W R 1961 Finite-difference methods for partial differential equations *AM Math. Mo* **68** (8) 826.
[10] Li J L, Ma H P and Li H Y 2018 A triangular spectral element method for elliptic and Stokes problems *Appl. Math. Comput* **321** 195-208.
[11] Khater A H, Temsah R S and Hassan M M 2008 A Chebyshev spectral collocation method for solving Burgers'-type equations *J. Comput. Appl. Math* **222** (2) 333-350.
[12] Yang X, Ge Y and Lan B 2021 A class of compact finite difference schemes for solving the 2D and 3D Burgers' equations *Math. Comput. Simulat* **185** (2) 510-524.

The Phase Transition of Calcium Sulfide Under High Pressure

Huidong Feng, Shuai Wang and Baojia Wu*

Department of physics, Yanbian University, Yanji, 133002, China
*Email: jbwuybdx@163.com

Abstract. We carried out the accurate in situ angle derisive x-ray diffraction experiments and the electrical resistivity in CaS under high pressure. The result shows the pressure-induced phase transition of CaS from Na Cl-type structure (the space group Fm3m, phase I) to Cs Cl-type structure (the space group Pm3m, phase II) is reversible and the transition point is around 40.0 GPa; in the meantime, the electrical resistivity and light permeability decrease as the pressure increases and occur obvious discontinuities at the corresponding phase transition point; the result of the first-principles calculations is in concordance with the measurement results.

1. Introduction

Calcium sulphide (CaS) is an important binary alkaline earth metal compound semiconductor crystallizing into typical face-centered cubic structure (Na Cl-type structure)[1] as shown in Figure 1(a) and (b).In the crystal structure of CaS, the octahedron complementary to every six Ca^{2+} ions surround one S^{2-}, and every six S^{2-} ions surrounds one Ca^{2+} ion. Calcium sulphide, with doping rare earth elements, has excellent luminescent properties and has been investigated and applied widely[2-4]. After the phase transition under high pressure, the crystal of calcium sulfate is transformed to a Cs Cl-type structure. Figure 1a,b show the structure of CaS before and after phase transition, respectively.

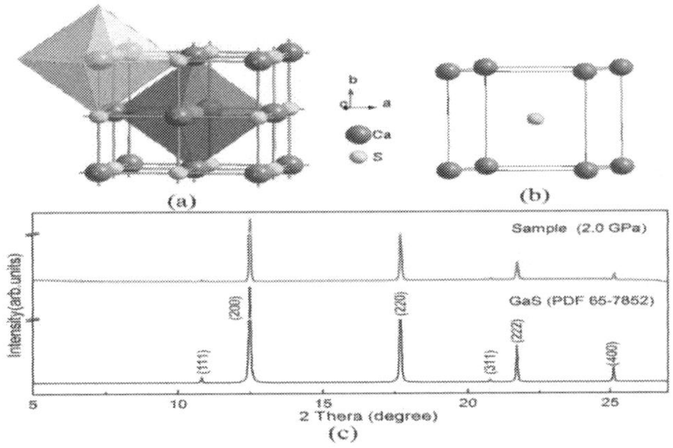

Figure 1. (a) the space group Fm3m of CaS (Na Cl-type structure,phase), (b) the space group Pm3m of CaS (Cs Cl-type structure,phaseII), (c)The comparison diagram of XRD diffraction spectra of CaS at 2.0 GPa and JCPDS 65-7852.

CaS is a insulator with the band gap rang from 2.5 to 4.5 eV at ambient pressure which has a feature in electric band structure is that there exist no d electrons in the valance band. Previously reported studies mainly focused on preparation or optimizing the performance of CaS at ambient pressure, but few reports involve on the structure and electronic characteristics under high pressure, an effective approach to tune the physical properties of materials. Recently published X-ray diffraction (XRD) studies on CaS demonstrated that phase Fm3m transformed to phase Pm3m to at 40.0 GPa. [3, 5] But Z.J.Chen forecaster that the Na Cl-type structure of CaS transformed to Cs Cl-type structure at 36.5 GPa by a method based on plane-wave potentiality density functional theory[4, 6-7]. There are some different between the results of theoretical and experimental. In the meantime, the structure transition under compression is often accompanied by the changing of physical properties. Therefore, it makes most sense to study physical characteristics of CaS under high pressure.

In this paper, by combing the XRD, the electrical resistivity measurements, the light permeability measurements and the first-principles calculations, we conducted a comprehensive study on the structural, electrical transport and optical characteristics of CaS powders under compression.

2. Experiments and Theoretical Calculation

The sample was powder CaS (99.999%, from Alfa Aesar Company). Figure 1(c) shows the structure of obtained sample matches well with previous publication (JCPDS 65-7852). A diamond anvil cell (DAC) was used, generating high pressure with the per-indented rhenium as the gasket. The pressure transmitting medium consists of a liquid mixture of CH3OH, C2H5OH and H2O (16:3:1). Pressure calibration was conducted by use of the wave number shift of the R1 line of a ruby (~5 μm)[5,8]. One-dimensional strength versus diffraction angle 2θ patterns was obtained by integrating Bragg diffraction images with FIT2D software [6, 9]. The XRD pattern obtained from the experiment was refined by using the software package of GSAS.

By using angular dispersion source (λ = 0.6199 Å), In situ high-pressure XRD experiments were conducted at beam line 15U1 of Shanghai Synchrotron Radiation Facility (SSRF) and beam line 4W2 of Beijing Synchrotron Radiation Facility (BSRF).

By using the *Van der Pauw* method, the electrical resistivity was measured by a four-electrode microcircuit. [7, 10, 11] The insulation method between gasket and film electrode and the fabrication of thin film microcircuit used in this experiment is consistent with the method reported previously.[8 12-14] A Keithley 2400 Source Meter provided the measuring current, generally 100mA, a Keithley 2700 multi meter measuring the voltage. A Keithley Kusb-488 interface adapter and a general purpose interface bus were utilized to connect all instruments with a computer. On the basis of the *Van der Pauw* method, the measurement process was automatically performed.

3. Results and Discussion

3.1. In Situ High-pressure XRD Measurements.

The high-pressure XRD experiments on CaS powders up to 50.2 GPa were firstly conducted to compare with previous studies conveniently. Figures2 (a) and (b) display the XRD patterns of CaS at compress and decompress processes, respectively. In the compression process, all diffraction peaks were moving to relatively large angles with increasing pressure, and the strength of the diffraction peaks was growing smaller. This trend indicates that the inter-planar spacing decreases with the pressure increases, and also that the crystal structure becomes denser under pressure. In this figure, the angle of the (400) diffraction peak exceeds the maximum measurement range of the experiment under a pressure of 21 GPa, and the (300) diffraction peak weakens with the pressure increases, but it always exists until a maximum pressure is reached. At 40.5 GPa, a new peak phaseII (110) appeared at 2θ=16.2°, which indicated that CaS began to transit from phase Fm3m to phase Pm3m. Under continuous pressure, both phaseIand phaseIIcoexist, but the phaseIcompletely transformed into the phaseIIat 45.5 GPa. After this point, there was no phase transition observed until the pressure increased to 50.2 GPa.Under decompression, starting with the maximum pressure 50.2 GPa, there was no significant change in the diffraction spectra of CaS, as shown in figure2(b). After the pressure was reduced to 34.9 GPa, the diffraction maximum at (110) of phaseIIdisappeared, which indicated a return to the Na Cl-type structure. The disappearing (400) diffraction peak of the phaseIin the compression process returns to its original position when the pressure is decreased continuously to 0.3 GPa (approximately atmospheric pressure). Thus, we confirm that the phase transition of CaS under high pressure from phaseIto phaseIIis reversible. Comparing to the compression curve, the decompression curve has an approximately 5.0 GPa hysteresis effect. This is because the symmetry of the crystal structure of CaS suffered some damage in the compression process. So CaS must overcome a barrier, if it wants to revert back to the original structure. Through the above analysis, the conclusions of the previous experiment are verified using a high-pressure XRD experiment. Specifically, it is observed that the phase transition of CaS from phaseIto phaseII is reversible, with the phase transition occurring around 40.0 GPa.

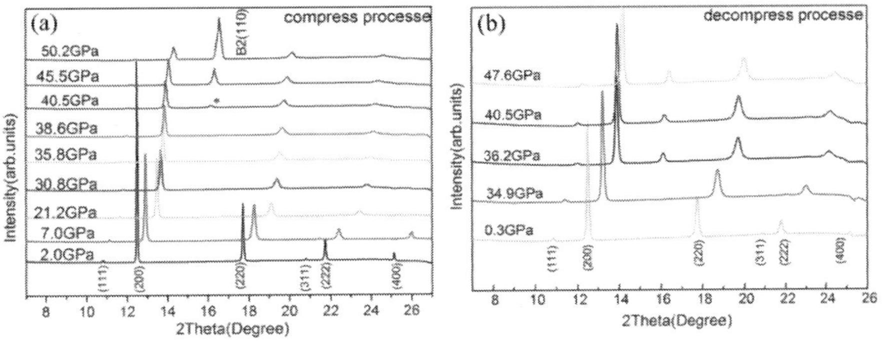

Figure 2. XRD diffraction spectra of CaS at a few representative pressures, (a) compress process, (b) decompress process.

3.2. In-situ Resistivity Measurement Under High Pressure

Figure 3 shows that during the compression, the pressure dependence of the electrical resistivity at room temperature. In the initial state, CaS has a very high resistance value greater than $1\times10^{11}\Omega$ and hence is demonstrated characteristics of an insulator. As the pressure increased, the resistance value of CaS declined slowly. Around 40.0 GPa, an abrupt resistivity decrease was observed, which reflects the structural transition from the phase I (Na Cl-type) to the phase II (Cs Cl-type). At this point, the stability in the energy band of CaS with Na Cl-structure was destroyed, and the band gap began to shrink. The phase transition of CaS from the six-coordinate structure (Na Cl-type) to the eight-coordinate structure (Cs Cl-type) was achieved when the pressure increased to 51.1 GPa. After that, the resistance value was reduced almost seven orders of magnitude compared to the previous phase change. In addition, during the decompression process we also detected what has changed in electrical transport properties of CaS , where the resistance values rose to the previous values when the pressure fell to 34.1 GPa from the maximum pressure 51.1 GPa. This illustrates once again that the phase transition is reversible.

4. Conclusions

In the present study, the in situ angle disperse x-ray diffraction experiments have been performed, the electrical resistivity measurements in CaS under high pressure. The data from these two methods are consistent with each other. The results of the two methods above confirmed that the pressure-induced phase transition of CaS from Fm3m (phase I) structure to a Pm3m (phase II) structure is reversible, and that the phase transition pressure of CaS transforming from phase I to phase II is around 40.0 GPa. In our electrical resistivity measurement experiment, the electrical resistivity of CaS decrease with the pressure increases. After the phase transition, the reduction of the electrical resistivity is exacerbated. These data provide some reference for future applications under high pressure.

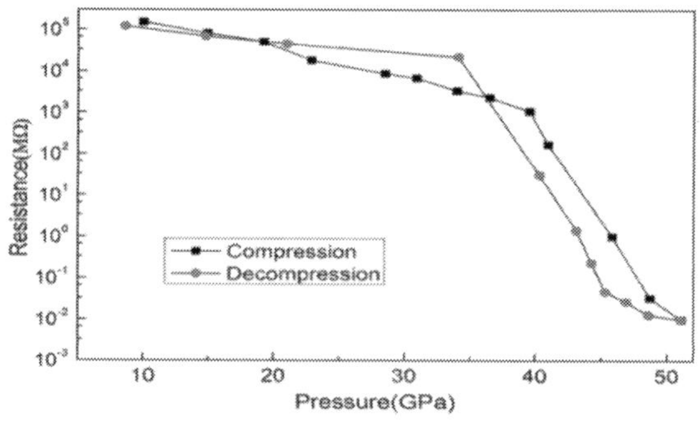

Figure 3. Measurement results of CaS electrical resistivity at room temperature under high pressure.

References

[1] Varshney D, Rathore V, Kinge R. 2009 High-pressure induced structural phase transition in alkaline earth CaX (X = S, Se and Te) semiconductors: NaCl-type (B1) to CsCl-type (B2) *Journal of Alloys and Compounds* **484** (1-2) 239-245.

[2] Ghosh P K and Pandey R. 1982 Thermoluminescence due to intrinsic point defects in polycrystalline calcium sulphide[J] *Journal of Physics C: Solid State Physics* **15** (28) 5875.

[3] Nakao Y. 1980 Luminescence Centers of MgS, CaS and CaSe Phosphors Activated with $Eu2^{+}$ Ion[J] *Journal of the Physical Society of Japan* **48** (2) 534.

[4] Tanaka K, Mikami A, Ogura T and Taniguchi K. 1986 High brightness red electroluminescence in CaS:Eu thin films[J] *Applied physics letters* **48** (25) 1730-1732.

[5] Luo H, Greene R G, Ghandehari K, Li T and Ruoff A. 1994 Structural phase transformations and the equations of state of calcium chalcogenides at high pressure[J] *Physical Review B* **50** (22)
16232.

[6] Charifi Z, Baaziz H and Hassan F E H. 2005 High pressure study of structural and electronic properties of calcium chalcogenides [J] *Journal of Physics: Condensed Matter* **17** (26) 4083-4092.

[7] Chen Z J,Xiao H Y and Zu X T. 2007 Structural and electronic properties of CaS Crystal: A density functional theory investigation Physica[J] *B: Condensed Matter* **391** (1) 193-198.

[8] Mao H, Bell P, Shaner J T and Steinberg D 1978 Specific volume measurements of Cu, Mo, Pd, and Ag and calibration of the ruby R1 fluorescence pressure gauge from 0.06 to 1 Mbar[J] *Journal of Applied Physics* **49** (6) 3276-3283.

[9] Hammersley A, Svensson S, Hanfland M, Fitch A and Hausermann D. 1996 Two-dimensional *detector* software: From real detector to idealised image or two-theta scan [J] *International Journal of High Pressure Research* **14** (4-6) 235-248.

[10] Mao H and Bell P. 1981 Electrical Resistivity Measurements of Conductors In The Diamond Window, High - Pressure Cell [J] *Review of Scientific Instruments* **52** (4) 615-616.

[11] Mao H, Xu J A and Bell P 1986 Calibration of the ruby pressure gauge to 800 kbar under qua-hydrostatic conditions [J] *Journal of Geophysical Research: Solid Earth* **91** (B5) 4673-4676.

[12] Han Y H, Gao C X, Ma Y Z, Liu H W, Pan Y W, Luo J F, Li M, He C Y, Huang X W and Zou G T. 2005 Integrated microcircuit on a diamond anvil for high-pressure electrical resistivity measurement[J] *Applied Physics Letters* **86** (6) 064104.

[13] Gao C X, Han Y H, Ma Y Z, White A, Liu H, Luo J, Li M, He C, Hao A and Huang X. 2005 Accurate measurements of high pressure resistivity in a diamond anvil cell[J] *Review of scientific instruments* **76** (8) 083912.

[14] Li M, Gao C, Ma Y, Wang D, Li Y and Liu J. 2007 In situelectrical conductivity measurement of high-pressure molten $(Mg_{0.875},Fe_{0.125})_2SiO_4$ *Applied physics letters* **90** (11) 113507.

Dispersion Dynamics of Clustered Particles by Centrifugal Force

Tiancong Feng

Shanghai Starriver Bilingual School, 2588 Jindu Road, Minhang, Shanghai, China
Email: alexfengtc@sina.com

Abstract. In this paper an approach is proposed to solve the problem of aggregation in nanomaterials through the mean of rotational separation aiming to quickly disperse clustered nanoparticles while not affecting their purity. If it is possible, this approach may replace the current mean of mechanical mixing, which may cause impurities issues. The hypothesis is that the centrifugal force due to rotational velocity acting on the nanoparticles can overcome the cohesive force between the nanoparticles, therefore dispersing the clustered nanoparticles. The experimental mean is to put different spheres connected by different types of glues imitating different nanoparticle clusters into centrifuges imitating the swivel plate. The results from both the theoretical model and the experiment show that for a cluster with a cohesive force of 1.75N, a rotational velocity of about 800rad./s is required to disperse the cluster. While for a cluster with a cohesive force of 0.25N and the same mass and position, a rotational velocity of about 150 rad./s is required to disperse the cluster. Except for the cohesive force, the mass and position of the nanoparticle on the swivel plate also have a large effect on the required rotational velocity. The observation of the physical mechanism of the dispersion has also shown that while using this way, the cluster is dispersed slowly with small parts separated from it. Therefore, this way can also eliminate re-clustering problems of nanoparticles.

1. Introduction

With the improvement of various properties such as mechanical strength, heat resistance and permeability, composite nanomaterials have been emerged as a new alternative to conventional nanomaterials. More and more kinds of nanoparticles are synthetized and mixed in order to tailor properties of the materials and to bring some specific properties to nanocomposites. However, the dispersing and homogenous mixing of fine particles into other matrix remains a challenge. Thus, dispersion of nanoparticles and powders has been a important requirement in various applications.

In all these industrial processes, high intensity of energy is required to overcome the adhesion forces of clustered nanoparticles or to increase interfacial surface area per unit volume, which was achieved by high-pressure homogenizers, ultrasonic-assisted devices, and high shear mixers [1]. Ultrasonic-assisted devices need longer time to make the clustered materials disperse, but the longer-time using of this device may make material become heat and new clustering could occur. High shear mixers, which are comprised of rotational and stationary parts, were found that the rotational part could produce friction with the materials to be mixed and thus some impurity may be brought into the mixed materials, leading to the deviation of properties of the material.

In this paper, a new approach is proposed to solve the problem of clustering and aggregation of nanomaterials by using rotational separation. It may replace the current mean of mechanical mixing, which may cause impurities issues. The approach is based on fact that the centrifugal force acting on

Journal of Physics: Conference Series **2148** (2022) 012016 doi:10.1088/1742-6596/2148/1/012016

the nanoparticles can overcome the adhesive force of the nanoparticles. The paper aims to quickly disperse the nanoparticles while not affecting the purity.

2. Theoretical Framework and Analysis

In this paper, small rigid metal spheres with mass and volume connected by a contact force F are adapted as the main subject of theoretical model and experimentation in replacement of nanoparticles. Such small rigid spheres have more controllable properties such as their mass, volume, cohesive force, and the number of spheres in a cluster. All of the following theoretical models are created based on these small rigid spheres, and most of the experiments are conducted based on them.

2.1. Clustered Double Spheres

To begin from the simplest form of cluster, given the drawing below, two rigid spheres with mass m_1, m_2 and radius r_1, r_2 have distance R_1, R_2 to the centre of a boundless swivel plate. There is a cohesive force F between the two spheres. The coefficient of friction (μ) between each sphere and plate is constant. The critical rotational velocity at which the two balls lose contact can be found.

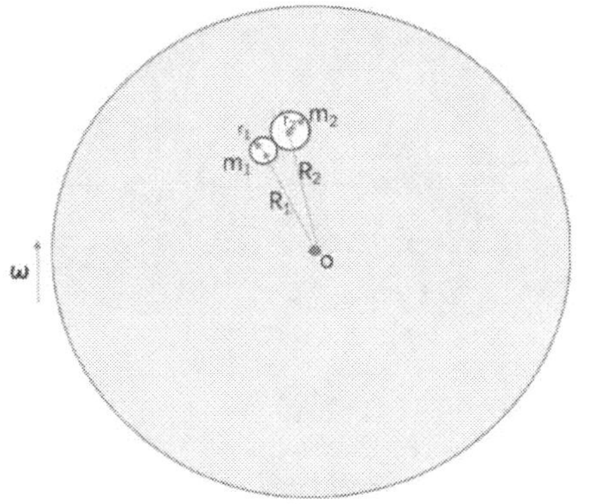

Figure1 (a). Position of clustered double spheres on a swivel plate.

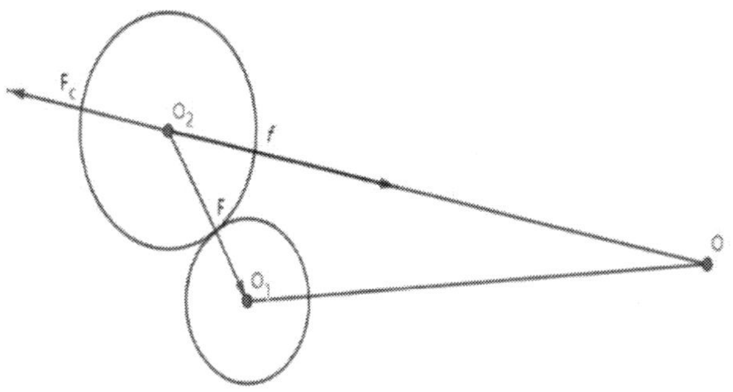

Figure 1(b). Force analysis of the sphere farther to the centre of the swivel plate.

In order to have the two spheres lose contact, there must be inadequate net force on the spheres to support the centripetal acceleration required for the spheres to reach the angular velocity. First, while the angular velocity has only a small value, the friction of both spheres is able to fully support the centripetal force. But as the velocity increases, the centrifugal force will exceed the friction.

Let the centripetal direction be the positive x direction. While the cohesive force F is not put into consideration,

$$\Sigma F_x = f = mR\omega^2 \tag{1}$$

$$\mu mg = mR\omega^2 \tag{2}$$

$$\mu g = R\omega^2 \tag{3}$$

Therefore, the greater the distance, the quicker the frictional force would not be able to support the centrifugal acceleration as the angular velocity increases. Thus, a force analysis on the sphere farther to the centre of the swivel plate can be done to derive the critical angular velocity required for the two spheres to lose contact.

Let sphere 2 to have greater distance from the centre of mass of the sphere to the centre of the disk. At the moment when the two blocks lose contact, it must be at the point when the net force is just about to exceed the centrifugal force required. For Sphere 2

$$\Sigma F_x = F_x + f = m_2 R_2 \omega_{critical}^2 \tag{4}$$

$$F\cos\alpha + \mu m_2 g = m_2 R_2 \omega_{critical}^2 \tag{5}$$

The expression of the $\cos\alpha$ can be found by solving the triangle OO_1O_2 in figure 1 using the law of cosines

$$R_1^2 = R_2^2 + (r_1 + r_2)^2 - 2(r_1 + r_2)R_2\cos\alpha \tag{6}$$

$$2(r_1 + r_2)R_2\cos\alpha = R_2^2 + (r_1 + r_2)^2 - R_1^2 \tag{7}$$

$$\cos\alpha = \frac{R_2^2 + (r_1 + r_2)^2 - R_1^2}{2(r_1 + r_2)R_2} \tag{8}$$

Then substituting the expression back to the equation

$$F\frac{R_2^2 + (r_1 + r_2)^2 - R_1^2}{2(r_1 + r_2)R_2} + \mu m_2 g = m_2 R_2 \omega_{critical}^2 \tag{9}$$

Yielding the critical rotational velocity to be

$$\omega_{critical} = \sqrt{F\frac{R_2^2 + (r_1 + r_2)^2 - R_1^2}{2(r_1 + r_2)R_2^2 m_2} + \frac{\mu g}{R_2}} \tag{10}$$

2.2. Clustered Triple Spheres

To complicate the situation by a little, one more sphere is added, creating a triple sphere cluster problem. Similarly, three rigid spheres with mass m_1, m_2, m_3 and radius r1 r2, r_3 have distances R_1, R_2, and R3 to the centre of a boundless swivel plate. There is a cohesive force F between the three spheres, the coefficient of friction μ between the spheres and the swivel plate. The critical angular velocity at which the sphere farthest to the centre of the swivel plate disperses from the cluster is to be found.

Using the same conclusion derived in the two-spheres cluster model, when the farthest sphere lose contact from the other two, its centrifugal force is exactly equal to the maximum net force.

The expression of the net force when the sphere is at its critical velocity can be found.

$$F\cos\alpha + F\cos\beta + \mu m_2 g = m_2 R_2 \omega_{critical}^2 \tag{11}$$

For $\cos\alpha$ and $\cos\beta$, they can be derived through the law of cosines

$$\cos\alpha = \frac{(r_1 + r_2)^2 + R_2^2 - R_1^2}{2(r_1 + r_2)R_2}, \qquad \cos\beta = \frac{(r_3 + r_2)^2 + R_2^2 - R_3^2}{2(r_3 + r_2)R_2} \tag{12}$$

Then substituting the expressions back to the equation

$$F\frac{(r_1 + r_2)^2 + R_2^2 - R_1^2}{2(r_1 + r_2)R_2} + F\frac{(r_3 + r_2)^2 + R_2^2 - R_3^2}{2(r_3 + r_2)R_2} + \mu m_2 g = m_2 R_2 \omega_{critical}^2 \tag{13}$$

Yielding the critical angular velocity

$$\omega_{critical} = \sqrt{F\frac{(r_1 + r_2)^2 + R_2^2 - R_1^2}{2(r_1 + r_2)R_2^2 m_2} + F\frac{(r_3 + r_2)^2 + R_2^2 - R_3^2}{2(r_3 + r_2)R_2^2 m_2} + \frac{\mu g}{R_2}} \tag{14}$$

2.3. Theoretical Results

2.3.1. Effect of Positions of Spheres on Critical Rotation Velocity. For double-sphere case, with the parameters: $m_1=m_2=1.0$mg, $r_1=r_2=1.0\mu$m, $\mu=0.1$, and F=0.1N, the equation (10) is used to calculate the critical rotational velocity of swivel plate when the farthest sphere lose contact.

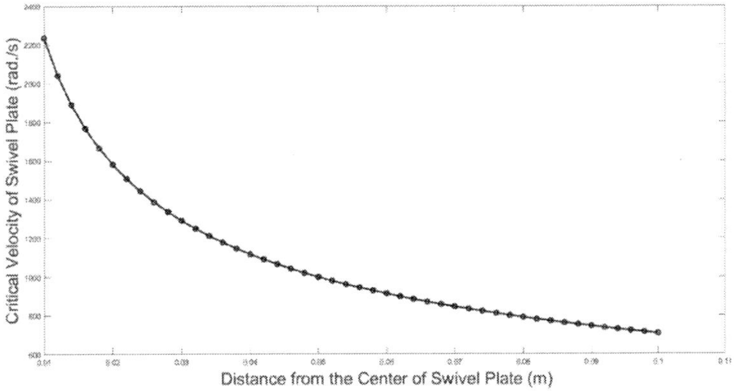

Figure 2. The variation of critical velocity of the swivel plate with the distance of the spheres from the centre of swivel plate.

Figure 2 shows the effect of the distance of the spheres from the centre of swivel plate on the critical velocity of the swivel plate. It is known from the figure that with the distance increasing from 0.01m to 0.1m, the critical velocity dramatically decreases from 2200 to 700 rad./s. Thus, it can be concluded that the initial position of the cluster has a large effect on the critical rotational velocity.

2.3.2. Effect of Radius of Spheres on Critical Rotation Velocity. With the parameters: $m_1=m_2=1$mg, $R_1=10$mm, $R_2=R_1+r_1$, $\mu=0.1$, and F=0.1N, the equation (10) is used to calculate the critical rotational velocity of swivel plate when the farthest sphere lose contact.

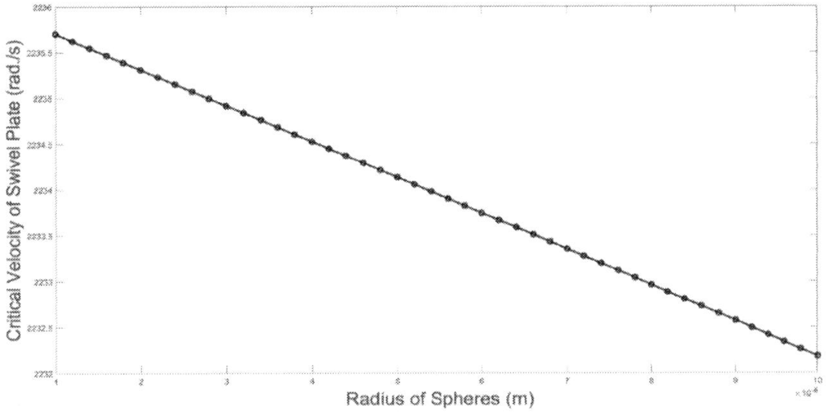

Figure 3. The variation of critical velocity of the swivel plate with the spheres radius.

Figure 3 shows the effect of the sphere radius on the critical velocity of the swivel plate. It is shown from the figure that with the radius increasing from 1μm to 10μm, the critical velocity slowly decreases from 2235.55 to 2232 rad./s. Thus, it can be concluded that the sphere radius has only a minor effect on the critical velocity.

2.3.3. Effect of Mass of Spheres on Critical Rotation Velocity. With the parameters: $r_1=r_2=1.0\mu m$, $R_1=10.0mm$, $R_2=10.001mm$, $\mu=0.1$, and $F=0.1N$, the equation (10) is used to calculate the critical rotational velocity of swivel plate when the farthest sphere lose contact.

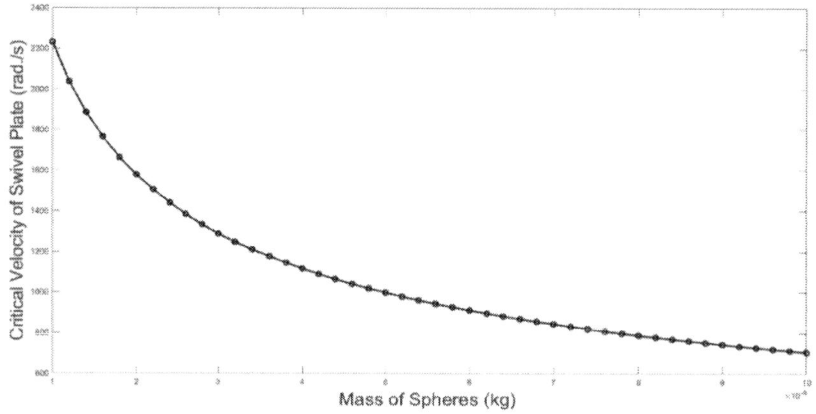

Figure 4. The variation of critical velocity of the swivel plate with the spheres mass.

Figure 4 shows the effect of the sphere mass on the critical velocity of the swivel plate. It is shown from the figure that with the mass increasing from 1.0mg to 10.0mg, the critical velocity dramatically decreases from 2200 to 700 rad./s. Thus, it can be concluded that the spheres mass has a large effect on the critical rotational velocity.

2.3.4. Effect of Cohesive Force of Spheres on Critical Rotation Velocity. With the parameters: $m_1=m_2=1.0mg$, $r_1=r_2=1.0\mu m$, $R_1=10.0mm$, $R_2=10.001mm$, and $\mu=0.1$, the equation (10) is used to calculate the critical rotational velocity of swivel plate when the farthest sphere lose contact.

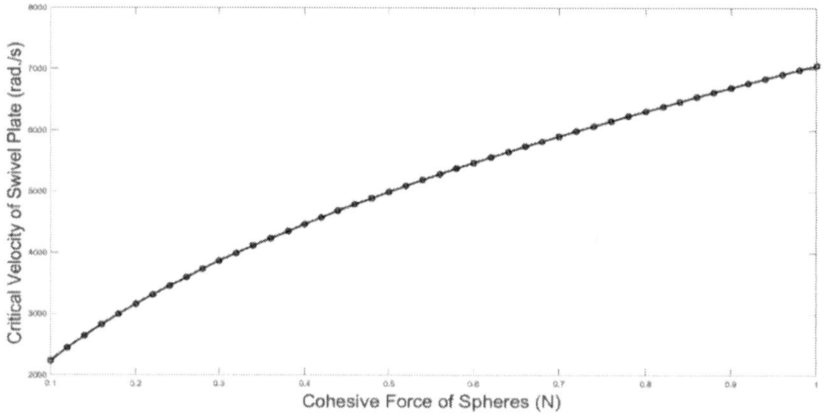

Figure 5. The variation of critical velocity of the swivel plate with the cohesive force of spheres.

Figure 5 shows the effect of the cohesive force of spheres on the critical velocity of the swivel plate. It is illustrated from the figure that with the force increasing from 0.1 to 1N, the critical velocity increases from 2000 to 7000 rad./s. Thus, it can be concluded that the cohesive force between the spheres has a deciding effect on the critical rotational velocity.

The analytical results show that for the clustered two-sphere case, with the increasing distance or/and particle mass, the critical velocity dramatically decreases. With the increasing radius of particles or/and increasing friction coefficient, the critical velocity nearly keep constant. With the increasing cohesive force, the critical velocity dramatically increases.

3. Experimental Results and Discussion

After the derivation and analysis of theoretical model, the experiment is conducted to verify and compare the results of the theoretical model with the experimental results.

3.1. Experiment Instruments

The key instrument is a small high-speed centrifuge JOANLAB type MS-12 Pro with rotational velocity of 1000-12000 rpm. In order to measure the cohesive force, a push-pull force probe is used. In order to imitate the cohesive force between each particle, three different types of glues are used: Liquid glue, PVA Solid glue, and 502 Cyanoacrylate Instant Adhesive Super Glue. To imitate the particles, two types of metal spheres with different mass and radius are used. The larger spheres have a diameter of 6mm and are made of 304 stainless steel. The smaller spheres have a diameter of 2.5mm and are made of 316 stainless steel. Besides the small metal spheres, since the ultimate goal of this work is to disperse the nanoparticles with the centrifugal force, SiO_2 nanoparticles are also prepared for observation on how the actual nanoparticles will disperse in the centrifuge. While using the SiO_2 nanoparticles, they are wrapped together with the weakest liquid glue imitating the cohesive force inside them while in use.

3.2. Experiment Parameters

Before finding the rotational velocity, the strength of the cohesive force from different glues is to be calculated. Through applying adequate amounts of different types of glue at the pull end of the force probe to connect to a 6mm sphere and then pulling the sphere, the value of the cohesive force that each glue provides is derived. The cohesive force of Liquid Glue, PVA Solid Glue and 502 Super Glue are <0.25, 0.27, 1.75N.

Then finding the mass of the two different spheres. Since the metal spheres are perfectly spherical, for the 6mm diameter spheres, the density for its 304 stainless steel material is 7930kg/m^3, while for the 2.5mm diameter spheres, the density for its 316 stainless steel material is 7980 kg/m^3. Therefore, their masses are found to be $m_{2.5} = 0.0653g$, $m_6 = 0.897g$.

With a ruler, the distance between the centre and the edge of the centrifuge's plate is measure to be 25.0mm. The coefficient of friction, since the swivel plate in the centrifuge is made of graphite, is 0.1.

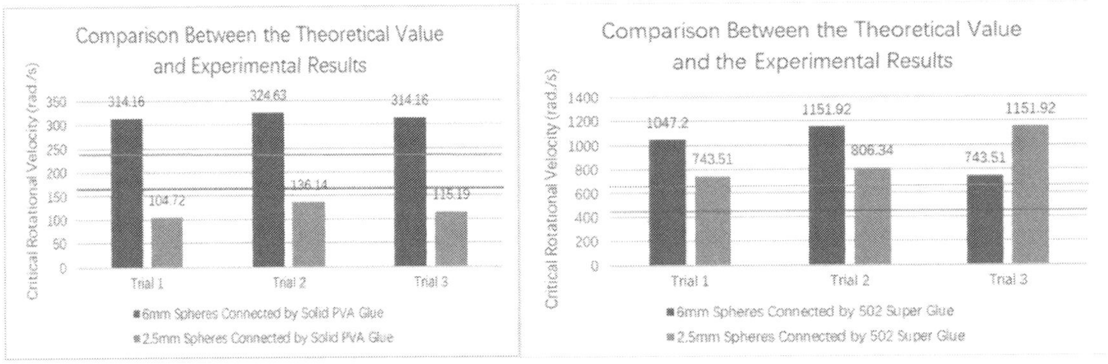

Figure 6. The comparison of theoretical and experimental results for the double sphere case.

3.3. Comparison of Experimental and Theoretical Results

Except the actual SiO_2 nanoparticles that are tested, all of the experiments are conducted in same situation as the theoretical model. For better comparison, the quantified recorded experiment results will be put into graphs along with the theoretical value.

For this section, the comparison will be done in the same order as the experiments are conducted. As shown in figure6, the theoretical results and the actual results for the double sphere cluster are analysed below.

The theoretical and the experimental values deviate a little. The bars are the experimental results, while the lines are the theoretical results. The analysis on cohesive force here appears to be applicable, as the critical rotational velocity is dramatically increased while increasing the cohesive force with the 502 Super Glue. However, comparing the experimental values for each graph with the theoretical values, it can be determined that the effect of the mass and radius are different for the two. In the theoretical analysis, it has been expected that the mass of the spheres has a deciding effect on the critical rotational velocity while the radius has only a little deciding effect. Therefore, it is expected that the critical rotational velocity for 2.5mm spheres will exceed the critical rotational velocity for 6mm spheres.

However, the experiment results show that the critical rotational velocity for 2.5mm spheres is actually less than the critical rotational velocity for 6mm spheres. The reason for this deviation is possibly due to the slight difference in the cohesive force between the spheres. As mentioned before, the cohesive force has a deciding effect on the critical rotational velocity. In this case, the larger spheres will form a larger contact surface with each other, therefore creating the larger cohesive force, which impacted on the results of experimentation. If this cohesive force is reasonably considered, it can be found that the results of the experiment does still fit the theoretical model.

4. Conclusion

As a summary, in this paper, the physical and mathematical models for clustered two-spheres, three-spheres, are deduced to model the relationship between the geometry, mass, mechanical parameters of particles as well as the angular velocity of a swivel plate for dispersing the clustered particles. The analytical results showed that for the clustered two spheres and three-sphere models, with the increasing distance from the centre of swivel plate, the critical velocity dramatically decreases. With the increasing radius of particles and friction coefficient, the critical velocity nearly keep constant. With the increasing mass of particles, the critical velocity dramatically decreases. With the increasing cohesive force, the critical velocity dramatically increase. Thus, the distance from the centre of disk, the mass of particles, cohesive force of the spheres have larger on the critical velocity of the disk, and the radius of particles, friction coefficient have smaller effect. The theoretical results are basically in agreement with the experiments results, and thus the approach proposed in this paper may separate the clustered particles and thus quickly disperse the nanoparticles while not affecting the purity.

References

[1] Nguyen S, Didier R, Brice V. 2014 Dispersion of nanoparticles: From organic solvents to polymer solutions [J] *Ultrasonics Sonochemistry* 21 149–153.

[2] Xie L, Rielly C, Eagles W, Zcan-Tas G. 2007 Dispersion of NanoParticle Clusters Using Mixed Flow and High Shear Impellers in Stirred Tanks[J] *Chemical Engineering Research and Design* **85** (A5) 676-684.

[3] Adam J, Stuart W. 2008 Dispersion of Nanoparticle Clusters by Ball Milling [J] *Journal of Dispersion Science and Technology* **29** (4) 600–604.

[4] Chia-Jung C, Lijuan X , Qiang H, Jianjun S. 2012 Quantitative characterization and modeling strategy of nanoparticle dispersion in polymer composites[J] *IIE Transactions* 44 523–533.

[5] Chenghu K, Jianlin X, Lei N, Chengsi L, Jiliang F. 2021 Effect of thermoplastic polyurethane elastomer on the properties of polybutylene terephthalate matrix flame retardant composites[J] *Polymer Composites* 42 3098–3113.

[6] Halen T, Laurence S. 2021 Combined combustion and heat-flux measurements on a supersonic jet-in-crossflow configuration using luminescent paint [J] *Exp Fluids* 62 70.

A Compatible and Efficient Fabrication of Bi-Continuous Nanoporous Copper Films

Sizhen Chen[1, 2], Chao Ren[1, 2], Penglei Zhang[1, 2], Yunna Sun[1] and Guifu Ding[1, *]

[1] National Key Laboratory of Science and Technology on Micro/Nano Fabrication, Shanghai Jiao Tong University, Shanghai, 200240, China

[2] School of Electronic Information and Electrical Engineering, Shanghai Jiao Tong University, Shanghai, 200240, China

Email: gfding@sjtu.edu.cn

Abstract. To cater to the application in various micro/nano devices, this paper reports a compatible and efficient approach combining electrodeposition and dealloying to fabricate bi-continuous nanoporous copper films. In the electrodeposition step, effect of concentration ratio of different ions, pH of the solution and cathode current density on the elementary composition and microscopic morphology of the deposited alloy are systematically investigated, obtaining an optimum condition with good stability and process compatibility. A uniform copper-zinc alloy with its zinc content improved to ~67 at.% is prepared under this condition. A uniform nanoporous copper with an average pore size of 150 nm is fabricated by ultrasonic assisted dealloying whose efficiency is significantly improved compared to free dealloying. This method is promising to be used in the mass production of nanoporous copper films, benefiting researches on various practical applications in micro/nano devices.

1. Introduction

Due to their high surface-to-volume ratio, low density and good electric and thermal conductivity, metallic nanoporous materials have attracted considerable attention in various fields, such as catalysis, sensors and actuators, fuel cells etc. [1-5]. As a typical metal nanoporous material, nanoporous copper (NPC) has become more popular than other precious metals, because of the advantages of abundant sources, low cost, and diverse preparation methods.

Till now, many physical and chemical methods have been proposed to fabricate nanoporous copper materials. Compared with the template and casting methods [6-7], preparation of nanoporous copper by the dealloying process has been widely studied because of its flexibility and low cost. The principle of the dealloying method is to selectively dissolve the less noble component from the alloy through chemical/electrochemical methods while the residual component rearranges to form a nanoporous structure by choosing suitable alloy precursors and controlling the dealloying conditions. Chen et al. [8] fabricated NPC with pore sizes ranging from ~15 nm to ~200 nm by selective corrosion of the melting single-phase Cu30Mn70 alloy in HCl solutions with different duration. A hierarchically structured micro/nanoporous copper [4] was formed in a 3 wt.% HCl solution at 50 °C with a multi-phase Cu-Al alloy synthesized by a pack cementation method as its precursor. In the study by Yang et al. [9], an enhanced bulk nanoporous copper with a three-dimensional bi-continuous porous structure was obtained by dealloying of Cu30Zn70 precursors which were synthesized through a casting process at 900 °C. It has been found that free corrosion in acid solution is a common method for dealloying. However, the hydrogen bubbles generated by the dissolution of Zn may remain in the nanopores formed by the early dealloying period, which would hinder the contact between the corrosive liquid and the undealloyed layer, delaying the further dealloying process. Besides, prior to dealloying, the preparation of the precursor alloys greatly affects the application of the nanoporous structures. Conventional methods such as casting and melting require a high working temperature and the prepared alloys are mostly bulk materials, which damages the process compatibility and limits the application of NPC in micro/nano devices. Meanwhile, manufacturing costs will be increased if we choose the sputtering process [10]. A compatible co-electrodeposition process [11-12] has been proposed to prepare a Cu-Zn alloy precursor, whose thickness can be tuned by controlling the deposition time. Whereas, to meet the component threshold of the less noble element (Zn), the concentration of Cu^{2+} and Zn^{2+} ions in the electrolyte solution differ largely. Given the consumed ions cannot be replenished from a suitable alloy anode, the bath system of the co-electrodeposition is erratic.

In this study, we propose a compatible and efficient two-step fabrication of bi-continuous nanoporous copper films, combining electrodeposition and dealloying. In the electrodeposition step, the Cu-Zn alloy precursors are obtained by co-electrodeposition at a constant current density. In addition, the effects of concentration ratio of different ions, pH and cathode current density on the composition and morphology of the deposited alloy layer are investigated systematically. In the dealloying step, the dealloying time is controlled in different concentrations of hydrochloric acid to obtain uniform and stable nanoporous copper, and the dealloying efficiency is significantly improved through the assistance of ultrasound during the process.

2. Methods

As is shown in figure 1, nanoporous copper films were fabricated by the electrodeposition of Cu-Zn alloys and a subsequent dealloying process.

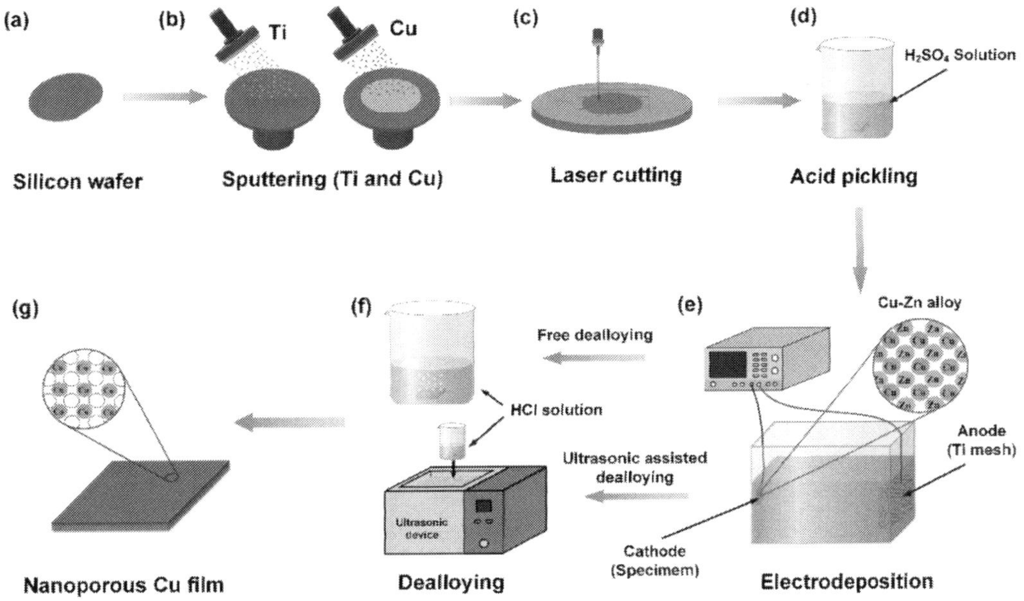

Figure 1. Schematic of the proposed compatible and efficient method to fabricate nanoporous Cu by electrodeposition of Cu-Zn alloys and a dealloying process.

2.1. Substrate Preparation

Polished monocrystalline silicon wafers coated with thin layers of titanium (Ti) and copper (Cu) by magnetron sputtering were prepared as the substrate to provide a super smooth surface for electrodeposition. It should be noted that the titanium film served as an adhesion layer and the copper film acted as a conductive layer respectively. After being coated with Ti/Cu, the wafer was cut into square pieces of 1.5×1.5 cm^2 by laser for use. Before electrodeposition, the substrate was immersed into 7 wt.% H_2SO_4 to remove the oxide layer on the surface and rinsed with deionized water.

2.2. Co-electrodeposition of Cu-Zn Alloy Precursors

The electrodeposition was performed at a constant current density in a two electrodes system, where a titanium mesh coated with platinum was chosen as the anode and the prepared substrate as the cathode. The working current was provided by an adjustable DC power supply (GPD-3303S, GWINSREK). The electrolyte solution for electrodeposition was composed of $CuSO_4 \cdot 5H_2O$, $ZnSO_4 \cdot 7H_2O$, with $Na_3C_6H_5O_7 \cdot 2H_2O$ as the complex agent. Concentration ratio of Cu^{2+} and Zn^{2+}, pH and cathode current density were studied to obtain an optimal condition for the co-electrodeposition of Cu-Zn alloy. Three kinds of baths with different concentration of $CuSO_4$ are listed in table 1. pH of the electrolyte solution was adjusted by the addition of H_2SO_4 and NaOH. Because temperature also has an important effect on electrodeposition, it was set at 25 °C in all deposition process. After deposition, the prepared Cu-Zn alloy precursors were rinsed with deionized water and dried at 65 °C for the following dealloying.

Table 1. Chemical composition of electrolyte.

Bath	$CuSO_4$ (M)	$ZnSO_4$ (M)	$Na_3C_6H_5O_7$ (M)
Bath1	0.02	0.3	0.5
Bath2	0.03	0.3	0.5
Bath3	0.06	0.3	0.5

2.3. Formation of Nanoporous Copper

Cu-Zn alloy precursors were dealloyed in HCl solutions with various concentrations (0.025M, 0.05M, 0.1M) to fabricate nanoporous copper films. The process temperature was set at 25 °C. Besides, to promote the dealloying process, the selective dissolution of Zn was performed in an ultrasonic cleaner (TES-1010, TIME) to help hydrogen bubbles get rid of the pores. The dealloyed specimens with nanoporous structure were then rinsed with deionized water and dried at 65 °C.

2.4. Characterization

The morphology and chemical composition of electrodeposited Cu-Zn alloy precursors and dealloyed nanoporous copper were characterized by a field emission scanning electron microscope (FESEM) (Ultra55, ZEISS) combined with an EDS analyzer.

3. Results and Discussion

3.1. Preparation of Cu-Zn Alloy Precursors by Electrodeposition

To obtain uniform bi-continuous nanoporous copper by dealloying, the alloy precursor should meet two requirements [13]: (1) There should be a large difference in the standard metal/metal-ion electrode potentials between copper and another less noble (LN) element (at least a few hundred millivolts). (2) The composition of the less noble element should surpass the component threshold. Therefore, Cu-Zn alloy is an excellent precursor for dealloying with the difference in standard electrode potential between Cu and Zn as 1.1 V. It was reported that [14] Cu-Zn alloys with Zn content of ~70 at.% are more likely to obtain bi-continuous nanoporous copper with open pores through dealloying.

In this study, Cu-Zn alloys were fabricated by electrodeposition, which is a process of depositing a thin layer of metal, alloy or composite material on the surface of a conductive solid by electrochemical methods. Based on electrodeposition theory, that the precipitation potential of the two metals are very close is necessary for co-electrodeposition, which can be adjusted by the addition of a complex agent to regulate the activity of metal ions. To obtain better process compatibility, trisodium citrate ($C_6H_5Na_3O_7$)

was chosen as the complexing agent, whose complexing ratio to Cu^{2+} and Zn^{2+} change with pH. For a co-electrodeposition system, several factors impact the preparation of the alloy. Herein, the effects of concentration ratio of Cu^{2+} and Zn^{2+}, pH of the solution and cathode current density on the composition and morphology of the alloy have been investigated step by step.

Electrolyte solutions with the concentration ratio of Cu^{2+} and Zn^{2+} as 1:15 (Bath1), 1:10 (Bath2) and 1:5 (Bath3) were prepared. Based on different baths, pH and cathode current density were regulated to obtain a uniform and stable Cu-Zn alloy with the composition of Zn at ~70 at.%.

3.1.1. The Effect of pH on Deposited Alloys. In Bath1, pH is firstly investigated. The raw electrolyte solution with 0.02M $CuSO_4·5H_2O$, 0.3M $ZnSO_4·7H_2O$ and 0.5M $Na_3C_6H_5O_7·2H_2O$ has the pH of 5.25. Cu-Zn alloys prepared in this solution have a brassy appearance (figure 2a) with microscopically unevenly spherical grains (figure 2g). EDS analysis manifests that the Zn content in this alloy is only 23.74 at.%. Figure 2b-c show that alloys obtained under pH of 6 and 6.5 are heterogeneous with a distinct color division on the surface, which indicates the instability of the electrolyte solution. When pH is regulated to 7, the deposited Cu-Zn alloy shows a uniform light gray surface macroscopically (Figure 2d) with a regular grain size (Figure 2h) and the Zn content is measured at 67 at.%, which is close to ideal. When the pH rises to 7.5, as is shown in Figure 2e, the alloy surface has brassy and green divisions. Continue to increase the pH, as is shown in Figure 2f, the alloy coating begins to show a uniform green color and large protruding grains can be observed on SEM images of the microscopic morphology. Besides, the Zn content of the alloy has been reduced to ~30 at.%. The complexing ability of citrate ion to Cu^{2+} is stronger than that to Zn^{2+}. When the pH is 7, the difference in complex ability reaches the maximum value, which significantly reduces the activity of Cu^{2+} in solution, favoring the deposition of Zn. The above experiments demonstrate that in Bath1, it is most likely to obtain a uniform Cu-Zn alloy with its Zn content of ~70 at.% when pH = 7.

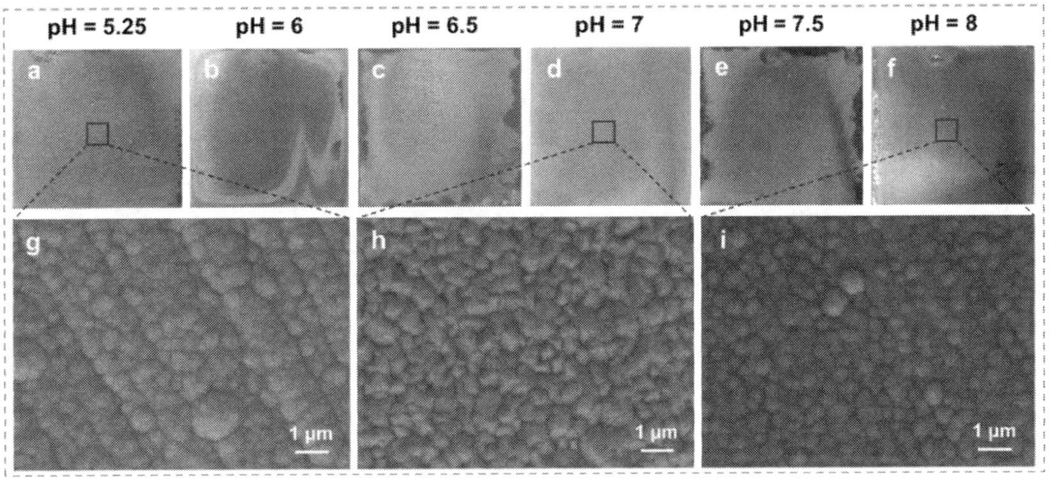

Figure 2. (a)-(f) Appearance of Cu-Zn alloys obtained in Bath1 under different pH. (g)-(i) SEM images of the microscopic morphology of the prepared alloys at the pH of 5.25, 7 and 8.

3.1.2. The Effect of Cathode Current Density on Deposited Alloys. With the pH of deposition solutions fixed at 7, the alloy *coatings* obtained under different cathode current density (J) are shown in figure 3. It is depicted in figure 3a-b that, Cu-Zn alloys deposited under the current density of 3 mA/cm^2 and 4 mA/cm^2 have macroscopically smooth and uniform surfaces and the regular grains can be observed by the SEM images in figure 3f-g. When J is larger than 5 mA/cm^2, the distribution of current density on the substrate is uneven, and the edge appears black due to the concentration of current density (figure3c-d). EDS analysis shows that the Zn content in the edge area is much higher than the central area. When J is 6 mA/cm^2, the Zn content even reaches 100%. For the central area, as shown in figure 3e, the Zn content in Cu-Zn alloys increases from 60% to 82% with the increase of current density. For a given cathode current density, the deposition rate of the more noble metal (Cu) is relatively closer to the critical current density. Therefore, increase of cathode current density is more likely to improve the deposition rate of the less noble metal (Zn), thereby increasing the content of Zn in the alloy. Besides, as J increases from 3 mA/cm^2 to 6 mA/cm^2, the microscopic morphology of the alloys (figure 3f-i) changes from regular square grains to sharp cone-shaped grains with fuzzy boundaries. Overall, although the increase of cathode current density is beneficial to obtain a higher Zn content, the uneven distribution of the current density will make the coating lose its uniformity, which makes it difficult to obtain uniform nanoporous copper by subsequent dealloying process. Therefore, the optimal condition in Bath1 is pH=7 and J=4 mA/cm^2 (**condition1**), in which the prepared Cu-Zn alloy shows a uniform appearance of light gray with regular and compact grains microscopically, and the Zn content is ~67 at.%.

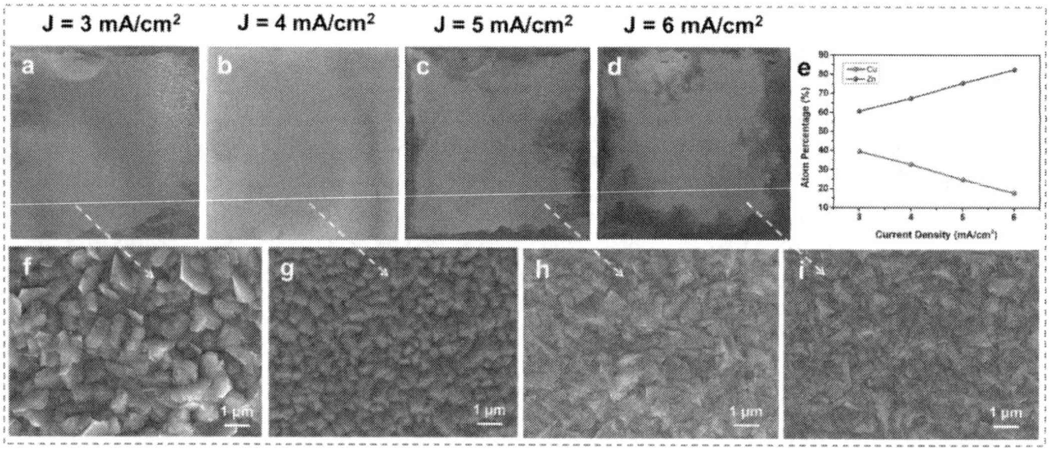

Figure 3. Appearance of Cu-Zn alloys obtained in Bath1 at the current density of (a) 3 mA/cm^2, (b) 4 mA/cm^2, (c) 5 mA/cm^2, (d) 6 mA/cm^2. (e) Change of the atom percentage of Cu and Zn alloys with the current density. SEM of the micro morphology of Cu-Zn alloys obtained in Bath1 at the current density of (f) 3 mA/cm^2, (g) 4 mA/cm^2, (h) 5 mA/cm^2, (i) 6 mA/cm^2.

3.1.3. Optimization of Concentration Ratio of Cu^{2+} and Zn^{2+}. The same method was applied to Bath2 and Bath3. The best condition in Bath2 is pH = 6, and J = 8 mA/cm^2 (**condition 2**) while that in Bath3 is pH = 6, and J = 15 mA/cm^2 (**condition 3**). As is shown in Figure 4, the surface of the alloy layer obtained under these two conditions has divisions in microscopic morphology, while the Zn content of the two structures is similar, being ~75 at.% under condition 2 and ~65 at.% under condition 3, respectively.

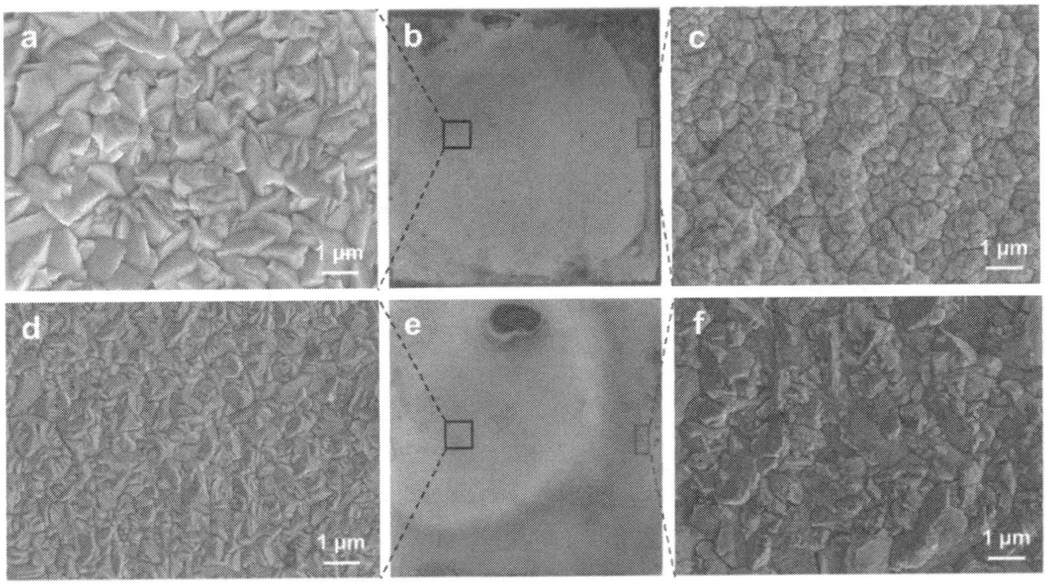

Figure 4. The appearance and morphology of Cu-Zn alloys obtained in (a)-(c) Bath2 with cathode current density of 8 mA/cm2 and pH of 6 and (d)-(f) Bath3 with cathode current density of 15 mA/cm2 and pH of 6.

Size of the substrate and volume of the electrolyte solution are changed to conduct further research on the above three conditions. Only the Cu-Zn alloy prepared under condition 1 can still maintain its uniformity and stability in elementary composition and microscopic morphology, as is shown in figure 5. Therefore, the optimal condition for the fabrication of Cu-Zn alloy precursor in this study is shown in table 2, with its Zn content at ~67 at.%. And the thickness can be tuned by controlling the deposition time.

Table 2. The optimal condition of electrodepostion for fabrication Cu-Zn alloy precursor.

$CuSO_4$ (M)	$ZnSO_4$ (M)	$Na_3C_6H_5O_7$ (M)	pH	Current Density (mA/cm^2)
0.02	0.3	0.5	7	4

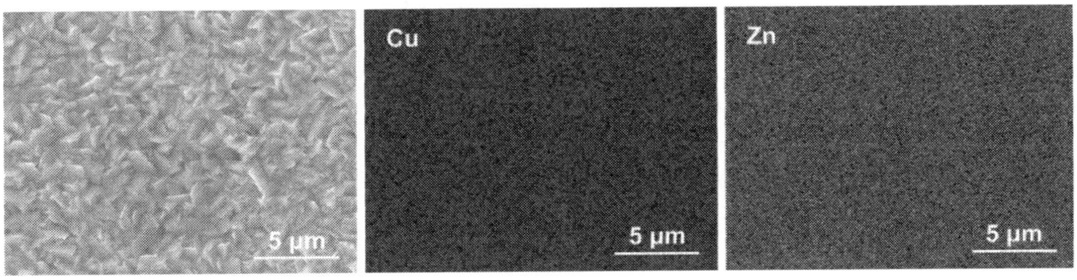

Figure 5. The SEM image and EDS mapping of the Cu-Zn alloy precursor obtained in Bath1 with the cathode current density of 4 mA/cm^2 and the pH of 7.

3.2. Fabrication of Nanoporous Copper by Dealloying

3.2.1. Free Dealloying. Hydrochloric acid is chosen as the electrolyte for dealloying to conduct the selective dissolution of Zn. The *acid* concentration has a critical effect on the evolution of the microstructure, which determines the uniformity and stability of the prepared nanoporous copper. As is shown in Figure 6, the dealloying process in HCl with concentration of 0.025M, 0.05M, 0.1M are investigated. Figure 6a-c depict that, in the initial stage of dealloying in 0.025M HCl, some small pores were formed on the surface of the film. With the progress of dealloying, the porous structure did not deepen into the bulk while a locally smooth surface began to appear. After 48 hours of dealloying, no continuous nanoporous structure was formed. A low acid concentration results in a low atom mobility rate at the solid/electrolyte interface [15]. After Zn atoms in the first layer are selectively dissolved, the coordinated copper atoms remain as adatoms on the surface, leading to the enrichment of Cu atoms, which blocks the dealloying channel, and hinders the further reaction. Finally, the copper atoms on the surface migrate slowly and join together to form a stable and smooth surface. In 0.05M HCl, a uniform bi-continuous nanoporous copper with an average pore diameter of 120 nm (figure 6e) was gradually formed within 36 h. Extending the dealloying to 48 h, the pore size was slightly coarsened while the nanoporous structure still exists. When the concentration of HCl was increased to 0.1M, as shown in figure 6g-i, a continuous NPC can be obtained in 24 h. However, with the further extension of the dealloying time, the NPC became heterogeneous and the structure collapsed. This indicates that NPC is unstable in concentrated acid solutions which promotes the mobility of Cu atoms at the solid/electrolyte interface to form a compact structure of nanoparticles with lower surface energy.

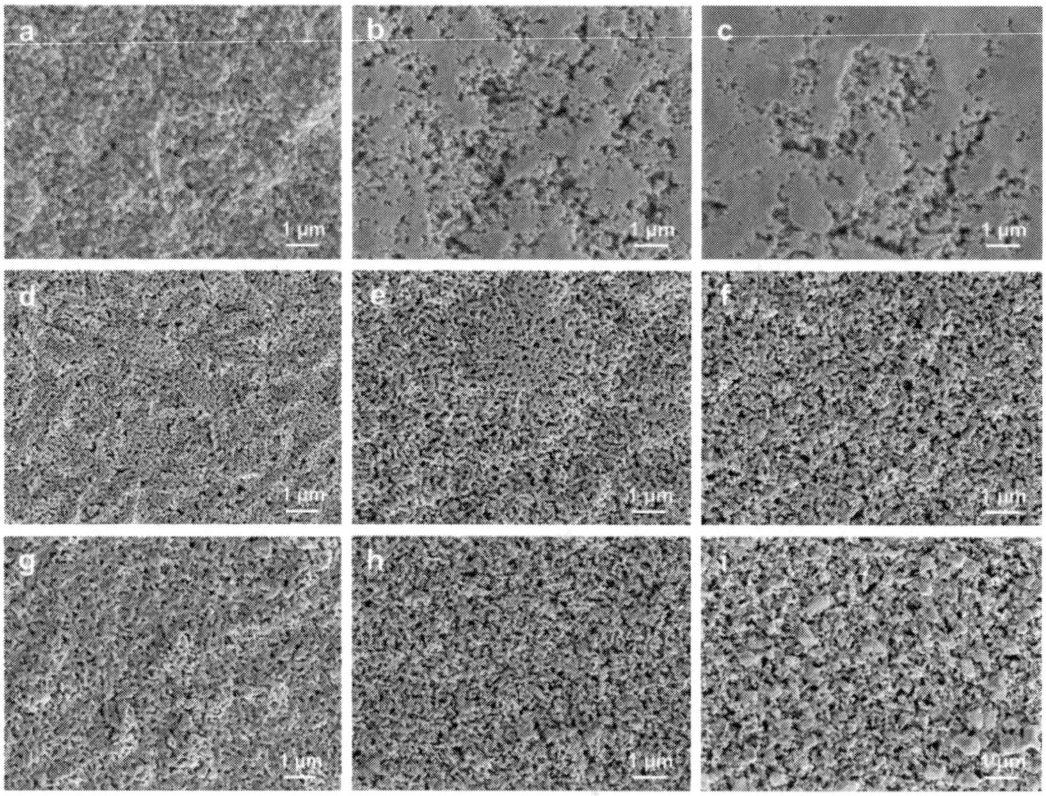

Figure 6. SEM images of NPC produced by dealloying Cu-Zn alloys in (a)-(c) 0.025 M HCl for 12 h, 24 h, 48 h; (d)-(f) 0.05M HCl for 24 h, 36 h, 48 h; (g)-(i) 0.1M HCl for 12 h, 24 h, 36 h.

The above experiments show that it is more difficult to form a bi-continuous NPC with open pores in HCl solution with low concentration. Although increase in acid concentration can improve the dealloying efficiency, the prepared NPC tends to evolve into nanoparticles with lower surface energy in concentrated solutions. Overall, in 0.05M HCl, a uniform bi-continuous NPC with a suitable pore size can be obtained by controlling the dealloying time.

3.2.2. Ultrasonic Assisted Dealloying. The hydrogen bubbles generated by the reaction of HCl and Zn may block the nanopores formed by dealloying, preventing the contact between electrolyte solution and the undealloyed alloy, which will delay the dealloying process. Therefore, to improve the efficiency of dealloying, ultrasonic assistance is applied to promote the escape of bubbles. As is shown in figure 7, in 0.05M HCl solution, with the aid of ultrasound, after 5 h of dealloying, a uniform and continuous NPC with an average pore diameter of 150 nm can be obtained. Compared with free dealloying, the efficiency is improved 7 times. Extending the dealloying time to 7 h, the nanoporous structure can still exist stably, which indicates its potential in the efficient fabrication of a stable and uniform NPC.

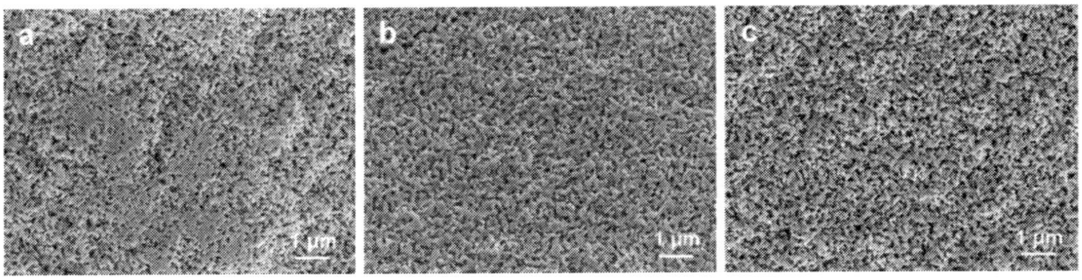

Figure 7. SEM images of NPC produced by dealloying Cu-Zn alloys with ultrasonic assisted 0.05M HCl for 2 h, 5 h, 7 h.

4. Conclusion

In summary, we have proposed a compatible and efficient two-step fabrication of bi-continuous nanoporous copper films by electrodeposition of Cu-Zn alloy and a dealloying process. Trisodium citrate is chosen as the complexing agent in the solution of electrodeposition for the difference in its complexing ability to Cu^{2+} and Zn^{2+}. In this study, concentration ratio of Cu^{2+} to Zn^{2+}, pH of the solution and cathode current density are regulated step by step to investigate their effect on the elementary composition and morphology of the deposited alloys. Compared to previous studies, we obtain a uniform Cu-Zn alloy with a higher atom percentage of Zn reaching 67 % at the same concentration ratio of Cu^{2+} to Zn^{2+} with pH = 7 and J = 4 mA/cm^2, which makes the process more flexible and compatible. 0.05M HCl is selected as the best concentration for dealloying, in which a stable and uniform bi-continuous NPC can be formed. Besides, ultrasonic assistance is applied to promote the escape of hydrogen bubbles generated during the dealloying process. Compared with free dealloying, the efficiency is improved 7 times. This compatible and efficient method with the characteristic of low cost is promising to be applied to the mass production of stable and uniform bi-continuous nanoporous copper films, which will promote researches on the application of NPC in catalysis, sensors, heat dissipation for micro/nano devices.

Acknowledgments

The authors would like to thank supports from the Shanghai Professional Technical Service Platform for Non-Silicon Micro-Nano Integrated Manufacturing. This work is supported by pre-research project (XXX3502) and the National Natural Science Foundation of China (No. 62104141).

References

[1] Zhang Z, Wang Y, Qi Z, Zhang W, Qin J and Frenzel J. 2009 Generalized fabrication of nanoporous metals (Au, Pd, Pt, Ag, and Cu) through chemical dealloying[J] *Phys. Chem. C* **113** (29) 12629-36.

[2] Wang R, Wang C, Cai W and Ding Y. 2010 Ultralow-platinum-loading high-performance nanoporous electrocatalysts with nanoengineered surface structures[J] *Cheminform* **22** (16) 1845-48.

[3] Jin H, Wang X, Parida S, Wang K, Seo M and Weissmüller J. 2010 Nanoporous au-pt alloys as large strain electrochemical actuators[J] *Nano Lett* **10** (1) 187-194.

[4] Han G, Um J H, Park H, Hong K, Yoon W and Choe H. 2019 Hierarchically structured nanoporous copper for use as lithium-ion battery anode[J] *Scr. Mater* **163** 9-13.
[5] Qi Z and Weissmüller J. 2013 Hierarchical nested-network nanostructure by dealloying[J] *ACS Nano* **7** (7) 5948-54.
[6] Li Y, Jia W, Song Y and Xia X. 2007 Superhydrophobicity of 3d porous copper films prepared using the hydrogen bubble dynamic template[J] *Chem. Mater* **19** (23) 5758-64.
[7] Jin Q, Li Zhenhua, Li Zaijiu, Yang T and Zhou R. 2011 Fabrication of lotus-type porous copper with a continuous casting technique[J] *Foundry Technol* **32** (3) 306-309.
[8] Chen L, Yu J, Fujita T and Chen M. 2010 Nanoporous copper with tunable nanoporosity for sers applications[J] *Adv. Funct. Mater* **19** (8) 1221-26.
[9] Yang Q, Liang S, Han B, Wang J and Mao R. 2012 Preparation and properties of enhanced bulk nanoporous coppers[J] *Mater. Lett* **73** 136-138.
[10] Diao F, Xiao X, Luo B, Sun H, Ding F, Ci L and Si P. 2018 Two-step fabrication of nanoporous copper films with tunable morphology for sers application[J] *Appl. Surf. Sci* **427** 1271-9.
[11] Yue H, Zhang C, Yang Z, Wang H, Ding G and Zhao X. 2013 A preparation method of patterned nanoporous copper[J] *Adv. Mater. Res* **663** 322-5.
[12] Tuan N T, Park J, Lee J, Gwak J and Lee D. 2014 Synthesis of nanoporous Cu films by dealloying of electrochemically deposited Cu–Zn alloy films[J] *Corros. Sci* **80** 7-11.
[13] Erlebacher J. 2004 An atomistic description of dealloying *J. Electrochem. Soc.* **151**(10) C614.
[14] Mao R, Liang S, Wang X, Yang Q and Han B. 2012 Effect of preparation conditions on morphology and thermal stability of nanoporous copper[J] *Corros. Sci* **60** (3) 231-7.
[15] Erlebacher J, Aziz M J, Karma A, Dimitrov N and Sieradzki K. 2001 Evolution of nanoporosity in dealloying[J] *Nature* **410** (6827) 450-3.

Punching Shear Failure Regulation of Reinforced Concrete Beams Subjected to Close-in Explosion

Hong Tang[1], Hongbo Zhai[1], Jianjun Su[1] and Teng Fu[2]

[1] Xi'an Modern Chemistry Research Institute, Xi'an 710065, China
[2] Fifth Military Representative Office of PLA Rocket Force Equipment Department in Xi'an Area, Xi'an 710065, China
Email: 1341435783@qq.com

Abstract. Aiming at the damage effect of reinforced concrete beams subjected to close-in explosion, numerical simulation and dimensional analysis are carried out to study the damage of beams under close-in explosion with spherical charge. The damage characteristics of beams were obtained by the methods of AUTODYN and DYNA. Using dimensional analysis, the relationship between the damage characteristics of reinforced concrete beams and the stand-off distance is deduced. The empirical equation of the beam punching shear failure characteristics is fitted on the basis of numerical simulation. The results show that the simulation method is more efficient than the fluid-solid coupling calculation; the proposed empirical equation is in good agreement with the experimental results. The relationship between failure width and the stand-off distance obtained by the fitting and the modelling method have reference value for the research on the damage effect of reinforced concrete beams.

1. Introduction

Accidental explosions and terrorist attacks occur from time to time. Reinforced concrete beams are the key load-bearing components of construction facilities. Studying its response and damage characteristics under explosions is of great significance to damage analysis and safety design. During a close-in explosion, the beam structure is in the detonation product area and is coupled by multiple field loads such as shock waves, detonation products, and heat. The evolution law of beam surface load and structural response is very complicated, and it is difficult to accurately describe it through analytical methods. Research is often carried out through experiments or numerical simulations.

Scholars have carried out a large number of simulations on reinforced concrete beams under explosions, but the simulation results are affected by material models and erosion. Under the condition of similar reinforcement ratio, material and size, 1.63kg TNT equivalent caused severe punching shear failure of the beam [1] while 4kg TNT equivalent caused slight damage [2], so numerical simulation needs to be established on the basis of experiments. Zhou [3] used DYNA to verify the experiment conducted by Wang [4]. The simulation obtained the beam deflection and the damage length of the mid-span area to be in good agreement with the experiment, and they used numerical simulation to study the damage of the beam by the span-to-height ratio. Gholipour [5] simulated the failure of reinforced concrete beams under the combination of close-in explosion and impact load by finite element, and studied the changing trend of the failure characteristics of the beams under different loading modes. Wang [2] found that the length of the beam concrete failure zone and the degree of deformation of the steel bar are approximately scaled to the 1/3 power of the explosive amount using numerical simulation. Wang [4] divided the beam damage criterion based on the scaled distance.

According to experiments, Wang [6] proposed that the local damage parameter and the charge amount are approximately linear. Yao [7] fitted the relationship between the flex-thickness ratio and scaled distance and stirrup ratio. reinforcement ratio[8], concrete strength, and explosive size all affect the failure mode of beams, but most of them were qualitative analysis. Related studies had analysed the influence of these parameters on the failure mode of beams, but no quantitative conclusions have been given.

Fluid-solid interaction is a common method for numerical simulation, but the solution has strict requirements on the grid size. In order to avoid leakage of explosive fluid into the concrete, the structural grid should be smaller than the fluid grid size or increase the coupling point, which will increase the calculation time. The keyword LOAD BLAST in DYNA is only applicable to where the scaled distance is greater than 0.2 m/kg$^{1/3}$. When the scaled distance or the standoff distance is small, the pressure curve calculated by CONVEP has a large error with the actual situation. In order to improve the solution speed and accuracy of the simulation results, Yu [9] and Yan[10] used the principle of equal impulse to simplify the explosion load into a uniformly distributed load applied to the reinforced concrete column. This method avoided modelling the air domain. Modelling had greatly improved the calculation efficiency. However, due to over-simplification [10], the simulated concrete column produced compressive failure at the support. This failure did not occur in the test, but this provides a way to calculate the reflection overpressure on the beam surface and solve the response of the beam separately, the key lies in the accurate description of the overpressure and the reasonable selection of the material model of the beam. In terms of overpressure, Wang [11] compared the time history curve of the target point reflection overpressure obtained by AUTODYN simulation and experiment. The peak size and arrival time of the reflection overpressure of the two are in good agreement, and AUTODYN has been used to obtain Overpressure of various explosives [12]. As a typical material, reinforced concrete has a material model suitable for explosive conditions in DYNA.

To sum up, in order to obtain the damage law when the beam is severely damaged and improve the efficiency of numerical simulation, this article uses a method to analyse the damage of the beam under close-in explosion. The method mainly includes the following two aspects: I Calculating the explosion load by AUTODYN; II simulating the response of the beam in DYNA; then verifies the validity of the numerical simulation according to the experiment; Uses dimensional analysis to deduce the relationship between the size of the damage area of the front and back explosion surface of the beam and standoff explosion. On the basis of numerical simulation, the damage area and standoff distance are fitted, and the empirical equation of the fitting is verified according to the experiments in other documents.

2. Numerical Simulation of Reinforced Concrete Beams

2.1. Calculation of Beam Surface Pressure Load

AUTODYN is used to simulate the spherical explosion of a 1kg TNT charge. The material parameters used are all taken from the material model library that comes with AUTODYN. The initial internal energy of air is 2.068*10^{2}kJ/g, explosive is TNT, central detonation of spherical charge, and the density is 1.63g/cm3. Symmetrical modelling is adopted, air and explosives are Euler grids, and unit length is 1mm. The reflection of the shock wave on the beam surface and the symmetrical boundary are simulated through total reflection boundary, and the other two sides are the outflow boundary conditions. The gauges are used to obtain the reflection overpressure time history curve of the beam surface under different standoff distances. Since the shock wave distribution on the beam is approximately exponential, the distance between gauges gradually increases. The numerical model and the distribution of gauges are shown in figure 1.

Figure 2 shows the calculated time history curve of reflected overpressure at the mid-span position of the standoff distance of 0.2m. When the shock wave reaches the surface of the beam, the reflected pressure rises to the peak instantaneously, then the pressure drops, attenuates to the ambient pressure and forms a negative pressure section, and gradually return to atmospheric pressure (Solid line). In order to verify the simulation method, it is agreed in the calculation: Ignore the rise time of the shock wave, that is, when the shock wave reaches the surface of the beam, it reaches the peak value, and the time when the reflected overpressure begins to change is taken as the time t_a when the shock wave reaches the surface of the beam; The time history curve is simplified to linear according to the principle of equal impulse and the negative pressure section is ignored (Dotted line);

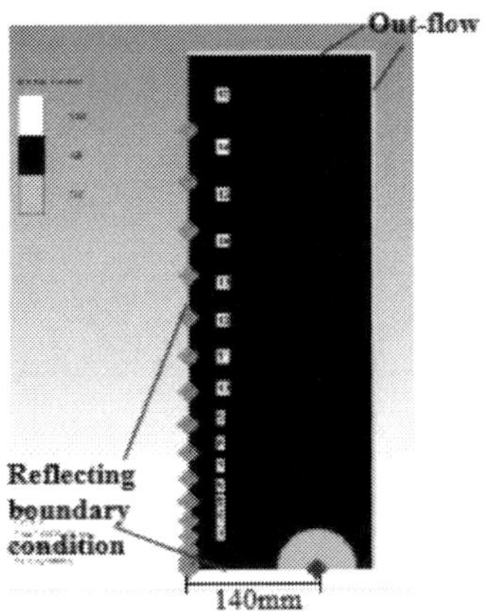

Figure 1. Computational model of reflection overpressure0.

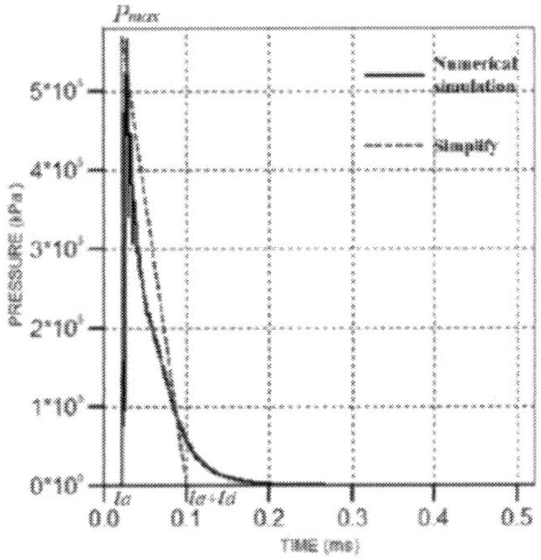

Figure 2. Overpressure time history curve of mid-span reflection0.

Using MATLAB to obtain the specific impulse i_j of gauges, the method is $i_j = \sum P_t dt$, reflected overpressure peak P_{max}, overpressure arrival time t_a, in which environmental pressure is reduced when calculating the specific impulse and reflected overpressure peak.

Write the standoff distance, specific impulse, and shock wave arrival time of each gauge into DYNA, and obtain the peak reflected overpressure, shock wave arrival time and specific impulse of each position on the beam through linear interpolation (1), and then use the principle of equal impulse to calculate the positive pressure action time t_d, see equation (2):

$$\varphi = \left(\varphi_{n+1} - \varphi_n\right)\left(Z - Z_n\right)\left(S_{n+1} - S_n\right)^{-1} + \varphi_n \tag{1}$$

$$t_{\mathrm{d}} = 2i / P_{\max} \tag{2}$$

Where φ is the parameters (P_{max}、i_j、t_a、t_d), S is the standoff distance, Z is the scaled distance.
The reflected pressure P at any time is calculated by the following equation:

$$\begin{cases} P = \left[P_m - P_m\left(t - t_{\mathrm{a}}\right)\left(t_{\mathrm{d}} - t_{\mathrm{a}}\right)^{-1}\right], & t_{\mathrm{a}} < t < t_{\mathrm{d}} \\ P = 0, & \text{others} \end{cases} \tag{3}$$

2.2. Numerical Simulation of Beam Response

Using DYNA to study the local failure characteristics of reinforced concrete beams under close-in explosions, using a separate modelling method. Due to the load and beam symmetry, a 1/2 model was established along the beam axis. According to the dimension and boundary condition of the beam in the experiment, the model is built. Concrete and fixing devices use SOLID165 elements, and steel bars use BEAM161 elements. The keyword BEAM IN SOLID is used to couple those. Defining the contact between the fixed devices and beam, and the symmetry boundary conditions are imposed on the symmetry plane. Since there is no air domain, the number of grids has dropped significantly, so a more

 2148 (2022) 012018 doi:10.1088/1742-6596/2148/1/012018

refined grid is used to solve the response of reinforced concrete beams, and the unit is 2.5mm, a total of 880,816 units. The simulation model is shown in figure 3.

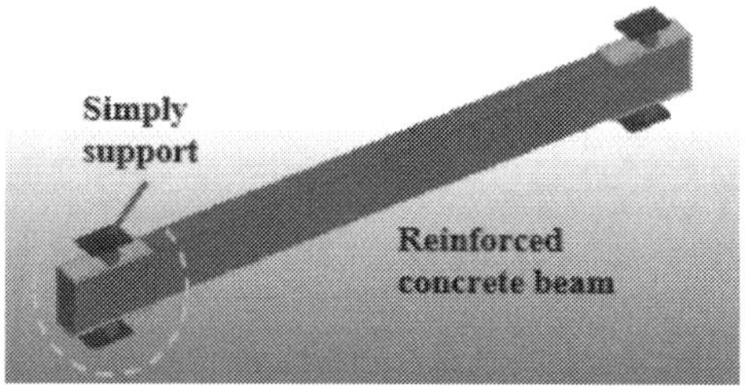

Figure 3. Simulation model of reinforced concrete beams.

The concrete adopts the K&C model, the steel bar adopts bilinear elastoplastic model, and fixing device adopts elastoplastic model. Relevant material parameters are shown in table 1.

Table 1. Material parameters.

	Density ρ (g/cm^3)	Poisson's ratio P_r	Yield strength F_c (MPa)	Elastic modulus E (MPa)
concrete	2.4	0.2	41.7	200
reinforced	7.8	0.26	435	200
support	7.8	0.26	700	200

The fracture and failure of concrete are simulated by erosion. Due to erosion effect on local failure characteristics, this paper adopts a relatively conservative erosion criterion [13] with a principal strain of 0.5; the concrete strain rate magnification factor is calculated according to equation 4 and 5[14, 15], magnification factor of concrete r_{fc} under compression is:

$$r_{\mathrm{fc}} = \begin{cases} (\dot{\varepsilon} / \dot{\varepsilon}_s)^{1.026\alpha} & \dot{\varepsilon} \le 30/s \\ \gamma\dot{\varepsilon} / \dot{\varepsilon}_s^{\ 1/3} & \dot{\varepsilon} > 30/s \end{cases} \quad (4)$$

Where $\dot{\varepsilon}$ is the strain rate; $\dot{\varepsilon}_s$ is the reference strain rate, $\dot{\varepsilon}_S = 3\times10^{-5}s^{-1}$, $\log\gamma = 6.156\alpha - 2$, f_C' is compressive strength of concrete.

Magnification factor of concrete r_{ft} under tension is:

$$r_{\mathrm{ft}} = \begin{cases} (\dot{\varepsilon} / \dot{\varepsilon}_s)^{1.026\alpha} & \dot{\varepsilon} \le 1.0/s \\ \beta(\dot{\varepsilon} / \dot{\varepsilon}_s)^{1/3} & \dot{\varepsilon} > 1.0/s \end{cases} \quad (5)$$

Where $\dot{\varepsilon}$ is the strain rate; $\dot{\varepsilon}_s$ is the reference strain rate, $\dot{\varepsilon}_S = 3\times10^{-5}s^{-1}$, $\log\beta = 6\delta - 2$, $\delta = 1/(1+0.8f_C')$, f_C' is compressive strength of concrete.

3. Validation of the Partial Method

3.1. Calculation Time of Different Simulation Methods

In order to show the improvement of the solution efficiency of this method, a reinforced concrete beam fluid-solid interaction model is established. As shown in figure 4, the 1/2 model is adopted, the air mesh element size is 10mm, the beam element mesh size is 5mm, and the total number of element is 154768, the material parameter setting of reinforced concrete beam is invariable. Explosives and air use multi-material ALE algorithm, explosives use JWL equation of state, and air use linear polynomial equation of state.

Figure 4. Fluid-solid interaction model.

The number of elements in the fluid-structure coupling model is 1/6 of the previous one, but the time required to solve the model is equivalent.

3.2. Comparison of Results Between Numerical Simulation and Experiment

In order to verify the validity of the numerical simulation results, a close-in explosion test of reinforced concrete beam 1# is carried out. The reinforced concrete beam is a rectangular section beam with a length of 160 cm, a width of 13 cm, and a height of 13 cm. The simply supported boundary is simulated by spacers and sticks, as shown in figure 5.

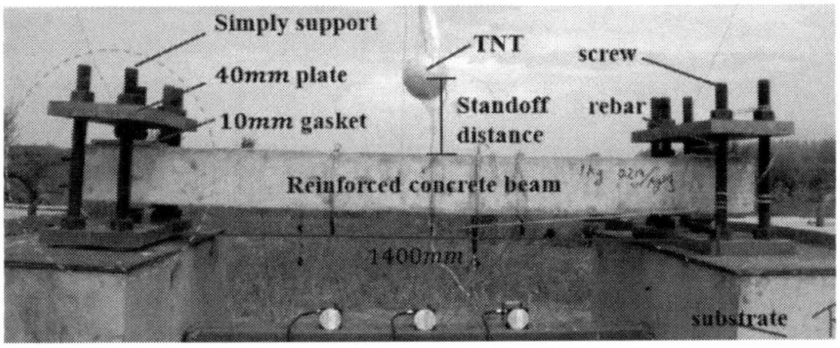

Figure 5. Explosion test.

The thickness of the protective layer is 2 cm, and the uniaxial test is carried out with prefabricated cement blocks. The measured strength is 41.7 MPa, the longitudinal reinforcement is 4Φ10 mm, the symmetrical reinforcement, the stirrup is Φ6 mm, and the spacing is 100 mm. The longitudinal steel bars and stirrups ultimate strength are about 625 MPa, and the yield strength is about 435 MPa. The test conditions are shown in table 2.

Table 2. Test condition.

Scaled distance Z (m/kg$^{1/3}$)	TNT equivalent W (kg)	Standoff distance S (m)
0.1	1	0.1

The whole response of the beam lags behind the local failure of the beam, so the deflection is not chosen as the damage parameter when verifying the damage characteristics of the numerical model. In this paper, the width of blasting area on the front and back face is selected as the damage parameter. Due to the irregular rectangle shape of the fracture zone, the fracture zone is calculated as the length of the beam *L* minus the length of the undamaged zone *D* as shown in figure 6.

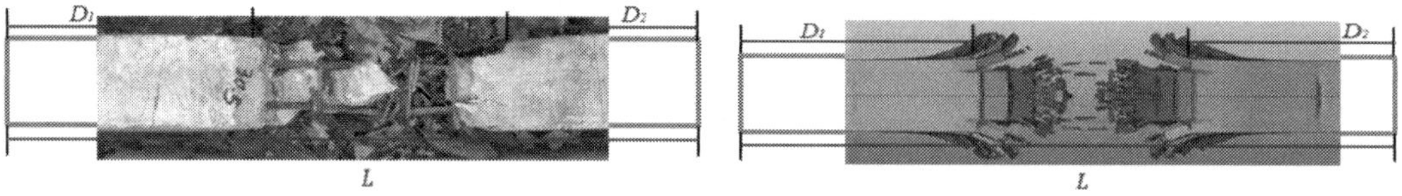

(a) test (b) simulation

Figure 6. Calculation method of fracture zone.

According to the numerical calculation results, the local damage of the beam has been completed within 1-2ms, so 3ms is selected as the time when the local damage parameter is measured. The damage result of the beam is shown in figure 7. The beam has severe punching and shear failure, forming a trapezoidal hollow area in the middle of the span.

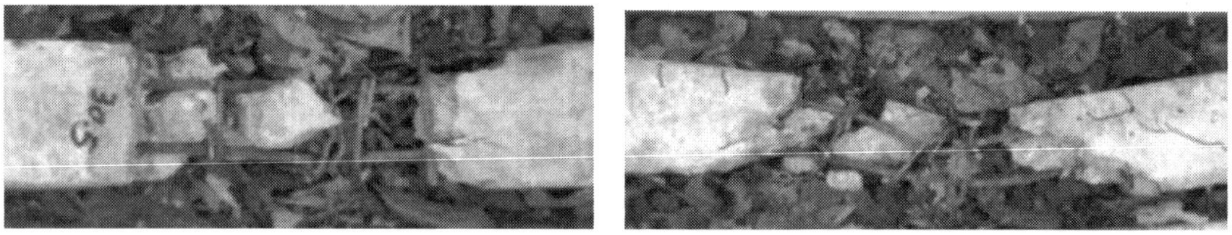

(a) frontal failure mode (b) lateral failure mode

Figure 7. Local failure mode.

According to the measurement, the test and simulation results are shown in table 3

Table 3. Failure dimension of concrete.

	Front face L_1 (cm)	Back face L_2 (cm)
test	36.5	52
simulation	37.8	51.7
error/%	3.5	0.6

The error between the two results is less than 5%, so the simulation method and model used in this paper can simulate the damage of the beam under the close-in explosion.

4. Punching Shear Failure Model Under Close-In Explosion

4.1. Dimensional Analysis

The damage law of the blast wave to the reinforced concrete beam involves various physical parameters, and the relationship between the various parameters can be found through dimensional analysis. The destruction width of the front and back explosion surface is the main parameter to measure the punching-shear failure of beams. Therefore, this paper mainly conducts dimensional analysis on the damage size of the punching-shear failure of reinforced concrete beams. Figure 8 shows a schematic diagram of a reinforced concrete beam under a spherical charge.

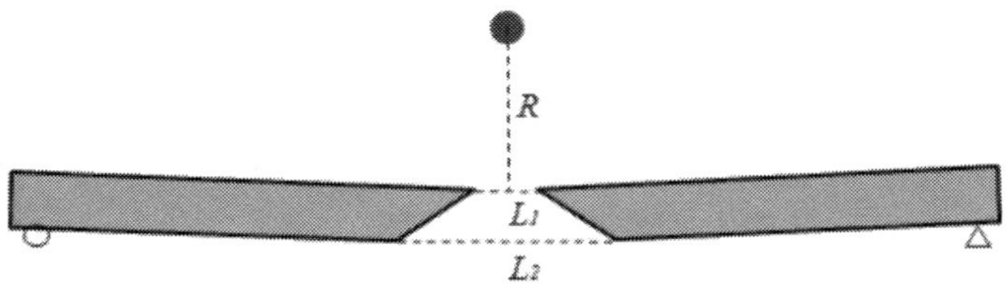

Figure 8. Schematic diagram of punching shear failure.

Among them, L_1 and L_2 are the broken widths of the front and back explosion surfaces. Through analysis, the main influencing factors that affect the beam damage parameters are:

1) TNT: charge radius r_0(L), density ρ_T($L^{-3}M$), standoff distance S(L), detonation velocity V_T(LT^{-1}), energy density e_T(L^2T^{-2}), product expansion coefficient γ(1);

2) Rebar: strength σ_S($L^{-1}MT^{-2}$), elastic modulus E_S($L^{-1}MT^{-2}$), shear modulus G_S($L^{-1}MT^{-2}$);

3) Concrete: thickness d(L), width b(L), strength σ_c($L^{-1}MT^{-2}$), elastic modulus E_c($L^{-1}MT^{-2}$), shear modulus G_c($L^{-1}MT^{-2}$), reinforcement ratio η(1)。

The basic dimensions of the above physical quantities are L, M, T, so take the charge radius r(L), density ρ_T($L^{-3}M$), detonation velocity V_T(LT^{-1}) as the basic parameters, and the size of the failure zone must satisfy a certain dimensionless function relationship:

$$\left(\frac{L_1}{r_0},\frac{L_2}{r_0}\right)=f\left(\frac{R}{r_0},\frac{d}{r_0},\frac{b}{r_0},\frac{e_T}{V_T^2},\frac{\sigma_S}{\rho_T V_T^2},\frac{E_S}{\rho_T V_T^2},\frac{G_S}{\rho_T V_T^2},\frac{\sigma_C}{\rho_T V_T^2},\frac{E_C}{\rho_T V_T^2},\frac{G_C}{\rho_T V_T^2},\gamma,\eta\right) \tag{6}$$

Dimensionless parameters can be approximated by power exponents within a certain range. The damage effect of beams is studied by changing the burst center distance, so the parameters of reinforced concrete beams and explosives can be regarded as invariants. The relevant quantities in equation (6) are all constants, so:

$$(r_0,d,b,e_T,V_T,\sigma_S,E_S,G_S,\sigma_C,E_C,G_C,\gamma,\eta)=\text{const} \tag{7}$$

equation (6) can be expressed as:

$$\begin{cases} L_1 = A' r_0 \left(S/r_0\right)^{\alpha_1} & S \ge r_0 \\ L_2 = B' r_0 \left(S/r_0\right)^{\beta_1} & S \ge r_0 \end{cases} \tag{8}$$

4.2. Similarity Rate of Beam Broken Area

Because this article considers that the charge is spherical, there is

$$r_0 = \left(3m/4\rho\right)^{1/3} \tag{9}$$

Substitution equation (8):

$$\begin{cases} L_1/m^{1/3} = A'\left(S/m^{1/3}\right)^{\alpha_1} & S \ge r_0 \\ L_2/m^{1/3} = B'\left(S/m^{1/3}\right)^{\beta_1} & S \ge r_0 \end{cases} \tag{10}$$

so the scaled crushing width ($L/m^{1/3}$) is related to the scaled distance.

When $S=r_0$, the standoff distance is equal to the charge radius, and contact explosion occurs at this time. When $S>r_1$, the failure mode of the beam changes, the equation (9) of the failure size remains unchanged, and the undetermined coefficient changes.

The following equation is obtained by comparing the two crushing widths in equation (10):

$$\frac{L_1}{L_2} = C\left(S / m^{1/3}\right)^{\tau_1} = CZ^{\tau_1} \quad S \geq r_0 \tag{11}$$

Where Z is the scaled distance, m·kg$^{-1/3}$.

For the same explosive, this ratio is only related to the scaled distance. If the broken size under different working conditions is compared, for the same explosive, the ratio of the broken size of the beam is

$$\frac{L / m^{1/3}}{L' / m'^{1/3}} = \frac{A\left(S / m^{1/3}\right)^{\alpha}}{A'\left(S / m^{1/3}\right)^{\alpha}} = D\left(\frac{Z}{Z'}\right)^{\alpha} \quad S \geq r_0 \tag{12}$$

In a certain range, the size and the proportion distance of concrete failure zone have similarity ratio.

4.3. Fitting the Relationship Between Broken Area and Standoff Distance

In order to study the influence of the standoff distance on the damage characteristics of the beam, the following conditions are set up. The charges are all spherical charges. The simulation method is to calculate the response of the reinforced concrete beam in the second section. The size of the broken zone, as shown in table 4, is the numerical simulation working condition and the beam failure size at 3ms. The simulation results of the beam at 3ms under each working condition are shown in figure 9.

Table 4. Simulation conditions and results.

	Standoff distance S (mm)	Scaled distance Z/(m·kg$^{-1/3}$)	Size of front surface L_1/(mm)	Size of back surface L_2/(mm)
1	100	0.1	378.21	517.34
2	110	0.11	362.76	501.77
3	120	0.12	332.32	489.69
4	130	0.13	295.53	454.95
5	140	0.14	284.83	434.83
6	150	0.15	285.35	414.59
7	160	0.16	279.26	411.84
8	170	0.17	277.57	394.11
9	180	0.18	272.28	336.43
10	190	0.19	261.19	346.74
11	200	0.20	254.94	329.32
12	210	0.21	247.26	

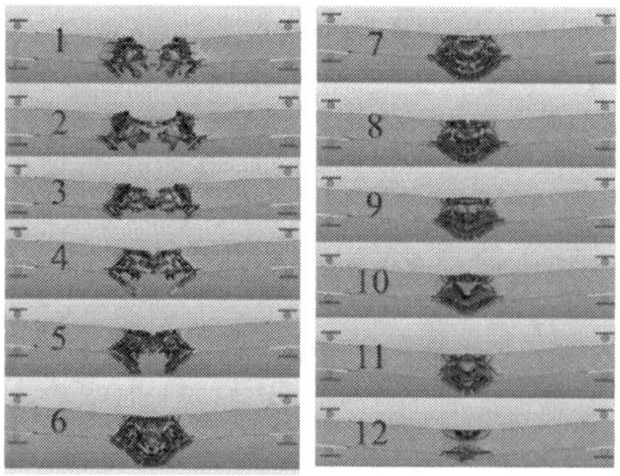

Figure 9. Failure mode of different beams.

According to the simulation results, it is found that under a small scaled distance, the punching shear failure of the beam occurs due to the local shear force exceeding the shear limit, forming a trapezoidal hollow area. When the scaled distance is greater than 0.16m·kg$^{-1/3}$, the failure mode of the beam changes to collapse. When the scaled distance is greater than 0.2 m·kg$^{-1/3}$, the beam does not penetrate, but there is a loss of concrete on the front and back explosion surfaces. In order to reduce the error, the simulation result is no longer used when the scaled distance exceeds 0.2 m·kg$^{-1/3}$.

According to the simulation results, the relationship between the width of the concrete broken zone and the standoff distance can be obtained as:

$$L_1 / r_0 = 9.65\left(S / r_0\right)^{-0.564} \quad S \geq r_0 \tag{13}$$

$$L_2 / r_0 = 14.85\left(S / r_0\right)^{-0.654} \quad S \geq r_0 \tag{14}$$

Where L_1 is the size of the broken area on the front surface, mm; L_2 is the size of the broken area on the back surface, mm; S is the standoff distance, mm; r_0 is the charge radius, mm;

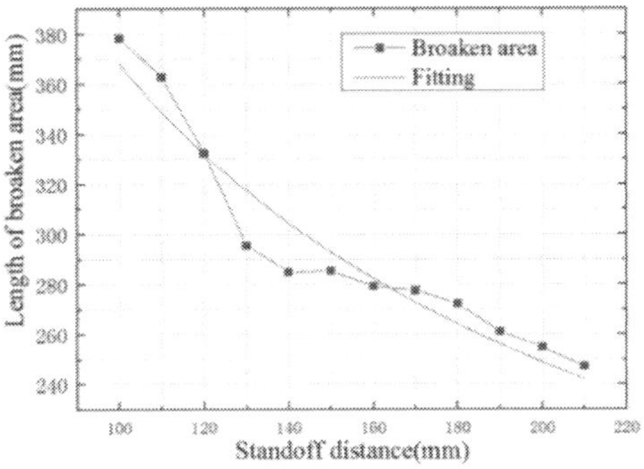

Figure 10. Fitting of broken area and standoff distance in front blasting face.

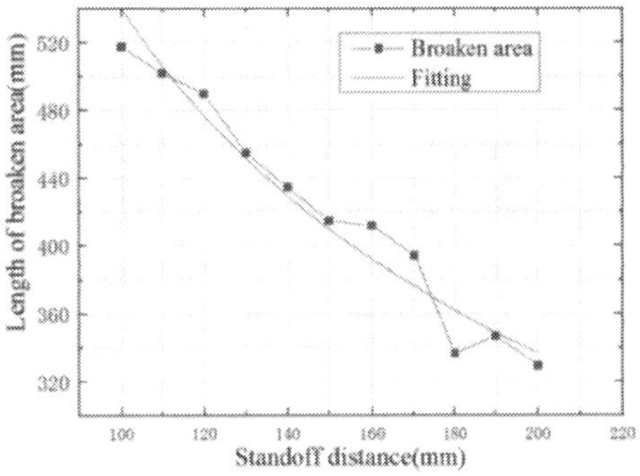

Figure 11. Fitting of broken area and standoff distance in back blasting face.

4.4. Empirical Equation Verification

Nagata [16] carried out 4 rounds of close-in explosion test, the charge is C4, the equivalent TNT mass is 470g, and the equivalent radius is 41.24mm. This article will combine its test with the test data carried out in this article to verify the crushing area fitting. The test conditions in reference [16] are shown in table 5.

Table 5. Test conditions and results.

	Standoff distance S/(mm)	Acaled diastance Z/(m·kg$^{-1/3}$)	Experimental results
n1	35	0.045(contact)	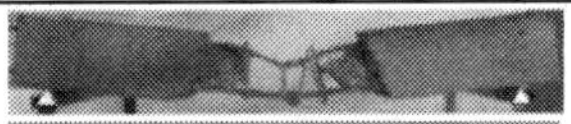
n2	77	0.1	
n3	154	0.2	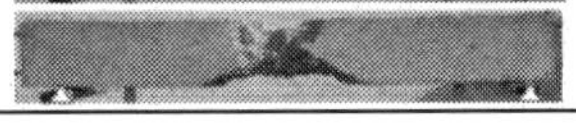

The results of test and equation (13) and (14) are shown in table 6.

Table 6. Comparison of test and calculation results.

	Test (mm)	Calculation (mm)	Error (%)
1# Front surface	36.50	36.81	0.9
n1Front surface	362.25	337.75	6.7
n2Front surface	305.32	279.88	8.3
n3Front surface	216.57	189.32	12.5
1#Back surface	52.00	53.50	2.9
n1Back surface	530.12	519.75	2.0
n2Back surface	449.75	407.17	9.5

The error is less than 15%, which may be due to the difference of charge shape and beam thickness as well as the change of failure mode.

5. Conclusion

(1) The method of using AUTODYN and DYNA to obtain the beam response is more accurate in the simulation of the local failure of the beam, which is suitable for the components with simple structure and easy to determine the explosion-bearing position. Compared with the fluid-structure coupling method, this method has higher efficiency and finer mesh.

(2) On the basis of numerical simulation and dimensional analysis, the relationship between the standoff distance and the beam damage parameters is proposed, and the calculation equation for the damage area of the front surface of the beam is obtained by fitting $L_1 / r_0 = 9.65\left(S / r_0\right)^{-0.564}$; the calculation equation of the back surface damage area for: $L_2 / r_0 = 14.85\left(S / r_0\right)^{-0.654}$.

(3) The empirical equation is used to calculate the broken area in the test, and the error is less than 15%. The modelling method, related equations and derivations used in this paper have reference value for the damage assessment of reinforced concrete beams.

Acknowledgments

The authors thank Dr Jianjun Su, professor of Xi'an Modern Chemistry Research Institute, for his guidance.

References

[1] Peng Y and Fu Y. 2016 Viability analysis and numerical calculation of damaged reinforced concrete beam [J] *Journal of Southwest University of Science and Technology* **31** 30.

[2] Wang J H, Lv L M, Shang W and Huang Z X. 2020 Study of damage characteristics of reinforced concrete beams under contact explosion load [J] *Protective Engineering* **42** 22.

[3] Zhou Q, Sun L and Liu H Y. 2019 Dynamic response and failure form of RC deep beams under blast load [J] *Special Structures* **36** 31.

[4] Wang W, Liu R C, Wu Biao, Wu B, Li L, Huang J R and Wang X. 2016 Damage criteria of reinforced concrete beams under blast loading [J] *Acta Armamentarii* **37** 1421.

[5] Gholipour G, Zhang C and Mousavi A A. 2019 Loading rate effects on the responses of simply supported RC beams subjected to the combination of impact and blast loads [J] *Engineering Structures* **201** 109837.

[6] Wang H M, Liu F, Yan L H, Wang J H, Shang W and Lv L M. 2020 Local damage effects of reinforced concrete beams under contact explosions [J] *Explosion and Shock Waves* **40** 34.

[7] Yao S J, Zhang D, Lu F Y, Wang W and Chen X G. 2016 Damage features and dynamic response of RC beams under blast [J] *Engineering Failure Analysis* **62** 103.

[8] Yan J B, Liu Y, Li Z, Xu Z X and Huang F L. 2020 Effects of transverse reinforcement radio and axial load on the blast response of steal reinforced concrete columns [J] *Acta Armamentarii* **41** 35.

[9] Yu R, Chen L, Fang Q and Huan Y. 2018 An improved nonlinear analytical approach to generate fragility curves of reinforced concrete columns subjected to blast loads [J] *Advances in Structural Engineering* **21** 396.
[10] Yan Q S, Chen T Y and Liao W Z. 2020 Comparison of numerical analysis loading methods of RC columns under near-field explosion [J] *Journal of Beijing University of Technology* **46** 154.
[11] Wang W. 2012 Study on damage effects and assessments method of reinforced concrete structural members under blast loading [D] *Chang Sha: National University of Defense Technology*.
[12] Liu H, Xu C D, Zhang J W, Kong D R. 2020 Numerical simulation of the influence of triple point in TNT explosion field [J] *Journal of Projectiles, Rockets, Missiles and Guidance* **40** 135.
[13] Yuan S J, Hao H, Zong Z H and Li J. 2017 A study of RC bridge columns under contact explosion [J] *International Journal of Impact Engineering* **109** 378.
[14] Malvar L J and Ross C A. 1998 A Review of Strain Rate Effects for Concrete in Tension [J] *Aci Materials Journal* **95** 735.
[15] CEB 1990 CEB-FIP Model Code 1990, First Draft [J] CEB Bulletin D'information **195**.
[16] Nagata M, Beppu M, Ichino H and Matsuzawa R. 2018 A fundamental investigation of reinforced concrete beams subjected to close-in explosion [J] *International Journal of* Protective *Structures* **9** 174.

Study on Correlation of Physical and Mechanical Properties Indexex of Cohesive Soil in Hilly and Plain Region along the Yangtze River in Anhui Province

Daoxiang Wu [1], Lei Ye [1], Huahong Zhao [2,3], Leilei Wu [2,3], Jiacheng Guo [1] and Bin Ji [1]

[1] Department of Geotechnical Engineering, School of Resources and Environmental Engineering, Hefei University of Technology, Hefei 230009, China
[2] Anhui Transport Consulting & Design Institute CO., Ltd., Hefei 230088, China
[3] Research and Development Center on Technology and Equipment for Energy Conservation and Environmental Protection of Highway Transport, Hefei 230088, China
Email: wdx@hfut.edu.cn

Abstract. In this paper, a large number of geotechnical engineering survey data are collected in hilly and plain region along the Yangtze River in Anhui Province. Based on the statistical analysis and calculation of the experimental data of physical and mechanical properties of cohesive soils (the main quaternary soil layers in the area), the correlation between the liquidity index and water content and other physical and mechanical properties indexes are analyzed, and the fitting regression is carried out respectively. The results show that the liquidity index (I_L)) and water content (w) are highly correlated with cohesion (C), compression modulus (E_s), compression coefficient (a), natural density (ρ), void ratio (e), and the regression equations have high goodness of fit and good fitting effect; In addition, the fitting regression equations of water content, void ratio and natural density are compared with the theoretical calculation formula, it is found that the calculation results are close, which proves that the fitting regression equations are reliable and can be used in engineering practice.
Keywords. Hilly and plain region along the Yangtze River in Anhui Province; Cohesive soil layer; Physical and mechanical property index; Correlation analysis; Empirical formula

1. Introduction

The physical and mechanical properties indexes of soil are the basis of analysis, evaluation and design calculation in geotechnical engineering. In engineering practice, the physical and mechanical properties indexes of soil can be obtained directly by indoor geotechnical tests, in-situ tests or back analysis. However, any test or analysis method needs a certain period, and some test methods have high cost. Especially in the preliminary survey and design stage of some small and medium-sized projects, sufficient effective data often can not be obtained in time. According to the relationship between the physical and mechanical property indexes, it is also an indirect method commonly used in engineering to determine the parameters which are difficult to obtain in time by using some easily obtained parameters. At the same time, it is of great significance to study the relationship between the physical and mechanical property indexes and master their internal laws for the reasonable selection of geotechnical property parameters and the study of geotechnical engineering reliability.

Domestic and foreign scholars have done a lot of research on the correlation between the physical and mechanical properties of soil. Azzouz A S et al.[1] used statistical methods to analyze and evaluate more than 700 consolidation test data of undisturbed soil, and established regression equation. Collotta T et al.[2] discussed the correlation between residual friction angle and physical and mechanical properties of cohesive soil. Anthony T. C[3] discussed the method of establishing the correlation relationship between physical and mechanical property indexes of soil by using neural network. Kang X et al.[4] established the relationship between undrained shear strength and plasticity index, and compared it with the existing empirical formula. Li Feifei et al.[5] and Wu Linian et al.[6] respectively conducted statistical analysis and regression on the cohesive soil in Hefei City, and discussed the correlation between the physical and mechanical property indexes of the soil. Jiang Jianping et al.[7,8] respectively analyzed and studied the correlation between the physical and mechanical properties of clay soil at the site of Nanjing Metro and Sutong Bridge. Liu Jia et al.[9] analyzed and studied the correlation between the compression index and physical property indexes of clay soils in Dalian City.

Due to the differences in sedimentary times, environment, material composition and structure of soil layers in different regions, their properties also vary greatly. Therefore, it is necessary to strengthen the research on the properties of regional soil layers[10]. The hilly and plain region along the Yangtze River in Anhui Province is a part of the middle-lower Yangtze Plain. The river terrace geomorphic units are developed in the region, and the Quaternary cohesive soil layers with different thickness and states are widely distributed on the surface[11,12]. This paper collects a large number of geotechnical engineering investigation reports in the study area, makes statistical analysis and calculation of physical and mechanical properties test data of cohesive soil layers in the area, and studies the correlation between the indexes.

It is found that the physical and mechanical properties indexes of cohesive soils in hilly and plain region along the Yangtze River in Anhui Province have different degrees of correlation. For cohesive soil, the moisture condition is the most important factor affecting its engineering properties, and the parameters representing the moisture condition of cohesive soil are mainly liquidity index and water content. Due to limited space, the paper focuses on the correlation between the liquidity index and water content of the cohesive soil in the study area and the commonly used soil property parameters such as the natural density, void ratio, cohesion, compression coefficient and compression modulus.

2. General Situation of the Study Area

The hilly and plain region along the Yangtze River in Anhui Province is located in the south of Anhui Province with an average altitude of about 20 m, including Wuhu City, Ma'anshan City, Anqing City and some areas of Chuzhou City, etc. It belongs to the humid monsoon climate in the northern subtropical, with clear four seasons and mild climate. The annual average precipitation is 1 200 ~ 1 400 mm.

The surface water system in the study area is well developed and belongs to the Yangtze River system. The types of microgeomorphology in the study area are floodplain, first class terrace, second class terrace and erosion residual hill. Except for erosion residual hill, the cohesive soil layers of different thickness and state are distributed on the surface of the other three types of microgeomorphic units, which mainly include Quaternary Holocene mucky silt, mucky soil, clay, silty clay, etc.[11,12]

3. Correlation Analysis and Fitting

3.1. Data Sources and Processing

The data used in the paper are all from several large survey and design institutes with decades of engineering experience in Anhui Province, and the data are accurate and reliable. Through the statistics of the data, a total of 125 projects are obtained for analysis and research, which basically covers the cities and counties under the jurisdiction of the hilly and plain region along the Yangtze River in Anhui Province, making the research results reliable and extensive.

In this paper, the main physical and mechanical properties of the cohesive soil layer in the study area are statistically calculated. After eliminating the data with large deviations, the correlation analysis between the liquidity index and water content and other common physical and mechanical properties indexes is carried out, and the linear and nonlinear methods are used to fit one by one to seek the optimal fitting regression equation.

3.2. Correlation of Liquidity Index, Water Content and Mechanical Properties Indexes

The commonly used mechanical property indexes of soil mainly include cohesion, angle of internal friction, compression modulus, compression index and compression coefficient, etc. which are the necessary soil property parameters for the design and calculation of geotechnical engineering. Therefore, how to obtain the mechanical properties of foundation soil economically and quickly is an important issue for engineering and technical personnel. In this paper, it is found through fitting that the liquidity index and moisture content of the cohesive soil layers in the study area have good correlations with cohesion, compression coefficient and compression modulus. The specific results are shown in table 1 and shown in figure 1, but the correlation with other mechanical indexes is poor, so those fitting results are not given.

Table 1. Fitting formulas of liquidity index, water content and mechanical properties of cohesive soil.

The independent variables	The dependent variable	Fitting equation	Good fit R^2	Dependent variable range
Liquidity index	Cohesion	$C=42.43(I_L)^2-122.12(I_L)+101.43$	0.769	5~139kPa
	Compression modulus	$E_s=4.15(I_L)^2-12.61(I_L)+13.32$	0.670	2.45~21.15MPa
	Compression coefficient	$a=0.36I_L+0.09$	0.740	0.021~0.89MPa^{-1}
Water content (%)	Cohesion	$C=237945w^{-2.55}$	0.658	5~139kPa
	Compression modulus	$E_s=4702.69w^{-1.92}$	0.671	2.45~21.15MPa
	Compression coefficient	$a=0.03w-0.49$	0.835	0.08~1.49MPa^{-1}

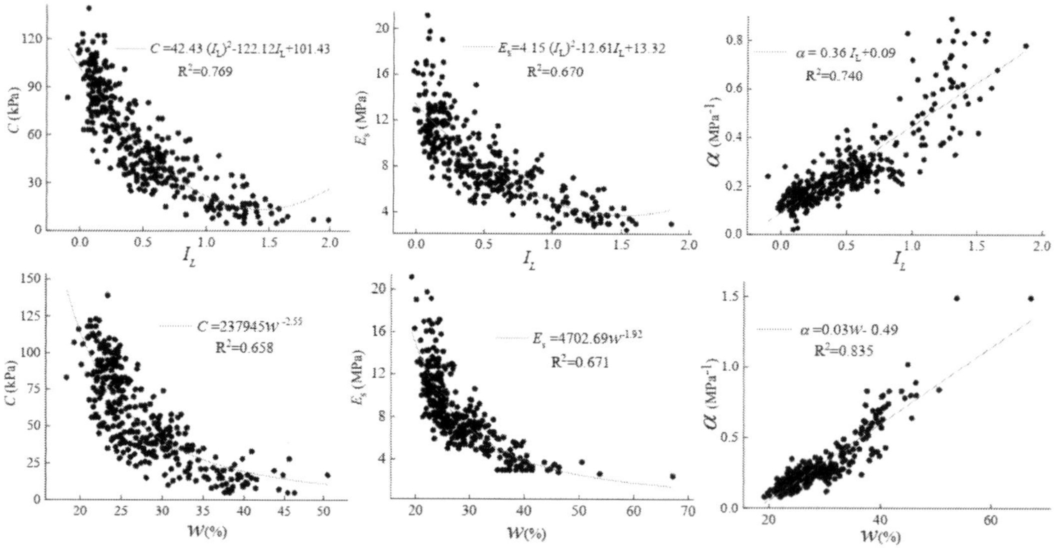

Figure 1. Scatter diagram and fitting curve of the relationship between liquidity index, water content and mechanical properties of cohesive soil in the study area.

The moisture condition of cohesive soil is the main factor affecting its mechanical properties. When the water content increases, the strength decreases while the compressibility increases; when the water content decreases, the strength increases while the compressibility decreases. All the fitting regression equations reflect this law. It can be seen from table 1 and figure 1, the goodness of fit R^2 of each empirical formula is relatively high, ranging from 0.67 to 0.835, indicating the good fitting effect.

3.3. Correlation of Liquidity Index, Water Content and Physical Properties Indexes

The commonly used physical properties indexes of cohesive soil mainly include water content, void ratio, liquid limit, plastic limit, natural density, liquidity index, plasticity index, etc. Through fitting, it is found that the liquidity index and water content of the cohesive soil layers in the study area have good correlation with natural density and void ratio. The specific results are listed in table 2 and shown in figure 2, but the correlation with other physical properties indexes is poor, so those fitting results are not given.

It can be seen from table 2 and figure 2 that the goodness of fit R^2 of each empirical formula is relatively high, ranging from 0.763 to 0.975, indicating that the correlation between the liquidity index and water content and the above physical properties is relatively high.

Table 2. Fitting formulas of liquidity index, water content and physical properties of cohesive soi.

The independent variables	The dependent variable	Fitting equation	Good fit R^2	Dependent variable range
Liquidity index	Water content	$\omega = 12.84I_L+21.24$	0.795	18.3~46.4%
	Natural density	$\rho = -0.17I_L+2.01$	0.768	1.64~2.06g/cm^3
	Void ratio	$e = 0.35I_L+0.61$	0.763	0.554~1.403
Water content (%)	Natural density	$\rho = -0.01w+2.26$	0.925	1.7~2.06g/cm^3
	Void ratio	$e = 0.03w+0.04$	0.975	0.554~1.819

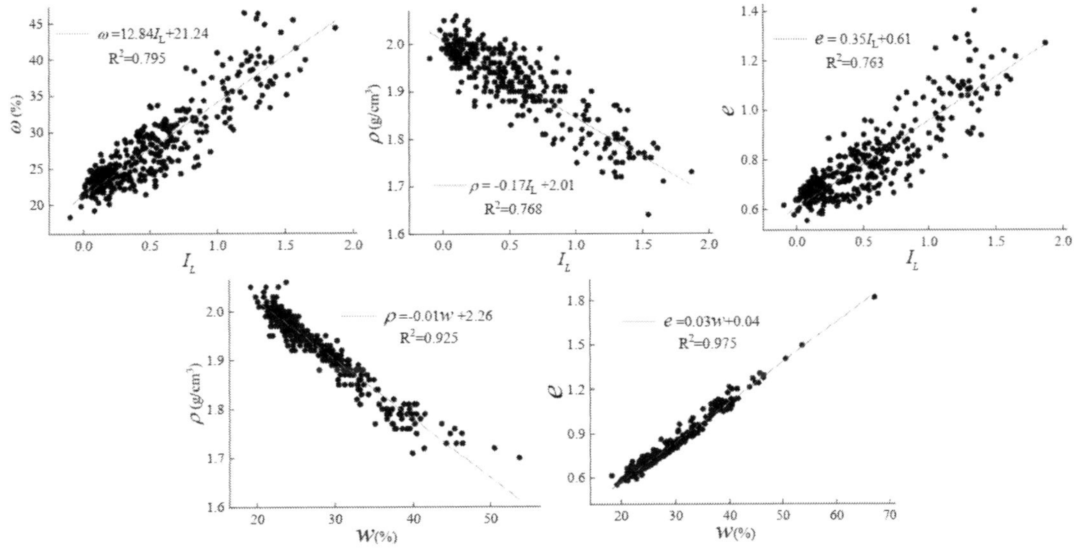

Figure 2. Scatter diagram and fitting curve of the relationship between liquidity index, water content and physical properties of cohesive soil in the study area.

4. Comparison Between Empirical Formula and Theoretical Formula

The theoretical conversion formula of void ratio and water content is as follows:

$$e = w \cdot G_s / S_r \tag{1}$$

According to the experimental data of cohesive soil collected from the study area, the specific gravity of soil particles (G_s) and soil saturation (S_r) are relatively stable, where G_s is generally about 2.73 and S_r is generally about 97%. It can be obtained by substituting them into Equation (1):

$$e = 0.028w \tag{2}$$

The theoretical conversion formula between the natural density of soil, water content and void ratio is as follows:

$$\rho = \frac{\rho_s + e\rho_w}{1+e} = \frac{G_s + wG_s/S_r}{1+wG_s/S_r} \tag{3}$$

Similarly, taking Gs = 2.73, Sr = 97 % into Equation (3), it can be obtained that:

$$\rho = \frac{264.8 + 2.73w}{97 + 2.73w} \tag{4}$$

In Equations (1)(2)(3)(4), ρ is the natural density of soil (g/cm^3); ρ_s is the soil particle density (g/cm^3); ρ_w is the density of water (g/cm^3); e is the void ratio; G_s is the specific gravity of soil particles; w is the water content (%); S_r is the saturation (%).

The regression equations of water content and soil void ratio and natural density in Table 2 are plotted with Eq. (2) & (4) respectively (as shown in figure 3 (a) & (b)). It can be seen from the figures that the regression equations obtained in the paper are close to the theoretical formula, indicating that the regression equations obtained in the paper are reliable and can be applied to engineering practice.

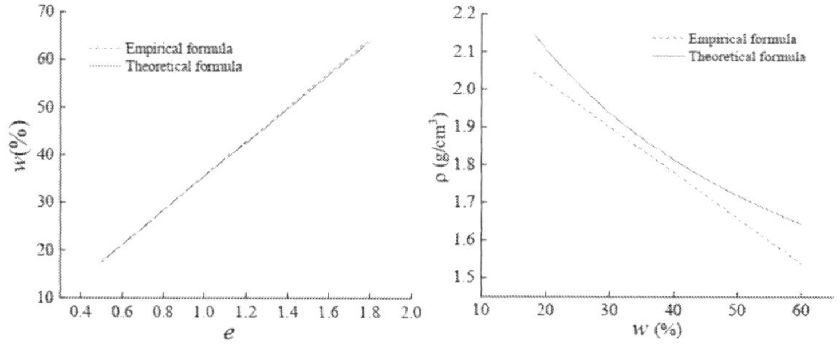

Figure 3. Comparison of fitting regression equations in table2 and theoretical formula Eq. (2) & (4)

5. Conclusions

(1) In this paper, the physical and mechanical properties indexes of cohesive soil layers in the study area are statistically analyzed and fitted. The results show that the correlation between the liquidity index, water content and cohesion, compression modulus, compression coefficient, natural density, void ratio and other common physical and mechanical properties indexes of cohesive soil layers in the study area is good, and the goodness of fit is up to 0.975;

(2) It is found that the goodness of fit between water content and physical and mechanical properties indexes is generally higher than that between liquidity index and physical and mechanical properties indexes. That is, the correlation between water content and physical and mechanical properties indexes is higher, which can more directly characterize the properties and state of cohesive soil;

(3) The regression equations of water content, void ratio and natural density are compared with the corresponding theoretical formulas. It is found that the regression equation is close to the theoretical formula, indicating that the regression equation is reliable and can be applied to engineering practice.

References

[1] Azzouz A S, Krizek R J, Corotis R B. 1976 *Soils and Foundations.* **16** 19-29.
[2] Collotta T, Cantoni R, Pavesi U, Ruberl E, Moretti P C. 1989 *Géotechnique* **39** 343-346.
[3] Anthony T C. 1995 *Journal of Computing in Civil Engineering* **9** 275-278.
[4] Kang X, Onyejekwe S, Ge L, Stephenson R. 2011 *Advances in Geothchnical Engineerin* (United States:Dallas, Texas) 263-2639.
[5] Li F F, Niu Z R, Wang T. 2020 *Railway Investigation and Surveying* **46** 91-95.
[6] Wu L N, Xie Q Q. 2002 *Hydrogeology & Engineering Geology* **29** 43-45.
[7] Jiang J P, Luo G Y. 2008 *Rock and Soil Mechanics* **29** 1669-1674.
[8] Jiang J P, Li X Z, Gao G Y, Luo G Y. 2007 *China Railway Science* **28** 17-24.
[9] Liu J, Zhu B, Duan L, Li S J. 2018 *Journal of Civil Engineering and Management* **35** 118-123.
[10] Gong X N. 2000 *Chinese Journal of Geotechnical Engineering* **22** 238-242.
[11] Cheng Y X, Zhang F S, Wang W R, Dong B L, Cai Y. 1996 *Geology of Anhui* **6** 63-69.
[12] Zhao H H, Wu D X, Ye L, Guo J C, Meng X L, Wu L L. 2020 *Journal* of *Hefei University of Technology(Natural Science)* **43** 72-79.

Study on Seismic Performance of UHV Transformer Based on Soil-structure Interaction

Yaodong Xue[1*], Yongfeng Cheng[1], Zhicheng Lu[1], Zhubing Zhu[1], Haibo Wang[1], Zhenliu Liu[1] and Shujun Zhang[2]

[1] China Electric Power Research Institute, Beijing, 100055, China
[2] State Grid Zhejiang Electric Power Co., Ltd., Hangzhou, 310007, China
Email: xuecumtb@163.com

Abstract. At present, the seismic performance of UHV transformers is mostly studied without considering the interaction between soil and superstructure. In practical engineering, the transformer is installed on the foundation slab buried in the soil. Under the action of earthquake, the interaction between the soil and the structure changes the earthquake response of the upper electrical structure. In order to study the influence of the interaction between soil and structure on the seismic performance of the transformer, the shaking table test method of simulated earthquake is used, and the shaking table test of UHV transformer with scale ratio of 1:4 is carried out in class I field conditions. The dynamic characteristics of the equipment and the seismic response of the bushing under different test conditions are obtained respectively. The test results show that when the peak acceleration is 1.2g, the acceleration response at each measuring point on the box is 1.63-1.92 times that when the peak acceleration is 0.4g. With the increase of seismic peak acceleration, the acceleration and strain increase of high voltage bushing are greater than that of medium and low voltage bushing, which has a great influence on the seismic response of high voltage bushing. The research conclusion can provide reference for substation engineering design.
Keywords. Soil-structure interaction; UHV transformer; shaking table test; seismic response.

1. Introduction

China is located in the Eurasian seismic zone, and earthquakes occur frequently. As an important part of the lifeline project, once the earthquake happens, the damaged equipment will affect the safe and stable operation of the power grid inevitably and cause economic losses extremely [1-2].

The UHV transformer is an important electrical equipment in the substation. The earthquake damage forms of the transformer are mainly as follows: the transformer body slips, falls, overturns and other phenomena; the high voltage bushing is fractured due to its material characteristics and relatively tall structure[3].In 1994 Northbridge earthquake, transformer bushing was damaged seriously [4]. In 1995, during the Kobe earthquake in Japan, a number of 770 and 275 kV substations were damaged. Post-disaster investigation found that the power interruption was mainly caused by the damage of transformer porcelain bushings in several substations [5]. Based on the above reasons, it is necessary to conduct seismic analysis for the transformer, especially in the early stage of UHV transformer. Gilani and other researchers[6] carried out shaking table tests on two kinds of 230 kV, 3000A and U-type transformer bushings, and took improvement measures for weak seismic positions. Filiatraul[7] conducted a dynamic response analysis of bushing mounted on the bushing, which showed that the bushing response was amplified when the frequency of bushing was tuned to the frequency of bushing.

Bender and others [8] showed that the dynamic amplification factor of the bushing might be greater than 2.0. Schiff[9] proposed a new method to identify the transformer and its bushing so that it could meet the requirements of IEEE-693.

The influence of soil on the upper transformer structure was not considered in the above research. Soil-structure interaction originated from Lamb's analysis of elastic foundation vibration in 1904 [10], and many scholars have carried out related research work on this problem from different perspectives. There are many researches on soil-structure interaction in the field of bridge and construction. But the seismic analysis of transformer has not been reported.

On the basis of the existing research, this paper uses the shaking table test method to carry out the test of UHV transformer considering soil-structure interaction. The seismic response of transformer box and bushing under different test conditions is compared and analysed to provide reference for subsequent seismic design.

2. Test Design

2.1. Prototype Size of UHV Transformer

The UHV transformer prototype used in this test is Baoding Tianwei 1000kV UHV transformer. The UHV transformer is mainly composed of a box, a body (containing iron core, winding resistance, etc.), a high voltage bushing, a medium voltage bushing, a low voltage bushing, a neutral point bushing and rising seats. Figure 1 shows the structure diagram of the UHV transformer. Among them, the size of the transformer box is (length × width × height): 10.9m×3.94m×4.84m. The thickness of the side plate of the transformer box body is 12mm, the thickness of the box cover is 30mm, and the thickness of the bottom of the box is 50mm. The total mass is about 396t. Figure 1 shows a schematic diagram of a transformer. Table 1 shows the geometric parameters of each bushing.

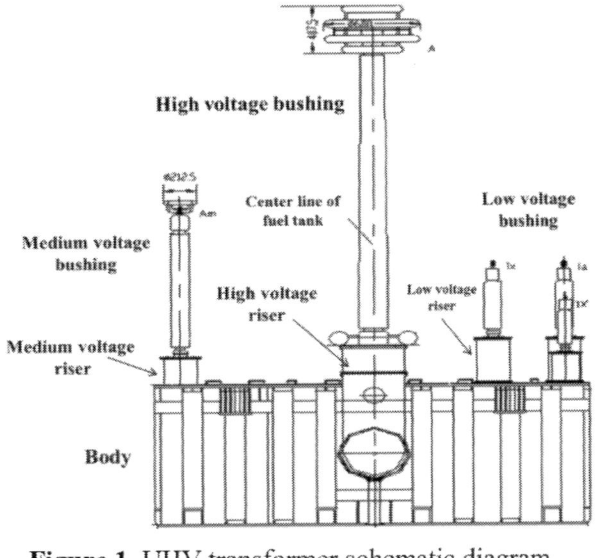

Figure 1. UHV transformer schematic diagram.

Table 1. Transformer bushing parameters.

Part	Inside diameter(mm)	Outer diameter (mm)	Height(m)	Weight(kg)
Low voltage bushing	220	275	2.12	602
Neutral point bushing	160	210	1.2	256
Medium voltage bushing	220	300	4.4	1536
Lower section of high voltage bushing	590	710	10.16	4544
Upper section of high voltage bushing	430	500		

2.2. Determination of Similarity

According to Buckingham's principle, for the earthquake response of the structure, the expression within the range of linear elasticity appears as shown below:

$$\sigma=f(l, E, \rho, t, r, v, a, g, \omega) \tag{1}$$

In this formula, σ is the structural response stress, l is the size of the member, E is the elastic modulus of the member, ρ is the density of the member, t is the time, r is the displacement, v is the velocity, a is the acceleration, g is the acceleration of gravity, and ω is the frequency.

S_n is defined as the similarity ratio between the model and the prototype. In order to enable the model test to simulate the earthquake response of the prototype structure, the similarity ratio of each variable shall meet the following requirements:

$S_\sigma = S_E$, $S_t = S_l\sqrt{S_\rho S_E}$, $S_r=S_l$, $S_v = \sqrt{S_E/S_\rho}$, $S_a = S_g = \frac{S_E}{S_l S_\rho}$, $S_\omega = \frac{\sqrt{S_E/S_\rho}}{S_l}$, in which $S_\sigma=\sigma_m/\sigma_p$, and the subscript m represents the model and p represents the prototype.

In general, the similarity constants of length, elastic modulus and acceleration can be selected as controllable similarity constants:

$$\frac{S_E}{S_\rho S_a S_l}=1 \tag{2}$$

Considering the performance parameters of the shaking table, the processing difficulty of the UHV transformer model and the boundary conditions of the soil-structure interaction, the geometric scale of the transformer model used in this model test is determined to be 1:4. The material of the model is the same as that of the prototype. According to expression (2), the elastic modulus is obtained, and the density similarity is 1:1. Then the acceleration ratio is 4:1. The similarity ratios of other parameters are shown in table 2:

Table 2. Similar parameters of model design.

Physical properties	Physical parameters	Similarity constant
Geometric performance	Length	1:4
Material performance	Strain	1:1
	Elastic Modulus	1:1
	Stress	1:1
	Density	1:1
	Mass	1:64
Power performance	Time	1:4
	Frequency	4:1
	Displacement	1:4
	Acceleration	4:1
Mass of model		6.2t

2.3. Model Size

The model size and mass of each part of the UHV transformer are determined by the design similarity constant and mass design similarity constant based on the geometric length determined by the similarity dimension relation.

The size of the transformer box is 2.7m×1.2m×1.0m (length × width × height). The total weight of the oil tank and accessories is about 1200kg; The mass of iron core and winding resistance in the box body is 4740kg; The total bushing weight was approximately 130kg; The weight of the lift seat is about 140kg. Figure 2 is diagram of the UHV transformer model after scaling, and figure3 is the three-dimensional model diagram.

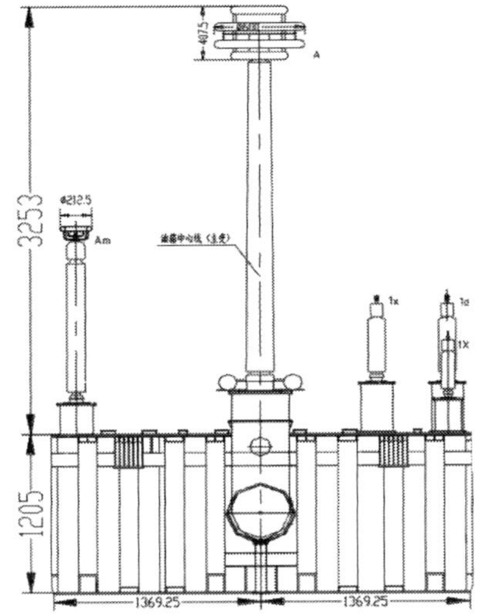

Figure 2. Model Size.

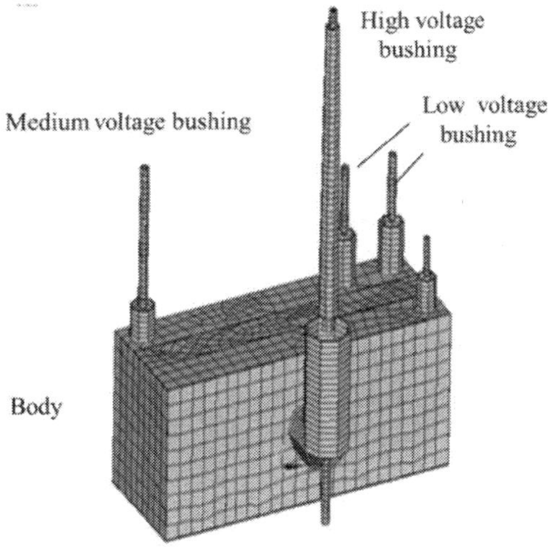

Figure 3. Three-dimensional model.

2.4. Determination of Soil Boundary Conditions

In soil-structure dynamic interaction, the determination of soil boundary conditions is very important. In practical engineering, the soil is in a semi-infinite domain, which is subject to the test conditions during the shaking table test. Therefore, it is necessary to put the soil in a limited range to simulate the earthquake process.

According to the research in literature [11], if a rigid soil box is used in the test, the length of the box body in the direction of vibration should be greater than 4 times of its height, which will exceed the size of the shaking table. The use of flexible soil box, in which the soil can slide in one direction or both directions in the plane of the box, can greatly reduce the plane size of the box and effectively reduce the reflection of seismic waves on the boundary surface of the box. Therefore, flexible soil box is selected for the test.

The plane of the soil box is rectangular. On the basis of satisfying the bearing capacity of the shaking table of 60t, the size of the shaking table of 6m×6m and the size and weight of the upper transformer, the length of the model box is selected as 4.6m and the width is 2.9m. The transformer is located in the middle of the model box, with the length of 2.7m and the width of 1.0m. The lower layer of the transformer is a concrete foundation with the length of 3.3m and the width of 1.6m. The transformer is 0.95m away from the edge of the model box. In order to increase the stability of the model box, scissors cross brace is set at the bottom of the model box, and two braces are set in the transverse direction of the top of the model box, which 0.3m away from the edge of the transformer. At the same time, in order to make the laminated structure stable, brace is set around the model box, and all meet the requirements of the design specification. Figure 4 shows the plane dimensions of the soil box.

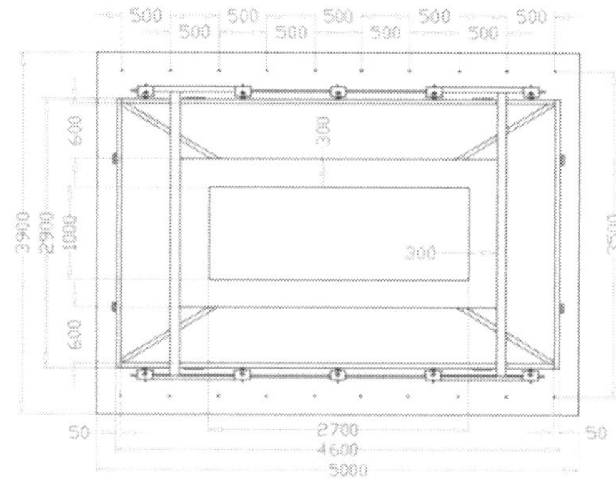

Figure 4. Soil box plane size.

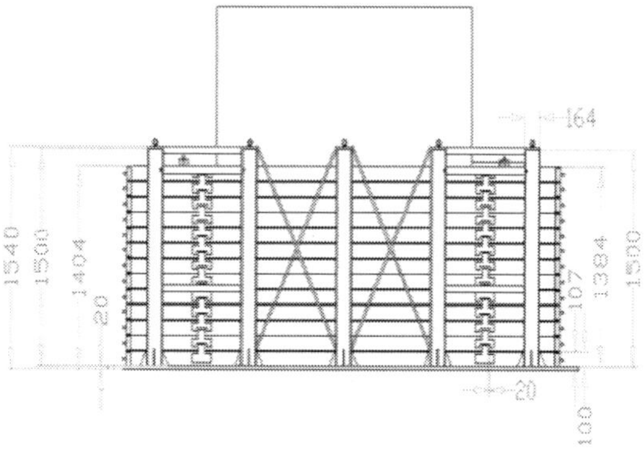

Figure 5. Front elevation of the soil box.

2.5. Simulation Method of Class I Site

According to the above design, the length similarity ratio of the UHV transformer is S_l=1:4. In the test, the length similarity ratio is the same as that of the UHV transformer, and the length similarity ratio of soil is also S_l=1:4.

S_E=1:1 and S_ρ=1:1 are used as the same cohesive soil as what geotechnical test used. The soil is the same as the UHV transformer model, and the similar model neglecting gravity is also adopted. Due to the test soil is cohesive soil and the shear wave velocity is relatively fixed, the method of controlling the thickness of soil layer is adopted to simulate different types of sites. According to the bearing capacity of the shaking table used in the test and the size of the soil box, the buried depth of the soil used in the test is 0.5m. According to the length similarity ratio of 1:4, the thickness of the prototype soil layer is respectively 2m. According to Code for Seismic Design of Buildings, when the thickness of soil layer is 0.5m during the test, the site condition can be considered as class I site.

3. Ground Motion Input and Measurement Point Layout

3.1. Test Input

In this paper, artificial synthetic ground motion and two measured ground motions—El-Centro wave and Taft wave are selected to carry out the shaking table tests of earthquake simulation. Synthetic ground vibration is generated by stipulating the response spectrum (hereinafter referred to as "the standard response spectrum") in Technical specification for seismic design of ultra-high voltage porcelain insulating equipment and installation/maintenance to energy dissipation devices (Q/GDW 11132-2013). The characteristic period of the standard response spectrum is 0.9s, which can almost envelope Class I_0~III sites in China. El-Centro wave and Taft wave are typical examples of ground motion in class II or class III site soils.

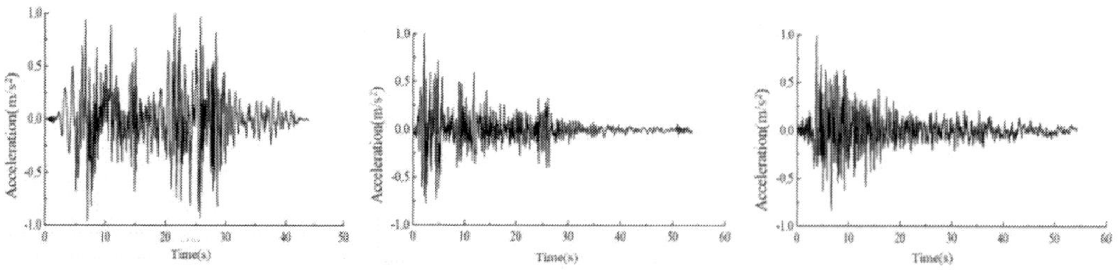

(a) Artificial wave (b) El-Centro wave (c) Taft wave

Figure 6. Seismic wave time history.

3.2. Arrangement of Measuring Points

To measure the acceleration response of each bushing, an acceleration sensor (X, Y, Z direction) is arranged on the top of each bushing. The acceleration sensor is arranged at the connection between the riser seat and the box body (X, Y, Z direction) and the acceleration sensor (X, Y, Z direction) is arranged at the connection between the riser seat and the bushing. In order to obtain the seismic response of the transformer box, the acceleration sensor (X and Y direction) is arranged in two top angles of the transformer body, the centre of the top cover and the three side centre of the box body (excluding the side centre of the high voltage bushing).

The strain gauges are pasted on the bottom of the four bushing tubes (high voltage bushing, medium voltage bushing, low voltage bushing 1 and low voltage bushing 2) along the opposite side of X and opposite side of Y respectively to measure the stress.

The XY direction displacement meter is arranged at the top of the high voltage bushing and the connection between the rising seat and the bushing. XY direction displacement is arranged at the top of the medium voltage bushing, the connection between the lifting seat and the bushing, and the connection between the lifting seat and the box body. Figure 7 shows the detailed diagram of measuring points.

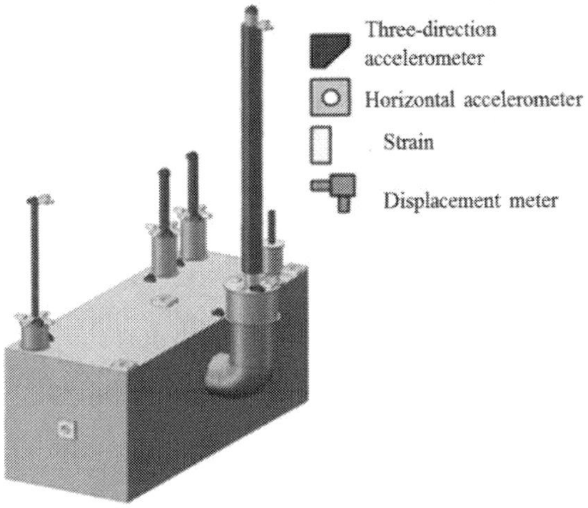

Figure 7. Layout of measuring point.

3.3. Table of Test Conditions

After the completion of the same seismic peak acceleration condition, the white noise test is carried out to detect whether the dynamic characteristics of the structure have changed. Transformer high voltage bushing belongs to long cantilever structure. It is necessary to carry out seismic response test in Z direction, so the seismic excitation direction is X direction and X+Z direction. The shear wave velocity of the soil was measured before and after the type I field test to ensure that the mechanical properties of the soil did not change significantly and the test conditions were consistent. Table 3 shows the table of test conditions.

Table 3. Class I field test condition.

	Test condition	direction
1	White noise	
2	0.4g artificial wave	X/XZ
3	0.4gEl-Centro wave	X/XZ
4	0.4gTaft wave	X/XZ
5	White noise	

6	0.8g artificial wave	X
7	0.8gEl-Centro wave	X
8	0.8gTaft wave	X
9	White noise	
10	1.2g artificial wave	X
11	White noise	

4. Analysis of Test Results

4.1. White Noise Result Analysis

Figure 8 shows the frequency response result obtained after the white noise sweep frequency test in test condition 1. It can be seen from the figure that the frequency in the X direction is f_x=10.75 Hz, and the frequency in the Y direction is f_y=7.25 Hz. Table 4 shows the natural frequency of the equipment under different test conditions. From Table 4, it can be seen that after the end of all test conditions, the X frequency of the equipment changes from 10.75 Hz to 10 Hz, with a decrease of 6.9%; the Y frequency changes from 7.25 Hz to 8.50 Hz, with an increase of 17.24%.

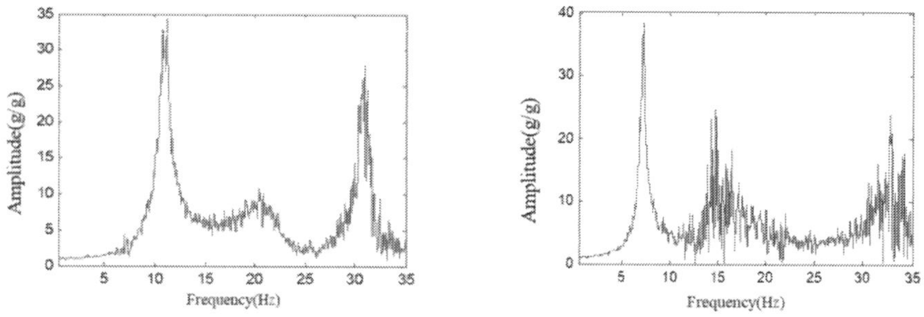

(a)X direction (a)Y direction

Figure 8. Equipment frequency response.

Table 4. Class I field test condition.

Test condition	X direction	Y direction
2	10.75	7.25
6	10.50	8.00
10	10.50	8.25
12	10.00	8.50

4.2. Analysis of Seismic Response of Transformer Box

The side wall (front) of the transformer box, the centre of the top plate of the box, the junction of the high-voltage riser-box, and the connection of the high-voltage riser-sleeve were selected as key measurement points to analyse the dynamic response change law of the transformer box. The acceleration response at the vibrating table surface propagates through the soil layer to the foundation floor, and the acceleration response is bound to change. Therefore, an accelerometer is also installed at the foundation floor to calculate the amplification effect of the soil on the acceleration. Figure 9 shows the acceleration time history under different test conditions. Table 5 shows the maximum acceleration of the box under different test conditions. It can be seen from the table that the acceleration response at each key measuring point gradually increases with the increase of seismic excitation. The acceleration response of the measuring point on the box is greater than the single X-direction input when the seismic wave is input in the X+Z direction.

For the top plate of the box, when the seismic excitation is 1.2g, the acceleration response is 1.92 times that of 0.4g. For the high voltage riser-box, when the seismic excitation is 1.2g, the acceleration response is 1.85 times that of 0.4g. For the high voltage riser-bushing, the acceleration response of 1.2g seismic excitation is 1.63 times that of 0.4g seismic excitation.

From the above analysis, it can be seen that when the seismic peak acceleration is 1.2g, the acceleration at each measuring point on the box is 1.63-1.92 times the acceleration response at 0.4g.

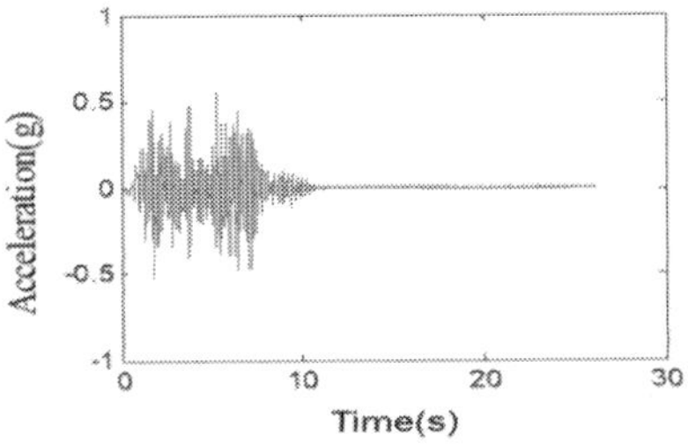

(a) Foundation

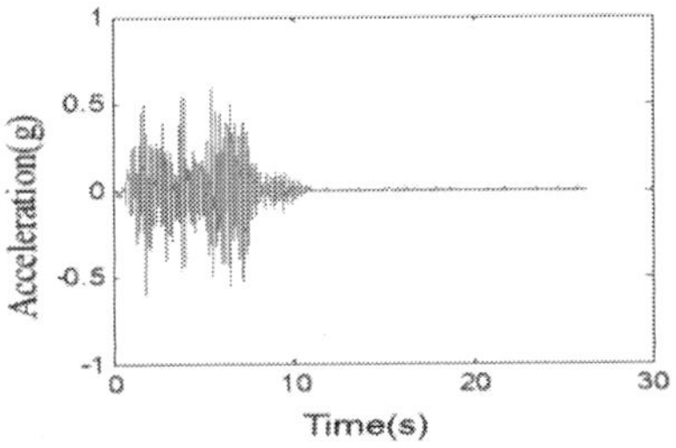

(b) Sidewall

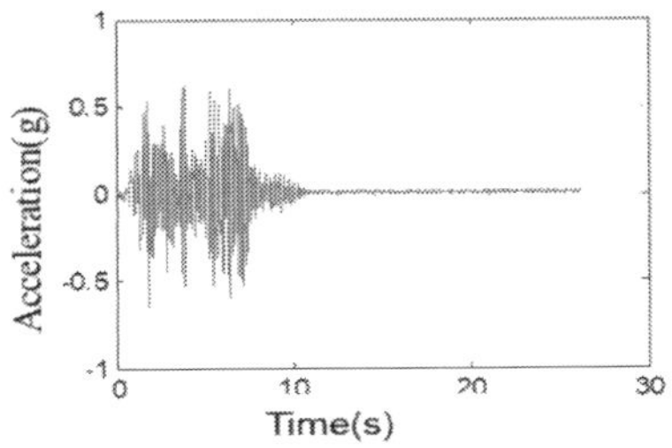

(c) top plate

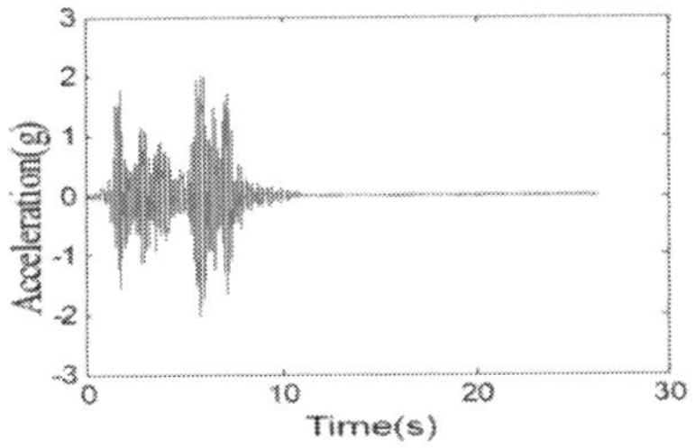

(d) high-voltage riser-bushing

Figure 9. Acceleration time history (0.4gX direction).

Table 5. Maximum acceleration.

Test condition	Maximum acceleration /g					direction
	Foundation	Sidewall	top plate	High voltage riser-box	High-voltage riser-bushing	
3	0.56	0.6	0.63	0.63	2.37	X
	0.54	0.6	0.68	0.72	3.10	X+Z
7	0.85	0.9	0.93	0.90	3.30	X
11	1.05	1.1	1.21	1.17	3.86	X

4.3. Bushing Seismic Response Analysis

In order to compare and analysis the seismic response of high, medium and low voltage bushings under different seismic excitations. Table 6 shows the strain response of the bushing under different seismic excitations. It can be seen from the table that the strain increase of the high-voltage bushing is greater than that of the medium and low-voltage bushing. When the seismic peak acceleration is 0.8g, the peak strain of the high-voltage bushing increases by 70% compared with the 0.4g test condition; the peak strain of the medium-voltage bushing and low-voltage bushing increases slightly with the increase of the peak acceleration, and when the peak acceleration is 0.8g, the peak strain of the medium and low voltage bushing increased by 6.6% and 50% respectively compared with 0.4g.

Table 6. Maximum strain.

Test condition	Maximum strain /με			direction
	High-voltage bushing	Medium-voltage bushing	Low-voltage bushing	
2	130	45	38	X
6	181	37.6	30.28	X
10	221	48	57.3	X

5. Conclusion

(1) The similarity relationship based on the soil-structure interaction shaking table test was studied, the similarity ratio of the model was determined to be 1:4, and the similarity relationship between the physical quantities was derived;

(2) The simulation method of the model foundation is determined, and the size of the soil box is determined, and the simulation method of the Class I site is determined

(3)The shaking table test based on soil-structure interaction was carried out, the X-direction frequency of the structure was f_x=10.75Hz, and the Y-direction frequency was f_y=7.25Hz. When the peak acceleration is 1.2g, the acceleration response at each measuring point is 1.63-1.92 times that of the response when the peak acceleration is 0.4g; with the increase of the seismic peak acceleration, the increase in acceleration and strain of the high-voltage bushing is greater than that of the medium and low-voltage bushing, which has a greater impact on the seismic response of the high-voltage bushing.

Acknowledgments
This study is funded by Science and Technology Project of State Grid Corporation of China: Research on Seismic Technology of UHV Transformers based on Soil-Structure Dynamic Interaction (Grant No: 521104180015).

References

[1] Li S, Tsang H H, Cheng Y, et al. 2016 Effects of sheds and cemented joints on seismic modelling of cylindrical porcelain electrical equipment in substations [J] *Earthquakes and Structures* **12** (1) 55-65.
[2] Xue Y , Cheng Y , Zhu Z , et al. 2021 Study on Seismic Performance of Porcelain Pillar Electrical Equipment Based on Nonlinear Dynamic Theory[J] *Advances in Civil Engineering* (6) 1-17.
[3] Xie Q, Zhu R, Qu W. Analysis on Seismic Failure Mechanism of 500 kV High-Power Transformer
[4] During Wenchuan Earthquake [J] *Power System Technology* 2011 **35** (3) 221-226.
[5] Xie Q, He C, Yang Z, et al. 2017 Tests and Analyses on Failure Mechanism of 1100 kV UHV Transformer Porcelain Bushing[J] *High Voltage Engineering* **43** (010) 3154-3162.
[6] Whittaker A S, Fenves G L, Gilani A S. Earthquake performance of porcelain transformer bushings[J] *Earthquake Spectra* **20** (1) 205-223.
[7] Gilani A S, Whittaker A S, Fenves G L. 2001 Seismic Evaluation and Retrofit of 230-kV Porcelain Transformer Bushings[J] *Earthquake Spectra* **17** (4) 597-616.
[8] Filiatrault, André, Matt H. 2006 Seismic Response of High Voltage Electrical Transformer–Bushing Systems [J] *Journal of Structural Engineering* **132** (2) 287-295.
[9] Bender J, Farid A. 2018 Seismic vulnerability of power transformer bushings: Complex structural dynamics and seismic amplification [J] *Engineering Structures* 162 1-10.
[10] Schiff A J, Torres-Cabrejos R E, Yao J T P. 2010 Evaluating the seismic reliability of electrical equipment containing ceramic structural members[J] *Earthquake Engineering & Structural Dynamics* **7** (1) 85-98.
[11] Shin C J. 1985 Dynamic Soil-Structure Interaction [M] Prentice-Hall.

[12] Shi X, Yue Q, Li J. 2007 Influence factor analysis of foundation modal in shaking table test considering soil-structure dynamic interaction [J] *Journal of architecture and civil engineering* (04) 54-57+63.

Mechanical Properties and Apparent Morphology of Sandstone under Chemical Corrosion Conditions

Jiangchun Hu[1, a], He Feng[1,*], Luge Sun[1, b], Zhipeng Liu[1, c] and Qin Wang[1, d]

[1]School of Civil & Architecture Engineering, Zhongyuan University of Technology, Zhengzhou 450007, China

[*]Corresponding author Email: 983155076@qq.com, [a]hujiangchun@126.com , [b]704238819@qq.com, [c]1449070521@qq.com, [d]511273880@qq.com

Abstract. The rise of the underground engineering provides more guarantee and convenience for human life, and the mechanical property of the surrounding rock is gradually lost, which has great harm to the long-term stability of the project. in that context of the environmental background of the project under the base of this article, firstly, a sandstone sample is taken at the site, a test sample of suitable size is made in the chamber, and then the test sample is arranged in a special device to simulate the simulated corrosion in the background of the simulation environment, Finally, the mechanical properties and apparent morphology of rock samples under different corrosion conditions were studied. The results show that the loss of the mechanical properties of the rock under different corrosion conditions is large, and the change of the acid and alkali of the solution is larger and the rock is The more obvious the damage difference of mechanical properties is, the more obvious the difference is that the pH value is from low to high, the peak strength loses 52%, 27.7%, 7%, 23%, 54% respectively. The failure morphology of corroded sandstone shows special conical morphology. Finally, the equivalent strain principle is used to interpret the corrosion of sandstone. The research results can be used for reference and reference for the long-term stability control of underground engineering based on water corrosion environment.

1. Introduction

With the rapid growth of the domestic economy, the "One Belt And One Road" policy implementation, people's underground exploration has also had a further development. In the absence of recognizing the characteristics of rocks in the underground environment, unreasonable and blind excavation and construction, there will be serious safety risks in the construction, so the study of rock characteristics is essential. The location of rock is more complex, different factors have different effects on its influence, and water chemical factor is one of the main factors affecting rock characteristics [1], it is urgent to study the characteristics of rock in water chemical environment.

Some research on rock in water environments is also common, Feng Xiaoting and others[2] have passed the tests under different stress states, The erosion of the chemical environment of the rock fracture characteristics was examined; Chen Sili and others[3] conducted a granite uniaxial compression and fracture test under chemical corrosion, Actual observation using digital microobservation system; Ding Wuxiu and others[4] studied the triaxial compression damage process under penetration and precast fracture sandstone through CT scanning test; Han Tielin and others[5] studied the physical and mechanical properties of sandstone after soaking in different aqueous chemical solutions, The corrosion degree of pH, concentration and hydrochemical composition of solution to sandstone were analyzed, The water chemical corrosion mechanism of sandstone specimen is discussed. So far, many scholars have compared and analyzed the corrosion degree of sandstone by solution pH value, concentration and hydrochemical composition, and its mechanical properties [6-8], and the characterization of rock and its mechanism in the hydration state [9-11]. Similarly, the study of pH values and microscopic particles is important for mineral flotation [12 -14].

However, the above results are the effects of mechanical or surface damage in a single study and are not organically combined. Based on the above research, indoor simulation test, study the mechanical characteristics of sandstone samples in five different hydration states, use data obtained from uniaxial compression test to calculate tensile and bending strength, and evaluate the damage form of sandstone samples under different water chemical states during the experiment, which provides important engineering significance for the safe construction of geotechnical engineering in hydrochemical corrosion environment.

2. Corrosion Environment of Sandstone Mechanical Properties Test

The blocks used in the test were quartz feldspar sandstone. The irregular blocks obtained on the site obtained the Φ50mm*60mm cylindrical specimen through the rock core rig, and then polished the sample surface, smooth and placed in the ventilation place for air dry treatment.

According to the actual situation of the field water chemical environment, the configured solution concentration [15] is shown in table 1.

Table 1. Design ratio of aqueous chemical solution[a].

Solution Number	Ingredient	Solution Concentration/mol/L	pH
1	HCL	0.1	2
2	HCL	0.1	5
3	Distilled Water	/	7
4	NaOH	0.1	10
5	NaOH	0.1	13

[a]Remove several samples every 30 days at the YR-2000 rock creep for uniaxial compression test See table 2 for the parameters.

Table 2. Main technical parameters of yr-2000 rock creep meter[b]

Maximum axial load	Axial force measurement accuracy	Dismeasurement range	Dismeasurement accuracy	Deformation limit value Axial and Radial direction	Demeasurement resolution
2000kN	<±1%	1-100mm	<±1%	5mm、2.5mm	0.001mm

[b]Using the displacement control method, the designed displacement load rate of 0.5mm/min, damage detection threshold is 50% , and it is believed that when the strain of the rock sample reaches the peak strength,the strain has weakened to 50% believes that the rock sample has damaged [16].

3. Mechanical Properties of Sandstone Under Corrosion Conditions

3.1. Stress - Strain Characteristics

The control variable method tests only the stress - strain curve under the same solution of pH. Sandstone stress - strain curve in different states is as shown in Figure.1-5.

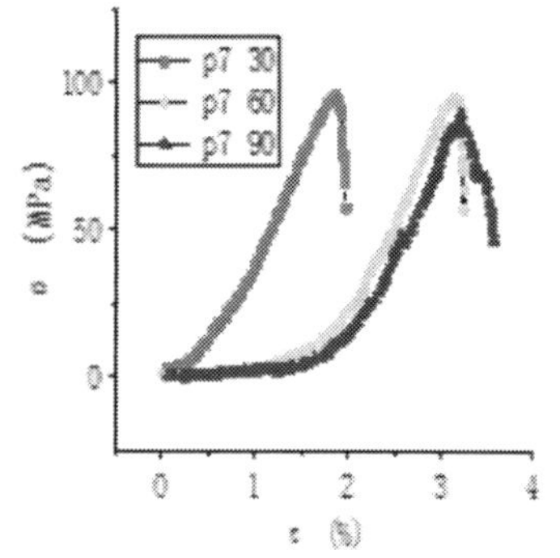

Figure 1. Stress-strain curve of sandstone in solution of pH=7.

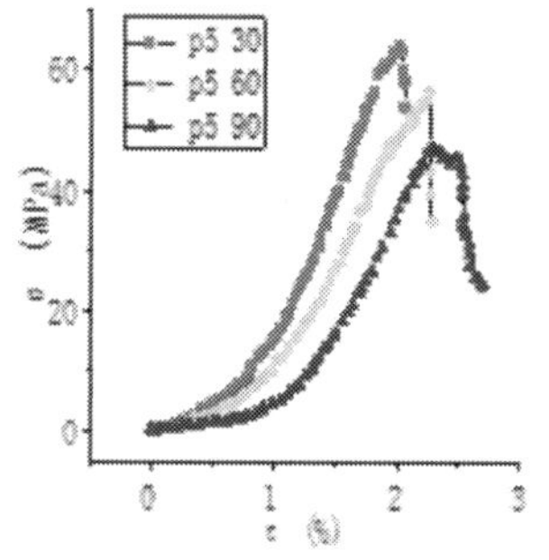

Figure 2. Stress-strain curve of sandstone in solution of pH=5.

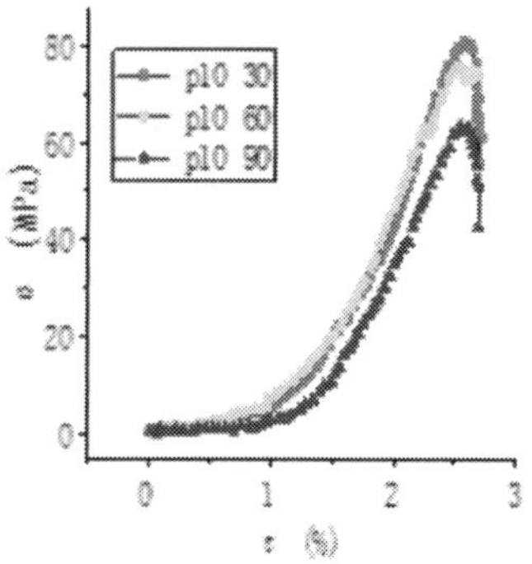

Figure 3. Stress-strain curve of sandstone in solution of pH=10.

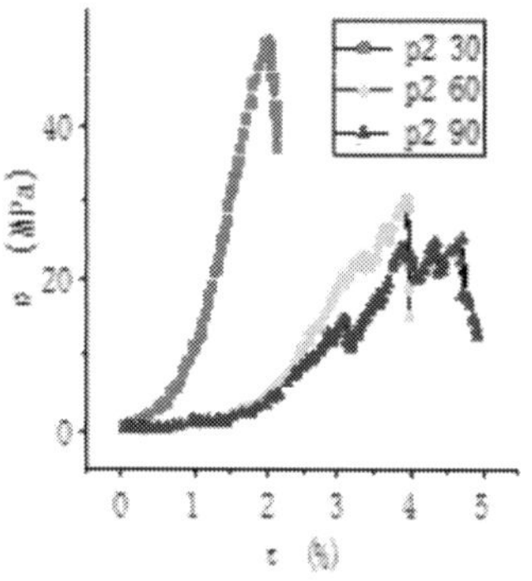

Figure 4. Stress-strain curve of sandstone in solution of pH=2.

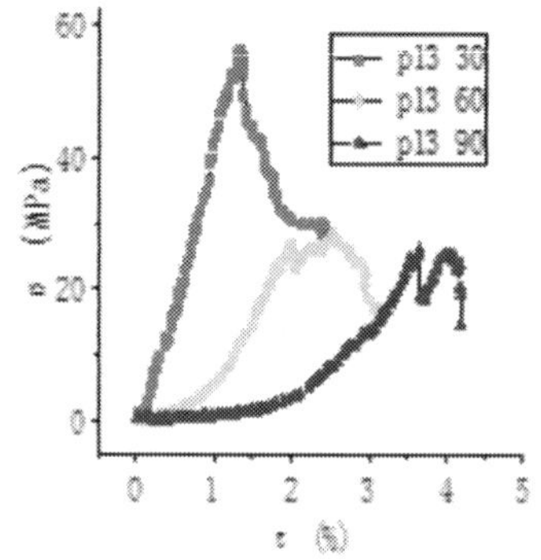

Figure 5. Stress-strain curve of sandstone in solution of pH=13.

In the process of stress start loading into the sample failure, the sample has experienced 5 stages: compression phase, elastic phase, yield phase, crack expansion phase, and failure phase.

The first stage is the compression stage, This phase is related to the microstructure of the rock, The gap inside the rock grew smaller under pressure, Until the press is to close, Overall, "U" shaped concave trend; The second phase is the elastic phase, The friction within the rock increases with stress, Make the interior of the rock even closer; The third phase is the yield phase, As the stress increases, According to Figure 1-5, no significant yield phase occurs except for pH=7, Other pH values all appear or have long or short yield stages; Stage IV is the crack expansion stage, The original cracks were significantly significantly, Mixed together with new cracks, Rock mass strength has decreased significantly, Stress growth rate is becoming significantly faster; Phase 5 is the destruction stage, The strain has no longer increased with the stress, The rock mass is seriously deformed and then damaged.

Figure 6. Rock fracture in pH=7 solution.

Figure 7. Rock breakdown in the remaining pH solutions.

Combined with figure.1-7, too large stress will reorganize the internal structure of the rock mass, thus losing the original strength, and causing damage when the stress in the rock mass exceeds the maximum strength of the rock mass. Rock fracture is the manifestation of failed rock resistance to external loads, and the fracture forms and characteristics shown by different rocks or the same rocks supporting different external load environments are different. A lot of classical mechanical tests show that rock fracture is conical or cylindrical fracture in most cases. In figure 1, only under the case of pH= 7, the stress - strain curve conforms to the conventional form, only the rock sample damage in pH= 7 solution can be clearly seen as the main shear fracture, and the shear fracture can be clearly observed, there is no obvious fracture in the remaining solution, corresponding to the above.

3.2.Changes of Mechanical Parameters of Sandstone Under Different Acid and Alkali Conditions

It is seen from Figure.8-10, there is a peak of 90d and greater than 60d and over 30d for the strain peak. Peak the lowest acidic and alkaline at pH= 5 and pH= 10, and the solution pH lower or higher than pH= 5 will increase the pH= 10. The peak elastic modulus of 30d is the highest, and similarly, the closer to pH= 7, the higher the peak elastic modulus of sandstone, and the overall trend presents a "convex" shape. The overall trend of the peak intensity is roughly the same as the peak elastic modulus.

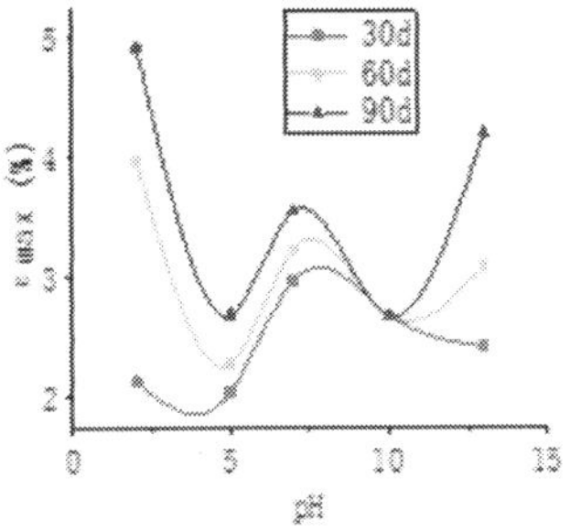

Figure 8. Relationship between peak strain and solution pH.

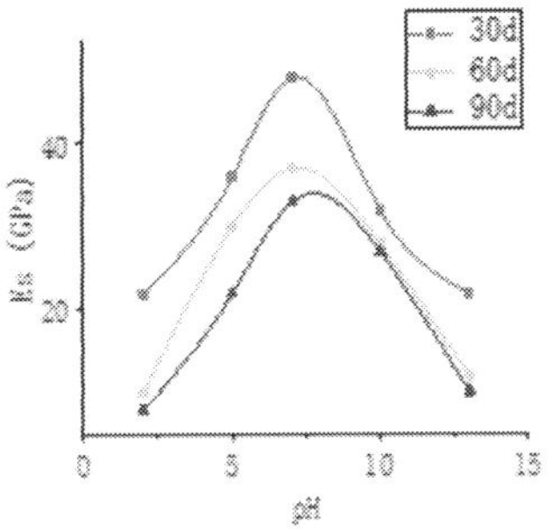

Figure 9. Relationship between elastic modulus and pH value of solution.

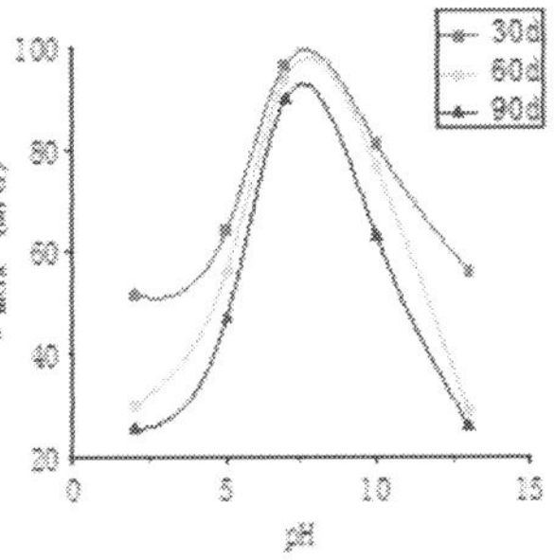

Figure 10. Relationship between peak intensity and pH value ofsolution.

3.3.Deformation Strength and Time Effect of Sandstone Under Different pH Degree

It is seen from figure.11-13 that in the solution with high acid base, the corresponding strain peak increase is significantly increased, with the largest difference between the rock peaks of pH= 2, 86% between 30d-60d, 25% between 60d-90d, and the difference between the rock peaks in the solution of pH= 13. The corresponding pH values of elastic modulus of rock samples in different aquochemical

2148 (2022) 012021 doi:10.1088/1742-6596/2148/1/012021

solutions decreased from large to small are pH= 2, pH= 5, pH= 13, pH= 7, pH= 10, and the calculated data are 63% , 54% , 38% , 31% and 15% , respectively. Meanwhile, at the same time, the initial peak intensity of the rock sample corrosion was significantly higher than that of the later stage, indicating that the initial chemical reaction of the rock sample is the most intense and then becomes chronic chemical reaction.

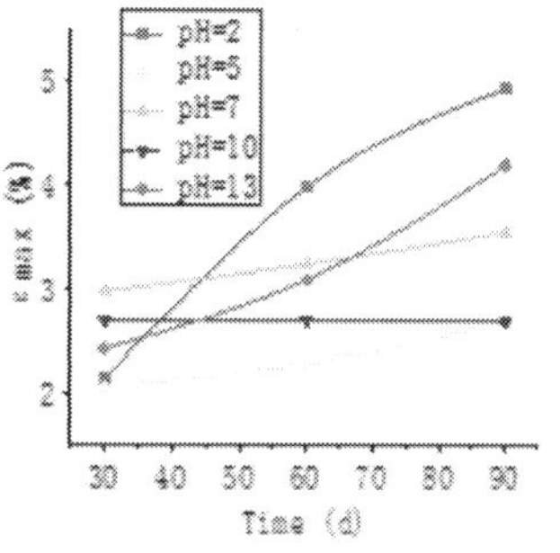

Figure 11. Relationship between peak intensity and pH value of solution.

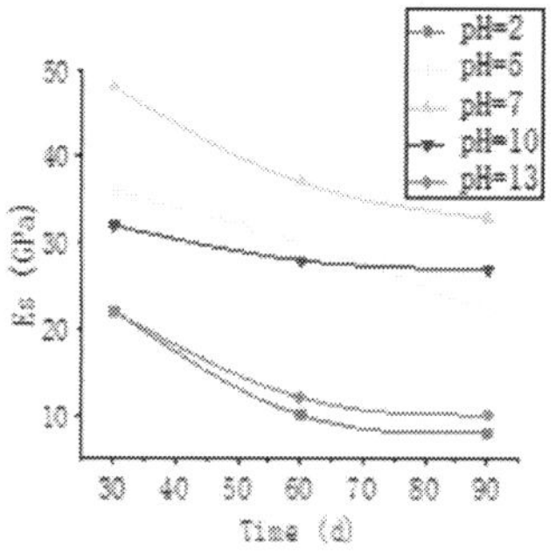

Figure 12. Relationship between elastic modulus and time.

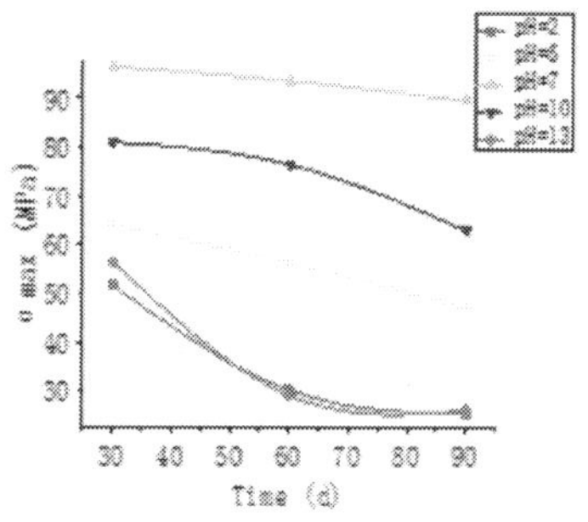

Figure 13. Relationship between peak intensity and time.

4. Theoretical Interpretation of Sandstone Corrorosion

Huo runke et[17]analyzed the damage constitutive model of rock.

If the initial damage state of sandstone is taken as the first state and the state after acid corrosion as the second state, the equivalent elastic modulus after acid corrosion can be expressed as:

$$E = E_0 (1 - D') \tag{1}$$

In formula: E Equielastic modulus after acid corrosion of sandstone; E_0 Flexible modulus of unacid - corroded sandstone.

Based on the principle of Lemaitre and damage mechanics theory, the damage constitutive relationship of sandstone under uniaxial compression is as follows:

$$\sigma = E\varepsilon(1 - D) \tag{2}$$

In formula: σ Stress; ε Strain; D damage variable.

Using formula (1) into formula (2), the damage constitutive relationship of acid-corroded sandstone under uniaxial compression is:

$$\sigma = (1 - D)(1 - D')E_0\varepsilon \tag{3}$$

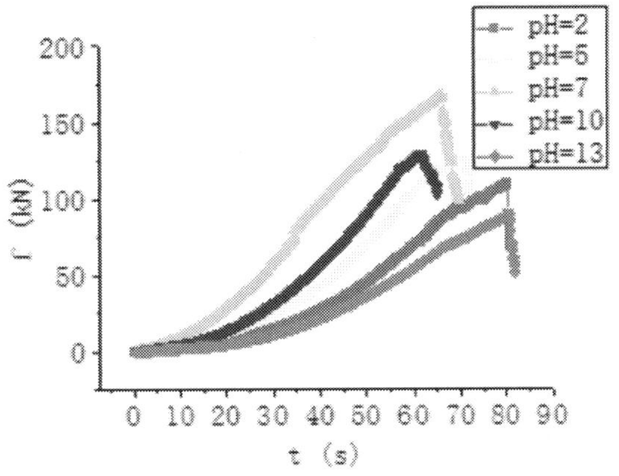

Figure 14. Stress-time curve of different PH values.

According to the previous data, the greater the acid and base degree of the solution, the more the H and OH-ions exist in the solution, the stronger the corrosion to the rock mass, the greater the damage is, the greater the D value. In the case of pH= 7, only a small amount of H ions and OH- ions are separately free in the solution, that is, the degree of sandstone damage in the solution of pH= 7 is less than that in the remaining pH solutions. According to formula (3), the stress - time curve form is a relatively high parabola, while with the increase of acid - base concentration, the sandstone stress - time curve form in its solution is a low parabola. This also corresponds with FIG. 14, with the sandstone stress - time curve form height in pH= 7 solution above the remaining pH and the lowest sandstone stress - time curve height in pH= 2 and pH= 13.

5. Conclusion

This paper observed the sandstone corrosion in different hydration states and drew the following conclusions:

(1) In sandstone, the stress - strain curve shows the front section and the back end.

(2) Sandstone in hydration state is more likely to destroy in the same force field environment. The sandstone failure form in pH= 7 solution is cylindrical fracture destruction, while the sandstone failure form in other pH solutions is mostly conical destruction.

(3) Significant mechanical performance loss of sandstone after corrosion of hydration solution. pH value ranged from low to high, with peak strength lost 52% , 27.7% , 7% , 7% , 23% and 54% ; after long corrosion, sandstone showed softening trend of 63% , 54% , 38% , 31% and 15% respectively. That is, different hydration solutions for different sandstone corrosion conditions, the greater the acidity and alkalinity, the more serious the corrosion situation.

References

[1] Cui Q, Feng X T, Xue Q, Zhou H, Zhang Z H. 2008 Mechanism study of porosity structure change of sandstone under chemical corrosion [J] *Journal of Rock Mechanics and Engineering* (06) 1209-1216.

[2] Feng X T, Lai H Z H. 2000 Rock fracturing behaviors under chemical corrosion-parti: experimental study [J] *Journal of Rock Mechanics and Engineering* (04) 402.

[3] Chen S L, Feng X T, Li S J. 2003 The fracturing behaviors of Three Gorges granite under chemical erosion [J] *Geotechnical Mechanics* (05) 817-821.

[4] Ding W X, Feng X T. 2008 CT experimental research of fractured rock failure process under chemical corrosion and permeation [J] *Journal of Rock Mechanics and Engineering* (09) 1865-1873.

[5] Han T L, Shi J P, Chen Y S, M W T. 2014 Experimental study of pHysical and mechanics characteristic of sandstone sample subjected to chemical erosion [J] *Journal of Xi'an University of Technology* **30** (01) 34-39.

[6] Liu Y S, Diao X H, Chen Z L, Yang M. 2014 Dynamic mechanical properties of surrounding rock in the circumstance of chemical corrosion [J] *Journal of Nanjing Forestry University* (Natural Science Edition) **38** (06) 175-178.

[7] Tan Z Y, Chai H B, Liu W J, Zhao X G. 2005Rock strength deterioration under acidified conditions and its static accelerated simulation [J] Journal of Rock Mechanics and Engineering (14) 2439-2448.

[8] Chen W L, Xie J, Sun Y. 2013 Effects of acid rain attack on uniaxial compression properties of marble [J] *Journal of Central South University* (Natural Science Edition) **44** (07) 2897-2902.

[9] Yue H W, Ma Z Z, Bao Y W. 2011 Influence of acid corrosion on contact deformation and damage of rock [J] *Journal of Central South University* (Natural Science Edition) **42** (05) 12821289.

[10] Chen Z C. 2001 Bao is also looking. Influence of Environmental Corrosion on Contact Deformation and Damage of Concrete Materials [J] Journal of Building Materials (03) 249.

[11] Yin Y C, Chen Y L, Meng X R. 2019 Comparison of mechanical properties and fracture toughness changes of sandstone and granite under freeze-thaw cycle and chemical corrosion [J]. Journal of Water Resources and Water Engineering **30** (04) 217-224.

[12] Liu L, Hao L Y, Li Q, Dai J P. 2016 Potential Controlled Preferential Flotation Test on Copper-lead-zinc Sulfide Ore from Liaoning [J] *Metal mines* (03) 82-85.

[13] Chen W. 2018 Experimental Study on Flotation of a Gold-bearing Copper Ore from Yunnan Province at Natural pH [J] *Metal mine* (12) 94-98.

[14] Xi Y, Dai S J, Zhang Z J. 2019 Effect of Ca2+ on Flotation Behavior of Magnesite [J] *Metal Mine* (09) 98-101.

[15] Jiang H, Sun F, Zhang X Y. 2010 Formation and prevention of the inferior groundwater in Zhengzhou area [J] *Engineering investigation* (09) 39-44.

[16] Shi C S, Li S Y, Shi L H, Wang C. 2019 Failure mechanisms under uniaxial compression and microscopic parameter inversion of frozen fractured coal rock [J/OL] *Glacier permafrost*: 1≤10 [2019×10:21]. Http://kns.cnki.net/kcms/detail/62. 1072. P. 20191018. 1146. 004. Html.

[17] Huo R K, Han F, Li S G, Wang G J. 2019 Experimental study on physical, chemical and mechanical properties of acid-corroded sandstone [J] *Journal of xi 'an university of building science and technology* (natural science edition) **51** (01) 21-26.

A Quantitative Index Representing Crack Shapes and Corresponding Techniques

Junhua Wu[1, 2], Tangliang Kuang[1], Fangyuan Fu[1] and Jiahao Li[1]

[1] College of Civil Engineering and Architecture, Nanchang Hangkong University, Nanchang, Jiangxi Province, China

[2] Key Laboratory of Failure Mechanism and Safety Control Techniques of Earth-Rock Dam of the Ministry of Water Resources, Nanjing, Jiangsu Province, China

E-mail: wjhnchu0791@126.com

Abstract. In order to quantificationally describe the soil cracks due to dry-wet cycles, the concept of gray level entropy is applied according to the physical significance of the information entropy to represent various shapes of cracks. Then a piece of simple and easy-to-use equipment for taking photos is used to monitor and record the crack propagation. A grayscale image and the corresponding gray level entropy are obtained automatically by a program. Test results showed that gray level entropy can quantificationally describe the shape of cracks reasonably well and evaluate the degree of crack development effectively.

Keywords. gray level entropy; crack; photograph; image processing.

1. Introduction

Desiccation cracking is an important phenomenon in soils. Bronswijk & Evers-Vermeer [1] cited several positive and negative effects of cracking on agricultural soils. The development of desiccation cracking is determined by the interaction of numerous factors, including soil mineralogy, rate of water loss, total drying time, the repeated rise and decline of underground water, repeated rain and evaporation, etc. The positive effects include improved infiltration[2], drainage[3], and soil structure[4]. Soil cracks can also help alter the pore size of soil through intermittent wetting[5], and are utilized during reclamation of high shrink-swell soils as pathways for water and salts[6]. The main negative effect of cracking on agricultural soils is stunting the growth of plants as the rapid transport of water through soil cracks results in the lack of water or higher osmotic pressure (named 'crop water stress' in pedology)[7]; or the rapid transport of solute through cracks results in the lack of nutrient (named 'crop nutrient stress' in pedology)[8]. Another negative effect of cracking on environment is that surface water contamination if transport is to a drainage system.

Desiccation soil cracking due to dry-wet cycles is commonly observed in expansive soils. As these soils dry, the collapsing layers of the clay micelles cause the soil volume to decrease. While soil swells upon imbibing water, it will result in the healing process of cracks. Soil surface cracks are formed through repeated dry-wet cycles. The shapes of cracks are influenced by numerous factors as mentioned above. In geotechnical engineering, desiccation cracks resulting from dry-wet cycles have significant effects on soil properties such as strength, deformation, permeability etc.[9]. Different crack shapes affect the soil properties differently. Therefore, it is necessary to investigate how to describe the shapes of cracks.

We know from images of cracks that the luminance at places where there are cracks is lower than that at places where there is no cracks. Due to this, RGB images are usually converted into grayscale or binary images. The main characteristic of cracks are preserved. In this study, grayscale images are used to analyze soil cracks. A testing method and a quantitative index for describing cracks are presented.

2. Theory and Techniques

2.1. Information Entropy

Information entropy is a mathematical nonrepresentational concept in information theory. Generally speaking, information entropy can be considered as the probability of specific information. An ordered system has lower information entropy than a disordered system. The more ordered of the system, the lower the information entropy, and vice versa. In other words, information entropy is an index to describe the degree of order of a system. The grayscale system can be used to represent image information effectively. A colored image can be converted into a 8-bit grayscale image as follows:

$$GS = (0.299 * R + 0.587 * G + 0.114 * B) / 255 \tag{1}$$

where
GS = gray level ranging from 0 to 255,
R = value of primary red color of a pixel in an image,
G = value of primary green color of a pixel in an image, and
B = value of primary blue color of a pixel in an image.

The gray level GS shows how close the color of a pixel in an 8-bit grayscale image to white or black. The greater gray level means the color is closer to white. When GS equals 255 it is white, while it's black when GS equals 0. Suppose the frequency of a certain gray level i is P_i, where $P_i = f_i / \sum_{i=0}^{N-1} f_i$, N is the total number of gray levels (256 for 8-bit grayscale image), and f_i is the number of pixels which GS equal i. The bigger the P_i, the more the number of pixels have gray level of i. The information entropy for the image is defined as following:

$$H = -\sum_{i=0}^{N-1} P_i \log_2 P_i \tag{2}$$

Where H is the information entropy of a gray level image, named as the gray level entropy.

We know from the physical significance of information entropy that the greater the information entropy, the more the average uncertainty of information source, and vice versa. Thus the gray level entropy reflects the gray level distribution in a discrete manner. When the image is relatively uniform, the range of gray level is smaller and consequently the average uncertainty is small. This leads to small gray level entropy. While the gray level at places where there are cracks is lower, the range of gray level is larger and the average uncertainty increases which results in the increased gray level entropy.

3. Equipments and Techniques

As mentioned before, photographic techniques and image processing techniques, which are already mature, provide us with a new method to research soil surface cracks. First the image of cracks is taken with a digital camera. Then the information contained in the image is obtained using image processing techniques. The main equipments needed are: a digital camera (the Nikon COOLPIX L16 in our case); and a computer to operate the image processing software or program developed in-house using Visual Basic in our case. The schematic of our photography setup is shown in figure 1. In order to eliminate the effect of artificial factors, the digital camera is fixed at some distance from the soil surface with the tripod. Additionally, the camera lens and the soil surface should be in parallel and the image should be captured in the middle visual field. The image can be obtained by putting the specimens on a pedestal. A code using Visual Basic is adopted to extract the RGB information from an image and convert to the grayscale image. Then the gray level of each pixel is calculated using Eq. [1]. Finally, the image gray level entropy can be obtained using Eq. [2] by counting the gray level distribution of the image.

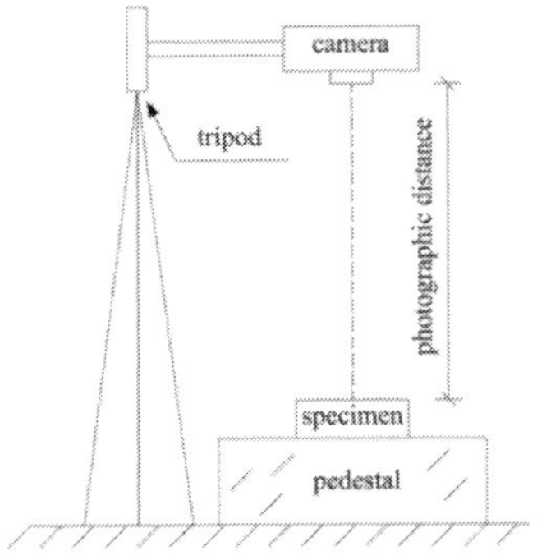

Figure 1. Schematic of the photography setup.

Table 1. Parameters of the expansive soil obtained from Xinxiang, Henan, China.

liquid limit	plastic limit	Grain size/%							
		sand/mm				silt/mm		clay/mm	colloidal/mm
W_L/%	W_P/%	0.5~ 2.0	0.25~ 0.5	0.1~ 0.25	0.075~ 0.1	0.005~ 0.075	0.005~ 0.05	<0.005	<0.002
29.5	15.1	8.3	1.0	1.9	1.4	1.2	40.3	45.9	37.7

Table 2. Continued.

Specific gravity	Maximum dry density	optimum water content	free swelling ratio
/	kg/m^3	%	%
2.72	1810	15.7	57.5

4. Test and Verification

In order to verify the feasibility of using gray level entropy as an index to describe the shape of cracks, a series of tests on soil cracking are carried out and images of a soil surface with cracks are obtained at regular intervals. In this study, soil was obtained from the South-to-North Water Diversion Project in Xinxiang, northern Henan, China. The soil parameters are given in table 1 and table 2.

First, all the specimens and equipments are placed in the room. The photographic distance (see figure 1) is 600 mm and the resolution is set to 800×600. Pictures are taken in natural light and the water content is obtained every time a picture is taken. The images we took and the corresponding histograph of gray level are drawn in figure 2 (a)-(f). Cracks are generated and propagated due to water loss. In time, the main cracks are widened and deepened, and new cracks are generated near the main cracks. It can be seen in the images that the luminance of the cracks is higher than the luminance of other places. In other words, the gray levels of the pixels forming the crack are lower than those of pixels forming the rest of the image. We can see that the frequency distribution of gray level becomes bimodal from the initial unimodal distribution. There is one main big crack in figure 2(a), but the area is small compared to the whole. This has limited influence on the gray level distribution, and the main values of gray level still depend on the area without cracking. That means the gray level is more concentrated in a smaller range. So a unimodal distribution is shown in figure 2(b). There are two main big cracks in figure 2(c). It can be seen that the width and length of cracks are wider than that one in figure 2(a). These areas cannot be ignored compared to the whole. That means there are two main distribution ranges of gray level here: the crack areas lower gray level contains and the rest of image higher gray level contains. The range of gray level distribution increases, mainly to the lower values. So a bimodal distribution is shown in figure 2(d). Similar results are obtained in Figure 2(f) and the range of gray level distribution increases further. Going from (a)/(b) to (c)/(d) and (e)/(f), the gray level distribution is evidently enlarged to include mainly the lower values. The peak distribution frequency reduces with crack propagation. The values of gray level entropy calculated using Eq. [3] are plotted against the corresponding water contents in figure 3. It is observed that gray level entropy increases with crack propagation. This shows that the gray level range of a soil image changes due to crack propagation. The uncertainty that one value of gray level has appeared increases greatly which results in the increased gray level entropy. From the above analysis, it can be concluded that gray level entropy is an effective index to represent and evaluate the shapes of cracks.

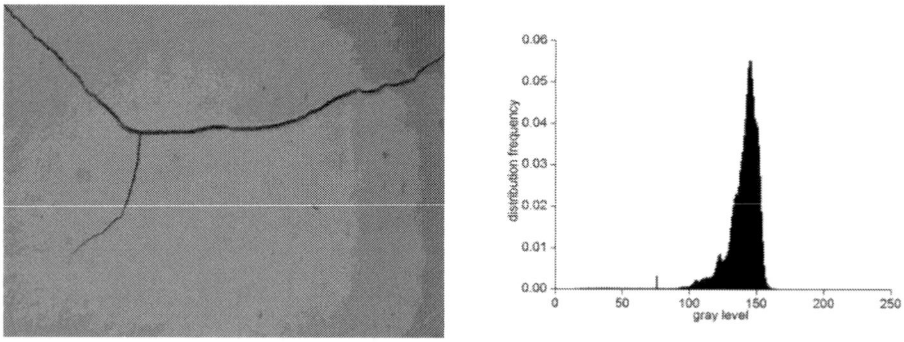

(a) Photo of cracks (b) Frequency distribution of gray level

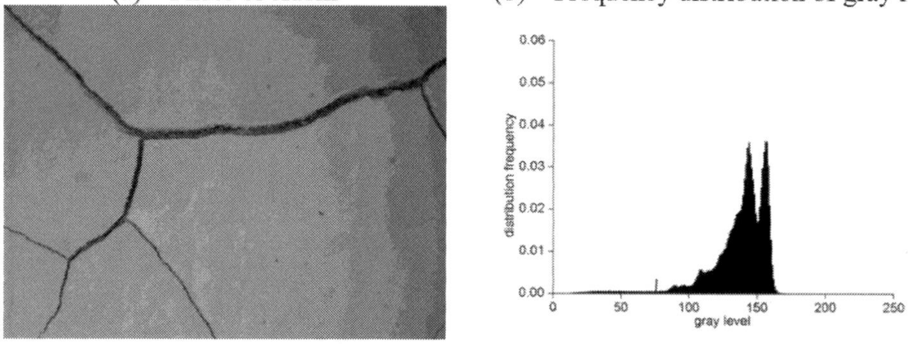

(c) Photo of cracks (d) Frequency distribution of gray level

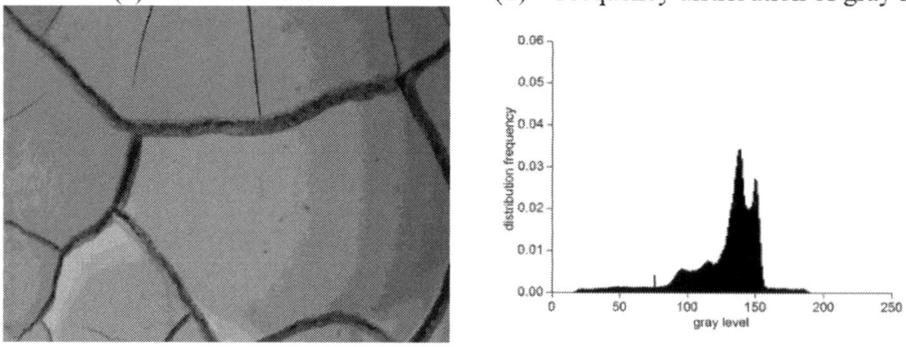

(e) Photo of cracks (f) Frequency distribution of gray level

Figure 2. The photos and corresponding frequency distribution of gray level.

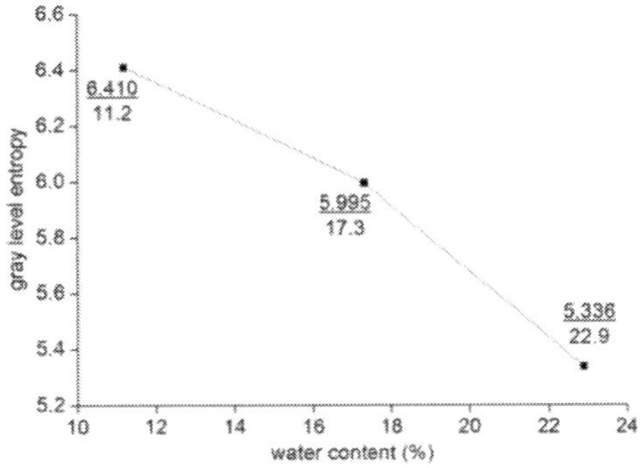

Figure 3. The gray level entropy-water content relationship.

5. Summary and Conclusions

1. In order to quantificationally describe soil cracks, a simple setup for taking photos of cracks is designed in this study. According to the physical significance of information entropy, crack images are converted from RGB model to grayscale model, and the concept of gray level entropy is applied to represent various shapes of cracks. Test results showed that gray level entropy can quantificationally describe the shape of cracks reasonably well and represent the degree of crack development effectively. A sensitivity analysis of influencing factors such as photographic distance, luminance, resolution etc. has also been carried out.
2. The luminance of cracks on a soil surface is lower and gray level is lower than that at places where there are no cracks. This leads the gray level range and therefore gray level entropy to increase. Thus the gray level entropy increases with crack propagation.

Acknowledgements
This work is supported by Project (51869013) of the National Natural Science Foundation of China and Project (YK321013) of Open Research Fund of Key Laboratory of Failure Mechanism and Safety Control Techniques of Earth-Rock Dam of the Ministry of Water Resources. All opinions, findings and conclusions in this work represent the views of the authors only.

References

[1] Bronswijk J J B, Evers-Vermeer J J. 1990 *Shrinkage of Dutch clay soil aggregates* [J] *Journal of Agricultural Science* 38 175-194.
[2] Swartz G L. 1966 Water entry into a black earth under flooding [J] *Journal of Agricultural Animal* Science 23 407–422.
[3] Wilding L P, Hallmark C T. 1984 Development of structural microfabric properties in shrinking and swelling clays, Proc. Symposium ISSS. International Institute for Land Reclamation and Improvement, Wageningen, Netherlands 1–22.
[4] Bronswijk J J B. 1989 Prediction of actual cracking and subsidence in clay soils [J] *Journal of Soil* Science 148 87–93.
[5] Jalai-Farahani H R, Heermann D F, Duke H R. 1993 *Physics of surge irrigation. II. Relationships between soil physical and hydraulic parameters* [J] *Transactions of the ASAE* 36 45-50.
[6] Malik M, Amrheim C, Letey, J. 1991 *Polyacrylamide to improve water flow and salt removal in a high shrink-swell soil* [J] *Soil Science Society of America Journal* 55 1664 - 1667.
[7] Thomas G W, Phillips R E. 1979 *Consequences of water movement in macropores* [J] *Journal of Environmental Quality* 8 149 - 152.

[8] Harris, G L, Nicholls P H, Bailey S W, et al. 1994 *Factors influencing the loss of pesticides in drainage from cracking clay soils* [J] *Journal of Hydrology* 159 235 - 253.
[9] Inoue H. 1993 *Lateral water flow in a clayey agricultural field with cracks* [J] *Geoderma* 59 311 - 325.

Vibration Effect Induced by Rock Breaking Technology Based on Dry Ice and Energy-gathered Agent in Trench Excavation

Wu Ye[1, 2], Yong Lu[1, 2], Lingzhi Xi[1, 2], Qiang Zhang[1, 2,*] and Shaobin Hu[3]

[1] Powerchina Huadong Engineering Corporation Limited, Hangzhou, Zhejiang, 311122, China
[2] Zhejiang Huadong Engineering Consulting Corporation Limited, Hangzhou, Zhejiang, 311122, China
[3] College of Civil Engineering and Transportation, Hohai University, Nanjing, 210098, China
Email: zhangqiang02016@163.com

Abstract. Rock breaking technology based on dry ice and energy-gathered agent has been developed and successfully applied in trench excavation for construction of oil pipeline. The vibration velocity waveform induced by this technology was monitored in site test to determine the attenuation law of vibration velocity with hypocentral distance. The results show that this rock breaking technology is effective method of trench excavation. It does not excessively damage the adjacent rock mass, ensuring the integrity of ditch walls. The vibration velocity induced by this technology is decay with the increase of hypocentral distance. At the hypocentral distance of 10m, the vibration velocity reduces to less than 20mm/s, which meets the requirements of the safety standard of blasting vibration in general buildings engineering. The results of this experiment have an important guiding effect on the field engineering practice and application of rock breaking technology based on dry ice and energy-gathered agent.
Keywords. Rock breaking technology; Dry ice; Energy-gathered agent; Vibration velocity.

1. Introduction

At present, explosive blasting, high energy gas fracturing (HEGF) and supercritical fluid fracturing technology are the typical technologies applied to breaking rock mass, which were used in mining and related engineering. Explosive blasting is a process of energy conversion and rapid expansion of gas volume in a very short time. It has the characteristics of high reaction heat (3-6 MJ/kg), fast reaction speed (within 10-6s) and large amount of gas generation. It is an efficient way to break rock and widely used in mining and urban infrastructure construction [1-3]. However, explosive blasting poses a threat to the ecological environment and the health of workers at different degrees [4-6]. It usually causes intolerable noise, strong earthquake, blasting shock wave and a large number of blasting stones, which are harm to the surrounding environment, buildings, and safety of life [7-9]. Meanwhile, explosive blasting produces toxic and harmful gases and waste residue [10-12], causing harm to the health of animals and plants [12-19].

In the 1980s-90s, carbon dioxide (CO_2) phase transition fracturing technology had been gradually promoting in the field of rock excavation and coal mining. This technology uses liquid CO_2 and heating agent as medium of rock breaking, which are encapsulated in an airtight container. The heating agent is excited to generate a high temperature above 800 °C within 0.1 s, and liquid CO_2 is instantly gasified, resulting in rapid volume expansion and high pressure [18]. Compared with compressed air, nitrogen and other media, liquid CO_2 is an excellent work medium because it is easy to compress and obtain high density at ambient temperature.

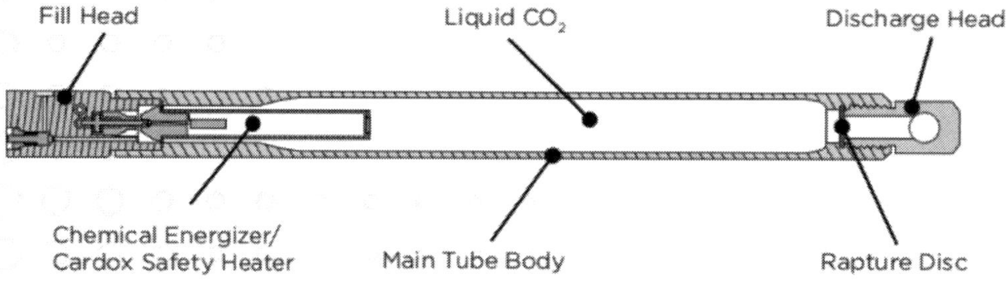

Figure 1. Structure diagram of CO_2 fracturing tube.

The CO_2 phase transition fracturing technology was first proposed by Cardox company in the 1950s, and the fracturing of rock was realized by CO_2 fracturing tube [20]. The fracturing apparatus are composed of main tube body, chemical energizer, rapture disc, fill head and discharge head as shown in figure 1. Since the technology was introduced into China in the 1990s, it had been widely studied by many researchers, and used in the field of coalbed methane exploitation to solve the problem of low permeability of coal seams. Fan et al. [21] studied the application of CO_2 phase transition fracturing technology in gas drainage. He et al. [22] established a numerical calculation method for the influence range of permeability of coal seam after CO_2 fracturing. Yan et al. [24] found that the coal fracture after CO_2 fracturing is more complex than that caused by hydraulic fracturing. At present, it hasn't been specifically studied for the mechanism of crack initiation and crack propagation using supercritical CO_2 phase transition fracturing [25-27]. Zhang et al. [28] established the cyclic dynamic model of rock fracture, and pointed out that the rock fracture using CO_2 phase change fracturing technology is result from the stress wave. Goodarzi et al. [29] concluded that the gas pressure has great influence on the crack length than the stress wave.

Generally speaking, fracture of intact rock is mainly caused by the impact effect of stress wave, while rock mass with relatively developed primary fractures is mainly caused by the expansion wedge effect of gas [30]. Because liquid CO_2 rock breaking is a physical process, the vibration caused by liquid CO_2 rock breaking is less than that caused by explosive blasting used the same charge amount. For the energy distribution of a single vibration waveform in different frequency bands, the study found that most of the energy is mainly distributed below 100Hz, and the frequency of the main vibration is generally below 20Hz [31-35]. In addition, the peak velocity of liquid CO_2 rock-breaking vibration is different in different directions. The peak velocity is affected by the direction and types of the fracture [36]. In general, vibration velocities in the vertical direction are higher than the horizontal direction, and the vibration velocities in the three directions gradually decrease and converge with increasing of distance [37].

Compared with explosive blasting, liquid CO_2 phase transformation process is slower than it, which is completed within milliseconds [23]. Therefore, it is more conducive to the generation and expansion of rock mass fractures, and the vibration damage and noise pollution are less. However, many shortcomings are revealed when this technology was widely used in engineering. Firstly, the high pressure gas was released in the lower end of the liquid storage pipe, and the breaking energy was limited. Secondly, the high pressure vessel was used to transport and filling the liquid CO_2, these processes pose a threat to workers; In addition, many accidental triggering events occurred during the filling process of the CO_2 cracker, such as pipe burst and fly-pipe accidents. Finally, the conventional CO_2 fracturing technology is limited by policies, as the chemicals used to heat liquid CO_2 is characteristic of explosive properties. Hu et al. [38] developed a new CO_2 fracturing technology, which adopted an intrinsically safe energy accumulating agent to realize the CO_2 phase change. As the energy accumulating agent is only a heating product rather than an explosive product, it improves the security of the CO_2 fracturing technology.

In this paper, the new rock breaking technology based on dry ice and energy-gathered agent (Hereafter, dry ice rock breaking technology) was applied to the excavation of trenches. The characteristics of vibration velocity waveform induced by this rock breaking and the attenuation law of vibration velocity with hypocentral distance were monitored and analysed.

2. Test Site and Test Method

2.1. Test Site

The site is located in Longhui No. 002-H5 Construction Project beside Xiangtangzi River, Wenchong Town, Dazhou City, Sichuan Province, China, as shown in figure 2. The depth of the trench for the construction pipeline is 1.2 m, the width is 0.8 m, and the top is 1.0 m. The lithology is mainly purple mudstone and grayish yellow sandstone. Because the test site is close to residential buildings, explosives blasting cannot be used to rock breaking. In order to improve the efficiency of rock breaking and to diminish the influence of vibration on the adjacent dwellings, dry ice rock breaking technology was carried out in this area. Microseismic monitoring system was used to monitor the vibration in whole process of rock breaking.

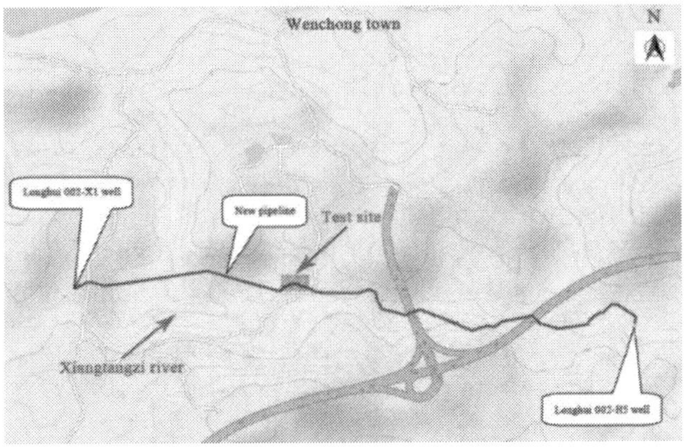

Figure 2. Location of filed test.

2.2. Test Methods and Procedures

Dry ice rock breaking device is composed of dry ice powder cabin, energy accumulating agent tube, heater, gas blocking system, safety relief valve, detonator and accumulator, etc. (see figure 3).

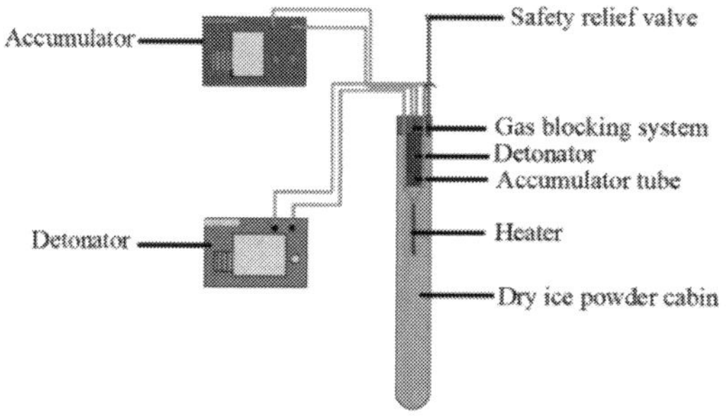

Figure 3. Schematic diagram of dry ice fracturing device.

The dry ice powder cabin used in this study were 600 mm and 400 mm in length, and 100mm in diameter, which could accommodate CO_2 of 1.5 kg and 1 kg, respectively. The energy accumulating agent tube is 400 mm in length, 50 mm in diameter and 0.4 kg in weight. It is a kind of functional material with porous micro nano structure, which has strong adsorption characteristics. This structure is conducive to the combustion exothermic reaction of the energy accumulating agent [38]. The physical and chemical parameters of the energy accumulating agent are shown in table 1.

Table 1. Basic physical and chemical parameters of the CO_2-EA [38].

Material	Apparent density (g/mL)	Porosity	Specific surface area (m^2/g)	Heat of combustion (kJ/g)	Thermal conductivity (W/m k)
CO_2-EA	0.12	0.92	240	21.33	0.016

As shown in figure 4, the test procedures were divided into drilling, cement sealing, sensor layout, and data collection. As shown in figure 5, a total of two holes were drilled at a 45 angle to the ground, the diameter was Φ110 mm, and the depth was 1.5 m. Dry ice powder cabin were inserted into the hole, and then sealed with quick-drying cement. Vibration sensors were arranged at different distance and different direction. Finally, microseismic data were collected after initiating the detonator.

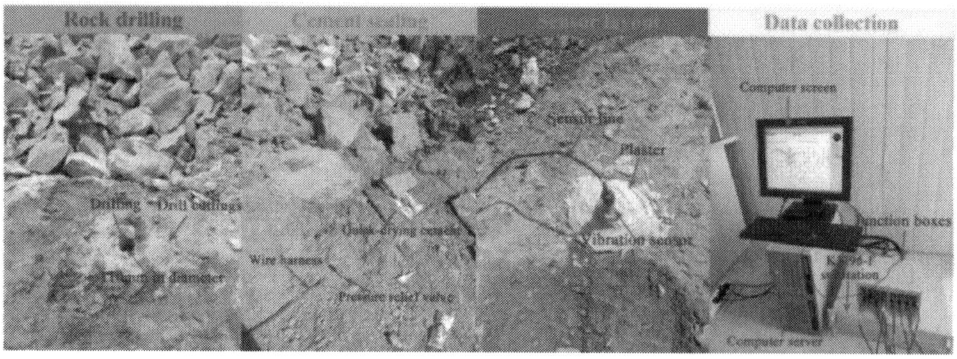

Figure 4. Schematic diagram of the test.

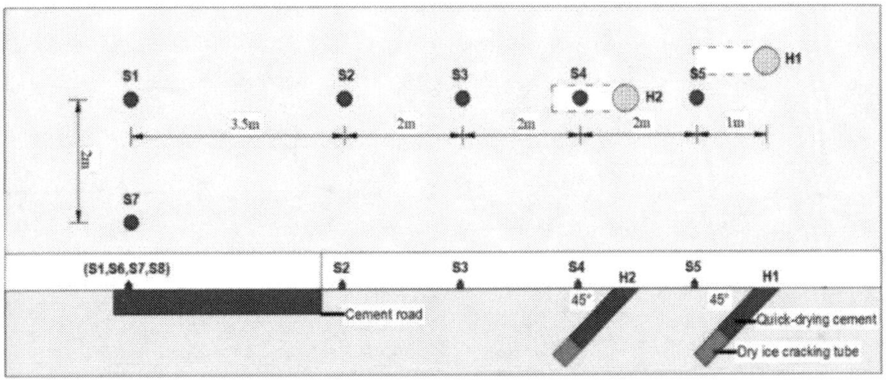

Figure 5. Schematic diagram of drilling and sensor layout.

2.3. Vibration Monitoring System and Layout Method

KJ522 microseismic monitoring system was used to monitor the vibration wave. This study focuses on the vibration attenuation law induced by dry ice rock breaking technology within 15 meters. The vibration sensor layout network was shown in figure 5. A total of 8 vibration sensors were arranged, S5 sensor was closest to H1 (1.05m) and S4 was closest to H2 (1m).S1, S6, S7 and S8 were installed along the left side of the rural road. S8 was the farthest from H1 and H2. A mixture of gypsum powder was used to fix the sensor, and the test was carried out after the gypsum hardened.

3. Test Results and Analysis

(a)H1 excited, H2 reserved (b) H1and H2 all excited

Figure 6. Schematic diagram of crack effect of single row of grooves.

 2148 (2022) 012023 doi:10.1088/1742-6596/2148/1/012023

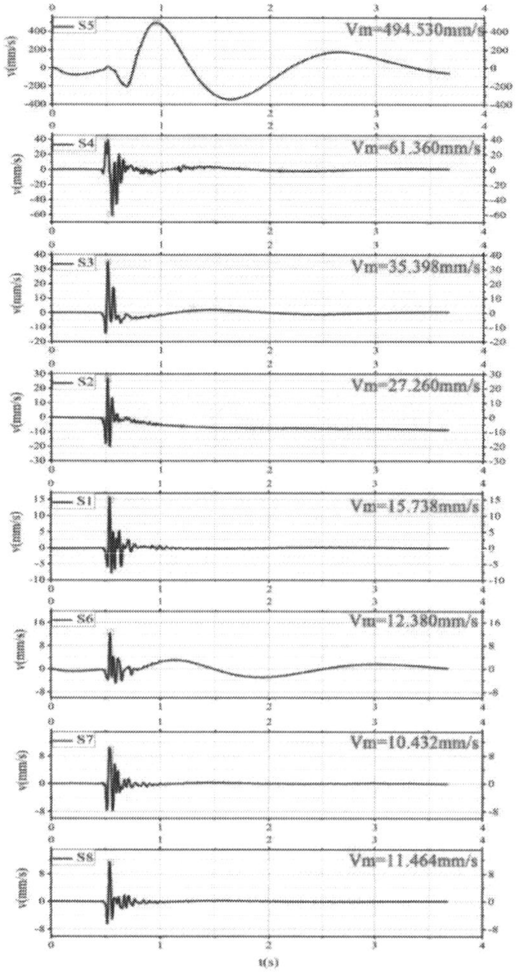

Figure 7. Schematic diagram of H1 vibration waveform.

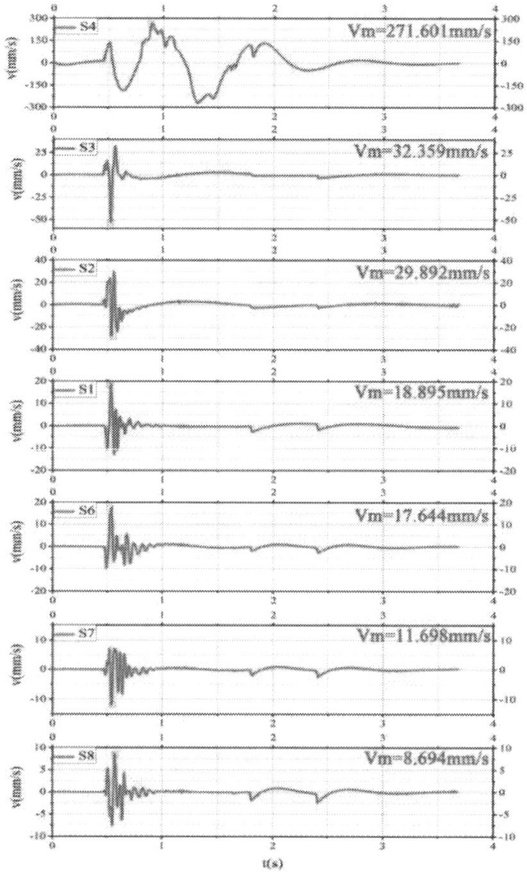

Figure 8. Schematic diagram of H2 vibration waveform.

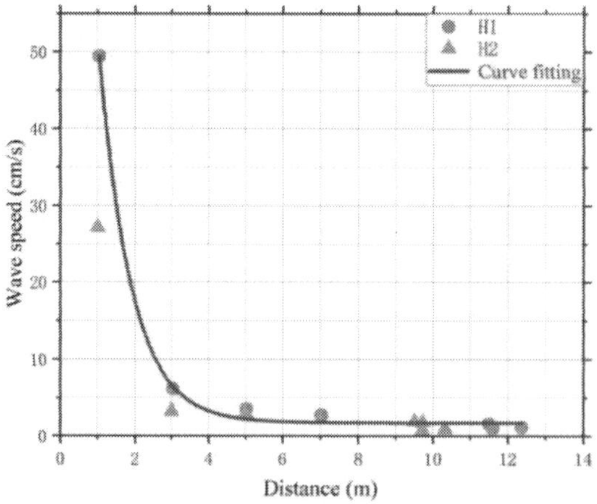

Figure 9. Vibration velocity attenuation diagram.

H1 and H2 were excited successively, and the result of rock breaking was shown in figure 6. Rock mass within trench was severely damaged by dry ice rock breaking technology, while side wall of trench remains integrity. It has achieved the desired purpose. The events of fly-pipe punching and fly-stone were not occurred. Compared with pick-axe machine, dry ice rock breaking technology exhibited higher construction efficiency in trench excavation. This result indicates that dry ice rock breaking technology is an effective method for trench excavation.

Figure 7 and figure 8 show the vibration waveform induced by two rock-breaking events. The amplitude of waveform velocity decrease with the increase of hypocentral distance. The first arrival time of different waveforms is obvious different.

Figure 9 shows the attenuation statistics of the vibration velocity induced by two rock breaking events with the hypocentral distance. The sensor S5 was the closest (1m) to the position of H1, and the maximum vibration velocity was 494.530 mm/s; the sensor S4 was the closest (1m) to the position of H2, and the maximum vibration velocity was 271.601 mm/s. The vibration velocity induced by dry ice rock breaking is attenuated with increasing of distance, and it conforms to the attenuation of power function, and the fitting accuracy is 0.99. Therefore, the vibration velocity of any seismic source distance can be predicted by the power function. According to figure 8, the vibration attenuation induced by dry ice rock breaking can also be divided into three regions, namely, the rapid reduction area (0~3m), the slow reduction area (3~10m) and the stable area (>10m).

4. Analysis and Discussion

The size of the dry ice fracturing cylinder in the H1 and H2 drilling holes used in this time is Φ100*600 mm and Φ100*400 mm respectively, and the fracture pressure obtained through the laboratory test is 60MPa.The energy calculation of CO_2 phase transformation cracking in the application of coal seam gas [39, 40], gas extraction [41], and exploration source are all calculated by the following equations [42]:

$$Eg = \frac{PV}{K-1}\left[1-\left(\frac{P_1}{P}\right)^{\frac{K-1}{K}}\right] \tag{1}$$

As in equation (1), ***Eg*** is the burst energy of gas, kJ; ***P*** is the absolute pressure of gas in the container, MPa; ***V*** is the volume of the container, m^3; ***K*** is the adiabatic index of the gas, 1.295. It can be obtained that blasting energies of the 60 mm and 40 mm tubes are 793.402 kJ and 508.952 kJ, respectively.

The approximate TNT equivalent of dry ice rock breaking device can be calculated by using equation (2):

$$W_{\mathrm{TNT}} = \frac{E_g}{Q_{TNT}} \tag{2}$$

As in equation (2), Q_{TNT} is 1 kg TNT explosion energy, 4250 kJ/kg. After calculation, the TNT equivalent of 60mm tube is 179.674g, and that of 40mm tube is 119.754g.

Sadovsky formula is an empirical formula based on a large number of measured data and the principle of similarity law, through studying the blasting seismic effect of concentrated charge [43-47].

$$v = K\left(\frac{\sqrt[3]{Q}}{R}\right)^{\alpha} \tag{3}$$

As in equation (3), ***V*** is the particle vibration velocity, cm·s^{-1}; ***K*** is the parameters related to blasting site conditions; ***Q*** is volume of tube, kg; ***R*** is the distance between the measuring point and the center of the charge, m; α is the coefficient related to geological conditions. According to equation (3), the calculating formula of dose is

$$Q = R^3 \left(\frac{v}{K} \right)^{\frac{3}{\alpha}} \tag{4}$$

Table 2. Relationship between ***K*** value and ***a*** value.

Lithology	***K***	***a***
Hard rock	50～150	1.3～1.5
Medium hard rock	150～250	1.5～1.8
Soft rock	250～350	1.8～2.0

According to the engineering report, the strength of highly weathered sandstone was about 20 MPa. Table 2 shows the relationship between ***K*** value and ***a*** value. The estimated dose range of 60mm tube was 173.244g ~ 236.157 g, and that of 40 mm tube was 94.179 ~ 129.681 g.

According to the calculation results, the TNT equivalent of 60 mm tube is 179.674 g, and that of 40 mm cracked pipe is 119.754 g, which is obviously within the charge range of vibration energy. It indicates that it is reasonable to use equation (1) to calculate the energy of dry ice tube. The effect on the building can be judged by estimating the vibration velocity of the dose.

Dry ice rock breaking technology is mainly through the rapid combustion of the energy accumulating agent under the excitation of instantaneous high voltage, releasing a large amount of heat to promote the rapid gasification of CO_2, so as to form a transient high pressure in the closed space to break the tube. The high-pressure CO_2 gas could rush from the tube rupture will impact on the borehole wall and generate stress waves. The stress state or dynamic stress field formed by stress wave disappears quickly due to the impact of the initial time is short. The range of cracks formed at the beginning of reaction is small. When CO_2 gas penetrates into impact cracks and primary cracks, it will produce an effect similar to an air wedge under static pressure, which will make the cracks continue to extend forward and cause rock fracture until the energy of high-pressure gas is exhausted.

In addition, the peak frequency of vibration induced by dry ice powder pneumatic rock breaking is mainly distributed in 0-50 Hz, which is similar to the peak frequency distribution of explosive blasting vibration. According to the vibration safety standard, the low frequency of the main frequency will cause more serious damage to the surrounding buildings. As the distance increases, the low frequency of dry ice rock breaking vibration decreases faster than the high frequency of that. When the distance far way 10m, the vibration wave velocity drops to less than 20mm/s, and the low frequency vibration wave attenuates more rapidly. Therefore, the vibration damage of buildings with a distance of more than 10m caused by dry ice rock breaking meets the safety allowable standards of blasting vibration. Compared with traditional blasting, dry ice rock breaking technology is a safer method of rock breaking.

5. Conclusion

(1)Dry ice rock breaking technology can be well applied to trench excavation, the cracking depth is about 1m, and it will only deepen the depth of the bottom groove, and protect the side. Compared with the pick machine, dry ice fracturing is more efficient and the noise is less;

(2)The vibration velocity induced by dry ice rock breaking technology will drop to less than 20mm/s the distance far way 10m. It is lower than the vibration velocity requirement of general buildings in the safety allowable standards of blasting vibration. Therefore, it is a more safe and reliable rock breaking technology.

(3)The TNT equivalent of cracking energy is applicable to Sadovsky formula, and this formula can be used to predict the wave velocity in field engineering, so as to carry out on-site safety assessment.

(4)The vibration velocity induced by dry ice rock breaking is conform to the attenuation of power function. The vibration attenuation induced by dry ice rock breaking can also be divided into three regions, including the rapid reduction area (0~3m), the slow reduction area (3~10m) and the stable area (>10m).

Acknowledgments
The work in this paper is supported by the grant from the Open Foundation of State Key Laboratories (No. HKLBEF202004), and the Natural Science Foundation of Jiangsu Province (No. BK20201313). In addition, special thanks to the Hohai University for its professional support and assistance.

References
[1] Sanchidrián J A. SLIM 2018 *Technology for Blasting to Improve Mining* (Sweden: International Symposium on Rock Fragmentation by Blasting).
[2] Dobratz B M, Crawford P C. 1985 LLNL *Explosives Handbook: Properties of Chemical Explosives and Explosive Stimulants* (America : Lawrence Livermore National Laboratory).
[3] Haugen S. Rock. 2018 *Fragmentation in Mining-past, Present and Future* (Sweden: 12th International Symposium on Rock Fragmentation by Blasting) 34-37.
[4] Attalla M I, DAY S J, Lange T, et al. 2008 NOx emissions from blasting operations in open cut coal mining *J. Atmospheric Environment* **42** (34) 7874- 7883.
[5] Bailey B L, Smith L J D, Blowes D W, et al. 2013 The Diavik Waste Rock Project: Persistence of contaminants from blasting agents in waste rock effluent [J] *Applied Geochemistry* **36** 256-270.
[6] Abdollahisharif J, Bakhtava E and Nouriza-Deh H. 2016 Green biocompatible approach to reduce the toxic gases and dust caused by the blasting in surface mining [J] *Environmental Earth Sciences* **75** (3) 191.
[7] Hajihassani M, Armaghani D J and MONJEZI M, et al. 2015 Blast-induced air and ground vibration prediction: a particle swarm optimization-based artificial neural network approach [J] *Environmental Earth Sciences* **74** (4) 2799-2817.
[8] Abdul-Karim N, Blackman C S and Gill P P, et al. 2014 Post-blast explosive residue-a review of formation and dispersion theories and experimental research *J.RSC Adv* **97** (4) 54354-54371.
[9] Banas A, Banas K and Bahou M, et al 2009 Post-blast detection of traces of explosives by means of Fourier transform infrared spectroscopy [J] *Vibspec* **51** (2) 168-176.
[10] He N, Ni Y C and Teng J, et al 2019 Identification of inorganic oxidizing salts in homemade explosives using Fourier trans- form infrared spectroscopy[J] *Spectrochimica Acta Part A.* **221** 117164.
[11] Hendry M J, Wassenaar L I and Barbour S L, et al. 2018 Assessing the fate of explosives derived nitrate in mine waste rock dumps using the stable isotopes of oxygen and nitrogen [J] *Science of The Total Environment.* 640-641 127-137.
[12] Mahmood F N, Barbour S L and Kennedy C, et al. 2017 Nitrate release from waste rock dumps in the Elk Valley, British Columbia, Canada [J] *Science of The Total Environment* 605-606:915-928.
[13] Villeneuve S A, Barbour S L and Hendry M J, et al. 2017 Estimates of water and solute release from a coal waste rock dump in the Elk Valley, British Columbia, Canada [J] *Science of The Total Environment* 601-602 543-555.
[14] Hendry M J, Wassenaar L I and Barbour S L, et al. 2018 Assessing the fate of explosives derived nitrate in mine waste rock dumps using the stable isotopes of oxygen and nitrogen [J] *Science of The Total Environment* 640-641 127-137.
[15] Mahmood F N, Barbours L and Kennedy C, et al. 2017 Nitrate release from waste rock dumps in the Elk Valley, British Columbia, Canada [J] *Science of The Total Environment* 605-606 915-928.

[16] Villeneuve S A, Barbour S L and Hendry M J, et al. 2017 Estimates of water and solute release from a coal waste rock dump in the Elk Valley, British Columbia, Canada [J] *Science of The Total Environment* 601-602: 543-555.
[17] Karlsson T and Kauppila T. 2015 *Release of Explosives Originated Nitrogen from the Waste Rocks of a Dimension Stone Quarry* (Chile: the 10th International Conference on Acid Rock Drainage & IMWA Annual Conference).
[18] Mattila K and Langwaldt G Z J. 2007 *Biological Removal of Nutrients from Mine Waters* (Finland: Finnish Forest Research Institute, Rovaniemi Unit Geological Survey of Finland).
[19] Karlsson T and Kauppila T. 2016 Explosives-originated nitrogen emissions from dimension stone quarrying in Varpaisjrvi, Finland [J] *Environmental Earth Sciences* **75** (9) 834.
[20] Bo K, Zhou K P, Ren G F, Shi J and Zhang Y N. 2019 Positive Phase Pressure Function and Pressure Attenuation Characteristic of a Liquid Carbon Dioxide Blasting System [J] *Energies* **12**.
[21] Fan, Y C, Qin B T, Zhou Q, Shi Q L and Wu J H. 2020 Liquid CO_2 phase transition fracturing technology and its application in enhancing gas drainage of coal mines [J] *Adsorption Science & Technology* **38** 393-412.
[22] He W R, He F L, Zhang K, Zhao Y Q and Zhu H Z. 2018 Increasing Permeability of Coal Seam and Improving Gas Drainage Using a Liquid Carbon Dioxide Phase Transition Explosive Technology [J] *Advances in Civil Engineering* **15**.
[23] Kang J H, Zhou F B, Qiang Z Y, Zhu S J. 2018 Evaluation of gas drainage and coal permeability improvement with liquid CO_2 gasification blasting [J] *Advances in Mechanical Engineering* **10**.
[24] Yan H, Zhang J X, Li B Y, Zhu C L. 2021 Crack propagation patterns and factors controlling complex crack network formation in coal bodies during triaxial supercritical carbon dioxide fracturing [J] *Fuel* **286**.
[25] Donze F V, Bouchez J and Magnier S A. 1997 Modeling fractures in rock blasting [J] *International Journal of Rock Mechanics and Mining Sciences* **34** (8) 1153–1163.
[26] Kutter H K, Fairhurst C. 1971 On the fracture process in blasting *J. International Journal of Rock Mechanics & Mining Sciences & Geomechanics Abstracts* **8** (3) 181IN1, 189–188, IN10, 202.
[27] Nilson R H, Proffer W J, Duff R E. 1985 Modelling of gas-driven fractures induced by propellant combustion within a borehole [J] *International Journal of Rock Mechanics & Mining Sciences & Geomechanics Abstracts* **22** (1) 3–19.
[28] Zhang, Y A, Deng J R, Deng W H, Ke B. 2019 Peridynamics simulation of rock fracturing under liquid carbon dioxide blasting [J] *International Journal of Damage Mechanics* **28** 1038-52.
[29] Goodarzi, M., S. Mohammadi, and A. Jafari 2015 Numerical analysis of rock fracturing by gas pressure using the extended finite element method [J] *Petroleum Science* **12** 304-15.
[30] Xie Xiaofeng, et al. 2018 Review of liquid CO_2 phase transformation rock breaking technology [J] *Journal of Railway Science and Engineering* **15** (6) 2018.
[31] Liu Xiaoxiong, et al. 2018 Study on vibrational energy distribution of rock-solid phase change in liquid carbon dioxide [J] *Mining and metallurgy engineering* **38** 181 (3) 5-10.
[32] Li Qiyue, et al. 2018 Application of liquid CO_2 phase change rock breaking technology in subway foundation pit excavation [J] *Journal of Railway Science and Engineering* **15** (1) p 163-169.
[33] Li Qi-Yue, Chen Guan, Luo Da-You, Ma Hai-Peng and Liu Yong 2020 An experimental study of a novel liquid carbon dioxide rock-breaking technology [J] *International Journal of Rock Mechanics and Mining Sciences* **128** 104244.
[34] Liu Xiaoxiong, et al. 2018 Study on vibrational energy distribution of rock-solid phase change in liquid carbon dioxide *J. Mining and metallurgy engineering* **38** 181(3) p 5-10

[35] Zhou S T, Jiang N, He X, Luo X D. 2020 Rock Breaking and Dynamic Response Characteristics of Carbon Dioxide Phase Transition Fracturing Considering the Gathering Energy Effect [J] *Energies*. 1336; doi:10.3390/en13061336.

[36] Chen G, Li Y Y, Liu X X, et al. 2018 Research on Energy Distribution Characters about Liquid CO_2 Phase-transition Broken Rock Vibration Signal [J] *Blasting*. 035(002):155-163.

[37] Huang J H, Luo Y, Zhang G, Zheng B W, Li X P, Xu M N, Xu P C. 2020 Numerical analysis on rock blasting damage in Xiluodu underground powerhouse using an improved constitutive model [J] *European Journal of* Environmental *and Civil Engineering*. DOI: 10.1080/19648189.2020.1780475.

[38] Hu S B, Pang S G, Yan Z Y. 2019 A new dynamic fracturing method: deflagration fracturing technology with carbon dioxide [J] *International Journal of Fracture* 220 99-111.

[39] Yang X L, Wen G C, Sun H T, et al. 2019 Environmentally friendly techniques for high gas content thick coal seam stimulation-multi-discharge CO_2 fracturing system [J] *Journal of Natural Gas Science and Engineering* **61** 71–82.

[40] Bai X, Zhang D M , Zeng S, et al. 2020 An enhanced coalbed methane recovery technique based on CO_2 phase transition jet coal-breaking behavior [J] *Fuel* **265** 116912.

[41] Hua G Z, He W R, Miao S N. 2018 Enhancing coal seam gas using liquid CO_2 phase transition blasting with cross-measure bore hole [J] *Journal of Natural Gas Science and Engineering* **60** 164–173.

[42] Wang B, Qiu W Y, Liu S D, Sun H C, et al. 2020 Supercritical CO_2 source for underground seismic exploration [J] *Journal of King Saud University-Science* **32** 1731–1737.

[43] Sheng X, Jones C and Petyt M. 1999 Ground vibration generated by a load moving along a railway track [J] *Journal of Sound and Vibration* **228** (1) 129.

[44] Krylov V V. 1999 Ground vibration boom from high-speed trains [J] *Journal of Low Frequency Noise* **18** (4) 251.

[45] Jones D V, Petyt M. 1991 Ground vibration in the vicinity of a strip load: two-dimensional half space model [J] *Journal of Sound and Vibration* **147** (1) 155.

[46] Yang Y B, Hung H H, C hang D W. 2003 Train induced wave propagation in layered soils using finite/infinite J element simulation [J] *Soil Dynamics and Earthquake Engineering* **23** 263.

[47] Marrara F and Suhadolc P. 1998 Site amplifications in the city of Benevento (Italy) comparison of observed and estimated ground motion from explosive sources [J] *Journal of Seismology* **2** 125-143.

Experimental Study on Creep of Sandstone under Hydrochemical Corrosion

Jiangchun Hu[a], Luge Sun[*], He Feng[b], Zhipeng Liu[c] and Qin Wang[d]

School of Civil & Architecture Engineering, Zhongyuan University of Technology, Zhengzhou 450007, China

[*]Corresponding author Email: 704238819@qq.com, [a]hujiangchun@126.com, [b]983155076@qq.com, [c]1449070521@qq.com, [d]511273880@qq.com

Abstract. Corrosion creep of sandstone has great influence on the safety and stability of geotechnical engineering, and it is of great significance to study the creep characteristics of sandstone under the action of hydrochemical corrosion for the long-term stability of geotechnical engineering. Through creep test of corroded rock, the characteristics of rock deformation and strength are analyzed. Combined with electrochemical impedance spectroscopy (EIS), the microcracks of corroded sandstone during creep process are detected, and the variation law of equivalent circuit parameters and the development trend of microcracks during creep process of corroded sandstone are analyzed. The test results show that the creep characteristics of rock samples are related to the hydrochemical environment; The medium-and long-term strength of the solution with high pH value is relatively small, and the difference with uniaxial compressive strength is large; Moreover, there is a certain correlation between creep of rock samples and equivalent circuit parameters. In this paper, the influence of creep deformation, mechanical properties and hydrochemical environment on sandstone creep is analyzed, and the research results can provide reference for stability analysis of related geotechnical engineering.

1. Introduction

Geotechnical engineering is an engineering relying on soil or rock mass, which mainly tends to foundation, slope and underground engineering. From design, construction, completion and use, and even later maintenance, the mechanical properties of rock mass change throughout the life cycle of geotechnical engineering. The effect of occurrence environment on rock mass is a long-term and cumulative process, which reflects the timeliness of deformation-failure. In-situ stress, groundwater, temperature and chemical environment are the four main factors forming complex geological environment. Rocks in groundwater seepage field and chemical environment for a long time, on the one hand, bear the static and dynamic water pressure of water, on the other hand, water acts as a carrier to bring chemical substances into rock pores, and some elements in rock minerals react with chemical substances, breaking the balance state between water and minerals, which makes the dissolution and migration of rock minerals change the mineral composition and structural composition of the rock itself, and weakens the cementation between rock mineral particles, resulting in the change of rock mineral composition and dissolution holes in the rock[1-2]. With the long-term action of the underground stress field, the rock is constantly deformed, the internal cracks are generated, the micro-cracks are penetrated and expanded, and the existing micro-cracks are gradually expanded under the action of chemical solution. Serious rock mass damage will eventually have an irreversible impact on the stability of geotechnical engineering [3-5].

At present, direct detection method and indirect detection method are often used to detect rock microcracks. Direct methods mainly include microscopy, scanning electron microscopy, and laser holographic interferometry. Most samples need to be prepared before direct detection technology, and the process of sample preparation also aggravates the damage of rock mass. Although it can visually display and detect the damage state of rock mass materials, it is limited to plane measurement, and there are still limitations in the application process. Indirect detection methods mainly include mass density change detection method, acoustic emission technology, mercury intrusion detection, nuclear magnetic resonance, ultrasonic wave, rock damage CT, etc[6]. Indirect detection technology mostly compares the test results of samples before and after or tests in specific indoor conditions. Some methods are limited by the conditions of rock samples, and the detection accuracy is not high, so most methods can not realize real-time detection.

Based on the study of the influence of hydrochemical environment on the instantaneous strength of rock, the creep test of sandstone under graded loading in hydrochemical environment is carried out in this paper, combined with the detection method of electrochemical impedance spectroscopy[7], in order to understand the creep deformation of sandstone, the mechanical properties and the influence of hydrochemical environment on sandstone creep, and also hope to provide reference for practical engineering.

2. Experimental Design

2.1. Test Materials and Equipment

The rock block used in the test is Shi Ying feldspar sandstone. Sandstone samples are mainly composed of detritus, matrix, accessory minerals and secondary minerals. After the rock block is taken from the construction site, the sample with diameter d of 50mm is drilled by the rock core drilling machine, and the height of the sample is intercepted by the rock cutting machine, and the sampling height h is 60mm; Grind both end surfaces of the sample on a rock grinding machine to ensure that the cross section of the sample is parallel, flat and vertical. After sampling, place the sandstone sample in a ventilated place for natural air drying. The design of hydrochemical solution is shown in table 1.

Table 1. Design ratio of aqueous chemical solution.

Solution Number	Ingredient	Solution Concentration/mol/L	pH
1	HCL	0.1	2
2	HCL	0.1	5
3	Distilled Water	/	7
4	NaOH	0.1	10
5	NaOH	0.1	13

Several samples were taken from each group every 30 days for rock creep test. Creep test of corroded rock is carried out on YR-2000 rock creep tester in Henan Mechanical and Engineering Structure Test Center. The creep tester is controlled by TestExpert software and can provide different loading modes and different loading rates. The main technical parameters are shown in table 2.

Table 2. Main technical parameters of yr-2000 rock creep meter.

Maximum axial load	Axial force measurement accuracy	Dismeasurement range	Dismeasurement accuracy	Deformation limit value Axial and Radial direction	Demeasurement resolution
2000kN	<±1%	1-100mm	<±1%	5mm、2.5mm	0.001mm

[b] Using the displacement control method, the designed displacement load rate of 0.5mm/min, damage detection threshold is 50% , and it is believed that when the strain of the rock sample reaches the peak strength,the strain has weakened to 50% believes that the rock sample has damaged [16].

2.2. *Experimental Process and Matters Needing Attention*

(1) According to the above, select samples from rock samples with different treatment methods. Boil distilled water for 30min, and dilute chemical solution remains.

(2) After the rock sample is naturally dried, it is put into an oven for drying at 105°C for 24 hours, and the dried sample is placed at room temperature.

(3) Add a rigid cushion block with a height of 30mm and a diameter of 50 on the lower bearing platform of YR-2000 rock creep instrument, and put the rock sample on the upper cushion block.

(4) Graded loading is set for the test. In this creep test of corroded rock, constant speed and step-by-step loading are adopted. The grading parameters of this test are set as follows: at the beginning of the first stage, displacement control is adopted, the displacement speed is controlled at 0.5mm/min, and the target value is set at 1mm or 5kN. When the receiving displacement of the rock creep gauge sensor reaches 1mm or the receiving axial stress reaches 5kN, it is considered that the first stage begins and remains there for 30min. Second stage: force control is adopted in this stage, the loading rate is controlled at 1kN/min, the target axial stress is 10kN, and this stage is maintained for 4h; In the last five stages, force control is adopted, the acceleration rate is controlled at 1kN/min, the target axial stress is 20, 30, 40, 50 and 60kN respectively, and it is maintained for 10h; hours in no stage.

(5) In the later stage of the test, because of the individual differences of sandstone samples, the number of segments is different, and the target axial stress in the last stage is taken according to the uniaxial compressive strength; For force majeure such as power failure during creep test, stress shall be loaded step by step, and shall be maintained in the accident section for corresponding time before entering the next stage.

(6) Protective measures should be taken in advance before the test. First, prevent the sudden destruction of rocks and the flying of cuttings.

3. Experimental Results and Analysis

3.1. *Effect of Hydrochemical Solution on Uniaxial Creep Curve of Sandstone*

Several sandstone samples were selected from different hydrochemical solutions for uniaxial compression creep test. The creep curves of rock samples under different hydration solutions and different corrosion time are integrated and compared. Considering that rock samples in different pH solutions are taken from different rock blocks and different rock blocks, only the creep stress-strain curves of rock samples in the same pH solution are characterized. As shown in figure 1-5.

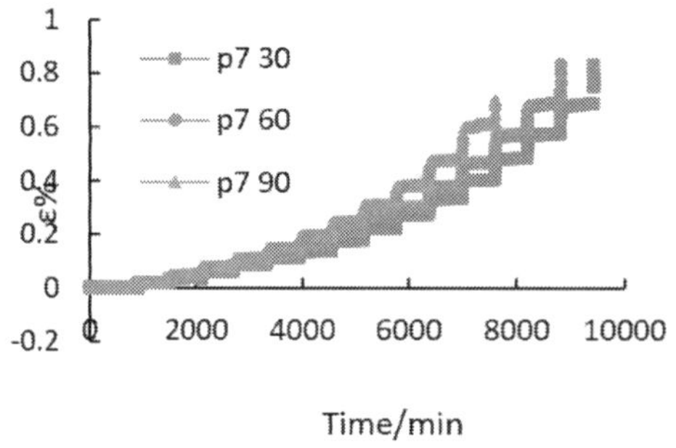

Figure 1. Creep curve of rock samples corroded by different time in pH=7 hydrochemical solution.

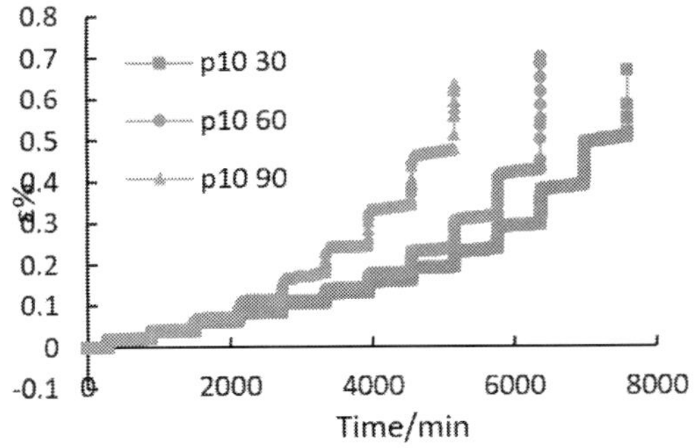

Figure 2. Creep curve of rock samples corroded by different time in pH=10 hydrochemical solution.

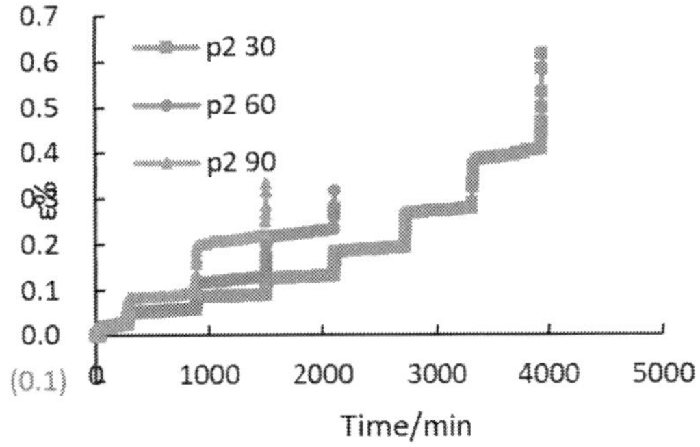

Figure 3. Creep curve of rock samples corroded by different time in pH=2 hydrochemical solution.

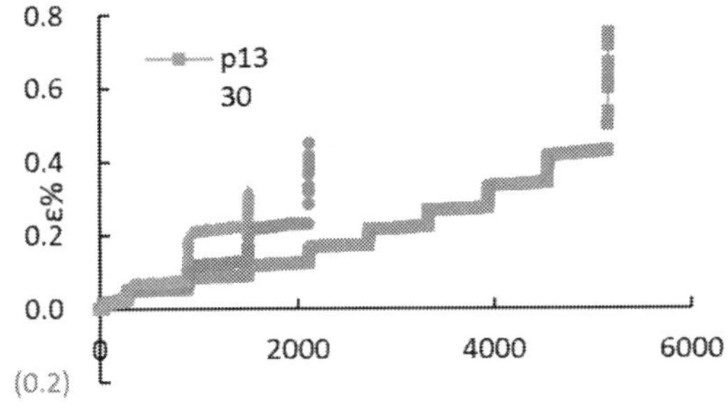

Figure 4.Creep curve of rock samples corroded by different time in pH=13 hydrochemical solution.

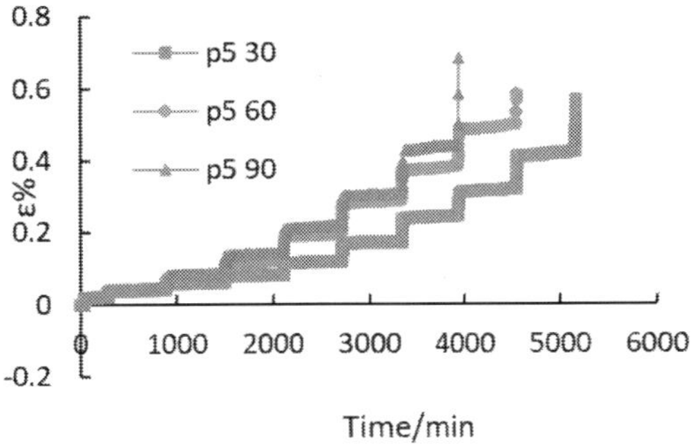

Figure 5. Creep curve of rock samples corroded by different time in pH=5 hydrochemical solution.

Figure 1-5 shows the creep curves of rock samples under various working conditions obtained from uniaxial compression creep test of corroded sandstone. Because the rock samples in different working conditions adopt the same loading mode, the creep curve shows the same trend, which mainly includes instantaneous stress loading deformation section and constant stress deformation section [8]. The main characteristics of creep curve of rock samples are as follows:

(1) Time effect. The creep end time of rock samples in the same hydrochemical solution is advanced with the increase of corrosion time. The longer the corrosion time is, the load instantaneous strain and creep of rock samples will increase correspondingly.

(2) Corrosion effect. Under different working conditions, the creep failure time of rock samples in the same corrosion time is gradually shortened with the increase of pH value of hydrochemical solution. Taking corrosion for 30 days as an example, the corresponding creep failure time of water chemical solution from low to high pH value is 65.75h, 86h, 157h, 126.65h and 85.8h, respectively.

(3) Instantaneous strain appears in each sample after each stress level is recorded. It can be seen that with the increase of loading stress, the deformation of rock sample generally increases. Under the action of low stress, the rock sample is mainly subjected to instantaneous deformation, and the creep strain is small. During this period, the creep of the rock sample presents a stable trend. When the stress increases gradually, the creep of the rock sample increases gradually, and the strain increases sharply during the last loading [9]. It can be considered that the rock sample has internal damage and cracks due to stress concentration.

3.2. Long-Term Strength Characteristics of Corroded Sandstone

According to the data, the isochronous stress-strain curve of corroded sandstone is drawn. Take the rock samples corroded for 30 days at pH=2 as an example. Firstly, the graded loading creep curve of rock sample is transformed into the separately loaded creep curve [10], as shown in figure 6, and then the isochronous stress-strain curve of rock sample is made according to the method in figure 7.

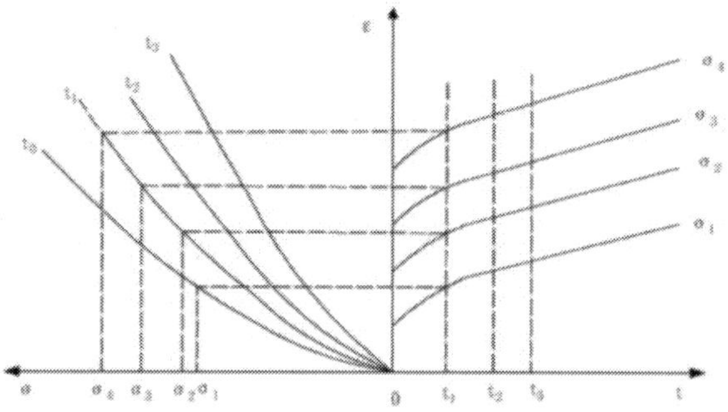

Figure 6. Creep curve under separate loading.

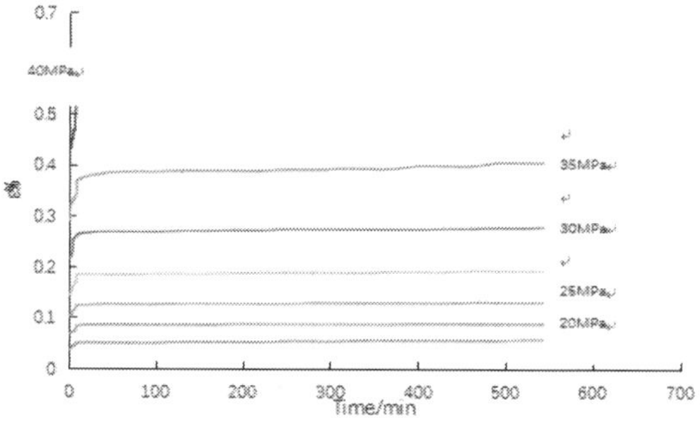

Figure 7. Method drawing of isochronous stress-strain curve.

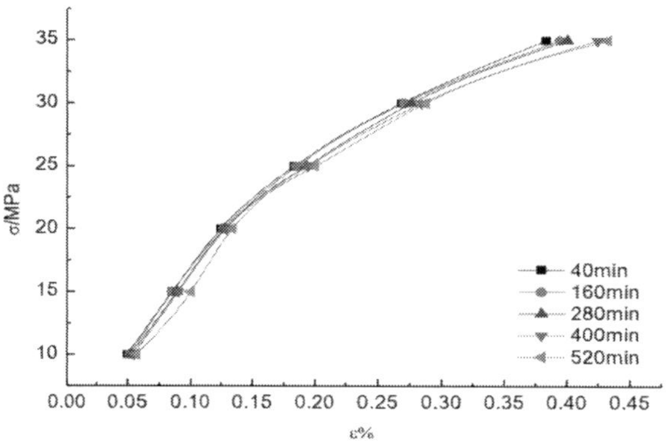

Figure 8. Isochronous stress-strain curve.

According to relevant research, the inflection point in the isochronous stress-strain curve can be expressed as the characteristic point of rock sample's transition from viscoelasticity stage to viscoplasticity, that is, the yield point of rock sample at each time point, which is reflected in the isochronous stress-strain curve as the characteristic point of curve trend from steep to slow [11]. It can be seen from fig. 8 that the inflection point appears approximately at 0.14% strain. According to the test data, the loading stress corresponding to 0.14% strain is 22.6MPa, and it can be considered that the long-term strength of rock sample is 22.6MPa.

According to the above method, the long-term strength of rock samples under various working conditions is calculated as shown in table 3.

Table 3. Long-term strength of rock samples under various working conditions.

Hydration environment	Time/d	Compressive strength/MPa	Long term strength/MPa	strength ratio%
pH=7	30	96.39	76.95	79.8
	60	93.57	69.45	74.2
	90	89.57	62.07	69.3
pH=5	30	64.232	40.864	63.6
	60	56.174	30.934	55
	90	47.095	22.058	46.9
pH=10	30	80.94	58.276	71.9
	60	76.434	50.43	65.9
	90	63	37.75	59.9
pH=2	30	51.53	22.6	44.3
	60	30.19		
	90	25.295		
pH=13	30	56.3	33.78	60
	60	29.063	15.46	53
	90	26.07		

It can be clearly seen from the test results in Table 3 that the long-term strength of rock samples in all working conditions is smaller than the uniaxial compressive strength of rock samples, with a drop of 40%-80%. The long-term strength of rock samples corroded by pH=2 for 30 days is the lowest, followed by rock samples corroded by pH=13 for 60 days. The long-term strength of rock samples in distilled water is relatively high, but the decrease is that the corrosion time increases by 30.7%. It can be seen that the long-term strength of rock samples changes with the properties of hydrochemical solution and corrosion time.

Where there is no data in table 3, the creep time of rock samples corroded by pH=2 water chemical solution at 60 and 90 and rock samples corroded by pH=13 water chemical solution for 90 days is shorter, and the isochronous stress-strain curve is linear without obvious inflection point.

3.3. Change Law of Electrochemical Impedance Spectroscopy in Creep Process of Corroded Sandstone

According to the application of electrochemical impedance spectroscopy (EIS) in microscopic description of rock samples, this paper attempts to apply EIS testing method to describe the development of microstructure in creep process of rock samples. Combined with the reaction of sensitivity of rock samples to hydrochemical solution in various working conditions summarized above, the rock samples corroded in hydrochemical solution with pH=2 for 60 days were selected for electrochemical detection of creep process. It is designed that the stress maintenance time of each level of loading is 10 hours, and the loading is divided into four levels, and the loading stress of each level is 10kN. The detection stage is only the creep stage within one hour of each level of stress loading.

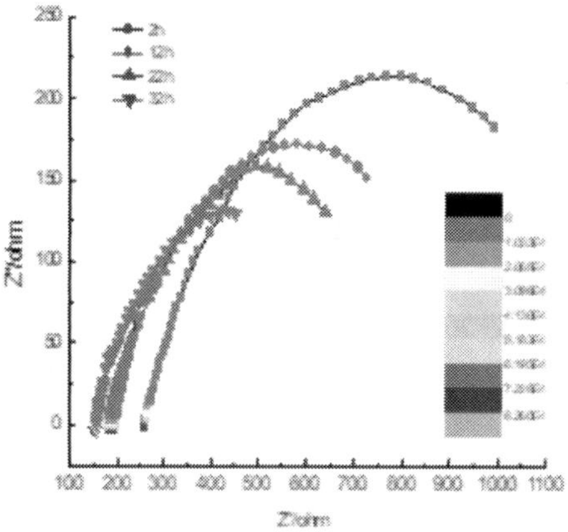

Figure 9. Nyquist diagram of rock sample creep.

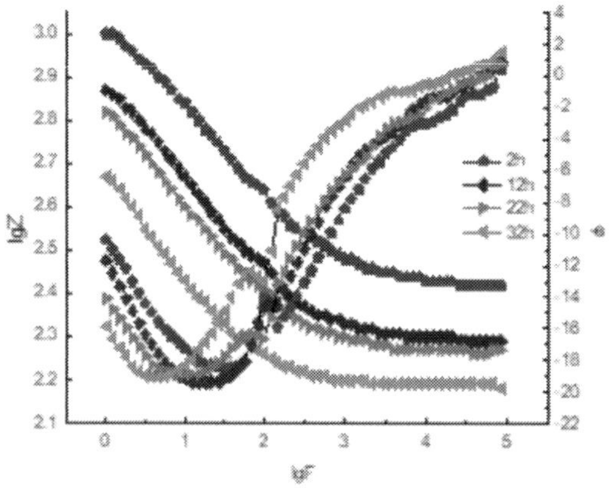

Figure 10. Bode diagram of rock sample creep.

Table 4. High-frequency technical parameters of rock creep Nyquist diagram.

Parameter / Time/h	Z_1	Z_2	$\alpha(^o)$	$\varphi_{Z_1}(rad)$	$\varphi_{Z_2}(rad)$	R_2-R_1
2	(230.63.9.86)	(247.6,31057)	4.57	0.03	0.13	16.79
4	(200.2,6.132)	(212.7,23.87)	3.93	0.03	0.11	12.5
6	(185.8,6.614)	(195.5,20.169)	3.69	0.04	0.10	9.7
8	(154.3,1.847)	(158.5,7.629)	1.47	0.01	0.05	4.2

In table 4, Z_1 and Z_2 represent the coordinates of the start and end segments of high-frequency arc in Nyquist diagram. α represents the average overlooking angle of high frequency band; φ_{Z_1} and φ_{Z_2} represent the top-down angles of the start and end sections of the high-frequency arc. R_1 and R_2 represent the projection of Z_1 and Z_2 on the real axis in Nyquist diagram.

The electrochemical impedance spectroscopy of rock samples during creep is described and analyzed as follows:

It can be seen from fig. 9 that the radius of frequency band arc in Nyquist diagram is 2h, 12h, 22h and 32h respectively, which indicates that the internal crack of rock sample begins to develop under load;

(1) According to Bode diagram, the value of high-frequency phase angle measured by stress loading for 8 hours is the smallest; Bode diagram, it can be clearly seen that the difference between the first stage and the second stage of loading is larger than that between the third stage and the fourth stage of loading, and the impedance value decreases gradually with the increase of time. It can be considered that the microcracks of rock samples expand rapidly during initial loading, and then the stable development stage of cracks appears. When loading the last stress, the cracks develop again and the microcracks expand completely.

(2) Combined with the high-frequency technical parameters in Nyquist diagram, it can be concluded that the variation law of rock microstructure is as follows: ① The change of high-frequency arc phase angle index in Nyquist diagram is 0.1, 0.08, 0.06 and 0.04 respectively according to the increase of corrosion time, which shows that the micro-cracks in rock samples increase obviously under long-time load. ② In ②Nyquist diagram, the changes of real impedance at the beginning and end of high frequency arc period with the increase of corrosion time are 16.97, 12.5, 9.7 and 4.2, respectively, indicating that the internal structure of rock samples changes obviously during creep process, and some voids gradually penetrate.

Similarly, the equivalent circuit is used to fit the creep electrochemical impedance spectrum of rock samples, and the parameters are shown in table 5.

Table 5. Fitting equivalent circuit parameters.

Parameter / Time/h	R_s	$CPE-T$	$CPE-P$	R_p
2h	219	2.0×10^{-4}	0.53	768
12h	192.5	2.2×10^{-4}	0.55	726.3
22h	179.1	3.1×10^{-4}	0.56	671.1
23h	151.6	5.2×10^{-4}	0.58	509

In table 5, R_s represents the resistance of solution in rock samples; $CPE-T$ represents electric double layer capacitance; $CPE-P$ is dispersion angle parameter, $\varphi=\pi(1-CPE-P)/2$ is dispersion angle; R_P represents the connection resistance.

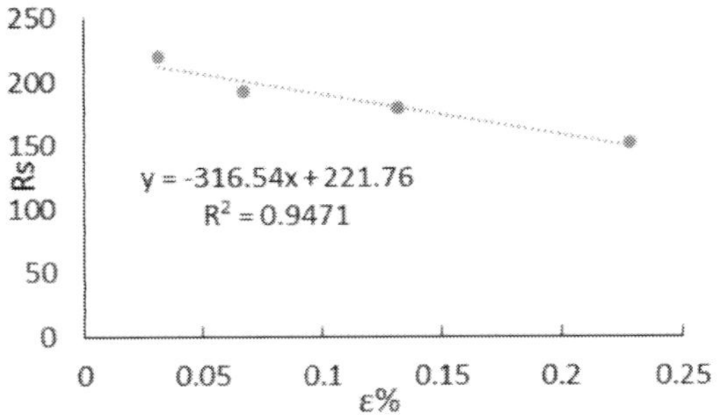

Figure 11. Relationship between rock sample strain and R_s.

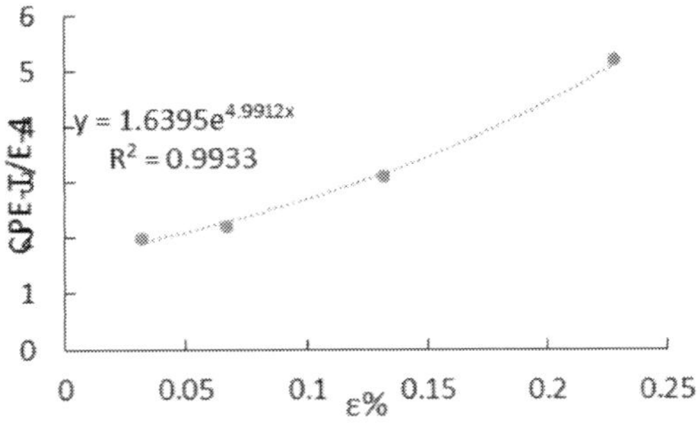

Figure 12. Relationship between rock sample strain and $CPE-T$.

According to the data in table 5, with the increase of stress loading grade, R_s decreases gradually as the creep time continues. When electrochemical test shows that the ion concentration of pore solution in the system remains unchanged, there is the following relationship between pore solution resistance and test substance particles [12]:

$$R_S = K\left(\frac{\rho_S}{1-\psi_S}\right) \tag{1}$$

In which ρ_s is the resistivity of pore solution, and $\psi_s = a\varphi_s$ is the relationship between solid fraction and volume fraction. The relationship between volume fraction of φ_s and porosity of P is $\varphi_s = 1-P$, so R_S can be expressed as:

$$R_S = \frac{K\rho_s}{1-(1+a)P} \tag{2}$$

It can be seen that is inversely proportional to. Therefore, it can be seen that the porosity of rock samples increases gradually during creep. The change trend of microstructure of rock samples can be directly reflected. When pores expand, electrolytic solution gradually fills the pores, and the corresponding capacitance increases. In addition, Burtman et al. analyzed the induced polarization mechanism of rocks and tested the IP response of sandstone, and found that the electric double layer is related to the size and shape of pores filled with electrolyte [13]. Compared with the discrete phenomenon in the third chapter, there is a certain regularity in creep electrochemical test, that is, with the development of creep deformation of rock samples, it shows a decreasing trend. It can be considered here that due to the load, the developed fissures make the electrolytic solution communicate and the current path open. Micro-cracks in rock samples are the main cause of deformation and fracture with time effect. Therefore, through the relationship between deformation and deformation of rock samples, it can be considered that creep deformation of rock samples has a certain correlation with them.

4. Conclusion

Through uniaxial creep test and electrochemical test of rock samples under different working conditions, the following conclusions are obtained:

(1) The creep characteristics of rock samples are related to the hydrochemical environment. The creep amount of rock sample increases gradually with the increase of pH value of hydrochemical solution, and its creep rate increases. At the same time, the creep of rock sample has corresponding time effect, and the longer the corrosion time, the greater the creep deformation.

(2) The isochronous stress-strain curve method is used to determine the long-term strength of rock samples under different working conditions. It is found that the long-term strength of the solution with high pH value is relatively small, and the difference with the uniaxial compressive strength is large

(3) Applying electrochemical impedance spectroscopy to detect creep deformation of rock samples, it is found that there is a certain correlation between creep amount of rock samples and equivalent circuit parameters.

5. References

[1] Li K G. 2016 Study on rock mechanical properties under water-rock physics [M] Beijing. *Metallurgical Industry Press*.

[2] Sun J. 2007 Some advances in rock rheological mechanics and its engineering application [J]. Chinese Journal of Rock Mechanics and Engineering **26** (6) 1081-1106.

[3] Feng X T, Lai H, Zheng H. 2000 Rock fracture characteristics under chemical environment erosion-Part I: Experimental study [J] *Chinese Journal of Rock Mechanics and Engineering* (04) 403-407.

[4] Chen S L, Feng X T, Li S J. 2003 Chemical corrosion effect of rock uniaxial compressive strength and fracture characteristics [J] Chinese Journal of Rock Mechanics and Engineering (04) 547-551.

[5] Liu Y S, Wang L, Dong X Y. 2015 Study on dynamic properties and constitutive model of rocks under chemical corrosion [J] *journal of yangtze river scientific research institute* **32** (5) 72-75.
[6] Peng R D, Liu H B, Wang J Q, et al. 2007 Research progress of rock damage detection technology [J] *Rock and Mineral Testing* (01) 61-66.
[7] Hu J C. 2014 Theoretical and experimental study on geotechnical electrochemistry [M] *Beijing. Coal Industry Press*.
[8] Wang X B. 2016 Study on creep experiment and constitutive model of rock materials [D] Chongqing. *Chongqing University*.
[9] Yang M M. 2014 Study on dynamic mechanical effect of rock under chemical corrosion [D] Nanchang. *East China Jiaotong University*.
[10] Jiang Z B. 2016 Experimental Study on Creep Characteristics of Rock under Multi-field Environment and Mechanical Model [D] Dalian. *Dalian Maritime University*.
[11] Xu D. 2016 Study on creep characteristics and nonlinear constitutive model of red bed rock [D] Chengdu. *Southwest Jiaotong University*.
[12] Fang W. 2013 Electrochemical impedance spectroscopy study on hydration process of concrete and cement [D] Dalian. *Dalian University of Technology*.
[13] Burtman V, Zhdanov M S. 2015 Induced polarization effect in reservoir rocks and its modeling based on generalized effective-medium theory [J] *Resource-efficient technologies* **1** (1) 34-48.

Numerical Simulation on the Effect of Rock Joint Roughness on the Stress Field

J Wang[1, *], J Liu[2] and Y Q Fu[3]

[1] Beijing Urban Construction Design & Development Group Co., Limited, Beijing, 100037, China
[2] YangZhou Architecture Design & Research Institute Co., Ltd., YangZhou, 225000, China
[3] China IPPR International Engineering Co., Ltd., Beijing, 100089, China
*2986141244@qq.com

Abstract. In view of the influence of Joint Roughness Coefficient (JRC), which is for quantitative description of the joint surface roughness, on the stress field of the rock mass, compression test and shear-compression test were simulated on models with different joint roughness. The photoelasticity technique is applied to examine the feasibility of numerical simulation. The results show that numerical simulation results are in agreement with the results of photoelastic experiments. The stress concentration area is distributed near the joint plane. Thus, the joint plane controls the shear strength of the rock. In compression test, the maximum shear stress of the model is proportional to JRC and the normal pressure. In shear-compression test, when the ratio of the axial shear to the normal pressure is small, the maximum shear stress is nonlinearly positively correlated with JRC. When the ratio of the axial shear to the normal pressure is relatively large, the relationship curve between the maximum shear stress and JRC is parabolic. When the JRC is small, as the ratio of the axial shear force to the normal pressure increases, the maximum shear stress changes abruptly, and the maximum shear stress after the mutation decreases significantly. The reason is that the upper and lower parts of the model have slipped, resulting in a redistribution of stress. In addition, when the JRC is 6 to 12, it is more likely to cause stress concentration.

1. Introduction

Rock discontinuities not only have impacts on the stability of geotechnical structures, but also effect the geotechnical operations like rock drilling, cutting and blasting [1]. Among discontinuities such as bedding planes, faults, joints and fractures, rock joints play the critical role in the effect on the mechanical behavior [2]. Considerable efforts have been devoted to study the effect of surface roughness on the stress field of the rock joints in the past several decades. Photoelasticity technique has been widely used to reveal the distribution of the stress and stress concentrations of geomaterials subjected to various loading conditions [3]. The combination use of photoelasticity and high-definition photography is a simple and effective method for the study of stress field, owing to that it can clearly see the full-field stress distribution [4]. Photoelastic samples under uniaxial compressive were made experiments to analyze the influence of the fractal dimension on the joint intensity and joint deformation [5]. Photoelasticity was applied to study the effect of the rock joints on the shearing property and the obtained results show that the rock joint roughness has vital impacts on the shear dilatation and contact property [6].Although photoelastic tests are significantly advantageous of stress field visualization, there are some shortcomings of experiment techniques. In some tests, it is inevitable that the presence of initial stress in complex models during specimen preparation will disturb the experimental results. Furthermore, there are a large number of other issues linked with the deviations between experimental observations and real results. Consequently, as the computer technology develops rapidly in recent years, numerical simulations also seem to be a powerful tool to characterize the stress field of models during loading [7]. Finite element methods could handle well the problems that experimental and theoretical methods cannot overcome.

This paper focuses on the effect study of rock joint on the stress field of the rock masses by using finite element method. Photoelasticity technique is applied to examine the feasibility of numerical simulation. Maximum shear stress is chosen to quantify the effect of joint roughness on the stress field, JRC is used to quantify the rock joint roughness. The shear strength of rock masses is much lower than their compressive strength in rock masses due to the existence of joints. Therefore, this paper concentrates on acquiring the relationship between the maximum shear stress and the joint roughness of the rock mass.

2. Method Verification

FEM simulation and photoelastic testing were conducted to obtain the mechanical response of the same loaded disc, and the obtained stresses of the specific points by two methods were compared to examine the validity of FEM simulation. Since the verification model is relatively simple, it can be considered that there is no interference from the initial stress in the photoelastic test. Therefore, it is effective to compare the numerical simulation results with the photoelastic test results.

2.1. Determination of Fringe Constant

The material parameter, fringe constant, was determined by disk radial loading test. As a critical design parameter, the fringe constant of a photosensitive material is of great importance on the stress field analysis based on the stress optical method [3]. Principal stress difference and fringe order are related by fringe constant. Thus, it is possible to compare the experimental results with other results such as simulation results and theoretical solutions according to this parameter. From formula (1), it can be seen that the fringe constant is determined by the wavelength of the light source and the stress-optical coefficient of the photoelastic material. Therefore, the fringe constant can be obtained from a sample with a theoretical solution of stress as shown in formula (2).

$$f = \frac{\lambda}{C} \tag{1}$$

Where f is fringe constant, λ is light source wavelength, C is stress-optical coefficient.

$$(\sigma_1 - \sigma_2) = \frac{f}{t} n \tag{2}$$

where n is the fringe order of interest, t is the thickness of the sample, $(\sigma_1 - \sigma_2)$is principle stress difference.

The photoelastic material used in this experiment is polycarbonate. The detail the specimen is shown in figure 1, and the experimental device is shown in figure 2. The light source is a monochromatic light with a wavelength of 532 nm. Figure 3 shows the dark field optical layout of photoelasticimeter.

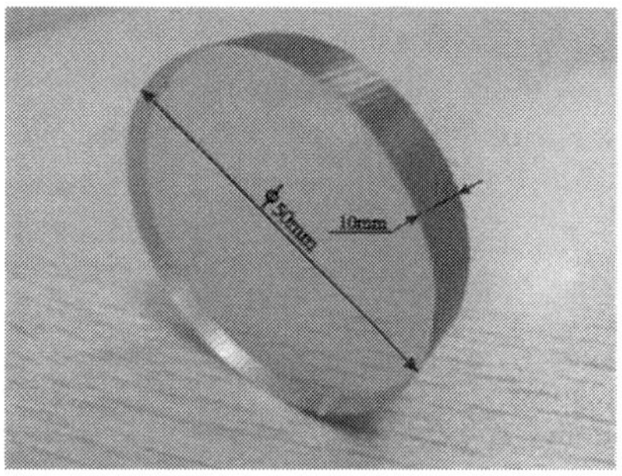

Figure 1. The circular disc used in testing.

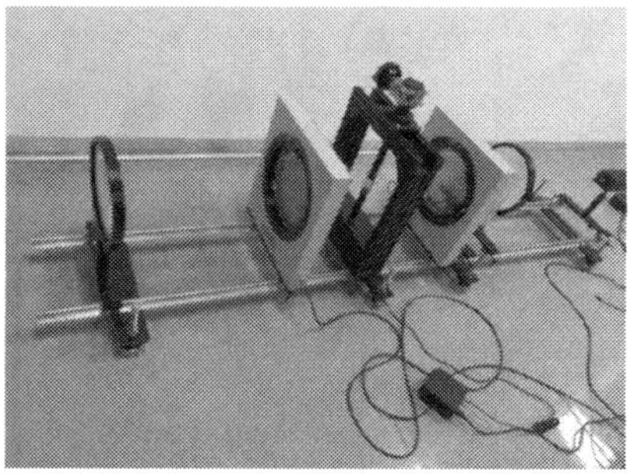

Figure 2. photoelasticimeter.

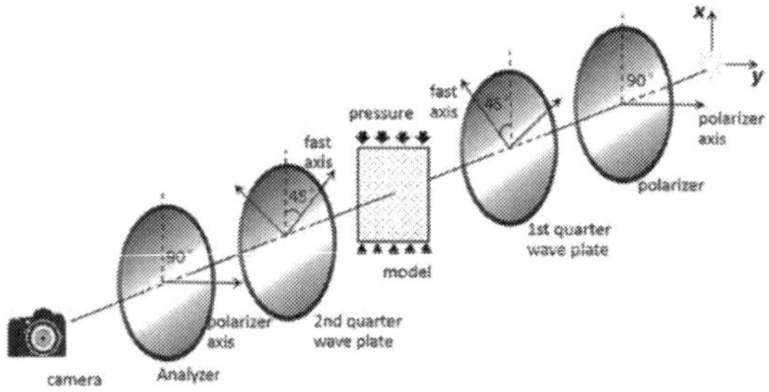

Figure 3. Schematic diagram of the dark field optical layout of photoelasticimeter.

Before loading, no fringe pattern is displayed on the disc. The fringes appear constantly from the upper and lower ends and move towards the side boundaries with the increase of load. The corresponding load was recorded under the circumstance of the integer fringes reached the center of the disk. The high definition camera captured the fringe pattern of the circular disc under load. Figure 4 shows the pictures of the radially loaded disk with one to four fringe orders. The final result of fringe constant f is obtained by averaging the four test values with a view to measurement error. The fringe constant of photosensitive material is determined according to formula (3) [8].

$$f = \frac{8P}{\pi n d} \tag{3}$$

where d is the diameter of the circular disc, n is the fringe order at the center of the circular disc, and P is the corresponding load.

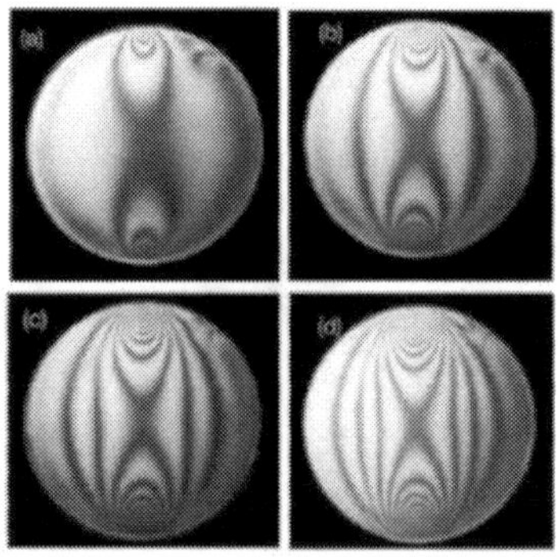

Figure 4. The pictures of the radially loaded disk with one to four fringes.

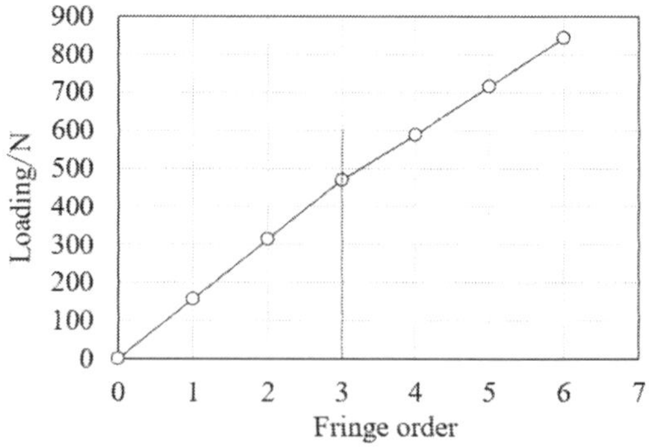

Figure 5. Evolution of the fringe order at the center of the disc according to the loading.

Figure 5 shows the relationship between the loading at the center of the specimen and fringe order, where the slope of the broken line decreases after the point A. This phenomenon is caused by the specimen having plasticity at the end point. The fringe constant is determined as 7.986N/mm.

2.2. Photoelastic Testing

The experiment to obtain fringe constant is also the experiment to obtain the principal stress difference at point A and point B as shown in *figure* 6. The load when the fringe passes through point A and point B is recorded, thus the experimental data corresponding to the fringe order and the load are obtained. The specific data are shown in table 1. According to formula (3), the principal stress differences at point A and point B under a specific load are obtained.

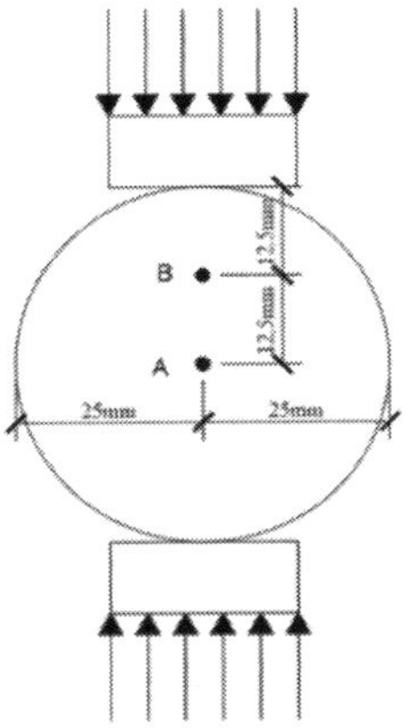

Figure 6. Schematic diagram of the test points.

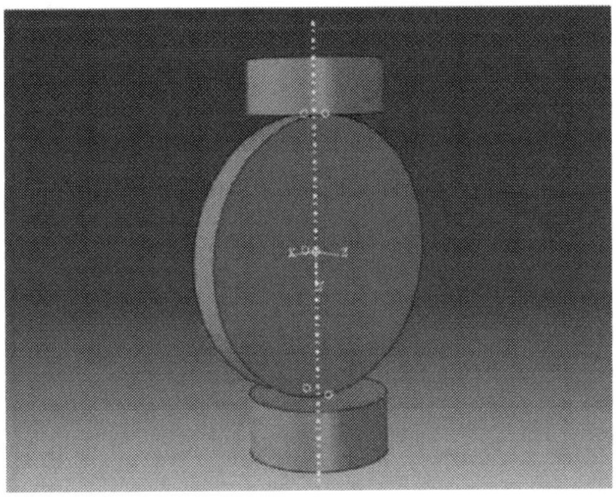

Figure 7. The FEM model.

2.3. Numerical Simulation

This model cannot be reduced to a two-dimensional model because the sizes of the model in three dimensions are close. Therefore, we set a three-dimensional model as shown in figure 7. To keep consistency with the photoelastic experiment, we set up a loading table with a relatively large elastic modulus to exert uniform load on the specimen. The elastic modulus and Poisson's ratio of the specimen are set to be the same as the experiment, which are 2.3 GPa and 0.39. The horizontal displacements of the upper and lower loading tables are constrained, only the vertical displacement of the loading tables is permitted, in order to simulate the actual loading process. Actually, principle stress difference cannot be obtained directly in the post-processor module. Consequently, data processing of σ_x, σ_y and τ_{xy} is shown as follows in order to convert into principle stress difference.

$$\begin{cases} \sigma_1 = \dfrac{\sigma_x + \sigma_y}{2} + \sqrt{\dfrac{(\sigma_x - \sigma_y)^2}{4} + \tau_{xy}^2} \\ \sigma_2 = \dfrac{\sigma_x + \sigma_y}{2} - \sqrt{\dfrac{(\sigma_x - \sigma_y)^2}{4} + \tau_{xy}^2} \\ \sigma_1 - \sigma_2 = \sqrt{(\sigma_x - \sigma_y)^2 + 4\tau_{xy}^2} \end{cases} \tag{4}$$

2.4. Results and Comparison

Table 1. Synthesis of principle stress difference of point A and point B.

point A				point B			
Load (N)	photoelastic testing (MPa)	Numerical simulation (MPa)	Relative error (%)	Load (N)	photoelastic testing (MPa)	Numerical simulation (MPa)	Relative error (%)
156.8	1.3	1.4	3.8	137.2	1.3	1.4	11.3
313.6	2.7	2.8	3.7	254.8	2.6	2.7	3.3
470.4	3.9	4.1	3.6	382.2	3.9	4.1	3.3
588	5.3	5.2	2.8	499.8	5.3	5.4	1.3
715.4	6.7	6.3	5.7	617.4	6.7	6.6	0.2
842.8	7.9	7.4	7.6	725.2	7.9	7.8	1.9

As shown in table 1, the data sets obtained by these two methods are closely matched. Most relative errors are lower than 4%, in addition to some outlier highlighted in red. In other words, the finite element method has been proved to provide a basis for reliable mechanical analysis.

3. Uniaxial Compression Testing

3.1. Modelling Procedure

A series of numerical models were set to study the mechanical response of rock joints under uniaxial compression so as to obtain the relationship between JRC and the maximum shear stress of the rock. The roughness of the rock joint is simulated through changing the shape of the contact surface. JRC is calculated by formula (5).

$$JRC = 49.2114e^{\frac{29L_0}{450L_n}} \arctan\left[\frac{8R_y}{L_n}\right] \tag{5}$$

where $l_0 = 10cm,\ R_y\ is$ tooth amplitude, $L_n$is the length of the profile curve. The detailed geometry property of the models is displayed in table 2.

Table 2. Geometry property of the models.

Model	Height (mm)	Width (mm)	R_y (mm)	JRC
1	100	100	0.07	3
2	100	100	0.14	6
3	100	100	0.22	9
4	100	100	0.29	12
5	100	100	0.37	15
6	100	100	0.45	18

Rock type varies in nature. In this study, we select granite, one of the most widely existed rock, as research object. However, the density, elastic modulus and Poisson's ratio of granite are not certain due to the environmental complexity, these values are in a range. For the convenience of research, we set the density of granite as 3×10^9 t/mm^3, the elastic modulus as 75GPa, and the Poisson's ratio as 0.2. Meanwhile, we only consider the relationship between JRC and the maximum shear stress in elastic phase. Considering that this case satisfies the plane strain principle, to eliminate the complexity of computation, these test models are set in 2D space.

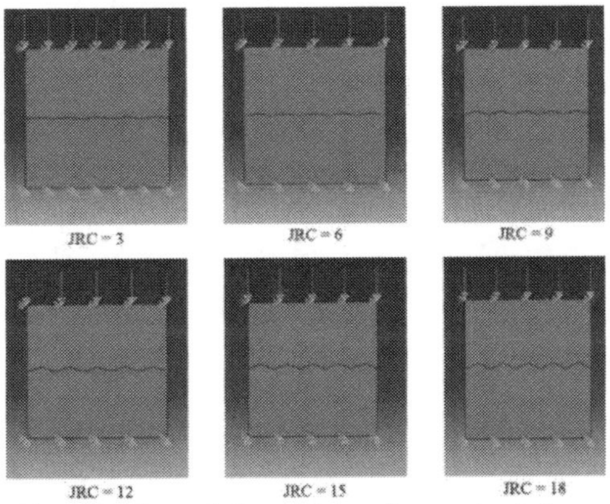

Figure 8. The FEM models subjected to uniaxial compression.

The accuracy of boundary conditions and contacts has a great impact on the reliability of numerical simulation. Complex boundary conditions and contacts are not easy to control in modeling, which may lead to the inaccuracy of numerical simulation [9]. In this study, the boundary conditions of the model are established to make the boundary conditions of the model as consistent as possible with the actual engineering onditions. The detailed boundary conditions are as follows:

- The maximum pressure applied on top and bottom surface of the model is 10MPa.
- The horizontal and vertical displacements of top surface and bottom surface of the specimen are both fixed to enable the uniaxial compression process.
- The type of sample surface interaction is defined as a surface-surface contact with finite sliding. Due to the roughness of the surface of the two specimens, the tangential behavior of contact interaction was set as punishment, and the friction coefficient was 0.4. The normal behavior of contact interaction is set to hard contact, and separation is not allowed after contact.

Due to the discontinuity of specimens, the mesh of finite element model is of great significance. It seems impossible to divide complex models in a high-quality structured way, in which case unstructured grids are a better alternative [10]. Considering the complexity of the model, we combine uniform grid with adaptive grid refinement to improve the performance of simulation.

3.2. Results and Discussion

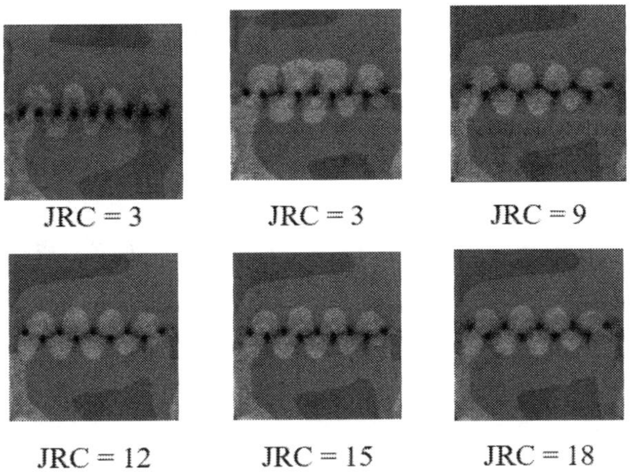

Figure 9. Mises stress plots of the FEM models under uniaxial compression.

As figure 9 shows, the element sizes are different for different regions. The reason is that the adaptive grid partitioning technology can subdivide the grid according to the stress level to meet the requirement of computational accuracy. In other words, stress concentration occurs in a dense grid. The simulation results show that the stress in joint surface area is higher than other areas, and the stress field distribution characteristics have not changed significantly with the increase of JRC.

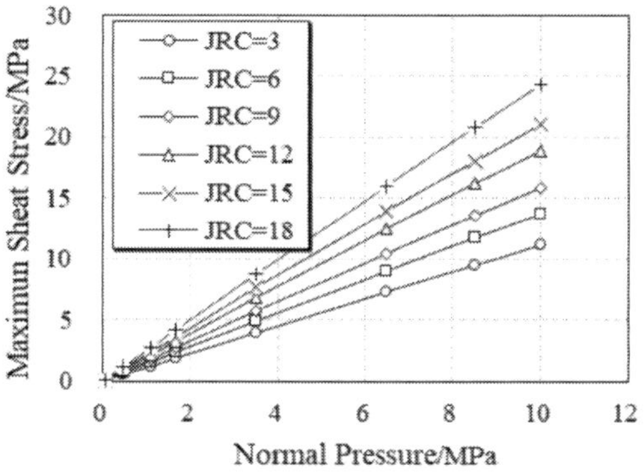

Figure 10. The curves of the maximum shear stress of the models with different JRC.

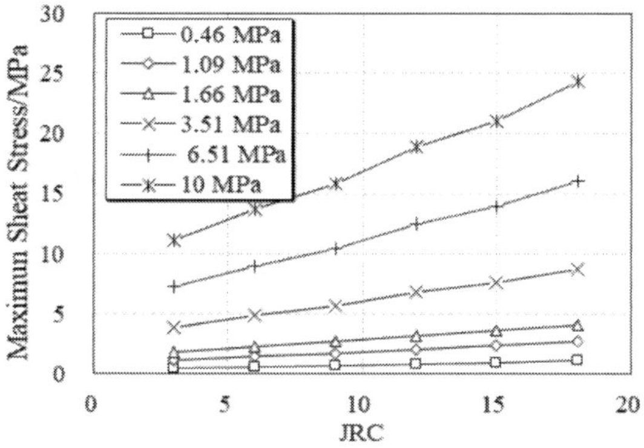

Figure 11. The curves of the maximum shear stress of the models with different normal pressure.

Firstly, it can be determined from figure 10 that the maximum shear stress is proportional to normal pressure. In particular, different roughness, different scale constant. Secondly, the relationship between the maximum shear stress and JRC will be discussed. Figure 11 illustrate the fitting curves of maximum shear stress of the models with different JRC under uniaxial compression. One surprising finding is that the maximum shear stress is proportional to JRC at a given load, although it is well known that the maximum shear stress increases with normal pressure.

4. Shear-Compression Testing

4.1. Modelling Procedure

To be consistency with the former simulation, there are not any changes in the geometry and material properties compared with former models. Considering the actual situation and the convergence of the model, the boundary conditions of the model under shear and compression are more complicated. The detailed boundary conditions are as follows:

- The maximum pressure exerted on the top surface of the specimen is 10MPa.
- Horizontal uniform loads of 20MPa were applied to the left and right sides of the specimen to obtain the mechanical response of rock joints under shear loads.
- The horizontal displacement of the top surface of the specimen is fixed to exert the axial compression.
- All the freedoms of the bottom surface of the specimen is constrained.
- The type of surface interaction between two specimens is also defined as surface-surface contact with finite sliding. The tangential behavior of contact interaction is set as penalty and the friction coefficient is 0.4 due to the roughness of two sample surfaces. The normal behavior of contact interaction is set to "hard" contact, and separation is not allowed after contact.

For the sake of improving the precision of the simulation results, adaptive mesh refinement methods are also applied in these models.

4.2. Results and Discussion

The simulation results are similar to the uniaxial compression testing, the stress in joint surface area is higher than other areas, and the stress field distribution characteristics have not changed significantly with the increase of JRC.To begin with, shear-compression ratio n is defined as follows.

$$n = \frac{shear\ pressure}{normal\ pressure} \tag{6}$$

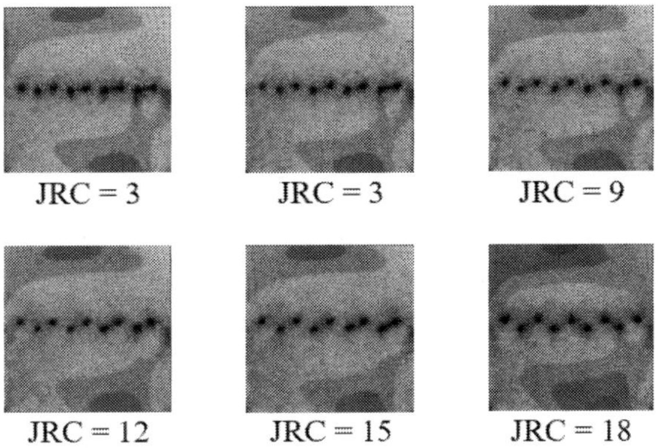

Figure 12. The Mises stress plots of the FEM models under shear- compression.

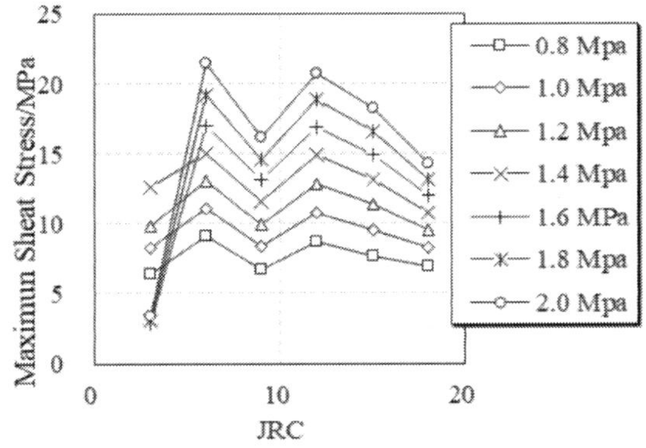

Figure 13. The curves of the maximum shear stress of the models with n > 0.6.

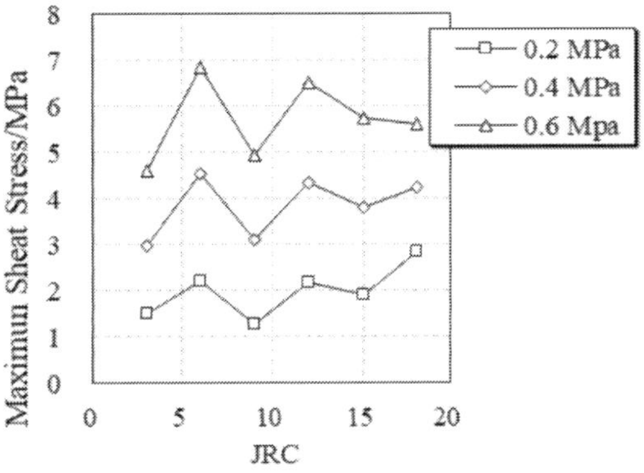

Figure 14. The curves of the maximum shear stress of the models $n \leq 0.6$.

Next, cases where n is more than 0.6 will be discussed. There are two maximal points for all curves and the maximal points are located at the position where JRC equal to 6 and 12. When JRC is less than 6, the maximum shear stress increases with the increase of JRC, and when JRC is greater than 12, the maximum shear stress decreases with the increase of JRC. It can be seen that when JRC is greater than 6 and less than 12, the joints are more prone to stress concentration. The tendency of the curves shown in figure 14 differ from the curves shown in Figure 13 when JRC is greater than 15. In uniaxial compression testing, we get a conclusion that the maximum shear stress is proportional to JRC under a given load. Thus, when n is quite small and JRC is greater than 15, the maximum shear stress is nonlinear proportional to JRC.

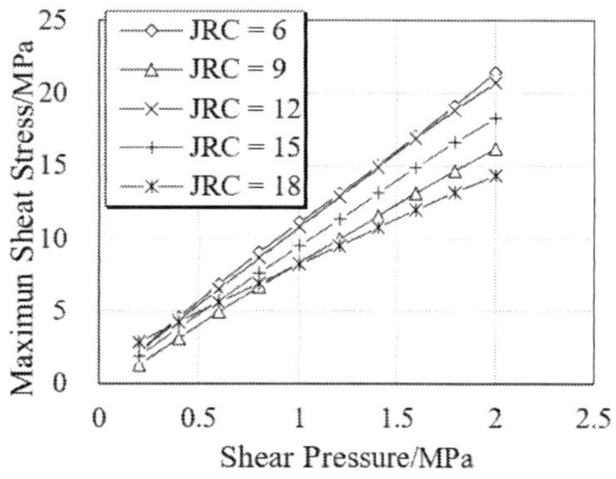

Figure 15. The curves of the maximum shear stress of the models with JRC > 3.

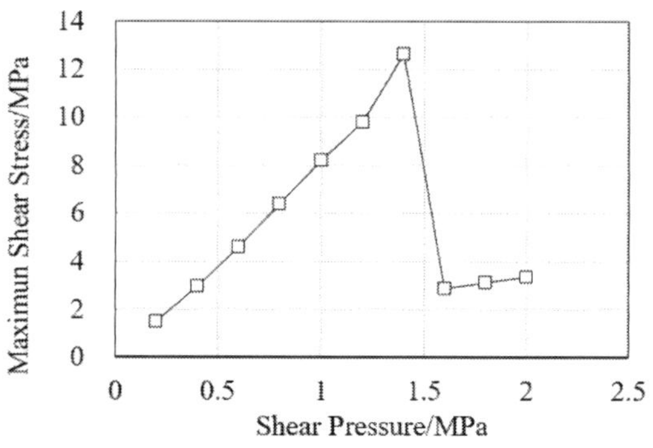

Figure 16. The curve of the maximum shear stress of the model with JRC = 3.

According to figure 15, we can determine that the maximum shear stress is proportional to shear pressure when JRC is greater than 3. The mutation of the maximum shear stress when JRC equals 3 and shear pressure equals 1.4 MPa is shown in figure 16. The reason is that there is a relative slipping between the top part and the bottom part of the model. Whether occurred slipping is determined by JRC and n. There is a threshold that slipping will happen when JRC is lower than the threshold and n is big enough. Conversely, slipping will not happen when JRC is bigger than the threshold. The phenomenon is referred as self-locking.

5. Conclusion

The following conclusions were drawn:

- The stress in joint surface area is higher than other areas, and the stress field distribution characteristics have not changed significantly with the increase of JRC. The shear strength of the joint controls the shear strength of the rock.
- Under uniaxial compression, the maximum shear stress of the model is proportional to JRC and the normal pressure, and the greater JRC, the greater the proportional constant.
- Under shear-compression, when n is relatively small, the maximum shear stress is positively correlated with the JRC nonlinearity. With the increase of JRC, the positive correlation gradually weakens. When n is relatively large, the relationship curve between the maximum shear stress and JRC is parabolic. When JRC is small, as n increases, the maximum shear stress changes abruptly, and the maximum shear stress after the abrupt change decreases significantly.
- Under the same loading conditions, when JRC is greater than 6 and less than 12, the joints are more prone to stress concentration.

References

[1] Chen S F, Liu D S, Wang S R. 2006 Static photo-elastic experiment study on maintenance of roadway with relieving blasting [J] *Journal of Liaoning Technical University* (Natural Science Edition) **25** (6) 873-875.

[2] Xie W H, Li H Q, Zhang Y Z, et al. 1997 A photoelastic investigation on the effect of roughness on the stress field of fractal rock joint [J] *International Journal of Rock Mechanics and Mining Sciences* **34** (3-4) 341. e1-341. e8.

[3] Ju Y, Wang L, Xie H, et al. 2017 Visualization and transparentization of the structure and stress field of aggregated geomaterials through 3D printing and photoelastic techniques [J]. Rock Mechanics and Rock Engineering **50** (6) 1383-1407.

[4] Lu Y S, Zhang H M, Lian Z Y. 2004 Experimental study on interaction of stress wave and interface using dynamic photoelasticity [J] *Explosion and Shock Waves* **24** (4) 352-355.

[5] Dan D Q, Konietzky H. 2014 Numerical simulations and interpretations of Brazilian tensile tests on transversely isotropic rocks [J] *International Journal of Rock Mechanics and Mining Sciences* **100** (71) 53-63.
[6] Weihong X, Xianghui G, Yu'an C. 1996 Photoelastic Study of Mechanical Properties of Fractal Joints [J] Hydrogeological Engineering Geology **23** (6) 11-13.
[7] Wu J B, Xie H P, Gao F. 2000 Photoelastic Analysis of Shearing Property of Rock Joints [J] *Journal of China University of Mining & Technology* **29** (6) 640-642.
[8] Lei Z, Yun H, Kang Y. 2009 Automatic evaluation of photoelastic fringe constant by the nonlinear least-squares method[J]. Optics & Laser Technology **41** (8) 985-989.
[9] Chalivendra V B, Hong S, Arias I, et al. 2009 Experimental validation of large-scale simulations of dynamic fracture along weak planes [J] *International Journal of Impact Engineering* **36** (7) 888-898.
[10] Jones M T, Plassmann P E. 1997 Adaptive refinement of unstructured finite-element meshes [J] *Finite Elements in Analysis & Design* **25** (96) 41-60.

Numerical Simulation Analysis of Deep Foundation Pit Engineering Based on HS-Small Constitutive Modell

Dawei Liu [1*], Xinsheng Li [1] and Dongzhou Xie [2]

[1] College of Geological Engineering and Geomatics, Chang'an University, Xi'an 710054, Shaanxi, China

[2] Institute of Rock and Soil Mechanics, Chinese Academy of Sciences, Wuhan, 430071, Hubei, China

Email: Liudw1103@163.com

Abstract. Numerical simulation has become one of the most important means of foundation pit engineering analysis, but there is a lack of numerical simulation of deep foundation pit engineering examples under the condition of adjacent buildings in Xi ' an. In this paper, MIDAS GTS NX software is used to simulate the deep foundation pit engineering of Xi ' an International Conference Center Phase II by using the Hardening Soil-Small constitutive model which is more suitable for foundation pit engineering. Combined with the measured data, the study on the characteristics of building settlement, the characteristics of surrounding surface settlement, the deformation characteristics of foundation pit retaining structure, the soil pressure at pile side and the variation characteristics of pile internal force is conducted under the condition of stepwise excavation- reduction of groundwater level of foundation pit. The results show that the top displacement of the retaining structure and surface settlement near the building are larger than those without load. The building settlement and surface settlement rate present the characteristics of medium-term slowdown. The lateral earth pressure and internal force of pile show different trends in free section and anchorage section, affected by building load. The simulation results of the constitutive parameters of HHS numerical simulation determined by parameter inversion are in good agreement with the measured data, which verifies the reliability of HHS constitutive model in the numerical simulation of foundation pit engineering in Xi ' an area, and provides reference for the numerical simulation of similar engineering cases in relevant areas.
Keywords. numerical simulation, deep foundation pit engineering, adjacent building, Hardening Soil-Small constitutive model.

1. Introduction

With the rapid development of urban construction, urban land is becoming increasingly tense, and the construction of foundation pits under complex conditions adjacent to buildings is gradually increasing. Under complex engineering conditions, while ensuring the stability of foundation pit, it is more necessary to ensure the normal use of buildings, which puts forward higher requirements for the surface settlement and building settlement control caused by foundation pit excavation.

In recent years, computer numerical simulation has been widely used in engineering practice. Numerical simulation method has become an important means to analyze the environmental impact of foundation pit excavation under sensitive environment. In geotechnical engineering analysis and calculation, the constitutive model and calculation parameters have a significant impact, and the key is to select the appropriate soil constitutive model and calculation parameters. Selecting appropriate soil constitutive model and calculation parameters is of great significance for the establishment of engineering practical constitutive model considering the influence of soil types and regional characteristics.

Relevant research [1-3] show that in the foundation pit engineering, except for a few areas where obvious plastic deformation occurs, most other areas are in a small strain state, and the small strain range is about 10^{-6}~10^{-3}. Moreover, the strain range of soil in the state of complete stiffness (initial small strain shear modulus G_0) is very small, and the soil has great stiffness when the strain is very small. However, with the increase of shear strain, the stiffness presents the characteristics of nonlinear sharp decrease.

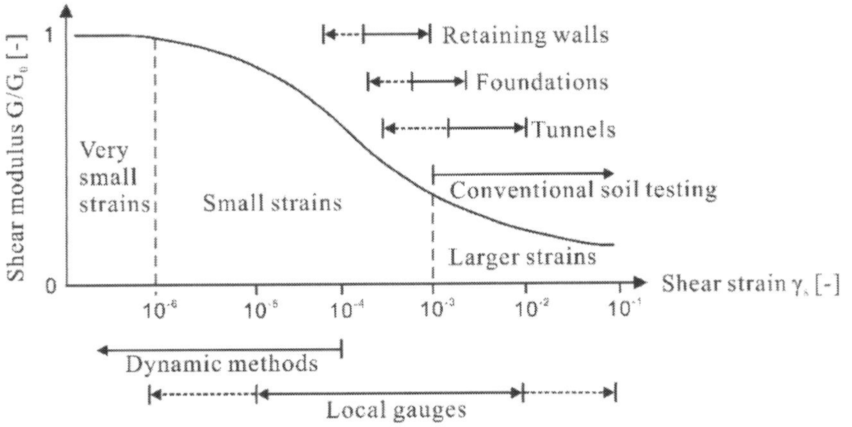

Figure 1. The variation characteristics of soil shear modulus ratio with shear strain (Benz [1]).

Hardening Soil-Small (hereinafter referred to as HHS constitutive model) is based on the Hardening Soil constitutive model proposed by Schanz [4] considering shear hardening and compression hardening of soil, and Benz [1] is based on the research results of Atkinson on small strain test of soil. The small strain shear modulus G_0^{ref} and the shear strain $\gamma_{0.7}$ of the soil stiffness modulus in the small strain range are increased. This model can be more suitable for the actual description of the stiffness characteristics of the soil in the small strain range, so it has good applicability in the numerical simulation of foundation pit engineering dominated by soil small strain. Shi Youzhi et all. [5] used HSS model to carry out sensitivity analysis of small strain parameters for typical soil layers in Xiamen, and studied the evolution law of parameters with excavation. Li Lianxiang et al. [6] used the small strain model of hardened soil for numerical simulation of foundation pit engineering on typical strata in Jinan area, and obtained the general selection method of HSS model parameters of soil under the typical strata according to the actual monitoring data and displacement back analysis technology. Wang Weidong et al. [7] preliminarily and completely obtained the HSS model parameters of typical soil layers in Shanghai by using numerical analysis of foundation pit excavation in Shanghai combined with laboratory tests and parameter sensitivity analysis. Zhang Weibing et al. [8] carried out a numerical analysis on the deep foundation pit excavation-reduction of groundwater level process of an underground garage in Yinchuan, which was supported by anchors. By comparing the simulation results with the measured data, the parameter values of HSS model of typical soil layers in Yinchuan were obtained.

There are few examples of numerical simulation of foundation pit engineering using HSS constitutive model in Xi ' an area. Taking the second phase project of Xi ' an High-tech International Conference Center as an example, the HSS constitutive model is used to simulate the excavation- reduction of groundwater level process of the foundation pit engineering. Combined with the monitoring data, the settlement law of the surface and buildings around the foundation pit, the deformation law of the retaining structure and the distribution law of the soil pressure and the internal force of the pile are discussed under the conditions of adjacent buildings and graded excavation- reduction of groundwater level. Verify the reliability of numerical simulation of foundation pit engineering based on HSS constitutive model and the applicability of model parameters selection in Xi ' an area.

2. Project Profile

2.1. Site Location and Engineering Conditions

The site of Phase II of Xi ' an High-tech International Conference Center is located in the southwest corner of the intersection of Tiangu 6 Road and Yunshui 1 Road in Xi ' an. The excavation shape of the foundation pit is irregular polygon. The maximum width in the east-west direction is 132.3 m, and the maximum width in the north-south direction is 134.1 m. The average depth of foundation pit excavation is 19.1 m. The left side of the excavation pit is close to the high-tech conference center building of the stage I project, and the spacing is 16.0 m. The height of the left building is 24.0 m, and the frame shear wall structure is in the form of pile foundation. The buried depth of foundation is − 7.0 m, and the design value of base load is 180 kPa.

2.2. Hydrogeological Conditions

(1) Artificial fill soil Q_4^{ml}: yellow-brown, mainly clay, and uneven soil. The average thickness of this layer is 3.9 m.

(2) Loess soil Q_4^{al+pl}: yellow-brown, slightly wet, plastic. The soil is uniform and the layer is distributed continuously in the field. The average thickness of the layer is 5.1 m.

(3) silty clay Q_4^{al+pl}: yellow-brown, slightly wet-saturated. The soil is homogeneous with an average thickness of 7.8 m.

(4) silty clay Q_3^{al+pl}: gray-brown gray, hard-plastic. The soil is relatively uniform, and the average thickness of this layer is 11.7 m.

(5) Powder clay Q_2^{al+pl}: light gray-gray yellow, plastic-hard plastic, saturated. The soil is uniform with an average thickness of 8.6 m.

The stable groundwater level depth of the site is 10.9 -14.2 m. The groundwater type belongs to phreatic water. The recharge mainly comes from atmospheric reduction of groundwater level and underground runoff. The discharge modes are mainly natural evaporation, artificial mining and underground seepage. The annual variation range of phreatic water level is 2.0 m.

2.3. Formatting Author Affiliations

The main retaining structure system of foundation pit adopts the combined support of bored pile and prestressed anchor cable. Bored cast-in-place piles are mainly of Φ900 @ 1800 and Φ800 @ 1600 sizes, with an average depth of 9.2 m. The excavation depth of the foundation pit is 3.5 m, and five prestressed anchor cables are set for each 3.0 m. The length is 15.0 -18.0 m, and the horizontal spacing is 1.6 -1.8 m with the center distance of bored pile. The stable groundwater level of the site is 10.9 -14.2 m. The dewatering of the foundation pit adopts the method of tube well dewatering, and the reduction of groundwater level is graded with the excavation of the foundation pit.

2.4. Monitoring Scheme

The west side of the foundation pit is adjacent to the built conference center of the phase I building, and the engineering conditions around the foundation pit are complex. The excavation- reduction of groundwater level work of the foundation pit needs to be alternately carried out, and the deformation needs to be strictly controlled. Therefore, the deformation monitoring and stress monitoring are carried out in the excavation-reduction of groundwater level project. The monitoring contents include settlement of surrounding buildings, settlement of surrounding surface, settlement of retaining structure, horizontal displacement of retaining structure, soil pressure of pile side and internal force of pile body. The layout of monitoring points is shown in figure 2.

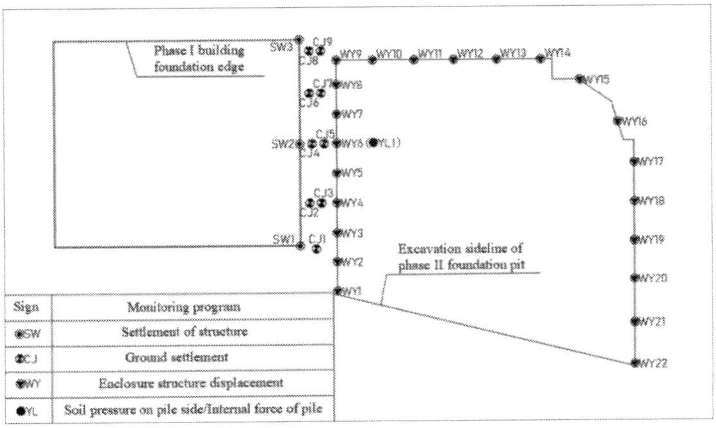

Figure 2. Layout Plane Diagram of Monitoring Points.

Figure 3. Model unit and grid division.

3. Numerical Model Establishment and Parameter Selection

3.1. Model Meshing and Determination of Boundary Conditions

In this paper, the finite element software MIDAS GTS NX is used to simulate the construction condition of foundation pit. In order to simulate the actual working condition more accurately, three-dimensional model is used to simulate.

According to the actual size of the second phase of the foundation pit project and the surrounding buildings, the plane size of the foundation pit is taken as the actual size, and the maximum range of the influence area of the foundation pit project is the possible influence area, that is, the range of 2.0- 3.0 H around the foundation pit. H is the design depth of foundation pit, and the average depth of foundation pit excavation in the second phase of the high-tech international conference center is 19.1 m, so the maximum range of foundation pit engineering is 3.0 H, 57.3 m. The overall size of the model is 350 × 250 × 60 (m). Fixed boundary is applied around and at the bottom of the model, and the initial node water head of the model is - 9.0 m. The foundation pit is divided into four times of reduction of groundwater level, 3.0 m each time, and the final node head in the pit is - 21.0 m. The mesh size of the foundation pit and the first-stage building is 6.0 m, and the mesh size of the model boundary is 12.0 m. The element and mesh division are shown in figure 3.

3.2. Selection of Constitutive Model Parameters

In this numerical simulation, the soil model adopts the HHS constitutive model to simulate the excavation- reduction of groundwater level of groundwater level process of foundation pit. The HHS constitutive model parameters include three strength parameters: effective cohesion c', effective internal friction angle φ', dilatancy angle ψ, and nine stiffness parameters: the reference tangent (compression) modulus E_{oed}^{ref} of consolidation test, the main loading reference tangent modulu sE_{50}^{ref} of triaxial consolidation drainage test, the unloading and reloading reference modulus E_{ur}^{ref}, the unloading and reloading Poisson's ratio v_{ur}, and the reference stress level p^{ref}. Reference stress level correlation power exponent m, failure ratio R_f static earth pressure coefficient K_0. Small strain parameters: refer to the initial small strain shear modulus G_0^{ref} and shear strain threshold $\gamma_{0.7}$, the stress level when the secant modulus decreases to 70%G_0^{ref}.

The five general parameters determined on the basis of geotechnical experiments reported by geotechnical investigation are bulk density γ, saturated bulk density γ_{sat}, void ratio e_0, permeability coefficient k, and average compression modulus E_s^{1-2} of clay under 100kPa- 200kPa pressure. The parameters determined by geotechnical test are effective cohesion c', effective internal friction angle φ 'and dilatancy angle ψ. According to the empirical relationships of reference tangent compression modulus E_{oed}^{ref}, reference secant modulus E_{50}^{ref} and unloading-reloading reference modulus E_{ur}^{ref} by Wang et al. [9], the reference tangent (compression) modulus $E_{oed}^{ref}=E_s^{1-2}$, reference secant modulus $E_{50}^{ref}=1.5E_{oed}^{ref}$, unloading-reloading reference modulus $E_{ur}^{ref}=3.0E_{50}^{ref}$ were preliminarily selected. The value range of unloading and reloading Poissons' ratio (0.10- 2.50) is taken as v_{ur}=0.2[9], the reference stress level p^{ref}=100kPa, and the value range of reference stress level correlation power index for cohesive soil *m* is (0- 1), taken as *m*= 0.5[10], the failure ratio R_f=0.9[11], and the static earth pressure coefficient $K_0=1\text{-}sin\varphi'$[12]. The reference initial small strain shear modulus G_0^{ref} in small strain parameters is determined according to the empirical formula $G_0^{ref}=33\times(2.97\text{-}e_0)^2/(1+e_0)$, by Hardin [13]. The shear strain threshold $\gamma_{0.7}=0.0001+5\times10^{-6}I_p\times(OCR)^3$ for clay.

On this basis, the model parameters are selected for parameter inversion analysis based on the measured foundation pit monitoring data, and the parameter values are determined, which are limited to space. The calculation parameters of soil HHS constitutive model are shown in table 1. Among the parameters of soil model, the Poisson's ratio of unloading and reloading v_{ur}=0.25, the power index of reference stress level correlation m=0.5, the reference stress level p^{ref}=100kPa, and the failure ratio R_f=0.9 are not listed in table 1. The calculation parameters of component materials are shown in Table 2. The retaining pile structure is converted by equal stiffness, and the underground continuous wall element is used for simulation.

Table 1. Soil calculation parameters of HHS constitutive model.

soil horizon	γ/ (kN·m^{-3})	γ_{sat}/ (kN·m^{-3})	e_0	K/ (m·d^{-1})	E_s^{1-2}/ (Mpa)	c'/ (kPa)	φ'/ (°)	ψ/ (°)	E_{oed}^{ref}/ (Mpa)	E_{50}^{ref}/ (Mpa)	E_{ur}^{ref}/ (Mpa)	K_0	G_0^{ref}/ (Mpa)	$\gamma_{0.7}$/ (×10^{-4})
Artificial fill soil①	17.8	19.1	0.818	6.0	15.0	8.0	13.0	0	15.0	22.5	67.5	0.775	84.1	1.52
Loess soil②	17.5	19.1	0.829	4.0	5.9	29.7	20.5	0	5.9	8.85	26.55	0.649	82.7	1.57
Silty clay③	18.0	19.2	0.797	5.0	6.9	31.7	21.7	0	6.9	10.35	31.05	0.630	86.7	1.59
Silty clay④	19.1	19.7	0.712	4.0	7.2	31.4	21.7	0	7.2	10.8	32.4	0.630	98.3	1.62
Powder clay⑤	19.3	19.8	0.702	8.0	6.9	31.9	22.8	0	6.9	10.35	31.5	0.612	99.7	1.64

Table 2. Calculation parameters of structural materials.

Construction materials	Unit type	γ/(kN·m^{-3})	E/(Mpa)	Sectional dimension/(mm)
Building basement wall	Plate	25	300	T=1000
Building raft	Plate	25	300	T=300
Building raft pile	Implanted beam	25	300	D=500
Crown beam	Beam	25	300	H=700 B=500
Anchor bar	Implantable truss	78	2060	D=300
Diaphragm wall	Plate	25	300	T_1=532 T_2=598

3.3. Construction Condition Definition

The construction conditions are set according to the actual sequence of engineering reduction of groundwater level-excavation, and the numerical simulation of construction conditions is shown in table 3.

Table 3. Numerical Simulation of Construction Conditions.

Stage number	Stage name	Stage setting	Stage number	Stage name	Stage setting
1	Initial condition Settings	Initial seepage field and stress field analysis	7	Reduction of groundwater level-2	Groundwater level decreased to -15.0 m
2	Construction before excavation	Construction of Phase I Buildings, Crown Beams, Underground Continuous Walls	8	Excavation of foundation pit-4	Excavation to -13.0 m and application of fourth anchor
3	Excavation of foundation pit-1	Excavation to - 4.0 m and application of first anchor	9	Reduction of groundwater level-3	Groundwater level decreased to -18.0 m
4	Excavation of foundation pit-2	Excavation to -7.0 m and application of second anchor	10	Excavation of foundation pit-5	Excavation to -16.0 m and application of fifth anchor
5	Reduction of groundwater level-1	Groundwater level decreased to -12.0 m	11	Reduction of groundwater level-4	Groundwater level decreased to -21.0 m
6	Excavation of foundation pit-3	Excavation to -10.0 m and application of third anchor	12	Excavation of foundation pit-6	Excavation to -19.1 m

4. Results Analysis

4.1. Analysis of Surrounding Building Settlement

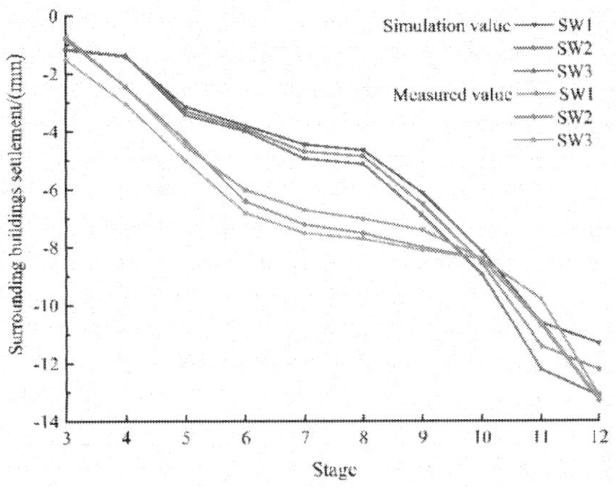

Figure 4 Comparison between measured and simulated settlement values of surrounding buildings.

Figure 4 is the measured data and numerical simulation results of the left building settlement under the condition of stepwise excavation- reduction of groundwater level. It can be seen that the monitoring values of building settlement under different working conditions are generally close to the simulation values. The settlement of adjacent buildings increases with the increase of excavation depth, and the difference between the data of the three settlement monitoring points is small, indicating that the differential settlement of the building is small, which is characterized by overall settlement. In addition, with the gradual increase of excavation depth, the settlement rate of buildings decreases first and then increases, which is mainly due to the lateral stress release of soil and the continuous decrease of groundwater level, and the consolidation and compaction of soil caused by Reduction of groundwater level.

4.2. Analysis of Surrounding Surface Subsidence

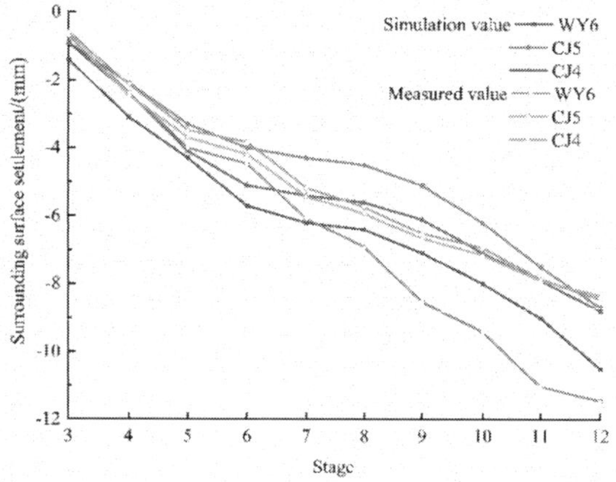

Figure 5. Curve diagram of surrounding ground settlement variation with distance from retaining structure.

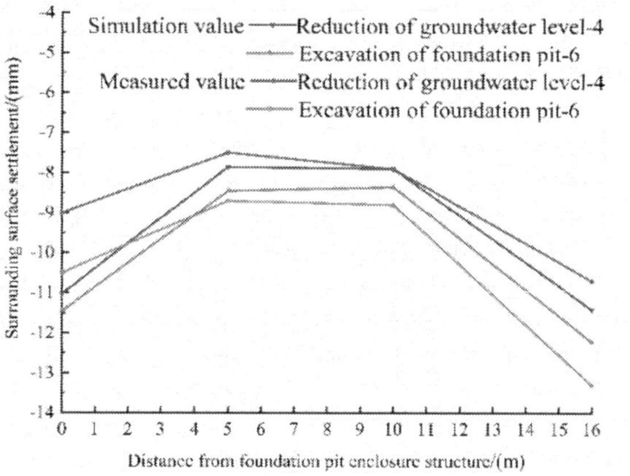

Figure 6. Comparison of measured and simulated values of surrounding ground settlement.

Figure 5 is the monitoring data and simulation results of WY6, CJ5, CJ4 and SW2 at the monitoring points of 0, 5, 10 and 16 m away from the foundation pit retaining structure under the conditions of Reduction of groundwater level -4 and excavation-6 stages. The results show that the surface settlement of the foundation pit retaining structure and the building is larger than that of the non-load. Under the influence of additional loads such as building load and retaining structure load, the compaction consolidation of soil is more obvious and the settlement is larger than that of the non-load.

Figure 6 is the measured data and numerical simulation of surface settlement at WY6, CJ5 and CJ4 monitoring points. The results show that the monitoring value and simulation value of building settlement are generally close under different working conditions. The surface settlement around the foundation pit increases gradually, and the settlement rate decreases first and then increases. By analyzing the influence of prestressed anchor supporting structure, the surrounding surface settlement rate slowed down, and the prestressed anchor had a positive effect on the surface deformation control. Then with the increase of excavation depth, the lateral unloading of soil is intensified, and the control effect of anchor on surface settlement is weakened, which makes the surface settlement rate increase. In the process of stepwise excavation-reduction of groundwater level, the reduction of groundwater level stage is more sensitive to the surrounding surface settlement than the excavation stage.

4.3. Analysis of Horizontal Displacement of Retaining Structure

Figure 7 is the comparison of the measured data and the simulated results of the horizontal displacement at the top of the retaining structure at WY (3-4-5-6-7) near the building side and WY (17-18-19-20-21) far from the building side under the conditions of reduction of groundwater level -4 and excavation-6 stages. It can be seen that the horizontal displacement of the top of the retaining structure of the foundation pit increases to the horizontal displacement of the inside of the foundation pit from the reduction of groundwater level -4 to the excavation-6. Under the action of building load, the lateral displacement from the top of the retaining structure near the building side to the inside of the foundation pit is significantly greater than that from the far building side. The horizontal displacement at the top of the enclosure structure near the building side is a convex curve of ' small at both ends and large in the middle', while there is no influence of building load on the far building side. The horizontal displacement at the top of the enclosure structure is nearly linear.

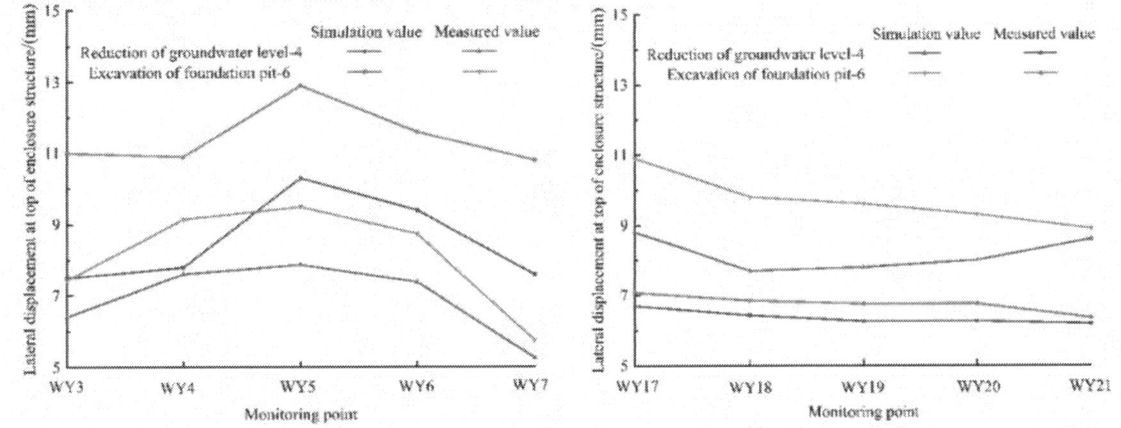

(a) Near the building side (b) Far from the building side

Figure 7. The comparison of the measured data and the simulated results of the horizontal displacement at the top of the retaining structure.

4.4. Analysis of Soil Pressure on Pile Side of Retaining Pile

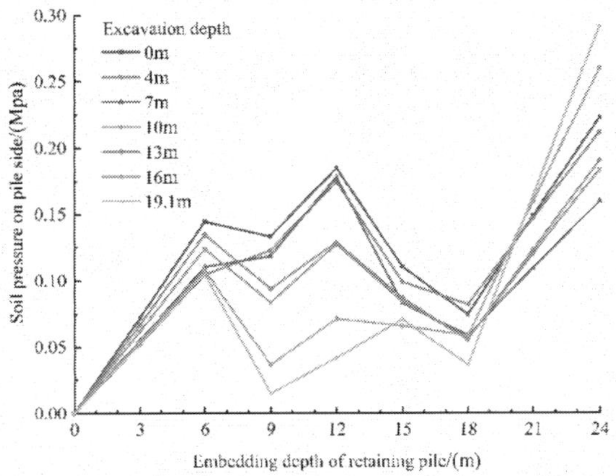

Figure 8. Curve of Soil Pressure on pile side with embedded depth.

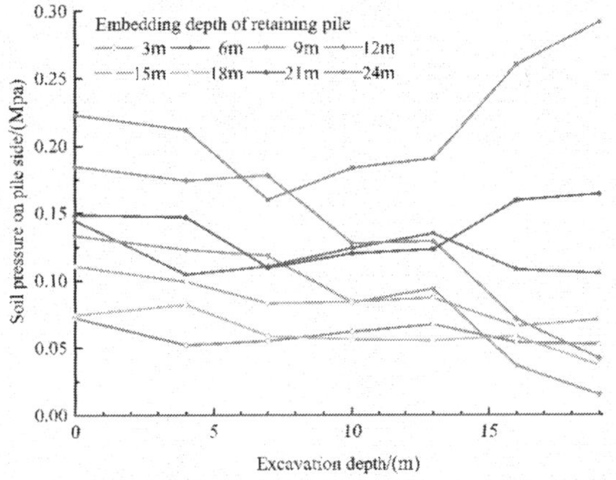

Figure 9. Curve of Soil Pressure on pile side with excavation depth.

Figure 8 shows the variation of lateral soil pressure of retaining pile with the increase of embedded depth under different excavation depths. The monitoring data of pile lateral soil pressure are the lateral soil pressure of foundation pit excavation non-lateral free face. With the increase of excavation depth, the soil pressure on the pile side of the free section of the retaining pile is smaller than the corresponding static soil pressure when the excavation depth is 0 m. With the increasing displacement of the free section of the retaining pile to the inner side of the foundation pit, the active soil pressure on the pile side gradually decreases.

With the increase of the displacement of the free section of the retaining pile to the inner side of the foundation pit, the active soil pressure on the pile side decreases gradually. The buried depth of the retaining pile is greater than that of the foundation pit, that is, the soil pressure on the pile side of the anchorage section of the retaining pile presents a different law from that of the free end. When the foundation pit is excavated in the range of 0-13m, the soil pressure on the pile side of the anchorage section of the retaining pile is the same as that of the free section, which is mainly dynamic soil pressure. When the foundation pit is excavated in the range of 13-19.1m, the earth pressure of anchorage section is greater than the static soil pressure when the excavation depth is0m, and the earth pressure of pile side is only passive soil pressure at this time.

Figure 9 is the curve of the lateral earth pressure of retaining pile with the increase of excavation depth under different buried depths. It can be seen that the lateral earth pressure of the retaining pile at the free end (0-18 m) decreases gradually with the excavation of the foundation pit. When the depth of the retaining pile is 21m and 24m, the soil pressure at the anchorage end decreases first and then increases with the increase of the excavation depth of the foundation pit. The distribution of soil pressure at each excavation stage is also verified in figure 10.

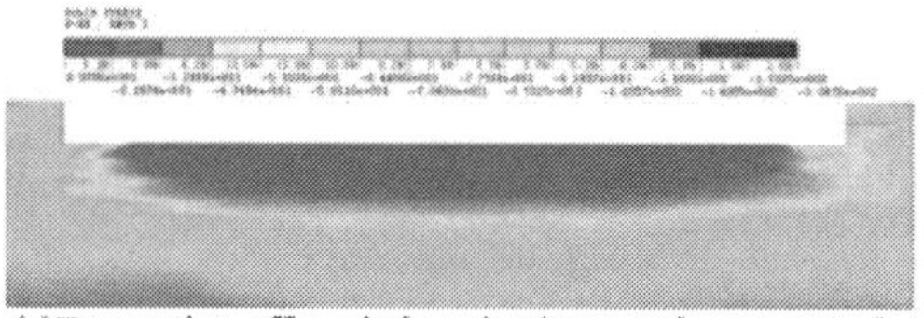

(a)Excavation offoundation pit-2(Excavation to -7.0 m)

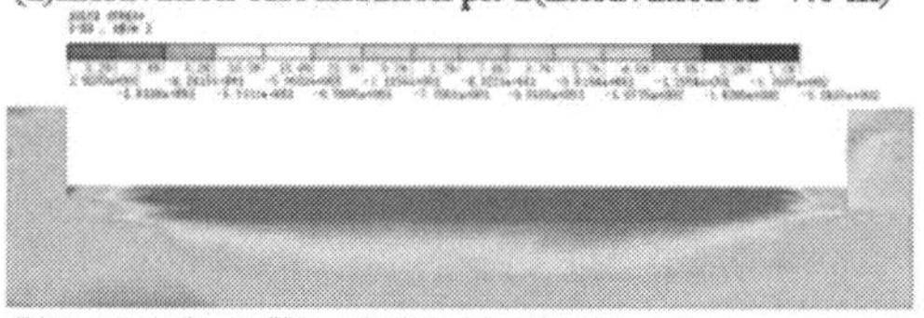

(b)Excavation offoundation pit-4(Excavation to -13.0 m)

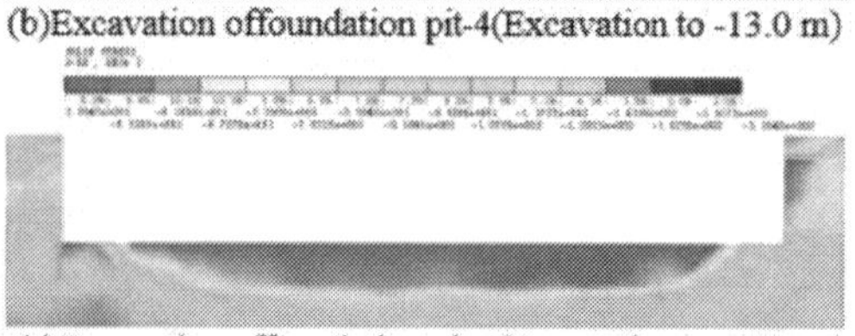

(c)Excavation offoundation pit-4(Excavation to -19.1 m)

Figure 10. Distribution nephogram of earth pressure changing in different excavation stage.

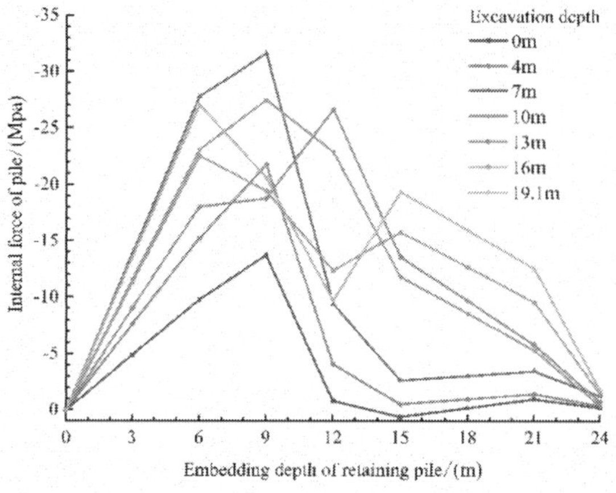

Figure 11. Curve diagram of pile internal force with embedded depth.

4.5. Analysis of Internal Force of Retaining Pile

Figure 11 shows the variation curve of the internal force of the pile on the side of the excavation face at the monitoring point WY6 along with the burial depth, where the internal force of the pile on the excavation side is positive under tensile state. In the measured data, only in the local range near the bottom of the foundation pit excavation, the excavation side is compressed, and the internal force of the retaining pile is characterized by small end and large middle. With the increase of foundation pit excavation depth, the influence range of pile internal force gradually expands from the free end of pile to the anchorage end. However, affected by the lateral free surface of foundation pit excavation, the variation range of internal force at the free end of pile is larger than that at the anchorage section. When the excavation depth is 0-13m, the curve shows a single peak curve, and the peak is concentrated in the range of 6-12m. And with the increase of excavation depth, the peak gradually shifts downward. When the excavation depth is16-19.1m, the internal force of the pile body declines at the depth of 12m due to the anchoring force of the prestressed anchor on the surface soil, and the curve shows a bimodal curve. The peak values appear at the depth of 6m and 15m respectively. With the increase of the distance from the bottom of foundation pit excavation, the internal force of the pile in the anchorage section gradually decreases to 0.

5. Conclusions

Taking the second phase project of Xi 'an High-tech International Conference Center as an example, the HHS constitutive model is used to numerically analyze excavation- the reduction of groundwater level of the deep foundation pit engineering. The following conclusions were reached:

(1) Through the numerical simulation of the excavation- the reduction of groundwater level of the deep foundation pit engineering in Xi ' an area, combined with the measured data, the model parameters are determined by parameter inversion analysis. It verifies the reliability of HHS constitutive model in numerical simulation of foundation pit engineering in Xi ' an area, and provides reference for numerical simulation of similar engineering cases in relevant areas.

(2) Under the influence of building load and retaining structure load, the surface settlement is more sensitive to the settlement caused by foundation pit excavation- reduction of groundwater level than that without load. The settlement rate of surrounding buildings and the surrounding surface first decreases and then increases.

(3) The horizontal displacement of the top of the enclosure structure near the building side is larger than that of the non-building side. Compared with the non-building side, under the influence of building load, the convex expansion curve of ' small at both ends and large in the middle ' is more obvious.

(4) Due to the influence of building load, the lateral earth pressure and internal force of retaining structure pile show different trends in the free section and anchorage section with the excavation of foundation pit.

References

[1] Benz T. 2006 Small-strain Stiffness of Soils and Its Numerical Consequences *University of Stuttgart*.

[2] Burland J B. 1989 Small is beautiful- The stiffness of soils at small strains *Revue Canadienne De Géotechnique* **26** (4) 499-516.

[3] Lanzo G, Vucetic M, Doroudian M, et al. 1997 Reduction of Shear Modulus at Small Strains in Simple Shear[J] *Journal of Geotechnical & Geoenvironmental Engineering* **123** (11) 1035-42.

[4] Schanz T, Vermeer A, Bonnier P, et al. 1999 the hardening soil model: formulation and verification *Beyond in Computational Geotechnics*.

[5] Shi Y Z, Lin S Z, Zhao H L, et al. 2016 Sensitivity analysis of the small strain parameters of the soil in the deep foundation pit excavation effect——Taking Xiamen as an example[J] *Journal of Engineering Geology* **24** (06) 1294-1301.

[6] Li L X, Liu J D, Li K J, et al. 2019 Study on selection and applicability of HSS parameters for typical strata in Jinan[J] *Rock and Soil Mechanics* **40** (10) 4021-29.

[7] Wang W D, Wang H R, Xu Z H, et al. 2013 Study on HS-Small model parameters of soil in numerical analysis of foundation pit excavation in Shanghai [J] *Rock and Soil Mechanics* **34** (06) 1766-74.

[8] Zhang W B, Zhang G Z, Han Z G, et al. 2020 Numerical analysis of deep foundation pit excavation- reduction of groundwater level process in Yinchuan based on HS-Small constitutive mode[J] *Journal of Water Resources and Water Engineering* **31** (02) 211-18.

[9] Wang W D, Wang H R, Xu Z H. 2013 Study on HS-Small model parameters of soil in numerical analysis of foundation pit excavation in Shanghai [J] *Rock and Soil Mechanics* **34** (06) 1766-74.

[10] Song E X, Kong Y F, Yang J. 2016 Discussion on Definition and Calculation Method of Safety Coefficient of Geotechnical Structure [J] *Engineering Mechanics* **33** (11) 1-10.

[11] Janbu N. 1963 Soil compressibility as determined by odometer and triaxial tests *Conf. SMFE* **1** 73-75.

[12] Brinkgreve R B J and BROERE W. 2006 *Proc. Europ. Plaxis material models manual.* (The Netherlands: Delft University od Technology and PLAXISB) 196-197.

[13] Hardin B O and Black W L. 1969 Closure to Vibration Modulus of Normally Consolidated Clays *Journal of the Soil Mechanics and Foundations Division* **95** (SM6) 1531-37.

Influence of Relative Molecular Mass of SBS on Low-Temperature Performance of Modified Asphalt

Ting Peng[1,2], Kesheng Yang[1], Taoyan Zhou[1] and Heng Wu[1]

[1] School of Highway, Chang'an University, Xi'an, Shaanxi, 710064, China
[2] Special Region Highway Engineering Lab, Chang'an University, Xi'an, Shaanxi, 710064, China
Email: 553063793@qq.com

Abstract. Styrene Butadiene Styrene (SBS) is widely used in pavements because it significantly improves the performance of the modified asphalt. The dosage of SBS has an important effect on the low-temperature performance of the modified asphalt. The difference in relative molecular weight influences the magnitude of the change in low-temperature performance. In this paper, low temperature bending beams rheology test of five different relative molecular weight SBS modified asphalt is conducted at -12°C, -18°C and -24°C to evaluate the low-temperature properties of modified asphalt. Then, the experimental results are analysed. The results show that there is a certain correlation between the relative molecular mass of SBS and the low-temperature performance of modified asphalt. The creep stiffness and creep rate of the modified asphalt are influenced by the relative molecular mass of SBS and temperature. As the relative molecular mass increases, the low-temperature performance of the modified asphalt first rises and then decreases.

1. Introduction

As a result of its better driving comfort and low traffic noise, asphalt concrete pavement has the characteristics of short construction period and easy repair, and it also has high reutilization value and becomes the main paving form of highway pavement. However, due to the characteristics of asphalt itself and asphalt concrete structure, under the joint action of vehicle load and climatic factors, asphalt pavement is prone to low temperature cracking and other damage behaviors, and low temperature stability cannot meet the practical requirements. In order to improve the performance of asphalt pavement, various modifiers such as rubber particles, SBS, nano-silica, fiber, etc. are added to the asphalt, and these modifiers can improve the low-temperature performance of the asphalt to some extent, in which SBS modifier can significantly improve the low temperature performance of asphalt, and the modification effect is ideal. With the advancement of scientific research methods, the SBS modified asphalt is gradually deepened, and the degradation law of SBS in SBS modified asphalt is studied [1]. The low temperature performance of SBS modified asphalt after aging is analysed by the rheological behavior of low temperature bending beam [2]. Cao liping [3] studied SBS modified asphalt using phase angle and viscoelastic modulus The low temperature rheological properties, analysis of the effect of modifier content on performance; asphalt temperature sensitivity study shows that when the modifier content is greater than 4%, the permeability index of SBS modified asphalt increases with the content of modifier Linear decline, The temperature sensitivity is increased [4]; the physical equation between the creep stiffness S(t) and the creep rate m(t) at any time t is established by mathematical derivation, and the modification is proposed to have better low temperature performance.

The m value of the agent is proposed to select the asphalt binder according to the S and m values [5]; the creep stiffness (S) and creep rate (m) of the SBS modified asphalt are combined with the raw material performance parameters, and the SBS modified asphalt is analysed. Correlation between raw material performance parameters and low temperature rheological properties [6]; BBR experiments were conducted to analyse the effect of SBS polymer type, SBS content, sulfur content and rubber processing oil addition on the low temperature performance [7].

SBS has great differences in structural composition and parameters, such as its relative molecular mass, Styrene/Butadiene block ratio, Styrene Butadiene diblock content, etc., the difference in structural parameters will be important for the physical and chemical properties of SBS. Influence, so when SBS is added as modifier to asphalt, it will have different effects on the low temperature performance of asphalt due to its own structure, but the research on the influence of SBS relative molecular mass on asphalt performance is rare. The relative molecular mass of SBS used for modification is also selected only by experience. The research on the influence of SBS relative molecular mass on asphalt performance remains in the aspect of basic index change, and its comprehensive influence on asphalt pavement performance and related mechanism are not discussed in depth.

What are the differences in the modification effect of SBS on asphalt with different relative molecular mass levels, and the different effects on the low temperature performance of asphalt? It has important research value for optimizing SBS for road modification and improving pavement performance. The relative molecular mass of linear SBS currently used for road asphalt modification is mainly around 80,000~120,000, and there are also SBS modifiers outside this range, but the relative use is relatively small. Therefore, five kinds of linear SBS modified asphalt with different relative molecular mass (79,000, 87,000, 106,000, 124,000, 138,000) were used to test the rheological properties of low temperature bending beam, and the creep stiffness S value and creep were analyzed. The relationship between the variable rate m value and the relative molecular mass and temperature of SBS, on the one hand, to explore the effect of SBS relative molecular mass on the low temperature performance of modified asphalt, and on the other hand, the preferred SBS for modified asphalt with better low temperature performance. Relative molecular quality levels provide an important reference.

2. Materials and Methods

2.1. Materials

2.1.1. Matrix Asphalt. The matrix asphalt used in this paper is Esso 70# asphalt. Refer to the "Test Procedure for Highway Engineering Asphalt and Asphalt Mixture" for the benchmark test of the matrix asphalt. The basic technical indexes of the asphalt are shown in table 1.

Table 1. Basic indicators of Esso 70# matrix asphalt.

Test index	Results of testing	Technical indicators
Penetration(25°C, 5s, 100g)(0.1mm)	62	60~80
Softening point(°C)	48.5	>45
Ductility(15°C, cm)	>100	>100
Solubility (%)	99.6	>99.5
Flash point(°C)	260	>260
Viscosity(135°C, Pa·s)	0.550	-

2.1.2. SBS Modifier. The modifiers used in this paper were synthesized by coupling polymerization process. All of them were linear SBS. Five kinds of SBS with different molecular weights were prepared by controlling the ratio of raw materials and polymerization process. The relatives were determined by Gel Permeation Chromatography (GPC). Molecular mass, the relative molecular mass of SBS modifier used in this paper and the main technical indicators are shown in table 2.

Table 2. SBS relative molecular mass difference and main parameter indicators.

SBS number	Weight average molecular weight(Mw)	Number average molecular weight(Mn)	Distribution index(Mw/Mn)	Styrene content (%)	SB content (%)
SBS-7.9	78826	75317	1.05	29.1	22.5
SBS-8.7	87451	81607	1.07	28.8	23.0
SBS-10.6	106008	98329	1.08	29.3	22.6
SBS-12.4	123652	116955	1.06	29.3	25.1
SBS-13.8	138325	122073	1.13	29.5	20.3

According to the main parameters of SBS, the content of five kinds of SBS styrene is relatively close, the block ratio is not much different, and the SB content is in the range of 20~25%. The main difference is the difference of relative molecular mass. Since the relative molecular mass of SBS used in road asphalt modifier mostly refers to its weight average relative molecular mass, therefore, the SBS of five different relative molecular masses is numbered by weight average molecular weight, such as SBS-7.9. The average relative molecular mass is 79,000 SBS. The relative molecular masses described herein also refer to their weight average relative molecular masses.

2.2. Standard Experimental Methods

In this paper, low temperature bending beam rheometer is used to test the modified asphalt pitch of matrix asphalt and different SBS relative molecular masses to evaluate the effect of SBS relative molecular mass on the low temperature performance of modified asphalt. During the test, each modified asphalt was poured into a standard mold to form a 127 mm × 12.7 mm × 6.35 mm girders test piece, which was placed in an instrument-controlled greenhouse for heat preservation after low-temperature demoulding, and 980mN±50 was applied to the asphalt trabecular test piece. For a constant force of 50 s, the deformation value is recorded, and the creep stiffness modulus S value and the creep rate m value at 60 s are calculated. The S value characterizes the ability of asphalt to resist deformation at low temperatures. The m value is the rate of change of creep stiffness modulus, indicating the stress relaxation capacity of asphalt. The larger the S value, the greater the asphalt stiffness and the worse the deformation ability at low temperature. The lower the temperature cracking occurs, the smaller the m value is, the lower the stress relaxation performance is. Therefore, in the low temperature environment, the smaller the S value, the larger the m value, indicating that the low temperature performance of the asphalt is better.

In this paper, the BBR test tests the low temperature performance of each modified asphalt sample at -12 °C, -18 °C and -24 °C, and obtains the S value and m value at each temperature to evaluate the relative molecular mass of different SBS. The difference in low temperature properties of asphalt.

3. Experimental Results

Bending Beam Rheometer (BBR) is used to evaluate low temperature performance of the modified asphalt. The low temperature performance evaluation test of modified asphalt with different SBS relative molecular mass was carried out at -12 °C, -18 °C and -24 °C to determine the S and m values of matrix asphalt and modified asphalt. The test results are shown in table 3.

Table 3. Test results of modified asphalt BBR.

Asphalt type	-12°C		-18°C		-24°C	
	S(MPa)	m	S(MPa)	m	S(MPa)	m
Matrix asphalt	163	0.404	425	0.285	805	0.171
SBS-7.9	105	0.406	285	0.305	593	0.196
SBS-8.7	105	0.409	287	0.311	573	0.208
SBS-10.6	107	0.406	296	0.309	580	0.203
SBS-12.4	111	0.400	301	0.301	597	0.198
SBS-13.8	125	0.395	330	0.281	632	0.185

With the decrease of temperature, the creep stiffness modulus S value of matrix asphalt and SBS modified asphalt increases rapidly, the asphalt becomes hard and brittle, and the S value of matrix asphalt is significantly larger than SBS modified asphalt at each temperature, and the m value of asphalt The decrease in temperature is drastically reduced, and the m value of the matrix pitch is reduced more significantly. It is indicated that the matrix asphalt has poor stress relaxation performance and low temperature performance at low temperature. The five different molecular weight SBS studied in this paper have a certain improvement on the low temperature performance of the matrix asphalt.

3.1. Creep Stiffness

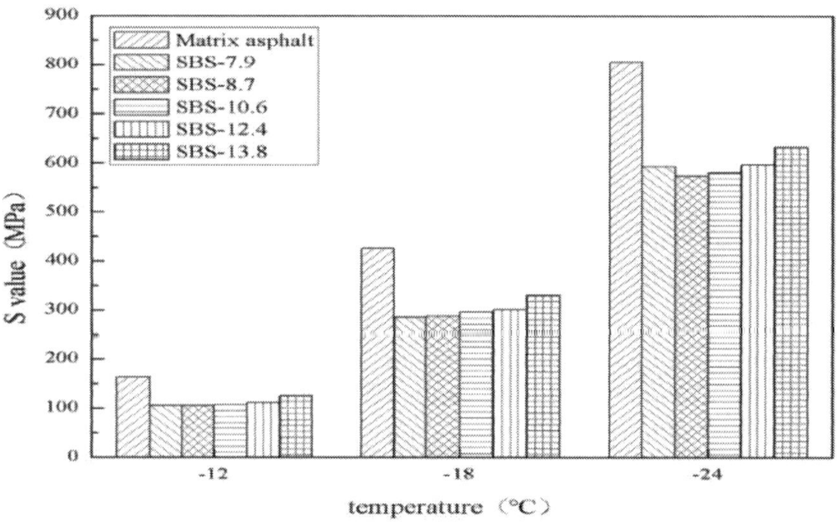

Figure 1. Creep Stiffness Modulus.

It can be seen from figure 1 that with the decrease of temperature, the creep stiffness modulus S value of matrix asphalt and SBS modified asphalt increases rapidly, the asphalt becomes hard and brittle, and the S value of matrix asphalt is significantly larger than SBS modified asphalt at each temperature. It is indicated that the SBS with five different relative molecular masses studied in this paper can improve the low temperature performance of matrix asphalt to a certain extent. Comparing the difference of modified low temperature properties of different SBS relative molecular masses, the S value of five kinds of SBS modified asphalts is not obvious at -12 °C, especially when the relative molecular mass of SBS modifier is less than 124,000, creep stiffness The modulus is relatively close, and when the relative molecular mass reaches 124,000 and above, there is a significant increase. The

SBS-13.8 modified asphalt has the largest stiffness modulus compared with other modified asphalts. When the temperature is lowered to -18 °C, the stiffness modulus of each modified asphalt increases, and the difference of the stiffness of the modified asphalt with different SBS relative molecular mass is slightly obvious. Overall, the modified asphalt creep. The stiffness modulus increases with the increase of the relative molecular mass of SBS modifier, and the creep stiffness modulus of SBS-13.8 modified asphalt is obviously larger. At -24 °C, the creep stiffness of the modified asphalt showed different trends. The stiffness modulus of the modified asphalt of SBS-7.9 increased significantly, which was significantly higher than that of SBS-8.7 and SBS-10.6 modified asphalt at the same temperature. The modifier used in SBS-7.9 modified asphalt has the lowest relative molecular mass, and the effect on the low temperature performance of the matrix asphalt is not obvious. When the temperature is low, the low temperature performance is poor. The creep stiffness modulus of SBS-8.7 modified asphalt is the smallest. When the relative molecular mass continues to increase, the stiffness modulus of SBS-12.4 and SBS-13.8 modified asphalt increases obviously.

3.2. m-Value

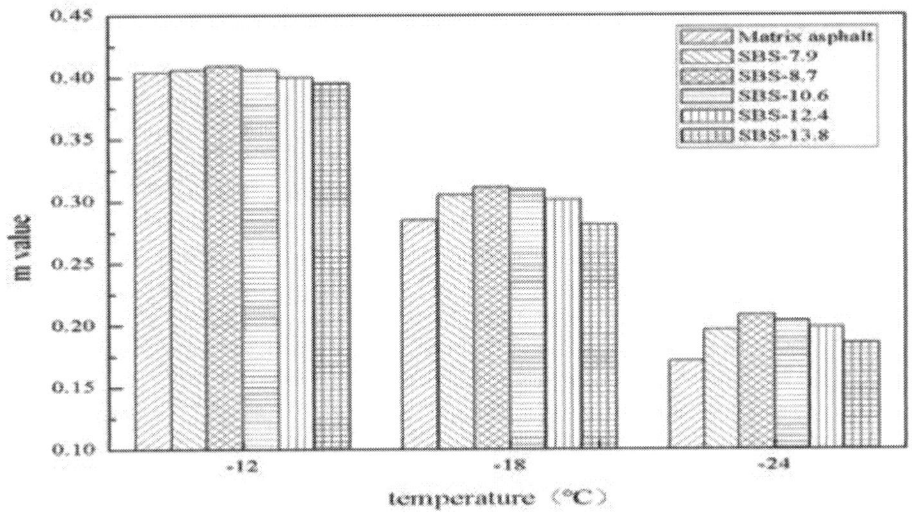

Figure 2. Creep Rate.

It can be seen from figure 2 that the m value of asphalt decreases sharply with the decrease of temperature, and the m value of the matrix asphalt decreases more obviously. At low temperature, the matrix asphalt has poor stress relaxation performance and poor low temperature performance. Under the condition of -12 °C, the m value of each modified asphalt is not much different, but it can be seen that except for SBS-7.9, the modified asphalt with large relative molecular mass has smaller m value and poorer low temperature deformation ability, and the increase in relative molecular mass is more pronounced as the value of m decreases. At -18 °C, there is a similar law. The creep rate of the modified asphalt with a relative molecular mass of more than 87,000 is more obvious. The modified asphalt with a relative molecular mass of only 79,000 has a small creep. Stiffness modulus, but m value is not as good as SBS-8.7 modified asphalt, again indicating that SBS-7.9 has poor modification effect on asphalt, and the effect of improving the low temperature performance of asphalt is not obvious, and when the temperature is lowered to -24 °C When the creep rate m value is reduced to a large extent, the creep rate is lower than the SBS modified asphalt with a relative molecular mass of 8,7000 to 124,000 at -24 °C.

4. Conclusion

In this work, the low temperature properties of SBS modified asphalt with different molecular weights are studied. According to the results, the following conclusions can be drawn:

(1) There is a certain correlation between the relative molecular mass of SBS and the low-temperature performance of modified asphalt.

(2) Temperature has a decisive influence on the Bending Beam Rheometer (BBR) test results, while both s-value and m-value are influenced by relative molecular mass of SBS, which is used as modifier of asphalt. The BBR test results show that when the molecular weight of SBS as a modifier increases, the s-value increases, and the m-value first increases and then decreases.

References

[1] Yan C Q, Xiao F P, Huang W D, Lv Q. 2018 Critical matters in using attenuated total reflectance fourier transform infrared to characterize the polymer degradation in styrene-butadiene-styrene-modified asphalt binders[J] *Polymer Testing* 70 289-296.

[2] Dong W L, Guan W Y, Huang W D. 2018 Low temperature performance analysis of sbs modified asphalt under different aging process [J] *Journal of Building Materials* 21 268-274.

[3] Cao L P, Dong Z J, Tan Y Q. 2009 Research on rheological performance of sbs modified asphalt at low temperature 190 9-14.

[4] Zhuang C Y, Li N, Zhao W, Cai C F. 2017 Effects of sbs content on the performance of modified asphalt[C] *IOP Conference Series: Materials Science and Engineering* 216 5.

[5] Liu S T, Cao W D, Shang S J, Qi H, Fang J G. 2010 Analysis and application of relationships between low-temperature rheological performance parameters of asphalt binders[J] *Construction and Building Materials* 24 471-478.

[6] Ma X Y, Chen H X, Zhang X Y, Xing M L, Yang P W, Wang Z L. 2018 An analysis over the correlation between low temperature rheological properties of sbs-modified asphalt and raw materials' performance parameters [J] *Materials Review* 32 3885-3890.

[7] Lin P, Huang W D, Li Y, Tang N P, Xiao F P. 2017 Investigation of influence factors on low temperature properties of sbs modified asphalt [J] *Construction and Building Materials* 154 609-622.

Simulation of Mechanical Properties of Special-Shaped Concrete-Filled Steel Tubular Column and Steel Beam Joints after Fire

Yi Zhao[1], Huihui Xu[1], Menghui Yan[1,2] and Shuai Wang[1,2]

[1] School of Civil Engineering and Architecture, Zhong Yuan University of Technology, Zhengzhou 451191, Henan, China
[2] The Sixth Design and Research Institute of Machinery Industry Co., LTD., Zhengzhou 450007, Henan, China
Email: zhaoyi091218@163.com

Abstract. To study fire after the mechanical performance of steel girder node special-shaped concrete-filled steel tube column, based on standard ISO - 834 litres of cooling curve, the node temperature field model was established based on finite element software ABAQUS, the compute nodes in the overall uniform temperature field under fire as a result, the reasonable choice of fire after the steel and concrete constitutive model, the temperature field results into the node stress model, considering the factors that influence the whole effect of fire loading in low cycle, the nodes of the finite element model, and contrast analysis of the temperature after the fire of the node and hysteretic performance and ultimate bearing capacity. The results show that the failure modes of special-shaped CFST column-steel beam joints at room temperature and after fire are the same, and the ultimate bearing capacity of the joints after fire decreases significantly by 14.88% compared with that at room temperature.
Keywords. Concrete filled steel tube; Cross node; fire; The temperature field.

1. Introduction

Fire is one of the common disasters in people's lives. Once a fire occurs, it will pose a huge threat to people's lives and property. Building fires occur most frequently, The losses caused are also the most serious. In recent years, concrete-filled steel tube structure has been used more and more widely in building structures due to its excellent mechanical properties. Its fire risk is also increasing gradually. The research on the properties of concrete-filled steel tube after fire will provide a basis for post-disaster repair and reinforcement [1]. As a force transfer hub of the frame structure, Transfer the bending moment and shear force of the beam and column. In order to further study the mechanical properties of special-shaped concrete-filled steel tubular structures after fire, In this paper, the temperature field and mechanical properties of concrete-filled steel tube cross column and steel beam joint after fire are simulated and analyzed. It provides a theoretical basis for the reinforcement and repair of such joints after disaster.

2. Model Design Scheme

The sample diagram of node model is shown in figure 1.The node attaches the side plate flat to the node domain, It is above the I-beam and extends a certain distance along its axis. The side panels strengthen the node domain obviously, The construction of the node is convenient, the force transmission is clear, This paper designs two models, Model size parameters are shown in table 1.JD1 is the node analysis model at room temperature.TJD1 is a thermodynamic coupling model after high temperature action.

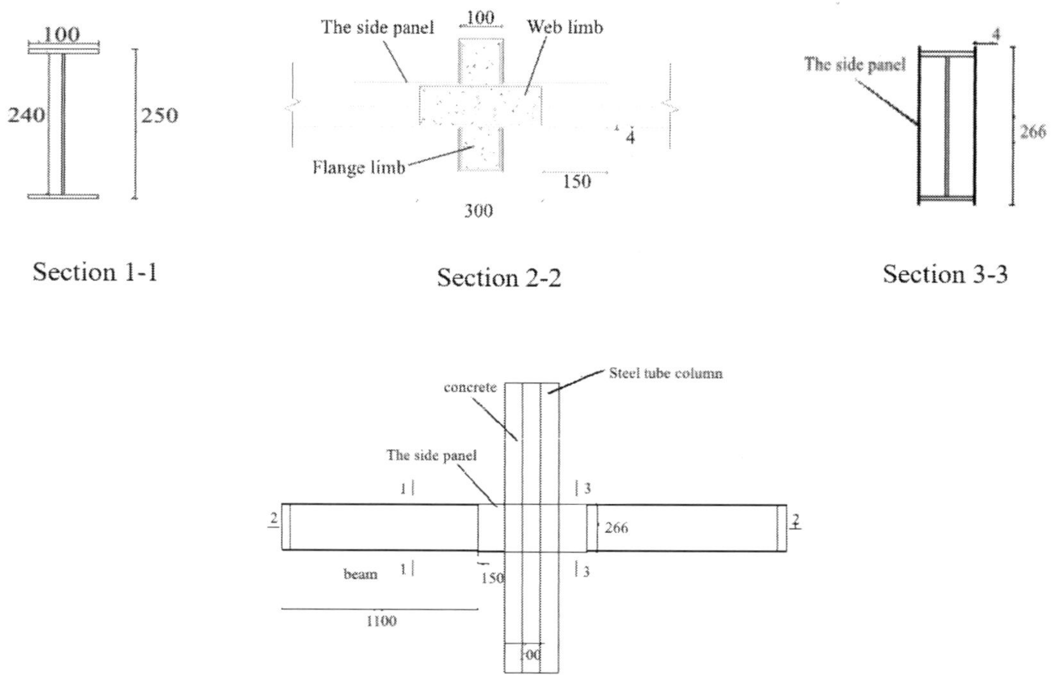

Figure 1. Node model construction diagram.

Specimen number	Section size of column	Frame beam dimensions	Axial compression ratio	th(min)	The working condition of the fire
JD1				—	From the fire
TJD1	Web limb:300×100 ×5;Flange limb:100×100 ×5	h×bf×tw×tf: 250×100×4×6	0.3	30	The nodal model is evenly exposed to fire

3. Fire Action Simulation of Node Model

3.1. Thermal Parameters of Node Model Materials

Under the action of fire, Heat is exchanged through heat conduction inside the cross - shaped concrete-filled steel tubular column-steel beam joints. In order to ensure the accuracy of heat transfer simulation, In this paper, the thermal conductivity of steel was calculated according to the formula (1) given by Lie [2] and the formula (2) for calculating the thermal conductivity of ordinary concrete suggested by the European Code [3].

$$K_s = \begin{cases} -0.022T + 48\dfrac{W}{m.^{\circ}C} & \\ {}^{\circ}C \le T \le 900^{\circ}C & \\ 28.2\dfrac{W}{m.^{\circ}C} & T > 900^{\circ}C \end{cases} \tag{1}$$

$$K_c = 0.012\left(\frac{T}{120}\right)^2 - 0.24\left(\frac{T}{120}\right) + 2 \qquad 20^{\circ}C \le T_s \le 1200^{\circ}C \tag{2}$$

According to the recommendation of European Code [3], The specific heat calculation formula (3) of steel and the specific heat calculation formula (4) of ordinary concrete are selected.

$$C_s = \begin{cases} 425 + 7.73 \times 10^{-1} T_s - 1.69 \times 10^{-3} \times T_s^2 + 2.22 \times 10^{-6} T_s^3 & 20^{\circ}C \le T_s \le 600^{\circ}C \\ 666 - \dfrac{13002}{T_s - 738} & 600^{\circ}C < T_s \le 735^{\circ}C \\ 545 + \dfrac{17820}{T_s - 731} & 735^{\circ}C < T_s \le 900^{\circ}C \\ 650 & 900^{\circ}C < T_s \le 1200^{\circ}C \end{cases} \tag{3}$$

$$C_c = 4\left(\frac{T}{120}\right)^2 + 80\left(\frac{T}{120}\right) + 900 \qquad 20^{\circ}C \le T_s \le 1200^{\circ}C \tag{4}$$

Where, T is temperature (°C), and the specific heat unit is J (kg·°C).

The mass density of steel and concrete is less affected by temperature, Take 2300kg/m³ and 7850 kg/m³, respectively.

3.2. Temperature-Time Curve of Composite Heating

In this paper, the rising and cooling curve suggested by the national standard ISO-834[4] is adopted to simulate the real situation of fire. The specific expression (5) is as follows:

Heating stage :($T \le t_h$)

$T = T_0 + 345\log_{10}(8t+1)$

Cooling period :($T > t_h$)

$$T = \begin{cases} T_h - 10.417(t - t_h) & t_h \le 30\,\text{min} \\ T_h - 4.167\left(3 - \dfrac{t_h}{60}\right)(t - t_h) & 30\,\text{min} \le t_h \le 120\,\text{min} \\ T_h - 4.167(t - t_h) & t_h > 120\,\text{min} \end{cases} \tag{5}$$

Where, the temperature unit is °C

t: total cooling time

T: temperature at time T

T_0: initial temperature

T_h: critical temperature for rising and cooling

3.3. Analysis of Cross Section Temperature Field

In order to better analyze the temperature variation rule at different locations of nodes, A feature point of each component at the horizontal midline section of the node is selected as the research object. The feature points of side plate, steel tube column, core concrete and steel beam are shown in figure 2.Corresponding to feature points 1, 2, 3 and 4, and extracting the temperature of feature points under the action of fire in the whole process of rising and cooling. Draw the time-temperature curve as shown in figure 3.

Figure 2. Location distribution of feature points.

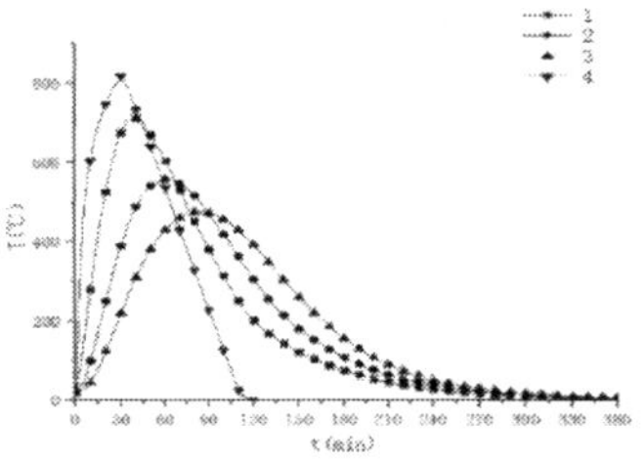

Figure 3. TJD1 section time (t) -temperature (T) simulation results.

As can be seen from figure 3, The time-temperature curve of each component of the node model is basically consistent with the law of the ISO-834 standard lift-temperature curve. The time-temperature curve of feature point 4 is the most consistent. The maximum temperature at feature points 1, 2 and 3 decreases in turn, It takes longer to reach the maximum temperature, In the heating process of feature point 3, The temperature of feature point 3 is always low, At the cooling stage of feature point 3, The temperature of feature point 3 is higher than that of other parts in the figure. This is due to the close contact between the side plate and the steel tube column and the core concrete, In the heating process, Evaporation of water from the core concrete consumes heat and cools the core concrete, The thermal inertia of the concrete will cause the temperature rise and drop of the cross concrete filled steel tube to lag behind that of the steel beam.

Based on reasonable setting of material thermal parameters, fire conditions and cell mesh, The heat transfer is simulated by integrating heat transfer coefficient and thermal radiation coefficient under the temperature load of the fire surface. The finite element analysis of the whole fire process of concrete-filled steel tube cross column and I-beam joint is calculated. According to the finite element calculation results, Finite element results of temperature field in TJD1 model at t_h time, 30min cooling and 60min cooling are shown in figure 4, the unit of temperature is °C.

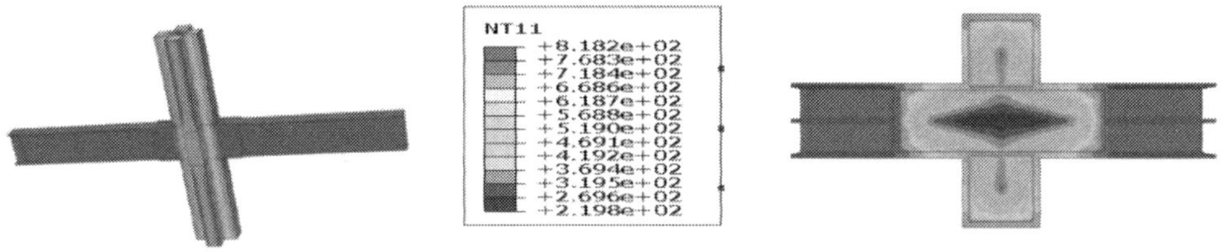

T_h moment

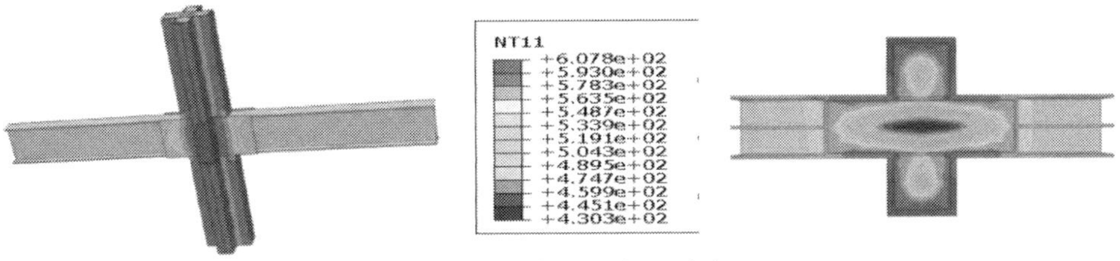

t=60min(cool30min)

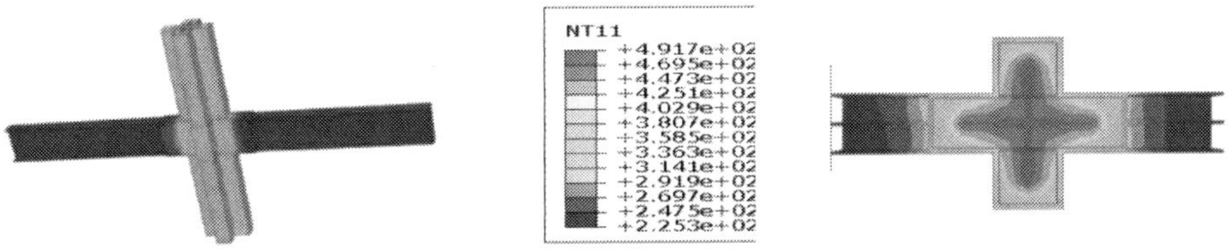

t=90min(cool 60min)

Figure 4. Finite element results of TJD1 temperature field.

As can be seen from the figure above, The temperature at different positions of the node model varies greatly. Mainly because of the steel pipe and concrete two kinds of material thermal performance is different. At the th moment of the ISO-834 standard lift-temperature curve, the steel beam reaches the highest temperature of 818.2°C, At this point, on the section of the CFST column, The side plate has the highest temperature, about 750 °C, The closer you get to the center of the column, The colder the core concrete gets, The temperature of the core concrete near the corner of the steel tube column is higher. This is because the concrete near the corner of the steel tube column has a large contact area with the steel tube column, It also transfers heat faster when exposed to fire, So the core concrete at the corner is warmer than the other places. Combined with Figure 3, it can be seen that, When the total time of exposure to fire is 60min,At this point, the temperature of the ISO-834 standard temperature curve has been lowered for 30min,By this time the steel columns, concrete and side slabs had begun to cool down, The core concrete is still heating up, When the node model enters the later cooling period (the total time t is greater than 120min),Core concrete is the hottest, It is the thermal inertia of concrete that leads to the difference of rising and cooling time of each component of the joint model.

3.4. Simulation of Mechanical Properties of Joints After Fire

3.4.1. Selection of Material Constitutive. Based on the research results in literature [1], This paper ignores the effect of the whole fire process, In other words, the effect of rising and cooling on the mechanical properties of the joints after fire is not considered. Cao Wenxian[5] was selected to study the constitutive relationship of steel in concretefilled steel tube at high temperature. Formula (6) is as follows:

$$\sigma = \begin{cases} E_s(T_{\max})\cdot\varepsilon & \varepsilon \le \varepsilon_y(T_{\max}) \\ f_y(T_{\max}) + E_s^{'}(T_{\max})\cdot[\varepsilon - \varepsilon_y(T_{\max})] & \varepsilon > \varepsilon_y(T_{\max}) \end{cases}$$

In Equation (6),
$E_S\ (T_{\max})$: the highest temperature in thehistory of the material,
$E_S\ (T_{\max}) = E_S$; $E_S\ (T_{\max}) = 0.01E_S\ (T_{\max})$;

$$\varepsilon(T_{\max}) = f_y(T_{\max}) / E_s(T_{\max})$$

Where, Es: elastic modulus of steel at room temperature

$E\ (T_{max})$: Elastic modulus of steel after high temperature, Es=2.06×103MPa;
$\varepsilon_y\ (T_{max})$: yield strain of steel at high temperature
$f_y\ (T_{max})$: yield strength of steel after high temperature.

$$f_y(T_{max}) = \begin{cases} f_y & T_{max} > 400^\circ C \\ f_y[1+223\times10^4(T_{max}-20)-588\times10^7(T_{max}-20)^2] & T_{max} > 400^\circ C \end{cases}$$

f_y: yield strength of steel at room temperature
The mechanical properties of concrete change after high temperature,Lin Xiaokang [6] modified the peak stress and peak strain of the concrete compression constitutive model.After modification, formula (7) is as follows:

$$\sigma_o = \frac{f_c^{'}}{1+2.4(T_{max}-20)^6\times10^{-17}}(N/mm^2)$$
$$\varepsilon_c = (1300+12.5\cdot f_c^{'})10^{-6}\times[1+(1500T_{max}+5T_{max}^2)\times10^{-6}] \qquad (7)$$

Guo Zhenhai and Li Wei [7] based on the study of concrete after high temperature,The peak tensile stress and peak strain of concrete are modified.Formula (8) is modified as follows:

$$\sigma_p = 0.26\cdot(1.25\cdot f_c)^{\frac{2}{3}}\cdot(1-0.001T_{max}) \qquad (8)$$

3.4.2. Establishment of Node Simulation Model. The joint model consists of a cross-shaped steel tube column, core concrete, side slabs and steel beams. It can save the calculation time and ensure the accuracy of calculation. All components are linear reduction integral element (C3D8R).The steel pipe and concrete use face to face contact, The tangential behavior adopts the Coulomb friction model provided by Abaqus, The friction coefficient at room temperature is 0.6,The friction coefficient after high temperature is 0.25[8], The normal behavior selects the ABAQUS default hard contact.

In this paper, the column end loading method recommended by JGJ101-2015 Code for Seismic Test Methods of Buildings [9] was selected. The loading diagram is shown in Figure 5.The top surface of the column is coupled to the reference point RP1, Constrain its translation in the y direction and its rotation in the y and z directions, By applying pressure on the top surface of the column to simulate

The axial force at the top of the column, A reciprocating horizontal displacement was applied to the reference point RP1 to simulate pseudo-static loading. The displacement increment is 5mm.The bottom of the column is coupled to the reference point RP2, It can only rotate about the Y-axis, The remaining degrees of freedom are all constrained. The left and right beam end faces are coupled to RP3 and RP4, respectively. Constraint y and z translation and x and z rotation.

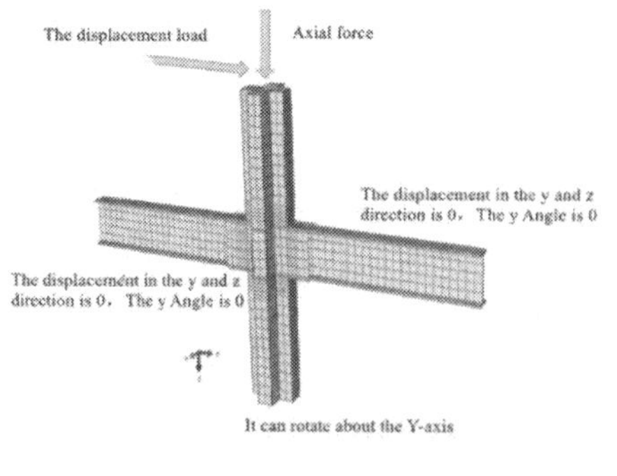

Figure 5. Schematic diagram of model loading.

3.4.3. Initial Temperature Conditions. Since the uncoupled thermal analysis is carried out in this paper,

That is, without considering the effect of high temperature on the initial stress and strain inside the component, The temperature at a time in the temperature field ODB file was imported into the stress analysis model. Then the program determines the temperature of each node, And give the corresponding material properties.

According to the material constitutive relation model after high temperature, The stress-strain model after high temperature is only related to the highest historical temperature of the material, Because of the thermal inertia of concrete, Each element of the node model cannot reach the maximum temperature at the same time. As the steel heats up, The temperature of concrete increases with the increase of the temperature of steel on the fire surface, The steel begins to enter the early cooling period, The concrete will still be warming up for a certain amount of time, When the steel has cooled to room temperature, Steel is still in the cooling section, And keep it at a certain temperature. The nodes selected in this paper are designed according to the principle of "strong nodes and weak components". The joint failure position is on the steel beam, Therefore, the material properties of steel have a greater impact on the stress analysis model. In this paper, the joint temperature when the steel beam temperature reaches the highest temperature is taken as the initial temperature condition. That is, t is the cross section temperature at the moment when 30min.

4. Comparison of Node Simulation Results at Normal Temperature and After Fire

4.1. Model Failure Process

Mises stress nephograph of JD1 model loading process is shown in Figure 6.As can be seen from figure 6 (a), When the horizontal displacement is loaded to 25.32mm, Beam-column connection parts, side plates and steel beam flanges begin to yield, At this point, the component enters the yield phase, The rest parts of the model were all in the elastic stage. When the horizontal displacement is pushed to 73.97mm, The ultimate load is 122.48kN.As the load continues, The flange deformation at the variable section position of the side plate overhanging steel beam is serious. When the horizontal displacement is loaded to 108.76mm, The load drops to 104.11kN, The capacity dropped to 85% of the ultimate capacity, Loading finished. At this point, the failure state is shown in figure 6 (b).

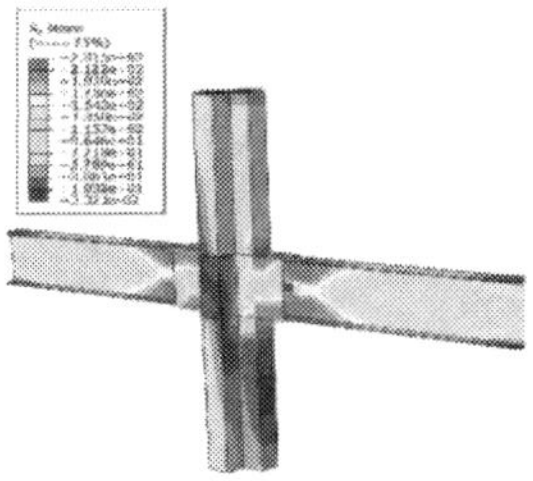

(a)The yield state

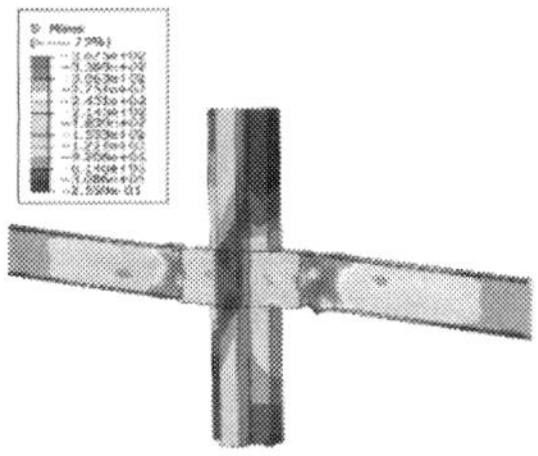

(b)Damage state

Figure 6. Stress nephogram of JD1Mises.

Mises stress nephograph during loading process of TJD1 model is shown in figure 7. As can be seen from figure 7 (a), When the horizontal displacement is loaded to 22.89mm (the same direction as the X coordinate in ABAQUS is positive, and the opposite direction is negative), The slope of skeleton curve changes obviously, The connecting parts of the upper and lower flanges of the steel beam and the steel tube column began to yield. TJD1 enters the yield phase, The steel tube column, core concrete and other parts of the steel beam were observed to be in the elastic stage. When this continues to load to 54.53mm, The maximum load is 104.25kN, The limit state is shown in figure 7 (b).

(a)The yield state (b)Damage state

Figure 7. Stress nephogram of TJD1Mises.

By observing the JD1 and TJD1 corruptions, Both JD1 and TJD1 are "beam hinge failure", That is, the flange of the steel beam at the outstretched end of the side plate yields first. As the displacement is further loaded, Deformation and bearing capacity gradually increase, Until a cross section plastic hinge is formed at this position, The node is broken. This suggests that the failure mode of the node model did not change after high temperature.

4.2. Skeleton Curve

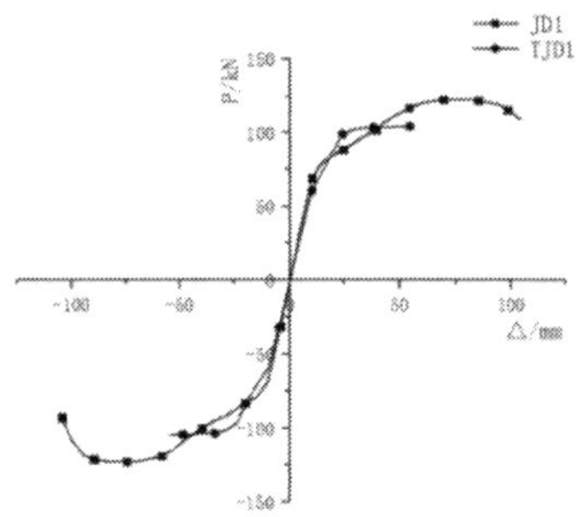

Figure 8. Skeleton curve.

The skeleton curve of the model is shown in figure 8.As can be seen from the figure, Compared with the finite element results of the nodal model at room temperature, The initial stiffness is basically unchanged, This is because the elastic modulus of steel and concrete is greatly reduced at high temperatures, After cooling to room temperature, The elastic modulus recovered obviously. The ultimate bearing capacity of T1JD1 decreased significantly compared with JD1.That's a 14.88 percent decrease. This means that when exposed to fire, This means that when exposed to fire, This is due to the effect of the whole fire, The mechanical properties of steel and concrete have been reduced to a certain extent, The extent of the drop depends on the record high temperature experienced by the material.

5. Conclusion

(1) After the whole process of rising and cooling is experienced by the side plate connecting the cross-shaped CFST to steel beam joints, The section temperature varies greatly in different positions. The time-temperature curve of steel beam characteristic points is in good agreement with the ISO-834 standard fire rise and drop curve. Because of the thermal inertia of concrete, The rising and cooling stages of each node component are not consistent. When the steel beam has cooled, The inside of the steel tube section may still be in the heating stage.

(2) Under normal temperature and high temperature, the failure mode of the side plate connecting the concrete-steel tubular cross beam joints is the same. That is, the flange of the steel beam at the

outstretched end of the side plate yields first. As the displacement is further loaded, Deformation and bearing capacity gradually increase, Until a cross section plastic hinge is formed at this position, The node is broken. This indicates that the inhomogeneity of temperature field does not lead to the change of failure mode of such nodes. However, its ultimate bearing capacity decreased significantly, The ultimate bearing capacity of T1JD1 is 14.88% lower than that of JD1.

Reference

[1] Zeng X. 2009 Experimental Study on Hysteretic Performance of CFST Columns after Whole FireProcess [D] *Hunan University.*
[2] Lie T T. 1994 Fire resistance of circular steel columns filled with bar-reinforced concrete [J] *Journal of structural Engineering* 120 (5) 1489-1509.
[3] CEN (European Committee for Standardisation) DAFT EVN 1993, Eurocode 3 Design of steel structures. British Standards Instituion.1995.7.
[4] International Standard ISO 834, Fire resistance tests-elements of building construction. Amendment 1. Amendment 2. 1980.
[5] Cao W Xi. 1998 Study on fire response of steel frame structures under damage accumulation [Ph. D. dissertation] Shanghai: Tongji University.
[6] Lin X K. 2006 Study on hysteretic behavior of concrete-filled steel tubular flexural members after fire [Ph. D. dissertation] Fuzhou: Fuzhou University.
[7] Li W, Guo Z H. 1993 Experimental study on strength and deformation behavior of concrete at high temperature [J] *Journal of Building Structures* **14** (1) 8-16.
[8] Huo J S, Qu H. 2007 Analysis of mechanical properties of concrete-filled steel tubular structures after fire based on ABAQUS. See the proceedings of the 4th National Symposium on Fire Prevention and Corrosion Technology of Steel Structure and the 2nd National Symposium on Fire Resistance of Structure. Shanghai: Tongji University Press 517-515.
[9] JGJ101-2015, Code for Seismic Test Method of Building [S] Beijing: China Building Industry Press 2015.

Study on Interaction of RC Pile-Soil with High Reinforcement Ratio of Integral Abutment Bridges under Quasi-Static Test

Ying Luo [1], Fuyun Huang [2], Zhifu Chen [3], Xinghua Liu [1], Zhengfeng Liu [2] and Chenxi Hu [2]

[1] China Construction Fifth Bureau Civil Engineering Co., Ltd, Hunan, China
[2] College of Civil Engineering, Fuzhou University, Fujian, China
[3] Quanzhou taiwanese investment zone urban construction development Co., Ltd , Fujian, China
Email: 52375495@qq.com

Abstract. In order to improve the ability of the reinforcement concrete (RC) pile foundation of integral abutment to absorb the horizontal reciprocating deformation under the action of temperature or earthquake, a pseudo-static low cycle test on interaction of pile-soil with high reinforcement ratio was carried out. The failure location, hysteresis curve, skeleton curve and horizontal deformation of three piles with different reinforcement ratios were compared. The test results show that, with the increase of the reinforcement ratio, the crack of the RC pile develops along the pile body to the depth, and the pile body failure area and the position where the maximum bending moment moves down, the crack resistance of the pile body is improved, and the effective interaction pile length increases; The test results also show that the hysteresis curve of the model pile becomes fuller with the increase of the reinforcement ratio, compared with the RCP-1 specimen with the lowest reinforcement ratio, the equivalent viscous damping ratio of the RCP-3 specimen is increased by 31.6%, and the energy dissipation capacity is improved. In addition, with the increase of the reinforcement ratio, the bearing capacity and deformation capacity of model piles are greatly improved. Compared with RCP-1 specimen, the ultimate bearing capacity of RCP-3 specimen increased by 150%, and the corresponding ultimate displacement increased by 153%. Increasing reinforcement ratio can significantly improve the mechanical properties and deformation capacity of RC pile.

1. Introduction

The integral bridge eliminates the expansion joints and expansion devices, and connects the superstructure, abutment and pile foundation as a whole, which has good seismic performance and life cycle, reduces the operation and maintenance costs, and has been widely popularized and applied in Europe, America and Japan[1-6]. China has gradually started to apply it. The latest standard for design of concrete structures durability [7] (GB/T 50476-2019) has clearly required that small and medium-span bridges should give priority to integral bridges. However, the reciprocating displacement of the main girder of the integral bridge under the action of ambient temperature or earthquake will drive the horizontal reciprocating motion of the abutment, which will lead to a large horizontal deformation of the pile foundation at the bottom of the abutment, and then cause the pile-soil interaction. Therefore, the integral bridge pile foundation has higher requirements for resisting horizontal deformation. The greater the temperature difference and bridge length, the greater the horizontal deformation, and the higher the requirement for the deformation ability of pile foundation.

In order to meet the needs of high horizontal deformation resistance, H-shaped steel pile foundation is mostly used for pile foundation at the bottom of integral abutment built abroad [8-11]. However, due to economy, engineering experience and habit, the common pile foundation for bridges in China is concrete pile (RC). RC piles are also used in the pile foundation of integral bridge in China. The research shows that [12-15], the horizontal deformation resistance of the traditional RC pile foundation reinforced only by the structure is low, so it is difficult to apply to the integral bridge, which limits the popularization of this type of bridge.

In order to improve the deformation resistance of RC pile foundation, scholars of domestic and foreign have carried out extensive research. Literature [16] improves the flexibility and deformation ability of pile body by reducing the diameter of section in a certain range at the top of pile, and studies the interaction between stepped pile and soil; However, due to the earth pressure around the pile, reducing the cross-section size can only improve its flexibility, but has little effect on improving its deformability. Literature [17] reduces the earth pressure around the pile and increases the horizontal deformation ability of the pile by reaming the pile top within a certain depth range and filling loose sand in the reaming, which is applied to the longest integral bridge in China-Yong-chun Shang-ban Bridge; However, due to the horizontal reciprocating interaction between pile and soil, loose sand will be gradually compacted, and the use effect will gradually decrease. Literature [18] has carried out the research of reaming depth and reaming material, and increased its horizontal deformation ability by reaming and filling rubber in a certain depth range of pile top; In addition, reference [19] improves the horizontal deformation ability of RC pile foundation by wrapping the pile body. These methods can improve the horizontal deformation ability of RC pile foundation, but the construction is inconvenient.

In addition, the pile-soil interaction test of PHC pipe pile in Reference [20] showed that the higher the reinforcement ratio and prestress degree were, the better the deformation performance was. Literature [21] has carried out numerical analysis on the parameters of concrete pile of integral bridge, and the influence of strong axis and weak axis of concrete pile on the mechanical performance of integral bridge is studied. It is considered that the horizontal deformation capacity of integral bridge can be improved by taking weak axis as the stress direction. Literature [22] thinks through finite element simulation that increasing the height of abutment can reduce the horizontal displacement of the top of pile foundation, but experimental research and in-depth finite element analysis have not been carried out.

Because the traditional concrete pile foundation has low reinforcement ratio and low horizontal deformation resistance, it is not suitable for the pile foundation of integral bridge. However, according to the above research summary, there are few researches on the anti-deformation ability of RC piles and pile-soil interaction of integral bridges with high reinforcement. Therefore, this paper has carried out the reciprocating low-cycle pseudo-static test of RC pile-soil interaction with high reinforcement to study the influence of reinforcement ratio on its horizontal deformation resistance, which provides reference for the popularization and application of this type of bridge in China.

2. Brief Introduction of Test

2.1. Model Pile Design

According to the different reinforcement ratio, one RC pile specimen with lower reinforcement ratio (0.8%) and two RC pile specimens with higher reinforcement ratio (1.6% and 3.2%, respectively) were designed and manufactured. The numbers of the three piles are RCP-1, RCP-2 and RCP-3 respectively. The pile length and diameter of each specimen are 3.5m and 155mm respectively.

The specimen is C40 concrete. The measured 28-day strength of concrete cube test block is 47.6MPa, and the elastic modulus is 3.25×104MPa. Longitudinal reinforcement adopts HRB335, and steel bars with diameters of 6mm, 8mm and 12mm are adopted according to different reinforcement ratios. The stirrup adopts HPB235, circular stirrup with diameter of 6mm, and the stirrup spacing in non-encrypted area is 120mm; The 600mm range of pile top and bottom is stirrup encryption area, and the encryption distance is 100mm; Thickness of protective layer is 20mm. Figure 1 only shows the reinforcement schematic of RCP-1 model pile, and other components are arranged similarly, so they are not given one by one due to the limitation of space.

2.2. Soil Box and Soil Parameters

The rectangular steel box with length of 3m, width of 2m and height of 4m is selected as the soil box, as shown in figure 2. The bottom of the soil box is welded with 4 angle steels with a length of 0.2m to fix the bottom of the pile.

The relevant parameters of the test sand are shown in table 1, according to code for investigation of geotechnical engineering (GB50021-2001), the test sand belongs to medium coarse sand. Before the test, the model pile is placed in the soil box, and then the sand is filled in the soil box and vibrating densification. After sand filling, the depth of the model pile is 3.1m, and the exposed part above the sand surface is 0.4m.

Table 1. Sand parameters in the experiment.

Types of soil	Moisture content ω [%]	Density ρ[g/cm^3]	Void ratio e	Angle of internal friction φ [°]	Modulus of compression [Mpa]	Average standard penetration number	Relative compaction
Medium coarse sand	0.8	1.50	0.80	35.86	28.9	11	53%

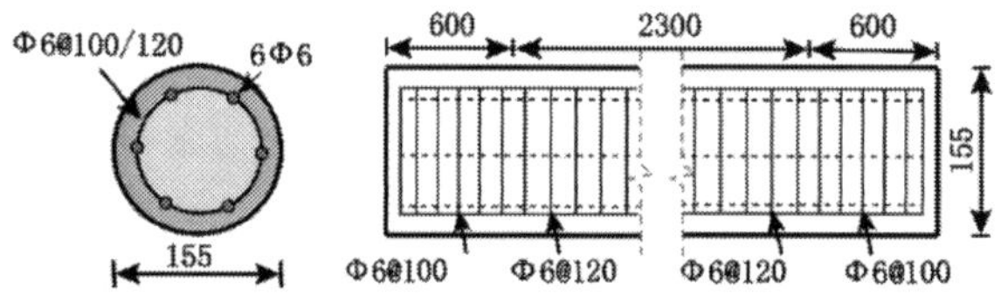

Figure 1. Reinforcement arrangement of RCP-1 specimen.

Figure 2. Soil-box.

2.3. Measuring Point Arrangement

Each side of the pile is arranged 10 strain gauges, symmetrically arranged on both sides, a total of 20. Strain gauge measuring points are arranged down at a distance of 350mm from the pile-soil junction, numbered from S1 to S20, as shown in figure 3a.

Eight pairs of soil pressure gauges (8 on each side, 16 in total) are symmetrically arranged along the stress direction of the pile. The layout method starts from 175mm down at the pile-soil junction and is arranged at a distance of 350mm, numbered from T1~T16, as shown in figure 3b.

In addition, 13 horizontal displacement meters (L1~L13) were arranged in this experiment. Among them, L1 is located at the top of the pile, which is used to compare the stroke of MTS actuator; other displacement meters are arranged at 200mm intervals from the pile-soil junction and become 600mm at the bottom, as shown in figure 3c.

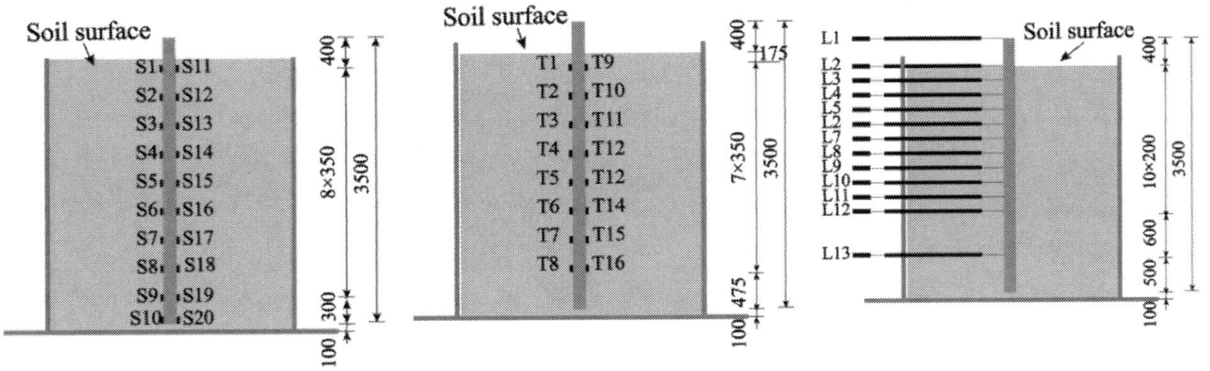

(a) Strain gauge layout (b) Earth-pressure cell layout (c) Displacement meter layout

Figure 3. Layout of the sensors (mm).

2.4. Loading Method

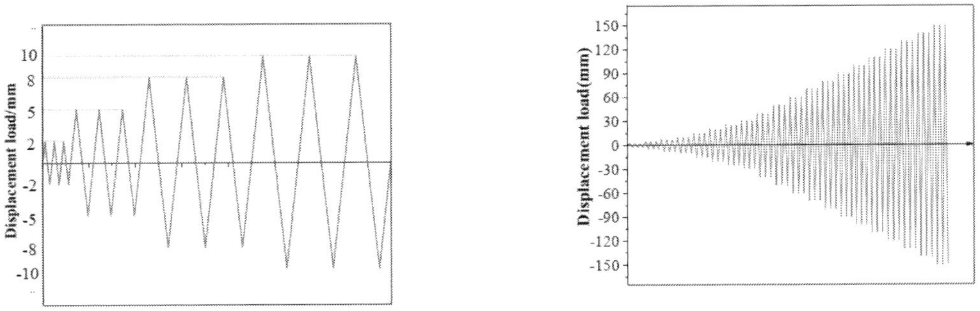

(a) The first stage loading scheme (b) whole process loading

Figure 4. Loading process diagram.

In this experiment, MTS electro-hydraulic servo loading system of Fuzhou University is used to apply reciprocating horizontal displacement to the top of model pile. The specific loading scheme is as follows: 1. Apply reciprocating displacement to the top of pile by 2mm, 5mm, 8mm and 10mm in the initial stage of loading; 2. When it is 10mm~30mm, it will increase according to the displacement increment of 5mm, that is, the loading displacement is 10mm, 15mm, 20mm, 25mm and 30 mm in turn; 3. When the displacement exceeds 30mm, make the displacement increment of each stage be 10mm, until the corresponding horizontal load is equal to or less than 0.85 of the maximum bearing capacity under a certain amplitude displacement, the experiment is stopped. In the experiment, the loading speed is 1.0mm/s, each load cycle is three times, and the load is held for 1 minute before the loading direction is changed. The loading history is shown in figure 4.

3. Test Result Analysis

3.1. Pile Failure Characteristics

After the experiment, the failure of model pile was observed, and it was found that the failure area of RCP-1 model pile was 2.8D~4.5D. The first slight crack appeared at 2.8D (D is the pile diameter). At 3.5D, the pile concrete was severely crushed and damaged, the steel bars were exposed, and some steel bars were broken. At 4.5D, there is obvious cracking of pile concrete, as shown in figure 5. The location and failure mode of RCP-2 model pile are similar to those of RCP-1 model pile, and the failure area is

2.9D to 4.6D. However, the failure area of RCP-3 model pile is 4.8D~6.8D, and there are four obvious cracks in the pile body.

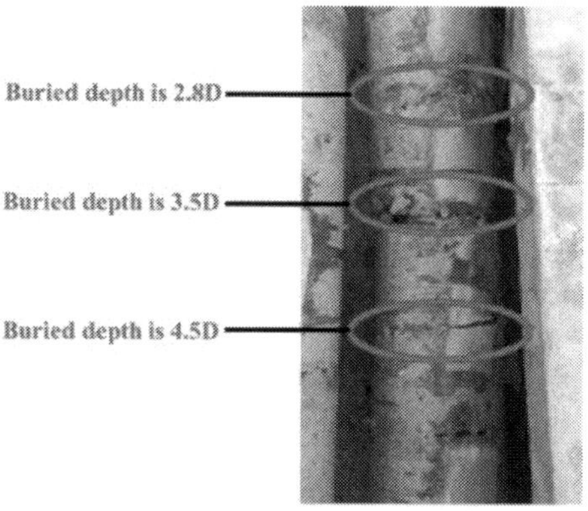

Figure 5. Damages and failure of RCP-1.

It can be seen from the comparison of each specimen that the crack position of RCP-3 is obviously lower than that of RCP-1 and RCP-2,and a crack appears at a deeper position, which indicates that with the further increase of reinforcement ratio, the depth of the failure position of RCP-3 pile body increases, the effective interaction length increases, and materials are effectively utilized. It shows that the larger reinforcement ratio has a great influence on the failure mode of RC piles.

3.2. Hysteresis Curve and Equivalent Viscous Damping Coefficient

The area enclosed by the hysteresis loop can be used to measure the energy dissipation capacity of pile-soil system in earthquake resistance. Figure 6 shows the hysteresis curve of RCP-1 model pile. It can be seen from Figure 6 that the hysteretic curve of the model pile is centrosymmetric as a whole, which shows that the mechanical properties of the model pile under forward and reverse loading are similar. At the initial stage of loading, the force-displacement curve of model pile is linear, the area of hysteresis loop is small, and the model pile is in elastic state With the increase of displacement load, the area of hysteresis loop increases, the energy dissipation capacity of pile-soil system increases, and the pile-soil interaction becomes more obvious.

Figure 7 shows hysteretic curves of three model piles under 50mm displacement load. It can be seen from the figure that under the same displacement load, the horizontal force of RCP-1 is the smallest, which is 7.2kN; The horizontal forces of RCP-2 and RCP-3 are 9.9kN and 13.2kN respectively, which are increased by 37.5% and 83.3% respectively compared with RCP-1. With the increase of reinforcement ratio, the hysteretic curve of model pile becomes fuller and the area surrounded by hysteretic loop becomes larger, which indicates that increasing reinforcement ratio of model pile can enhance the anti-push ability and pile-soil interaction ability of model pile.

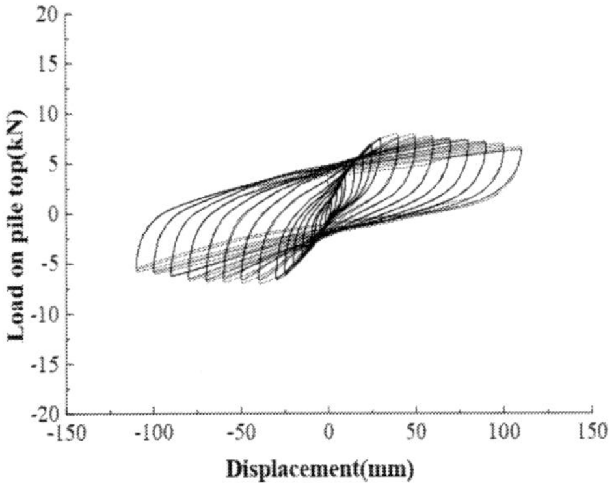

Figure 6. Hysteretic curves of RCP-1.

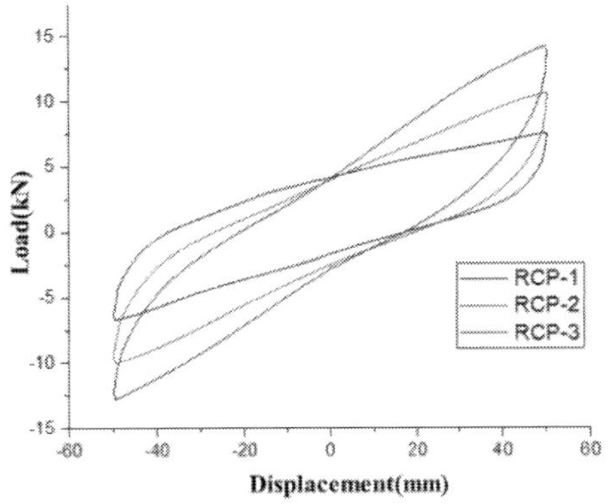

Figure 7. Hysteretic loops of three piles under 50mm displacement load.

The equivalent viscous damping ratio ξ_e can be used to measure the energy dissipation capacity of the pile-soil system. Figure 8 shows a hysteresis loop, and ξ_e can be calculated according to Eq. (1):

$$\xi_e = \frac{A_{\mathrm{ABCD}}}{2\pi(A_{\mathrm{OFD}} + A_{\mathrm{OBE}})} . \tag{1}$$

In this formula, A_{ABCD} is the hysteresis loop area, A_{OFD} and A_{OBE} are the areas of triangle OFD and triangle OEB respectively.

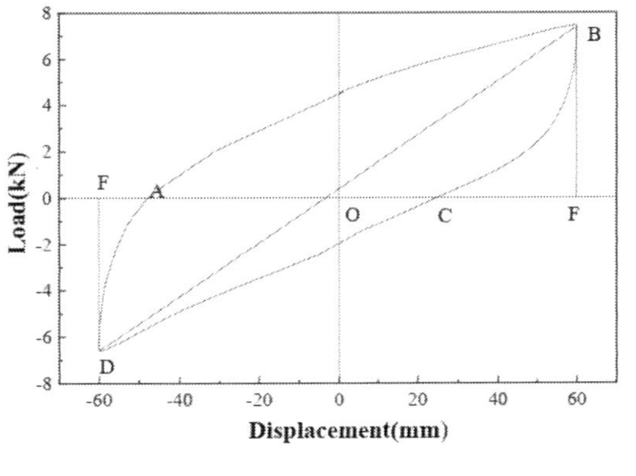

Figure 8. Hysteretic loop.

Figure 9 shows the equivalent viscous damping ratio of three model piles under various loads. It can be seen from Figure 9 that the equivalent viscous damping ratio of the three model piles has the same change trend in the loading process. For RCP-1 model pile, when the displacement is less than 40mm, ξe increases with the increase of displacement load, and reaches the maximum value when then displacement is 40mm, then decreases with the increase of displacement. This is because, in the initial stage of loading, the deformation of pile is small, which has not been fully transmitted to the lower part of pile body, and the interaction between pile and soil is not obvious, so the equivalent viscous damping ratio is small. With the increase of displacement load, the pile-soil interaction increases, the energy dissipation of pile-soil system increases, and the damping ratio also increases. However, when the displacement reaches 40mm and the concrete is crushed , the energy dissipation of the system decreases and the damping ratio begins to decrease. The RCP-2 and RCP-3 reached the maximum at 50mm displacement.

Figure 9 also shows that under the same displacement load, the higher the reinforcement ratio, the greater the ξe of the model pile. Comparing the peak values of three model piles, it can be seen that the peak values of RCP-1 ~ RCP-3 are 0.19, 0.22 and 0.25 respectively. Compared with RCP-1, RCP-2 and RCP-3 increase by 15.7 % and 31.6 %, respectively, indicating that increasing the reinforcement ratio can improve the energy dissipation capacity of the pile-soil system and the seismic capacity of the model pile.

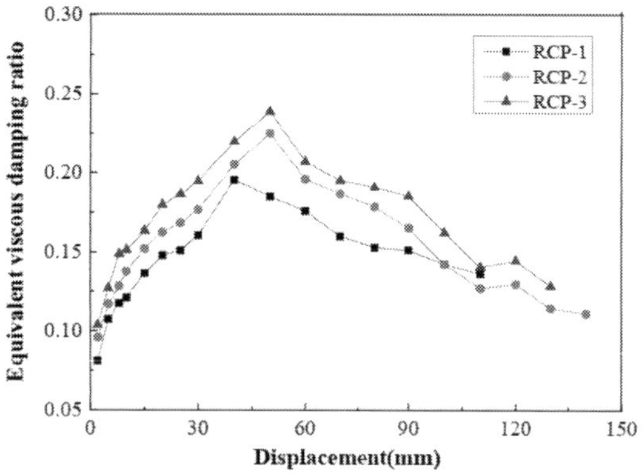

Figure 9. Equivalent viscous damping ratio of piles.

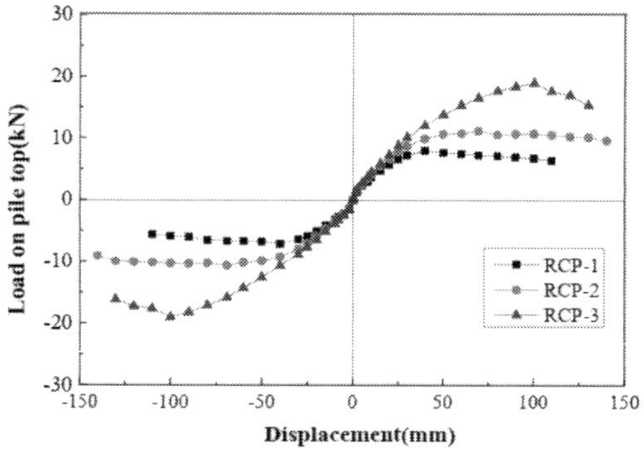

Figure 10. Skeleton curves.

3.3. Skeleton Curve

According to the skeleton curve, the bearing capacity and deformation capacity of the pile-soil system can be evaluated. Figure 10 shows the skeleton curves of three model piles. It can be seen from Figure 10 that the skeleton curve is symmetrical in the forward and reverse directions. The skeleton curve can be divided into three parts: elastic stage, elastic-plastic stage and failure stage. It can be seen from Figure 10 that with the increase of reinforcement ratio, the deformation capacity of the model pile is significantly improved, the ultimate bearing capacity is correspondingly improved, and the corresponding displacement when reaching the ultimate bearing capacity is also significantly increased. Therefore, increasing the reinforcement ratio of model piles can improve the horizontal deformation capacity of piles, that is, RC piles with high reinforcement ratio have better horizontal deformation resistance.

Figure 11 compares the load and displacement of three model piles under elastic limit state. The ultimate elastic loads of RCP-1~RCP-3 are 2.8kN, 4.3kN and 6.0kN respectively, and the corresponding ultimate elastic displacements are 8.0mm, 10.0mm and 15.0mm respectively. Compared with RCP-1, the ultimate elastic load of RCP-2 and RCP-3 increased by 42.3% and 110.7% respectively, and the ultimate elastic displacement increased by 25.0% and 87.5% respectively. It shows that increasing reinforcement ratio can significantly improve the deformation capacity and bearing capacity of model piles in elastic state.

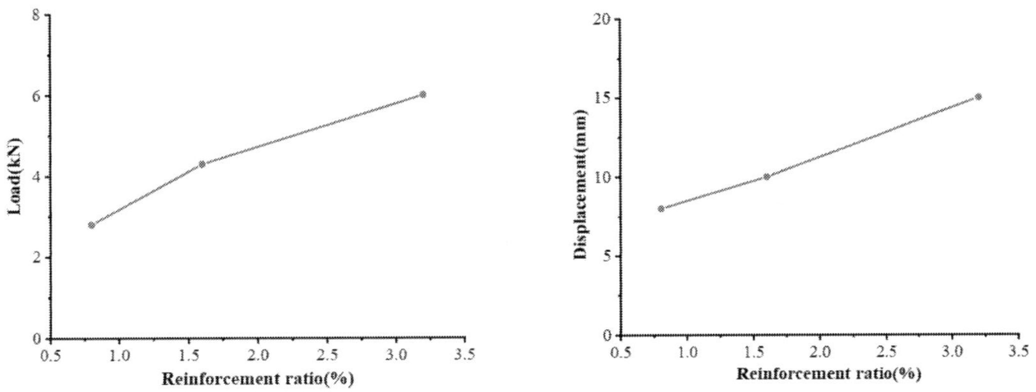

(a) Elastic ultimate load of each pile (b) Elastic ultimate displacement of each pile

Figure 11. Curves on elastic limit state.

Figure 12 shows the load and displacement of three model piles under the ultimate limit state of bearing capacity. It can be seen from figure 12 that the ultimate loads of RCP-1~RCP-3 are 7.5kN, 11.1kN and 19.0kN respectively, and their corresponding ultimate displacements are 40.0mm, 70.0mm and 100.0mm respectively. Compared with RCP-1, the peak load of RCP-2 and RCP-3 increased by 48.0% and 153.3% respectively, and the peak displacement increased by 75.0% and 150.0% respectively. It shows that increasing reinforcement ratio can significantly improve the bearing capacity and deformation capacity of model piles in ultimate state.

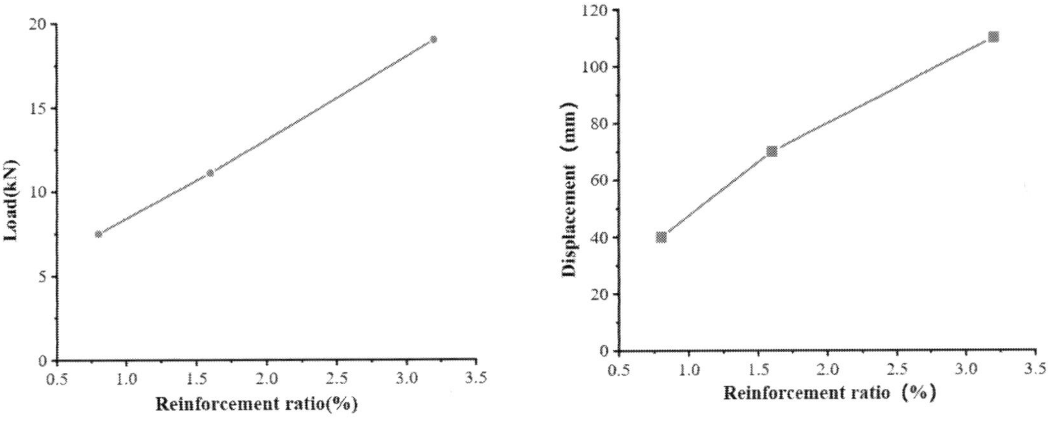

(a) Ultimate load of bearing capacity of each pile (b) Ultimate displacement of bearing capacity of each pile

Figure 12. Curves on ultimate limit state.

3.4. Distribution Rule of Horizontal Deformation of Pile Along Depth Direction

Figure 13 shows the distribution law of horizontal deformation along the depth direction in the whole process of model pile test. Because the three model piles have similar horizontal deformation distribution rules under displacement load, only RCP-1 model piles are given. It can be seen from figure 13 that the distribution law of horizontal deformation of pile body gradually decreases with the increase of depth, and horizontal deformation which is obviously consistent with the loading side appears at the measuring points L1~L7 of pile body (buried depth 0.0~1.0m); And there is obvious reverse direction at the measuring point L8 of pile body (buried depth of 1.2m), as shown in figure 13. Starting from L10 measuring point (buried depth: 1.6m), the horizontal displacement of pile body is very small and basically zero.

Figure 14 is a comparison of horizontal deformation of three model piles under 20mm displacement load. It can be seen from figure 14 that the horizontal deformation distribution laws of the three piles are basically the same. It shows that the change of reinforcement ratio of model pile has little effect on the global bending stiffness of pile-soil system. However, under the same displacement load, the deformation of RCP-1 in shallow layer soil is too large, which shows that the integrity of pile foundation with low reinforcement ratio is lower than that with high reinforcement ratio.

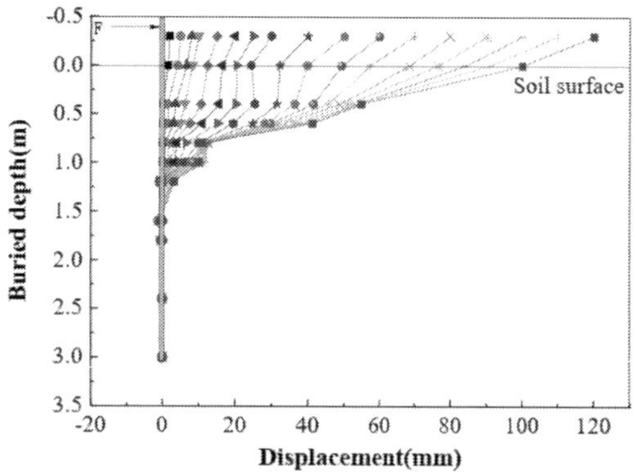

Figure 13. Distribution of horizontal deformation of RCP-1.

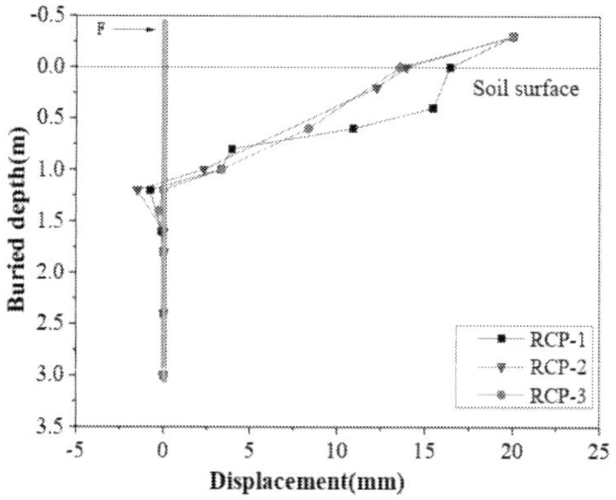

Figure 14. Horizontal deformation of piles under 20mm displacement load.

4. Conclusion

In this paper, the low-cycle quasi-static test of high-reinforced RC pile-soil interaction is carried out, and the following conclusions are obtained:

(1) By comparing the failure position, hysteretic curve, skeleton curve, equivalent viscous damping coefficient and deformation of three model piles, it basically shows that increasing the reinforcement ratio of RC piles can improve the pile-soil interaction effect, horizontal deformation, bearing capacity and energy dissipation capacity.

(2) The damaged area of the pile body is 2.8D～6.8D. Increasing reinforcement ratio has great influence on the failure of model pile, and the range of pile failure increases and develops along the buried depth to the deep; The effective length of pile-soil interaction increases.

(3) In the early stage of loading, the skeleton curve shows linearity; after concrete cracks, the load increases and slows down, which shows nonlinearity. Increasing the reinforcement ratio can improve the horizontal deformation capacity and bearing capacity of model piles. When the reinforcement ratio increases from 0.8% to 3.2%, the load and displacement corresponding to the elastic limit state of the model pile increase by 110.7% and 87.5% respectively; The load and displacement corresponding to the ultimate state of bearing capacity increased by 153.3% and 150% respectively.

(4) The equivalent viscous damping coefficient increases first and then decreases with the increase of displacement load. Compared with RCP-1, the maximum equivalent viscous damping ratio of RCP-3 is increased by 31.6%, its hysteretic curve is fuller, and the energy consumption of pile-soil system is stronger.

In a word, increasing the reinforcement ratio can significantly improve the mechanical performance and deformation capacity of RC pile foundation. However, only RC piles with reinforcement ratio of 0.8%, 1.6% and 3.2% are tested in this paper, and the reinforcement of RC piles can be optimized by finite element simulation and in-depth analysis. In addition, RC piles with high reinforcement can be combined with reaming at the top of piles to study their horizontal deformation ability.

References

[1] Huang Fu Y, Zhuang Y Z, Fu C, Qian H M. Chen B C. 2015 Review on the seismic performance and simplified design method of jointless bridge [J] *Earthquake Engineering and Engineering Dynamics* **35** (5) 15-22(in Chinese).

[2] Wolde Tinsae A M, Klinger J E, Whitr E J. 1988 Performance of jointless bridges [J] *ASCE Journal of Performance of Constructed Facilities* **2** (2) 111-125.

[3] White II H, Collin P. 2010 Integral abutment bridges: the European way [J] *ASCE Practice Periodical on Structural Design and Construction* **15** (3) 201-208.

[4] Civjan S A, Bonczar C, Brena S F. 2007 Integral abutment bridge behavior: Parametric analysis of a Massachusetts bridge [J] *Journal of Bridge Engineering* **12** (1) 64-71.

[5] Kong B, Cai C S, Zhang Y. 2016 Parametric Study of an integral abutment bridge supported by prestressed precast concrete piles [J] *Engineering Structures* 120 37-48.

[6] Chen B C, Zhuang Y Z, Huang F Y, Bruno Briseghella. 2019 *Jointless Bridges* [M] Beijing: China Communication Press (in Chinese).

[7] GB/T 50476-2019, Standard for design of concrete structure durability[S].

[8] Davids W G, Sandford T, Delano J, Ashley S, Delano J, Lyons C. 2010 Field-measured response of an integral abutment bridge with short steel H-piles [J] *Journal of Bridge Engineering* **15** (1) 32-43.

[9] Razmi J, Ladani l, Aggour M S. 2013 Fatigue crack initiation and propagation in piles of integral abutment bridges [J] *Comput-aided Civil and Infrastructure Engineering* **28** (5) 389-402.

[10] Far N E, Maleki S, Barghian M. 2015 Design of Integral abutment bridges for combined thermal and seismic loads [J]. Earthquakes and Structures **9** (2) 415-430.

[11] Zhao Q H, Qi Z Y, An Z Y, Chen B C. 2018 Parametric analysis on seismic behavior of integral abutment steel bridge considering SSI [J] *Journal of Transportation Engineering* **18** (5) 35-46(in Chinese).

[12] Gama D, Almeida J F. 2014 Concrete integral abutment bridges with reinforced concrete piles [J] *Structural Concrete* **15** (3) 292-304.

[13] Hong J X, Peng D W. 2002 Research on the loaded property of integral abutment bridges with flexible piles [J] *China Journal of Highway and Transport* **15** (4) 43-48(in Chinese).

[14] Jin X Q, Shao X D. 2002 Study and practice on integral abutment bridge [J] *Journal of Chongqin Jiaotong University* **3** (21) 7-10(in Chinese).

[15] Chen B C, Fu C, Zhuang Y Z, Bruno B. 2018 Application status and development strategies of jointless bridge in china[J] *Journal of China & Foreign Highway* **38** (1) 87-95(in Chinese).

[16] Chen B C, Chen G D, Su J Z, Lin S S, Xue J Q, Tabatabai H. 2018 Trial-design Study on integral abutment bridge by using UHPC-RC stagewise piles[J] *Journal of architecture and Civil Engineering* **35** (1) 1-8(in Chinese).

[17] Peng D W, Chen C W, Hong J X. 2005 Study of loaded property of abutment node of integral abutment bridges[J] *China Journal of Highway and Transport* (01) 50-54(in Chinese).

[18] Cheng L X, Cheng J F, Wang J J, Wang S Z. 2018 Experimental study on p-y curve of dynamic interaction between micro-pile with predrilled oversize hole and soil [J] *Journal of Water Resources and Architectural Engineering* **16** (1) 156-163(in Chinese).

[19] Dunker K F, Liu D. 2007 Foundations for integral abutments [J] *Practice Periodical on Structural Design and Construction* **12** (1) 22-30.

[20] Zhuang Y Z, Huang F Y, Qian H M, Xiong Z H. 2017 Pseudo-static test on mechanic behavior of PHC piles with soil-pile interaction [J] *China Journal of Highway and Transport* **31** (4) 42-51(in Chinese).

[21] Huang J, Shield C K, French C E W. 2015 Parametric study of concrete integral abutment bridges [J] *Journal of Bridge Engineering* **13** (5) 511-526.

[22] Wang X Q, Guo X Y, Yan G Q. 2018 Parametric study on the mechanical properties of integral bridges[J] *Journal of Railway Science and Engineering* **15** (9) 2276-2284(in Chinese).

The Numerical Analysis of Tunnel Traversing Piles and Its Seismic Response in Deep Soft Soil

Xiaomin Liu[1], Liwei Song[1], Songbao Cai [1], Xin Zhao[2], Peng Xie[3]
[1] China Construction Sixth Engineering Bureau Co,.Ltd., No.219 Liuwei Road, Hedong Dist., Tianjin, P.R. China
[2] College of Civil Engineering, Hebei University of Engineering, No.19 Taiji Road, Handan Economic and Technological Development Zone, Hebei, P.R. China
[3] Tianjin Saiying Engineering Construction Consultancy Management Co., Ltd, No.239 Yingkou Road, Heping Dist., Tianjin, P.R. China
Email: 56370328@qq.com

Abstract. The Ou River estuary deposits deep soft silt and silt clay, the pipe-jacking tunnel departures from Qidu island, then passes underneath north levee road, viewing platform, Ou River and Ou River road, finally reaches the received well. The net distance between tunnel segment and the viewing platform piles is only 230mm. In order to analyze the negative influence of large diameter pipe-jacking tunnel traversing piles of viewing platform, the numerical model is established, and the levee road settlement value is found which matches with observed settlement very well. The dynamic analysis is also conducted by EI Centro seismic wave, and the result reveals that the max displacement occurs on the segment between the rear row piles of viewing platform, and the max displacement reaches 68.6mm in the earthquake.

1. Introduction

The soft soil layers such as silt and silty clay distributes in Tianjin, Shanghai, Wenzhou and some other coastal cities extensively, and the soft soil layers are higher water content and less deformation modulus, so the large displacement and deformation occur obviously when the soft soil was disturbed. Underground engineering Construction is difficult in deep soft soil layers, especially in the area where is close to structures, underground utilities. Construction work is easy to bring adverse effect on the adjoining structures, so the engineering analysis is necessary before the project construction, and then take some measures to reduce adverse effect for nearby structures.

2. Project Profile

The power tunnel crosses Ou River in Wenzhou city, and the beginning well sunk in Qidu island while the receiving well sunk in Ou River road treelawn. The total pipe-jacking distance is 1202m. The power tunnel starts from Qidu island, then passes underneath levee crown road, Ou River and Ou River road consecutively, and reaches the receiving well at last. The power tunnel's external diameter is 4.14m while the distance between adjacent piles is 5.4m. The concrete piles' diameter is 800mm. The net distance between tunnel and piles is only 0.23m.

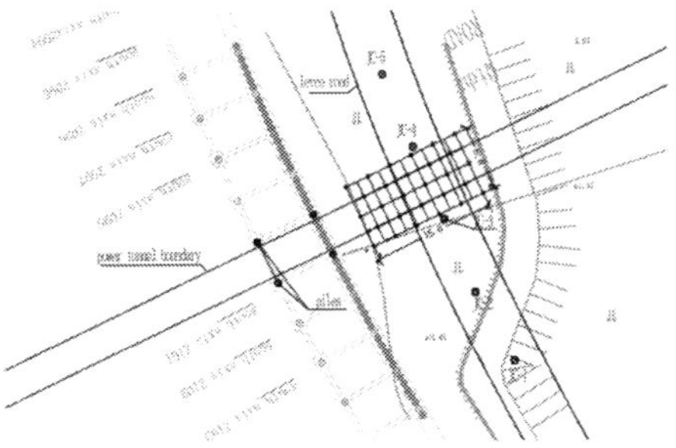

Figure 1. The relative position of tunnel and piles.

3. Project Geological Condition

The geological exploration reveals 5 layers by physical & mechanical characters, lithology & hidden rule. Some geological layers are subdivided into secondary layers and sandwich layers. The soil layers is ①$_1$ miscellaneous fill, ②clay, ③$_1$ silty sand,③$_1$ silt, ③$_2$ sandy silt, ③$_3$ silt, ③$_4$ sandy silt, ④$_1$ silty clay, ⑤$_1$ clayey silt, ⑤$_{11}$ clay,⑤$_{12}$ silty clay, ⑤$_2$ pebble respectively from top to bottom. The power tunnel is constructed in silt layer, and the soil's parameters are listed in table1:

Table1. The physical & mechanical parameters of soils.

Serial number	layer	depth/m	Gravity /kN/m³	c/kPa	φ/°	Es/ MPa
1	clay	1.2	16.7	19.98	12.33	2.42
2	Sandy silt	4.3	16.9	4.68	18.63	2.11
3	silt	21	17	12.1	10	3.16
4	Silty sand	1	18.6	10	18	6.08
5	Silty clay	9.1	17.5	17	13.8	3.29
6	Clayey silt	2.5	19.9	13.01	23.22	8.55
7	pebble	20.9	22	0	32	30

4. Pipe Jacking Passes Through Closely to the Viewing Platform Piles

The viewing platform basement includes double row piles. The diameter of front row piles is 800mm whose axial distance is 5.4m. The pile length reaches 35m, which is constructed by C25 concrete. The length of back row pile is 15m, and other parameters are same to front row pile. The relative position of tunnel and piles is shown in the figure.2.

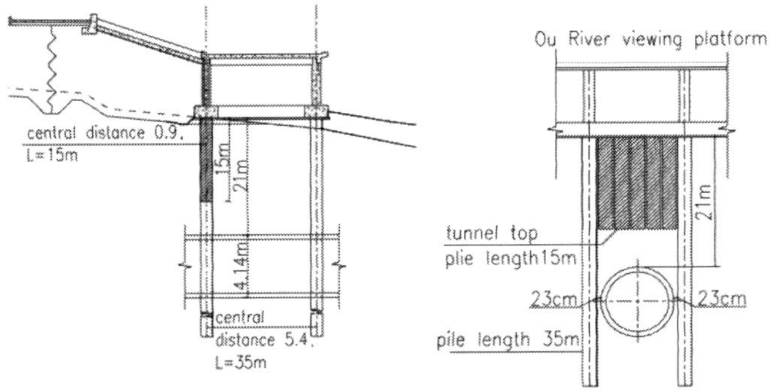

(a)The longitudinal profile(b)The cross section profile

Figure 2. The relative position of tunnel and pile.

In order to survey the concrete pile's real position and quality, the piles are need to be measured before the tunnel passes through. The survey method is in figure3.

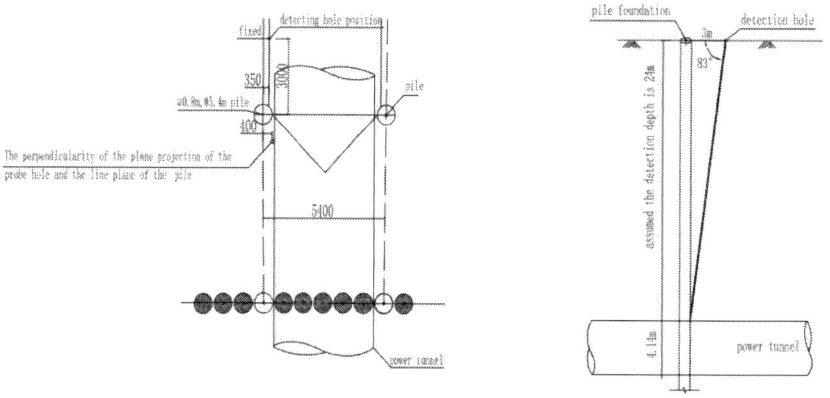

(a)The planform of piles and sound-pipes (b)The profile of piles and sound-pipes

Figure 3. The net distance surveying between piles.

The surveying is conducted as follow:

1) Place the piles and sound-pipes exactly, and vertical plan of sound-pipes parallels the vertical plane of electric power tunnel axis.

2)Drilling vertical hole in the site of surveying inclined hole, and the axis is projected on the levee slope, thus control the precision of vertical plan of surveying inclined hole.

3) Drill stem drills into soil slantways at a specific angle, and the aiguille diameter is 20mm. In the middle of exploration hole, the hole diameter expands to 100mm to set oriented pipe.

4) Stop drilling when the drill stem touches pile surface, and records the drilling length, then puts the inclinometer into the hole. In order to guarantees the cross oriented slot parallels the vertical plane, the cross slots posture should be control accurately.

5) Each measuring point was measured by inclinometer along two orthogonal planes of the cross guide groove of the surveying pipe. Within the height range of the tunnel pipe, at least three measuring points were given for each pile to give a more accurate net distance between piles. The detection results show that the theoretical value of net distance between piles in the front row adjacent river is 4600mm, and the measured value is 4564mm. The theoretical value of net distance between rear piles is 4730mm, and the measured result is 4693mm. Both front and rear piles have small protruding, but it can meet the requirements of pipe jacking passing through the piles.

5. Adverse Effect of Pipe Jacking on the Embankment Top Road

In order to analyze the adverse effects of pipe jacking on the top road of Ou River embankment, settlement observation points are uniformly arranged in the direction perpendicular to the axis of the power tunnel on the top road of the embankment. The six monitoring points close to the tunnel is located as shown in figure 4:

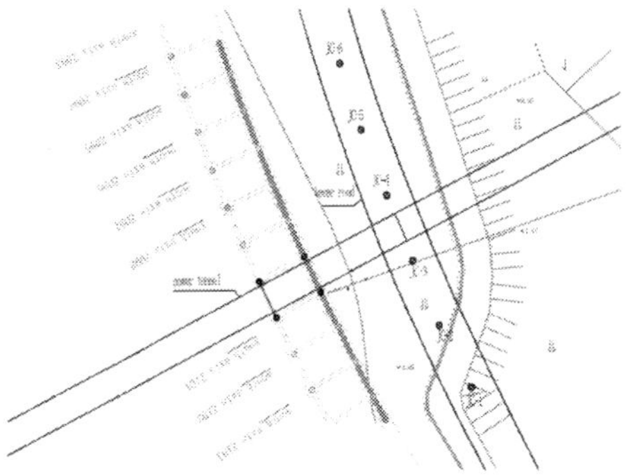

Figure 4. The service point of levee road displacement.

The corresponding relationship between the observation point settlement and the horizontal distance to the tunnel central axis is shown in Figure 5. The monitoring points' settlement law is obtained, as shown in Formula (1):

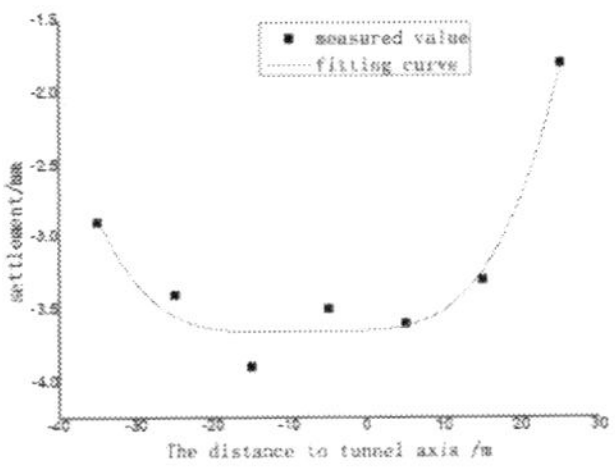

Figure 5. The settlement of observation points.

$$s = 2 \times 10^{-6} \cdot D^4 + 5 \times 10^{-5} \cdot D^3 + 5 \times 10^{-4} \cdot D^2 + 2.1 \times 10^{-3} \cdot D - 3.6485 \quad (1)$$

where:
$R^2 = 0.961$
s — settlement, mm;
D—horizontal distance to the tunnel central axis, mm.

6. Numerical Simulation of the Impact of Pipe-Jacking on the Environment Near Viewing Platform Pile Foundation

According to the soil layers revealed in the geological exploration report, a calculation unit with a width of 6m is selected. The numerical model is shown in figure 6:

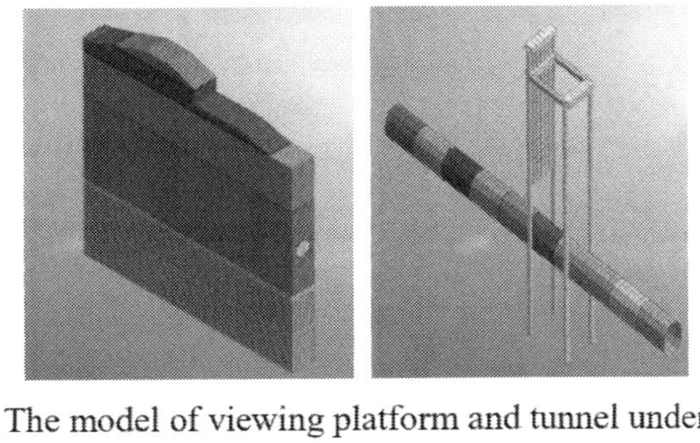

Figure 6. The model of viewing platform and tunnel under the levee.

The design elevation of the water level of Ou River is 0.8m, and the elevation of the XK-13 hole near the viewing platform is -6.12m. Considering that the water level reaches the design level in the process of pipe jacking construction, 7m water pressure is loaded on the bottom of the river. As the pushing force of the pipe jacking cutter head is balanced with the active earth pressure on the face of the tunnel, the pushing force is acted on the face of the tunnel during the pipe jacking process in the numerical model. When the pipe-jacking passes through the embankment and the pile foundation of viewing platform, the maximum calculated settlement of the embankment is 2.0mm, and the actual observed average value of the observation points is 3.2mm. The difference between the calculated value and the measured value is 1.2mm. This difference may be caused by the change of the swelling mud pressure around the segment after the pipe-jacking cutter head cutting the soil. When the pipe is pushed through, the settlements of embankment top road and the piles are shown in figure 7.

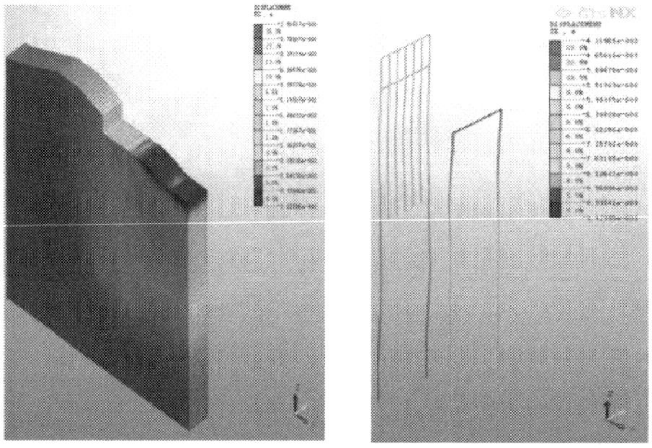

Figure 7. The soil and piles displacement when machine jacking through the levee road.

After the pipe jacking through the embankment and viewing platform, the distribution of von-Mises stress on the segment structure is shown in figure 8:

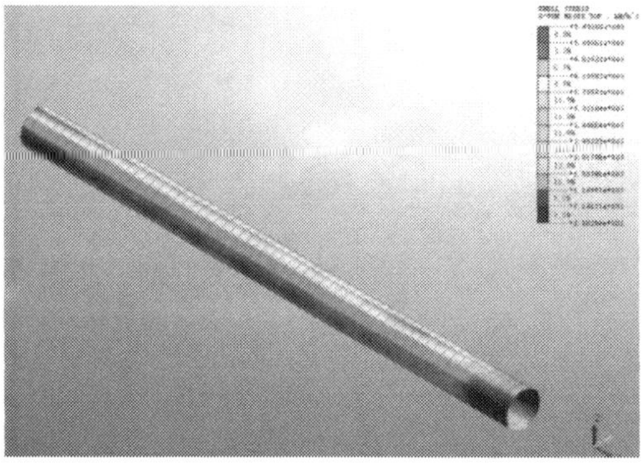

Figure 8. The Von-Mises distribution on segments.

7. Study on Seismic Dynamic Response of Tunnel Structure and Soil

The 1940 EI Centro Site seismic wave was adopted in numerical model. The peak acceleration is 0.3569g and the duration of the seismic wave is 53.72 seconds. The acceleration curve of the seismic wave is shown in figure 9.

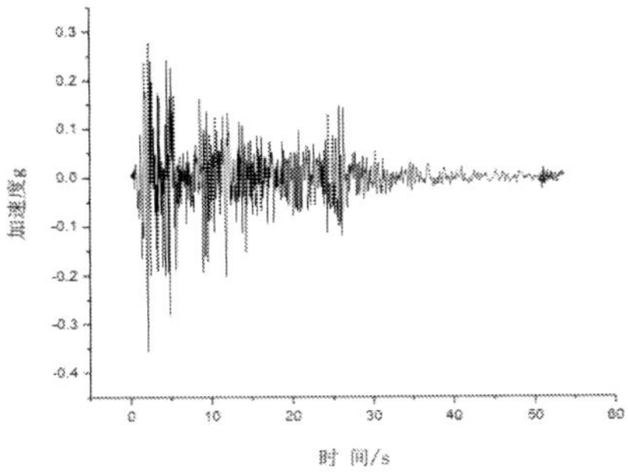

Figure 9. The curve of earthquake acceleration.

The top six vibration modes of piles and tunnel are shown in figure 10:

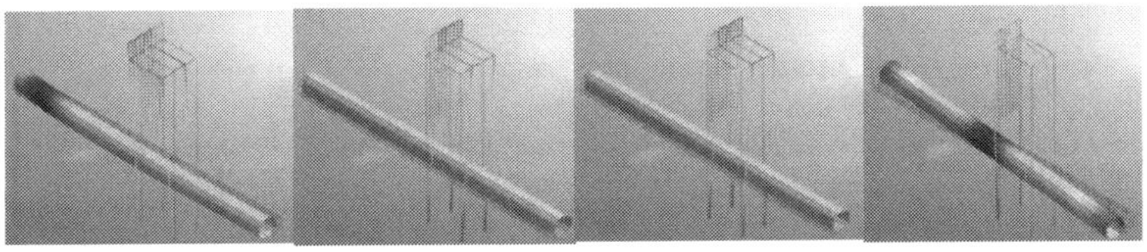

(a)F=0.2635Hz (b)F=0.3016Hz (c)F=0.3032Hz (d) F=0.7327Hz

(e)F=1.3359Hz (f)F=1.3437Hz

Figure 10. The first six modes of piles and tunnel.

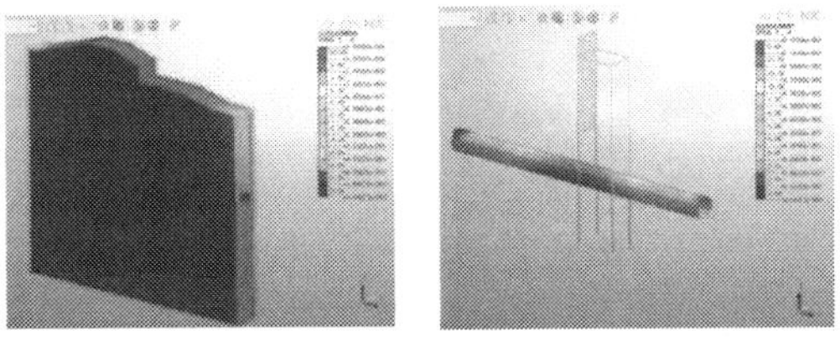

Figure 11. The max displacement of tunnel and piles in the earthquake.

As can be seen from figure 11, under the action of EI Centro seismic wave, the maximum displacement of the tunnel structure is 68.6mm. The maximum displacement pipe is between the rear row piles underneath the viewing platform. With the increase of the distance to the pile foundation, the maximum displacement of the pipe segments decreases gradually.

8. Conclusion

The Ou River power tunnel has an outer diameter of 4.14m and passes between two 35m piles on the north landscape platform. The net distance between the surface of the segment and the piles is only 230mm. By the accurate detection of pile foundation and the precisely control of pipe-jacking during construction, the pipe-jacking tunnel traversed the pile foundation successfully, and the settlement of embankment top road was controlled in the limitation, and the average settlement of measuring points is only 3.2mm.

Acknowledgment
Supported by: CSCEC-2020-Z-21.

References

[1] Hu X C, Zhang G Q, Ye J R. 2021 Research on construction technology of large section and long distance slurry balance rectangular pipe jacking [J] *Construction Technology* 50 105-107.
[2] Luo G, Zhang Y L, Pan S K, Ren Yi. 2021 Dynamic response analysis of submerged floating tunnels to coupled wave-seismic action [J] *Engineering Mechanics* 38 211-220.
[3] Zhang Ji, Tan C X, Huang Y, Huang L H, Xu K C. 2020 Dynamic response analyses of soft soil-tunnel-aboveground frame systems under earthquake [J] *Journal of Vibration Andshock* 39 278- 284.
[4] Ai S J, Liang Q G, Zhao T, Qiao X J, Wang L L, Zhou W D. 2020 Shaking table experimental study on seismic dynamic response of slope-tunnel system at the portal section of loess tunnel [J] *Earthquake Engineering and Engineering Dynamics* 40 168 -176.
[5] Yang H, Liu C, Jiang X L, Shi H T, Sun G C. 2020 Study on seismic response law of shallow-buried bias double-arch tunnel [J] *Journal of Natural Disasters* 29 161- 172.
[6] Zhang C M, Zhong Z L , Zhen L B, Shen Y Y, Zhao M. 2021 Seismic intensity measures for the damage evaluation of circular tunnels [J] *Engineering Mechanics* 38 100- 106.

Settlement Analysis of Sand Compaction Pile Composite Foundation in Transition Section of Immersed Tube Tunnel

Jinfang Hou[1, 2, 3, 4] and Jian Yu[1, 3, 4]

[1] CCCC Tianjin Port Engineering Institute Co., Ltd., Tianjin 300222, China.
[2] CCCC First Harbor Engineering Company Ltd., Tianjin 300461, China.
[3] Key Laboratory of Port Geotechnical Engineering of Tianjin, Tianjin 300222, China.
[4] Tianjin Key Laboratory of Construction, Operation and Maintenance Technologies for Underwater Tunnels, Tianjin 300461, China.
Email: jinfanghou@163.com

Abstract. The foundation of the immersed tube transition section of the Hong Kong-Zhu Hai-Macao Bridge Project is distributed with thick silty soil. The method of sand compaction pile + surcharge preloading is used to treat the soft soil ground. In order to determine the foundation consolidation settlement and subsequent residual settlement in the transition section of the immersed tube tunnel, a monitoring system using long wires combined with wireless transmission and long-distance data collection was developed to obtain the measured foundation settlements during the surcharge period. After comparing the measured value with the calculated value, the formula for the composite foundation of the sand compaction pile was revised to obtain a more reasonable residual settlement of the foundation, which could guide the design and construction of the immersed tube tunnel.

1. Introduction

The immersed tube tunnel of the Hong Kong-Zhuhai-Macao Bridge is 5664m long and consists of 33 sections of immersed tubes and 1 final joint. It is the first offshore immersed tube tunnel in China. As it is located in the open sea, the foundation is mainly soft soil, and the soft foundation needs to be treated before it can be used as the foundation of an immersed tube tunnel. Especially in the transition section, the amount of foundation excavation is small, so the soft soil is deep, which is prone to excessive residual settlement, resulting in tube joint cracking and water inflow. The foundation treatment design of the immersed tube transition section of the west artificial island adopts the method of sand compaction pile + surcharge preloading. The sand compaction pile are arranged in a square shape, the spacing is 1.8m-2.2m, the pile diameter is 1.5m-1.7m, and the replacement rate is respectively 42%, 55% and 70%, the pile length is 15.6m-25m. After the construction of the sand compaction pile is completed, the foundation is preloaded by the application of upper riprap, and the load is about 6MPa, and when the consolidation degree of foundation reaches more than 90%, and the residual settlement of foundation is reduced through the surcharge preloading. In view of the situation where the water depth of the transition section is more than 20m, the offshore is more than 200m, and the environment is harsh, it is of great significance to develop appropriate monitoring methods to obtain effective measured data during the stacking period, and correct the settlement calculation method of sand compaction pile composite foundation and determine the reasonable residual settlement of foundation through the comparative analysis of calculated data and measured data, so as to ensure the safety of immersed tube tunnel.

2. Engineering Geology

The elevation of the original mud surface of the transition section of the immersed tube tunnel is about -8.0m~-10.0m, the surface layer is silt and silty clay layer, and the lower part is the super-consolidated soil layer $(3)_1$ clay layer. The silt layer has high water content and high compressibility; the super-consolidated soil layer has greater strength and strong resistance to deformation. The foundation treatment penetrates the silt layer and enters the bottom of the super-consolidated soil layer $(3)_1$ clay layer. The parameters of the formation to be treated are shown in table 1.

Table 1. Stratum parameters.

Stratum number and name	Bottom elevation (m)	Unit weight(kN/cm^3)	Porosity ratio	Undrained strength (kPa)
$(1)_1$Silt and silty clay	(-28.3) - (-31.0)	16.2	1.732	23.0
$(2)_1$Clay	(-29.4) - (-31.78)	18.7	0.866	42.7
$(3)_1$Clay	(-34.92) - (-36.02)	18.1	1.052	124.0

3. Settlement Analysis of Sand Compaction Pile Composite Foundation

3.1. Settlement Monitoring of Sand Compaction Pile Composite Foundation during Surcharge Preloading Period

The sand compaction pile are constructed by sand pile ships, and the top of the foundation is pre-loaded after the construction is completed. As the water depth of the surcharge preloading area exceeds 20m and the distance from the artificial island is more than 200m, it is very difficult to obtain foundation settlement data during the preloading period. The traditional manual measurement does not have the conditions for implementation, and it is necessary to use automated monitoring methods to obtain actual measurement data. Therefore, after the long wire is guided along the axis of the immersed tube tunnel to the steel cylinder of the artificial island, the method of automatic acquisition and wireless transmission is carried out. This method also avoids the underwater sealing treatment of automatic acquisition and transmission equipment, is convenient for maintenance, reduces the risk of the entire monitoring system, ensures the continuity and accuracy of data, and saves the costs. In this method, the protection of the wire is very important, so the wire is protected by a protective hose after passing through the hole, and fixed on the ground with a "U"-shaped clip, with a sand bag on it, so that the wire will not be protected from water flow and construction before being stacked, so as to ensure the acquisition of measured data.

The surface settlement under water is measured by a liquid differential pressure settlement instrument, and the layered settlement of the foundation is observed by embedding a layered single-point displacement meter. A total of 6 monitoring sections are set up, of which the sections A, D, E, and F are embedded with 3 sets of liquid differential pressure sedimentation meters and a group of layered single-point sedimentation meters. Sections B and C are only embedded with liquid differential pressure sedimentation meters. The 3 sets of liquid differential pressure sedimentation meters are respectively installed at the edges and the middle of the immersed tube tunnel; a single-point displacement meter is placed every 3m along the depth direction to form a layered sedimentation meter, which is placed in the middle of the immersed tube tunnel. The layout of monitoring points is shown in figure 1.

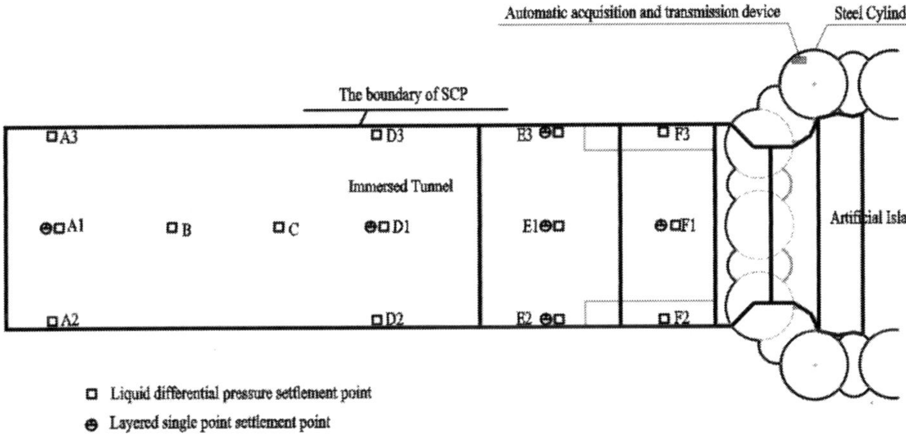

Figure 1. Layout of monitoring points.

During the monitoring process, the sensor at point C was damaged due to the impact of the stacking construction, and the other measuring points were intact, and real-time, effective and continuous monitoring data were obtained.

3.2. Settlement Calculation of Sand Compaction Pile Composite Foundation

The existing domestic settlement calculations for sand compaction pile composite foundations are mostly estimated by the calculation method recommended by the Japanese code [1], and the calculation formula is shown in equation (1).

$$S' = \beta \sum_{i=1}^{k} m_{vi} \Delta P_i h_i \tag{1}$$

In the formula, S', settlement of composite foundation, mm;β, settlement reduction rate, and β=1/(1+(n-1)m), %, n is the pile-soil stress ratio, m is the replacement rate of sand pile area,%; m_{vi}, the volume compressibility of the soil layer i;ΔP_i, the average additional stress of the soil layer i, kPa。

According to this method, the calculated settlement results of each monitoring point of the sand compaction pile composite foundation in the transition section during the preloading period are shown in table 2.

According to the calculation results in table 2, it follows that due to the installation of sand compaction pile on the soft foundation, the settlement of the foundation is greatly reduced relative to the natural foundation, from the 274mm-407mm of settlement of natural foundation to about 52mm-109mm at 90% degree of consolidation, the settlement is greatly reduced, indicating that the use of sand compaction pile and surcharge preloading can effectively reduce the settlement of the foundation.

Table 2. Settlement calculation value of sand compaction pile composite foundation during preloading period

Monitoring point	Replacement rate (%)	Height of surcharge preloding (m)	Calculated settlement of foundation (mm)	Calculated settlement of 90% consolidation degree (mm)
A1	42	14.5	308.7	78.8
A3	42	14.5	268.8	52.3
B	42	15.5	274.5	65.7
D	42	13.7	278.5	69.6
D2	42	13.5	340.6	79.8
D3	42	13.3	274.5	68.6
E1	55	14.4	382.6	100.6
E2	55	14.4	314.1	86.1
E3	55	14.4	407.0	109.4
F1	70	15.6	314.1	64.8
F2	70	16	407.0	101.8
F3	70	15.5	382.6	94.2

4. Result Analysis

4.1. Comparison of Calculated Value and Measured Data

The calculated value of foundation settlement at each monitoring point in table 2 is compared with the measured value, as shown in figure 2. It can be seen that the calculated settlement of the area where the replacement rate of sand compaction pile is 42% is close to the measured settlement. In areas with higher replacement rates of 55% and 70%, the calculated values differ greatly from the measured values. When the replacement rate is 55%, the difference between the theoretical calculation value and the measured value of the local position is nearly double. Therefore, when the high replacement rate of sand compaction pile is used for foundation treatment, the calculation result of the settlement formula of the sand compaction pile composite foundation is too large to truly calculate the consolidation settlement and residual settlement of the foundation, so the calculation formula needs to be modified[2,3].

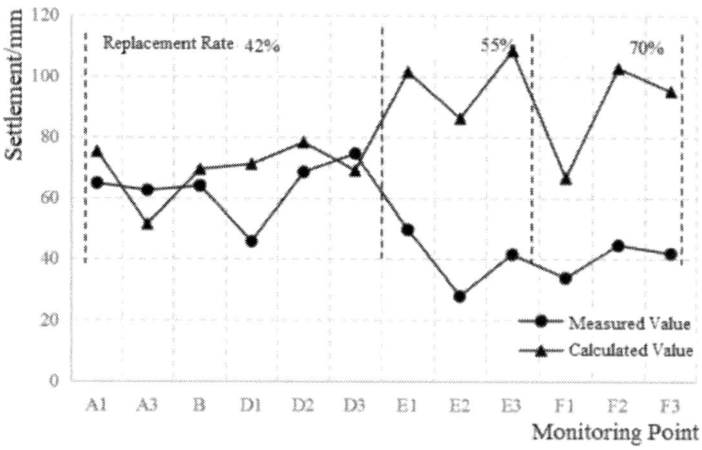

Figure 2. Comparison between calculated value and measured value.

4.2. Modification of Calculation Formula

In view of the large deviation of the settlement calculation results of the sand compaction pile foundation with high replacement rate, a fitting correction method to the original formula is proposed based on the field measured data. In order to further clarify the settlement law of the sand compaction pile foundation and eliminate the discrete influence of the different stratum distribution of each section on the settlement law of the foundation, the settlement reduction ratio coefficient of sand compaction pile composite foundation is introduced [2, 4], which is the ratio of settlement of sand compaction pile composite foundation to that of natural foundation.

In figure 3, it shows the distribution of the measured settlement reduction ratio under different replacement ratios for the composite foundation of sand compaction pile in the transition section of the immersed tube tunnel. The curve in the figure is the calculated settlement reduction ratio curve. Compared with the measured data, it can be seen that as the replacement rate increases, the settlement reduction ratio has a significant decreasing trend. However, after the replacement rate is greater than 50%, the overall distribution trajectory of the measured data is quite different from the theoretical calculation formula (η =1 in figure 3), and the distribution is more consistent with the curve characteristics of η =2.5.

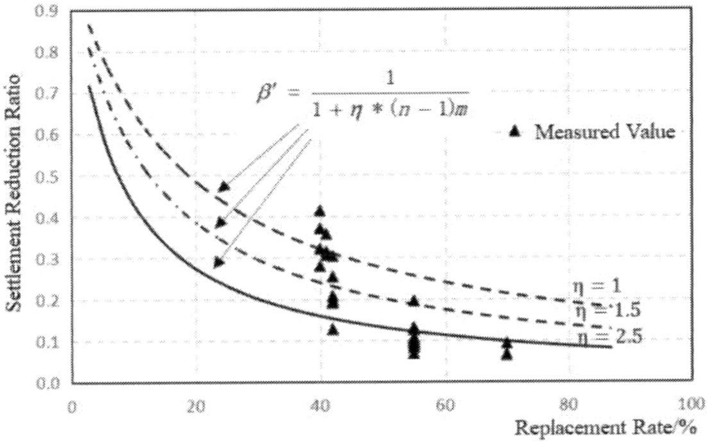

Figure 3. Settlement reduction ratio coefficient.

Therefore, the settlement reduction rate β in the formula (1) for the settlement calculation formula of the sand compaction pile composite foundation is corrected to, and the calculation formula is equation (2).

$$\beta' = \frac{1}{1+\eta(n-1)m} \tag{2}$$

In the formula, η , empirical correction coefficient, when the replacement rate m<0.5, take 1~1.5, when m≥0.5, take 2.5.

This modified formula is used to calculate the settlement of the composite foundation with high replacement rate sand compaction pile in other areas, as shown in figure 4. It can be seen that except for individual points and factors affecting the complex situation of measured data, the both are relatively consistent, indicating that the modified formula has good applicability and can be used as the prediction, calculation and analysis of settlement of high replacement rate sand compaction pile composite foundation.

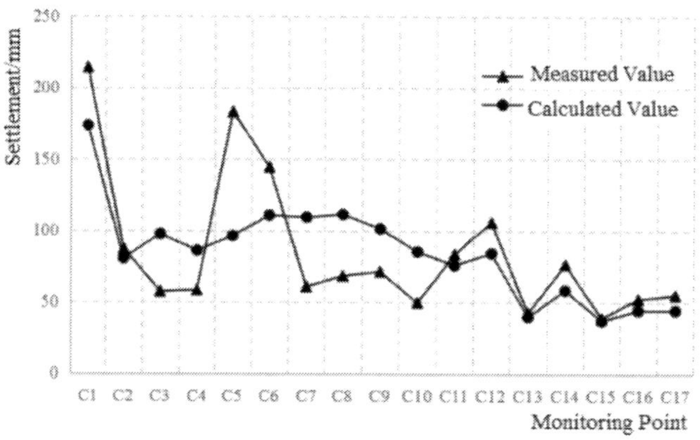

Figure 4. Comparison of the calculated value and the measured value.

Using this formula, the residual settlement of the foundation in the transition section of the immersed tube is calculated, and the residual settlement is less than 10mm, indicating that the soft foundation has been well strengthened.

5. Conclusion

According to the above analysis, the conclusions are as follows:

(1) The settlement of the composite foundation of high-replacement sand compaction piles calculated by the commonly used formula is quite different from the actual situation, and the calculated settlement is too large. When the replacement rate of sand compaction pile reaches more than 50%, the reduction rate should be corrected before the settlement can be more accurately estimated.

(2) In engineering application, appropriate monitoring technology should be adopted according to actual engineering conditions. This time, the long wire is guided along the axis of the immersed tube tunnel to the steel cylinder of the artificial island, and then automatically collected and wirelessly transmitted on land, which reduces the risk of the entire monitoring system for the transition section of the immersed tube tunnel of the Hong Kong-Zhuhai-Macao Bridge and ensures the continuity and accuracy of the data.

(3) The sand compaction pile + surcharge preloading treatment method makes the soft foundation of the immersed tube transition section better improved, and the residual settlement is small, which can meet the service requirements of the immersed tube transition section.

References

[1] Mo J Y, Huang J S. 2009 Application of sand compacted pile for ground treatment of oceanic shore protection engineering works [J] *Port & Waterway Engineering* 87-92.

[2] Xu Z H. 2011 Calculation of composite foundation's bearing capacity by high replacement ratio sand compacted pile in soft clay [J] *Port & Waterway Engineering* 131-5.

[3] Zhang D. 1999 Analysis and computation of the deformation modulus of the pile in the composite foundation [J] *Chinese Journal of Geotechnical Engineering* 205-8.

[4] Xu Y, Xu G P, Fu B Y, He X. 2020 Research on foundation settlement control of immersed tunnel at the joint of offshore artificial island and island tunnel [J] *Tunnel Construction* 88.

Journal of Physics: Conference Series **2148** (2022) 012032 doi:10.1088/1742-6596/2148/1/012032

Experimental Study on Shear Behavior of Reactive Powder Concrete Beam

Yuexia Li [1], Huijun Yang [1,*] and Chao Liu[2]

[1] Nanning College of Technology, Faculty of Civil Engineering, 317 Yanshan Street, Yanshan Town, Yanshan District, Guilin, Guangxi, China

[2] College of Civil Engineering and Architecture,Guilin University of Technology, 319 Yanshan Street, Yanshan Town, Yanshan District, Guilin, Guangxi, China;

*Corresponding author, Email:1114521419@qq.com

Abstract. In order to study the shear behavior of high-strength reinforced Reactive Powder Concrete (RPC) beams, eight test beams were designed and fabricated for the shear test under symmetrical concentrated load. By observing the development and failure mode of diagonal cracks, the influence of shear span ratio, stirrup ratio, and longitudinal reinforcement ratio on the cracking load, shear capacity, and deflection of the test beam is analyzed. The results show that: in a specific range, the shear capacity increases with the increase of stirrup ratio and longitudinal reinforcement ratio and decreases with the increase of shear span ratio. The shear span ratio has the most significant influence on the component's failure mode and deformation capacity. The increase of the stirrup ratio can improve the deformation capacity of the component in a specific range. It is conservative to use the code to design concrete structures to calculate the shear capacity of high-strength reinforced reactive powder concrete beams. It is suggested that the shear calculation formula suitable for high-strength reinforced reactive powder concrete should be adopted to make the theoretical calculation results closer to the measured values.

1. Introduction

With the development of society, ordinary concrete cannot meet the needs of some engineering construction due to the lack of strength and durability. Reactive Powder Concrete (RPC) is a high-performance concrete with long-term stable mechanical properties [1], high early strength, high toughness, good volume stability, and long service life under adverse conditions. It is the development direction of building concrete materials [2]. HRB500 steel bar is a kind of steel bar with high strength and good ductility. In recent years, there have been some researches on the shear behavior of high-strength reinforced RPC members at home and abroad. Through a series of shear tests of prestressed RPC beams without web reinforcement, Australian scholars Vool and Foster[3] summarized the effects of prestressing size, steel fiber type, and steel fiber content on shear capacity. Based on the softened truss theory model, Cladera [4] proposed a shear model and shear formula considering the contribution of tensile strength to concrete shear capacity. Zheng Hui [5] carried out the load test of prestressed RPC thin-walled box girder, summarized the influence of stirrup ratio, shear span ratio, and other parameters on the components, and put forward that the tensile strength of RPC can be improved by adding 2% steel fiber in volume, and the cracks in the test beam present the characteristics of "fine and dense".

2. General Situation

2.1. Design and Manufacture of Specimen

Eight RPC rectangular beams with high-strength steel bars are designed, and their section sizes are all B × h=150mm × 250mm, beam length is 2200mm. the parameters of each test beam are shown in table 1.

Table 1. Main design parameters of test beam.

Specimen number	Volume fraction of steel fiber	Shear span ratio λ	Reinforcement			
			Stirrup	Stirrup ratio %	longitudinal reinforcement	Longitudinal reinforcement ratio %
L1-1	2%	1.51	C6@450	≈0	4D25	6.58
L1-2	2%	2.26	C6@450	≈0	4D25	6.58
L1-3	2%	3.02	C6@450	≈0	4D25	6.58
L2-1	2%	2.26	C6@450	≈0	3D25	4.43
L2-2	2%	2.26	C6@450	≈0	5D25	8.04
L3-1	2%	2.26	C6@225	0.17	4D25	6.58
L3-2	2%	2.26	C6@150	0.25	4D25	6.58
L3-3	2%	2.26	C6@65	0.58	4D25	6.58

2.2. Loading Mode and Measurement Scheme

The static loading method is adopted in this test, and two equivalent and synchronous concentrated forces are obtained by distributing the symmetrical position of the steel beam on the test beam. The measurement contents include deflection, steel strain, crack, and load. See figure 1 for details.

Figure 1. Loading device diagram.

3. Test Results and Analysis

3.1. Experimental Phenomena

The failure characteristics of RPC beams are similar to those of ordinary beams. When the load is small, the tensile strength of RPC mainly resists the shear force, and there is no crack on the concrete surface. When the load is about 200kN, small shear cracks appear at the bottom of the beam in the shear span area. With the increase of the load, the cracks develop into bending inclined cracks. With the continuous increase of load, the inclined crack extends along the beam height direction, the crack width increases, and new inclined cracks appear at the same time. Finally, the bearing capacity of the beam decreased rapidly due to the concrete being pressed and peeled off.

3.2. Failure Mode

Adding an appropriate amount of steel fiber into RPC can improve the failure characteristics of RPC, bridge cracks and hinder the development of cracks. The failure forms of RPC inclined section of high-strength reinforcement are mainly divided into shear compression failure, oblique compression failure, and bending shear failure, as shown in figure 2 to figure 4.

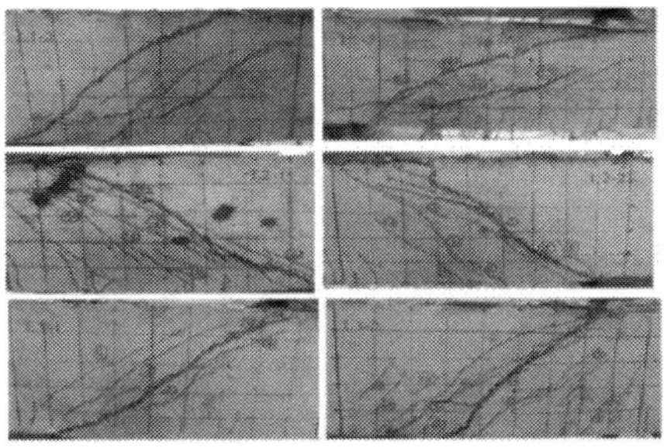

Figure 2. Shear damage.

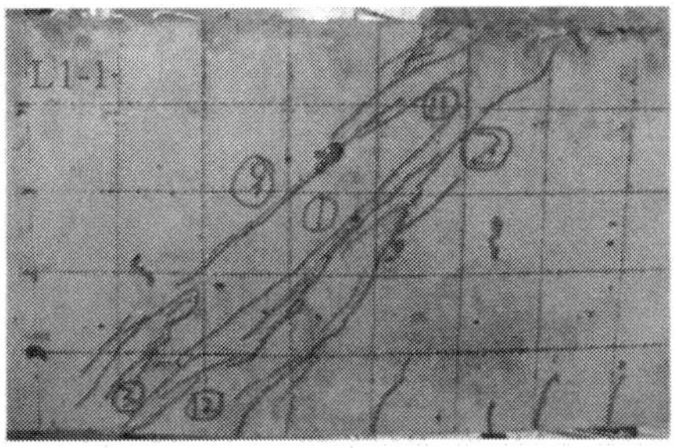

Figure 3. L1-1 diagonal compression damage.

Figure 4. L3-3 bending shear damage.

- Shear compression failure. It can be seen from Figure 2 that the shear compression failure of beams without web reinforcement is L1-2, L1-3, L2-1, L2-2, and beams with web reinforcement L3-1, L3-2, L3-3. During the loading process, the bending cracks first appear in the middle of the test beam span, then the inclined cracks appear in the shear span area, and finally, the inclined cracks are destroyed due to the excessive width of the inclined cracks. In the beam with web reinforcement, when entering the descending section, the stirrups yield, but the longitudinal bars do not yield, and the concrete in the compression area is crushed, which declares that the whole beam loses its bearing capacity and is damaged.
- Baroclinic failure. In this test, the shear span ratio of L1-1 is 1.51, and there are many small cracks in the concrete between the inclined cracks in the abdomen of the test beam, showing a certain characteristic of oblique compression failure, as shown in figure 3.
- Bending shear failure. Because L3-3 is equipped with more stirrups, bending shear failure occurs, not only part of the stirrups yield but also the longitudinal bars near the loading point in the shear span region yield. When the test beam is about to fail, the load is basically unchanged, and the beam deflection continues to increase significantly, showing the characteristics of bending failure, as shown in figure 4.

3.3. Bearing Capacity Analysis

It can be seen from figure 5 that the shear bearing capacity decreases with the increase of shear span ratio. It can be seen from Figure 6 that longitudinal reinforcement can effectively control the width and height of bending cracks, increase the area of concrete contributing to shear resistance, and thus improve the shear bearing capacity, which is called the pin bolt effect of longitudinal reinforcement. The higher the ratio of longitudinal reinforcement is, the greater the pin bolt effect is, and the higher the shear bearing capacity is. It can be seen from Figure 7 that the shear capacity of the inclined section increases with the increase of stirrup ratio, but beyond a certain range, the increase is not obvious. L3-3 has a high stirrup ratio, but the bearing capacity of the inclined section decreases slightly. This is due to the shear failure and bending failure. The stirrups yield individually, and the longitudinal bars yield, which leads to the reduction of the bearing capacity of the beam.

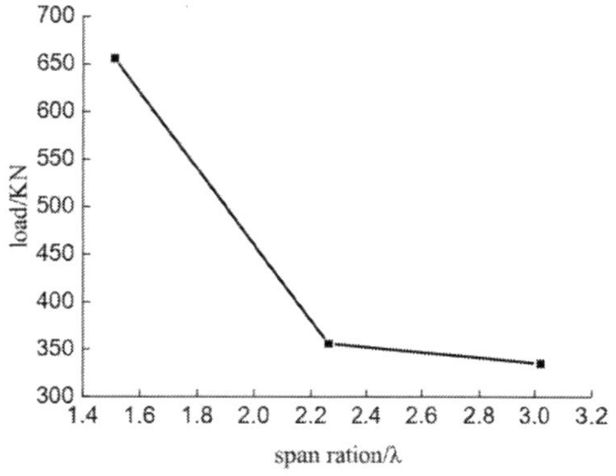

Figure 5. The curve of shear span ratio and bearing capacity.

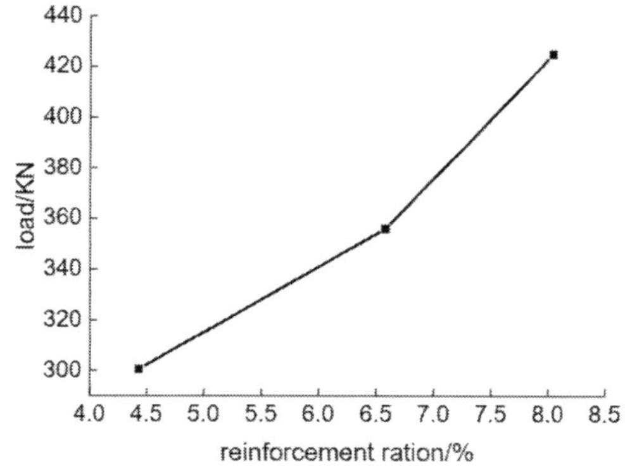

Figure 6. The curve of reinforcement ratio and bearing capacity.

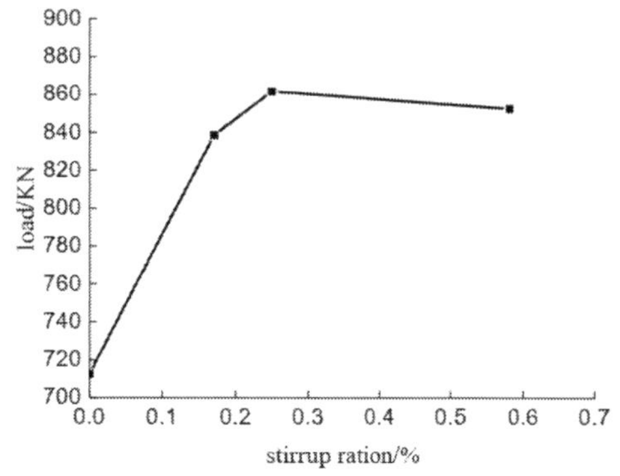

Figure 7. The curve of stirrup ratio and bearing capacity.

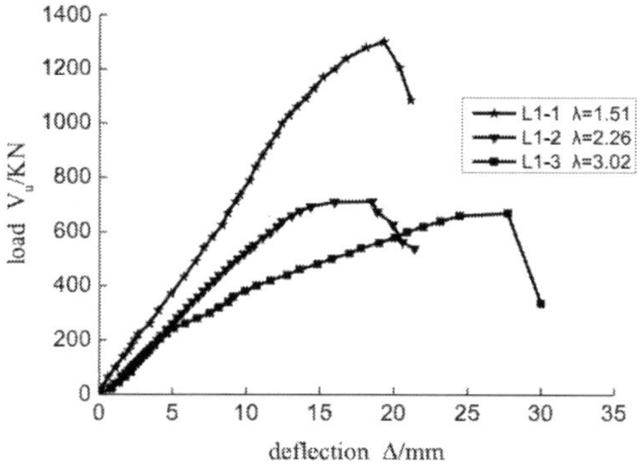

Figure 8. the curve of shear span ratio on load – deflection.

4. Deformation Behavior of Test Beam

4.1. Deformation Law of Test Beam

It can be seen from figure 8- figure 10 that when the load is relatively small, the load and deformation basically increase in a linear relationship. With the continuous increase of the load, the RPC beam enters the inelastic working stage, and the curve slope slows down, but there is no obvious breakpoint. Near the failure, the deformation increases suddenly. After the failure, the load decreases, the deformation still increases, and the maximum displacement is more than 14mm. This is due to the bridge function of steel fiber in RPC. After the concrete cracks, it effectively prevents the concrete from cracking on the inclined crack surface, and the concrete can still bear part of the force.

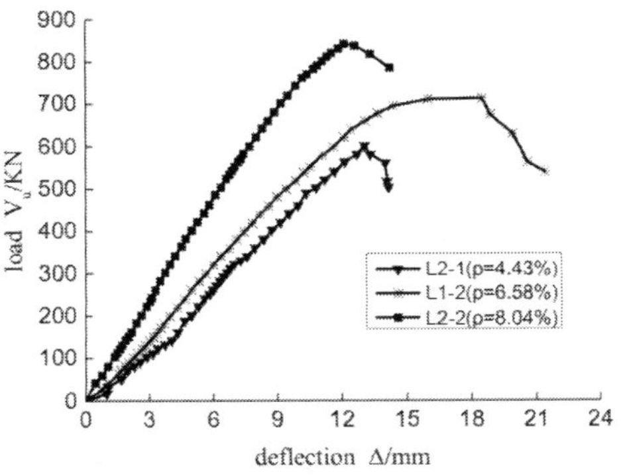

Figure 9. The curve of reinforcement ratio on load – deflection.

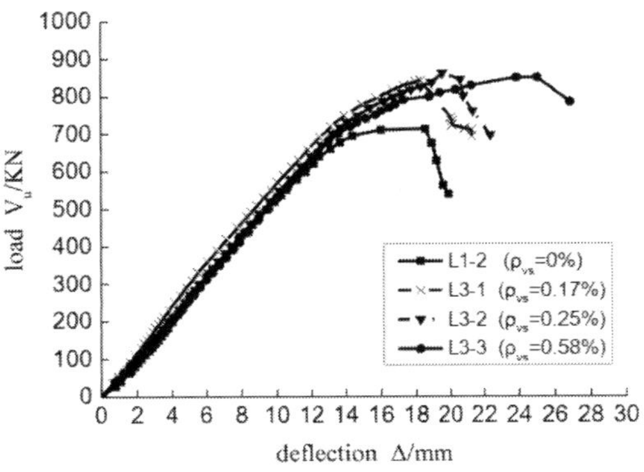

Figure 10. The curve of stirrup ratio on load – deflection.

4.2. Factors Affecting the Deformation Performance of Test Beams

- Shear span ratio. According to figure 8, the mid-span deformation and deformation speed increase with the increase of shear span ratio. On the one hand, the beam with a large shear span ratio produces a greater bending moment under the same conditions. On the other hand, the beam with a large shear span ratio has smaller stiffness. Therefore, after RPC cracking, the deformation of the beam with a large shear span ratio increases faster, and the deformation will be larger when it reaches failure.
- Longitudinal reinforcement ratio. It can be seen from figure 9 that the longitudinal reinforcement ratio is within a certain range, and the deformation of the RPC beam increases with the increase of longitudinal reinforcement ratio, but the improvement effect is not obvious when the longitudinal reinforcement ratio is too high.
- Stirrup ratio. It can be seen from figure 10 that although the high stirrup ratio restricts the development of oblique cracks, it has no obvious effect on the stiffness of the test beam. Therefore, the development trend of the load midspan deflection curve of the RPC beam is basically the same. In the later stage of loading, the deflection curve of the RPC beam with stirrups did not drop suddenly because the stirrups yielded successively in the load falling section. In particular, the curve of l3-3 is near the ultimate load, the load increases little or basically does not increase, but the deflection increases continuously, which is similar to the characteristics of bending shear failure of beams.

5. Shear Calculation Formula of RPC Beam

In this paper, the shear capacity of the oblique section of the RPC beam adopts the literature[6]:

$$V_u = \frac{2.5}{\lambda+1}\sqrt{0.8f_c} \times 5.4\sqrt{\rho}bh_0 + 1.25f_{yv}\frac{A_{sv}}{s}h_0 \tag{1}$$

In formula (1), V_u is the shear capacity of the beam, λ Is the shear span ratio, f_c is the compressive strength of concrete, ρ is the reinforcement ratio of longitudinal reinforcement, b is the width of concrete section, h_0 is the effective height of concrete section, f_{yv} is the tensile strength of stirrup, A_{sv} is the stirrup area, and s is the stirrup spacing.

Table 2. RPC recommended formula calculation value and experimental value.

Specimen number	Shear span ratio/λ	Stirrup ratio/%	Reinforcement ratio/%	f_c /N/mm²	h_o /mm	V_{cal} /kN	V_{exp} /kN	V_{cal}/ V_{exp}
L1-1	1.51	0	6.58	117.2	205	410.80	656	0.626
L1-2	2.26	0	6.58	117.2	205	316.29	356	0.888
L1-3	3.02	0	6.58	117.2	205	256.49	335	0.766
L2-1	2.26	0	4.43	117.2	221.5	280.41	300.5	0.933
L2-2	2.26	0	8.04	117.2	199	339.39	425	0.799
L3-1	2.26	0.17	6.58	117.2	205	357.57	419.25	0.853
L3-2	2.26	0.25	6.58	117.2	205	378.20	430.85	0.878
L3-3	2.26	0.58	6.58	117.2	205	459.17	426.25	1.077
Mean value								0.852
Standard deviation								0.123
Coefficient of variation								0.144

Through comparison, it can be seen that the average value of the ratio between the calculated value and the test value adopted by the formula is 0.852, the standard deviation is 0.123, and the coefficient of variation is 0.144. The calculated value is close to the measured results, and the shear force of RPC members can be better estimated according to the adopted formula of RPC shear capacity.

6. Conclusion

- The steel fiber added into RPC can improve the failure mode of non-web reinforced beams with a large shear span ratio. With the increase of shear span ratio, the shear capacity of RPC beams decreases, and the deformation accelerates; With the increase of longitudinal reinforcement ratio, the shear capacity and deformation of RPC beams increase, but too high longitudinal reinforcement ratio has no obvious effect on improving the deformation capacity of RPC beams; The shear capacity of RPC beam increases with the increase of stirrup ratio, but the bearing capacity decreases with too many stirrups.
- The shear bearing capacity of high-strength reinforced RPC beams calculated according to the current code for design of concrete structures (gb50010-2010) is quite different from the measured values in the test, but according to the recommended formula of semi-empirical and semi theoretical shear bearing capacity of high-strength reinforced RPC members, the calculated value is close to the theoretical value.

Acknowledgments
Foundation item: the Ability Improvementing Project for Young Teachers· scientific research in Guangxi University 2021(2021KY1668).

Reference

[1] Li J P. 2011 Study on the properties of reactive powder concrete with ultrafine slag [D] *Harbin Institute of Technology*.
[2] Zheng W Z, Lv X Y. 2015 Literature review of reactive powder concrete [J] *Journal of Building Structures* 10 44-58.
[3] Vool J Y L, Foster S J, Gilbert R I. Shear Strength of Fiber Reinforced Reactive Powder.
[4] Concrete Girders without Stirrups. Australia: The University of New South Wales 2003.
[5] Cladera A, Mari A R. 2005 Experimental Study on High—Strength Concrete Beams Failing in Shear [J] *Engineering Structures* 27 1519-1 527.
[6] Zheng H, Fang Z, Liu M. Experimental study on shear behavior of prestressed reactive.
[7] Fang Z, Liu M, Zheng H. 2015Experiment on Flexural Behaviors of Prestressed Reactive Powder Concrete Box Girders [J] *Journal of Architecture and Civil Engineering* 06 51-63.

Journal of Physics: Conference Series **2148** (2022) 012033 doi:10.1088/1742-6596/2148/1/012033

Damage Assessment of Concrete under Wind-Sand Load

Ying Jia[1, 2] and Zhiwei Xie[1,*]

[1] School of Civil and Architectural Engineering, Beijing Jiaotong University, Beijing 100044, China

[2] Professor, School of Civil and Architectural Engineering, Beijing Jiaotong University, Beijing 100044, China

Email: 20121129@bjtu.edu.cn

Abstract. In this paper, a numerical model of wind-sand flow is established to explore the influence of wind-sand flow parameters on the stress state of the concrete. According to the contact mechanics theory, the wind erosion critical pressure of concrete members and the corresponding stress state of concrete are analyzed. A wind erosion damage assessment index based on stress fatigue theory is proposed.

1. Introduction

The erosion wear of wind on solid surface is called wind erosion. Long-term wind erosion will accelerate the durability of concrete damage, and reduce service life, which result in reduced structural safety. Many scholars have studied the material abrasion behavior under wind erosion through wind erosion tests [1–5] and numerical models [6–10] to explore the mechanism of wind erosion. However, there are few studies on material stress under wind-sand load. In this paper, the stress state of concrete under wind-sand load is analyzed and studied by numerical model, and a wind erosion damage evaluation index based on fatigue theory is proposed, which provides a reference for the design of concrete structure in dust storm areas.

2. Numerical Model

The concrete model is a cube with side length of 150mm , and the calculation domain size is $150\text{m}\times150\text{m}\times150\text{mm}(X\times Y\times Z)$. Considering the gravity of sand particles, the inlet boundary condition is velocity inlet, the outlet boundary condition is pressure outlet, the side and top of the calculation domain are symmetrical boundary conditions, and the bottom of the calculation domain and the wall of the concrete model are non-slip wall. Numerical calculation uses standard k-ε model, turbulent kinetic energy k and dissipation rate ε equations are shown as in equation (1) and equation (2). The coupling algorithm of pressure and velocity is Phase Coupled SIMPLE. The numerical model was used to simulate the Han's test [5], and the numerical results were consistent with the experimental results, which verified the accuracy of the model.

$$\rho\frac{\mathrm{dk}}{\mathrm{dt}}=\frac{\partial}{\partial x_{\mathrm{i}}}\left[\left(\mu+\frac{\mu_{\mathrm{t}}}{\sigma_{\mathrm{k}}}\right)\frac{\partial}{\partial x_{\mathrm{i}}}\right]+G_{\mathrm{k}}+G_{\mathrm{b}}-\rho\varepsilon-Y_{\mathrm{M}} \tag{1}$$

$$\rho\frac{\mathrm{d}\varepsilon}{\mathrm{dt}}=\frac{\partial}{\partial x_{\mathrm{i}}}\left[\left(\mu+\frac{\mu_{\mathrm{t}}}{\sigma_{\varepsilon}}\right)\frac{\partial}{\partial x_{\mathrm{i}}}\right]+C_{1\varepsilon}\frac{\varepsilon}{k}\left(G_{\mathrm{k}}+C_{3\varepsilon}G_{\mathrm{b}}\right)-C_{2\varepsilon}\frac{\varepsilon^{2}}{k} \tag{2}$$

Where, G_k, G_b are the represent turbulent kinetic energy caused by the average velocity gradient and buoyancy respectively; Y_M is the effect of pulsation expansion of compressible turbulence on total dissipation rate; μ_t is the eddy viscosity; adjustable constants $C_{1\varepsilon}=1.44$, $C_{2\varepsilon}=1.92$, $C_{3\varepsilon}=0.09$, $\sigma_k=1.0$, $\sigma_z=1.3$.

3. Stress Analysis of Concrete Under Wind-Sand Load

Wind erosion is related to wind-sand parameters such as wind-sand velocity, sand content and sand particle size. In order to explore the influence of wind-sand parameters on wind erosion, this paper mainly studies and analyzes the variation law of principal stress of concrete under different sand wind velocities. The working conditions of wind-sand velocities are set as follows: $18\text{m}\cdot\text{s}^{-1}$, $22\text{m}\cdot\text{s}^{-1}$, $26\text{m}\cdot\text{s}^{-1}$, $30\text{m}\cdot\text{s}^{-1}$.

3.1. Critical Principal Stress Analysis of Concrete Wind Erosion

When the concrete just happens wind erosion, the corresponding wind-sand velocity is called the critical wind speed of wind erosion. As Han's test shown, wind erosion on concrete happens when sand wind continuous action for 3min with sand content of 2% and sand particle size of 0.3mm. The corresponding wind erosion critical wind speed is $13.78\text{m}\cdot\text{s}^{-1}$. In order to explore the stress state of concrete under different wind-sand loads, it is necessary to analyze and study the critical principal stress of concrete under the wind-erosion critical wind-sand load.

The load of wind-sand flow on the structure can be regarded as the net wind pressure caused by wind with sand and the impact pressure of sand particles on the surface of the objec[t] [11]. According to Load Code for Building Structures (GB50009-2012) and contact mechanics theory, the net wind pressure on the windward surface and the sand impact pressure are calculated under the wind erosion critical wind speed of $13.78\text{m}\cdot\text{s}^{-1}$, respectively. The sum of the two can be regarded as wind-sand load, which is 2879Pa. However, this method of calculation does not fully consider the parameters such as wind vibration coefficient, body shape coefficient and the coupling between wind and sand at all times. In the calculation process, the sand impact pressure and net wind pressure are both average pressure. In this paper, the numerical model adopts the fluid-solid coupling method to obtain the distribution of wind-blown sand pressure at the critical wind-sand velocity of wind erosion, $13.78\text{m}\cdot\text{s}^{-1}$.

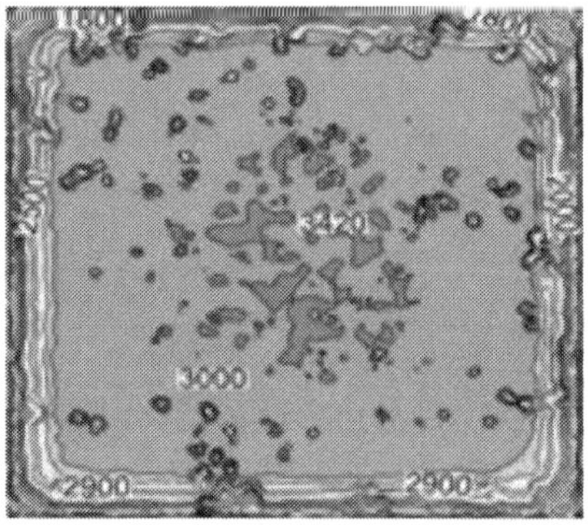

Figure 1. Critical pressure of wind erosion on the concrete model windward side.

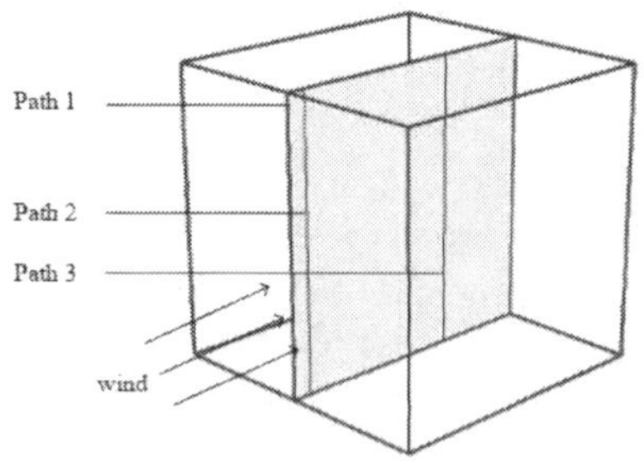

Figure 2. Path diagram.

As shown in figure1, the maximum wind-sand pressure appears in the center area of the concrete specimen, which is 3.4kPa, and the distribution is that the wind-sand pressure decreases from the inside to the outside, which is more in line with the actual situation. It is of certain reference significance to give the maximum wind-sand pressure by using the numerical simulation method.

The stress state of concrete under critical wind erosion pressure is further analyzed, and three paths are selected (shown in figure 2) to study the change of principal stress along the path: path 1 is the surface of the concrete windward surface, path 2 is located in the range of 0-0.08m from the surface of windward surface, and path 3 is located in the range of 0.08m-0.1m from the surface of windward surface. As shown in figure 3, the stress elements on paths 1 and 2 are in the state of tension-tension-compression stress. The elements on both sides of path 3 are in the state of tension-tension-compression stress, and the intermediate element is in the state of three-dimensional tension.

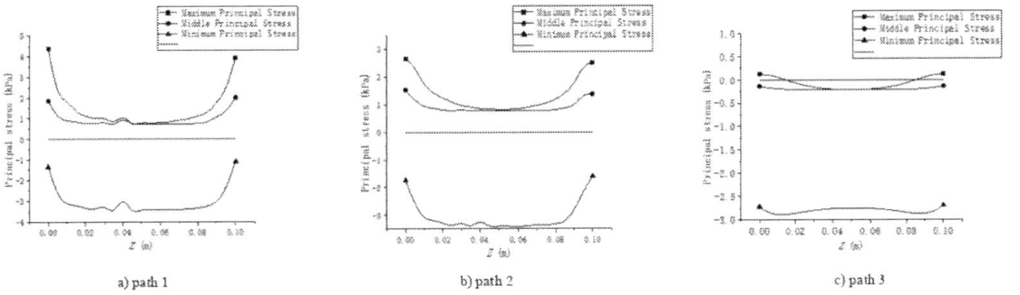

Figure 3. Principal stress history of different path.

3.2. Stress Analysis of Concrete Under Wind-Sand Load

The stress of concrete structure in wind-sand environment is complex, and the concrete units are in three-dimensional stress state. The principal stress value of the maximum tensile stress unit on the windward surface is extracted from the numerical model when the sand content is 2 %, the sand particle size is 0.3mm, and the wind-sand flow velocity is different (shown in table 1). The unit on the windward surface is in the three-dimensional stress state of tension-tension-compression, which is prone to damage. The maximum value of the first principal stress on the windward surface of the concrete increases with the increase of wind-sand velocity.

Table 1. Element principal stress value of maximum tensile stress on windward surface.

Wind-sand flow parameters		σ_1 (Pa)	σ_2 (Pa)	σ_3 (Pa)	$\lvert\sigma_1/\sigma_2\rvert$
Velocity	18m/s	8210.3	3918.8	-2256.8	2.12
	22m/s	11983.0	5293.5	-3245.4	2.26
	26m/s	17033.0	8047.5	-3847.7	2.12
	30m/s	22733.0	10738	-5132.6	2.12

4. Damage Analysis of Concrete Under Wind-Sand Load

High-cycle stress fatigue generally occurs in the concrete under wind-sand load [12]. With the increase of load, the cracks generated by sand particles impacting on the concrete surface expand and develop from the contact area to the interior in a tortuous path. With the generation of radial cracks, serious damage occurs in the contact area. All cracks interact with each other in an unsteady state to produce a sub-surface of the material. When the crack reaches a certain position, the expansion path turns upward. When the crack reaches the surface of the material, material spalling occurs around the damage area.

4.1. Wind Erosion Damage Criterion

Combined with Miner linear criterion, the damage of concrete model in wind erosion fatigue process is defined as damage variable, shown as equation (3), which is used as the judgment basis of fatigue damage of concrete model in wind and sand environment.

$$D = 1 - \frac{p}{p_0} = \frac{n}{N} \tag{3}$$

Where, p is the residual strength of concrete after wind erosion; p_0 is the initial strength of concrete; n is the number of cycles corresponding to the stress amplitude, N is the number of cycles corresponding to the fatigue life of the concrete, $N = 2\times10^6$; Damage factor D is a variable to measure the degree of material damage: $D = 1$ indicates that the material has been destroyed; $0 < D < 1$ indicates that the material has different degrees of damage, the smaller the D value is, the less the material damage is.

About the fatigue theory, Wholer equation gives the relationship between stress level S and fatigue life or cycle number N:

$$S = A - B\lg N \tag{4}$$

The current *S-N* relationship is essentially determined by the uniaxial fatigue test. However, in the actual analysis of wind erosion damage of the concrete, the randomness of wind load and the stability of wind speed distribution in the whole year are considered, and the concrete model is in a multi-axial stress state. Therefore, in the analysis of wind erosion fatigue damage of concrete model, the wholer formula needs to be modified. According to the numerical simulation results, the maximum and minimum stress amplitude of the wind-sand load of the concrete is determined according to Equation (5), so as to calculate the coefficients A and B related to the stress ratio and loading mode, and give the modified *S-N* curve of concrete wind erosion.

$$S_{\max} = \frac{\sigma_{\max}}{f_t} \tag{5}$$

Where: $S_{\max}$ — maximum stress level;

f_t —tensile strength of concrete, unit is MPa;

$\sigma_{\max}$ — maximum stress of structural section in fatigue analysis.

In this paper, tensile strength design value 1.71MPa is adopted in all relative calculations. According to the wind erosion test of the concrete, the wind erosion phenomenon of the concrete specimen occurs when the wind-sand with velocity of $14\mathrm{m\cdot s^{-1}}$ sustains for 3 min. The wind-sand with velocity of $26\mathrm{m\cdot s^{-1}}$ sustains 1 min, and the wind erosion occurred on the concrete specimens. The strength loss rates were 0.01 and 0.06, respectively. The number of wind-sand loads corresponding to the velocity of wind-sand can be obtained from equation 3. In the numerical simulation results, the maximum cyclic stress $\sigma_{\max}$ is 1.7kPa at $26\mathrm{m\cdot s^{-1}}$, and the maximum cyclic stress $\sigma_{\max}$ is 6.3kPa at $14\mathrm{m\cdot s^{-1}}$. Taking the maximum stress amplitude into equation 5, it can be determined that the correlation coefficient A of the modified *S-N* curve is 0.0502, and B is 0.007.

Based on the modified *S-N* relationship, the damage of the windward surface of the concrete model under wind-sand loads with different wind-sand velocities is studied. The relationship curve between the wind-erosion fatigue life and the velocity is shown in the figure 4. The natural logarithm of the wind-erosion fatigue life and the velocity are in line with the linear relationship.

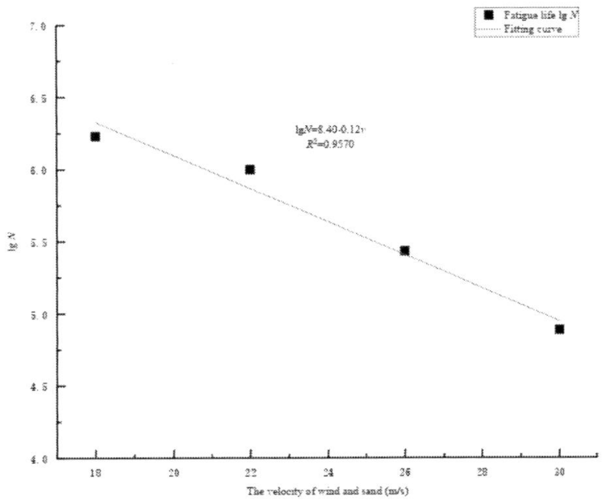

Figure 4. Relationship between the fatigure life N and the velocity of wind-sand.

4.2. Wind Erosion Damage Analysis

If the long-term cyclic loading of wind-sand will lead to wind-induced fatigue, long-term cumulative fatigue damage will degrade the performance of components. In order to improve the safety of structural design, it is necessary to carry out quantitative research on material performance degradation of concrete materials during the whole life.

According to Miner linear criterion and damage mechanics theory, the concrete damage model is obtained, such as equation (6):

$$\begin{cases} E_{\mathrm{D}} = \left(1-\sum_{i}^{m} D_{\mathrm{i}}\right) E \\ f_{\mathrm{tD}} = \left(1-\sum_{i}^{m} D_{\mathrm{i}}\right) f_{\mathrm{t}} \\ f_{\mathrm{cD}} = \left(1-\sum_{i}^{m} D_{\mathrm{i}}\right) f_{\mathrm{c}} \end{cases} \tag{6}$$

Where, E , E_{D} are the initial elastic modulus of concrete and the elastic modulus after *n* years,unit is GPa ; f_{t}, f_{c}, f_{tD}, f_{cD} are the initial tensile strength, compressive strength, tensile strength after *n* year and compressive strength after *n* year respectively, unit is GPa ; D_{i} is the loss variable in year i, $D_{\mathrm{i}} = d_{\mathrm{i}} \times n \times f\left(v_{\mathrm{i}}\right)$; d_{i} is the cumulative damage value in year i; $f\left(v_{\mathrm{i}}\right)$ is the probability density function of wind velocity at 90° wind attack angle.

According to the measured data of Zha Xiaopeng [13] and equation (5), it is calculated that the elastic modulus of the concrete decreases to 30.1GPa, the tensile strength decreases to 1.58MPa, and the compressive strength decreases to 17.6MPa after 50 years, which have a certain degree of stiffness degradation. Therefore, the wind erosion fatigue should be considered in the structural design process to improve the safety of the structure.

5. Conclusion

This paper studies the mechanism of wind erosion of the concrete under wind-sand load through numerical simulation analysis based on the existing test data, and the results are as follows:

1) When the wind-sand velocity is greater than the critical wind speed, the windward surface element of the concrete is in the tension-tension-compression three-dimensional stress state. The maximum

value of the first principal stress increases with the increase of wind sand velocity, which is easy to damage.

2) The wind erosion damage is essentially a kind of fatigue damage, and the occurrence of fatigue damage can be judged by the damage factor D. The natural logarithm of the wind erosion fatigue life satisfies a linear relationship with the wind-sand velocity.
3) Under the long-term action of wind-sand load, the elastic modulus, tensile strength, and the compressive strength dissipates, and a certain degree of stiffness degradation occurs. A stiffness degradation calculation model is proposed, which provides a reference for the structural design in dust storm areas.

Acknowledgments
Thanks to the project [51778051] of the National Natural Science Foundation of China for supporting this research.

References
[1] Yunhong H, Yongming X and Shiting Y. 2010 *Tribology* **30** 26-31.
[2] Chi L and Yu G. 2011 *Rock and Soil Mechanics* **32** 33-38.
[3] Yujiang F. 2015 the Study of Damage Mechanism and Evaluation of Concrete Under the Erosion of Wind-blown Sand Evironment (Hohhot: Hohhot Inner Mongolia University of Technology).
[4] Xiaofei F. 2019 Wind Tunnel Test Study on Wind-sand Flow Field and Wind-sand Resistance of buildings (Changsha: Hunan University).
[5] Chaochao H 2020 Experimental Research on Wind-sand Load and Wind Erosion of Concrete in Aeolian Environment (Beijing: Beijing Jiaotong University).
[6] Xin L. 2013 Experimental Research on Transmission Line Wron in Wind-sand Two-phase Flow (Baodong: North China Electric Power University).
[7] Lin Y C. 2015 Resaerch on LGJ-400/50 ACSR Surface Erosion Characteristics in Wind Sand Environment (Baodong: North China Electric Power University).
[8] Chaochao H. Experimental Research on Wind-sand Load and Wind Erosion of Concrete in Aeolian Environment (Beijing: Beijing Jiaotong University).
[9] LI S, Zheng S. 2017 Aerodynamic performance analysis of wind-sand flow on riding-type hangers of suspension bridges Journal of Vibroengineering 19.
[10] Mei Z. 2019 Wind-blown Sand Load and Wind Erosion of Bridge Under Air-borne Sand Environment (Beijing: Beijing Jiaotong University).
[11] Cheng Z. 2019 Resaerch on Coupled Fluid-Solid Theorem and Dynamic Damage Mechanism of Wind Turbine Based on Data Exchange Methods (Baodong: North China Electric Power University).
[12] Yingjian T. 2020 the Analysis Method of Wind-induced Fatigue of High-rise Reinforced concrete Structure in the life cycle (Dalia: Dalian University of Technology).
[13] Xiaopeng Z. 2008 Theory and Method of Wind-induced Fatigue Safety Prewarning for High-rise Structure (Wuhan: Wuhan University of Technology).

Journal of Physics: Conference Series **2148** (2022) 012034 doi:10.1088/1742-6596/2148/1/012034

Finite Element Analysis on Reinforced Concrete Beams Strengthened with CFRP

Yihong Hong

Digital Construction Institute, Shanghai Urban Construction Vocational College, 200438, Shanghai, China
Email: hyhzju@126.com

Abstract. Reinforced concrete structure is widely used in building structure because of its unique physical and mechanic properties, but with the increase of service life, there will be different degrees of damage in the structures. In this paper, combined with the test beam, a model of reinforced concrete beam strengthened with CFRP is established by Using ANSYS finite element software, nonlinear finite element analysis is carried out on the whole process of yield, cracking and destruction of the test beam under secondary load, while different working states of CFRP sheets were simulated by the life and death unit. The results show that the bending performance of reinforced concrete (RC) beams strengthened with CFRP can be predicted by selecting the finite element analysis model rationally.

1. Introduction

The technology of strengthening concrete beams with CFRP sheets is widely used at home and abroad in recent years.[1-2] In theory, sufficient strain is needed to exert the strong tensile properties of CFRP, as a brittle material, the strain range of concrete is very limited, so the effect and ultimate bearing capacity of CFRP reinforced concrete, especially considering the secondary load, has been the focus of attention and discussion in this field.[3-5] However, the structures that need to be strengthened in the actual engineering are those of the existing engineering with insufficient bearing capacity and other conditions, so an analysis on the strengthening of RC beams under secondary load is necessary. In this paper, the finite element analysis software ANSYS is used to simulate 3 test beams, the finite element analysis values and the test values are compared, the results show that they are in good agreement with each other. It is a feasible method to simulate CFRP reinforced concrete beams.

2. Profile of Model Beam

According to the Code for Design of Concrete Structures (GB 50010-2010), three beams of the same size (the sectional size was 120 mm × 250 mm, and the length was 2500 mm) were designed fabricated, as shown in figure1. In this experiment, unreinforced beam B1 is taken as control beam, the mechanical properties of intact reinforced concrete (RC) beam B2 and the secondary load beam B3 with a load of 0.6Py were studied. Parameters such as the amount of CFRP and the sustained load point of each specimen are shown in table 1. The main experimental results are shown in table 2

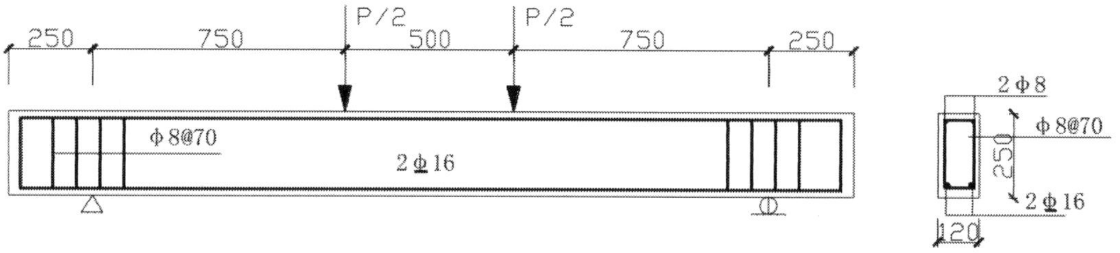

Figure 1. Diagram of test beam reinforcement.

Table 1. Specimen parameters.

Specimen No.	Ratio of longitudinal reinforcement	Amount of CFRP (Sheets)	Sustained load point	Remark
B1	1.52%	0	—	Not strengthened
B2	1.52%	1	—	Intact beam strengthened
B3	1.52%	1	$0.6P_y$	—

Table 2. Main experimental results.

Specimens	Load (kN)			Deflection (mm)			Rigidity (N/mm, $\times 10^4$)			Failure modes
	P_{cr}	P_y	P_u	cr	y	u	B_{cr}	B_y	B_u	
B1	10.0	65.2	68.7	1.0	7.6	17.2	0.971	0.840	0.037	Bending failure, concrete compression failure
B2	10.1	73.0	82.4	0.8	7.9	21.0	1.217	0.887	0.072	Protective layer in the bending segment peeled off
B3	12.1	80.1	92.1	0.7	7.3	23.7	1.862	1.026	0.073	CFRP 30% tensile failure, concrete compression failure

The experimental results show that the reinforcement technology of using CFRP to improve the bearing capacity of RC beams is effective, and the ultimate bearing capacity of the beam strengthened with CFRP sheets was greatly improved. The ultimate state of the RC beams strengthened with CFRP sheets was delayed under a certain load (such as $0.6P_y$). The rigidity of the RC beams strengthened with CFRP was significantly improved, compared with the ordinary RC beams, and the CFRP sheets could improve the rigidity and control the deflection of the components mainly resistant to bending.

3. Finite Element Analysis

3.1. Analysis Model

In this paper, the FEA software ANSYS is used to analyse the whole process of RC beams strengthened with CFRP, and the analytical beam model and material properties are the same as the test definition. The control beam B1 and the test beams B2 and B3 with one CFRP sheet are analysed. In the analysis, concrete is simulated by SOLID65 of 3D 8-node 6-hedron, and the stirrup is realized by the integral model, also called smeared element, which is directly defined by real number in SOLID65. Its advantage is that the modelling is simple, but the internal force of reinforcement cannot be obtained. A separate model LINK8 is used to simulate the tension and compression reinforcement for two reasons. The two reasons are: 1) they are not evenly distributed in the simple girder, if the integral model is used, it does not accord with the actual situation. 2) The strain value of the tensile steel bar needs to be read to investigate the structural performance of the beam in detail. CFRP is a material with only intra-plane stress and minimal external stiffness, and ANSYS provides SHELL41 element for simulation.

3.2. The Life and Death Unit Simulates Different Working States
In the practical application of CFRP reinforcement, due to various limitations, the load can't be completely removed before the reinforcement. CFRP is usually directly pasted on the surface of the member without unloading, so that the initial strain of CFRP is 0, and the initial working stress of CFRP is also 0. CFRP can only play a role with the further development of the deformation of the strengthened member, which is the so-called "secondary load" problem. ANSYS requires that the model must be built all at once in the front processor. Therefore, in the first load step, the unit life and death technology is used to kill all Shell41 units and then apply the load, which is equivalent to the component working without reinforcement. These units will be activated after the predetermined load is applied, so as to simulate the secondary load problem while the Shell41 units are never killed can simulate intact reinforced beam.

3.3. Compare with the Test Results
The comparison of ANSYS finite element analysis and test results are shown in figure 2 (a, b, c) which is about control beam B1, intact strengthened beam B2 and secondary load strengthened beam B3. The FEA results are consistent with the test results to a certain extent. Because of the difference between the constitutive relation used in the model and the actual test, the horizontal section in the test curve could not be clearly simulated. Figure 2 (b) shows that before B2 enters the yield stage of reinforcement, ANSYS finite element analysis can simulate the load-deflection curve of concrete beam very well. When the longitudinal reinforcement reaches yield, the bearing capacity of the test beam is slightly better than that of the finite element model. Since B3 is strengthened with CFRP sheets under a certain load point (such as $0.6P_y$), Figure 2(c) shows that there is a certain difference between the test value and the analysis value, which is related to the difference between the material constitutive relation used in the model and the actual specimen, and these issues need further study.

The comparison of analysis values of load deflection curves of the three test beams is shown in figure3. There is no significant difference betweenB1, B2 and B3 at the initial loading stage, when the load is increased to the holding point, the load deflection curves of B2 and B3 are separated. Under the same external force, the deflection of B2 is greater than that of B3, for B3 is strengthened when certain damage occurs to the member. In this way, the tensile properties of CFRP can be fully utilized and the ultimate bearing capacity of the member can be improved.

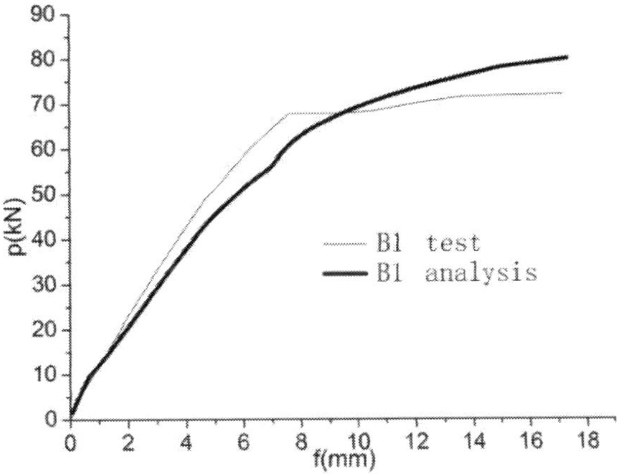

(a)Comparison of load-deflection curves of B1 beam

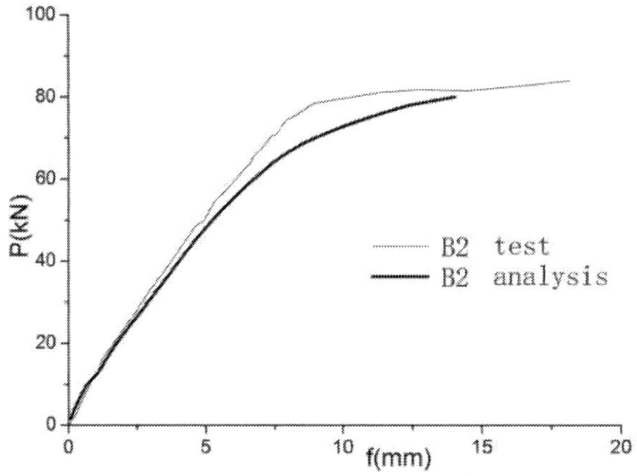

(b)Comparison of load-deflection curves of B2 beam

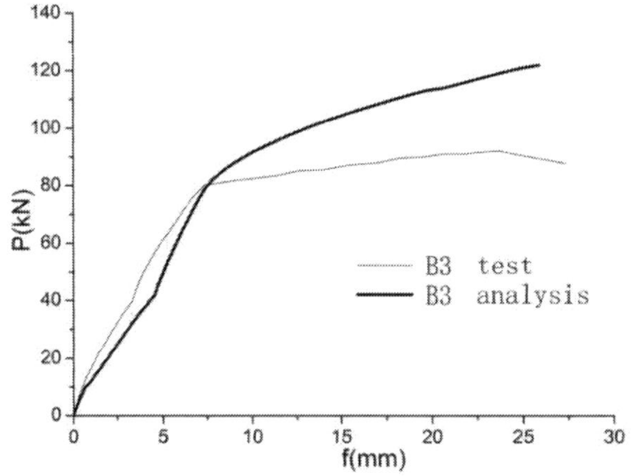

(c)Comparison of load-deflection curves of B3 beam

Figure 2. Comparison of Load- Deflection curve.

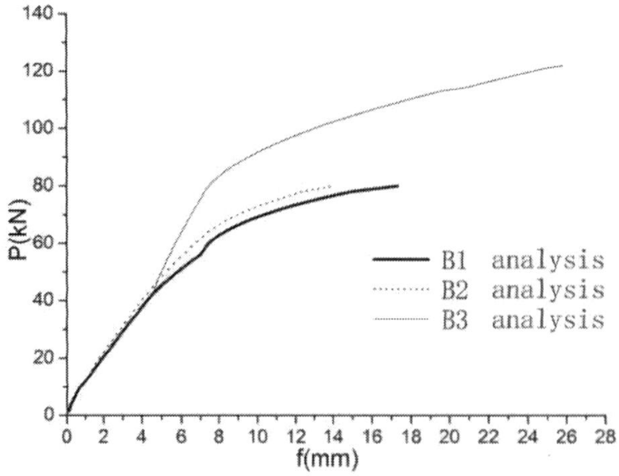

Figure 3. B1, B2 and B3 Load- Deflection curve.

4. Conclusion

In this paper, an experimental study of one control beam and two text beams pasted with one CFRP sheets is carried out and compared with the finite element analysis. From the comparison results, the two fit very well. The finite element analysis can save the test fund, and it is worth promoting in this field. The ultimate bearing capacity of strengthened beams under secondary load is higher than that of intact strengthened beams. Which indicates that the strengthened beams under secondary load can exert the strong tensile resistance of CFRP and conform to the engineering practice. In the finite element analysis, the life and death element is used to simulate the strengthened beams under secondary load. The analysis results agree well with the test results, which illustrates that the life and death technology is feasible.

Acknowledgement

We are grateful for the support from construction of first-class engineering cost major in Shanghai project (Z106-0603-20-001).

References

[1] Chen X B. 2009 Guide for the Design and Construction of Fiber Reinforced Plastics in Civil Engineering [M] Beijing: China Building Industry Press 1-4.

[2] Yue Q R. 2000 Present status of research and application of strengthening and repairing technology with CFRP and its outlook in China [J] *Industrial Construction* **30** (10) 23-26.

[3] Zhu Y P, Liao Y S, Zheng J J. 2010 Experimental study on RC beams' bending capacity strengthened with CFRP considering secondary load [J] *Building structure* **40** (6) 379-381.

[4] Wang Ch M, Chen Y Z. 2006 A Practical method of ultimate flexural capacity of RC beams strengthened with CFRP plates [J] *Industrial Construction* **36** (3) 1032-1034.

[5] Bu L T, Xiao X P. 2017 Study on the flexural behavior of RC beams strengthened with prestressed CFRP under secondary load [J] *Journal of Railway Science and Engineering* **14** (1) 126-134.

Mechanical Mechanism Analysis of Warping and Bulging Of Track Slab of High-Speed Railway in Service

Yali Liu[1], Jun Dong[1,*], Guohua Li[1], Xi Chen[2] and Yunkai Zhang[1]

[1] Beijing Higher Institution Engineering Research Center of Structural Engineering and New Materials, Beijing University of Civil Engineering and Architecture, Beijing, 102616, China;
[2] School of Humanity and Law, Beijing University of Civil Engineering and Architecture, Beijing 102616, China
Email: jdongcg@bucea.edu.cn

Abstract. Due to warpage and ballooning of high-speed rail track slab caused by debonding and environmental temperature changes during the deterioration of CA mortar layer being extremely concerned in practical engineering, it is based on numerical simulation of typical working conditions and analysis of test model that the deformation dynamics mechanism of track slab of high-speed rail in service is studied in this paper. Firstly, three kinds of damage conditions of CA mortar layer are designed to simulate the partial stress state of track slab under normal, warping and bulging conditions, and the results of model test are compared with those of finite element analysis so that the accuracy and credibility of the numerical simulation method and results are verified. Then, through finite element numerical simulation, the dynamical mechanism of actual full-scale high-speed rail track slab under vibration load is studied. The results show that the warping deformation around the track slab and the bulging deformation in the middle part under the action of positive and negative temperature gradient load caused by environmental temperature change will have a great impact on the structural performance of itself and CA mortar layer; Bulging deformation of track slab is more destructive to its structure than warping deformation. It is of great practical significance to further study the critical position of track plate warpage and bulging deformation, and to optimize and strengthen the structure of this part; The research results are of great significance to further study the deterioration.

1. Introduction

When CA mortar layer is repeatedly soaked by rainwater and underground seepage for a long time, the bonding effect between CA mortar layer and track plate will fail. Then, under the action of positive and negative temperature gradient load caused by ambient temperature change, the track plate that loses the bonding effect of CA mortar layer will have large surrounding warping deformation and bulging deformation in the middle [1-4]. CA mortar layer is constantly damaged under the alternating action of warping and bulging deformation of track plate, resulting in the loss, falling off and deformation of CA mortar, and the corresponding track plate will have diseases such as void and joint separation.

In this paper, three working conditions will be designed to study the stress characteristics of track plate under normal, maximum warpage and maximum bulging. In this paper, the scaled experimental model is established, the vibration load is applied by WS-Z30 shaking table, and the Abaqus finite element numerical simulation is used to compare the two results to verify the accuracy of the numerical simulation results. Then, the stress mechanism of actual full-scale high-speed rail track slab beam under vibration load is analyzed by finite element method.

2. Verify the Accuracy of Numerical Simulation Results

2.1. Model Experiment

Model test is mainly used to verify the accuracy of finite element numerical simulation. According to the similarity principle, when the main variables are similar, the secondary variables can be appropriately ignored. In this test, plexiglass plate is used to simulate track plate and support layer, and rubber pad is used to simulate CA mortar layer. According to the research results of Yilong Liu and Junlin Liu [5-10], the test vibration load frequency is 12.5Hz in the longitudinal direction and 80Hz in the vertical direction. Due to the limited conditions of the shaking table selected in this test, only the longitudinal vibration load is applied to verify the accuracy of the numerical simulation results.

WS-Z30 small precision simulation shaking table system is selected as the test instrument, as shown in figure 1 below. The system is composed of shaking table, electromagnetic exciter, power amplifier, shaking table control sensor, model, shaking table controller, computer and software. The test model is shown in figure 2 below. Three working conditions are designed for the test: working condition 1 is non-destructive, as shown in Figure 3 below; Working condition 2 is that the middle of CA mortar layer is not stressed due to the bulging of track plate, as shown in figure 4 below; Condition 3 is that the edge around the CA mortar layer is not stressed due to the warping of the track plate, as shown in figure 5 below. The parameters of test materials are shown in table 1 below.

Figure 1. Vibration table system.

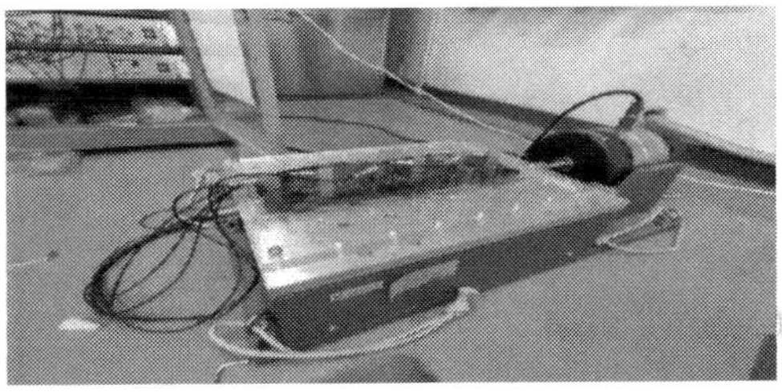

Figure 2. Test model.

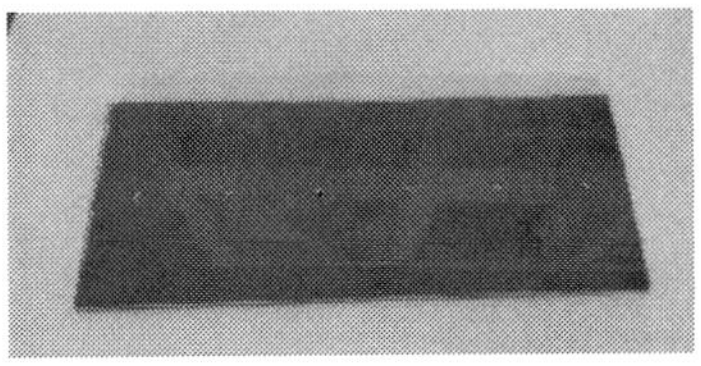

Figure 3. Working condition I.

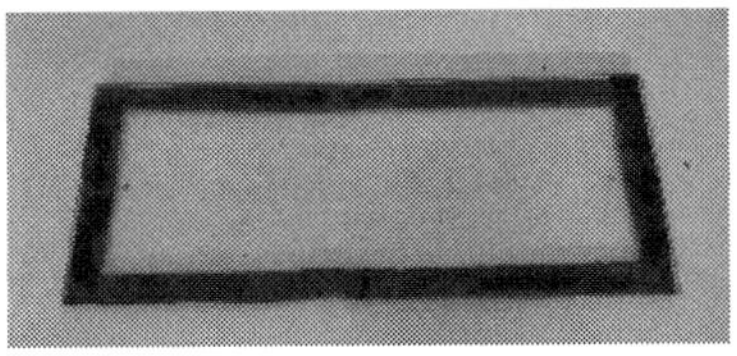

Figure 4. Working condition II.

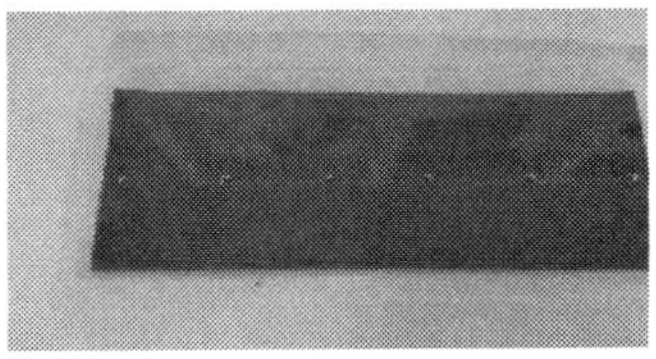

Figure 5. Working condition III.

Table 1. Axperimental material parameters.

Structure	Material	Size(mm)	Elastic modulus (MPa)	Poisson ratio	Density(kg/m^3)
Track llate	Organic glass	$420 \times 170 \times 1$	2.5×10^3	0.37	1200
		$420 \times 170 \times 5$			
CA Mortar layer	Rubber	$1180 \times 20 \times 5$	15	0.47	1300
		$400 \times 150 \times 5$			
Supporting layer	Organic glass	$450 \times 90 \times 20$	2.5×10^3	0.37	1200

2.2. Model Finite Element Analysis

Using Abaqus finite element analysis software, the finite element model of equal size and equal material which is exactly the same as the model test is established. Apply the same vibration load as the model test. The finite element model is shown in figure 6-8 below.

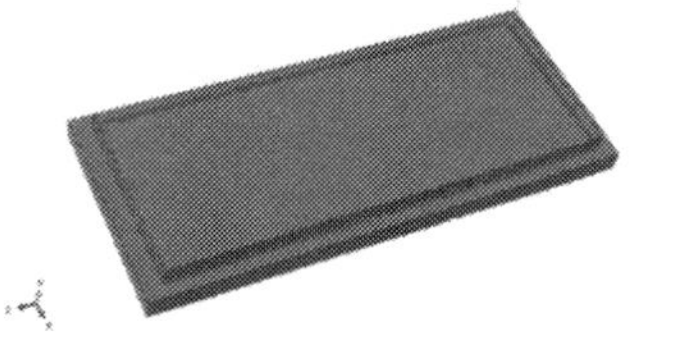

Figure 6. Working condition I.

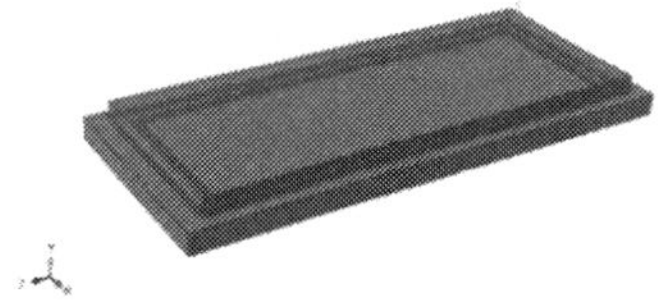

Figure 7. Working condition II.

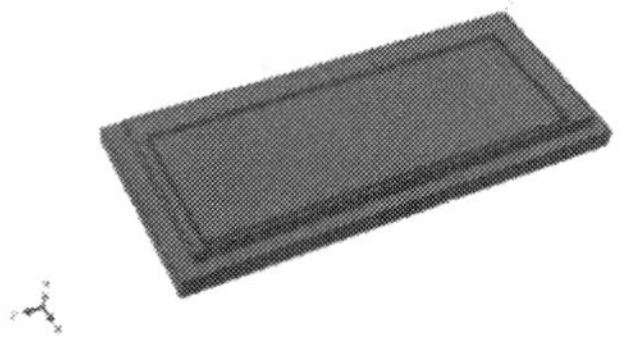

Figure 8. Working condition III.

2.3. Comparative Analysis of Model Test Results and Finite Element Analysis Results

In this study, the maximum acceleration of track plate in the process of model test and finite element analysis is selected as the index to evaluate the vibration level. The following figure 9 (a) - (c) respectively shows the maximum acceleration change curve of track slab beam during model test and finite element analysis from condition 1 to condition 3, and Figure 9 (d) shows the comparison of the maximum acceleration difference of track slab from condition 1 to condition 3 during model test, Figure 9 (e) shows the comparison of the maximum acceleration difference of the track plate from condition 1 to condition 3 during the finite element analysis.

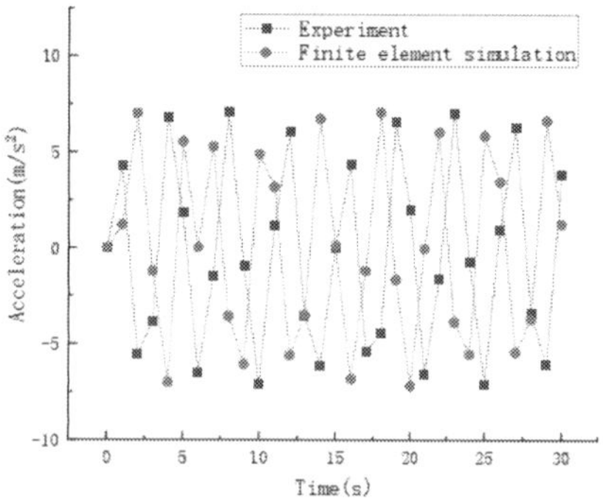

(a) Comparison diagram of working condition I

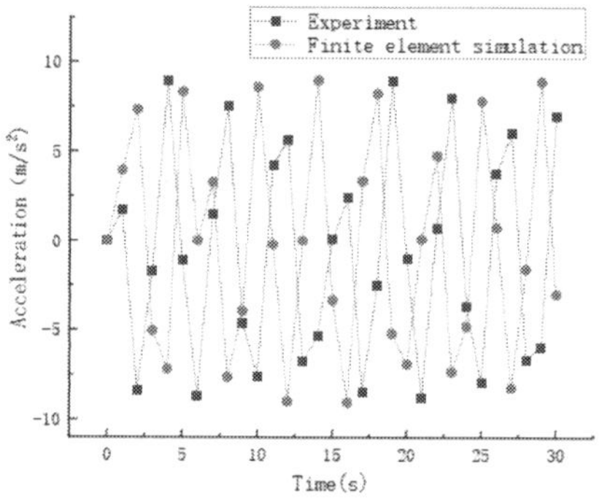

(b) Comparison diagram of working condition II

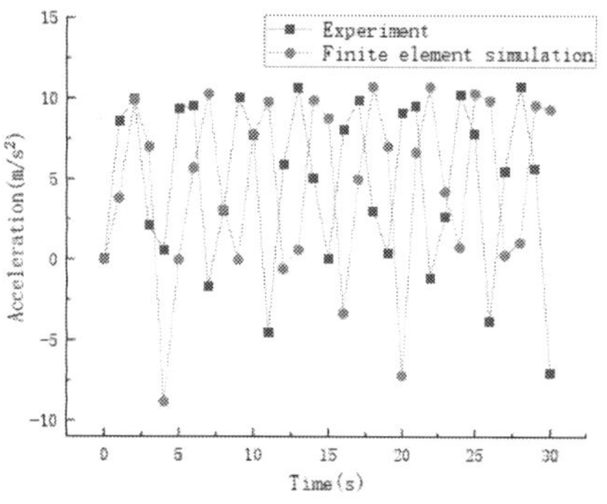

(c) Comparison diagram of working condition III

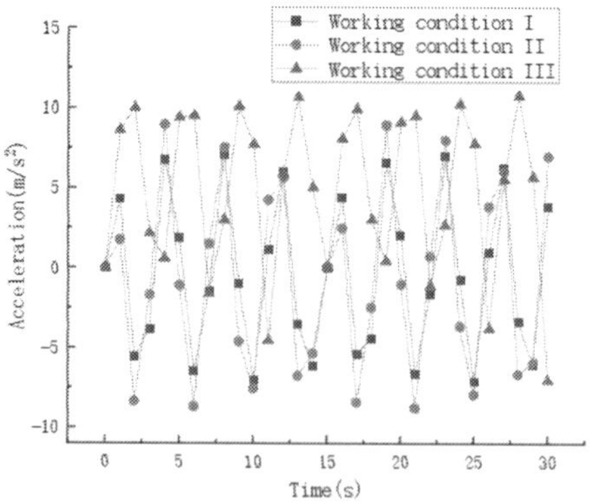

(d) Comparison diagram of test working conditions I - III

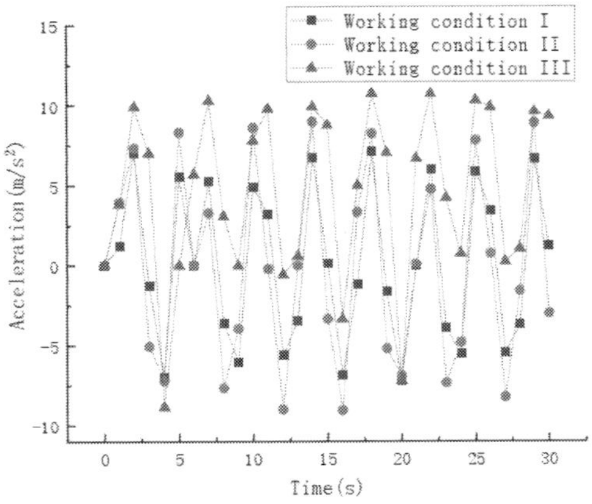

(e) Comparison diagram of finite element working conditions I - III

Figure 9. Comparison between model test results and finite element analysis results.

As can be seen from figure 9 (a) - (c), in the process of simulating the partial stress state of the track plate during normal, warping and bulging, the test results are relatively consistent with the finite element analysis results, and the inconsistency of some secondary conditions leads to small differences in some results. As can be seen from Figure 9 (d) and (e), in the model test, the maximum acceleration of the track plate in condition 1 is $7.377m/s^2$, $9.206m/s^2$ in condition 2 and $11.525m/s^2$ in condition 3; In the finite element simulation, the maximum acceleration of the track plate in condition 1 is$7.396m/s^2$, that in condition 2 is $9.335m/s^2$, and that in condition 3 is $11.375m/s^2$. It shows that the warpage and bulging of track plate will have a great impact on its mechanical properties, which is in line with the law. Through the comparative analysis between the model test results and the finite element analysis results, the accuracy of the numerical simulation results is fully verified, which provides a strong basis for the next finite element analysis of the stress of the actual full-scale high-speed rail track slab beam under vibration load.

3. Finite Element Analysis Actual Equal Size Model

Using Abaqus finite element analysis software, the finite element model of high-speed rail track system with actual equal size and equal material is established. The frequency of vibration load applied to it is: longitudinal 12.5Hz, vertical 80Hz. The finite element model of track system is basically similar to figure 6-8 except for different dimensions, which will not be shown here. See Table 2 below for actual material parameters of track system.

Table 2 Actual material parameters of track system.

Structure	Size(mm)	Elastic modulus (MPa)	Poisson ratio	Density(kg/m^3)
Track plate	$6450 \times 2550 \times 20($	35.5×10^3	0.167	2450
	$6450 \times 2550 \times 50$			
CA mortar layer	$17500 \times 250 \times 50$	7×10^3	0.160	1800
	$5950 \times 2050 \times 50$			
Supporting layer	$7000 \times 2950 \times 30($	30×10^3	0.2	2500

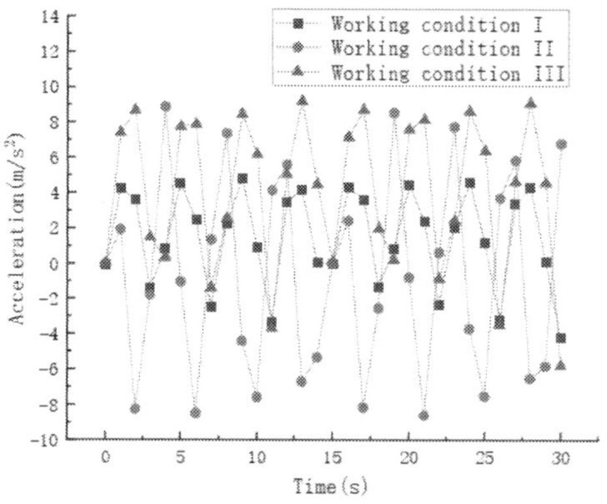

Figure 10. Acceleration variation diagram of track plate under working conditions I – III.

It can be seen from figure 10 that in the actual equal size finite element simulation, the maximum acceleration of the track plate in condition I is$5.120 m/s^2$, that in condition II is$9.045 m/s^2$, and that in condition III is$9.681 m/s^2$. It is consistent with the law of the above test. It shows that the warpage and bulging of track plate will have a great impact on its mechanical properties.

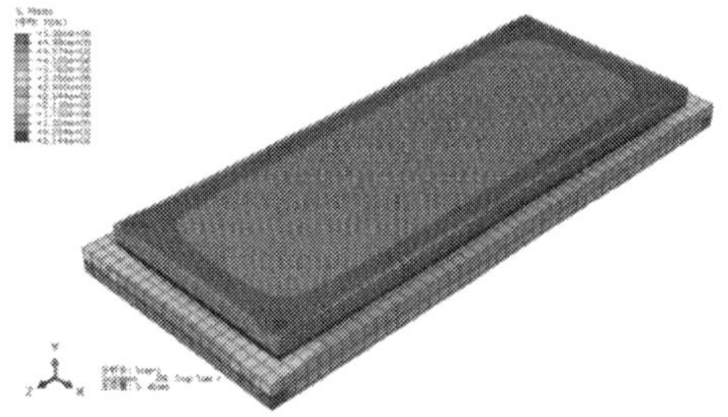

Figure 11. Stress nephogram of working condition I.

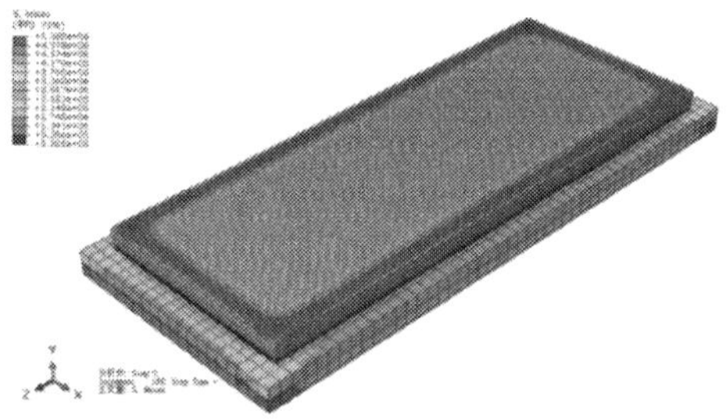

Figure 12. Stress nephogram of working condition II.

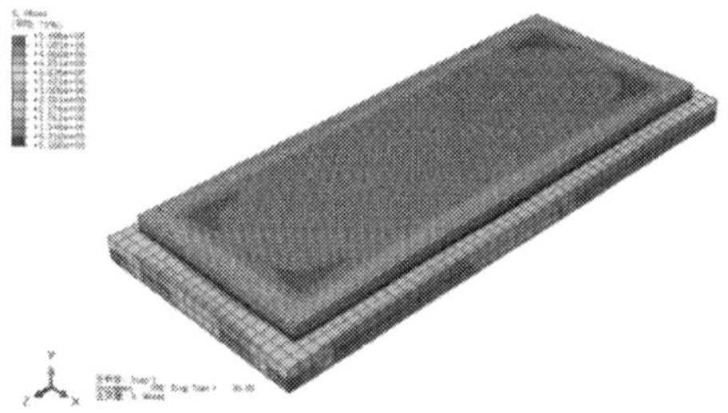

Figure 13. Stress nephogram of working condition III.

It can be seen from figure 11-13 that the maximum longitudinal stress of track plate under condition I is$1.040 MPa$, $5.035 MPa$ under condition II and $1.037 MPa$ under condition III; Most of the longitudinal stress of CA mortar layer under working condition I is $0.111 MPa$, $0.131 MPa$ under working condition II and $0.643 MPa$ under working condition III. It shows that the bulging of track plate is more destructive than warping. Through the stress nephogram, it can also be found that when the track plate warps and bulges, the stress concentration will occur at the contact edge between the track plate and the CA mortar layer. Under the alternating action of track plate warping and bulge, this part of the area will be destroyed first, and vice versa.

4. Conclusion

(1) Under the action of positive and negative temperature gradient loads caused by the change of ambient temperature, the warping deformation around the track plate and the bulging deformation in the middle part will have a great impact on the structural performance of itself and CA mortar layer;

(2) The bulging deformation of track plate is more destructive to its structure than warping deformation;

(3) It is of great practical significance to further study the critical position of track plate warpage and bulging deformation, and to optimize and strengthen the structure of this part;

(4) The research results are of great significance to further study the deterioration mechanism of CA mortar layer and the loss identification and monitoring of track slab.

5. Acknowledgments

Authors wishing to acknowledge assistance or encouragement from BUCEA, and financial support from the Project of Beijing Natural Science Foundation (No.8202012), the Teaching and Research Project of Beijing University of Civil Engineering and Architecture (201802), the Changjiang Scholars and Innovation Team Development Plan (IRT-17R06) and Beijing Advanced Innovation Center for Future Urban Design, Beijing University of Civil Engineering and Architecture (UDC2019021424).

Reference

[1] Xu H, Cao X G. 2019 Analysis and Research on warpage deformation of CRTs III track slab in severe cold area [J] *Journal of railway engineering* **36** (05) 20-24 + 84.

[2] Yi X. 2015 Study on warpage deformation of embedded track plate in hardened Section [D] Southwest Jiaotong University.

[3] Chen J. 2020 Study on temperature and stress evolution law of CRTs III ballastless track slab [D] East China Jiaotong University.

[4] Yang J J, Zhang N, Gao M, Sun J L. 2016 Temperature warping deformation of CRTSII ballastless track and its influence on dynamic response of train track [J] *Engineering mechanics* **33** (04) 210-217.

[5] Ma L X, Liu W N, Liu W F. 2012 Dynamic response of floating slab track under moving resonant load [J] *Engineering mechanics* **29** (12) 334-341.

[6] Wang T Y, Ding J M, Lou M L. 2010 Load and analysis method of site vibration caused by subway operation [J] *Engineering mechanics* **27** (01) 195-201.

[7] Liu Y L. 2017 Study on vibration propagation law of high speed railway transportation [D] Shijiazhuang Railway University.

[8] Lou P. 2007 Vertical vibration analysis of train track (bridge) system [D] Central South University.

[9] Tao L. 2015 Vibration analysis of station overpass caused by high-speed train [J] Railway architecture (05) 34-37.

[10] Liu J L. 2019 Study on dynamic characteristics of saturated sand foundation under high speed railway vibration load [D] China University of mining and technology.

Analysis on Stop-hole Parameters for Fatigue Cracks at Arc Notch in Steel Bridge Deck

Liangping Feng[1, a], Lipeng Ling[1, b], Cheng Meng[2, c] and Bohai Ji[2, d]

[1] China Communications Construction Company Highway Bridges National Engineering Research Center Co., Ltd., Huangsi Street A23#, Beijing, China
[2] College of Civil and Transportation Engineering, Hohai University, No.1 Xikang Road, Nanjing, China
Email: [a]fengliangping@bnerc.com, [b]linglipeng@163.com, [c]ycmc1995@163.com, [d]bhji@hhu.edu.cn

Abstract. Two types of fatigue cracks at arc notch in steel bridge deck were repaired by drilling stop-holes. The effect of stop-holes with different diameters and positions was considered. Based on finite element models, the variation laws of stress distribution and the effects of stress concentration were compared for different stop-hole diameters and positions. Analysis results indicated that stop-hole can effectively improve the stress concentration at crack tip and the fatigue life of components can be considerably increased. The crack-stopping performance enhances with the increase of stop-hole diameter, but large stop-hole cannot effectively retard crack growth. The stop-hole performs well with the location at -0.5*D*~0.5*D*. The maximum stress point still appears at crack tip when the stop-hole is outside or inside the crack. The stop-hole diameter has no effect on the stop-hole location.

1. Introduction

Orthotropic steel deck has been widely used in the construction of long-span bridges because of its advantages, such as light weight, good stability and strong spanning capacity [1]. Bearing the weight loading of cars, with welding residual tensile stress and welding defects, it is easy to cause fatigue cracks in orthotropic steel deck. The arc notch between diaphragm and longitudinal rib is one of the typical fatigue crack details because of its special structure [2].

With the continuous emergence of fatigue problems, the research on the cause mechanism, damage detection and maintenance of fatigue cracks has been carried out. There are many effective maintenance methods to repair fatigue cracks. Drilling stop-holes is one of the most common maintenance methods to delay or hinder the crack propagation now. The principle of drilling stop-hole is to drill a smooth hole, and relieve the stress concentration [3, 4].

Current researches on drilling stop-holes mainly focus on the parameters and technology of stop-holes. Wu et al. [5] predicted the fatigue crack initiation lives by employing classical concepts properly modified by short crack theory to model the stop-hale effect. Liu et al. [6] measured residual stress after drilling cold expansion hole and carried out a series of uniaxial fatigue tests. Fu et al. [7] proposed drilling advice and conducted fatigue testing of the cracks in steel bridge deck with stop-hole. Castro et al. [8] analysed fatigue cracks after drilling stop-holes and proposed a calculation method of crack fatigue life after drilling stop-holes.

In this study, fatigue crack details at arc notch of the diaphragm were studied. Based on the data analysis of a real bridge and the results of finite element models, the reasonable position and diameter of stop-hole were proposed to optimise the drilling technology in real bridges.

2. Finite Element Models

2.1. Model Size

Annual check records of a long -span suspension bridge with steel box girder shows that there are many fatigue cracks grow in the arc notch, which are mainly divided into two types of fatigue details:(1) cracks grow obliquely upward from the bottom of weld between diaphragm and U-rib, named as crack C1; (2) cracks grow horizontally from the arc notch, named as crack C2.

According to the size of the real bridge, Q345qD steel was used as the bridge material with the Young's modulus of 2.06×10^5MPa and Poisson's ratio of 0.3. The thickness of roof and diaphragm were 12mm and 10mm respectively, and the section size of U-rib was 324mm×262mm×6mm, the specific model was shown in figure 1.

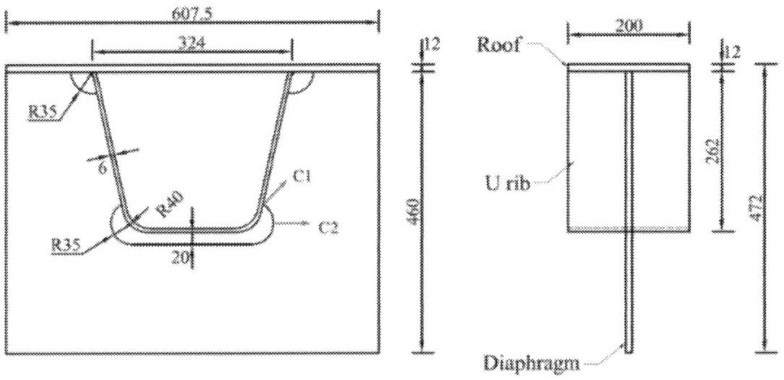

Figure 1. Model size and fatigue detail (dimensions in mm).

2.2. Finite Element Model

Local finite element models of diaphragm-to-rib weld were established by ABAQUS, as shown in figure 2. The roof was fixed constraints with full degrees of freedom, the loading area was set on the surface of the diaphragm, directly below the arc notch. The reference point was set at 10mm from the weld toe both in the horizontal and vertical directions with the nominal stress (σ_{nom}) of 100 MPa.

According to the actual statistics of fatigue cracks in real bridge, the average length of crack C1 and C2 were 35mm and 120mm respectively. The angle of 45° and 0° between the cracks and weld toe were adopted as the crack parameters in the finite element models. The cracks were set with lengths of 35mm and 120mm respectively, and width of 0.1mm both.

The models were meshed with hexagonal solid element type C3D8R and tetrahedral solid element type C3D10 and the meshes of crack regions were refined.

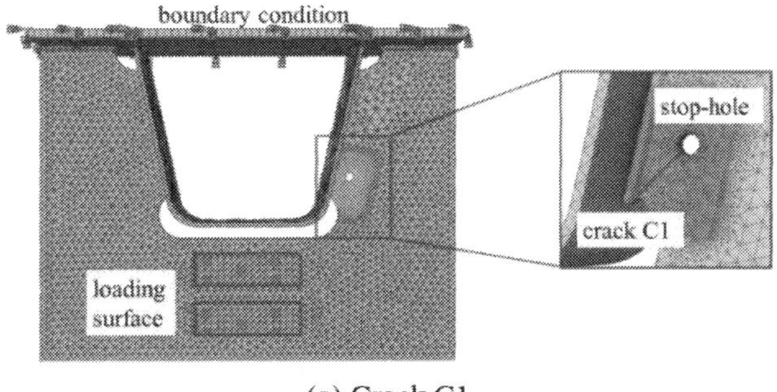

(a) Crack C1

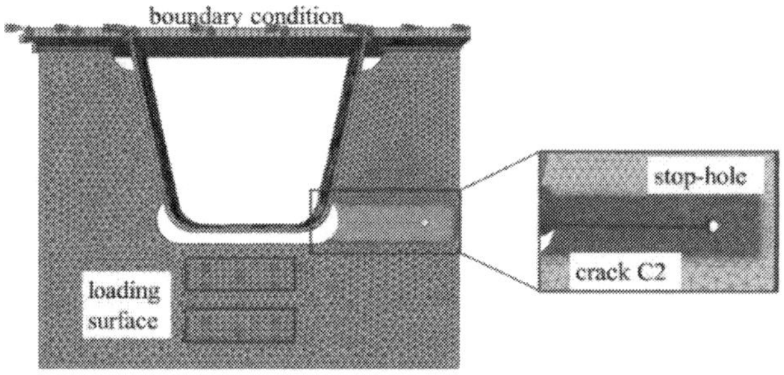

(b) Crack C2

Figure 2. Finite element model.

2.3. Determination of Cracks

Based on Extended Finite Element Method (XFEM) analysis, stress intensity factors (SIFs) of two cracks were calculated and the results were shown in figure 3.

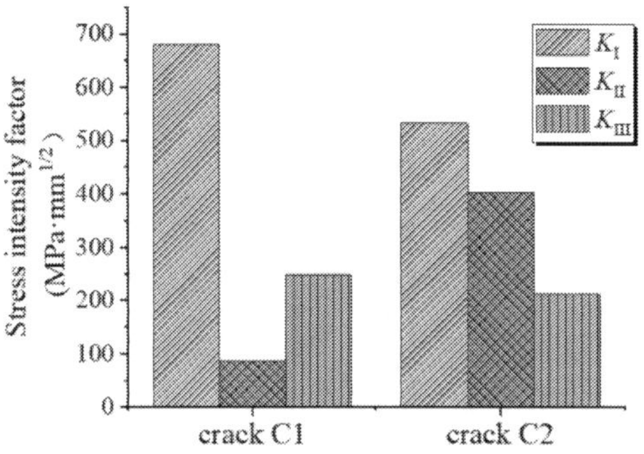

Figure 3. Stress intensity factors of two cracks.

According to the calculations of SIFs, K_{II} of crack C1was smaller than K_I and K_{III}, which showed crack C1 was a I-III mixed mode fatigue crack dominated by type I. In addition, crack C2 was a I-II-III mixed mode fatigue crack.

3. Influence of Stop-Hole Parameters

The stresses of two fatigue details under different stop-hole diameter and location were calculated. The stress concentration factor (K_t) and the fatigue notch factor (K_f) were compared to evaluate the fatigue performance. The solutions for K_t and K_f are given in equation (1) and equation (2).

$$K_t = \frac{\sigma_{max}}{\sigma_{nom}} \tag{1}$$

$$K_f = 1 + \frac{K_t - 1}{1 + \frac{c}{\rho}} \tag{2}$$

Where σ_{max} is obtained from the finite element model. And σ_{nom} is the nominal stress, here the value is 100MPa. c is a constant related with material, here the value is equal to 0.45. Then ρ is the radius of stop-hole.

3.1. Effects of Stop-Hole Diameter

Circular stop-holes with different diameters were drilled at the tip of crack C1 and C2, of which the center located at the extend line of crack tip. The stress concentration factor (K_t) and the fatigue notch factor (K_f) under different diameters were shown in table 1.

Table 1. Calculation results of different diameters.

D (mm)	σ_{nom} (MPa)	Crack C1			Crack C2		
		σ_{max} (MPa)	K_t	K_f	σ_{max} (MPa)	K_t	K_f
0	100	1583.5	15.835	/	758.4	7.584	/
8		555.8	5.558	4.963	402.4	4.024	3.630
10		489.6	4.896	4.502	363.5	3.635	3.369
12		420.2	4.202	3.938	336.7	3.367	3.172
14		403.5	4.035	3.823	316.7	3.167	3.016
16		343.4	3.434	3.287	300.9	3.009	2.888
18		313.7	3.137	3.023	287.7	2.877	2.777
20		298.4	2.984	2.890	277.6	2.776	2.692
22		292.3	2.923	2.840	269.3	2.693	2.620
24		278.1	2.781	2.711	262.8	2.628	2.564

As shown in table 1, the maximum stress of crack C1and C2 before drilling stop-holes were 1583.5 MPa and 758.4 MPa respectively, which means if left untreated, the cracks would continue to grow and cause structure destruction. After drilling stop-holes at the crack tips, the maximum stress, K_t and K_f decreased significantly. The stop-hole performance enhanced with the increase of stop-hole diameter.

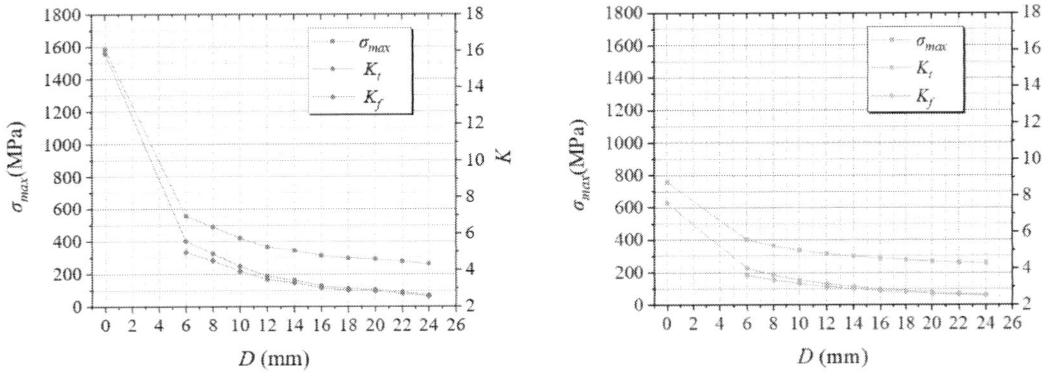

(a) Crack C1 (b) Crack C2

Figure 4. Parameter variation under different diameters.

Figure 4 shows that with the increase of stop-hole diameter D, the change of K_t and K_f got slower. Besides, stop-hole could weaken the stiffness and strength of structure. Therefore, the reasonable diameter for crack C1 and crack C2 should be controlled at 12-16mm.

3.2. Effects of Distance Between Stop-Hole and Crack Tip

The influence of stop-hole position on crack arrest effect was studied under the condition that the stop-hole was not eccentric, namely the center of stop-hole and the crack were on a straight line. The sketch of hole location was shown in figure 5, X is the horizontal distance between the center of stop-hole and crack tip.

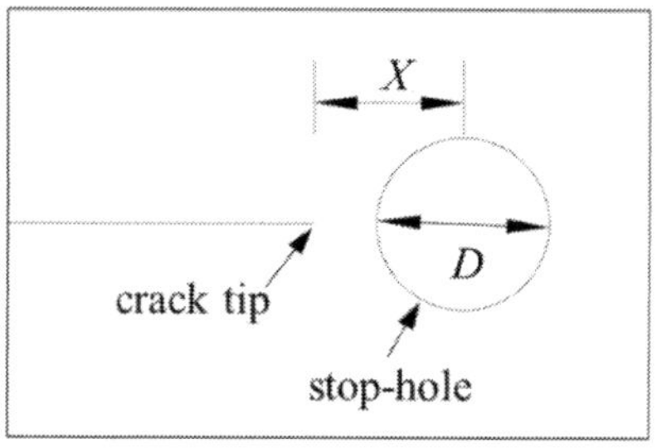

Figure 5. Sketch of hole location.

The hole diameters of stop-hole were 12mm, 14mm and 16mm in model calculation. Calculation results of the maximum stress around stop-hole or at crack tip were shown in figure 5.

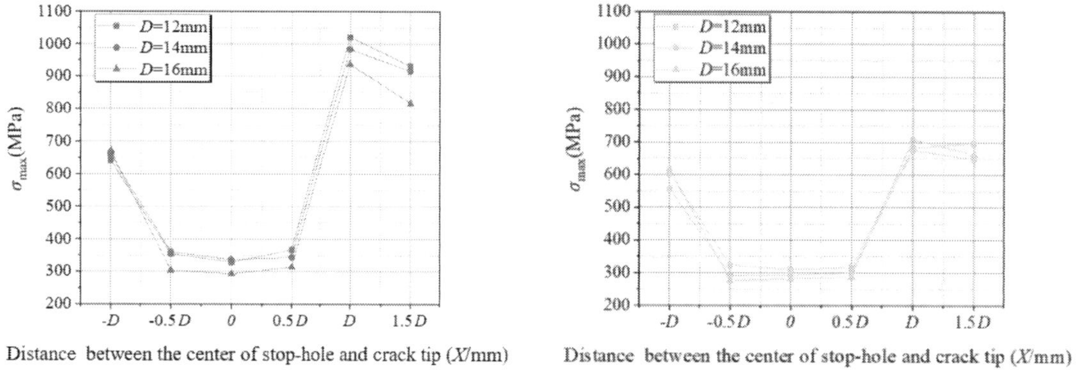

(a) Crack C1 (b) Crack C2

Figure 6. Variation law of the maximum stress.

As shown in figure 6, when the center of stop-hole was inside the crack, namely $-D \leq X < -0.5D$, although the maximum stress reduced slightly, the stress concentration still could not be improved well.

When $-0.5D \leq X \leq 0.5D$, the stress concentration improved substantially, and the crack stopping performances of different diameters were close in this range. When the stop-hole was completely outside the crack ($X > 0.5D$), stress concentration still existed at crack tip and would cause crack propagation.

The center of stop-hole should be set in range of $-0.5D \sim 0.5D$ from crack tip for these two types of fatigue cracks. In practical maintenances, crack tip is hard to locate accurately, therefore the visible tip should be completely covered by the stop-hole.

4. Conclusions

(1) These two typical types of fatigue cracks at arc notch are both mixed mode fatigue cracks. Drilling stop-holes can effectively relieve the stress concentration at the tips of crack C1 and crack C2, and stop crack propagation.

(2) For these two types of fatigue cracks, the crack-stopping performance improves with the increase of stop-hole diameter. And the reasonable stop-hole diameter should be in range of 12-16mm considering the weakening effect of the stop-hole on structure.

(3) The most efficient crack stopping performance occurs when the distance between the center of stop-hole and crack tip is in range of $-0.5D \sim 0.5D$. The stress concentration cannot be improved when

the stop-hole is drilled completely inside or outside the crack. And the stop-hole diameter has little effect on the stop-hole location.

Acknowledgments
The research reported herein has been conducted as part of the research projects granted by Guangdong Key Areas Research and Development Project (2019B111106002), Academician Project Foundation of CCCC (YSZX-03-2020-01-B) and National Key Research and Development Project (2017YFE0128700). The assistances are gratefully acknowledged.

References

[1] Ya S, Yamada K, Ishikawa T. 2011 Fatigue Evaluation of Rib-to-Deck Welded Joints of Orthotropic Steel Bridge Deck [J] *Bridge. Eng* **18** 492-499.
[2] Kolstein M H. 2007 Fatigue classification of welded joints in orthotropic steel bridge decks *Publication of Transport & Road Research Laboratory*.
[3] Song P S, Shieh Y L. 2004 Stop drilling procedure for fatigue life improvement *Int. J. Fatigue.* **26** 1333-1339
[4] Yao Y, Ji B, Fu Z, Zhou J, Wang Y. 2019 Optimization of stop-hole parameters for cracks at diaphragm-to-rib weld in steel bridges [J] *Constr. Steel. Res* **162** 105747
[5] Hao W, Imad A, Benseddiq N, Maeda K. 2010 On the prediction of the residual fatigue life of cracked structures repaired by the stop-hole method *Int. J. Fatigue* **32** 670-677
[6] Liu J, Shao X, Liu Y, Yue Z. 2008 Effect of cold expansion on fatigue performance of open holes [J] *Mat. Sci. Eng. A-Struct.* **477** 271-276
[7] Fu Z, Ji B, Xie S 2017 Crack stop holes in steel bridge decks: Drilling method and effects [J] *Cent. South. Univ* **24** 2372–2381.
[8] Castro J, Meggiolaro M, Miranda A, Hao W, Imad A, Benseddiq N. 2012 Prediction of fatigue crack initiation lives at elongated notch roots using short crack concepts *Int. J. Fatigue* **42** 172-182.

FLUENT Simulation Design and Optimization of Biological Aerated Filter

Shuqin Wang [1, 2, *], Zhiqiang Zhang [1], Ning Wang [1], Wenqi Zhao [1] and Chungang Yuan [1, 2]

[1] Department of Environmental Science and Engineering, North China Electric Power University, Baoding 071003, China;
[2] Hebei Provincial Key Laboratory of Collaborative Control of Multiple Pollutants in Flue Gas of Coal-fired Power Plants, North China Electric Power University, Baoding 071003, China
Email:wsqhg@163.com

Abstract. In this paper, a small biological aerated filter for experimental use was designed, and a method was explored to optimize the nitrogen removal efficiency by using FLUENT software to simulate the particle size of the filler, the amount of the filler, the initial concentration of ammonia nitrogen, dissolved oxygen and other operating parameters. Through the simulation experiment, the optimal design parameters of the particle size of filler, the amount of filler, the initial concentration of ammonia nitrogen and the dissolved oxygen of the biological aerated filter are 4mm, 60%, 15% and 1.5%, respectively, when the removal efficiency of ammonia nitrogen exceeds 30% reported in the literature. It provides a reference for the experimental research and practical application of biological aerated filter (BAF) denitrification.
Keywords. BAF; FLUENT simulation; Nitrogen removal; Optimized design.

1. Introduction

As a representative process of the third generation of biofilm, biological aerated filter (BAF) has the advantages of compact structure, convenient operation and high treatment efficiency, and has been widely used in municipal and industrial wastewater [1-2]. However, the biological aerated filter is often exposed to the disadvantage of low nitrogen removal efficiency in practical application. The study of Zhang Ruijing et al. [3] showed that the nitrogen removal efficiency of a biological aerated filter in a sewage treatment plant was 58.11%, while the ammonia nitrogen removal efficiency of Yue Zhifang et al. [4] was only 37.5%. And in the design and calculation of biological aerated filter is often used in the empirical or semi-empirical formula, these formulas lack of rigorous theoretical basis, resulting in the formula varies from person to person and various, so in the actual design selection and calculation of the work is large, and it is difficult to ensure accuracy. It is urgent to improve the nitrogen removal efficiency of biological aerated filter and explore a more simple and accurate design method.

Many studies show that the numerical simulation results of FLUENT software are very similar to the experimental results. Liao Tengyi [5] simulated the Biostyr filter, J. Ridgeman [6] simulated the distribution and reaction of anaerobic bacteria and activated sludge in small-scale experimental equipment, and Ding Rui et al. [7] simulated the hydrodynamic characteristics of self-made water treatment reactor, all of which were consistent with the results of verification experiments. Using FLUENT for reactor design can effectively reduce the workload of repeated experiments and calculation, and save cost. At present, there are few reports on the application of FLUENT in the denitrification of biological aerated filter. Wang Lijie [8] simulated the influence of water retention time and water vapor ratio on the nitrogen removal efficiency of BAF, but there is no further study on the key design parameters of the filter, such as filler, initial ammonia nitrogen concentration and dissolved oxygen.

In this study, the FLUENT nitrogen removal simulation of the experimental device of biological aerated filter was conducted to explore the particle size of filler, initial ammonia nitrogen concentration, dissolved oxygen and the amount of filler to achieve the highest nitrogen removal efficiency of the small biological aerated filter, and explore the method of numerical simulation to design the biological aerated filter and optimize the nitrogen removal efficiency.

2. Numerical Simulation

2.1. Simulation Object

The simulation object is a rectangular filter with a scale of 70*60*190mm. The height of the lower packing area is 70mm, the height of the upper packing area is 90mm, and the design height of the aeration area is 30mm at the bottom. The two-layer packing zone can effectively slow down the wall flow phenomenon, make the liquid phase more evenly distributed in the tower, and improve the processing efficiency. The packing adopts hollow spherical packing, which has large specific surface area and high space utilization rate, and the regular spherical packing is convenient to simulate. The lower layer is 20mm more than the upper layer to expand the capacity of sewage. See figure 1 for the three views of the tower body. After considering the size and cost of the aeration device, the aeration device was optimized, and the method of aquarium microhole aeration was chosen instead of the herring-shaped guide plate and perforated tube aeration commonly used in BAF. Because this method can not only meet the oxygen demand, but also have a good disturbance effect on the substrate, which is helpful to improve the efficiency of nitrogen removal.

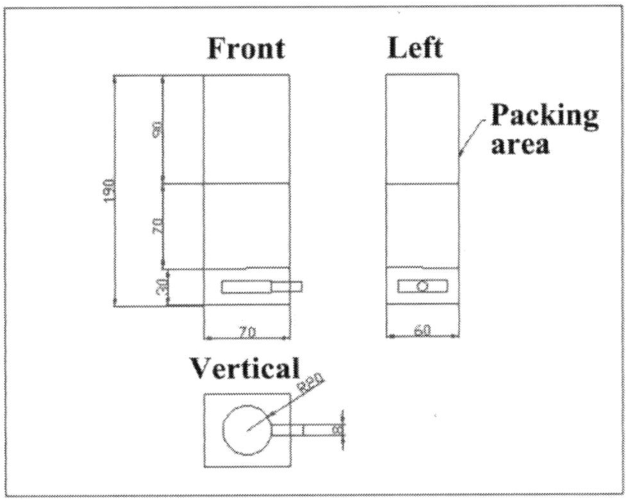

Figure 1. Design of biological aerated filter experimental device.

2.2. Simulation Method

FLUENT is a commonly used CFD software package, which has strong grid support ability. The finite volume method based on completely unstructured grid is used for simulation calculation. The software contains a wealth of turbulence models, which can accurately simulate inviscous flow, laminar flow, turbulence, etc. ICEM was used to read the three-dimensional filter model designed by CAD, and the triangle mesh was used to divide it. Boundary layer conditions such as inlet, outlet and fluid domain were defined, and the output was MSH file. The mesh file was read by FLUENT, and the Mixture model of multiphase flow was used to calculate the fluid dynamics. The nebulae of ammonia concentration distribution were output after changing the parameters. The reaction design was the transformation of NH_4^+ to N_2. The spherical filler was simulated by porous media model. The porosity of porous media was designed to be 0.5, and the size of the filler was characterized by the drag coefficient.

2.3. Simulation Experiment

In this simulation process, a total of 7 groups of conditions were designed. Two surface areas X and Y were designed in the direction parallel to and perpendicular to the wall where the water inlet was located. By observing the distribution of ammonia nitrogen concentration on the two surface areas, the nitrogen removal efficiency was calculated and analyzed. The first four groups of simulation tests respectively simulated the distribution of ammonia nitrogen concentration on two surfaces under different packing particle sizes, and determined the optimal packing size. In the last three groups of simulation experiments, the group conditions of the optimal particle size obtained in the previous four groups of experiments were taken as reference, and the values of ammonia nitrogen concentration, filler amount and dissolved oxygen in the intake were changed to explore the best parameter conditions. The design inlet and outlet water flow rate is 1m/s. The packing amount is expressed as the proportion of the volume occupied by the packing area.

Table 1. Experimental design parameter.

Group	Packing size (mm)	Inlet ammonia nitrogen concentration (%)	Proportion of packing area occupied (%)	Dissolved oxygen content (%)
1	10	10.00	60	1.0
2	8	10.00	60	1.0
3	6	10.00	60	1.0
4	4	10.00	60	1.0
5	required value	15.00	60	1.0
6	required value	15.00	60	1.5
7	required value	15.00	70	1.5

3. Results and Discussion

Table 2 lists the numerical simulation calculation results of ammonia nitrogen removal efficiency under 7 working conditions.

Table 2. Ammonia nitrogen removal efficiency calculation table.

Group	Packing size (mm)	Inlet ammonia nitrogen concentration (%)	Outlet ammonia nitrogen concentration (%)	Ammonia nitrogen removal efficiency (%)
1	10	10.00	7.65	23.50
2	8	10.00	7.16	28.40
3	6	10.00	6.80	32.00
4	4	10.00	5.75	42.50
5	4	15.00	10.44	45.60
6	4	15.00	4.90	67.33
7	4	15.00	4.66	68.93

3.1. Optimum Packing Size

It can be seen from the first four working conditions in table 2 that the smaller the packing size of the spherical filler, the higher the ammonia nitrogen removal efficiency, and the optimal particle size is 4mm. It is concluded that the smaller the packing size is, the larger the specific surface area is, which is conducive to increasing the mass transfer contact area and prolonging the reaction time, and is conducive to hanging the film in the practical application. Literature reports [9-10] also show that fillers with smaller packing size have better ammonia nitrogen removal performance. But too small packing size will increase the production cost, after comprehensive consideration, this design uses 4mm spherical packing.

3.2. Effect of Influent Ammonia Nitrogen Concentration on Denitrification
The nebulae of concentration distribution in X and Y areas of Group 4 (figure 2) and Group 5 (figure 3) were compared and analyzed. It was found that the content of ammonia nitrogen in the filter decreased significantly after the concentration of ammonia nitrogen was increased, and the overall distribution showed that the content in the upper layer decreased significantly while the concentration in the lower layer was higher. The concentration of nitrogen on the wall side where the sewage inlet is located is high, which may be due to the poor fluidity of the liquid phase, which is harmful to the mass transfer and reaction of oxygen. Table 2 shows that the nitrogen removal efficiency is slightly improved after increasing the concentration of influant ammonia nitrogen. The analysis reason may be that the nitrogen content set in the simulation is within the range that microorganisms can oxidize, and higher concentration of ammonia nitrogen is conducive to the growth of nitrification bacteria. There are also reports in the literature that increasing the nitrogen load within a certain range will increase the nitrogen removal efficiency [11].

3.3. Effect of Dissolved Oxygen on Denitrification
By comparing and analyzing the cloud map of concentration distribution in X and Y plane of Group 5 (figure 3) and Group 6 (figure 4), it is found that: it was found that when dissolved. oxygen was appropriately increased, the concentration of ammonia nitrogen at the outlet was significantly reduced. According to the calculation results in table 2, compared with Group 5, the dissolved oxygen content in group 6 only increased by 0.5%, and the nitrogen removal efficiency increased by 21.73%, indicating that the change of dissolved oxygen had a significant impact on the nitrogen removal efficiency. This is because dissolved oxygen will not only affect the growth of nitrification bacteria, but also affect the gas-liquid two-phase flow field of the BAF [12] and the shear action on the surface of the biofilm [13]. This design uses 1.5% oxygen content with high nitrogen removal efficiency.

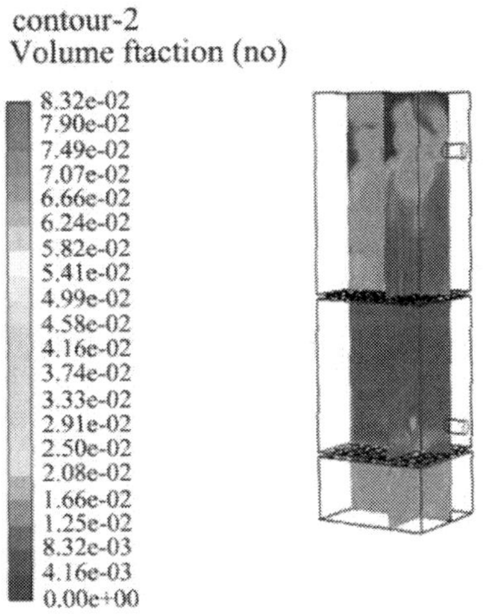

Figure 2. Nebulogram of concentration distribution in X and Y plane of group 4.

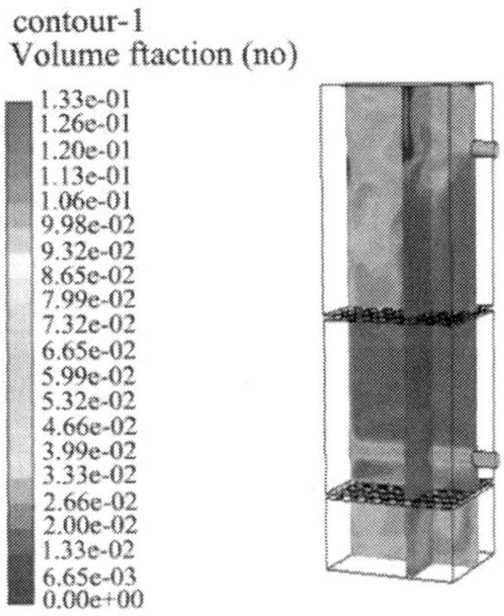

Figure 3. Nebulogram of concentration distribution in X and Y plane of group 5.

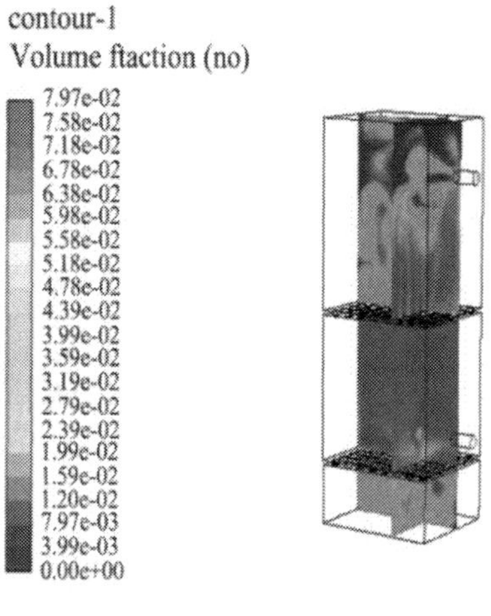

Figure 4. Nebulogram of concentration distribution in X and Y plane of group 6.

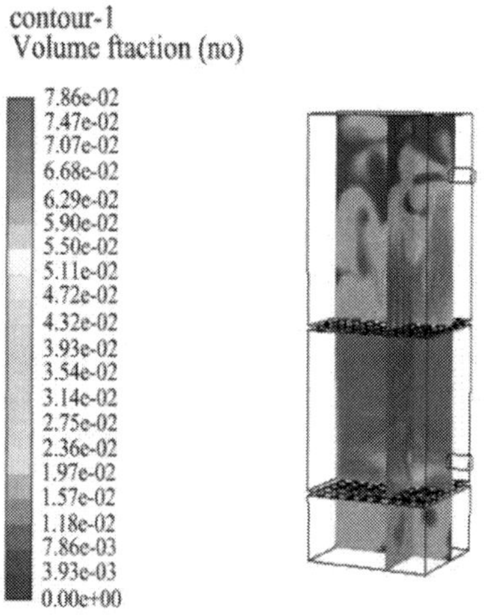

Figure 5. Nebulogram of concentration distribution in X and Y plane of group 7.

3.4. Effect of Packing Quantity on Denitrification

Group 7 increased the packing amount by 10% on the basis of Group 6. According to the concentration distribution nebulograph (figure 3 and figure 4), the ammonia nitrogen concentration at the outlet of Group 6 and Group 7 was significantly reduced, but there was no significant difference between them. According to the calculation results, the nitrogen removal efficiency of Group 7 only increased by 1.6% when the filler was increased by 10%, which indicated that increasing the filler amount within the allowable range of the filter load was beneficial to increase the ammonia nitrogen conversion efficiency. The analytical reason may be that increasing the filler amount properly could provide more space for microbial attachment and growth. However, the improvement of nitrogen removal efficiency is not very obvious. Due to cost considerations, the BAF is designed to take up 60 percent of the total packing area volume when running.

3. Conclusion

After optimizing the packing area and the aeration unit, the final design of the small biological aerated filter was a tower BAF with two layers of packing and aquarium microporous aeration. The numerical simulation of self-made small biological aerated filter by FLUENT shows that the decarburization efficiency is higher with the smaller particle size of filler. Dissolved oxygen concentration has great influence on the BAF operation and appropriate increase of dissolved oxygen can significantly improve the effect of nitrogen removal. With the increase of influent ammonia nitrogen concentration and filler amount, the nitrogen removal efficiency is increased but the effect is not obvious. The optimal design parameters of the small biological aerated filter in terms of packing particle size, packing quantity, initial ammonia nitrogen concentration and dissolved oxygen are 4mm, 60%, 15% and 1.5% respectively. Under these parameters, the ammonia nitrogen removal efficiency of 67.33% can be achieved, which is about 30% higher than that reported in the literature [4]. The FLUENT numerical simulation method was used to design the biological aerated filter and optimize the nitrogen removal efficiency, which provided a reference for the experimental research and practical application of nitrogen removal in biological aerated filter.

Acknowledgements

The project was supported by the Joint Funds of Education Reform Project of Hebei Province

(2019GJJG416) and The Education Reform Program of the Ministry of Education (E-HGZY20202003).

Reference

[1] Yu Q S, Huang H, Ren H Q, et al. 2016 in Situ Activity Recovery of Aging Biofilm in Biological Aerated Filter: Surfactants Treatment and Mechanisms Study [J] *Bioresource Technology* 219.
[2] Cai Q Q, Wu X P, Shen H C, et al. 2017 Degradation of Microcystin-LR in drinking water source with biological aerated filter enhanced by bacteria agent [J] *Environmental Chemistry* **36** (9) 2015-2024.
[3] Zhang R J, Zhang J S, Qian Y L. 2020 Application of EBAF@ Biological Active Filters in Guangxi A WWTP [J] *Guangdong Chemical Industry* **47** (11) 147-148.
[4] Yue Z F, Yang H Y, Yan Y L, et al. 2019 Treatment of Wastewater from a Development Zone by Hydrolysis Acidification/Multi-stage and Multi-level AO/Biological Aerated Filter Process[J] *China Water & Wastewater* **35** (18) 107-111.
[5] Liao T Y. 2014 Fluent Simulation in Biological Aerated Filter Based on Stratified Biofilm [D] Hunan University.
[6] Bridgeman J. 2011 Computational Fluid Dynamics Modelling of Sewage Sludge Mixing in an Anaerobic Digester [J] *Advances in Engineering Software* **44** (1).
[7] Ding R, Mao Z Y, Wang J L. 2017 CFD simulation and optimization of the water treatment reactor by electron beam [J] *China Environmental Science* **37** (3) 980-988.
[8] Wang L J. 2014 Numerical Simulation of Gas-liquid Two Phase Flow of BAF and Denitrification Experimental Study [D] Harbin Institute of Technology University.
[9] Li P, Wang Q H, Zheng T L, et al. 2012 Organic matter and ammonia removal and adsorption ability of three kinds of carriers [J] *Technology of Water Treatment* **38** (7) 17-20.
[10] Wang H T. 2012 Multiplc-scalc Analysis and CFD Simulation in Biological Aerated Filters [D] Hunan University.
[11] Lu L. 2011 Study on the Craft and Process Control of Nitrate Wastewater Treatment with Biological Aerated Filter [D] South China University of Technology.
[12] Ren J H, Cheng W, Jiao M, et al. 2020 Simulation of gas-liquid two-phases flow characteristics in the biological aerated filter [J] *Chinese Journal of Hydrodynamics* **35** (3) 306-313.
[13] Duan Y, Lu S M, Jiang D. 2018 Biofilm Formation and Start-up of Suspended Ceramic Biological Aerated Filter in Wastewater Treatment [J] *Technology of Water Treatment* **44** (2) 88-92.

Stiffness Degradation of Shear Wall with Frame Column and its Influence to Structural Performance

Ming Wen, Zipeng Huang, Xinfang Wang, Baokui Chen, Yu Wan

School of Civil Engineering and Architecture, Nanchang University, Nanchang 330031, China
Email: 734525570@qq.com

Abstract. The shear wall with frame column has a wide application prospect in practical engineering for it has better ductility, bearing capacity and stiffness degradation than the shear wall with unframed column. In this paper, the finite element model of shear wall with frame column for an experiment was established in ABAQUS software firstly, Then on the base of proving the model's precision, the axial compression ratio and the stiffness ratio of shear wall with frame column were adjusted, the influence of different axial compression ratio and stiffness ratio on the stiffness degradation coefficient was analyzed and the corresponding formulas were fitted. At last, by changing the axial compression ratio and the stiffness ratio of shear wall with frame column, The influence of stiffness degradation on the maximum displacement amplification coefficient and the bottom shear amplification coefficient were studied and the relevant formulas were fitted.

1. Introduction

With the development of science, technology and the improvement of people's living standard, the development and utilization of high-rise buildings with Shear walls are more and more extensive. Shear wall is generally divided into two types: shear wall with unframed column and shear wall with frame column. References [1] [2] show that the ductility, bearing capacity and stiffness degradation of shear wall with frame column are much better than the wall with unframed column. It is of great significance to study the performance of shear wall with frame column deeply.

In rare earthquakes, the lateral stiffness of the shear wall degrades seriously when entering the elastic-plastic stage, and the internal force will redistribute in the whole structure, so the stiffness degradation of the shear wall has an important influence on the elastic-plastic performance of the structure. The stiffness degradation characteristics of shear wall had been studied by many scholars in the world. By simulation calculating, the conclusion that the shear ratio of the shear wall decreases with the increasing of the horizontal load in the nonlinear stage was found by Jia Yigang et Al [3]. The conclusion that the stiffness degradation of joints under strong earthquakes is dynamic was found by Ma Lei et Al [4]. Based on building the model of degradation of bearing capacity and considering degradation of fatigue stiffness of concrete beam, the relationship between bearing capacity and stiffness degradation was studied by Liu Fangping et AL [5]. By using ABAQUS nonlinear finite element analysis software, it was found that the frame column and frame beam of steel plate shear wall have influence on plastic energy dissipation, lateral stiffness and bearing capacity by Fan Zhong et al [6].Through the ANSYS software, the performance of rectangular shear wall with concealed column, shear wall with frame column at one end and shear wall with frame column at two ends were compared by He chihe [7]. 10 groups high and low shear-span ratios of concrete shear walls with framed columns were tested under cyclic loading by Sun jinchi [8]. After analyzing the mechanical behavior of masonry shear wall with whole section and masonry wall with opening and border frame, the relevant calculation formulas were put forward by Liu Xianming et Al [9]. Through the experimental data of multi-group and multi-layer lightweight aggregate concrete shear walls with frame columns, it was found that the stiffness degradation of shear wall is related to inter-layer displacement angle, axial compression ratio and shear-span ratio by Zhao Shifeng et Al [10].

From the studies that had been done, it can be found that the research on stiffness degradation of shear wall with frame column was mostly confined to experiment, and systematic analysis was seldom done. In this paper, the concept of stiffness ratio of shear wall with frame column was proposed firstly, Then a finite element model for an experiment was established in ABAQUS software, after proving the accuracy of the model, the influence of axial compression ratio and stiffness ratio on the stiffness degradation of shear wall with frame column was analyzed and the corresponding calculation formulas were fitted .At last, for an actual frame-shear wall with frame column structure, the axial compression ratio and the stiffness ratio of shear wall with frame column were changed ,and the influence of stiffness degradation of shear wall with frame column on the magnification coefficient of maximum displacement and the magnification coefficient of bottom shear was studied. The conclusions have certain theoretical and practical value.

2. Establishment of Finite Element Model of Shear Wall with Frame Column

According to one of the experiments made by Sun Jinchi on shear wall with frame column [8], a finite element model was established in ABAQUS software. The parameters of the model were described as follows: the reinforcement ratio of frame column is 0.448% (4Φ4), the shear span ratio of shear wall is 3.37, and the reinforcement ratio of wall is 0.593% (3.5@35):The exact dimensions of the model were shown in figure 1.

The steel bar is grade 1 steel, the yield strength is 235 N/mm^2 , Poisson's ratio is 0.3, and the elastic modulus is 206 GPa. The concrete is C30, Poisson's ratio is 0.2, the elastic modulus is 30 GPa. The concrete constitutive model is plastic damage model, and the steel constitutive model is two-fold line follow-up hardening model. The finite element model established in ABAQUS was shown in figure 2.

Let $\beta = k_t / k_0$, Where β is stiffness Degradation coefficient ; k_0 is initial stiffness of the shear wall with frame column; k_t is the secant stiffness of the shear wall with frame column.

Using the same loading mode as the experiment [8], the comparison of β got by calculation and got by experiment was shown in table 1,

From table 1, it can be concluded that the results are almost identical and the errors are within the allowable range,the finite element model established in ABAQUS software is effective.

[illegible]

3. Study on Influence Factors of Stiffness Degradation of Shear Wall with Frame Column

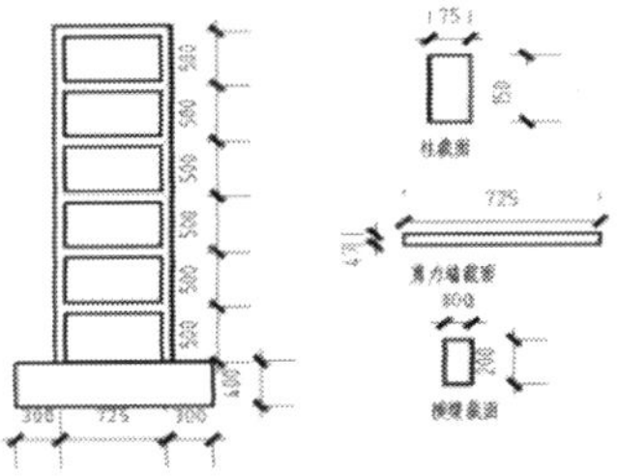

Figure 1. Overall dimension of shear wall with frame column specimen.

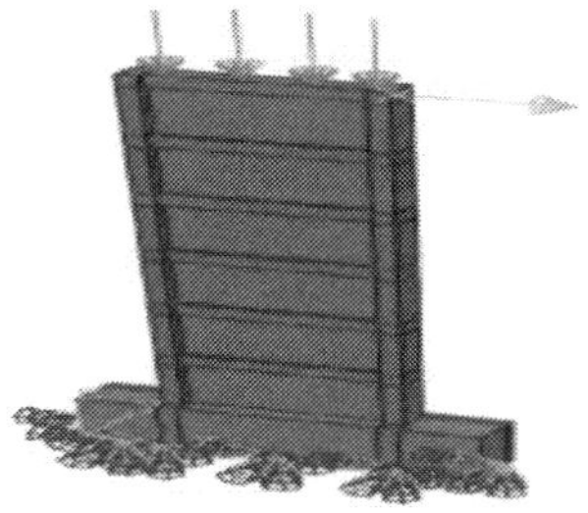

Figure 2. Loading model of shear wall with frame column structure.

Table 1. Comparison of β got by calculation and got by experiment.

Inter-layer displacement angle	1/3000	1/2000	1/1000	1/750	1/500	1/300	1/250	1/120
Experimental results	1.00	0.86	0.67	0.57	0.46	0.35	0.30	0.20
Calculation result	0.996	0.855	0.657	0.554	0.453	0.342	0.289	0.188

The stiffness degradation performance of shear wall with unframed column is related to axial compression ratio and shear-span ratio, while the stiffness degradation performance of shear wall with frame column is related to the stiffness ratio of frame column and shear wall besides axial compression ratio. The stiffness ratio of frame column to shear wall is defined as

Where C_f is the lateral stiffness of frame column; H is the height of shear wall and EI_w is the stiffness of shear wall .The effect of axial compression ratio and stiffness ratio on the stiffness degradation performance of shear wall with frame column was analyzed as follow in detail.

3.1. Influence of Axial Compression Ratio on Stiffness Degradation Performance of Shear Wall with Frame Column

The stiffness ratio m remained unchanged at 1.4, the axial compression ratio u was set as 0.15, 0.2, 0.25, 0.3, 0.35, 0.4, 0.45, 0.5, 0.55 and 0.6 respectively. The corresponding models were established and calculated. Let $\alpha = \theta / 3000$ (θ is inter-layer displacement angle), The relationship between inter-layer displacement angle and stiffness degradation coefficient under different axial compression ratio was shown in figure 3. As can be seen in figure 3

(1)The stiffness of shear wall with frame column decreases with the increasing of inter -layer displacement angle, and the degradation curve decreases obviously when α is in the range of 1.5-6.

(2)The axial compression ratio of shear wall has obvious influence on the stiffness degradation coefficient of shear wall with frame column. When the inter-layer displacement angle is the same, the stiffness degradation coefficient increases and the stiffness degradation degree decreases with the increasing of axial compression ratio.

(3)When the axial compression ratio changes between 0.1- 0.25, the stiffness degradation coefficient of the shear wall changes greatly; and when the axial compression ratio changes between 0.25 -0.6, the stiffness degradation coefficient of the shear wall changes little.

3.2. Influence of Stiffness Ratio on Stiffness Degradation Coefficient of Shear Wall with Frame Column

The axial compression ratio remained unchanged at 0.25, the stiffness ratio was set as 1.0, 1.2, 1.4, 1.6, 1.8, 2.0, 2.2, 2.4, 2.6, 2.8, 3.0, 6.0 respectively. The corresponding models were established and calculated. Let $\alpha = \theta / 3000$ (θ is inter -layer displacement angle), the relationship between inter-layer displacement angle and stiffness degradation coefficient under different stiffness ratio was shown in figure 4.

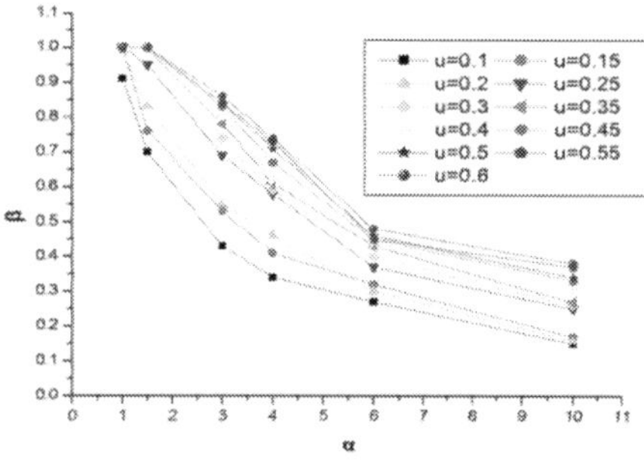

Figure 3. Relationship curves between Inter-layer displacement angle and stiffness degradation coefficient under different axial compression ratio.

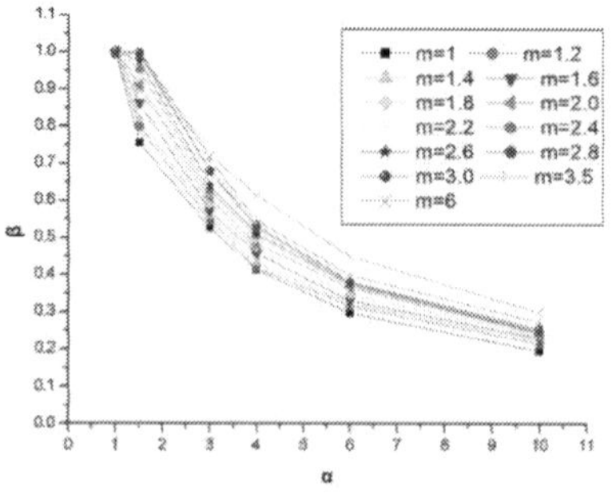

Figure 4. Relationship curves between Inter-layer displacement angle and stiffness degradation coefficient under different stiffness ratios.

As can be seen in figure 4

(1)With the increasing of stiffness ratio between frame column and shear wall, the stiffness degradation degree of shear wall decreases when reaching the same inter-layer displacement angle

(2)The stiffness degradation curve of shear wall decreases rapidly when α is in the range of 1-3, and the stiffness degradation coefficient is obviously affected by the inter-layer displacement angle. The stiffness degradation curve is relatively gentle when α is in the range of 3-10.

(3)With the increasing of the stiffness ratio between frame column and shear wall, the stiffness degradation velocity of shear wall decreases after elastic-plastic deformation. The stiffness ratio has an obvious effect on the stiffness degradation velocity when stiffness ratio is in the range of 1 - 2.8, and when the stiffness ratio is between 2.8 - 6.0, the stiffness degradation of shear wall is not obvious, the stiffness degradation performance of shear wall has been improved obviously.

3.3. Analysis of Stiffness Degradation of Shear Wall with Frame Column under Comprehensive Factors

The stiffness ratio was set as 1.0, 1.2, 1.4, 1.6, 1.8, 2.0, 2.2, 2.4, 2.6, 2.8, 3.0, 3.5, 6.0 respectively; axial compression ratio was set as 0.25, 0.3, 0.35, 0.4 respectively, 52 sets of shear wall with frame column models were established, and the stiffness degradation performance of shear wall with frame column under different stiffness ratio and axial compression ratio was analyzed separately. Due to the limitation of space, the stiffness degradation coefficients were shown in table 2 only when the stiffness ratio is 1 and the axial compression ratio is 0.25

Table 2. Relationship between inter-layer displacement angle and stiffness degradation coefficient of shear wall when m = 1, μ=0.25.

Inter-layer displacement angle	1/3000	1/2000	1/1000	**1/750**	**1/500**	**1/300**
Stiffness Degradation Coefficient	0.992	0.755	0.527	0.413	0.298	0.196

After fitting all the data of 52 models, the formula of stiffness degradation coefficient can be denoted as:

$$\beta = \sin\left[\frac{1+e^{p_1 m}+e^{p_2\theta}+e^{p_3 u}}{p_4+e^{p_5 m}+e^{p_6\theta}+e^{p_7 u}} - \frac{(p_8-u)^2}{(p_9-\theta)^3-(p_{10}-m)}\right] \quad (1)$$

Where β is stiffness degradation coefficient, values range from 0-1; m is stiffness ratio of frame column to shear wall, values range from 1-6; u is axial compression ratio, values range from 0.25-0.4; θ is Inter-layer displacement angle, values range from 1/3000~1/300; p1=0.17; p2=137.54; p3=-115.91; p4=-2.99; p5=0.00016; p6=19.60; p7=-0.0074; p8=-1.0; p9=0.66; p10=-218.03;

The comparison between data got from fitting formula and got by the finite element calculation (the target value) was shown in figure 5. It was shown that the fitting formula has higher precision.

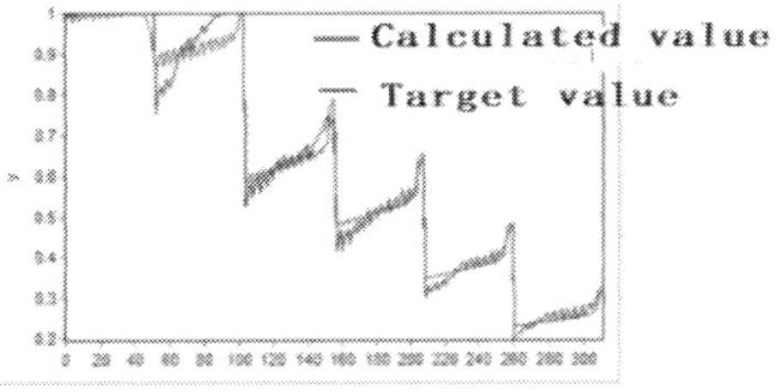

Figure 5. Comparison between data got from fitting formula and got by the finite element calculation.

4. Influence of Stiffness Degradation of Shear Wall with Frame Column on Seismic Behavior of Frame-Shear Wall Structure

For a frame-shear wall with frame column structure, the load on the structure was adjusted to change the axial compression ratio of the shear wall, and the section size of the frame-column was adjusted to change the stiffness ratio of the frame-column to the shear wall, elastic time-history and elastic-plastic time-history analysis of structure with different axial compression ratio and stiffness ratio were carried out in SAP software.

The ratio between maximum floor displacement got in elastic-plastic time-history condition and that got in elastic time-history is defined as magnification coefficient of maximum displacement, and the ratio between the bottom shear got in elastic-plastic time-history condition and that got in elastic time-history is defined as magnification coefficient of bottom shear. The maximum inter-layer displacement angle got by elastic-plastic time-history analysis and the corresponding axial compression ratio and stiffness ratio of the shear wall with frame column were taken into formula (1) to obtain the stiffness degradation coefficient. The influence of stiffness degradation coefficient of shear wall with frame column on the displacement magnification coefficient and the bottom shear magnification coefficient were analyzed and the related formulas were fitted.

The structure is a 14-story frame-shear wall with frame-column structure. The height of bottom floor is 3.6 m, and the height of the other floors is 3.3 m. The layout of the structure is shown in figure 6. The seismic fortification intensity is 7 degree, the site type is II, the design earthquake grouping is the second group, the site characteristic period is 0.4 s.

The concrete strength of wall and column is C40 from 1 to 4 layers,and is C35 from 5 to 11 layers. The concrete strength of beam and plate is C30 from 1 to 11 layers. The concrete strength of each member from 12 layers to top layer is C25. The live load of floors is 2 kN/m^2. The reinforcement grade of column, beam and wall is HRB400, the size of frame column from 1-4 layers is 750×750, and is 600×600 from 5-14 layers, the thickness of shear wall is 200.

According to the norm, one artificial seismic wave RH1TG040 and two natural seismic waves TH003TG040 and TH005TG040 were selected for time history analysis. The bottom shear force calculated from each time history curve is more than 65% result calculated from the formation decomposition response spectrum method, the average value of the bottom shear force calculated by several time history curves is more than 80% result calculated from the formation decomposition response spectrum method, which means that the specifications are met. The formulas were fitted according to the seismic wave data with the largest amplification coefficient.

By adjusting the load value, the axial compression ratio can be adjusted to 0.28,0.3,0.32.0.34,and by adjusting the frame column size to 600x600, 650x650, 600x700, 750x750, the stiffness ratio of shear wall with frame column can be adjusted to 1.965, 2.234, 2.469, 2.685. 16 groups of models were established and analyzed.

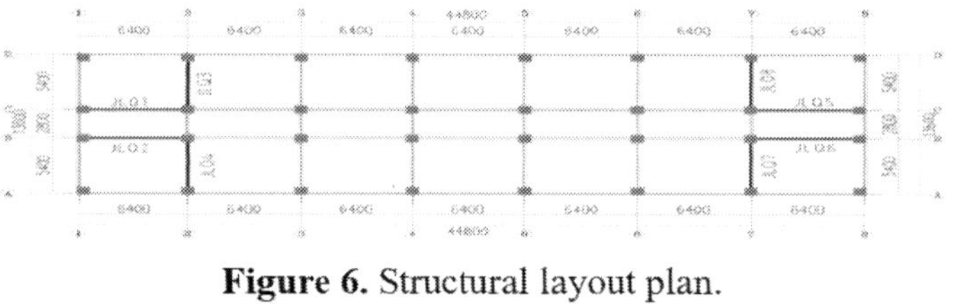

Figure 6. Structural layout plan.

4.1. The Influence of Axial Compression Ratio Variation on Structural Performance with Constant Stiffness Ratio

The stiffness ratio remained unchanged at 1.965, the axial compression ratio was set as 0.28, 0.3, 0.32, 0.34 respectively, the corresponding stiffness degradation coefficients of shear wall were 0.264, 0.303, 0.316 and 0.326. The relationship between the stiffness degradation coefficient and the bottom shear force magnification coefficient and the maximum displacement magnification coefficient was shown in figure 7.

As can be seen in figure 7

(1)When the stiffness ratio of frame column to shear wall is constant, the stiffness degradation coefficient of shear wall on the maximum displacement angle floor increases with the increasing of axial compression ratio in the limit range

(2)The maximum displacement amplification coefficient and bottom shear amplification coefficient of the structure decrease with the increasing of stiffness degradation coefficient of shear wall.

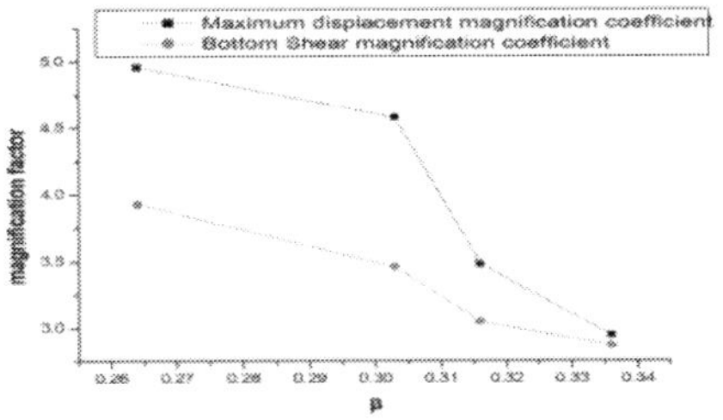

Figure 7. Effect of shear wall stiffness degradation coefficient on bottom shear force magnification coefficient and maximum displacement magnification coefficient (constant stiffness ratio)

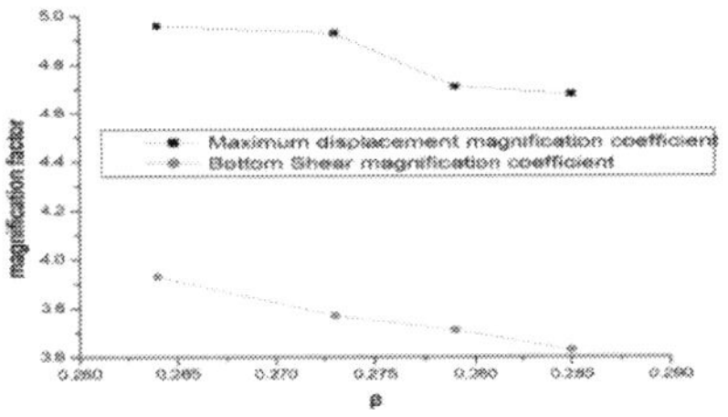

Figure 8. Effect of shear wall stiffness degradation coefficient on bottom shear force magnification coefficient and maximum displacement magnification coefficient(constant axial compression ratio)

4.2. The Influence of Stiffness Ratio Variation on Structural Performance with Constant Axial Compression Ratio
The axial compression ratio remained unchanged at 0.28, the stiffness ratio was set as 1.965, 2.234, 2.469, 2.685 respectively, the corresponding stiffness degradation coefficients of shear wall were 0.264, 0.273, 0.279 and 0.285. The relationship between the stiffness degradation coefficient and the bottom shear force magnification coefficient and the maximum displacement magnification coefficient was shown in figure 8. As can be seen in figure 8:

(1)When the axial compression ratio of shear wall remains constant, the stiffness degradation coefficient of shear wall increases with the increasing of the stiffness ratio of frame column to shear wall

(2)The maximum displacement magnification coefficient and bottom shear magnification coefficient decrease with the increasing of stiffness degradation coefficient of shear wall.

4.3. Influence to Structural Performance with Change of Axial Compression Ratio and Stiffness Ratio
Due to space limitation, the stiffness degradation coefficients of the layer with the maximum inter-layer displacement angle, the maximum displacement magnification coefficient and the bottom shear magnification coefficient were listed in table 3 only when the stiffness ratio was 1.965 and the axial compression ratio was 0.32.

Table 3. Elastoplastic analysis results when m = 1.965, μ=0.32.

coefficient Analysis category	Elastoplastic time-history analysis
Coefficient of stiffness degradation	0.316
Displacement magnification coefficient	3.48
Bottom Shear magnification coefficient	3.05

From calculation data of 16 models, the relationship between the stiffness degradation coefficient and the maximum displacement magnification coefficient can be fitted into formulas (2), and the relationship between the stiffness degradation coefficient and the bottom shear magnification coefficient can be fitted into formulas (3).

$$X_d=\cos(p_1u)+\frac{1}{p_2+u}-\frac{1+p_3\beta+p\,u+p\,m}{p_6+p_7\beta+p_8u+p_9m}+\frac{p_{10}}{(1-\beta)^2} \quad (2)$$

Where X_d is maximum displacement magnification coefficient; m is stiffness ratio of frame column to shear wall, Values range from 1.965~2.685; u is axial compression ratio,values range from 0.28~0.34;β is stiffness degradation coefficient of shear wall,p1=0.043; p2=-0.23; p3=190.39; p4=-64.87; p5=-7.22; p6=-11.793; p7=15.30; p8=36.82; p9=-0.60; p10=-0.34;

$$X_v=p_1+p_2mu+p_3m\beta+p_4u\beta+p_5m+p_6u+p_7m^2+p_8u^2 \quad (3)$$

Where Xv is bottom shear magnification coefficient; m is stiffness ratio of frame column to shear Wall, values range from 1.965~2.685; u is axial compression ratio, values range from 0.28~0.34; β is stiffness degradation coefficient; p1=28.48; p2=-19.42; p3=21.11; p4=-134.33; p5=1.11; p6=-149.72.

The comparison between the result calculated by fitting formula and the result calculated by finite element (target value in figure) were shown in figure 9, which shows that the fitting formulas have enough precision.

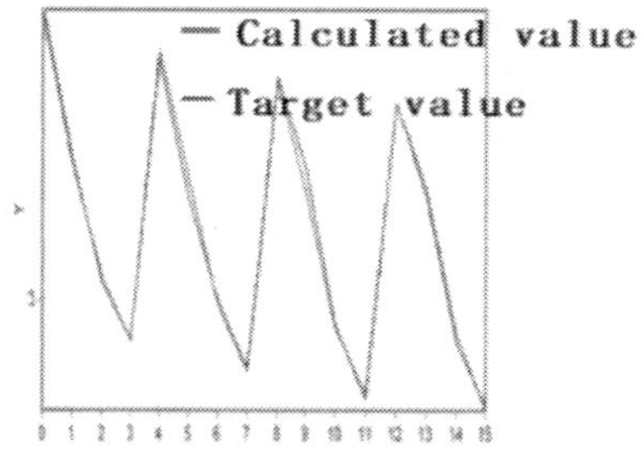

(a) Error of maximum displacement magnification coefficient between formula and target value
(b) Error of bottom shear magnification coefficient between formula and target value

Figure 9. Comparison between fitting formula calculation results and finite element calculation results.

5. Conclusion

The following conclusions can be drawn from previous analysis

(1) When the stiffness ratio remains constant, the stiffness degradation of the shear wall with frame column under the same displacement angle is more obvious with the decreasing of the axial compression ratio; and when the axial compression ratio remains constant, the stiffness degradation of the shear wall with frame column under the same displacement angle is more obvious with the decreasing of the stiffness ratio.

(2) The relationship between axial compression ratio, the stiffness ratio of the frame column to the shear wall, the displacement angle and the stiffness degradation coefficient can be fitted as formula (1), the formula has high precision.

(3) When the stiffness ratio of frame column to shear wall is constant, the shear wall stiffness degradation coefficient of the layer with the maximum inter-layer displacement angle increases with the increasing of axial compression ratio in the limit range, and the displacement magnification coefficient and bottom shear magnification coefficient decreases with the increasing of stiffness degradation coefficient.

(4) When the axial compression ratio is constant, the shear wall stiffness degradation coefficient of the floor with the maximum inter-floor displacement angle increases with the increasing of stiffness ratio of the frame column to the shear wall, and the displacement magnification coefficient and bottom shear magnification coefficient decrease with the increasing of stiffness degradation coefficient.

(5) The relationship between axial compression ratio, the stiffness ratio, stiffness degradation coefficient and the maximum displacement magnification coefficient can be fitted as formula (2),the relationship between axial compression ratio, the stiffness ratio , stiffness degradation coefficient and the bottom shear magnification coefficient can be fitted as formula (3),the formulas have high precision.

References

[1] Nader A. 2020 Khaled Galal. Effect of ductile shear wall ratio and cross-section configuration onseismic behavior of reinforced concrete masonry shear wall buildings [J] *Journal of Structural Engineering* **146** (4).

[2] Shen Z W, Wang S L, Wang J, et al. 2007 Study on anti-seismic behavior of r.c shear wall with frame column, Architectural Technique (05) 357-360.

[3] Jia Y , Wu G Y, Yang C B X et, al. 2016 Simulation analysis on lateral stiffness degradation and storey shear force redistribution of RC frame-shear wall structure with different axial compression ratios, Building Structure **46** (24) 50-55.

[4] Ma L, Huang S J. 2020 Discussion on stiffness reduction factor of semi-rigid joint of preloaded assembly structure, Engineering and Construction **34** (06) 1035-1037.

[5] Liu F P, Yi W T. 2020 Research on bearing capacity degradation model of reinforced concrete beams based on fatigue stiffness degradation *Building Structure* **34** (06) 1035-1037.

[6] Fan Z, Li Y Y, Li W, Tian Y J. 2020 Influence of frame stiffness on performance of buckling restrained steel plate shear walls *Engineering Mechanics* **37** (04) 30-40.

[7] He C H, Chen Z D. 2015 Seismic behavior analysis of shear walls with frame columns of different types *Engineering and Construction* **29** (02) 213-215.

[8] Sun J C, Guan Q X. 1988 Experimental research on strength and deformability of reinforced concrete farmed shear wall [J] *Journal of Building Structure* (06) 17-29.

[9] Liu X M, Li A Q, Ye J H. 2001 Calculation Analysis of Bearing Capacity and deformation of masonry shear wall with reinforced frame [J] *Journal of Southeast University* (Natural Science Edition) (01) 62-68.

[10] Zhao S F, Cheng W X, Wang F M. 1999 Study on Cross beam action of lightweight aggregate concrete Shear Wall with frame *Engineering Mechanics* 6.

Long-Term Performance of Seawater Sea-sand Concrete (SSC) Columns Wrapped with CFRP Strips Under Simulated Marine Environment

Junlong Yang, Ziru Wang and Jizhong Wang*

State Key Laboratory of Coastal and Offshore Engineering, Dalian University of Technology, Dalian, Liaoning 116024, China
Email: wangjz@dlut.edu.cn

Abstract. This article presents a durability study of carbon fiber-reinforced polymer (CFRP) partially wrapped seawater sea-sand concrete (SSC) columns exposed to natural seawater to explore the effect of exposure duration on the long-term performance of the specimens. Thirty-two cylinders were wrapped with CFRP jackets and exposed to different times of wet-dry cycles (up to 360 days) in an outdoor simulated marine environment. Test results indicate that exposure has no obvious influence on the failure process and ultimate strains of specimens, but the compressive strengths of confined columns (f_{cc}) increase with the increment of exposure time, especially for the partially confined concrete specimens. Moreover, due to the significant variation of unconfined concrete strength (f_{co}^*), the retentions of f_{cc} and f_{cc}/f_{co}^* exhibit an opposite trend. Therefore, the increase of f_{co}^* should be considered when using the parameter of the confined-to-unconfined ratio of strength to evaluate the long-term performance of the specimens.
Keywords. Partial wrapping; Durability; Seawater sea-sand concrete; Natural exposure; Ultimate condition.

1. Introduction

In recent years, fiber-reinforced polymer (FRP) composites have been widely used in strengthening existing concrete structures and new constructions due to their extraordinary properties of high strength-to-weight ratio and excellent corrosion resistance [1,2]. On the other hand, to develop more environmental-friendly construction in coastal regions, attempts of the direct use of seawater and sea-sand for concrete mixing have been made by many studies [3,4] and the usage of seawater sea-sand concrete (SSC) can significantly alleviate the resource shortage problems and reduce the material transportation costs. However, considering the high chloride contents in seawater and sea-sand, the conventional steel reinforcement cannot be employed, and the FRP can be taken as an attractive replacement of steel, leading to the innovative combination of FRP and SSC, i.e., FRP-SSC structures [5,6]. As the FRP is effective when it is acted as the confinement for concrete members, many investigations have been conducted on the performance of the FRP confined SSC columns [7-9]. In the above-mentioned studies, most of them are only focused on the short-term structural behavior of the columns, but the long-term performance may seem more important for the practical applications for FRP-SSC structures. More importantly, only the mechanical properties of FRP fully wrapped SSC columns have been discussed without considering other FRP wrapping schemes. According to the study by Hollaway and Teng [10], it is crucial to choose the suitable wrapping strategy for the structural members based on their specific locations in the marine environment to improve their long-term behavior. So the usage of partial wrapping is more reasonable for the columns not submerged in seawater since the evaporation of the moisture can be available in the unwrapped regions [11].

In addition, it is common to use the accelerated degradation tests by increasing the exposure temperature or(/and) salt concentration of the solution to examine the durability of the FRP confined concrete columns. However, Li et al. [12] investigated the mechanical performance degradation of the SSC-filled FRP tubular columns under the natural indoor environment for up to 2.5 years. The results reveal that the natural environment may not be as severe as the accelerated test environment, which may result in the failure of the existing degradation models obtained by accelerated tests in predicting the real performance of the columns in a natural environment.

Against this background, this paper presents an experimental study of the durability of carbon fiber-reinforced polymer (CFRP) partially wrapped SSC columns under a wet-dry condition in a natural seawater environment. The effects of clear spacing ratios and conditioned time on the long-term behavior of the specimens were investigated through the test. The experimental program and the test results are presented and discussed in the following sections.

2. Experimental Program

2.1. Specimens Preparation

A total of 44 standard seawater sea-sand concrete cylinders were fabricated in this research, including 12 unconfined control specimens, 24 partially CFRP confined SSC columns and 8 fully CFRP confined SSC columns. The main test parameters are listed in table 1, and figure1 shows the wrapping schemes of CFRP wrapped specimens. Besides, five conditioned ages (0, 60 days, 120 days, 180 days, 360 days) were set to explore the exposure time on the mechanical performance of the specimens. It should be noted that the unconditioned specimens were derived from the authors' previous study [13].

Table 1. Test program.

Specimen ID	*s/D*	Wrapping schemes	Control [a]	Aging time (Days)			
				60	120	180	360
R	/	/	3	3	3	3	3
W4A2	0.6	Partial	2	2	2	2	2
W5A2	0.5	Partial	2	2	2	2	2
W6A2	0.4	Partial	2	2	2	2	2
F1	0	Full	2	2	2	2	2

Note: (a) The control specimens were derived from the existing studies conducted by the authors' group [13].

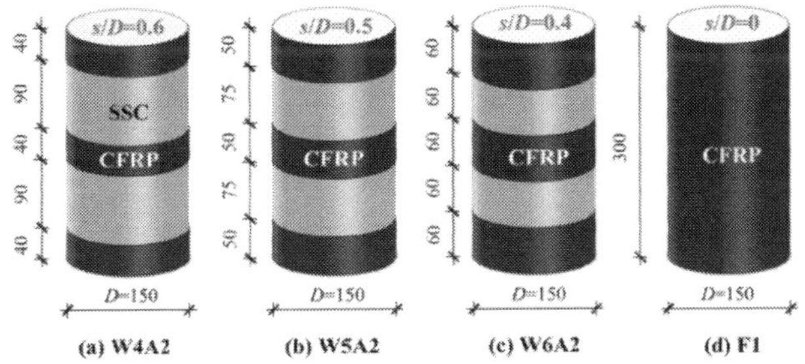

Figure 1. Wrapping schemes of the tested specimens (unit: mm).

The same concrete mix proportion was used for all specimens in this study. In addition, the unidirectional CFRP sheets were wrapped in the circumferential direction by the traditional wet-layup technique. To prevent the premature debonding failure, an additional 150-mm overlap zone was provided. For ease of the following discussions, each specimen was given a name. R represents the control specimens. W4A2, W5A2 and W6A2 stand for the CFRP partially wrapped columns with the width of the FRP strips equal to 40 mm, 50 mm and 60 mm, respectively. F1 means the one-layer CFRP fully wrapped SSC columns.

2.2. Exposure Conditions

As shown in figure 2, a simulation tank was constructed in the outdoor environment in Dalian, China to simulate the natural marine environment. The sea tidal regime was simulated by the wet-dry cycling condition, which was achieved by 12 h of immersion in natural seawater and 12 h of drying with the help of the water pump in each chamber (figure 2). Each complete wet-dry cycle consisted of 24 h and the conditioned specimens were tested after 60, 120, 180 and 360 days, respectively.

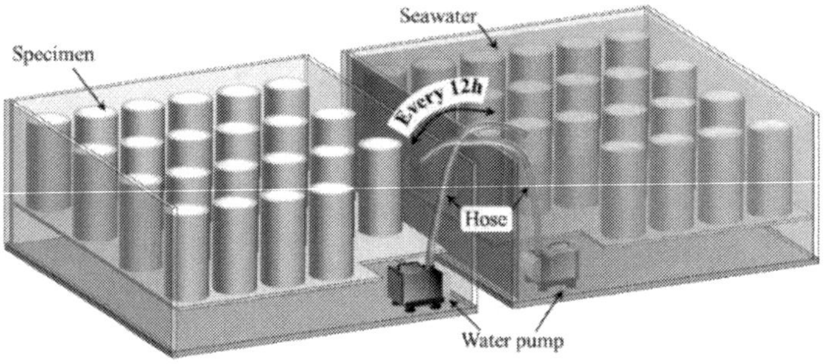

Figure 2. Simulation tank for seawater wet-dry cycling.

2.3. Material Properties and Test Setup

The same raw materials of concrete, CFRP, test setup and instrumentations adopted in the previous study [13] were also used in this test, and both the unconditioned and conditioned specimens were fabricated at the same time to make the results more reliable. The readers may refer to the literature [13] for more details.

3. Results and Discussions

3.1. Failure Modes

Figure3 shows the failure modes of some typical specimens at different aging times. It can be found that the surfaces of unconfined specimens and the unwrapped regions of partially wrapped specimens exhibited more obvious color change from whitish gray to dark gray with the increase of the aging time, which may result from the salt deposition on the surfaces after exposure in seawater. Besides, salt crystallization was also observed on the surfaces of CFRP jackets as shown in figure 3.

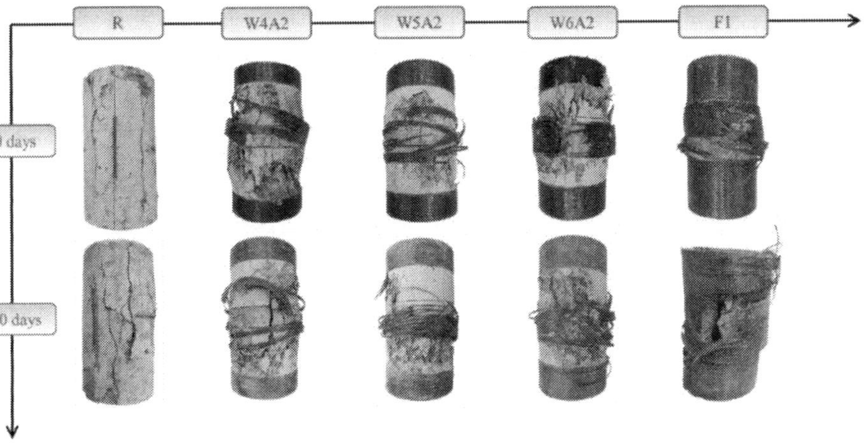

Figure 3. Failure modes of typical specimens.

However, apart from the above-mentioned visual changes, the conditioned ages had no significant influence on the final failure modes of specimens. The unconfined specimens failed in a classic manner with the crushing of SSC, and the interior part of the specimens retained their original color. It is indicated that only the outer side of the unwrapped regions of the columns was affected by the seawater action, which is consistent with the observation from the study conducted by Islam et al. [14].

For the unconditioned CFRP confined specimens, the fully wrapped specimens showed no changes before the applied load reached the ultimate load of unconfined specimens, while some slight vertical cracks were observed in the unwrapped regions of the partially wrapped columns. Then, with the increase of the applied load, the number and the width of the cracks increased, and the sporadic popping sounds could be heard when the load was approaching the ultimate value. Finally, the fully and partially wrapped specimens failed by the rupture of the CFRP jackets and the CFRP strips at the midheight, respectively. A similar failure process was detected for the conditioned columns, but their failure became more brittle, which may result from the embrittlement of the CFRP system after the wet-dry cycles. Moreover, less concrete debris was found on the internal surface of the CFRP jackets after the test, so the concrete-resin interface was also degraded to some extent.

3.2. Stress-Strain Relationships

Figure4 shows the axial stress-axial (lateral) strain relationships of the confined specimens. It should be noted that both the stress and strain in figure4 are normalized by the unconfined strength f_{co}^* and peak strain ε_{co}^* at the corresponding aging time to exclude the effect of material properties of concrete.

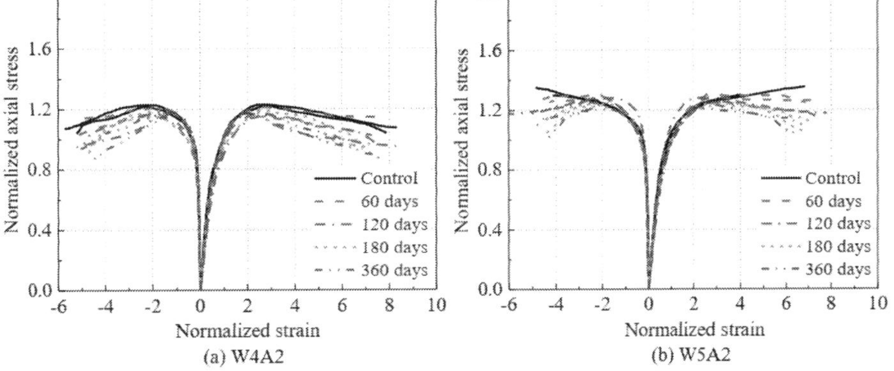

(a) W4A2 (b) W5A2

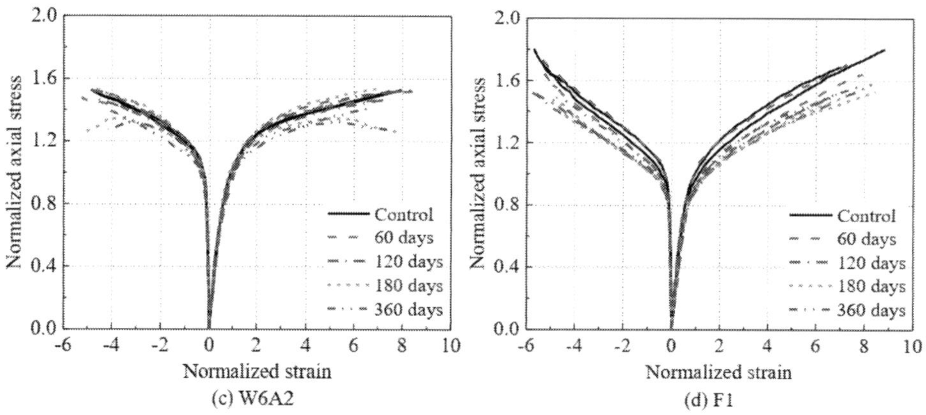

Figure 4. Normalized axial stress-strain curves of tested specimens.

As depicted in figure4, the natural wet-dry cycles have various impacts on the shape and key points of the normalized stress-strain curves for the specimens with different wrapping schemes. In specific, the slope of the post-peak segment on stress-strain curves of the partially wrapped specimens gradually degrades with the increase of the aging time, especially for the specimen W4A2 with the largest clear spacing ratio, which may be attributed to the plasticization of the resin matrix and the fiber-resin interface damage induced by the seawater action during wet-dry cycles, leading to the degradation of the confinement stiffness of CFRP strips. However, it is noteworthy that the locations of the transition points on the curves of the partially wrapped specimens are almost unchanged under different exposure times, illustrating that the exposure may have a similar degradation effect on the unconfined and partially confined cylinders. On the other hand, for the fully wrapped specimens, although the slope of the second linear portion decreases as a result of the wet-dry cycles, all columns possess a typical bi-linear stress-strain response regardless of the aging time. Moreover, the normalized stress-strain curves of unconditioned specimens are obviously higher than that of the conditioned columns, which manifests that the effect of the f_{co}^* on the fully wrapped specimens is more significant compared to the columns wrapped with CFRP strips.

3.3. Effects of Conditioned Ages on Compressive Strength

3.3.1. Unconfined Specimens. It is worth mentioning that the average strength and strain of the nominally identical specimens will be used in the following discussions unless otherwise specified. It can be seen from figure5 that the ultimate strains of the unconfined specimens only fluctuate slightly during the aging period, while more obvious enhancement is detected for the compressive strength with the increase of aging time, especially in the early stage, which is consistent with the observations found by Li et al. [12]. This may result from the following two reasons. Firstly, the seawater in the tank can promote further hydration of the unhydrated cement, leading to the increase of the compactness of the concrete. Besides, the ettringite and Friedel salt produced from the reactions of the ions (e.g., chloride and sulphate ions) in the seawater, calcium hydroxide, unhydrated and hydrated calcium aluminate can improve the pore structure of concrete, which is also beneficial to the compressive strength development of SSC.

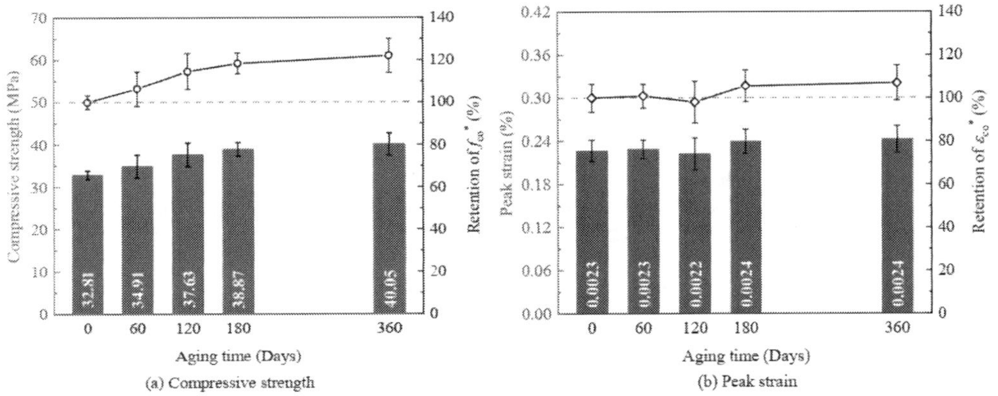

Figure 5. Effect of exposure duration on the material properties of SSC.

3.3.2. Confined Specimens. Figure 6 (a) presents the variations of the compressive strength f_{cc} of confined specimens during the aging period. The overall ascending trend is detected with the increase of the aging time for a single column. Furthermore, due to the difference in FRP volumetric ratio, specimen F1 possesses the highest compressive strength at each conditioned time among all specimens. To exclude the influence of volumetric ratio on the compressive strength, the retentions of f_{cc} and f_{cc}/f_{co}^* are given in figure6, where the retention of unconfined compressive strength f_{co}^* is also depicted for comparison.

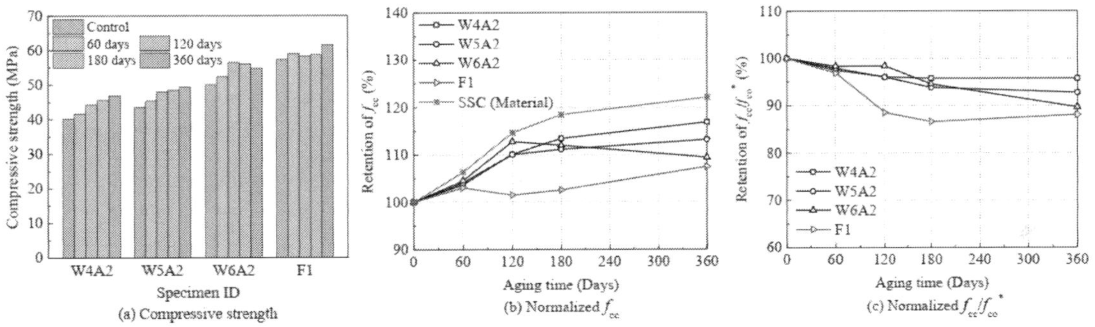

Figure 6. Variation of compressive strength with the increase of aging time.

As shown in figure6 (b), although the f_{cc} retentions of all confined specimens increase with the aging time, the retention of f_{co}^* is still larger than that of f_{cc} at every conditioned age. This is because apart from the SSC, both the fiber and resin matrix may also be degraded during the conditioned period. Moreover, due to the different thermal coefficients of fibers and matrix, the thermal variation may also cause the interface damage between fibers and matrix. However, the above-mentioned speculation needs further verification at the microscopic level in the future. Thus, the degradation mechanism seems more complicated for CFRP confined SSC compared to the unconfined specimens, leading to the more obvious decrease of f_{cc} retentions. Nevertheless, it should be noted that the partially wrapped concrete specimens exhibit larger f_{cc} retentions, which may result from the higher compactness of concrete since the aforementioned ettringite and Friedel salt can be produced more effectively due to the existence of unconfined regions in the partially wrapped columns. Besides, the possibility of matrix degradation and interface damage is also reduced because of the less usage of CFRP materials and smaller contact interface between CFRP and SSC.

However, the retention of the normalized f_{cc}/f_{co}^* shows a completely different trend as depicted in figure 6 (c). The f_{cc}/f_{co}^* retentions of all specimens are smaller than 100% and decrease with the increment of aging time, especially for the fully wrapped columns. This is because that the effect of improved pore structures of concrete on the compressive strength of f_{co}^* is more significant than the influence of the conditioned ages on the strength f_{cc}, which is consistent with the observations as shown in figure6 (b).

3.4. Effects of Conditioned Ages on Ultimate Strain

It should be noted that the ultimate strain of specimen W5A2 is much smaller than the other partially wrapped specimens due to the abnormal value of W5A2-2. Thus, to make the comparisons more reasonable, the ultimate strain of W5A2-1 instead of the average value will be used in the following discussions.

As shown in figure7 (a), the variations of ultimate strains are more scattered compared with those of the compressive strengths and similar results were also detected by Xie et al. [15], which means that the ultimate strain is more sensitive to the conditioned environment. Moreover, the variations seem to have no relationship with the wrapping schemes. On the other hand, since the peak strain of unconfined concrete exhibits insignificant changes with the aging time, a similar trend of ε_{cu} and $\varepsilon_{cu}/\varepsilon_{co}^*$ retentions is observed as depicted in figure7 (b) and figure7 (c). In specific, in the later stage of the conditioned period (180 days), the retention of $\varepsilon_{cu}/\varepsilon_{co}^*$ shows an obvious descending trend.

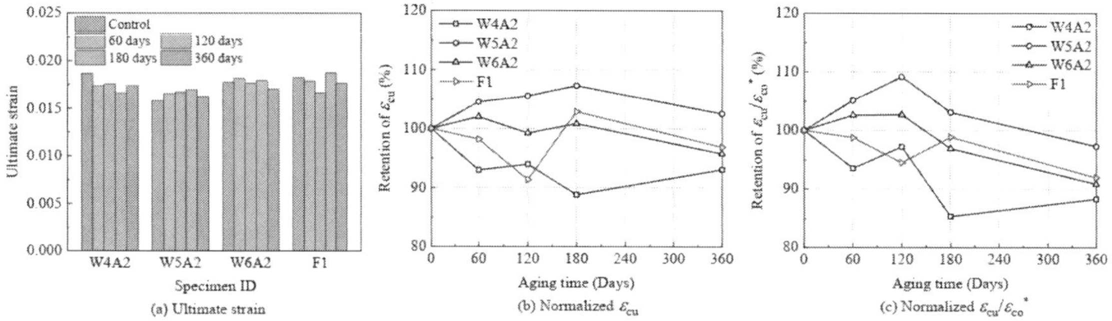

Figure 7. Variation of ultimate strain with the increase of aging time.

4. Conclusions

(1) Due to the seawater wet-dry cycling effect, the surfaces of concrete change into dark surfaces because of the salt deposits and the salt crystallization is also observed on the CFRP surfaces. However, the conditioned environment has little impact on the failure modes of the confined specimens, which is still controlled by the rupture of the CFRP jackets.

(2) The post-peak behavior of the CFRP partially wrapped specimens is significantly influenced by the aging time, while all fully wrapped columns possess typical bi-linear stress-strain responses regardless of the exposure duration. But the slope of the second linear segment is degraded to some extent.

(3) Under the conditioned environment in the present study, the compressive strength of unconfined specimens increases with the aging time, which leads to the different trend of the variations of f_{cc} and f_{cc}/f_{co}^* retentions.

(4) Compared to the compressive strength, the conditioned environment has no clear effect on the ultimate strain of confined specimens. Besides, the strain variations seem to have no relationship with the wrapping schemes.

References

[1] Elsanadedy H M, Al-Salloum Y A, Alrubaidi M A, Almusallam T H, Siddiqui N A, Abbas H 2020 Upgrading of precast RC beam-column joints using innovative FRP/steel hybrid technique for progressive collapse prevention *Constr Build Mater* **268** 121130.

[2] Ali A H, Mohamed H M, Benmokrane B 2016 Shear behavior of circular concrete members reinforced with GFRP bars and spirals at shear span-to-depth ratios between 1.5 and 3.0. *J Compos Constr* **20** (6) 04016055.
[3] Teng J G, Xiang Y, Yu T, Fang Z 2019 Development and mechanical behaviour of ultra-high-performance seawater sea-sand concrete. *Adv Struct Eng* **22** (14) 3100-3120.
[4] He X, Zhou J 2021 Mechanical characteristics of sea-sand concrete in simulated marine environment. *Constr Build Mater* **274** 122098.
[5] Ahmed A, Guo S, Zhang Z, Shi C, Zhu D 2020 A review on durability of fiber reinforced polymer (FRP) bars reinforced seawater sea sand concrete. *Constr Build Mater* **256** 119484.
[6] Dong Z, Wu G, Zhao X-L, Zhu H, Shao X 2019 Behaviors of hybrid beams composed of seawater sea-sand concrete (SWSSC) and a prefabricated UHPC shell reinforced with FRP bars. *Constr Build Mater* **213** 32-42.
[7] Yang J, Lu S, Wang J, Wang Z 2020 Behavior of CFRP partially wrapped square seawater sea-sand concrete columns under axial compression. *Eng Struct* **222** 111119.
[8] Yang J, Wang J, Wang Z 2021 Behavior and modeling of CFRP nonuniformly wrapped circular seawater sea-sand concrete (SSC) columns under axial compression. *Constr Build Mater* **299** 123887.
[9] Bazli M, Li Y L, Zhao X L, Raman RKS, Bai Y, Al-Saadi S, Haque, A. 2020 Durability of seawater and sea sand concrete filled filament wound FRP tubes under seawater environments. *Compos Part B-Eng* **202** 108409.
[10] Hollaway L C, Teng J G. 2008 *Strengthening and rehabilitation of civil infrastructures using fibre-reinforced polymer (FRP) composites*. Amsterdam: Elsevier.
[11] Belarbi A, Bae S-W 2007 An experimental study on the effect of environmental exposures and corrosion on RC columns with FRP composite jackets. *Compos Part B-Eng* **38** (5) 674-684.
[12] Li Y L, Zhao X L, Raman RKS 2021 Durability of seawater and sea sand concrete and seawater and sea sand concrete–filled fibre-reinforced polymer/stainless steel tubular stub columns. *Adv Struct En.* **24** (6) 1074-1089.
[13] Yang J, Wang J, Wang Z 2020 Axial compressive behavior of partially CFRP confined seawater sea-sand concrete in circular columns - Part I: Experimental study. *Compos Struct.* **246** 112373.
[14] Islam MM, Islam MS, Al-Amin M, Islam MM 2012 Suitability of sea water on curing and compressive strength of structural concrete *J Civ Eng* **40** (1) 37-45.
[15] Xie Z, Xie J, Guo Y, Huang Y 2018 Durability of CFRP-wrapped concrete exposed to hydrothermal environment. *Int J Civ Eng* **16** (5) 527-541.

Force Analysis of Supporting System during Form-Work Construction of Box Culvert without Tie Rod

Guangming Zhai [1], Yuwei Wang[1], Liu Liu[2] and Pan Guo[3,*]

[1] CCCC-SHB Fourth Engineering Co., Ltd., Luoyang, Henan, 471013
[2] Henan Anluo Expressway Co., Ltd., Zhengzhou, Henan 450061
[3] School of Civil Engineering, Zhengzhou University, Zhengzhou, Henan 450001
Email: guopan@zzu.edu.cn

Abstract. In the construction process of box culvert form-work without tie rod, the stress state of its supporting system is of great importance. Based on a construction project for the box culvert without tie rod form-work, this paper studied the force characteristics of diagonal braces during the concrete pouring of the box culvert. The force calculation theory of the support system was analyzed and the force was analyzed using the finite element software Midas Civil. The results showed that when the distance between the connection point of the diagonal brace and the foundation was 1 m, the diagonal brace cannot provide effective support for the template, and when the distance between the connection point of the diagonal brace and the template is 2 m, 3 m, and 4 m, the diagonal brace can provide effective support. It was recommended that the diagonal brace connection point is 3 m away from the template as the optimal diagonal brace layout position. It is recommended to strengthen the area of increased local deformation of diagonal braces to a certain extent according to the calculation results of this finite element analysis to improve the safety performance of the entire structure. The research results of this paper can provide reference for follow-up similar construction problems.

1. Introduction

Culvert construction is an important part of highway construction in China [1, 2]. Among many culvert structures, box culvert has become the most commonly used culvert structure in expressway construction in China because of its good overall mechanical performance and strong foundation adaptability [3, 4]. In the past, when the wall body of cast-in-situ slab culvert was constructed, the wall body formwork was often reinforced by pull rod, which would lead to dense pull rod holes on the wall body surface after formwork removal. In the construction process, in order to prevent the water after abutment back backfilling from flowing into the main wall through the tie rod hole and polluting the wall surface, the tie rod hole is generally blocked, but the effect is not ideal, and the tunnel surface needs secondary coating, which not only prolongs the construction period, but also increases the construction measure cost. The construction technology of box culvert without tie rod formwork can solve this problem well.

The concrete will produce certain lateral pressure on the formwork and auxiliary supporting structure during construction [5]. Liu et al [6] studied the lateral pressure of the newly poured concrete on the formwork and the stress characteristics of the support. The results show that the newly poured concrete will produce unbalanced force on the top formwork and bottom formwork of the normal section, and additional force should be applied to support the formwork system. Deng et al [7] studied the economic design of cast-in-situ concrete one-sided formwork and support system, and analyzed the economic layout distance of secondary ridge, main ridge and support in the economic design of one-sided formwork and support system, as well as the anchor bolt diameter in the economic layout distance of support, so as to save materials and reduce construction cost. Hu et al [8] analyzed the synergy between the formwork support system and the concrete floor by using the finite element software ANSYS, and proved the necessity of the formwork support system under the middle plate during the construction of the structural roof. During the construction of box culvert without tie rod formwork, due to the lack of the role of tie rod, the stress of its auxiliary diagonal brace is very important, so it is necessary to calculate and study it.

Based on the construction project of a box culvert without tie rod formwork, the mechanical characteristics of diagonal bracing in the process of box culvert concrete pouring are studied in this paper. The stress calculation theory of the support system is analyzed, and its stress is analyzed by using the finite element software MIDAS civil.

2. Project Overview

The schematic diagram of box culvert non tie rod formwork construction is shown in Figure 1. The inner formwork is a steel plate with a height of 6m and a thickness of 1cm. Since there is no internal tie rod, the diagonal brace plays a role in bearing the lateral pressure of concrete. One end of the diagonal brace is connected with the concrete foundation, and the other end is connected with the internal formwork. The diagonal brace is composed of vertical beams placed in parallel at intervals, and the vertical beams are connected through beams. All vertical beams and beams shall be made of steel pipes with an outer diameter of 45 mm and a thickness of 4 mm. Q345 section steel shall be used for formwork and diagonal bracing.

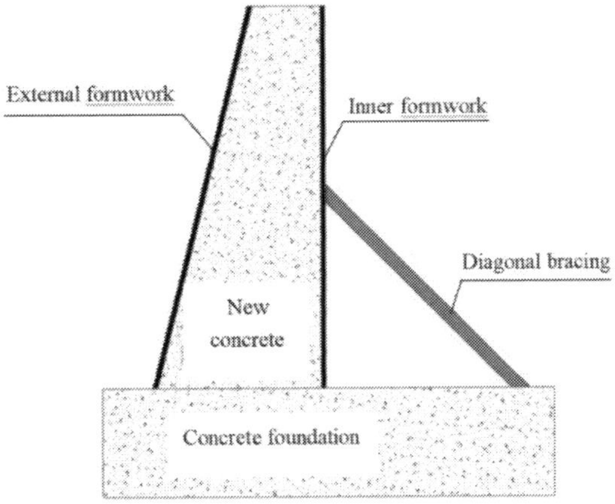

Figure 1. Diagram of construction of culvert without tie rod formwork.

3. Calculation Theory and Model Establishment

3.1. Calculation Theory

During the concrete pouring of box culvert, the concrete will produce lateral pressure on the inner form-work, and then transfer the force to the diagonal brace. According to relevant theoretical research [9], The lateral pressure can be calculated according to the following two formulas, and the minimum value is taken as the lateral pressure value of the form-work:

$$F = 0.22\gamma_c t_0 \beta_1 \beta_2 V^{1/2} \tag{1}$$

$$F=\gamma_c H \tag{2}$$

Where, γ_c is the unit weight of concrete, γ_c =25 kN/m^3; t_0 is the initial setting time of newly poured concrete, t_0=6h; V is the initial setting time of newly poured concrete and the pouring speed of concrete, V=2 m/h; β_1 is the influence correction coefficient of admixture, β_1=1.0; β_2 is the influence correction coefficient of concrete slump, β_2= 1.15; H is the formwork height, H= 6 m.

The lateral pressure of formwork is calculated according to the above two formulas, and the results are 53.67 kn/m^2 and 150 kn/m^2 respectively. Therefore, in this calculation, the lateral pressure of newly poured concrete on the formwork is taken as 53.67 kn / m^2.

3.2. Finite Element Model and Calculation Conditions

As shown in figure 2, the finite element analysis model established in Midas civil software. Plate element is selected for internal steel formwork and truss element is selected for diagonal bracing. The dimensions and materials of all elements are consistent with the actual project. In the actual construction process, the placement angle of the diagonal brace will have a great impact on the stress of the diagonal brace. Therefore, in this calculation, the analysis and calculation are carried out according to the four working conditions that the distance between the diagonal brace and the foundation connection point is 1 m, 2 m, 3 m and 4 m respectively.

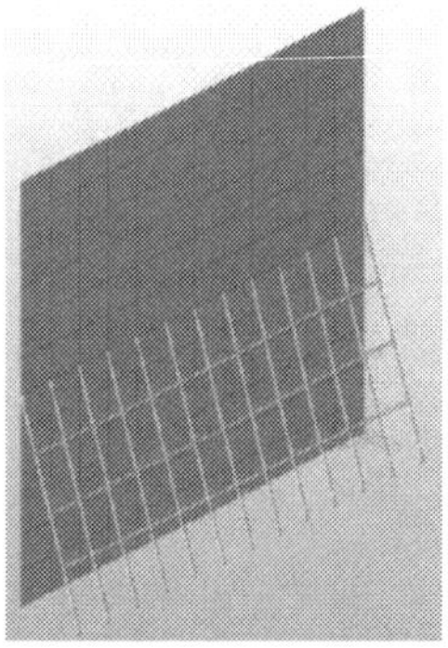

(a) The connection point of diagonal bracing is 1m from the formwork

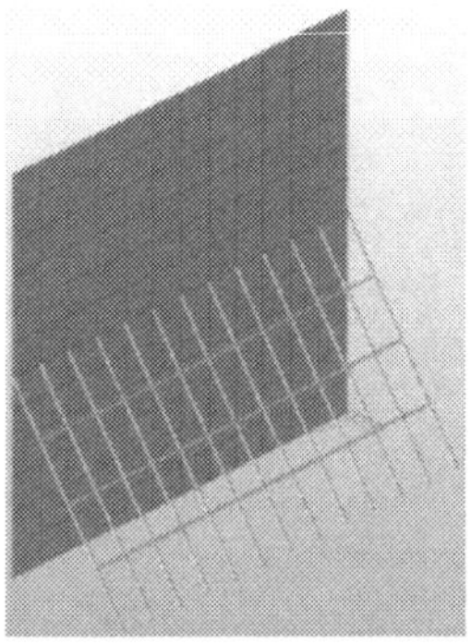

(b) The connection point of diagonal bracing is 2m away from the formwork

(c) The connection point of diagonal bracing is 3M from the formwork

(d) The connection point of diagonal bracing is 4m away from the formwork

Figure 2. Finite element analysis model for different calculation conditions.

4. Results and Analysis

4.1. Stress Analysis of Diagonal Bracing

The cloud diagram of diagonal brace stress under different calculation conditions is shown in figure 3. The farther the diagonal brace connection point is from the formwork, the smaller the stress of the diagonal brace. Under the working conditions of 1 m, 2 m, 3 m and 4 m, the maximum stress of the diagonal brace is 452 MPa, 270 MPa, 214 MPa and 189 MPa respectively. Considering that the diagonal brace is made of Q345 steel and its yield strength is 345MPa, when the diagonal brace connection point is 1m away from the formwork, the diagonal brace stress has exceeded the yield strength of the material. Therefore, it can be considered that the diagonal brace cannot provide effective support for the concrete formwork under this working condition. When the distance between the connection point of the diagonal brace and the formwork is 2 m, 3 m and 4 m respectively, the stress of the diagonal brace does not exceed the limit.

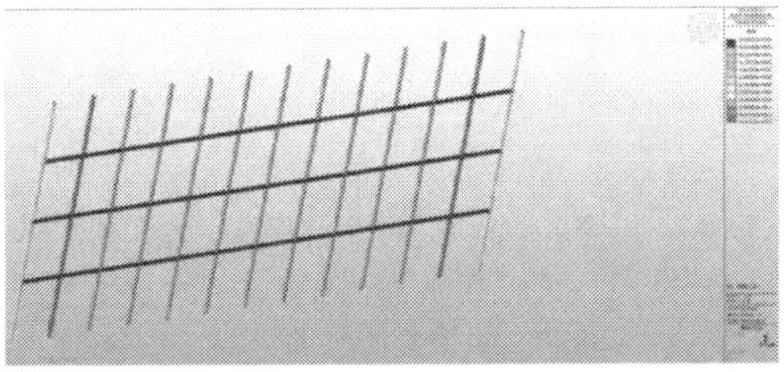

(a) The connection point of diagonal bracing is 1m from the formwork

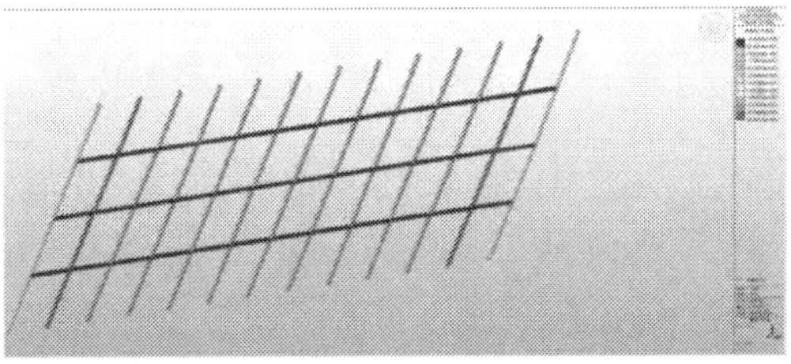

(b) The connection point of diagonal bracing is 2m from the formwork

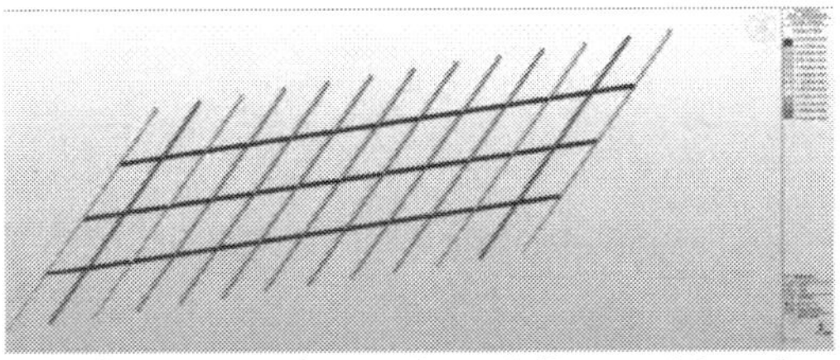

(c) The connection point of diagonal bracing is 3m from the formwork

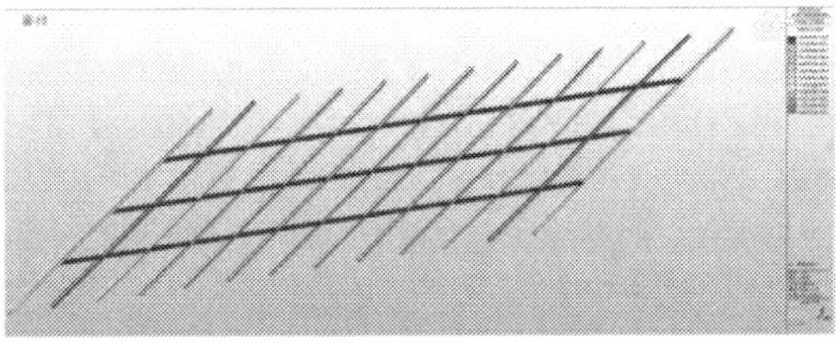

(d) The connection point of diagonal bracing is 4m from the formwork

Figure 3. Stress cloud diagram of diagonal braces under different calculation conditions.

4.2. Deformation Analysis of Diagonal Bracing

The displacement nephogram of diagonal bracing under different calculation conditions is shown in figure 4. The displacement of diagonal bracing is small in other areas except that it shows large deformation in individual areas on the whole. Under the working conditions of 1 m, 2 m, 3 m and 4 m, the maximum deformation of diagonal bracing is 914 mm, 18.1 mm, 16.7 mM and 31.6 mm respectively. When the connection point between the diagonal brace and the foundation is 1m away from the formwork, serious large deformation occurs in the diagonal brace, indicating that the diagonal brace cannot provide effective support under this working condition, which is consistent with the results of stress analysis. Under the condition that the distance between the connection point of diagonal bracing and the formwork is 2 m, 3 m and 4 m respectively, the maximum deformation of diagonal bracing is small.

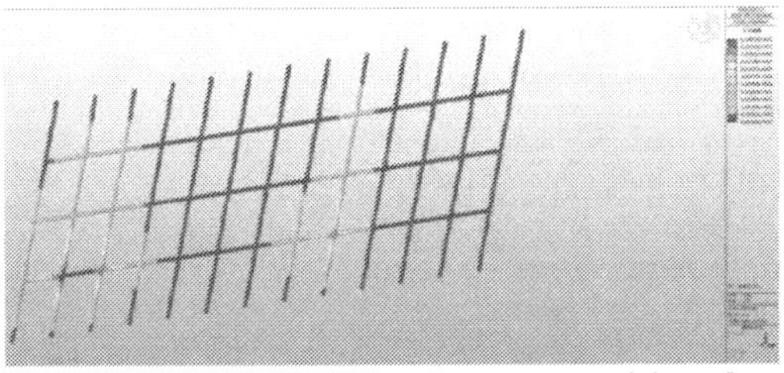

(a) The connection point of diagonal bracing is 1m from the formwork

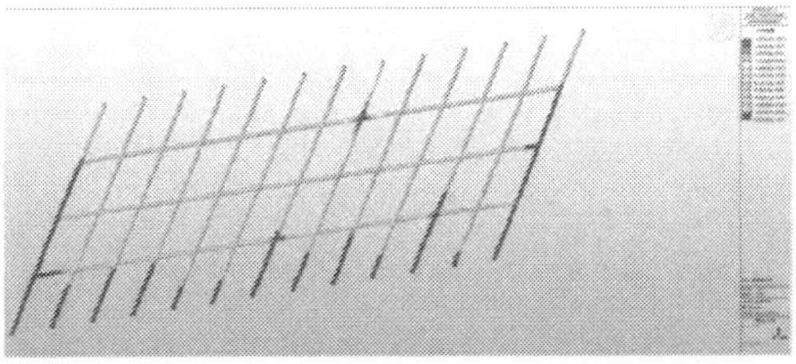

(b) The connection point of diagonal bracing is 2m from the formwork

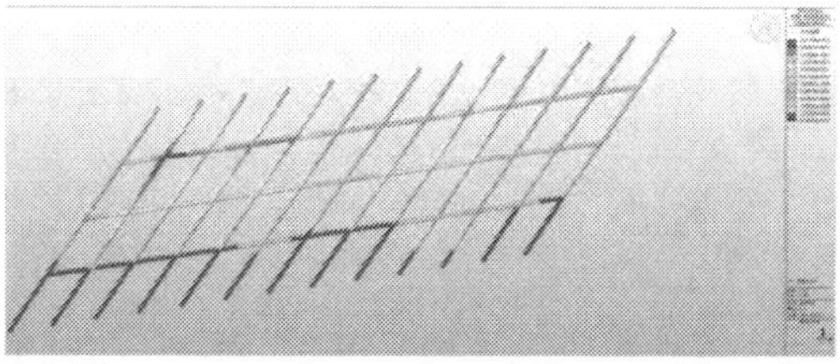

(c) The connection point of diagonal bracing is 3m from the formwork

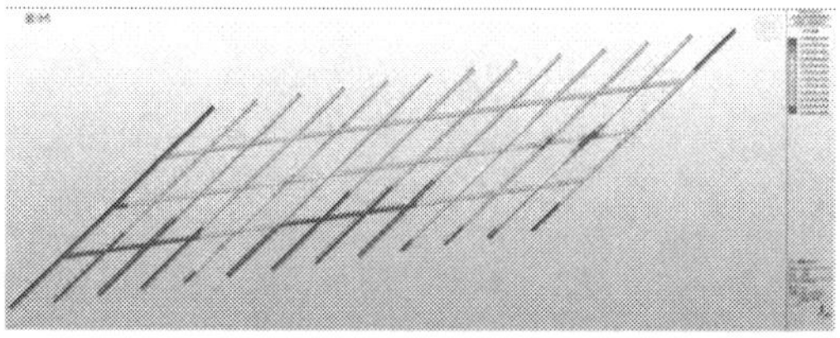

(d) The connection point of diagonal bracing is 4m from the formwork

Figure 4. Deformation cloud diagram of diagonal braces under different calculation conditions.

4.3. Optimal Diagonal Brace Layout Position

According to the comprehensive analysis of the force and deformation of the diagonal brace, when the connecting point between the diagonal brace and the foundation is 1m away from the formwork, the force and local deformation of the diagonal brace exceed the limit, indicating that the diagonal brace cannot provide effective support under this working condition. When the distance between the connection point of the diagonal brace and the formwork is 2 m, 3 m and 4 m respectively, the stress and local deformation of the diagonal brace are maintained at a low level, indicating that the diagonal brace can provide effective support for the concrete formwork. Considering the maximum stress and local maximum deformation of diagonal bracing under these three working conditions, 3M from the connection point between diagonal bracing and foundation to the formwork can be selected as the optimal layout position of diagonal bracing.

In addition, according to the finite element analysis and calculation results, the area with increased local deformation of diagonal bracing can be strengthened to a certain extent, so as to improve the safety performance of the whole structure.

5. Conclusions

In this paper, MIDAS / civil finite element analysis software is used to analyze the stress of the support system during the construction of culvert tie free formwork, and calculate the stress and deformation of the diagonal brace under the working conditions of different diagonal brace layout positions. The main conclusions and relevant suggestions are as follows:

(1) When the connection point between the diagonal brace and the foundation is 1m away from the formwork, the stress and local deformation of the diagonal brace exceed the limit, indicating that the diagonal brace cannot provide effective support for the concrete formwork under this working condition, while the diagonal brace can provide effective support under the other three working conditions.

(2) Considering the maximum stress and local maximum deformation of the diagonal brace under the three working conditions of 2m, 3M and 4m from the connecting point between the diagonal brace and the foundation to the formwork, it is recommended to select 3M from the connecting point between the diagonal brace and the foundation to the formwork as the optimal layout position of the diagonal brace.

(3) It is suggested to strengthen the area with increased local deformation of diagonal bracing to a certain extent according to the finite element analysis and calculation results, so as to improve the safety performance of the whole structure.

Acknowledgments

Authors wishing to acknowledge assistance or encouragement from colleagues.

References

[1] Ding L T, Wang X C, Fu L J, et al. 2021 Differential settlement technology of culvert subgrade treated by geogrid based on driving comfort *Material guide* **35** (4) 79-87.

[2] Zhao Y H, Mi W J, Sun R D, et al. 2017 Study on the effect of highway culvert foundation treatment measures in collapsible loess area [J] *Journal of railway engineering* **34** (1) 6-10 74.

[3] Zhou G Y, Li C, Hu Y, et al. 2020 Numerical simulation analysis of pavement settlement during box culvert jacking construction *Highway engineering* **45** (3) 29-37.

[4] Wang S H, Alepujiang J, Wang P Y, et al. 2018 Mechanical performance simulation and potential failure mode of prefabricated rectangular box culvert. Journal of Northeast University (Natural Science Edition) **39** (2) 260-265.

[5] He X P, Zhang Y S, Cheng Q, et al. 2014 Study on force transfer law of interlayer formwork support frame of cast-in-situ concrete structure *Industrial architecture* (1) 107-111.

[6] Liu Y T, Huang J, Jiang J S, et al. 2012 Analysis of lateral pressure of newly poured concrete on inclined formwork and stress of support *Construction technology* **41** (11) 85-87.

[7] Deng T J, Zhu M. 2015 Study on economic design of cast-in-situ concrete unilateral formwork and support system [J] *Journal of Hunan University* (Natural Science Edition) **42** (11) 133-138.

[8] Hu C M, Zhao Y B, Wang J, et al. 2014 Finite element analysis of the synergy between formwork support system and concrete floor [J] *Industrial architecture* (5) 106-110.

[9] Cao Y G, Tan X D, Lv S, et al. 2020 Structure and stress analysis of mass concrete construction formwork [J] *Ceramics* (8) 101-103.

Journal of Physics: Conference Series **2148** (2022) 012041 doi:10.1088/1742-6596/2148/1/012041

Calculation and Analysis of Truss Internal Force Based on Beam Element

Yao Zhang[1], Jun Dong[2,*], Guohua Li[2] and Xiufang Wang[2]

[1] Department of Bridge and Tunnel Engineering, Beijing University of Civil Engineering and Architecture, School of Civil and Transportation Engineering
[2] Beijing Higher Institution Engineering Research Center of Structural Engineering and New Materials, 15 Yongyuan Road, Huangcun Town, Daxing District, Beijing, 102616, China.
Email: jdongcg@bucea.edu.cn

Abstract. For plane truss structure, starting from the analysis of ideal truss model, the influence of tangential deformation and angular deformation on the secondary internal force of the truss is fully considered through Python program. It is obtained through analysis that: in the ideal truss model, the *P-δ* second-order effect causes the member to produce tangential deformation and angular deformation, resulting in secondary internal forces. Numerical analysis shows that due to the influence of secondary internal force, the axial force error of ideal truss model can reach 19.731% and the secondary shear force is almost all the members of the truss, and the secondary moment only appears at the support. The research results have important reference value for the engineering design and high-precision internal force analysis of truss structures.

1. Introduction

Truss structure has the advantages of simple structure and low steel consumption. It is used for roof truss, bridge, TV tower and other structures [4~6, 8~11, 13]. In engineering design, it is usually regarded as an ideal truss model for analysis in order to simplify calculation. The ideal truss model has the following characteristics:1) All members are smooth hinged.2) All bars are straight bars.3) All external loads imposed on nodes, and there is no non-nodal load.4) The members intersect axially at common hinge points.5) The internal force of the bar is only axial force without shear force and bending moment, and the truss only considers axial deformation.

Compared with the ideal truss model, the actual truss structure has some flexural deformation. Nair[1], Huixian Liu[2], Yongtao Bao[3], Yuehua Li[4], Hao Liu[5], Huijun Yan et all[6], Yue Cao et all[7], Hui Wang[8], Fanrong Liu[9], Rui Jiang[10], Xiaoqing Liu et all[11], Schmidt et all[12], Shundong Qiu[13] and so son blamed the difference to the secondary internal force or stress of the truss.And created anayltical methods including truss model, rigid frame model and non-bar element model, which is used to establishing finite element model.

Nair [1] saw the shear force and bending moment of the truss as secondary internal force, and the stress produced by secondary internal force is called secondary stress. When Huixian Liu [2] analyzed the secondary stress of the truss, he believed that the connecting plate at the node had an effect on the stiffening (stiffness strengthening) of the bar. And based on the stability principle of the pressure bar, the deflection caused by axial force is explained.

At present, it is generally believed that the main reasons for the secondary stress of the truss are: node rigidity, non-coincidence between the axis of the connected bar and the hinge node, and non-node load [4~6, 9, 12, 13]. Based on the finite element numerical analysis and experimental verification, it is considered that the main factors affecting the secondary internal force are: the line stiffness of the bar, the defect of the bar, the depth-span ratio of the bar, the connection mode of the node and the built-in function, the shape of the truss and the dead weight of the truss[1~12]. A large number of research results show that secondary internal forces are common in truss structures, and the analysis of secondary internal forces of truss structures is relatively mature. However, there are few literatures about the secondary internal force analysis of truss models. That is, does the ideal truss model itself have secondary internal forces?

Therefore, the secondary internal force analysis of truss finite element model will be carried out in this paper. Through Python finite element self-programmed program (Python program for short in this paper), the rigid frame model with discontinuous displacement is established, and the mechanism and influence factors of secondary internal force of truss model are explored, which provides a certain reference value for engineering design and finite element research.

2. Python Finite Element Program Introduction

Python program is based on the displacement method, can analyze bar structure, solid structure and composite structure, has the ability to analyze plane problems or space problems, and the loading process can be simulated once or multiple times through Python program. Figure 1 shows the block diagram of the program flow for the multi-condition mechanical model.

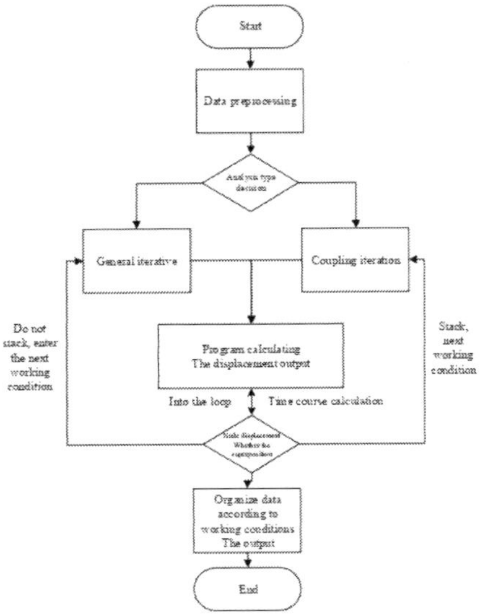

Figure 1. The block diagram of the Python programmer for multi-condition.

There are two types of iteration built into Python programs: general iteration and coupled iteration. General iteration means that there is no sequence between adjacent working conditions, and the initial data of the next working condition is the same as that of the previous working condition. Coupling iteration refers to the sequence of adjacent working conditions, and the solution result of the previous working condition will be loaded into the next working condition as the initial condition.

In Python programs, each degree of freedom has a corresponding serial number. When the total stiffness matrix is reduced (cut out row and column), the program will assign a new number to the degree of freedom to participate in the solution, which indicates the row and column order of the degree of freedom in the reduction stiffness matrix. The program is provided with a coupling element to simulate the functions of master-slave nodes. The coupling element can associate the corresponding degrees of freedom of any two nodes. The associated degrees of freedom have the same number in the reduced stiffness matrix, and have the same displacement value in the process of solving linear equations.

3. Analysis of Secondary Internal Force of Truss

3.1. The Basic Assumptions

In order to facilitate the analysis of the secondary internal forces of the ideal truss model itself, the model should meet the following basic assumptions:1) The members are hinged and have no built-in function.2) The axes of adjacent bar are intersected at common hinge points, and all the bar are straight bar with uniform cross sections and the same material properties.3) The load type is only node load, and all external loads imposed on the center of hinge point.4).The model satisfies the assumption of plane section and small deformation.

3.2. Establishing Finite Element Model

When there is secondary internal force in the truss, not only the axial displacement, but also tangential displacement and angular displacement will occur. In the ideal truss model, only the axial deformation of the member is considered, and the tangential and angular deformation of the member are ignored. When the total stiffness is reduced, the rows and columns irrelevant to the axis deformation are removed, so that the tangential and rotational degrees of freedom of the member do not participate in the solution of the equations.

Therefore, based on Bernoulli-Euler elementary beam theory, a rigid frame model with discontinuous displacement as shown in figure 2 was established. In order to fully consider the influence of tangential deformation and angular deformation on the secondary internal forces of the truss, the model relates the corresponding degrees of freedom through coupling elements, and makes tangential and angular degrees of freedom participate in the solution of linear equations.

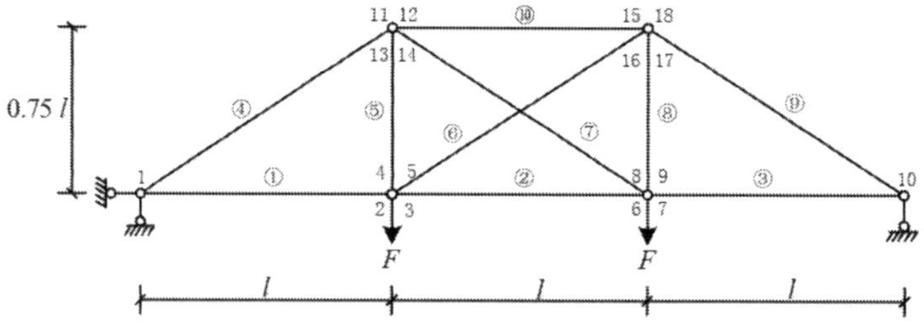

Figure 2. Mechanical model of truss.

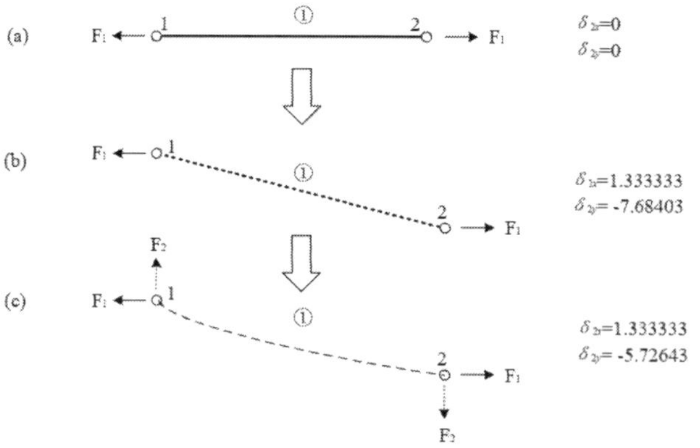

Figure 3. Deformation decomposition diagram of truss model.

Figure 2 shows the mechanical model diagram of the truss in this paper. The node serial number and element serial number are marked in the figure. Nodes 2, 3, 4 and 5 are overlapped nodes, which are used to establish coupling elements and associate corresponding translational degrees of freedom. Similarly, the same method is used to establish coupling elements at other locations. Discontinuous displacement also exists at node 1 and node 10, but through analyzing maximum linearly independent group of Stiffness matrix, it is found that the rotation variables at node 1 and node 10 are not independent. Therefore, in order to make the stiffness matrix having being reduced non-singular, no coupling element is set at node 1 and node 10.

In order to facilitate data analysis, the units of model parameters are all SI units and set as follows: bar element length L=1, elastic modulus E=1, cross-sectional area A=1, moment of inertia I_z=1, shear coefficient k_s=0. In the boundary conditions, the degree of freedom of horizontal and vertical translational of node 1 is constrained, and the degree of freedom of vertical translational of node 10 is constrained.The following four analysis conditions are established:1) Working condition 1: open axial deformation, close tangential deformation and angular deformation.2) Working condition 2: open the axial deformation and tangential deformation, close the angular deformation.3) Working condition 3: open axial deformation and angular deformation, close tangential deformation.4).Working condition 4: open axial deformation, tangential deformation and angular deformation.

Working condition 1 is equivalent to the ideal truss model, working condition 4 is rigid frame model, and working condition 2 and 3 are only used for comparative analysis. The Python program stores the model data in Excel(.xlsx) file.

3.3. Model Validation and Result Analysis

The solved displacement result of working condition 1 is the same as the example in literature [14], and the calculation accuracy is much higher than the example, which verifies the feasibility of the truss model in figure 2 and the reliability of the Python calculation results.

The iteration type of the above four working conditions was set as general iteration, and the above four working conditions were analyzed simultaneously by Python program.

Table 1. The summary table of bar end results.

Project	Working condition 1	Working condition 2	Working condition 3	Working condition 4
Maximum axial force	1.666666667	0.041727838	1.666666667	1.337822671
Minimum axial force	-1.666666667	-0.041727838	-1.666666667	-1.337822671
Maximum tangential force	1.80411e-16	0.890015913	5.2134e-15	1.026936027
Minimum tangential force	-1.80411e-16	-0.89005913	-5.2134e-15	-1.026936027
Maximum moment	0	0.445007957	9.53589e-16	1.026936027
Minimum moment	0	-0.445007957	-4.26091e-15	-1.026936027
Maximum displacement of X	3.833333333	0	3.833333333	4.0044893
Minimum displacement of X	0	-0.041679	0	0
Maximum displacement of Y	0	0	0	0
Minimum displacement of Y	-7.684028	-0.074168	-7.684028	-5.744108
Maximum Angle	0	0	4.34837e-16	5.897586981
Minimum Angle	0	0	-1.26137e-15	-5.897586981

The bar end displacement and bar end force of the four working conditions are compared, and the analysis results are as follows:1) In working conditions 1 and 3, bar end displacement and bar end force of bar element are exactly the same.2) In the working conditions 1, 2 and 3, the bar element has no angular deformation. In working 4, except bar element 2 and bar element 10, all other bar elements have angular deformation.3) In the working conditions 1 and 3, the bar end torque of the bar element is zero. In working condition 2, bar end torques exist except element 2 and element 10. In the 4 working condition, the bar end torque only appears at the support.4) In the working conditions 1 and 3, the bar ends of the bar element have no tangential component force. In the working condition 2, except bar element 2 and bar element 10, there is tangential component force at the bar end of the bar element. In the working condition 4, the tangential force of bar end of bar element only appears in bar element 1, bar element 3, bar element 4 and bar element 9.5)As shown in table 1, compared with working condition 1, the concentrated force at the bar end of the bar element decreases, the torque at the bar end increases, and the rotation Angle at the bar end increases. The relative errors of axial force, X displacement and Y displacement are 19.731%, 4.465% and 25.246% respectively.

When the bar has tangential deformation (lateral deformation) or angular deformation, due to the existence of certain lateral stiffness or bending stiffness of the bar, the corresponding secondary shear force or secondary bending moment will be generated after the bar moves sideways or turns. Based on Bernoulli-Euler elementary beam theory, the relation between shear force and tangential equation (1) and the relation between bending moment and angular equation (2) were obtained:

$$F_s = K_s \delta_s = \zeta \frac{EI_z}{L^3} \delta_s \tag{1}$$

$$M = K_\theta \theta = \eta \frac{EI_z}{L} \theta \tag{2}$$

In equations 1 and 2:

F_s: Shear force. M: Bending moment.

K_s: Lateral stiffness. K_θ: Rotational stiffness.

δ_s: Lateral deformation θ: Angular deformation.

ζ: The lateral stiffness coefficient, related to the restraint form at both ends of the bar

η: The rotational stiffness coefficient, related to the restraint form at both ends of the bar.

Based on equations (1) and (2), when the lateral stiffness K_s and rotational stiffness K_θ remain unchanged, the secondary internal force is proportional to the corresponding member displacement. Therefore, in the ideal truss model, the tangential (lateral) and angular displacements are the main causes of secondary internal forces.

3.4. Mechanism of Tangential Deformation and Angular Deformation

It can be concluded that there are differences between the ideal truss model and the discontinuous rigid frame model. The main reason for this difference is that the *P-δ* second-order effect of ideal truss model is ignored in internal force calculation, that is, the calculation of internal force precedes the occurrence of deformation. As shown in figure. 3, the force and deformation process of beam element ① in working condition 4 is decomposed into the following three stages:1) Loading stage (a): at this time, the beam element is in a balanced state of force, and the resultant force F_1 of the bar end is parallel to the beam axis, but it is considered that the beam element is not deformed at this time.2) Deformation stage (b): only axial deformation occurs in the beam element, and the final state, which is the same as working condition 1, is reached.3) Adjustment stage (c): The beam element is in equilibrium under the joint action of bar end forces F1 and F_2.

Loading stage (a) and deformation stage (b) are the deformation processes of the ideal truss model. However, the second order effect of *P-δ* is ignored, so the truss can not keep the equilibrium state after the deformation stage (b). Under the action of unbalanced external forces, the bar produces tangential deformation and axial deformation. According to the relation between equations (1) and (2), secondary shear force and secondary bending moment are generated inside the bar, and the resultant force of secondary shear and secondary bending moment forms the "additional" bar end force F_2.

4. Conclusion and Prospect

Based on Python program, this paper analyzes the secondary internal force of truss model and draws the following conclusions:1) The secondary internal forces of the truss model are generated by the tangential and angular displacements of the members. The secondary internal force is proportional to the corresponding displacement.2) The main reason why the ideal truss model cannot calculate the secondary internal forces is that the *P-δ* second-order effect is neglected in the calculation of internal forces.3) The angular displacement of the truss model only appears at the support, and the angular displacement at the non-support does not affect the calculation of other degrees of freedom. Tangential displacement is the main reason affecting the secondary internal force of truss non-support member. The secondary moment occurs only at the support, and the secondary shear force is almost all the members.4) When the ideal truss model is used for analysis, the axial force error, vertical settlement error and lateral displacement error of the model are at least 19.731%, 25.246% and 4.465% respectively.

Based on the research results of this paper, the following suggestions are given for the analysis and design of truss structures:1) The secondary internal force should be decomposed into the secondary internal force of truss model and the secondary internal force of truss structure. The secondary internal force of truss model is caused by neglecting *P-δ* second-order effect in internal force calculation. The secondary internal force of truss structure is caused by node rigidity, non-coincidence between the axis of connected bar and hinge node, non-node load and so on.2) Strengthen the bending stiffness of the member at the support, strengthen the shear stiffness of all the member.

Although the analysis conclusions of this paper confirm the research results of relevant literature, the analysis model of this paper cannot *contain* all possibilities. The influence of load forms and truss structure forms on the secondary internal forces of the truss model will be further advanced in subsequent studies.

Acknowledgments

Authors wishing to acknowledge assistance or encouragement from BUCEA, and financial support from the Teaching and Research Project of Beijing University of Civil Engineering and Architecture (201802), the Changjiang Scholars and Innovation Team Development Plan (IRT-17R06)

Reference

[1] Nair, R. 2008 Shankar. Secondary stresses in trusses [J] *Engineering Journal of the American Institude of Steel Constryction* **25** (4) 144-145.
[2] Liu H X. 1956 On moment distribution, shear distribution and truss secondary stress [J] *China Civil Engineering Journal* (01) 69-93 (in chinese).
[3] Bao Y T, Lu P. 2003 Accurate analysis of ideal truss calculation mode [J] *Mechanics in Engineering* (06) 68-69 (in chinese).
[4] Li Y H. 2005 Influence of secondary stress on steel truss structure [D] Zhengzhou University.(in chinese).
[5] Liu H. 2007 Secondary stress analysis of steel tube truss [D] Wuhan University Of Technology (in chinese).
[6] Yan H J. 2007 Analysis the secondary stress of joint rigidity and the method of orthotropic steel plate of steel truss bridge [D] Hefei University of Technology (in chinese).
[7] Cao L, Ma J. 2008 The influence analysis of spatial truss secondary stress [J] Jiangsu Construction (01) 40-42(in chinese).
[8] Wang H, Chen R J. 2008 Finite element analysis on secondary stress for two-span continuous square steel-pipe truss[J] *Steel Construction* (04) 4-7(in chinese).

[9] Liu F R. 2009 Study on secondary stress and ultimate bearing capacity of triangular steel roof truss with hot-rolled cut-T-section [D] Xi'an University of Architecture and Technology.(in chinese).

[10] Jiang R. 2009 Study on secondary stress and numerical analysis of parallel chord steel truss Bridge [D] Northeastern University. (In chinese).

[11] Liu X Q, Tong D G. 2010 Research on joint stiffness of steel truss continuous beam [J] *Sichuan Architecture* **30** (01) 129-131(in chinese).

[12] Schmidt L C, Morqan P R, Coulthard B R. 2010 Influence of eccentricity and continuity on the inelastic behaviour of a space truss [J] *American Society of Mechamcal Engineers* AMD 274-281.

[13] Qiu S D. 2014 Influence of secondary stress on long Span steel truss arch Bridge [A] Computer Application Branch of China Civil Engineering Society, Computer Application Committee of Building Structure Branch, Architectural Society of China, Proceedings of the 17th National Conference on Computer Applications in Engineering construction[C] *Computer Application Branch of China Civil Engineering Society, Computer Application Committee of Building Structure Branch, Architectural Society of China* 4 (in chinese).

[14] Li L K. 2017 Structural Mechanics (Vol. Ⅰ, 6th Edition) [M] Beijing: Higher Education Press. (in Chinese)

Study on Self-Cleaning Time and Suspended Particle Distribution in Medical Clean Room

Long Chen[1], Enyan Wang[1], Yang Li[2], Miaocheng Weng[1, 3, 4,*] and Fang Liu[1, 3, 4,*]

[1] School of Civil Engineering, Chongqing University, Chongqing 400045, PR China
[2] Hainan Tianjie Commercial Management Co., LTD, Hainan 570125, PR China
[3] Key Laboratory of New Technology for Construction of Cities in Mountain Area of Ministry of Education (Chongqing University), Chongqing 400045, PR China
[4] Joint International Research Laboratory of Green Buildings & Built Environments, Chongqing 400045, PR China
B Campus, Chongqing University, 83 Shabei Street, Shapingba District, Chongqing.
Email: chlchonglong@qq.com

Abstract. CFD numerical simulation of clean room in Class D medical factory was carried out and compared with the actual measurement to verify the feasibility of the simulation method. On this basis, four typical air flow organizations were simulated and compared by changing air change rate from two directions of self-cleaning time and suspended particle concentration field. According to the simulation results, in order to meet the self-cleaning time within 20 min, the best air change rate should be between 15/h and 25/h. Different air flow organizations have different self-cleaning capacity, and the value of air change rate can be relatively small in the form of single-side supply same-side down return. Different airflow organizations have different suspended particle distribution characteristics, and there are differences in the applicable scenarios, and the applicability of the top supply down return is the best.
Keywords. Clean room; CFD numerical simulation; Self-cleaning time; Suspended particles; Air flow organizations; Air change rate.

1. Introduction

To ensure the quality of drugs, the creation of indoor environment in medical workshop is particularly important, which puts forward a certain test for the design of clean room, not only to meet the requirements of relevant specifications, but also to achieve energy conservation and environmental protection, reasonable air flow organization and air change rate is the key. By comparing relevant domestic and foreign specifications, it can be found from the requirements of air change rate for class D clean rooms and ISO class 8 clean rooms that the span ranges from 3.5/h to 25/h[1][2][3], which brings difficulties to the actual design. Meanwhile, by comparing the changes of the old and new GMP in China and Code for design of pharmaceutical industry clean room, it can be found that the current specifications have removed the specific provisions on air change rate and instead require designers to select appropriate air change rate according to the actual situation to meet the requirements of indoor environment [3][4].

As an important performance indicator of non-unidirectional flow clean room, self-cleaning time is required by current GMP to be within 20 min [4] [5]. A lower air change rate can also meet the static standard of Class D clean room, but it may not meet the self-cleaning time requirement. Therefore, an appropriate air change rate is of great significance. Meanwhile, there are a variety of air flow organizations for this level of clean room to choose from, and the difference between each air flow organization should also be considered. However, most of the current researches are confined to the analysis and evaluation of the indoor flow field. Some scholars have studied the concentration field of suspended particles, but there is little horizontal comparison between different air flow organizations, and there is a lack of studies on the self-cleaning time of Class D clean room[6][7][8]. In this paper, the single-side supply same-side down return, double-side supply double-side down return, top supply down return and top supply top return the four kinds of typical air flow organizations, use the method of CFD numerical simulation in combination with the measured contrast, change the air change rate to simulate the distribution of suspended particles under self-cleaning process and steady state, and the optimal air change rate to meet the self-cleaning time is given, and the applicable characteristics of different air flow organizations are analyzed.

2. Physical Model Building and CFD Simulation

The pharmaceutical plant is located in Suihua City, Heilongjiang Province. In this paper, a clean room of the fluidized bed granulation process in the plant is selected to simulate the self-cleaning time test under as-built.

2.1. Clean Room Model

One side of the clean room has a local suspended ceiling, two side air supply outlets, a top air supply outlet, and five return air inlets, and two staff members are standing inside. Taking into account factors such as computing resources and computing time, a simplified physical model was set up as shown in figure 1, with the parameters shown in table 1.

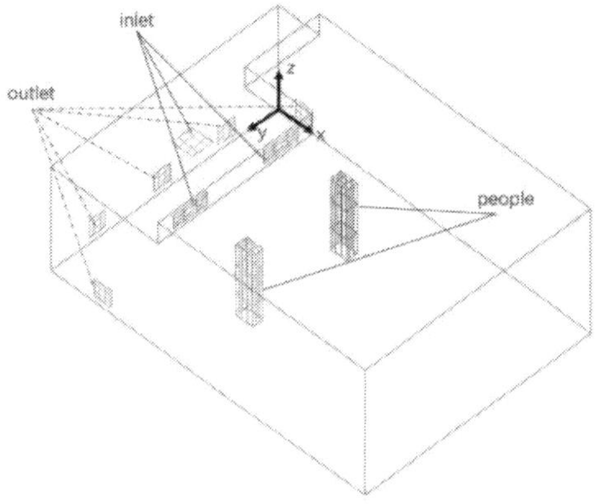

Figure 1. Clean room physical model.

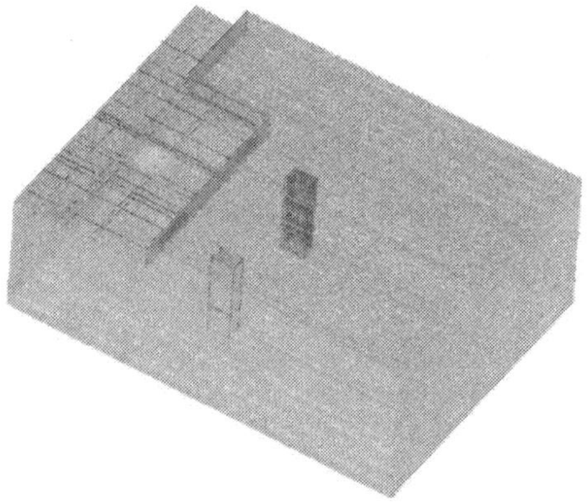

Figure 2. Schematic diagram of mesh division.

Table 1. Main parameters of clean room model.

Room size (m)	Room volume (m^3)	side air supply outlet (m)	top air supply outlet (m)	return air inlet (m)
7.6×5.5×3.0	122.9	0.83×0.30	0.50×0.50	0.40×0.40

2.2. Mesh Division

As shown in figure 2, since the model is composed of regular geometric shapes, it is divided into structured meshes. In addition, the quality of the meshes was checked and it was found that more than 99.5% meshes of the determinant 2×2×2 and determinant 3×3×3 were above 0.95, and no mesh below 0.75 was found, which proved the high quality of the meshing.

2.3. Computational Model Selection
To facilitate the calculation, the following assumptions and simplifications were made based on the characteristics of indoor air flow:

- Indoor air as incompressible Newtonian fluid.
- Indoor air is a viscous fluid, and the air flow is a steady flow.
- Indoor air leakage and outdoor air infiltration are not considered.
- Energy loss caused by air flow is not considered.

Meanwhile the following assumptions and simplifications are made for suspended particles:

- The movement of suspended particles has no influence on indoor air flow field.
- There is no merging and breaking of suspended particles.
- All suspended particles are spherical.

According to the above assumptions and simplifications, the standard k-ε model, which has been verified by most scholars to be feasible, has been chosen for the indoor air turbulence model. k and ε are the basic unknown quantities, and the corresponding equations are as follows:

$$\frac{\partial}{\partial t}(\rho k)+\frac{\partial}{\partial x_i}(\rho k u_i)=\frac{\partial}{\partial x_j}\left[\left(\mu+\frac{\mu_t}{\sigma_k}\right)\frac{\partial k}{\partial x_j}\right]+\mu_t\frac{\partial u_i}{\partial x_j}\left(\frac{\partial u_i}{\partial x_j}+\frac{\partial u_j}{\partial x_i}\right)-\rho\varepsilon \quad (1)$$

$$\frac{\partial}{\partial t}(\rho\varepsilon)+\frac{\partial}{\sigma x_i}(\rho\varepsilon u_i)=\frac{\partial}{\partial x_j}\left[\left(\mu+\frac{\mu_t}{\sigma_\varepsilon}\right)\frac{\partial\varepsilon}{\partial x_j}\right]-\rho c_2\frac{\varepsilon^2}{k}+c_1\frac{\varepsilon}{k}\mu_t\frac{\partial u_i}{\partial x_j}\left(\frac{\partial u_i}{\partial x_j}+\frac{\partial u_j}{\partial x_i}\right) \quad (2)$$

For the flow of suspended particles in air, the DPM model was chosen to simulate the discrete phase, which is suitable for situations where the volume fraction of the discrete phase is small, such as in a clean room.

2.4. Boundary Condition Setting
The air change rate in this clean room is 26/h, and the air supply volume is set according to the actual measurement. Table 2 shows the boundary condition settings for the air flow simulation.

Table 2. Air flow simulation boundary conditions.

Items	Boundary conditions	Velocity or pressure
Side air supply outlet	velocity	1.271 m/s
Top air supply outlet	velocity	0.979 m/s
Return air inlet	pressure	0 pa

This paper discusses the concentration of suspended particles with a diameter greater than or equal to 0.5μm, which can be approximated by a single particle with a diameter of 0.5μm instead, based on the experience of previous scholars. For the setting of the dust generation, it was determined after comparing the simulation results with the actual measurements based on the review of relevant information [9]. Table 3 shows the relevant boundary condition settings for the DPM model.

Table 3. DMP model boundary conditions.

Items	Dust generation rate	Discrete phase contact type
Staff	2765 pc·(s·per person)$^{-1}$	reflect
Floor	388 pc·(s·m^2)$^{-1}$	reflect
Wall	236 pc· (s·m^2)$^{-1}$	reflect
Ceiling	0	reflect
Supply/return air outlet	/	escape

2.5. Simulation and Measurement
In order to verify the reliability of the simulation, unsteady-state simulation and steady-state simulation were conducted to compare with the measured data of the self-cleaning process and the

suspended particles after stabilization. The location of the measured suspended particle concentration is shown in figure 3. The height of the arrangement is 0.8m. The specific location of the plane is shown in table 4.

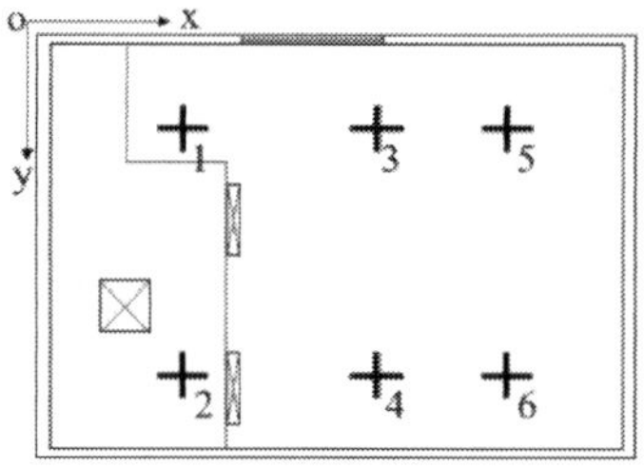

Figure 3. Layout of measurement points.

Table 4. Location of measurement points.

number	1	2	3	4	5	6
(x, y)(m)	(1.5, 1.2)	(1.5, 4.3)	(4.0, 1.2)	(4.0, 4.3)	(6.0, 1.2)	(6.0, 4.3)

Figure 4 shows the comparison between the simulation results and the actual measurement. It can be seen that the simulation of the self-cleaning process has the same trend as the actual measurement, and the data do not differ much, and the simulation of the concentration field of suspended particles after stabilization has the same trend as the measured concentration distribution, so the simulation can be considered reliable.

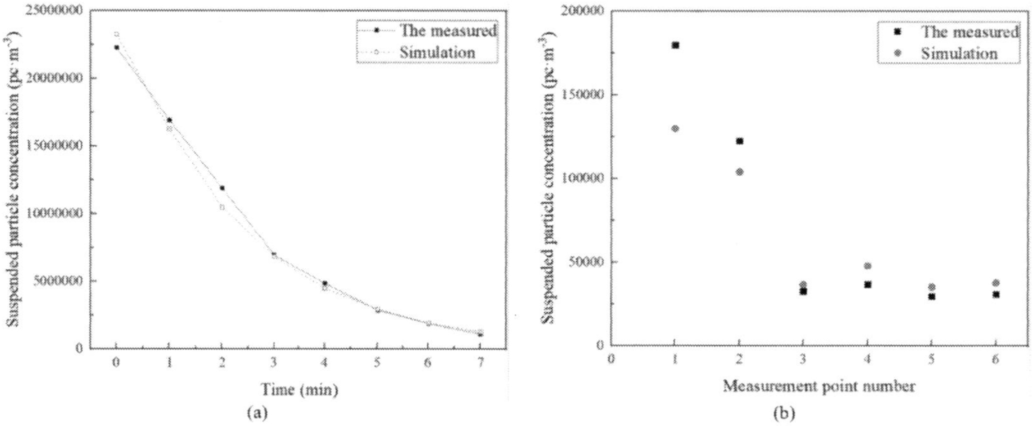

Figure 4. Simulated versus measured (a) Variation of suspended particle concentration during self-cleaning process; (b) Post-stabilization suspended particle concentration.

3. CFD Simulation Conditions and Results Analysis

3.1. Clean Room Model

In order to investigate the effect of air change rate on the self-cleaning time, and to compare the characteristics of each air flow organization, this paper conducted unsteady state simulation and steady state simulation for four typical air flow organizations[3], namely, single-side supply same-side down return, double-side supply double-side down return, top supply down return, top supply top return, and the respective physical models are shown in figure 5. With reference to the design range of air change rate in class D clean rooms, eight types of air change rate were set for each air flow organization, 11/h,

14/h, 17/h, 20/h, 23/h, 26/h, 29/h, 32/h, and the suspended particle concentration at 0.8m height was counted.

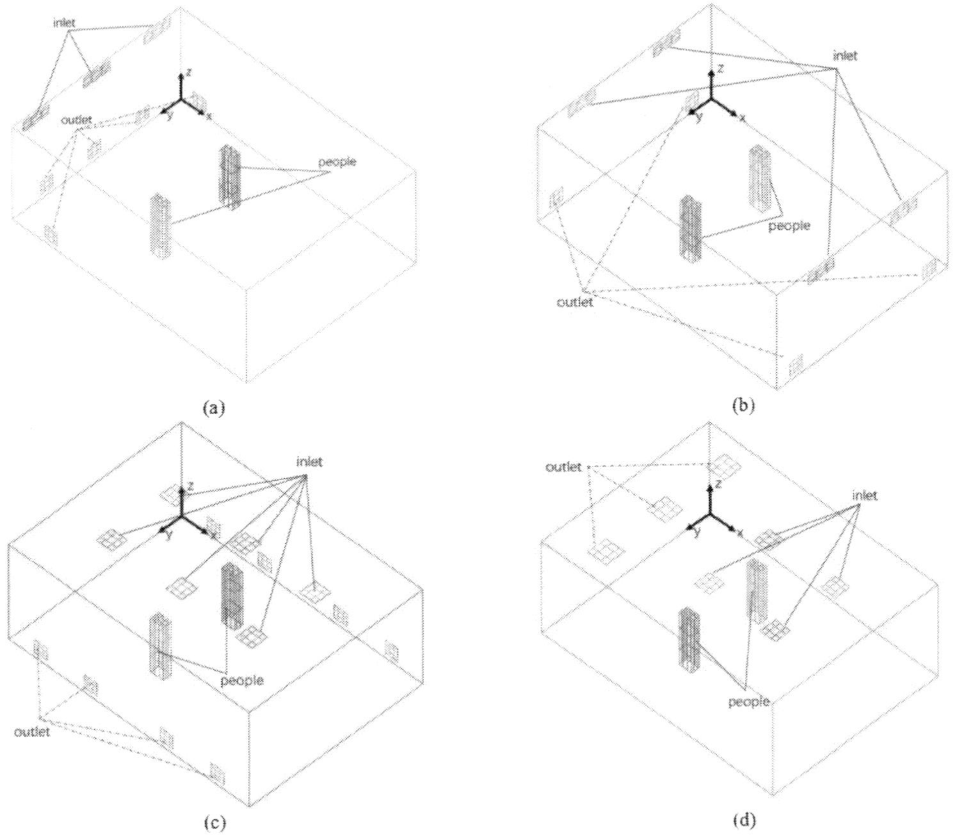

Figure 5. CFD physical model of typical air flow organization (a) Single-side supply same-side down return;(b) Double-side supply double-side down return;(c) Top supply down return;(d) Top supply top return.

3.2. Evaluation Criteria

The pharmaceutical workshop belongs to class D clean room, according to the GMP[4], the concentration of suspended particles greater than or equal to 0.5 μm does not exceed 3520000 pc·m^{-3} under as-rest, and it is recommended that the self-cleaning time does not exceed 20 min[5], where the self-cleaning time is measured according to 100:1 suspended particle concentration, which means that the clean room in the contaminated state is reduced to 0.01 times the suspended particle concentration after the air conditioning system is turned on[10]. The initial suspended particle concentration in the cleanroom in this simulation was 23135766 pc·m^{-3} and the 100:1 suspended particle concentration was 231358 pc·m^{-3}.

3.3. Analysis of Results

3.3.1. Self-cleaning Time. Through CFD simulation, the self-cleaning of each air flow organization with different air change rate is obtained, and the variation of suspended particle concentration with time is shown in figure 6. The specific values and trend of self-cleaning time are shown in table 5 and figure 7 respectively.

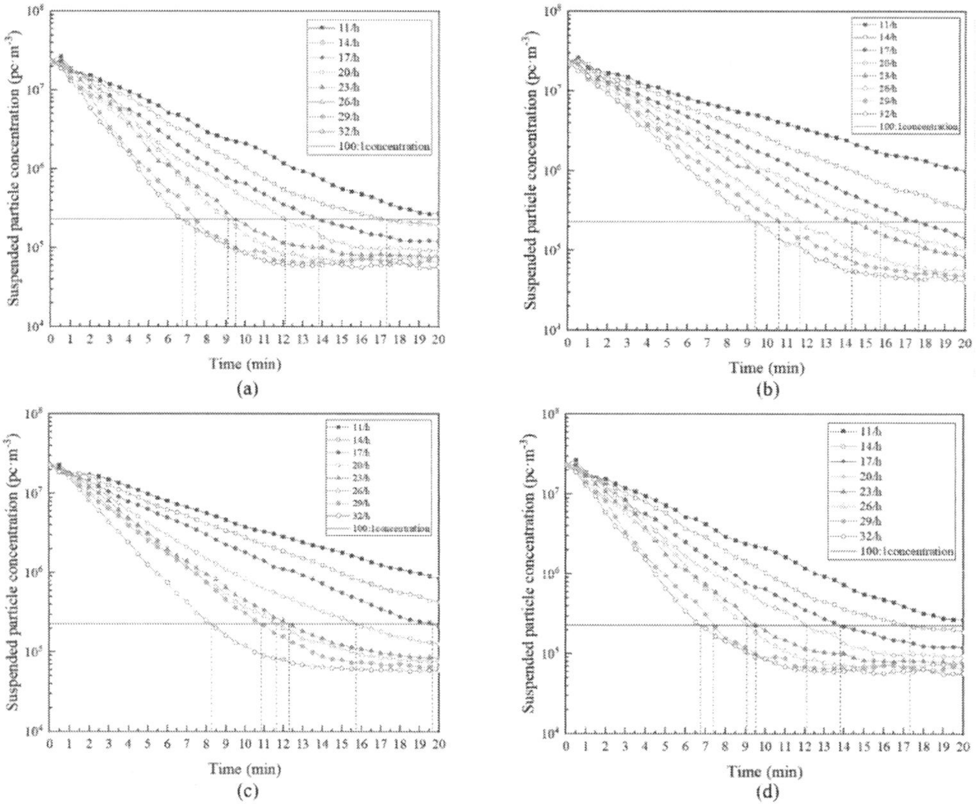

Figure 6. Variation of suspended particle concentration with time under different air exchange rate of each airflow organization(a) Single-side supply same-side down return;(b) Double-side supply double-side down return;(c) Top supply down return;(d) Top supply top return.

As shown in figure 6, the logarithm of the concentration of suspended particles at the beginning of the self-cleaning process under each air flow organization and time is approximately linear, with the increase of time, the greater air change rate, the earlier the concentration of suspended particles into a stable state, and when air change rate is about 25/h or more, increase air change rate the concentration change of suspended particles after stabilization is no longer obvious, indicating that for these types of air flow organizations, simply by increasing the air change rate to improve cleanliness needs to be under a certain air change rate to have a better effect.

Table 5. Self-cleaning time (min) under different air change rate of each air flow organization.

Air change rate	11/h	14/h	17/h	20/h	23/h	26/h	29/h	32/h
Single-side supply same-side down return	21.8	16.6	13.7	12.3	10.4	9.6	8.6	7.7
Double-side supply double-side down return	45.1	26.2	17.6	15.7	14.3	11.6	10.6	9.6
Top supply down return	43.8	27.4	19.6	15.7	12.3	11.4	10.6	8.3
Top supply top return	67.2	17.4	13.4	11.9	9.6	9.1	7.4	6.6

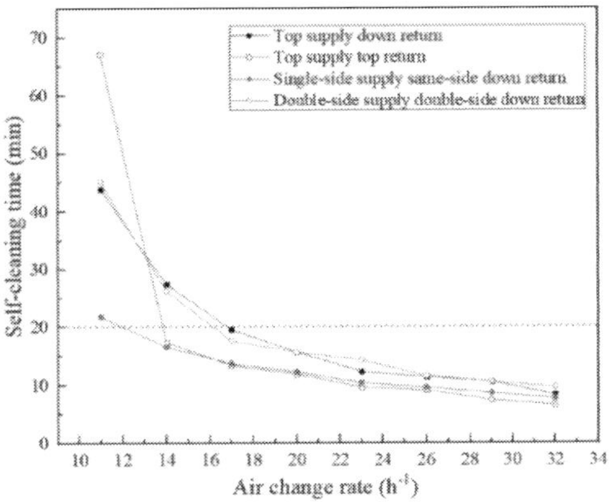

Figure 7. Self-cleaning time under different air change rate of each air flow organization.

Comparing each air flow organization, combined with table 5 and figure 7, it can be seen that the self-cleaning time of the four types of air flow organizations shows the same trend with the change of the air change rate, and the self-cleaning time of the single-side supply same-side down return and top supply top return is shorter than that of the double-side supply double-side down return and top supply down return with the same air change rate, among which it can be seen from figure 6 that the concentration of suspended particles of the top supply top return decreases more rapidly than other air flow organizations at the early stage of self-cleaning, but the concentration is relatively higher after stabilization. Meanwhile, it can also be seen that increasing the number of air change rate below 23/h has an obvious effect on shortening the self-cleaning time, and when air change rate is above 25/h, the change trend of self-cleaning time becomes smaller, and the difference between the airflow organizations is not big; when air change rate is below 15/h, reducing the air change rate may result in excessive self-cleaning time, in addition, it should be especially noted that even if air change rate in about 10/h, the concentration of suspended particles after stabilization is far below the requirements of the class D clean room (not more than 3520000 pc/m^3), but at this time the self-cleaning time has been too large. For these types of air flow organization, to meet the self-cleaning time within 20min, the air change rate in 15/h to 25/h between the best, where single-side supply same-side to the required air change rate can take a smaller value.

3.3.2. Suspended Particle Distribution. In order to further compare and analyze the characteristics of each airflow organization, z=0.8m and y=2.75m cross sections were selected for each air flow organization at 20/h air change rate to analyze the concentration fields of suspended particles after stabilization, as shown in figure 8 and figure 9.

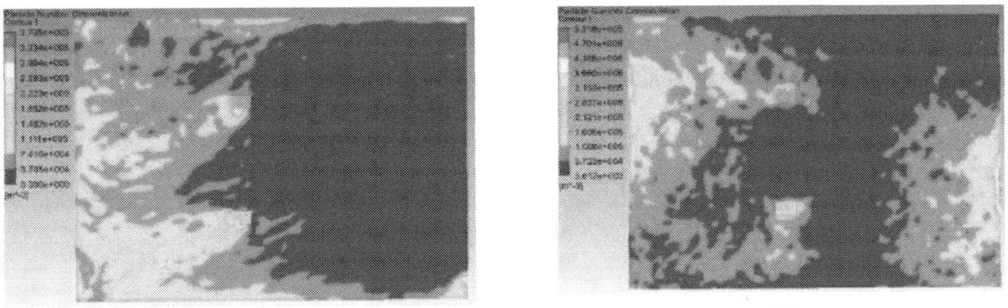

(a) (b)

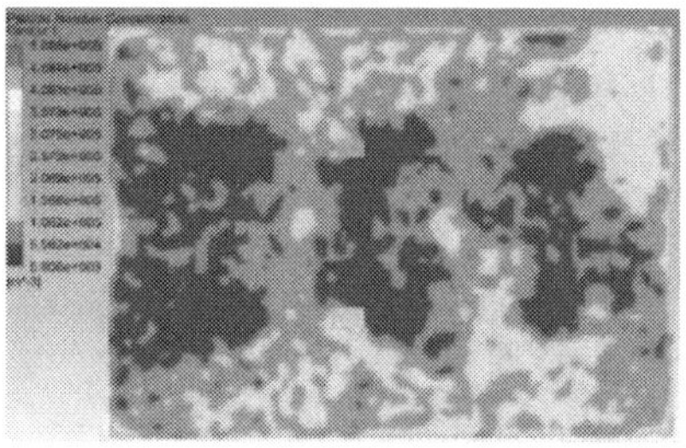
(c)

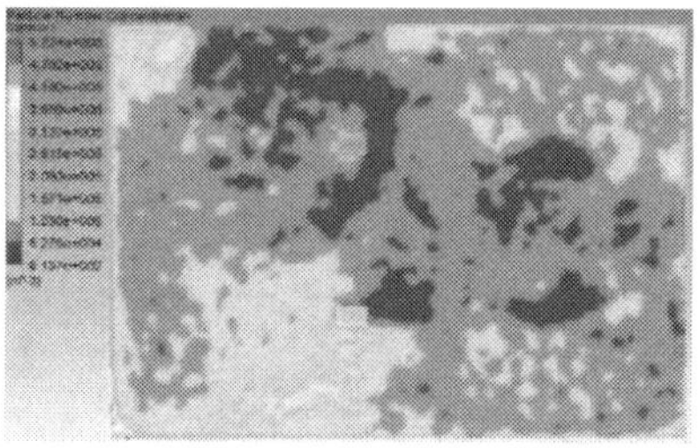
(d)

Figure 8. Each air flow organization under 20/h air change rate z = 0.8m section particle.concentration distribution(a) Single-side supply same-side down return;(b) Double-side supply double-side down return;(c) Top supply down return;(d) Top supply top return.

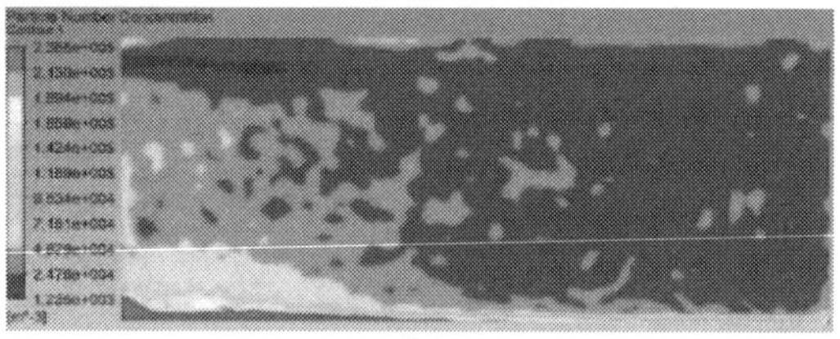
(a)

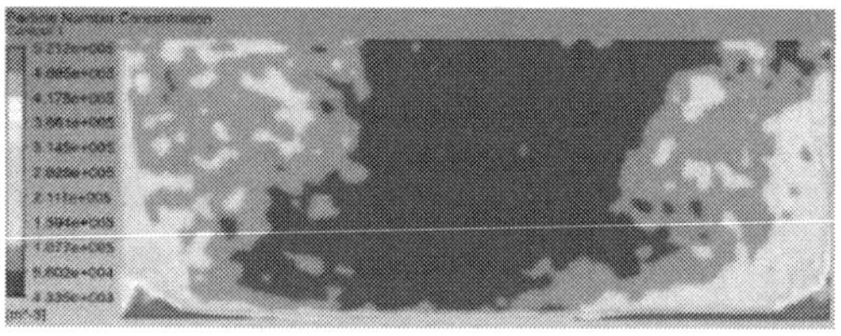
(b)

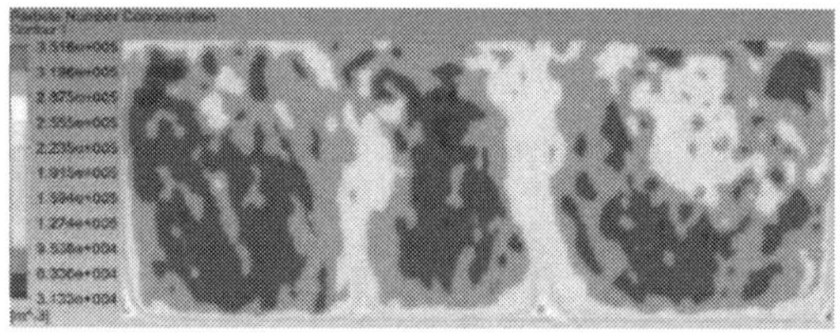
(c)

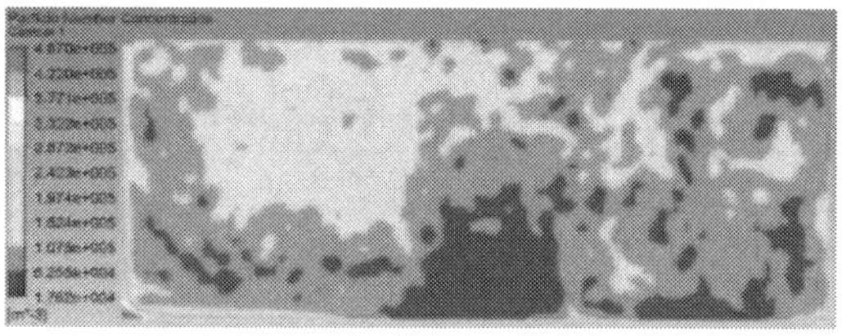
(d)

Figure 9. Each air flow organization under 20/h air change rate y = 2.75m section particle concentration distribution(a) Single-side supply same-side down return;(b) Double-side supply double-side down return;(c) Top supply down return;(d) Top supply top return.

As can be seen from figure 8, at the working surface height, for the two side supply side return forms, the concentration of suspended particles is high near the return side; for the single-side supply same-side down return, the concentration of suspended particles is lower in the area on the other side of the air outlet; for the double-side supply double-side down return, the concentration of suspended particles is lower in the middle of the room. On the working surface height, for the two top supply forms, the concentration of suspended particles is high near the four sides of the building enclosure, where the top supply top return also has a large area of high concentration of suspended particles on the return side; for the top supply down return, the concentration of suspended particles is low within the area about 1m from the building enclosure; for the top supply top return form, the concentration of suspended particles is low in the area below the air outlet.

As shown in figure 9, in the middle of the room, for the two side supply side return forms, the concentration of suspended particles is high near the return side, while the phenomenon of suspended particles gathering near the floor; for the top supply down return, the concentration of suspended particles is more uniform in most areas, and the phenomenon of high concentration of suspended particles appears in the lower area between the air supply outlets; for the top supply top return, the phenomenon of high concentration of suspended particles appears in a large area in the upper part of the room on the return side.

3.3.3. Summary. In a comprehensive analysis, the top supply down return is applicable to most cases, while the side supply and top supply top return form are applicable to the case where only part of the room is used as the working area, among which the side supply form should try to avoid the dust on the floor caused by indoor activities, while the top supply top return is best to stay away from the return side during production activities because of the obvious gathering of suspended particles in the space above the working height on the return side.

4. Conclusion

CFD method was used to perform unsteady-state simulation and steady-state simulation for a pharmaceutical plant of class D cleanliness, and compared with the measured data, and then four typical air flow organizations were simulated and compared, and the following main conclusions were drawn.

To meet the requirements of self-cleaning time within 20 min, according to different forms of air flow organization can be designed for different air change rate, but generally in 15 / h to 25 / h for the best, where the single-side supply same-side down return can be relatively small value. For the top supply top return, although the concentration of suspended particles in the early stage of the self-cleaning process decreases more quickly, air change rate above 15/h, the self-cleaning time is relatively shorter than other forms, but also faster to enter the steady state, while the concentration of suspended particles is relatively higher, it is not appropriate to take a lower air change rate.

When air change rate is below 15/h, reducing the air change rate will increase the self-cleaning time and the concentration of suspended particles after stabilization; when air change rate is above 25/h, increasing air change rate will not reduce the self-cleaning time significantly, and the concentration of suspended particles after stabilization will not change significantly. Therefore, when air change rate is lower than 15/h, the appropriate increase in air change rate can bring a greater improvement to the cleaning effect; when the air change rate is higher than 25/h, the increase in air change rate will not improve the cleaning effect much, but increase the energy consumption in vain.

Different forms of air flow organization have different characteristics of suspended particle distribution, the top supply down return does not have a particularly concentrated area of suspended particles, and is applicable to most scenes; the side supply form is applicable in part of the area for the work area, and when setting up the work area as far as possible in the middle of the room, which for the room length is relatively small room, single side supply same side down return than double side double side down return better performance, this is because The former in most of the space can form a similar one-way flow of parallel airflow and the latter will form more turbulent area; the top supply top return will form suspended particles gathering above the working height of the air return side, workbench arrangement should be far from the return side, applicable to workshops with low dust generation and low cleanliness requirements.

References

[1] European Committee for Standardization ISO 14644-4. 2001 *Cleanrooms and Associated Controlled Environments* Part 4: Design, construction and start up. p 18.
[2] Domestic - National standard - State Administration for Market Regulation CN-GB. GB 50073-2013 *Code for Design of Clean Room*. p 21.
[3] Domestic - National standard - State Administration for Market Regulation CN-GB. GB 50457-2019 *Standard for Design of Pharmaceutical Industry Clean Room*. pp 30–38.
[4] State Food and Drug Administration. 2011 *Good Manufacture Practice of Medical Products.2010*. Beijing: China Medical Science Press. p 74.
[5] Xu Z L. 2011 *Pharmaceutical Clean Room Design, Operation and GMP Certification*. Version 2. Tongji University Press. pp 23–24.
[6] Huang W H, Wei Q, Lu Q M. 2006 *Application of CFD in Airflow Design for Cleanroom* (Contamination Control & Air-conditioning Technology (02)) pp 7–11.

[7] Khoo C Y, Lee C C, Hu S C. 2012 *An Experimental Study on the Influences of Air Change Rate and* Free *Area Ratio of Raised-floor on Cleanroom Particle Concentrations* (Building & Environment, 2012, 48(Feb.) pp 84–88.

[8] Zhao J N, Zhang X S. 2008 *Clean Down Capability Prediction in Clean Operation Room Based on Technology of CFD and Reliability Test* (Journal of Harbin University of Commerce(Natural Sciences Edition),2008(02)) pp 227–232.

[9] Xu Z L. 2014 *Fundamentals of Air Cleaning Technology And Its Application In Cleanrooms.* Version 4. Science Press. pp 385–390.

[10] European Committee for Standardization ISO 14644-3. 2019 *Cleanrooms and associated* controlled *environments* Part 3: Test methods. pp 34–38.

Design and Analysis of the Location of Cold and Heat Sources for the Payload's Active Thermal Control System on the Lunar Surface

Hongyu Zhang[1], Yajing Li [2], Yifei Wang [1], Miaocheng Weng[1,3,4,*], Fang Liu[1,3,4,*]

[1] School of Civil Engineering, Chongqing University, Chongqing 400045, PR China
[2] Zhuhai Singyes Green Building Technology Co.Ltd
[3] Key Laboratory of New Technology for Construction of Cities in Mountain Area of Ministry of Education (Chonging Univrsity), Chongqing 400045, PR China
[4] Joint International Research Laboratory of Green Buildings & Built Environments, Chongqing 400045, PR China
B Campus, Chongqing University, 83 Shabei Street, Shapingba District, Chongqing.
Email: 1418238756@qq.com

Abstract. The payload of the Chang'e-4 biological experiment is used as an object for designing and analyzing the location of cold and heat sources. The research compares and analyzes the energy consumption and temperature uniformity of cooling and heating sources mounted on different surfaces using Thermal Desktop/Sinda Fluint, which may be used to guide the design and operation of active thermal control systems. The results indicate that when the hot and cold sources are mounted on the payload's top surface, the total energy consumption of the active thermal control system is minimized and temperature uniformity is improved.
Keywords. Active thermal control , energy consumption , temperature uniformity , cold and heat sources.

1. Introduction

The thermal control system regulates heat transfer from the payload's inside to its outside. Thermal control systems may be passive or aggressive. The thermal condition of the load is basically unadjustable once detected during the passive thermal control procedure [1]. Active heat control offers the benefits of being extremely flexible and precise. As a result of this, the payload's capacity to adapt and adjust to the external environment and operate is improved.

Domestic and international research on active thermal control numerical simulation computations has largely focused on extensive analyses of active thermal control systems in recent years. Zhou et al. [2] summarized the research on active thermal control systems by demonstrating that the research hotspots were primarily focused on spacecraft thermal control system performance analysis and equipment system optimization, and did not include the design of active thermal control systems during the development phase of the payload model. R. Nadalini [3] et al. presented active thermal control mechanisms for heat dissipation using variable thermal conductivity heat pipes and thermal switches, and validated their logic and dependability using ground thermal simulation tests. Derek W et al. [4] provided an overview and generalization of contemporary spaceship subsystem temperature management technologies, as well as a detailed assessment of their possible applications and use scenarios. Shao et al. [5] proposed a thermal control design for the Chang'e-3 probe, utilizing a radiative heat sink during the lunar day phase and a two-phase fluid loop and isotopic heat source during the lunar night phase, and analyzed the thermal control effect using numerical simulations; Li et al. proposed a joint heat dissipation active thermal control method, and the simulation results indicated that this thermal control method performed well. In summary, ongoing thermal control research at the moment is mostly focused on the design and development of thermal control system schemes for actual engineering applications.

However, using peak cold and heat loads to build active thermal control systems wastes energy and overcools or overheats the payload. Simultaneously, to maintain the payload's internal temperature regulation while responding to exterior climatic changes, the ground must feed the appropriate heat and cold in real time [6]. Since cold and heat sources are positioned on various surfaces, the paper concentrates on the payload model. This data helps design and operate the active thermal control system.

2. Governing Equations and Numerical Simulation

2.1. Governing Equations

The control target parameter of the payload model is temperature, and this paper will establish the heat transfer mathematical model by establishing the heat balance conditions in the space thermal environment where payload model is located. The total energy balance equation of the lunar surface load model is established based on the principle of energy conservation.

$$Q_1 + Q_2 + Q_3 + Q_4 = Q_5 + Q_6$$

Where Q_1-Q_4 are the heat gains from solar radiation, reflected solar radiation, lunar infrared radiation, and the heat gain from the heat source inside the spacecraft. Q_5 is the heat gain from spacecraft radiation, while Q_6 is the change in the spacecraft's internal energy.

The total energy balance calculation assumes the payload as a whole. However, in actuality, the payload model's surfaces have varying azimuthal angles to the Sun and Earth space, and hence different temperatures on each surface. Because of this, each instrument, component, or portion of the payload model has its own heat balancing equation [7]. Currently, the nodal network method is used for spacecraft thermal analysis. The nodal network approach emerged from Oppenhiem's analogous electrical and thermal modeling principles [8]. Analytic objects are divided into micro-element control bodies of specified dimensions, each considered as an isothermal object with similar thermophysical properties, and labeled as nodes. The heat balance equation 1.9 for each node can be listed.

$$Q_{sj} + Q_{pj} + \sum_{i=1}^{m} B_{i,j} A_i \varepsilon_i \sigma T_i^4 + \sum_{i=1}^{n} D_{i,j}(T_j - T_i) = (cm)_j \frac{dT_j}{d\tau} + A_j \varepsilon_j \sigma T_j^4$$

Where Q_{sj} is the heat flow outside the space absorbed by node j. Q_{pj} is the thermal power consumption of node j. $B_{i,j}$ is the Gebhardt coefficient. $D_{i,j}$ is the heat transfer coefficient between node i and j. $(cm)_j$ is the heat capacity of node j.m is the number of nodes with radiative heat transfer at node j. n is number of nodes with conduction heat exchange with node j.τ is the time.

The Monte-Carlo approach is now frequently used to complete computations, considerably enhancing the speed and accuracy of thermal simulations. The solver software SINDA/FLUINT, which is utilized throughout the paper, also adopts this approach.

2.2. Numerical Simulation

2.2.1. Modeling Geometry. Due to the limitation of using the data of Chang'e-4 "Biological Science Experiment Load", the payload model in the initial sample stage was chosen as the object of study. The model is divided into six surfaces, taking into account the different azimuthal angles between the payload model and the Sun and the Moon during one lunar day cycle. The model is shown in figure 1.

2.2.2. Boundary Conditions. The preliminary payload model of the lunar surface loading model at The Moon's Farside Von Kármán Crater is studied. The model is furnished with animal and plant life, as well as supporting instruments and equipment, and is exposed to external space environments such as high vacuum, microgravity, and extremely cold lunar surface temperatures (-180°C). Therefore, to maintain the survival of biological organisms and normal equipment operation, the inside temperature must be kept between 20°C and 30°C. However, due to space and budget constraints, the internal thermal control system is fitted with a small-scale, lightweight thermal control system. The active thermal control measures of semiconductor refrigeration and electric heating, which are inexpensive and straightforward to install and operate, are adopted for temperature management in this study. Table 1 summarizes the lunar surface load model's environmental conditions.

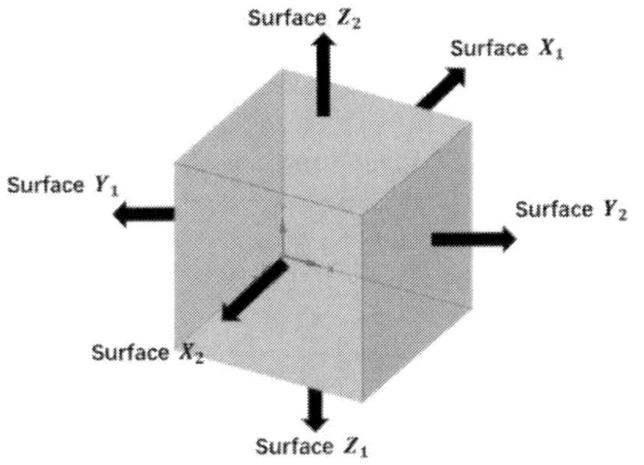

Figure 1. Schematic diagram of the payload model.

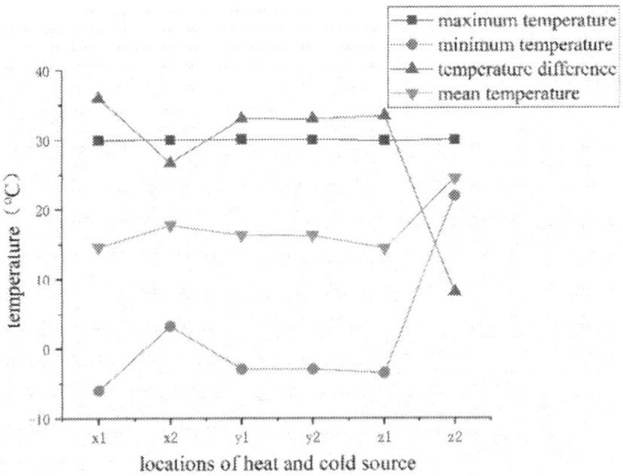

Figure 2. Payload internal extreme temperature and peak cold load curves.

Table 1. Environmental condition table of payload model.

Type		Parameter
Model	Temperature control target	20~30°C
Space environment	Location	Von Karman crater (177.6°E, 45.5°S)
	One lunar daily cycle	655.722 hours
Thermal control measures	Passive thermal control measures	MLI+ coating
		Inner surface black paint (0.9, 0.9)
		Outer surface Sio_2-Al (0.9, 0.1)
	Active thermal control measures	Semiconductor refrigeration
		Electric heating

3. Design and Analysis of Cold and Heat Sources Installation Position

3.1. Comparison of Mounting Surfaces of Cold and Heat Sources

During cooling and heating, the temperature of the payload model nodes must be near to the temperature control interval boundary. As a result, the internal cooling and heat calculation is only possible via software simulation, constantly changing the internal cooling and heat size. Figure 2 shows the highest temperature and peak cooling load inside the payload model at the extreme hot condition of 281.028.

From figure 2, The predicted peak cooling load under very hot conditions (281.028) of the payload model was Z_1, X_1, Y_1, X_2, Y_2, X_2 and Z_2 in descending order of surface position. The peak cooling load calculation value is -30.15W when the heat source is situated on the Z_2 surface of the payload model. So the payload model's bottom surface Z_1 is the least suitable for installing cold and heat sources. The heat source is mounted on the Z_2 surface at the top of the payload model, reducing the peak cooling demand and ensuring the most uniform temperature distribution. Thus, Z_2 on the payload model's top surface is the best place to install cold and heat sources.

In practice, the payload model is generally linked to the mothership. So the payload model's bottom surface Z_1 is the least favorable mounting surface for the cold and heat sources, and its top surface Z_2 is the most favorable. The other five surfaces except Z_1 are installed together as the mixed mounting surface of the cold and heat source.

3.2. Energy Consumption Analysis

In a lunar day operating cycle, the peak heat load Q_h is 3.05 W; the peak cooling load Q_c is -31.04 W. The no operation strategy condition is cooling at the peak cooling load during the cooling period and heating at the peak heat load during the heating period. The cooling and heating sources are installed on different surfaces of the payload model, and the time-by-time cooling and heating loads are calculated as shown in the table 2.

Table 2. Calculation result table of total power consumption of load in one track period.

Working condition	Cooling energy consumption /KJ	Heating energy consumption /KJ	Total energy consumption /KJ	Energy-saving efficiency
No operation strategy	41870.88	3085.69	44956.57	/
The most unfavorable surface	28896.37	3692.70	32589.07	27.50%
The most favorable surface	28219.94	3654.76	31874.70	29.09%
Mixed surfaces	28360.62	3619.35	31979.97	28.86%

From table 2, it can be obtained that the payload model consumes the most total energy under one lunar day operation cycle with no operation strategy. The principal heat transport modalities in microgravity are conduction and radiation. This is because the top surface faces the Sun and thus receives more heat from space. The bottom surface faces the Moon, which is less affected by heat flow from outer space. Heat and cold sources on top of the payload model can better resist heat conduction and radiation, resulting in the lowest overall energy consumption and the highest efficiency, 29.09 %. In this case, the heat and cold sources are on the payload model's bottom surface, causing higher overall energy consumption and the lowest energy saving efficiency of 27.50 %. It is recommended that the cold and heat sources be placed on the payload model's top surface Z_2.

3.3. Temperature Field Analysis

To compare and analyze the temperature field uniformity of nodes within the payload model when different cold and heat sources installation wall surfaces are used (the most unfavorable surface, the most favorable surface, and a mixed surface), the interior of the payload model is divided into six layers along the Z-axis direction, from bottom to top as layer I, layer II, layer III, and layer IV. As shown in figure 3.

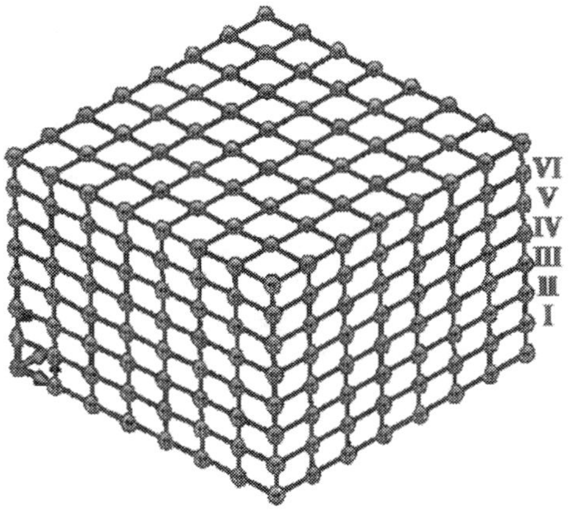

Figure 3. Temperature field stratification diagram.

Based on the payload model of von Karman crater, the mean temperature variation curves of each layer in the Z direction within the payload during a lunar daily cycle are analyzed. From figure 4, it can be seen that the payload model has the smallest difference in the mean value interval of all nodes temperature in the interior for the mixed surfaces during a lunar day operation cycle, whether in the cooling or heating time. From the total mean value interval of temperature of all nodes in the interior compared to the design temperature requirement (20-30°C), the mixed surfaces are the closest, and from the mean value of temperature of all nodes in the interior layers compared to the design temperature requirement (20-30°C), the mix surfaces reache the temperature control target for the most periods.

The average temperature interval difference of all internal nodes is the smallest for mixed surfaces, followed by the most favorable surface, and the highest is the least favorable surface in a lunar day operation cycle of the payload model. The mixed surfaces were the closest to the design temperature (20-30 °C), followed by the most favorable and the least favorable. In terms of time required to reach the temperature control objective (20-30 °C), the mixing surface takes the greatest time, followed by the most favorable surface, and finally the least unfavorable.

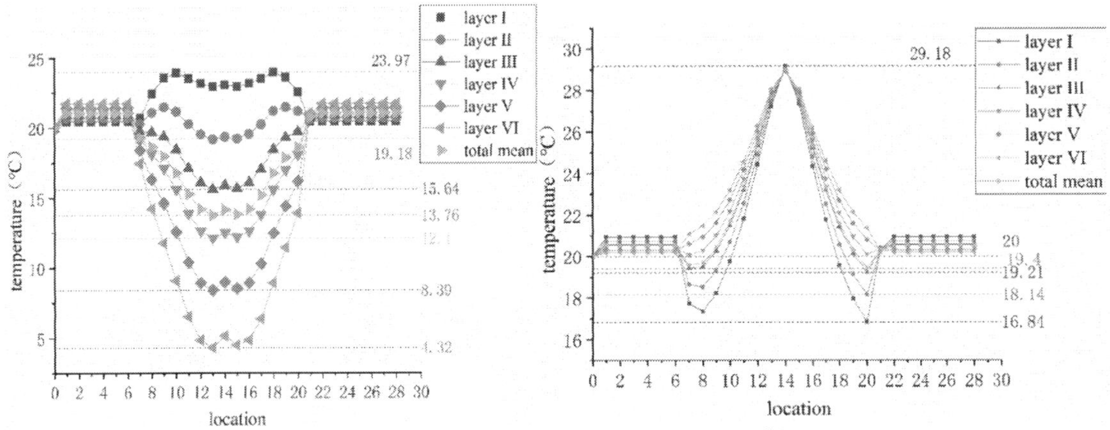

(a) (b)

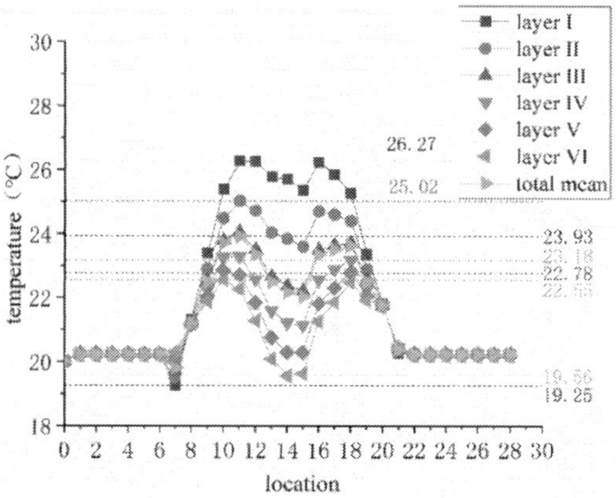

(c)

Figure 4. The cold and heat sources are installed on (a) surface Z_2 (b) surface Z_1 and (c) mix surfaces and the average temperature diagram of each layer in Z direction inside the payload.

Comparison of cold and heat source surfaces (unfavorable, favorable, and mixed) yielded the average temperature of each layer in the payload model's Z direction. A more pronounced internal temperature stratification was seen when the cold and heat sources were put on the payload model's surface Z_1 (the least favorable surface). The interior temperature stratification is more consistent when the heat source is situated on the Z_2 surface (the most favorable surface) of the payload model. The internal temperature stratification was most uniform and the effect was highest when both heat and cold sources were positioned on the other five surfaces (mixed surface) except for surface Z_1.

4. Conclusion

The cold and heat sources are installed on the top surface Z_2 of the payload model as the most favorable surface. When the cold and heat sources are installed on face Z_2, the greatest temperature and smallest temperature span interval are the smallest, and the temperature distribution is the most uniform, compared to the other five cold and heat sources installation faces.

To reduce the active thermal control system's energy consumption, the cold and heat source should be installed on the top surface Z_2 of the payload model, that is, the most favorable surface for installing the cold and heat source. The model's total energy consumption was the lowest and its energy saving efficiency was the highest, at 29.09 %. The mixed surface cooling and heating source has a slightly lower energy saving efficiency of 28.86 %. The source of cooling and heating installed on the least favorable surface saves the least energy at 27.50%.

In terms of payload interior temperature field uniformity, the best results were obtained by concurrently installing cold and heat sources on five surfaces without Z_1 (mixed installation surfaces). This is followed by Z_2 and Z_1.

Finally, considering the operability of the active thermal control system's installation and the analysis results for the active thermal control system's energy consumption, it is recommended that the cold and heat source be installed on the top surface Z_2 of the payload model, that is, the most favorable surface for the cold and heat source's action.

Acknowledgments

Funding: This work was also supported by the Third Pre-research Projects of the Civil Space Project from China National Space Administration(CNSA), under the project of "The Key Technology for the Construction of Micro-ecospheres Adapted to the Lunar Environment".

References

[1] Clou, S D, et al. 2015 "Performance and Control of a Pulse Thermal Loop Heat Transport System." *Journal of Thermophysics and Heat Transfer* **29** 4(2015) pp 1-9.

[2] Zhao C Y, et al. 2018 "Analysis of Development Trend of Lentinan Research Based on the Literature of Web of Science." *Edible Fungi of China*.

[3] Nadalini R, Bodendieck F. 2006 "The thermal control system for a network mission on Mars: The experience of the Netlander mission." *Acta Astronautica* **58** 11(2006) pp 564-575.

[4] *Hengeveld* D, et al. 2010 "Review of Modern Spacecraft Thermal Control Technologies." *HVAC&R RESEARCH* **16** 2(2010) pp 189-220.

[5] Shao X. "Heat Pipe Applications and Test in Chang'e-1 Satellite." *Spacecraft Engineering*.

[6] Jia, Y, et al. 2018 "Scientific Objectives and Payloads of Chang'E-4 Mission." *Chinese Journal of Space Science*.

[7] Hou Z. et al. 2007 *Spacecraft thermal control technology*. Press of University of Science and Technology of China.

[8] Oppenheim A V, and Lim J S. 1981 "The importance of phase in signals." Proc IEEE 69.5(1981): pp *529-541*.

Seismic Finite Element Analysis of Lightning Arrester Combination by Casing and Pillar Insulator

Xiaojun Zhang and Zhenlin Liu

China Electric Power Research Institute Co. , No. 33, Nanbinhe Road, Xicheng District, Beijing, China
Email: zhangxiaojun@epri.sgcc.com.cn

Abstract. The seismic performance of electrical equipment in substations has a great impact on the normal operation of the whole substation. The results of the modal analysis show that the fundamental frequency of the three devices is in the range of 0.9Hz~1.1Hz. The maximum stress of the casing for the three devices is respectively 55.43MPa, 45.39MPa, 35.26MPa, when the peak acceleration 0.4g seismic action is verified. The maximum stress of insulator is respectively 47.01MPa, 62.72MPa and 30.85MPa, and the maximum relative displacement of the top for the equipment is 617.2mm.

1. Introduction

As an important part of the lifeline project, once the power system fails or is damaged by earthquake, it will cause serious disasters and incalculable economic losses. Power interruption not only seriously affects normal production and earthquake relief work, but also may cause secondary disasters such as fire, which seriously threatens people's life and property safety.

Since the mid-1970s, with the frequent occurrence of seismic disasters, the seismic damage to electric power facilities has been reported throughout almost every strong earthquake. Among them, the 2008 Wenchuan earthquake that occurred in China, the 2011 earthquake in the Pacific Ocean in northeastern Japan, and the 2019 California earthquake in the United States all caused serious damage to electric power facilities, which directly caused serious rupture, displacement and oil leakage of electrical equipment such as lightning arresters and mutual transformers, thus it will affect the normal operation of the power system.

For the seismic safety of substation power facilities, scholars at home and abroad have conducted a lot of seismic research by using post-earthquake research, theoretical analysis and tests, and other technical means.

Ersoy S [1] et al. analyzed the coaction of high-voltage transformers and porcelain bushings during earthquakes by finite element simulation and obtained the relationship between the mutual influence of transformers and porcelain bushings; Filiatrault A [2] et al. performed finite element analysis on various types of high-voltage transformers and obtained the acceleration spectrum of porcelain bushings under the coaction with the box, which provided an analytical reference for the seismic study of transformers. Gilani A S [3] et al. conducted shaking table tests on several different sizes of high-voltage transformer bushings, and the test results showed that the bushings had better seismic performance with larger connection stiffness, and point washers were set at the bottom of the bushings, it effectively prevent leakage of transformer oil due to bushing displacement; Filiatrault A [4] et al. conducted a shaking table test on a 525 kV transformer and successfully obtained the dynamic characteristics of the transformer tank and bushing as well as other important components, and it provided valuable data for the seismic study of large transformers; in 2009, Knight B T [5] et al. proposed a "SVRP design method" to improve the seismic weakness of substations and gave several recommendations for the seismic design of substations. M. Fahad [6] conducted a seismic table test on electrical equipment which made of porcelain material, and the test results showed that the strength of the material is the controlling factor for the seismic performance of this type of equipment.

Chinese scholars have also studied the seismic performance of substation equipment gradually. Zhou Shiping [7] analyzed the damage to transformers in the Wenchuan earthquake, and the analysis results showed that the strong seismic resistance of the transformer body often led to the damage of accessories such as the emergence of bushings; Fan Rongquan [8] studied the mechanical properties and operation of synthetic insulators, and the results showed that the characteristics of composite insulators under dynamic loads were essentially different from those under static loads; in 2011, Zhang Jun et al [9] conducted a study on the seismic test and finite element analysis, which conducted on electric porcelain-type high-voltage electrical equipment, and the test data and simulations concluded that the bending resistance of electric porcelain-type high-voltage electrical equipment is poor, when it used as insulating sleeves or load-bearing columns. The reason is the fine and high body shape and the brittle porcelain parts.

This paper mainly uses the finite element software ANSYS to numerically analyze three kinds of composite surge arrester and insulator combination equipment, through the modal calculation and seismic calculation, the frequency, stress and displacement response of the equipment is compared and analyzed, and the seismic performance of the equipment is evaluated.

2. Finite Element Numerical Simulation

2.1. Equipment Overview

A domestic manufacturer's 750kV composite casing and insulator combination of lightning arrester electrical equipment was selected for the study, which consists of three upper sections of casing and one lower section of insulator. A total of three types of equipment are selected and represented by equipment 1, equipment 2 and equipment 3 respectively, with the structural arrangement shown in figure 1 and the specific structural parameters shown in table 1. The main difference between equipment 1 and equipment 3 is that the weight of each section for the casing, the radius of the insulator indexing circle and the size of the insulator diameter are different, but the arrangement form is the same, and the three insulators are evenly arranged on the circular plate. The main difference between equipment 2 and the other two types of equipment is that the arrangement form is not the same, i.e. equipment 2 uses single pillar insulators.

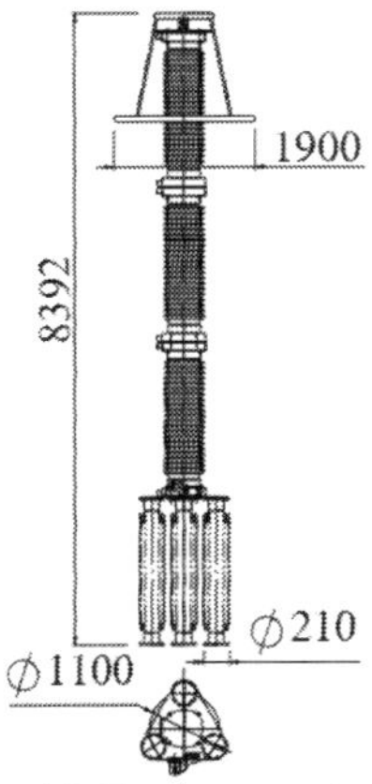

(a) Equipment 1

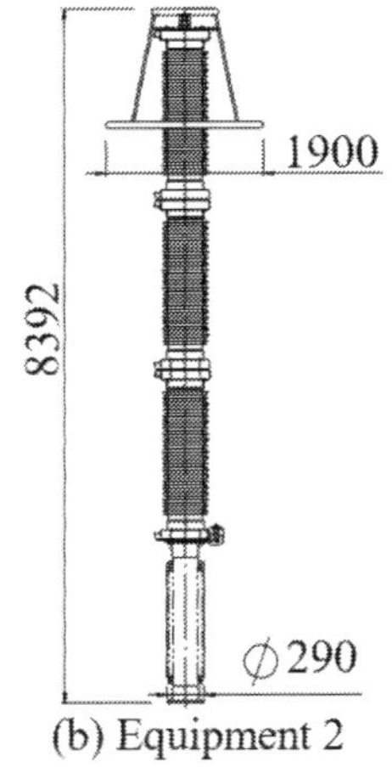

(b) Equipment 2

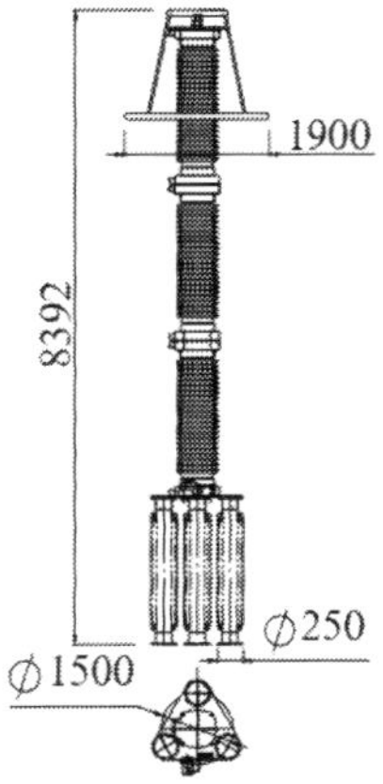

(c) Equipment 3

Figure 1. Equipment structure diagram.

Table 1. Structural parameter of equipment.

Type of equipment	Equipment Weight.(kg)	Young's modulus (×109Pa)	Siméon Denis Poisson	Equipment Height (m)
Equipment 1	2705	16/43	0.3	8.392
Equipment 2	2280	16/43		
Equipment 3	3600	20/43		

2.2. Finite Element Modeling

The simulation model of the equipment is established according to the structural scheme and related parameters. In establishing the finite element model of the structure, the beam unit (BEAM189) is usually used to simulate the electrical equipment, and the mean pressure ring and electromagnetic unit are simulated by the concentrated mass unit (MASS21). In the finite element mesh division, the cell size should be uniform and moderate to ensure sufficient calculation accuracy. The finite element models of the three types of equipment are shown in figure 2.

(a) Equipment 1 (b) Equipment 2 (c) Equipment 3

Figure 2. Finite element model of equipment.

2.3. Modal Analysis

Through the modal analysis of the three types of equipment, the order frequencies of the equipment structure are obtained and the dynamic characteristics of the structure are initially understood, and the modal analysis is also the basis for the seismic acceleration response spectrum analysis using the modal superposition method. Table 2 shows the calculated first 6 order frequencies of each equipment.

Table 2. The results of the first 6 modes calculation for each equipment.

Type of equipment	Frequency/Hz					
	order1	order2	order3	order4	order5	order6
Equipment 1	1.066	1.090	6.292	6.362	11.286	11.380
Equipment 2	0.903	5.279	14.400	16.817	27.461	37.461
Equipment 3	1.107	1.114	7.181	7.195	12.779	12.796

From table 2, it can be seen that among the three types of equipment, the fundamental frequency of Equipment 2 is the smallest and the fundamental frequency of Equipment 3 is the largest, and the fundamental frequencies of the three types for equipment are between 0.903 Hz and 1.107 Hz.

2.4. Seismic Load Calculation

The peak seismic input acceleration is taken as 0.4g, and the standard response spectrum of acceleration input for model analysis is shown in figure 3, and the seismic calculation results of some of the schemes are shown in figure 4~6.

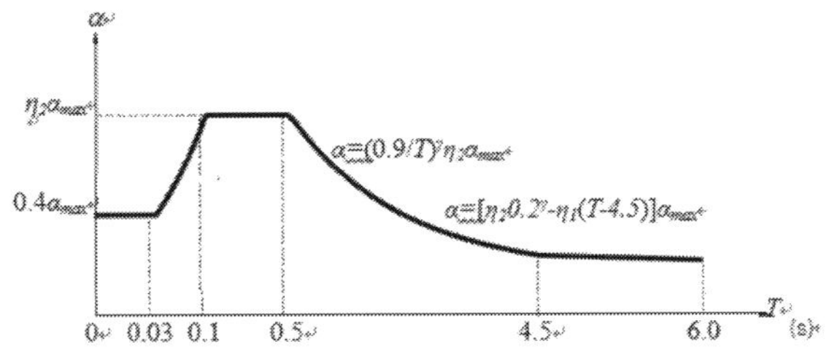

Figure 3. Input response spectrum.

Table 3. The result of earthquake resistance calculation of equipment.

Type of equipment	Compose	Direction of excitation	Maximum stress (MPa)	Maximum displacement (mm)
Equipment 1	lightning arrester	X	55.43	509.2
		Y	54.46	521.1
	insulator	X	43.23	509.2
		Y	47.01	521.1
Equipment 2	lightning arrester	X/Y	45.39	617.2
	insulator	X/Y	62.72	617.2
Equipment 3	lightning arrester	X	35.26	515.6
		Y	35.15	520.5
	insulator	X	28.65	515.6
		Y	30.85	520.5

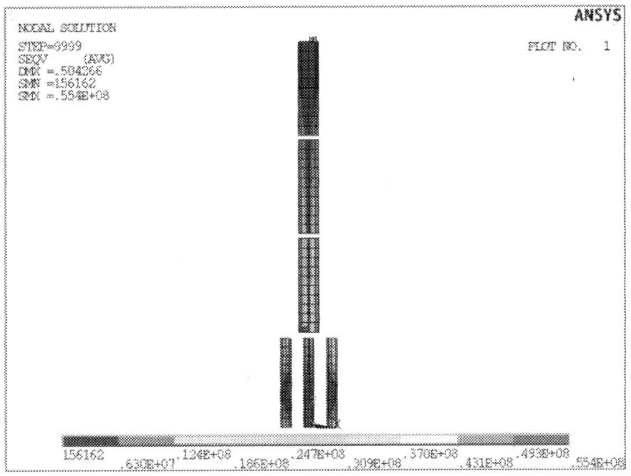

Figure 4. X-direction stress Nephogram of equipment 1.

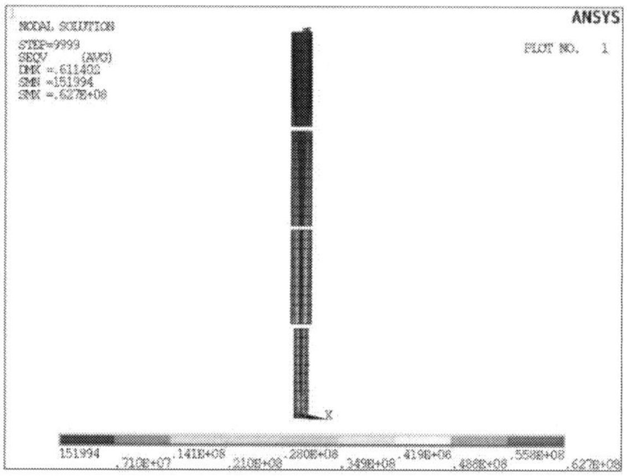

Figure 5. Stress Nephogram of equipment 2.

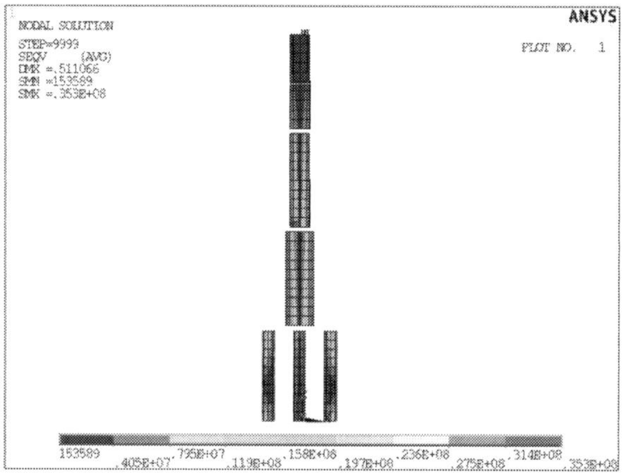

Figure 6. X-direction stress Nephogram of equipment 3.

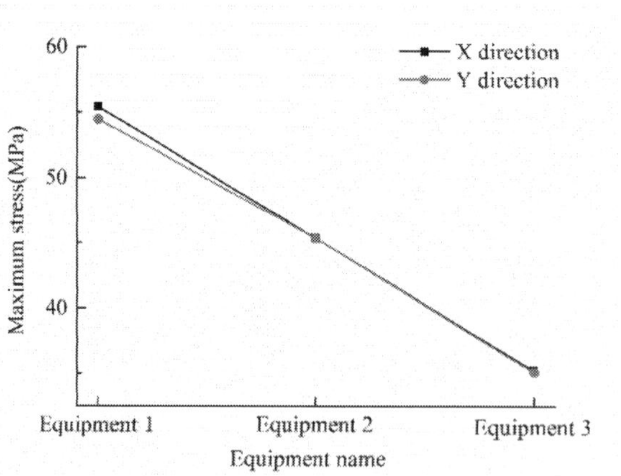

Figure 7. Maximum stress value of lightning arrester.

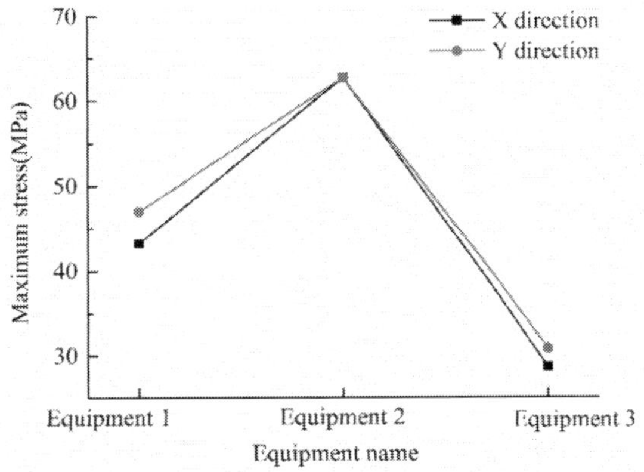

Figure 8. Maximum stress value of insulator.

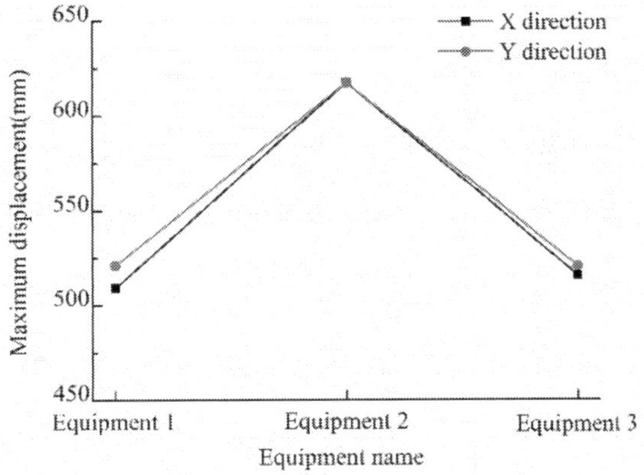

Figure 9. Maximum displacement value of top of the equipment.

From table 3 and figure 4 to Figure 9, it can be found that the maximum stress values of the top casing for equipment 1, equipment 2 and equipment 3 are 55.43 MPa, 45.39 MPa and 35.26 MPa, and the maximum stresses of equipment 1 and equipment 2 appear at the root of the third casing section (from top to bottom).The maximum stress of the casing in equipment 3 under X-direction excitation appeared at the root of the third casing section, and the maximum stress of the casing under Y-direction excitation appeared at the root of the second casing section; the maximum stress values of the insulators in the lower part of the three devices were 47.01 MPa, 62.72 MPa and 30.85 MPa, the maximum stresses appeared at the top of the insulators, except for the maximum stress at the root of equipment 2.

In the seismic displacement response, the maximum displacement of the top of the casing for equipment 2 is 617.2 mm, and the minimum displacement at the top of the casing for equipment 1 is 509.2mm. The difference between the maximum displacement at the top of the casing for equipment 1 and equipment 3 is very small, and the difference is less than 10 mm, i.e., the same arrangement has little effect on the displacement for the top of the equipment.

The main reason for the difference in seismic performance of each equipment is that Equipment 1 and Equipment 3 use three insulators in parallel, which increases the stiffness of the bottom of the equipment, and therefore it has better seismic performance than the structure of Equipment 2. The weight of each section of the casing for Equipment 3 is greater than the weight of the corresponding casing for Equipment 1, and Equipment 3 increases the size of the insulators, thus improving the seismic performance of the equipment on the whole. Equipment 3 has best seismic performance from the stress aspect.

3. Conclusion

For the numerical model of 750kV lightning arrester combination by casing and pillar insulator, modal calculation and seismic calculation are carried out, and the following conclusions can be obtained by analyzing the calculation results of the three devices.

(1) Among the three devices, the fundamental frequency of equipment 1 is the smallest, and the fundamental frequency value is 0.903 Hz, the fundamental frequency of equipment 2 is 1.066 Hz, and the fundamental frequency of equipment 3 is 1.107 Hz.

(2) The maximum stress values for the casing roots of the three devices is respectively 55.43 MPa, 45.39 MPa, and 35.26 MPa for equipment 1, equipment 2, and equipment 3, The maximum stress values of insulators respectively is 47.01 MPa, 62.72 MPa, and 30.85 MPa for equipment 1, equipment 2, and equipment 3. The equipment 3 has the best seismic performance.

(3) The top displacement of the casing for equipment 2 is the largest, and the top displacement of the casing for equipment 1 is the smallest. The difference between the maximum displacement at the top of the casing for equipment 1 and equipment 3 is very small, the difference is less than 10mm, the top displacement is not greatly affected by using the same arrangement form.

References

[1] Ersoy S, Saadeghvaziri M A. 2004 Seismic response of transformer bushing systems. *IEEE Transactions on* Power *Delivery* **19** (1) 131-137.
[2] Filiatrault A, Matt H. 2006 Seismic response of high voltage electrical transformer bushing systems [J] *Journal of Structural Engineering* **132** (2) 287-295.
[3] Gilani A S, Whittaker A S. 2001 Seismic evaluation and retrofit of 230kV porcelain transformer bushings *Earthquake Spectra* **17** (4) 597-616.
[4] Filiatrault A, Matt H. 2005 Experimental seismic response of high-voltage transformer bushing system *Earthquake Spectra* **21** (4) 1009-1025.
[5] Knight B T, Kempner L J. 2009 Vulnerabilities and retrofit of high-voitage electrical substation facilities Lifeline Earthquake Engineering in a Multihazard Environment, California, United States 232-243.
[6] Muhammad F. 2013 Seismic evaluation and qualification of transformer bushings. New York, USA: The State University of New York at Buffalo 20-48.
[7] Zhou S P. 2008 Transformer damage analysis in the 5.12 Wenchuan Earthuake. Hubei Electric Power **32** (4) 1-4.
[8] Fan R Q. 2009 Study of the characteristics of composite insulators. *Electric Power* **32** (2) 60-62.
[9] Zhang J, Qi L Z, Li K W, et al. 2011 Shock test and finite element analysis of porcelain High-voltage electrical equipment *Electric Power Construction* **32** (7) 6-10.

Creep Behaviour of Reinforced Concrete Beam under Long-Term Bending

Jianda Xin[1,2,3], Yi Liu[1,2,3], Guoxin Zhang[3], Juan Wang[3], Zhenhong Wang[1,2,3] and Wenqian Hou[3]

[1] State Key Laboratory of Simulation and Regulation of Water Cycle in River Basin,China Institute of Water Resources and Hydropower Research, Beijing 100038, China;
[2] Key Laboratory of Construction and Safety of Hydraulic Engineering of Ministry of Water Resources, China Institute of Water Resources and Hydropower Research, Beijing100038, China;
[3] Department of Structure and Materials, China Institute of Water Resources and Hydropower Research, Beijing 100038,China
Email: xinjd@iwhr.com

Abstract. Concrete structure after casting will be subjected to external load, peripheral components and other constraints, the overall performance of the structure under a long-term loading state will change continuously. In practical engineering, reinforcement is usually calculated and configured based on external load, but the role of reinforcement in the continuous loading process of concrete structure is not fully considered. Based on the elastic creep theory, this paper deduces the concrete creep calculation formula considering the influence of reinforcement, and analyses the law of the influence of reinforcement ratio and loading age on the strain and stress of concrete. The results show that the reinforcement can effectively restrain the deformation of concrete and reduce the load level of concrete.
Keywords. Reinforced concrete; Creep; Age; Reinforcement ratio.

1. Introduction

Concrete as a kind of structural material, has a very rapid application development and has become an indispensable part of construction engineering. Concrete will have creep effect under continuous load [1]. This effect will affect the service life and performance of concrete structures: on the one hand, creep can cause stress relaxation and reduce the risk of concrete cracking; On the other hand, creep will cause the loss of profit of prestressed concrete structure, and adversely affect the normal use of the structure. Therefore, it is particularly necessary to study the mechanical properties of concrete under the long-term loading.

In recent years, many scholars have carried out theoretical analysis and experimental research on concrete creep characteristics, and achieved a lot of results [2-4]. Mu et al. [5] conducted creep analysis on steel-reinforced concrete beam members and studied the variation rules of parameters such as strain and section bending moment. Xue et al. [6] studied the creep of prestressed reinforced concrete beam members, taking into account the influence of concrete types and prestressed tendons. Pan et al. [7] studied the influence of reinforcement on the shrinkage and creep properties of HPC. Koga et al. [8] studied the beam deflection and crack width of prestressed reinforced concrete beams under long-term load. Mari. Antonio et al. [9] analyzed the variation rule of deflection of cracked reinforced concrete beams under long-term load in tension zone.

Generally, the role of reinforcement is to limit crack width and improve bending deflection [10], but its "crack resistance" role is not obvious [11], so it is difficult to give full play to the advantages of reinforcement. In China's existing specifications [12], the configuration of steel bars is only to consider its "crack limit" function, with a certain number of temperature reinforcement. As a matter of fact, the influence of creep cannot be ignored because the concrete is in the state of loading for a long time after casting. If the influence of reinforcement is considered, the stress - strain state of concrete is more concerned.

Therefore, based on the basic theory of elastic creep, this paper deduces the calculation formula of concrete beam creep considering the influence of reinforcement, analyzes the variation law of strain and load of reinforcement and concrete under different loading, and provides a theoretical basis for the application of reinforcement in engineering.

2. Creep Theory of Concrete

The creep characteristics of concrete were discovered in the early 20th century, and since then, relevant computational theories have made great progress, including effective modulus method, aging theory and elastic creep theory, etc. [1].

Elastic creep theory is based on the assumption that there is a linear relationship between deformation and stress. When stress changes, the total creep deformation $\varepsilon_{total}(t)$ at age t can be calculated according to the sum of the creep deformation caused by the corresponding stress increment, which is the so-called superposition principle

$$\varepsilon_{total}(t)=\sigma(t_0)[\frac{1}{E(t_0)}+C(t,t_0)]+\int_{t_0}^{t}[\frac{1}{E(\tau)}+C(t,\tau)]d\sigma(\tau) \tag{1}$$

where $\sigma(t_0)$ is the initial stress of concrete; $E(t_0)$ is the initial elastic modulus of concrete; $C(t,t_0)$ is the specific creep of concrete at age of t with a loading age of t_0.

3. Creep of Concrete Considering Effect of Reinforcement

3.1. Basic Assumption

The basic condition that reinforcement and concrete can form a composite structural material is that there is a reliable bond between them. For newly cast reinforced concrete structures, the deformation of reinforcement and concrete coordinated under the early age loading is an important factor to be considered. Mimura et al. [13] conducted pull tests on reinforced concrete specimens at different ages, and found that under low load levels, the strains of reinforcement and concrete were basically the same, and the bond force at the initial stage of pouring could ensure the two to work together. Therefore, the basic assumptions of formula derivation in this paper are as follows:

(1) The section of the specimen conforms to the assumption of plane section;
(2) There is no slip between reinforcement and concrete;
(3) The creep properties of concrete are consistent;
(4) Concrete does not crack.

3.2. Deformation Behavior of Reinforced Concrete

After the load is applied, the load will be transferred to the reinforcement with time due to the creep effect of concrete. By dividing the whole calculation time into small periods, the final strain value of concrete can be obtained through continuous cyclic calculation. The elastic strain increment and creep strain increment of concrete in each time period are [14]

$$\Delta\varepsilon_n^e=\int_{t_0}^{t}\frac{1}{E(\tau)}\frac{d\sigma}{d\tau}d\tau=\frac{\Delta\sigma_n}{E(\bar{\tau}_n)} \tag{2}$$

$$\Delta\varepsilon_n^c = \eta_n + \Delta\sigma_n C(t_n, \bar{\tau}_n) \quad (3)$$

$$\Delta\varepsilon_n = \Delta\varepsilon_n^e + \Delta\varepsilon_n^c \quad (4)$$

where

$$\Delta\sigma_n = \bar{E}_n(\Delta\varepsilon_n - \eta_n) \quad (5)$$

$$\Delta\bar{E}_n = \frac{E(\bar{\tau}_n)}{1 + E(\bar{\tau}_n)C(t_n, \bar{\tau}_n)} \quad (6)$$

$$\eta_n = \sum (1 - e^{-r_s \Delta t_n})\omega_{sn} \quad (7)$$

$$\omega_{sn} = \omega_{s,n-1} e^{-\gamma_s \Delta t_{n-1}} + \Delta\sigma_{n-1}\psi_s(\bar{\tau}_{n-1})e^{-0.5\gamma_s \Delta t_{n-1}} \quad (8)$$

$$\omega_{s1} = \Delta\sigma_0 \psi_s(\tau_0) \quad (9)$$

$$\psi_s = \sum f_i + g_i \tau^{-p_i} \quad (10)$$

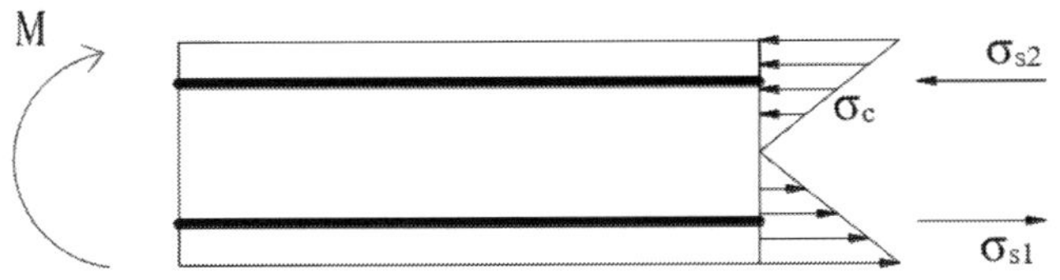

Figure 1. Stress distribution of reinforced concrete beam.

Figure 1 shows the stress distribution of reinforced concrete beam. The moment M together undertaken by concrete and reinforcement can be expressed by

$$M = [\frac{bh^2}{6} - \frac{\pi D^4}{8h} - \frac{\pi D^2(h-2a_s)^2}{2h}]\sigma_c + A_s'\sigma_s'(\frac{h}{2} - a_s) + A_s\sigma_s(\frac{h}{2} - a_s) \quad (11)$$

where b is width of beam; h is height of beam; D is diameter of reinforcement; σ_c is the concrete stress at outermost edge of compression zone; σ_s', σ_s, A_s' and A_s is stress and area of reinforcement at compression and tension zone, respectively. a_s is the distance between the gravity center of reinforcement and outermost edge of concrete.

The stress and strain of reinforcement at compression and tension zone is same due to symmetrical reinforcement

$$\varepsilon_s' = \varepsilon_s \quad (12)$$

$$\sigma_s' = \sigma_s \quad (13)$$

$$\varepsilon_s' = \frac{h - 2a_s}{h}\varepsilon_c' \quad (14)$$

Therefore, Equation (11) can be rewritten as

$$M = [\frac{bh^2}{6} - \frac{\pi D^4}{8h} - \frac{\pi D^2(h-2a_s)^2}{2h}]\sigma_c + A_s'\sigma_s'(h - 2a_s) \quad (15)$$

The increment of moment ΔM is

$$\Delta M = [\frac{bh^2}{6} - \frac{\pi D^4}{8h} - \frac{\pi D^2 (h-2a_s)^2}{2h}]\Delta\sigma_c + A_s'\Delta\sigma_s'(h-2a_s) \tag{16}$$

The increment of reinforcement strain is

$$\Delta\varepsilon_s = \frac{h-2a_s}{h}\Delta\varepsilon_c \tag{17}$$

Combining Equations (3), (16) and (17), one can obtain

$$\Delta\sigma_c = \frac{\Delta M - A_s E_s \eta \frac{h-2a_s}{h}}{[\frac{bh^2}{6} - \frac{\pi D^4}{8h} - \frac{\pi D^2 (h-2a_s)^2}{2h}] + \frac{A_s E_s}{E_n}\frac{(h-2a_s)^2}{h}} \tag{18}$$

$$\Delta\sigma_s' = \frac{\Delta M + [\frac{bh^2}{6} - \frac{\pi D^4}{8h} - \frac{\pi D^2 (h-2a_s)^2}{2h}]\overline{E}_n \eta}{[\frac{bh^2}{6} - \frac{\pi D^4}{8h} - \frac{\pi D^2 (h-2a_s)^2}{2h}]\overline{E}_n + A_s E_s \frac{(h-2a_s)^2}{h}} \frac{E_s(h-2a_s)}{h} \tag{19}$$

Based on the above formula, the variation rules of the parameters such as the strain and stress of reinforcement and concrete at different times can be calculated by programming.

4. Case Analysis

Figure 2 shows the dimensions of reinforced concrete beam. $b = 140mm$, $h = 160mm$, $a_s = 37mm$, $D = 14mm$, and the reinforcement ratio is 2%. According to CEB-FIP [15], the evolution of elastic modulus $E_c = 26.51\sqrt{e^{0.25(1-\sqrt{\frac{28}{t}})}}\ GPa$, the elastic modulus of reinforcement $E_s = 200GPa$. Based on the design code for hydraulic concrete structures [12], the cracking force of reinforced concrete N_{cr}=10.5kN, so the loading force is 0.5 N_{cr} , and the applied moment M_{cr}=1.136$kN \cdot m$. In the current case, the loading age is 28d, and the loading duration is 200d. The specific creep is[14]:

$$C(t,\tau) = \sum_{j=1}^{2}(f_i + g_i \tau^{-p_i})[1 - e^{-r_i(t-\tau)}] \tag{20}$$

Parameters in Equation (20) take the specific value shown in table 1.

Table 1. Parameters for specific creep of concrete.

Parameter	$f_i\,(10^{-4})$	$g_i\,(10^{-4})$	p_i	r_i
i=1	0.1045	0.9614	0.45	0.3
i=2	0.2360	0.4012	0.45	0.005

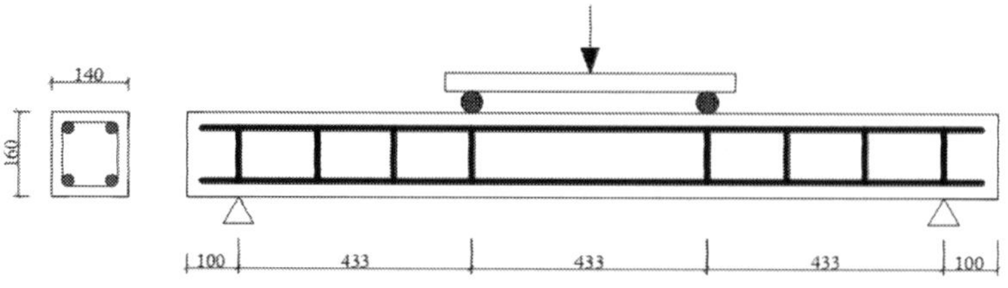

Figure 2. Dimensions of reinforced concrete beam.

Based on the assumed conditions of the case, the strain and stress of concrete and reinforcement in the compression zone and tension zone of reinforced concrete specimen are equal, but the signs are opposite. Therefore, the compression zone of the specimen is taken as the analysis object below.

4.1. Strain Evolution

Figure 3 shows the strain curves of the outermost concrete and adjacent reinforcement in the compression zone. It can be seen that with the growth of age, the strain of reinforcement and concrete increases rapidly in the early stage and gradually slows down in the later stage, which is consistent with the creep change law of concrete. Component separation of concrete strain is shown in figure 4. Due to the creep of concrete, the elastic strain of concrete strain generated by initial load is 58με, which increases to 109με after 200d of loading duration.

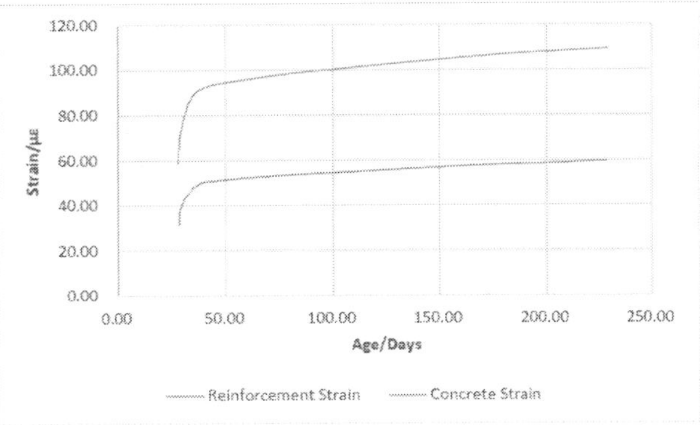

Figure 3. Strain curves of the outermost concrete and adjacent reinforcement in the compression zone.

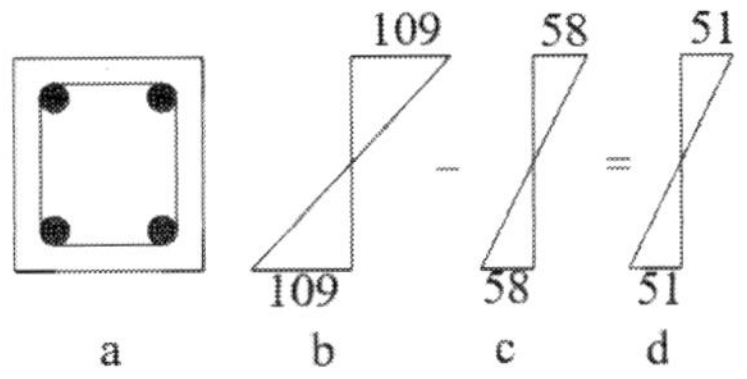

Figure 4. (a) Section of reinforced concrete beam, (b) Strain distribution after 200d loading, (c) Instantaneous elastic strain distribution, and (d) Creep strain distribution.

4.2. Stress Evolution

The creep characteristic of concrete causes stress redistribution of reinforced concrete specimens after loading. The stress of concrete decreases continuously, while the stress level of reinforcement increases continuously, as shown in figure 5. Similar to the variation curve of strain, the stress curve of reinforcement and concrete also shows the characteristics of rapid growth in the early stage and gradual growth in the late stage. The stress of concrete decreases from 1.54 to 1.17MPa, and the decrease ratio is about 23.7%. The stress of the reinforcement increases by nearly two times compared with that at the initial loading.

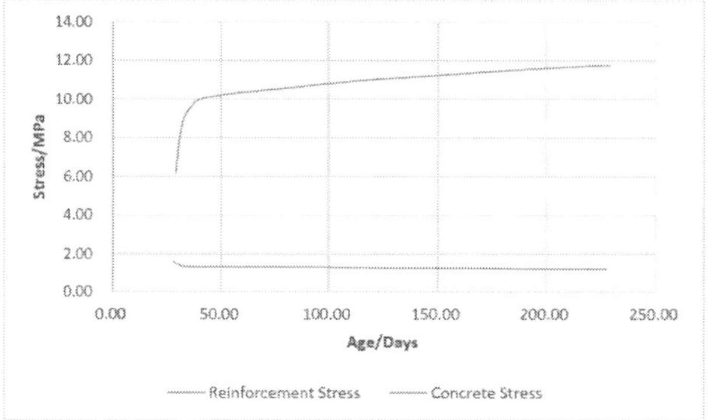

Figure 5. Stress curves of the outermost concrete and adjacent reinforcement in the compression zone.

4.3. Stress Transfer Ratio

Figure 6 shows the change curve of concrete stress transfer ratio with age. With the decrease of concrete stress level, the bending moment of concrete itself is gradually transferred to the reinforcement, and the stress transfer ratio tends to moderate with the increase of loading time. After 200 d loading duration, the transfer ratio of concrete stress is about 23.7%.

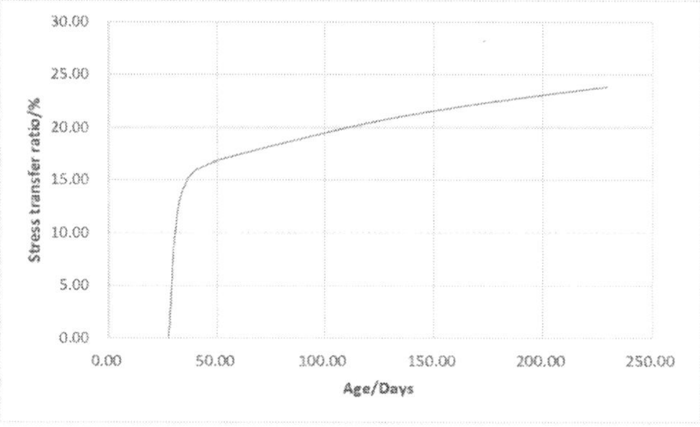

Figure 6. Concrete stress transfer ratio.

5. Conclusions

In this paper, formula for calculating the stress change of reinforced concrete beams considering creep is derived. The calculation results show that, due to creep characteristics of concrete, the stress transfer ratio of concrete increases with the loading duration, which increases rapidly in the early stage and tends to be gentle in the later stage. After 200 d loading duration, the transfer ratio of concrete stress is about 23.7%.

References

[1] Huang G X, Hui R Y, Wang X J. 2011 Creep and shrinkage of concrete. Beijing: China Electric Power Press.

[2] Ranaivomanana N, Multon S, Turatsinze A. 2013 Basic creep of concrete under compression, tension and bending. Construction and Building Materials 38 173–180.

[3] Rossi P, Tailhan J L, Maou F L, Gaillet L, Martin E. 2012 Basic creep of concretes investigation of the physical mechanisms by using acoustic emission. Cement and Concrete Research **42** 61–73.

[4] Reinhardt H W, Rinder T. 2006 Tensile creep of high-strength concrete. *Journal of Advanced Concrete Technology* **4** (2) 277-283.

[5] Mu G B, Wang L G, Wang C. 2007 Creep effect analysis of steel reinforced concrete beam. Journal of Northeastern University (Natural Science) **28** (5) 733-736.
[6] Xue W C, Hu Y M, Wang W. 2008 Experiment on creep behaviors of prestressed concrete beams *China Journal of Highway and Transport* **21** (4) 61-66.
[7] Pan Z F, Lv Z T, Meng S P. 2009 Experimental study on the influence of steel on high-strength concrete creep and shrinkage *China Civil Engineering Journal* **42** (2) 11-16.
[8] Koga H, Watanabe H, Takeuchi Y, Aoyama H , Kitano Y. 2009 Experimental study on the time dependent flexural behavior of prestressed reinforced concrete beams. Creep, Shrinkage and Durability Mechanics of Concrete and Concrete Structures-Proceedings of the 8th Int. Conference on Creep, Shrinkage and Durability Mechanics of Concrete and Concrete Structures 1 781-786.
[9] Antonio R, Bairán J M, Duarte N. 2010 Long-term deflections in cracked reinforced concrete flexural members *Engineering Structures* **32** (3) 829-842.
[10] Piyasena R, Loo Y C, Fragomeni S. 2003 Determination of crack spacing and crack width in reinforced concrete beams. Structural Engineering and Mechanics **15** (2) 159-180.
[11] Song W, Yuan Y, Gong J. 2002 Experimental research on tensile performance of reinforced concrete *Journal of Southeast University* (Natural Science Edition) 32 Sup.:98-101.
[12] SL 191-2008. Design code for hydraulic concrete structures. Beijing: China Water & Power Press, 2009.
[13] Mimura Y, Yoshitake I, Zhang W. 2011 Uniaxial tension test of slender reinforced early age concrete members. Materials 4:1345-1359.
[14] Zhu B F. 2009 Thermal stresses and temperature control of mass concrete. Beijing: China Water & Power Press.
[15] CEB-FIP. International recommendations for the design and construction of concrete structures. London: Cement and Concrete Association, 1970.

Analysis on Dynamic Characteristics and Seismic Response of Wooden Structure of South Gate Building in Jiangzhang Town

Xiancai Ren[1], Zhaobo Meng[1*], Xin Wang[2], Feifei Gao[1] and Ting Zhang[1]

[1] School of Civil Engineering, Liaocheng University, Shandong Province, China
[2] School of Civil Engineering, Tianshui Normal University, Gansu Province, China
Email: mengzhaobo@lcu.edu.cn

Abstract. In order to study the seismic performance of ancient wooden structures with single eaves and beam lifting in China, the finite element model of the upper floor of the south gate of Jiangzhang town in Shanxi Province was established by using ANSYS. Through modal analysis, the main frequencies and modes of the south gate of Jiangduan were obtained. Through the seismic response analysis of the south gate tower model, the displacement and acceleration response curves of the top nodes of the outer eaves column, golden column and through column under various working conditions of the South gate tower are obtained. The results show that the first and second order frequencies of the South Gate tower model are 1.830Hz and 1.855Hz, and the first two order modes are mainly transitional. With the increase of seismic excitation, the displacement and acceleration response of the top joints of the outer eave column, golden column and through column increase.
Keywords. South Gate Tower; Dynamic characteristics; Seismic response.

1. Introduction

The South Gate Tower in Jiangzhang Town, Shaanxi Province is a masonry high platform structure with a wooden structure "Sanyi Temple" on it. It is a key cultural relic protection object in my country and has important research significance.

In recent years, scholars at home and abroad have done a lot of research on the dynamic characteristics and seismic response of ancient timber structures.Zhaobo Meng [1] conducted field measurement of traffic vibration of Xi 'an bell tower, used ITD method to obtain the main dynamic parameters of the bell tower, and performed the stiffness of mortise-and-tenon joints of the wooden structure of the bell tower.Suzuki [2] made the overall model of Japanese traditional wood structure and conducted shaking table test, and the study showed that wood structure had good seismic performance.In the aspect of numerical simulation, the whole finite element model of wooden structure is established by large finite element software, and its seismic performance is analyzed. Xueqi Chen [3] studied the influence of external rammed earth wall and internal wood frame on the overall seismic performance of Fujian earth buildings by using the method of finite element numerical simulation, and discussed the difference between square and round earth buildings in seismic performance.

In order to accurately evaluate the seismic performance of ancient wooden structure with single eaves and beam, this paper takes shanxi province town south gate upstairs born of wood structure as the research object, is firstly established by ANSYS finite element software of south gate upstairs wood structure finite element model for modal analysis for model fundamental frequency compared with standard formula to prove the validity of the model, finally carries on the seismic response analysis, The results can provide theoretical basis for earthquake protection of the south Gate building and the ancient buildings in the same area.

2. Numerical Model Establishment

2.1. Introduction To South Gate Tower

Nanmen Tower was built in the twelfth year of Daoguang in the Qing Dynasty and rebuilt in 1998. It is located in Jiangzhang Town, Fufeng County, Shaanxi Province. The south gate building is a masonry high platform base structure, on which there is a wooden structure "Sanyi Temple". The "Sanyi Temple" is a single-eave raised beam wooden structure, the size is 13.3m×8.2m. There are 18 outer eaves columns, inner gold There are 10 columns with a diameter of 0.4m and a height of 4.1m. The main beams are evenly distributed, and the wooden structure is divided into three.

2.2. Unit and Material Parameter Selection

The beam-column frame is simulated by BEAM188 beam element. Tenon and tenon joints and bucket arch joints are simulated by COMBIN14 and COMBIN39 spring elements. With reference to related researches on ancient buildings in similar areas, the stiffness values of tenon-mortise joints and bucket arch joints are based on the existing research results in literature [5]. The roof is simulated by mass unit MASS21. Column foot COMBIN39 unit to simulate, it stiffness value refers to the data of the study on the mechanical model of the column foot joints by Yi Pan [6]. The wood used in this paper is poplar, and the density of the wood is 410kg/m3. Refer to Professor Maohong Yu [7] the measured values of Xi'an Jianlou wood as the material parameters of the upper timber structure of the South Gate Tower. The specific wood material parameters are shown in table 1.

Table 1. Wood physical parameters.

Elastic Modulus （MPa）			Poisson's ratio			Shear modulus （MPa）		
E_L	E_R	E_T	μ_{LR}	μ_{LT}	μ_{RT}	G_{LR}	G_{LT}	G_{RT}
8301	538	269	0.25	0.035	0.02	623	498	149

3. Dynamic Characteristics and Seismic Response Analysis

3.1. Dynamic Characteristics

The Block Lanczos algorithm is used to extract the first 6-order natural frequency and natural vibration period of the upper wooden structure model of the South Gate Tower, as shown in table 2. In this analysis, the x-direction is horizontal, y-direction is vertical, and z-direction is vertical.

Table 2. Modal analysis results.

Mode shape	1	2	3	4	5	6
Natural frequency (Hz)	1.830	1.855	2.314	2.842	2.857	3.356
Natural vibration period (S)	0.546	0.539	0.432	0.352	0.350	0.298

As shown in figure 1, the first two modes of the upper wooden structure of the South Gate Tower are horizontal vibration in the X and Z directions; the third mode is torsional vibration around the Y axis. It can be seen from Table 2 that the basic period of the upper wooden structure of the South Gate Building obtained by finite element analysis is 0.546s, which is slightly higher than 0.44s calculated by the empirical formula provided by literature [8], but the error is within an acceptable range. It can be seen from the period calculation formula $T = 2\pi / \sqrt{k/m}$, The main factor affecting the natural vibration period of the structure is the stiffness and quality of the structure. The main reason is that the influence of the wall is not considered in the modeling process. In practice, the wall also provides a certain lateral stiffness to the wooden structure itself. It can be seen from figure 1 that the changes in the first three-order modes are in line with reality. Therefore, the above analysis shows that the finite element model is more reasonable and accurate, and the seismic time history analysis can be performed on it.

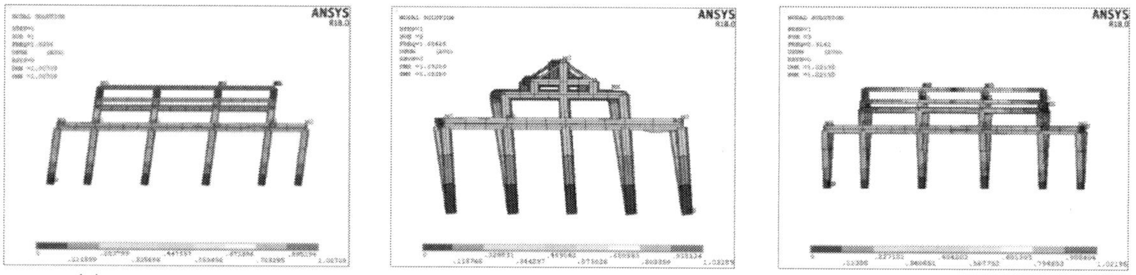

(a) First-order mode (b) Second-order mode (c) Third-order mode

Figure 1. The first three-order mode of the Nanmen Tower.

3.2. Seismic Wave Selection and Damping Determination

According to the "Code for Seismic Design of Buildings" (GB 5001-2010), the location of the South Gate Tower is located at an 8 degree seismic fortification, the basic acceleration is 0.2g, the site category is Class II, the design earthquake is divided into the second group, and the characteristic period is 0.40s . In this paper, EL-Centro waves, Taft waves and artificial waves are selected as the ground motion input. During the time history analysis, adjust the peak acceleration of the seismic wave to 70gal (for frequent earthquakes), 200gal (fortified earthquakes), and 400gal (rare earthquakes).

Using Rayleigh damping as the structural damping of the finite element model of the South Gate Tower [1]. Because the low-order mode shape contributes a lot, it is generally confirmed by the first two modes. In the seismic analysis of the structure, often take $\xi_i = \xi_j = 0.05$,determine the constant α、β according to the first two-order natural circular frequency of the structure, get $\alpha = 0.576$, $\beta = 0.00434$.

3.3. Acceleration Response Analysis

The seismic response of the upper wooden structure of the South Gate Building can be obtained by inputting the three kinds of seismic wave ex citations with different acceleration peaks into the finite element model. Figure 2 shows the acceleration response time history curves of the outer eave pillar top, the gold pillar top, and the through pillar top under the excitation of EL-Centro wave. See Table 3 for the acceleration peaks and dynamic amplification coefficients of the outer eave pillar tops, gold pillar tops, and through pillar tops under different acceleration peak seismic waves. In Table 3, E, T, and R correspond to EL-Centro waves, Taft waves, and artificial waves, respectively. 70, 200, and 400 correspond to seismic waves with acceleration peaks of 70 gal, 200 gal, and 400 gal.

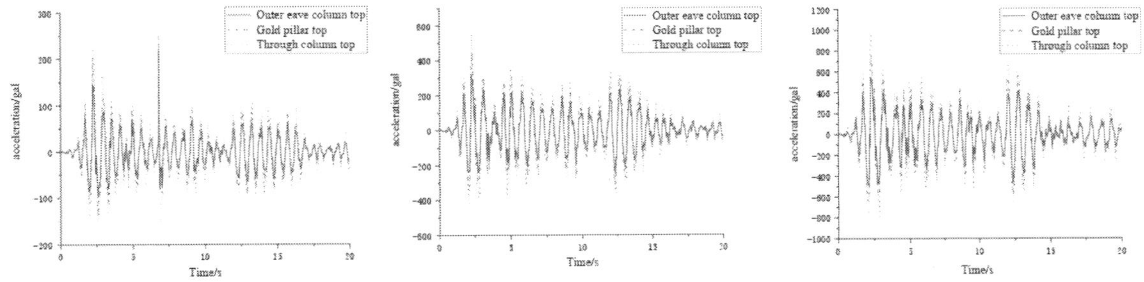

(a)70gal peak acceleration seismic wave excitation (b)200gal peak acceleration seismic wave excitation (c)200gal peak acceleration seismic wave excitation

Figure 2. Acceleration response of outer eave pillar top, gold pillar top, and through pillar top under the excitation of EL-Centro wave.

Table 3. Acceleration response peak values and dynamic amplification factors of outer eave pillar tops, gold pillar tops, and through pillar tops under different seismic waves.

		E70	E200	E400	T70	T200	T400	R70	R200	R400
Outer eaves column top	Acceleration(cm/s^2)	220.8	321.4	546.5	181.3	410.1	736.2	187.5	306.3	666.1
	amplification factor	3.16	1.61	1.37	2.59	2.05	1.84	2.68	1.53	1.67
Golden pillar top	Acceleration(cm/s^2)	247.9	438.0	757.6	236.3	530.2	974.4	228.1	399.5	869.1
	amplification factor	3.54	2.19	1.89	3.38	2.65	2.44	3.26	1.99	2.17
Column top	Acceleration(cm/s^2)	260.8	551.3	996.7	270.7	634.7	1210.5	233.8	479.3	1037.9
	amplification factor	3.73	2.76	2.49	3.87	3.17	3.03	3.34	2.40	2.60

It can be seen from figure 4 that under the action of the same seismic wave, the shape of the acceleration time history curve of each node is very close, and the time point of the acceleration peak is basically the same. It can be seen from Table 3 that with the increase of seismic excitation, the acceleration peak value of the same floor height becomes larger and larger; under the same intensity seismic excitation, as the height increases, the acceleration peak value and the amplification factor become larger and larger. The acceleration amplification factor of the top is between 1.37-3.16, the top of the gold column is between 1.89-3.54, and the top of the through column is between 2.40-3.87. Among them, the peak acceleration response of the top of the column under the action of the EL-Centro wave, the Taft wave and the artificial wave with the acceleration peak of 400gal are 996.70gal, 1210.53gal, 1037.96gal, and the amplification factors are 2.49, 3.03, and 2.60 respectively. Waves have the greatest impact on the South Gate Tower, followed by artificial waves, followed by EL-Centro waves, which may be related to the different spectral characteristics of seismic waves; under different seismic wave excitation, the acceleration amplification factor gradually decreases with the increase of the seismic wave peak value. Because the upper wooden structure of the South Gate Tower increases with the peak acceleration of the seismic wave, the energy consumption and shock absorption of the tenon and tenon joints and the bucket arch joints increase.

3.4. Displacement Response Analysis

Figure 3 shows the displacement time history curve of each node under the action of EL-Centro wave.

The peak displacement response and maximum inter-layer displacement angle of each node are shown in table 4.

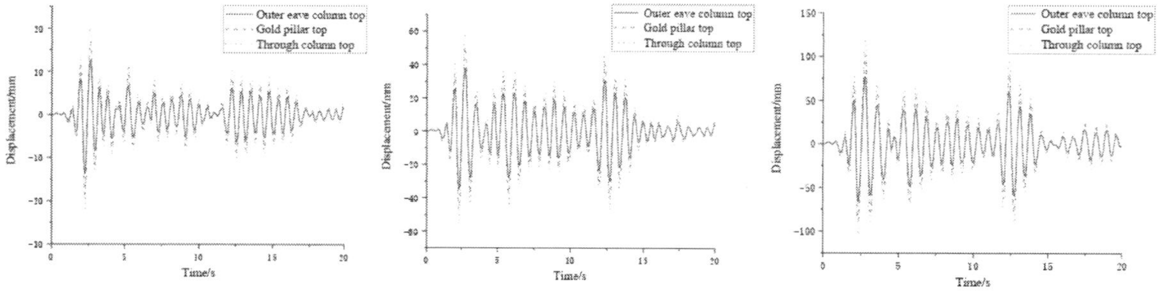

(a)70gal peak acceleration seismic wave excitation (b)200gal peak acceleration seismic wave excitation (c)200gal peak acceleration seismic wave excitation

Figure 3. Displacement response of outer eave column top, gold column top, and through column top under EL-Centro wave excitation.

Table 4. Displacement response peak values and inter-layer displacement angles of outer eave pillar tops, gold pillar tops, and through pillar tops under different seismic waves.

		E70	E200	E400	T70	T200	T400	R70	R200	R400
Outer eaves column top	Displacement (mm)	13.8	37.9	77.3	19.1	49.8	79.02	16.4	31.8	68.9
	Layer displace -ment angle	1/217	1/79	1/39	1/157	1/60	1/38	1/183	1/94	1/44
Golden pillar top	Displacement (mm)	18.6	50.3	102.1	25.5	66.1	104.87	21.9	42.2	90.9
	Layer displace -ment angle	1/231	1/90	1/45	1/173	1/68	1/43	1/202	1/107	1/51
Column top	Displacement (mm)	22.3	58.9	120.4	29.7	77.7	123.67	25.4	49.4	107.9
	Layer displace -ment angle	1/299	1/129	1/60	1/263	1/95	1/59	1/316	1/153	1/65

It can be seen from Figure 5 that under the action of the same seismic wave, the shape of the displacement time history curve of each node is very close, and the time point of the displacement peak is basically the same. It can be seen from table 4 that under the action of seismic waves of different amplitudes of the upper wooden structure of the South Gate Building, the displacement peaks generally show a gradually increasing trend from bottom to top. Under the action of 400gal rare earthquake, the maximum inter-story displacement angle of the upper wooden structure column frame of the South Gate Tower is 6/353, which meets GB 50165-2020, "Technical Specification for Maintenance and Reinforcement of Ancient Building Wood Structures" [12] for the maximum floor The displacement angle does not exceed the limit of 1/30.

4. Conclusion

(1) The modal analysis of the numerical model of the Nanmen Tower is carried out, and it is determined that the main vibration modes of the Nanmen Tower are mainly transitional, and the first and second order frequencies are 1.830 Hz and 1.855 Hz.

(2) Under different seismic wave excitation, the acceleration amplification factor gradually decreases with the increase of the peak acceleration of the seismic wave, mainly because the upper

wood structure of the south gate building increases with the increase of the peak acceleration of the seismic wave, which increases the energy consumption and damping of the tenon-and-mortise joints and the bucket arch joints.

(3) Under the action of rare earthquakes, the maximum inter-story displacement angle of the upper wooden structure column frame of the South Gate Tower is 1/38, which meets the requirement of the code not exceeding 1/30.

Acknowledgement
The writers appreciate the financial support provided by National Natural Science Foundation of China (No.52068063), Graduate Natural Science Foundation of Shandong Province (ZR2020ME240).

References

[1] Meng Z B. 2009 Analysis and Assessment of the Vibration Responds Traffic-Induced of Xi'an Bell Tower [D] Xi'an University of Architecture.
[2] Suzuki Y, Maeno M. 2006 Structural mechanism of traditional wooden frames by dynamic and static tests [J] *Structural Control and Health Monitoring* **13** (1) 508-522.
[3] Chen X Q, Luo W, Hu H S, et al. 2021 Numerical Simulation Analysis on Seismic Performance of Fujian Tulou Mixed Structure[J] Journal of Huaqiao University(Natural Science) **42** (02) 165-171.
[4] Zhao S J, Yang Y Q, Dai J W. 2020 Study on dynamic characteristics and seismic response of Yuhua Pavilion in the Forbidden City[J] *World Earthquake Engineering* **36** (01) 85-92.
[5] Su J. 2008 Study on the Seismic Performance of Ancient Chinese Timber Structure [D] Xi'an University of Architecture.
[6] Pan Y, An R B, Chen J, et al. Study on mechanical model of column footing joint in ancient timber structure based on rocking column [J/OL] *Journal of Building Structures* 1-11[2021-08-12].
[7] Yu M H, Wang Y, Yu T, et al. 2009 Xi'an's ancient city walls and bells and drum towers: history, art, science[M] Xi'an: Xi'an Jiaotong University Press.
[8] GB 50165-2020, Technical specifications for the maintenance and reinforcement of ancient building wood structures[S] *Beijing: China Construction Industry Press* 2020.

Joint Inversion of Rayleigh and Love Dispersion Curves Extracted from Ambient Seismic Noise Based on Secular Function

Feng Gong[1, 2], Xiaofei Chen[1, 5,*], Youhua Fan[2], Xuefeng Liu[3] and Haibing Tang[4]

[1] Shenzhen Key Laboratory of Deep Offshore Oil and Gas Exploration Technology, Southern University of Science and Technology, Shenzhen, 518055, China
[2] School of Science, Harbin Institute of Technology, Shenzhen, 518055, China
[3] Aviation Engineering Institute, Civil Aviation University of China, Tianjin, 300300, China
[4] Guangming Administration, Shenzhen Municipal Transportation Bureau, Shenzhen, 518055, China
[5] Department of Earth and Space Sciences, Southern University of Science and Technology, Shenzhen, 518055, China
[*] Corresponding author, E-mail: chenxf@sustech.edu.cn

Abstract. Traditional multi-mode dispersion curve inversion requires correct mode discrimination. However, when the stratum contains complex structures such as low-speed soft interlayer or high-speed hard interlayer, the dispersion curve may show phenomena such as "mode kissing" and "mode jumping", which can easily cause mode misjudgment and lead to erroneous inversion results. Based on the "secular function", this paper constructs a new type of objective function applied to the inversion of dispersion curve. This objective function does not require prior mode discrimination, which effectively solves the "mode misjudgment" problem of multi-mode dispersion curve inversion. The joint inversion of Rayleigh and Love dispersion curves extracted from ambient seismic noise is used to improve the constraint of the inversion and avoid the inversion falling into a local minimum in the case of a large-scale search of parameters. Finally, a numerical simulation was performed to verify the feasibility of the new inversion method.

1. Introduction

The ambient seismic noise exploration is an important method in passive source exploration. This method does not require seismic source, without any damage to the environment, and data collection is extremely simple. The flow is to collect the ambient seismic noise of the surrounding environment and extract the dispersion curve from it. The existing noise tomography is mainly for Rayleigh wave, and there are relatively few ambient seismic noise data tomography about Love wave. The advantage of Love wave over Rayleigh is that it is more sensitive to the shallow shear wave velocity structure and can better constrain the shallow velocity structure.

In 2019, Wang proposed the Frequency-Bessel transform method for extracting the Rayleigh wave dispersion curve from the ambient seismic noise data. Compared with the traditional method, it has a clearer image of the higher-order mode [1]. In 2020, Hu used the multi-component cross-correlation coefficient to extract the Love wave dispersion curve from the ambient seismic noise data [3]. These two research results have greatly promoted the development of ambient seismic noise exploration, and the dispersion curves of Love waves and Rayleigh waves can be extracted from seismic records at the same time.

In the inversion of multi-mode dispersion curve, the high-order mode dispersion curve is prone to mode misjudgment. If it is not handled correctly, the inversion result will be greatly deviated from the real situation. In 2010, Maraschini proposed to use the dispersion function as the objective function to avoid the mode misjudgment problem in the inversion of the multi-mode dispersion curve [4]. This new objective function also greatly accelerates the calculation speed, but at the same time, compared with the traditional objective function, the new inversion objective function lacks the constraints caused by the corresponding order mode, and has more local minima. The joint inversion of Love wave and Rayleigh wave will greatly reduce the ambiguity of the inversion.

2. Secular Function of Generalized E/R Coefficients Method

2.1. Basic Theory

In 1993, Chen proposed the generalized reflection and transmission coefficient method to calculate the dispersion curve of surface waves. This method subtly eliminates the e-exponential growth term that causes the instability of numerical calculations, and completely solve the problem of high frequency instability in the calculation of the theoretical dispersion curve [5]. He proposed to improve the original algorithm of the secular function family in 2006, which can still accurately and efficiently give the surface wave calculation results when there is a low-velocity layer in the stratum model [6]. The form of the secular function family is as follows:

$$det\left\{\boldsymbol{I}-\widehat{\boldsymbol{R}}_{ud}^{(j-1)}\widehat{\boldsymbol{R}}_{du}^{(j)}\right\}=0 \quad j=1,2,\cdots,N \tag{1}$$

where I is the identity matrix; $\widehat{\boldsymbol{R}}_{ud}^{(j-1)}$ and $\widehat{\boldsymbol{R}}_{du}^{(j)}$ are the generalized reflection coefficients of the corresponding stratum, which can be determined by the generalized transmission coefficients $\widehat{\boldsymbol{T}}_u$, $\widehat{\boldsymbol{T}}_d$and the modified reflection-transmission coefficient$\boldsymbol{R}_{du}$, $\boldsymbol{R}_{ud}$, $\boldsymbol{T}_u$, $\boldsymbol{T}_d$are deduced to obtain:

$$\begin{cases}\widehat{\boldsymbol{R}}_{du}^{(N+1)}=0\\ \widehat{\boldsymbol{T}}_d^{(j)}=\left[\boldsymbol{I}-\boldsymbol{R}_{ud}^{(j)}\widehat{\boldsymbol{R}}_{du}^{(j+1)}\right]^{-1}\boldsymbol{T}_d^{(j)}\\ \widehat{\boldsymbol{R}}_{du}^{(j)}=\boldsymbol{R}_{du}^{(j)}+\boldsymbol{T}_u^{(j)}\widehat{\boldsymbol{R}}_{du}^{(j+1)}\widehat{\boldsymbol{T}}_d^{(j)}\end{cases} \qquad \begin{cases}\widehat{\boldsymbol{R}}_{ud}^{(0)}=\boldsymbol{R}_{ud}^{(0)}\\ \widehat{\boldsymbol{T}}_u^{(j)}=\left[\boldsymbol{I}-\boldsymbol{R}_{du}^{(j)}\widehat{\boldsymbol{R}}_{ud}^{(j-1)}\right]^{-1}\boldsymbol{T}_u^{(j)}\\ \widehat{\boldsymbol{R}}_{ud}^{(j)}=\boldsymbol{R}_{ud}^{(j)}+\boldsymbol{T}_d^{(j)}\widehat{\boldsymbol{R}}_{ud}^{(j-1)}\widehat{\boldsymbol{T}}_u^{(j)}\end{cases} \tag{2}$$

$$j=N,N-1,\cdots,2,1 \qquad j=1,2,3,\cdots,N$$

The calculation formula of the modified reflection-transmission coefficient is as follows:

$$\begin{bmatrix}\boldsymbol{T}_{\mathrm{d}}^{(j)} & \boldsymbol{R}_{\mathrm{ud}}^{(j)}\\ \boldsymbol{R}_{\mathrm{du}}^{(j)} & \boldsymbol{T}_{\mathrm{u}}^{(j)}\end{bmatrix}=\begin{bmatrix}\boldsymbol{E}_{11}^{(j+1)} & -\boldsymbol{E}_{12}^{(j)}\\ \boldsymbol{E}_{21}^{(j+1)} & -\boldsymbol{E}_{22}^{(j)}\end{bmatrix}^{-1}\begin{bmatrix}\boldsymbol{E}_{11}^{(j)} & -\boldsymbol{E}_{12}^{(j+1)}\\ \boldsymbol{E}_{21}^{(j)} & -\boldsymbol{E}_{22}^{(j+1)}\end{bmatrix}\times\begin{bmatrix}\Lambda_{\mathrm{d}}^{(j)}\left(z^{(j)}\right) & 0\\ 0 & \Lambda_{\mathrm{u}}^{(j+1)}\left(z^{(j)}\right)\end{bmatrix}$$

$$j=1,2,\cdots,N-1$$

$$\begin{bmatrix}\boldsymbol{T}_{\mathrm{d}}^{(N)}\\ \boldsymbol{R}_{\mathrm{du}}^{(N)}\end{bmatrix}=\begin{bmatrix}\boldsymbol{E}_{11}^{(N+1)} & -\boldsymbol{E}_{12}^{(N)}\\ \boldsymbol{E}_{21}^{(N+1)} & -\boldsymbol{E}_{22}^{(N)}\end{bmatrix}^{-1}\begin{bmatrix}\boldsymbol{E}_{11}^{(N)}\Lambda_{\mathrm{d}}^{(N)}\left(z^{(N)}\right)\\ \boldsymbol{E}_{21}^{(N)}\Lambda_{\mathrm{d}}^{(N)}\left(z^{(N)}\right)\end{bmatrix} \tag{3}$$

From the free boundary conditions of the ground surface, we get:

$$\boldsymbol{R}_{\mathrm{ud}}^{(0)}=-\left(\boldsymbol{E}_{21}^{(1)}\right)^{-1}\boldsymbol{E}_{22}^{(1)}\boldsymbol{\Lambda}_{\mathrm{u}}^{(1)}(0) \tag{4}$$

For P-SV wave:

$$\mathbf{E}^{(j)}=\frac{1}{\omega}\begin{bmatrix}\alpha^{(j)}k & \beta^{(j)}\nu^{(j)} & \alpha^{(j)}k & \beta^{(j)}\nu^{(j)}\\ \alpha^{(j)}\gamma^{(j)} & \beta^{(j)}k & -\alpha^{(j)}\gamma^{(j)} & -\beta^{(j)}k\\ -2\alpha^{(j)}\mu^{(j)}k\gamma^{(j)} & -\beta^{(j)}\mu^{(j)}\chi^{(j)} & 2\alpha^{(j)}\mu^{(j)}k\gamma^{(j)} & \beta^{(j)}\mu^{(j)}\chi^{(j)}\\ -\alpha^{(j)}\mu^{(j)}\chi^{(j)} & -2\beta^{(j)}\mu^{(j)}k\nu^{(j)} & -\alpha^{(j)}\mu^{(j)}\chi^{(j)} & -2\beta^{(j)}\mu^{(j)}k\nu^{(j)}\end{bmatrix} \quad (5)$$

For SH wave:

$$\mathbf{E}^{(j)}=\begin{bmatrix}1 & 1\\ -\mu^{(j)}\nu^{(j)} & \mu^{(j)}\nu^{(j)}\end{bmatrix} \quad (6)$$

For the Rayleigh roots whose modal velocity is close to the Rayleigh wave velocity in half space, it is difficult to obtain using equation (1). To solve this problem, the following surface secular function is obtained from the free boundary conditions of the surface:

$$det\left\{\mathbf{E}_{21}^{(1)}+\mathbf{E}_{22}^{(1)}\Lambda_u^{(1)}(0)\hat{\mathbf{R}}_{\mathrm{du}}^{(1)}\right\}=0 \quad (7)$$

Equations (1) and (8) are collectively called the secular function family.

2.2. Inversion Objective Function Based on Secular Function

When only using the surface secular function to search the root of the dispersion curve, some modes with strong energy in the low-velocity layer will be missed. However, since the seismic source of the existing surface wave exploration is on the surface and the signal reception is also on the surface, so the dispersion curves extracted from the measured data is also unable to visualize certain modes that only have strong energy in the low-velocity layer. Therefore, in this paper, the objective function is constructed based on the surface secular function. Since the surface secular function has different orders of magnitude at different frequencies, its maximum value is proportional to the square of the frequency. Therefore, when summing the different dispersion points, it needs to be divided by corresponding value, the basic form of the inversion objective function is shown in equation (8):

$$obf(\boldsymbol{m})=\{\sum_{i=1}^{nr}[S_r(\boldsymbol{m},f_{ri},c_{ri})w_i]^k+\sum_{i=1}^{nl}[S_l(\boldsymbol{m},f_{li},c_{li})w_i]^k\}^{1/k} \quad (8)$$

where: $\boldsymbol{m}$ is the stratum model; (f_{ri},c_{ri}) and (f_{li},c_{li}) are the dispersion curves points of Rayleigh wave and Love wave; $S_r(\boldsymbol{m},f_{ri},c_{ri}),S_l(\boldsymbol{m},f_{li},c_{li})$ are the surface secular function values, $S(\boldsymbol{m},f_i,c_i)=\frac{det\left\{E_{21}^{(1)}+E_{22}^{(1)}\Lambda_u^{(1)}(0)\hat{R}_{du}^{(1)}\right\}}{f_i^2}$; w_i is the weight of different dispersion points; k is the norm, which is 2 in this paper.

3. Tests With Synthetic Data

This paper is based on the discrete wavenumber method to synthesize theoretical seismic records of ambient seismic noise data [7]. A total of 1000 seismic sources are randomly distributed in an annular area ranging from 500 m to 1500 m. The Ricker wavelet is selected as the source time function, the center frequency is randomly distributed between 6 Hz and 10 Hz, the onset direction is (1, 1, 1), the takeoff time is randomly distributed from -5 s to 60 s, and the intensity is randomly distributed between 0.001 and 1. A total of 100 observation stations are randomly distributed within a 100 m center range. Three-component background noise theoretical seismic records are synthesized, the frequency range is between 0.5 Hz and 28 Hz, the time length is 60 s, and the sampling rate is 68.23 Hz. The observation array is shown in figure 1 and three-component noise data is shown in figure 2.

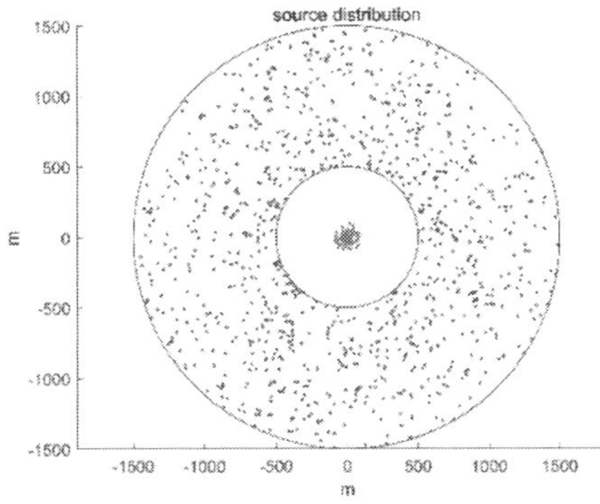

Figure 1. Distribution of source and stations.

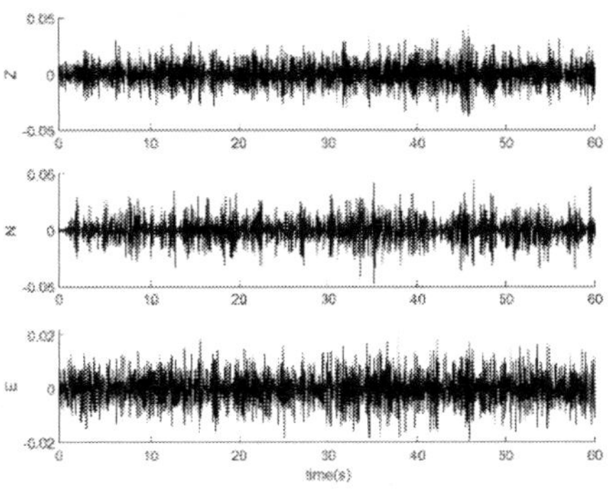

Figure 2. Synthesized ambient seismic noise data.

In practical applications, the appearance of low-velocity layers, underground cavities or silt layers stratum will lead to "mode jumping" and "mode kissing" of the dispersion curve, which can easily lead to misjudgment of the model in the inversion of multi-order modes. This paper uses four-layer model with low-velocity layer to perform numerical simulation, and the parameters and search range are shown in table 1 and figure 3. In this paper, a joint inversion of the multi-mode dispersion curve will be performed, the density and longitudinal wave velocity are set as the real values of the theoretical model, and the shear wave velocity and layer thickness are used as the target parameters of the inversion. A total of 50 inversion calculations are carried out using the particle swarm algorithm, and the anomalous solutions are eliminated and then the numerical average is performed to obtain final result.

The dispersion curve extracted from the synthetic seismic record through Frequency-Bessel transform method and based on the inversion result are shown in figure 5[10]. The fundamental-order mode of the Rayleigh wave dispersion curve is missing at 5~8 Hz, where the energy of the first higher-order mode dominates. A mode kissing occurs between the first high-order mode and the second high-order mode at 20 Hz. In this case, it is extremely difficult for the traditional inversion objective function to make a correct mode discrimination of the dispersion curve, resulting in wrong inversion results. The new type of inversion objective function based on secular function does not require model judgment, each order mode is automatically matched, and the dispersion curve fits well. It can still obtain more accurate inversion results under a wide range of parameters search. The Inversion results for shear wave structure is shown in Figure 4, and the layering error is less than 10%, the shear wave velocity error of the first three layers is less than 4%, and the error of the fourth layer increases to 10%. The main reason for error is that there are fewer dispersion points extracted at low frequencies, and the phase velocity has a certain deviation from the theoretical dispersion curve.

Table 1. The parameters and inversion search range for four-layer stratum model with low velocity interlayer.

Layer number	Model parameters			Search range		
	Z(m)	ρ(g·cm^{-3})	Vs(m·s^{-1})	Vp (m·s^{-1})	Vs(m·s^{-1})	Z(m)
1	10.0	1.78	180	1500	90~400	5~15
2	20.0	1.85	350	1700	150~700	10~30
3	40.0	1.80	250	1600	125~800	25~55
half space	∞	1.94	600	2000	200~1000	∞

Table 2. The inversion results of four-layer stratum model.

Parameters	Actual value	Rayleigh wave inversion		Joint inversion	
		Result	Relative error	Result	Relative error
Z1(m)	10.00	7.70	23.00%	10.60	6.00%
Z2(m)	10.00	8.40	16.00%	9.60	4.00%
Z3(m)	20.00	22.60	13.00%	21.90	9.50%
Vs1(m/s)	180.00	90.00	50.00%	182.70	1.50%
Vs2(m/s)	350.00	333.50	4.71%	361.00	3.14%
Vs3(m/s)	250.00	293.40	17.36%	248.60	0.56%
Vs4(m/s)	600.00	392.90	34.52%	422.05	11.06%

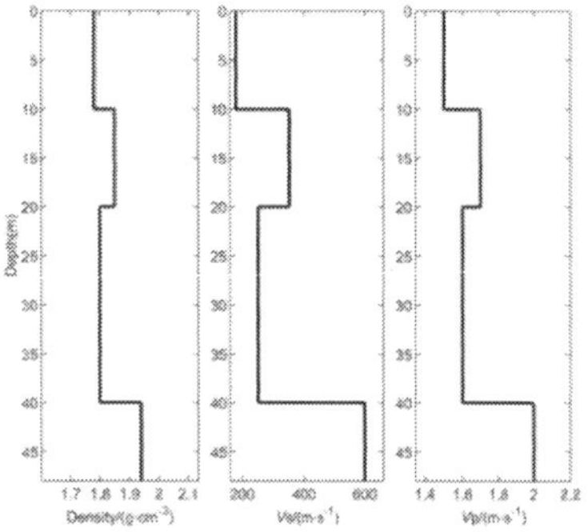

Figure 3. Horizontally layered stratum model.

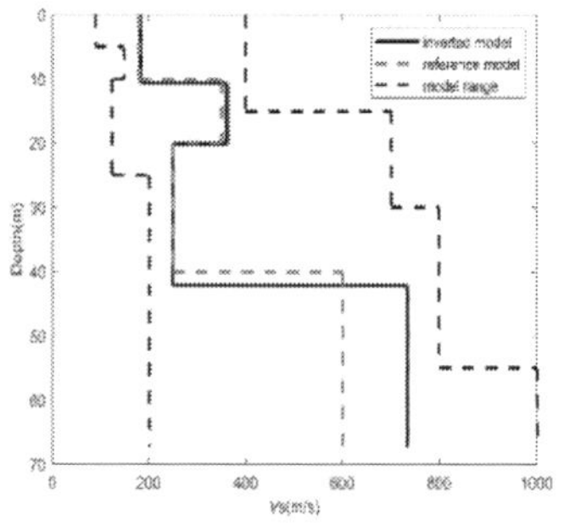

Figure 4. The Inversion results for Vs structure.

4. Conclusion

In this paper, a new type of inversion objective function is constructed based on the secular function. There is no need to judge the mode of the dispersion curve before the inversion calculation, which effectively solves the problem of pattern misjudgment of the traditional inversion objective function. The joint inversion of Rayleigh wave and Love wave is used to improve the constraint of the inversion and avoid the inversion falling into a local minimum in the case of a large-scale search of parameters. Finally, the ambient seismic noise data is synthesized by the vector wavenumber method and the dispersion curve is extracted by the Frequency-Bessel transform for inversion. There is a good fit for the dispersion curves and a more accurate shear wave velocity structure in the inversion result, which proves the new inversion method feasible.

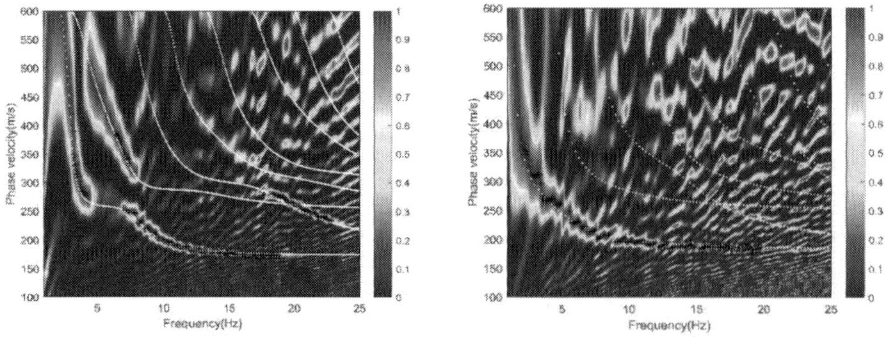

(a)Rayleigh dispersion curves (b) Love dispersion curves

Figure 5. The Inversion results for dispersion curve, Inversion results (solid white line); Dispersion curve for inversion (black plus)

Acknowledgements
This work is supported by Shenzhen Key Laboratory of Deep Offshore Oil and Gas Exploration Technology (Grant No.ZDSYS20190902093007855).

References

[1] Aki K. 1957 Space and Time Spectra of Stationary Stochastic Waves, with Special Reference to Microtremors [J] *Bull.earthq.res.inst.*

[2] Wang J, Wu G, Chen X. 2019 Frequency-Bessel Transform Method for Effective Imaging of Higher-Mode Rayleigh Dispersion Curves From Ambient Seismic Noise Data [J] *Journal of Geophysical Research: Solid Earth* **124** (4) 3708-3723.

[3] Hu S, Luo S, Yao H. 2020 The frequency - Bessel spectrograms of multi - component cross - correlation functions from seismic ambient noise[J] *Journal of Geophysical Research: Solid Earth* **125** (8) 1271-1280.

[4] Margherita M, Fabian E, Sebastiano F et al. 2010 A new misfit function for multimodal inversion of surface waves [J] *GEOPHYSICS* **75** (4) 31–43.

[5] Chen X F. 1993 A Systematic and Efficient Method of Computing Normal Modes for Multi-Layered Half-Space [J] *Geophysical Journal International* (115) 391-409.

[6] He Y F, Chen W T, Chen X F. 2006 Normal mode computation by the generalized reflection_transmission coefficient method in planar layered half space. Chinese J .Gophys. (In Chinese) **49** (4) 1074-1081.

[7] Bonnefoy C S, Cornou C, Kristek J, et al. 2004 Simulation of seismic ambient noise: I. Results of H/V and array techniques on canonical models 1120.

[8] Zhang H M, Chen X F, Chang S. 2003 An Efficient Numerical Method for Computing Synthetic Seismograms for a Layered Half-space with Sources and Receivers at Close or Same Depths [J] **160** (3-4) 467-486.

[9] Chen X F. 1999 Seismogram synthesis in multi-layered half-space Part I. Theoretical formulations. Earthquake Research in China **13** (2) 149–174.

[10] Wu B, Chen. X. 2016 Stable, accurate and efficient computation of normal modes for horizontal stratified models [J] *Geophysical Journal International* **206** (2) 1281-1300.

Quantitative Identification Cracks of Heritage Rock Based on Digital Image Technology

Xiufang Wang [1, 2*], Jingyuan Li[2], Ming Bai[1] and Yan Pei [2]

[1] Beijing Key Laboratory of Functional Materials for Building Structures and Environmental Remediation, School of Science, Beijing University of Civil Engineering and Architecture, Beijing, 102616, P.R. China
[2] School of Civil and Transportation Engineering, Beijing University of Civil Engineering and Architecture, Beijing, 102616, P.R. China
Email: wangxiufang@bucea.edu.cn

Abstract. Digital image processing technologies are used to extract and evaluate the cracks of heritage rock in this paper. Firstly, the image needs to go through a series of image preprocessing operations such as graying, enhancement, filtering and binaryzation to filter out a large part of the noise. Then, in order to achieve the requirements of accurately extracting the crack area, the image is again divided into the crack area and morphological filtering. After evaluation, the obtained fracture area can provide data support for the restoration and protection of heritage rock. In this paper, the cracks of heritage rock are extracted in three different locations.The results show that the three groups of rock fractures have different effects on the rocks, but they all need to be repaired to maintain the appearance of the heritage rock.

1. Introduction

As the most intuitive carrier in cultural traditions, heritage rock have left precious wealth for the brilliant history of the Chinese nation. It contains a large number of precious heritage rock, but with historical changes, social development, and the fate of heritage rock have also undergone tremendous changes with the development of the times. Rock is a valuable cultural relic because of its own characteristics, but with the change of dynasties and histories, these very meaningful rocks have been damaged to varying degrees. The reasons for the damage are various. Summary includes two main factors: natural environment and human influence.

Therefore, this paper will provide some ideas for the protection, rehabilitation and restoration of heritage rock through the exploration of digital image processing technology in the detection of rock cracks in heritage rock. The reason why this paper studies the application of digital image processing technology in the detection of heritage rock cracks is because it is found that image technology has obvious advantages in the protection and restoration of heritage rock, such as high accuracy and reliability, and Non-contact damage to rock surface and high detection efficiency.

With the development of computer science, the technology of image processing is also used by more people and gradually improved. Some scholars have applied computer image processing methods to the identification and research of road cracks [1], and some scholars have proposed a method for concrete apparent crack detection [2]. Fang Mengyang et al. used MATLAB image processing technology to identify the contours of particles in the clastic rock, and obtained various morphological parameter information of all particles in the clastic rock [3]. Some scholars used the crack detection embedded system to diagnose the bridge concrete cracks and recognition [4]. Some scholars use digital image processing technology to perform polarization imaging on stone carvings, which enhances the detection of carving features[5].

Based on the research of many scholars, this paper uses MATLAB to realize the function of digital image technology in the detection of heritage rock cracks. The system used in this paper is GUI design. The interface of the entire system is shown in figure 1. GUI is a graphical user interface. It has the advantages of convenient and friendly interface for users and easy operation [6].

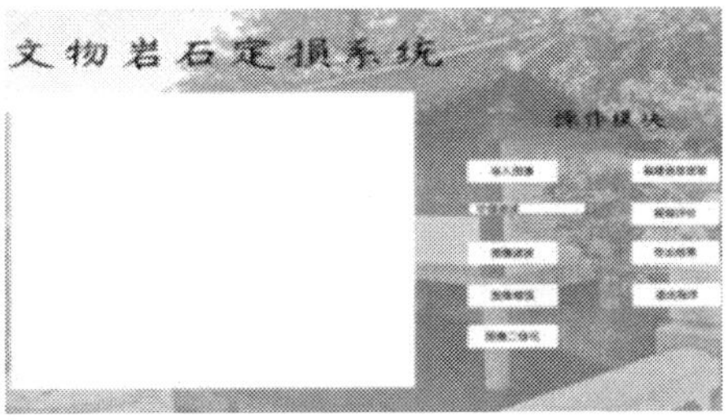

Figure 1. GUI interface.

2. Image Preprocessing

Through the preprocessing of the image, the influence of the noise caused by various factors on the experimental results can be minimized. Therefore, in order to better reduce the impact of noise and improve the efficiency of the algorithm, this paper will mainly introduce several image preprocessing methods:

2.1. Image Grayscale

From the spectrum theory, we can know that all visible light is mixed by the three colors of red (R), green (G), and blue (B). The pictures used in this experiment are all made by mobile phones, and these pictures are also described in the RGB color space. However, the processing calculation for the three components of R, G, and B is relatively large, so grayscale processing is required to ensure that the operation of image processing is easier and the result is more accurate.

In the RGB color space, there are four main methods for image gray-scale processing: component method, maximum method, average method, and weighted average method. This paper will give a brief introduction to these four methods one by one.

2.1.1. Component Method:

The component method refers to the gray value obtained after conversion of the brightness values of the three components of R, G, and B in the image as the gray value of the three gray images. This processing process can be expressed as formula (1)

$$\begin{aligned} f_1(i,j) &= R(i,j) \\ f_2(i,j) &= G(i,j) \\ f_3(i,j) &= B(i,j) \end{aligned} \tag{1}$$

Where $f_k(i,j)$ (k=1, 2, 3) represents the gray value of the image after gray-scale conversion processing.

2.1.2. Maximum Method

This method refers to selecting the largest amount among the three components of R, G, and B of the image as the gray value of the entire image. This process can be expressed as formula (2)

$$f(i,j) = \max\{R(i,j), G(i,j), B(i,j)\} \tag{2}$$

2.1.3. Average Method

Since there are three components of R, G, and B in the image, the average value method, as the name implies, is to use the average value of these three components in the image as the gray value of the entire image. This method can be expressed by the following formula (3).

$$f(i,j) = \frac{R(i,j)+G(i,j)+B(i,j)}{3} \tag{3}$$

2.1.4. Weighted Average Method

Because the human eye is not very sensitive to all colors, the sensitivity of the human eye to different colors is different, so if you want to get a gray value of the image, you need to deal with it according to the sensitivity of the human eye to the color. Different colors are given different weights, and then the values given with different weights are weighted and averaged. This method is the weighted average method. For the pictures in this paper, they are described in the three color environments of R, G, and B. Because human eye has different sensitivity to these three colors, it is most sensitive to green, followed by red, and for blue the sensitivity is the lowest [4], so the weight of 0.59 for green, 0.3 for red and 0.11 for blue is given. With the weights, the gray value of the image can be expressed as the following formula (4)

$$f(i,j) = 0.3R(i,j) + 0.59G(i,j) + 0.11B(i,j) \tag{4}$$

Choosing the weighted average method to preprocess the rock cracks can get more accurate results, and can reduce the difficulty and improve the efficiency for the subsequent processing. Therefore, this paper adopts the weighted average method to gray-scale the graphics.

2.2. Image Filtering

In actual images, there are often a variety of noises, and the presence of these noises will affect the processing of the image to varying degrees, and at the same time will also have a corresponding impact on the rendering effect of the processed image. By reading the literature, we know that most of the energy in the image information is concentrated in the middle and low frequency parts, and the high frequency parts will be covered by noise due to weak energy. Therefore, this part of the high-frequency noise needs to be filtered out in the experiment, and this is image filtering. Through filtering operations, the extraction of image features is more recognizable and more operable, and to achieve image noise reduction the goal of. This paper will introduce three filtering methods, namely mean filtering, Gaussian filtering and median filtering.

1) Mean filtering: this method first needs to select a target pixel, then according to this pixel as the center of the circle, the pixel values of the pixels within a certain radius are averaged, replace pixel values of target pixels with average values, this method is a linear filtering method can be expressed as formula (5):

$$g(x,y) = \frac{1}{M}\sum f(x,y) \tag{5}$$

$g(x,y)$ Represents the pixel value after the pixel point $(x,y)$is filtered, $f(x,y)$ represents the pixel value of the pixel point, and M represents the size of the range selected during the filtering process.

2) Gaussian filtering: It is also a linear filtering method. This method is similar to mean filtering, the difference is that the pixels in a certain radius are calculated according to the weight of a certain normal distribution. One-dimensional expressions such as formula (6), two-dimensional expressions such as formula (7):

$$G(x) = \frac{1}{\sqrt{2\pi}} e^{\frac{x^2}{2\sigma^2}} \tag{6}$$

$$G(x,y) = \frac{1}{2\pi\sigma^2} e^{-\frac{x^2+y^2}{2\sigma^2}} \tag{7}$$

σ^2 Is the variance, which represents the width of the Gaussian curve.

3) Median filtering: this method also needs to select a target pixel first, then determine the pixels within a certain radius, and then sort the pixel values of these pixels from low to high, select the middle value of this sequence, and use this median value replaces the pixel value of the target pixel. This method is a non-linear filtering method. Expressed as formula (8):

$$g(x,y) = \mathrm{mid}\{f(x,y)\} \tag{8}$$

$g(x,y)$ Represents the pixel value of the pixel (x,y) after median filtering, and $f(x,y)$ represents the pixel (x,y) value of the pixel.

2.3. Image Enhancement

Image enhancement refers to the operation of enhancing the useful information needed in the image. After image enhancement, useful features are enhanced, while less useful features will be suppressed. After image enhancement, the feature distribution of the image is more obvious, which is more conducive to the recognition of image features by human eyes and machines.

There are many methods for image enhancement. This paper uses one of many methods: spatial domain image enhancement technology. This technology is simple and easy to understand and has strong operability, this technique enhances images mainly through linear or nonlinear transformations. The neighborhood denoising method is a commonly used method in this technology. Determine a target pixel, select the pixel in the corresponding area in a specific direction, and enhance the pixel in this area according to certain rules. The definition of the operator can be expressed as formula (9):

$$A(x,y) = T[B(x,y)] \tag{9}$$

$A(x,y)$ Represents the output image after the enhancement operator is processed, $B(x,y)$ represents the original image, and T represents the enhancement operation in a certain direction of the pixel(x,y). The schematic diagram of the operation is shown in the figure.

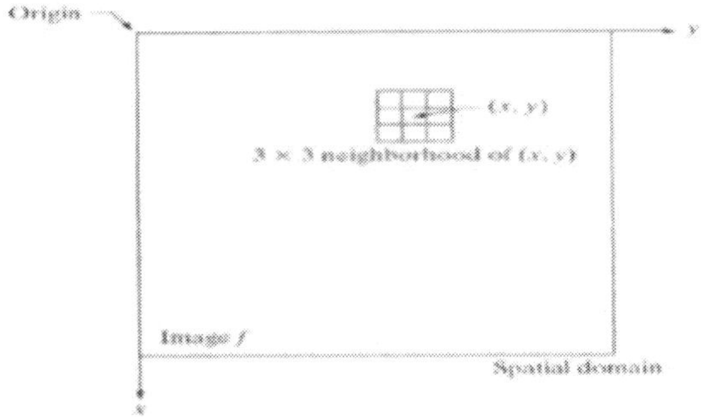

Figure 2. Description of spatial domain enhancement.

2.4. Image Binarization

The so-called image binarization means that each pixel value in the pixel matrix assigned to the image is only two values, which are 0 (representing black) or 255 (representing white), that is, the entire image presents only black and white effects. In this paper, the threshold method is used to binarize the image, that is, a fixed value V is set according to the need. When the pixel value in the grayscale image of the original image is less than or equal to this fixed value V, then these in the binarized image, the value of all pixels is set to 0, otherwise it is 1, and the expression is shown in formula (10)

$$f = \begin{cases} 1, W \leq V \\ 0, \ W > V \end{cases} \tag{10}$$

In the formula, V represents a fixed value, W represents the pixel value in the original image grayscale image, and f represents the pixel value of the binarized image.

3. Extraction of Crack Information

The image pre-processed in the previous paper has filtered out most of the noise that will interfere with the experiment, retains and enhances some features that are beneficial to subsequent operations. This paper aims to extract the cracks of the heritage rock, so the pre-processed image needs to extract the crack information.

3.1. Fracture Segmentation

Because most of the rock's cracks are not evenly distributed in the area and the distribution is wide, in order to make the extraction of the image cracks go smoothly, the image needs to be segmented first, and the image is divided into several small areas, analysis and processing of each small area, the commonly used image segmentation methods are based on threshold, region and edge, and this article mainly uses the region-based image segmentation method.

The region growing method [2] is to operate on pixels with the same properties. Before the operation, it is necessary to determine the classification rules. Commonly used rules include the gray value, the texture characteristics of the image content, and the color of the image. According to the determined rules, the pixels in the image are classified, and the same types of pixels or pixel blocks are grouped together as a category of image judgment standards. Therefore, this method can be understood as classifying pixels according to a certain rule, using pixels with a certain property as seeds as needed, and then selecting pixels with the same properties as the seeds among all the pixels in the image. Then these pixels with the same properties are combined, and finally displayed according to the combined situation of the pixels. In this paper, gray value is selected as the classification rule, point A is used as the seed, and then a certain neighborhood B of the crack is determined as the area to be executed by the region growth method, and then iterated according to formula (11)

$$|f(A) - f(B_i)| \leq T \qquad f(B_i) \in [T_1, T_2] \tag{11}$$

In the formula, $f(A)$ represents the gray value of point A, $f(B_i)$ represents the gray value of any point in the area B, and T, T_1, T_2 represents the gray value range.

3.2. Morphological Processing

For images that have undergone the area growth method, the recognition effect of the crack area has been greatly improved, but there may still be some problems that the crack shape is not ideal. In order to avoid these problems from affecting the subsequent operations, so Morphological processing of the image is also required. Morphological processing has four basic operations: corrosion, expansion, opening operation and closing operation [4]. Corrosion is the morphological treatment method used in this paper.

Morphological corrosion can eliminate the boundary points of the image and shrink the boundary to the inside. This method can be used to eliminate small and meaningless objects in the image and that will affect the image recognition. The target pixels can be processed by corrosion to achieve refinement purpose. This method has a vivid analogy, In the cell living environment, 0 is used to represent cells with viruses, and 1 is used to represent healthy cells. The process of viruses eroding healthy cells is the corrosion operation. As shown in figure 3, for a 3*3 pixel matrix, as long as there is a virus (0), healthy cells (1) will be corroded.

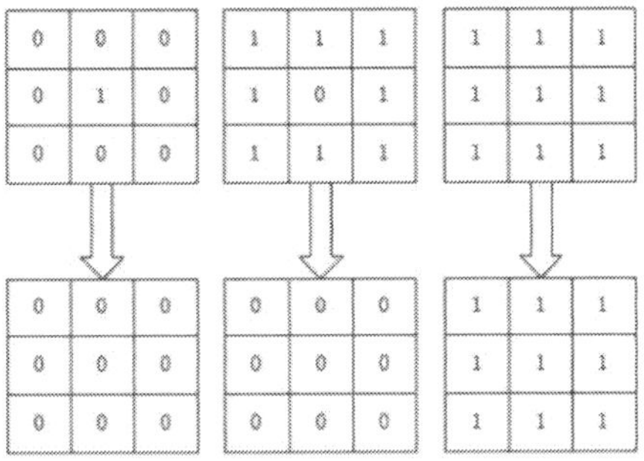

Figure 3. Schematic diagram of morphological corrosion.

3.3. Evaluation of Crack Information

After the above-mentioned fracture segmentation and morphological processing, the fracture area of the image is already obvious. Next, a certain evaluation of the extracted fractures is needed to determine the degree of impact of the fractures on the rock. Since there is no universal standard for evaluating rock cracks in stony heritage rocks [7], this paper formulates the judgment rules on heritage rock based on the actual situation and the research of other scholars, as shown in table 1 and table 2.

Table 1. Evaluation standard of crack length of heritage rock.

Crack length(mm)	influence degree
0-20	mild
20-80	moderation
>80	serious

Table 2. Evaluation criteria of crack width of heritage rocks rock.

Crack length(mm)	Influence degree
0	No influence
0-1.5	mild
1.5-10	moderation
>10	serious

4. Experimental Test

This paper selects three rock cracks in different locations for research, uses the mobile phone to take pictures of the cracks, and imports the taken pictures into the system. After a series of operations, the information of the cracks is obtained, and the severity of the corresponding rock cracks is analyzed. The protection and restoration of the rocks provide favorable data support.

(1) The first group is the stone road between the Hall of Prayer for Good Harvests in the Temple of Heaven and the Echo Wall. The picture taken is as shown in figure 4, where there is a longitudinal crack in the target picture, and the shape of the crack is relatively regular. There are large waves of water ripples in the lower part of the image, but the color is lighter, and the contrast with the surrounding environment is not large, and the cracked part is in the middle. The crack image extracted after processing is shown in figure 5.

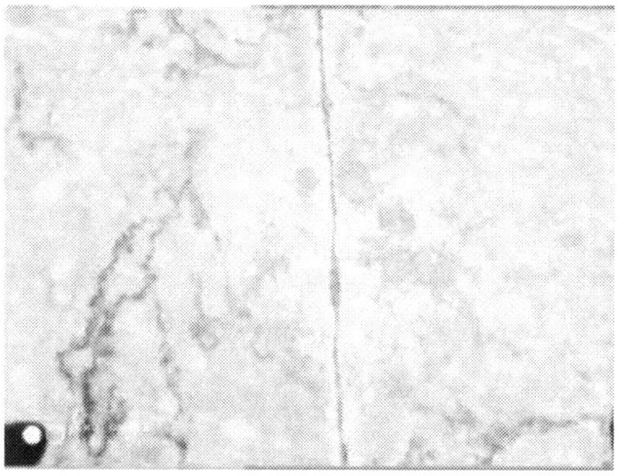

Figure 4. The first group of photos taken.

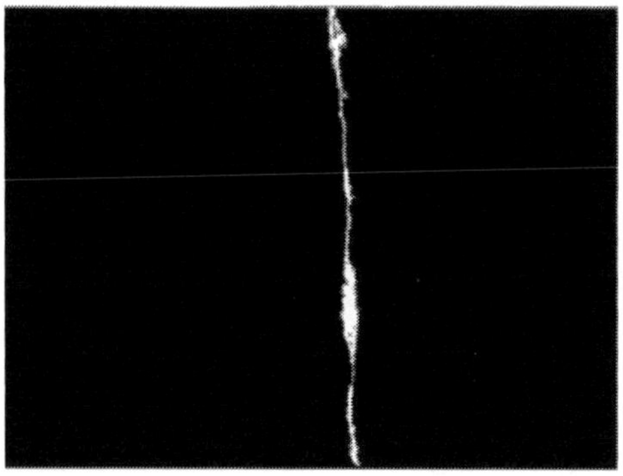

Figure 5. The first group of crack extraction effect diagram.

(2) The second group is the moat fence on the north side of the Temple of Heaven. The photo taken by the mobile phone is shown in figure 6. Although the cracks in the target picture are irregular, the overall shape is relatively smooth. The image around the crack has little contrast with the surrounding environment, the upper part is generally white, and the crack part is in the middle. The image of the crack area extracted after processing is shown in figure 7

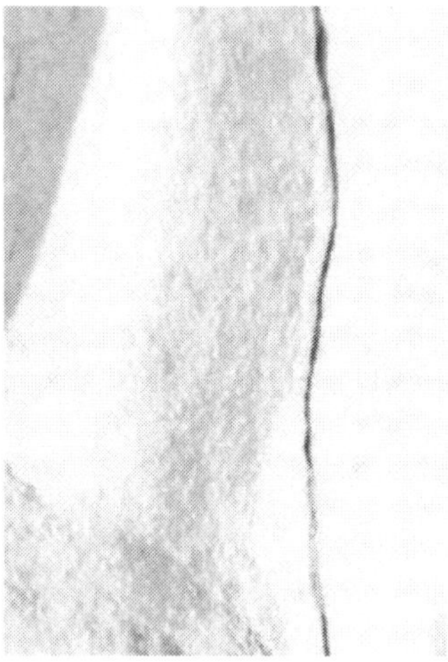

Figure 6. The second group of photos taken.

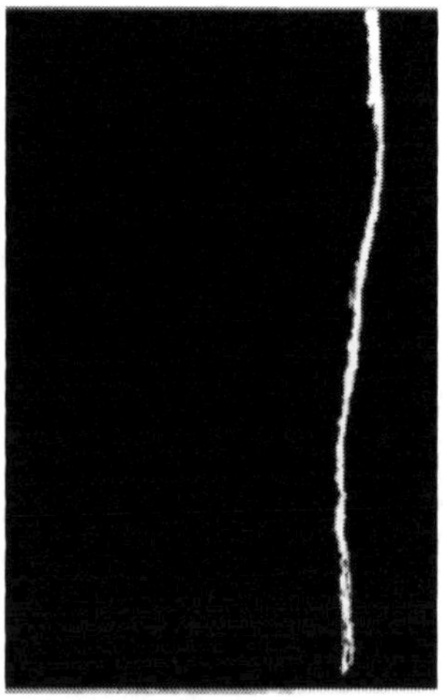

Figure 7. The second group of crack extraction effect diagram.

(3) The third group is the steps near the clerical pavilion in Zhengjue Temple of Yuanmingyuan garden. The picture taken by mobile phone is shown in figure 8. The cracks in the target picture are not regular but the overall shape is relatively smooth. There is little contrast between the image cracks and the surrounding environment. The crack area image extracted after processing is shown in figure 9.

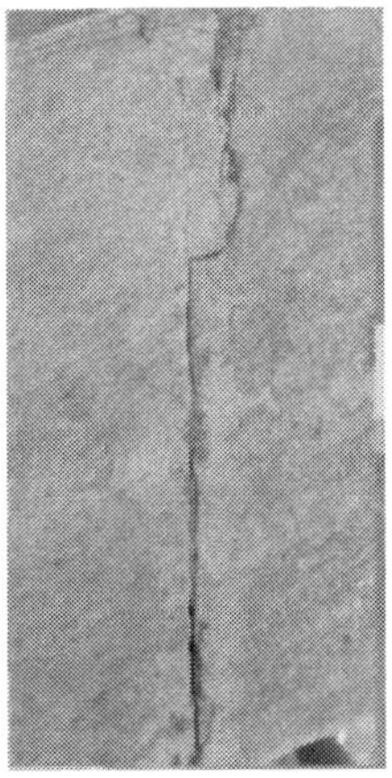

Figure 8. The third group of shots.

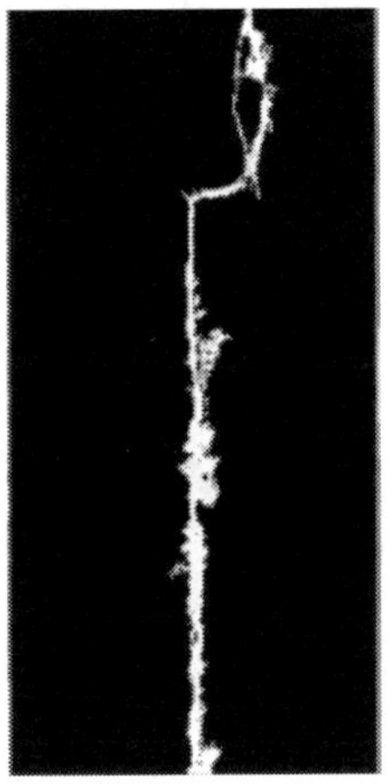

Figure 9. The effect of the third group of cracks.

Summarize the processed data of the above three groups of pictures. Due to the uneven distribution of the width of the entire crack, there is the widest part and the narrowest part in a crack. Therefore, the evaluation of the crack width in this paper uses the average value of the maximum width and the minimum width.

According to the evaluation rules on the cracks of cultural relics made in the previous paper, it can be concluded that the cracks in experimental groups 1 and 3 have a moderate impact on the rock, while the rock cracks in the experimental group 2 have a serious impact on the rock. Therefore, the three groups of rocks need to be repaired and protected, but because the impact of the second group is more serious, the second group of rocks needs to be repaired urgently to avoid more serious impacts on the rocks.

Table 3. Summary of experimental results.

Experimental group	Crack length (mm)	maximum crack width (mm)	Minimum crack width (mm)	Average crack width (mm)	Influence degree
1	200.30	8.10	1.00	4.55	moderation
2	71.10	25.00	1.00	13.00	serious
3	128.10	10.75	2.50	6.63	moderation

5. Conclusion

(1) The application of digital image technology in the extraction of heritage rock fractures first needs to preprocess the image, and first filter most of the noise that will affect the extraction of fractures, which is beneficial to improve the efficiency of subsequent operations. Then the pre-processed image is subjected to crack segmentation and morphological processing, and a part of the small noise is filtered out again, so that the crack area obtained after the operation is more accurate. The evaluation of cultural heritage rock according to the established rules provides favorable data support for the pre-protection and restoration of cultural relics.

(2) This paper uses mobile phone to take pictures of cracks and import the pictures into the heritage rock damage system to identify and evaluate the crack areas. From the evaluation results, it is concluded that the three groups of rocks have different degrees of cracks. The average width of the cracks show that the first group is 4.55 mm, the second group is 13.00 mm, and the third group is 6.63 mm, so these three groups of rocks all need to be repaired later. But in terms of the degree of influence, the cracks in the second group have a deeper impact on the rocks than the first and third groups. The average width of the cracks in the second group is about three times that of the first group and twice that of the third group. So the rocks in the second group need to be repaired urgently.

Acknowledgments

This work was supported by the Natural Sciences Foundation of China under Grant (NO.51808025 and NO.51978033), the Pyramid Talent Training Project of BUCEA (NO.JDYC20200329), the Scientific Research Fund Project of BUCEA (NO.ZF16085), Beijing Social Science Foundation under Grant (NO.17LSC008), and Beijing High School's high-level talent cross-cultivation and training program support project.

References

[1] Kang S Y, Nie W. 2020 Research on road crack identification based on MATLAB[J]*Computer knowledge and technology* **16** (30) 191-193.
[2] Xu G Z, Liao C C , Chen J J, Dong B and Zhou Y. 2020 Extraction of concrete apparent crack information based on HU-ResNet [J] *Computer Engineering* **46** (11) 279-285.
[3] Fang M Y, He J N, Wang W H and Li M Z. 2020 A particle size analysis method of clastic rock based on MATLAB [J] *Rock mineralogy journal* **39** (04) 504-510.
[4] Sun Y C. 2020 Research on image recognition method of concrete cracks in elevated bridges [D] *Beijing University of Chemical Technology.*
[5] Wang H F, Luo Y Q, Caidan An, Chu S N, Shen Z X, Huang L X and Zhang D Y. 2019 Application of imaging polarimeters to enhanced detection of stone carving [J] *Journal of Cultural Heritage* (40) 92-98.
[6] Zhang G C, Wan S P and He J R. 2019 MATLAB GUI digital image processing system development and design [J] *Software* **40** (11) 123-127.
[7] Li J. 2013 Research and application of health evaluation technology of ancient building stone components [D] *Beijing University of Chemical Technology.*

Impact of Lightning Protection Grounding System on the Ground Surface Potential of Substations

Tingji Chen [1], Lian Yang [1], Weibing Gu [1], Haiyang Gao [1], Junchi Zhou [2] and Xiaodong Fu [3]

[1] Nanjing Electric Power Design and Research Institute Co. LTD, Nanjing 210028, China
[2] Jiangsu Meteorological Disaster Protection Center, Nanjing 210019, China
[3] School of Mathematics and Physics, North China Electric Power University, Baoding 071003, China
Email: zhoujunchi1985@163.com

Abstract. Grounding device is an indispensable facility for lightning protection of buildings. Nowadays, SGCC (State Grid Corporation of China) is promoting steel structure substations, which are made of metal as a whole including the roof. There are now several grounding approaches when the roof was struck by a lightning flash, including external grounding, nearby grounding, separate grounding and common grounding. This paper took a metal structure substation in Nanjing as an example and calculated its ground potential in case of different grounding system. We came to such conclusions: 1) For substations of separate grounding system, the ground potential after a lightning strike could reach as high as 743.5kV and 230kV with a single earthing electrode and multiple electrodes respectively. 1000μs after the strike, the ground potential is 91.57 kV, which is still a significant threat to humans and equipment inside. 2) Nearby grounding and external grounding are both common grounding system. The peak of ground potential after a lightning strike is 101.4kV and 109kV respectively, much lower than that of separate grounding system. They also have similar waveform and peak time. 3) 3500μs after the lightning strike, the ground potential all over the grid is around 36V. 4) Separate grounding is not a sound choice of grounding system for steel structure substations. From the perspective of cost and discharging capacity, nearby grounding is the most reasonable scheme for a steel structure substation.

1. Introduction

Grounding device is an indispensable facility for lightning protection of buildings. Grounding network of power system has the function of lightning protection in addition to its system operation. Taking the substation as an example, the equipment in the substation are connected to the grounding network underground through the lead wire. Grounding network serves as a public potential reference ground. Meanwhile, in case of a lightning strike or short-circuit fault, it could also discharge the fault current quickly, reducing the ground potential, and thereby protect the staff and electrical equipment inside. Therefore, the design of grounding network, grounding parameters as well as grounding performance deserves highlight of relevant researchers and operation of substations [1-3]. At present, grounding impedance of grounding networks has been discussed widely by scholars. But ground potential rise and potential difference within the grounding network are often neglected, which is a real problem caused by formation of grounding networks. Lightning strikes impose significant impact on the time characteristics of the potential. As a result, the issue of evaluating the safety performance of large grounding networks scientifically has also attracted more researchers' attention [4,5].

Lightning is one of major threats of power system because of its high voltage, intense current, transiency and uncertainty. Theoretical and experimental analysis of the characteristics of lightning waves, the process of lightning intrusion into the power system is necessary for effective lightning protection [6-8]. With application of new materials and improving environmental awareness, SGCC (State Grid Corporation of China) is now promoting design and construction of steel substations, which are made of metal as a whole including the roof. This raises new issues about their lightning characteristics and lightning protection [9, 10]. There are now several grounding means of a metal substation when its roof was struck by a lightning flash, which includes external grounding, nearby grounding, separate grounding and common grounding. In the first type of grounding system, the down lead is wrapped with insulating material. It is then connected with the grounding grid conductor on the outer edge of the grounding grid. In the nearby grounding system, the downlead is directly connected with the nearest grounding grid conductor. The grounding system above are both common grounding, which means lightning protection grounding and substation structure share the grounding system. As for separate grounding, the lightning protection device is completely insulated from the substation structure. In this case, the metal roof is insulated from the steel structure wall, and the lightning protection downlead is wrapped with insulating materials and separately set with grounding electrode. Through modeling and simulation, this paper analyzed the variation characteristics and surface potential distribution caused by lightning strike on the roof with the four grounding modes mentioned above separately to find out the most appropriate grounding mode. This work provides a reference for the design and safety performance evaluation of lightning protection and grounding mode of metal structure substations.

2. Modeling Establishment

2.1. Modeling of the Substation and Grounding System

Figure 1 shows the calculation model of a steel structure substation under construction in Nanjing. The grounding grid of the substation has a span of 5.5m-6m, total length of about 84.5m, width of about 43.5m and underground depth of 0.8m. Due to the edge effect of lightning current, we set the metal roof lightning point at the edge of the building. The downlead is connected to the grounding grid at 12 points. The grounding grid and the downlead is made of $40\times4mm^2$ copper coated steel. Resistivity of grounding conductor was set as ρ=1.78×10-7Ωm and relative permeability was set as μ=200. Earth resistivity of the substation was set as 100 Ωm.

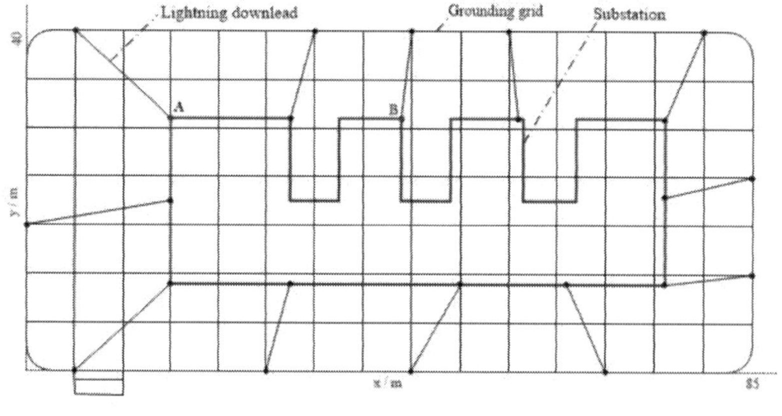

Figure 1. substation and grounding system model.

2.2. Lightning Current Model

According to GB50057-2010 Code for design of lightning protection of buildings, the substation was considered as a Class II lightning protection building. So this paper adopts the standard lightning wave with specifications of 10/350μs and 100kA. Its waveform is expressed by double exponential function as follow.

$$I(t) = I_m k(e^{-\alpha t} - e^{-\beta t})$$

In the equation above, lightning current peak I_m=100kA, waveform correction coefficient k=1.051, wavefront attenuation coefficient $\alpha=2.127\times10^3 s^{-1}$, wave tail attenuation coefficient β=2.461×105s-1. Sampling time, sampling frequency and time domain truncation window width were set as 0.6104s, 1.638MHz and 2500us separately. The time domain waveform is shown in figure 2, and the corresponding spectrum is shown in figure 3. We can see that the lightning current is mainly distributed in the frequency band of 0-10kHz, much within 0-1khz. The current distribution at each frequency point doesn't change dramatically. On the other hand, within the spectrum of 1-10kHz, the current distribution decreases rapidly with increase of frequency.

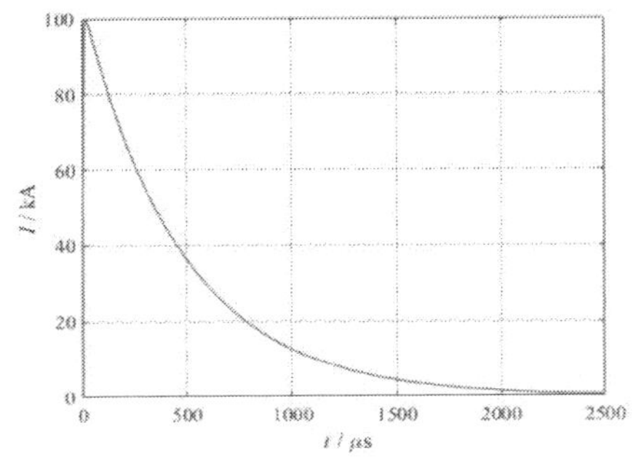

Figure 2. Time domain waveform of lightning current.

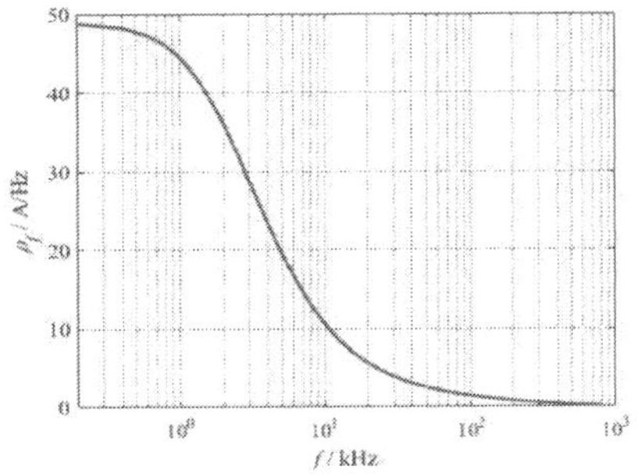

Figure 3. Frequency spectrum of lightning current.

3. Simulation and Calculation

3.1. Ground Potential of Nearby Grounding and External Grounding System

Many steel structure substations have lightning downleads connected to the grounding gird of the substation structure. But grounding modes are not the same. For some substations, the downlead is wrapped with insulating material and connected to the outer edge of the grounding grid. Except for the grounding point, the downlead is insulated with the substation. This approach is considered as external grounding. In another approach, the downlead is directly conducts lightning current to the nearest grounding grid conductor, which is called nearby grounding. We carried out simulation calculation by using CDEGS (a commonly accepted grounding performance analysis package) to analyze characteristics of surface potential distribution when the substation roof was struck by a strike. The lightning-struck point on the metal roof was assumed as A (17.25m, 29.1m, 8.4m). In case of external grounding and nearby grounding, the distribution characteristics of surface potential at the point(x=0-85m, y=0-40m) at different time are shown in figure 4 (a-Surface potential in case of external grounding, t=5.495μs; b-Surface potential in case of nearby grounding, t=7.326μs; c-Surface potential in case of external grounding, t=15.873μs; d-Surface potential in case of nearby grounding, t=15.873μs).

 2148 (2022) 012049 doi:10.1088/1742-6596/2148/1/012049

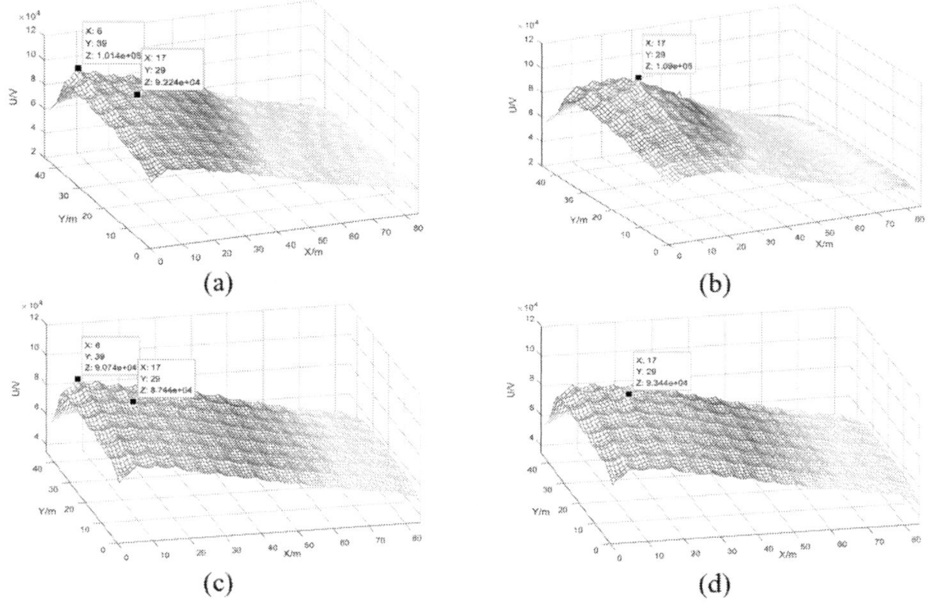

Figure 4. Surface potential at the point(x=0-85m, y=0-40m) at different time in nearby grounding system and external grounding system.

As is shown in figure 4(a) and figure 4(b), in the external grounding system the grounding potential reached its peak at 101.4kV 7.326μs after a lightning strike on the roof. In the nearby grounding system, the corresponding value was 109kV and 5.495μs. It is also seen that before the surface potential reaches its peak, the surface potential of nearby grounding rises faster than that of external grounding, and the distribution of the former system is steeper. The rise of surface potential is slightly slow and the distribution is relatively moderate in the external grounding system. Figure 4(c) and figure 4(d) shows the grounding potential 15.873μs after a lightning strike, when the two system has similar potential distribution. Besides, it is the connecting point of the downlead and ground grid where the potential reaches its peak.

3.2. Ground Potential of Seperate Grounding and Common Grounding System

3.2.1. Separate Grounding with Multiple Earthing Points. The external grounding and nearby grounding discussed in 3.1 are both common grounding system. In this chapter, we take single grounding into consideration. That is to say, downlead grounding electrode is insulated with the substation structure. The lightning-struck point on the metal roof was assumed as B (44.6m, 29.1m, 8.4m). The distribution characteristics of surface potential at the point(x=0-85m, y=0-40m) at different time in different grounding system are shown in figure 5 (a-Surface potential in case of separate grounding with multiple points, t=15.873μs; b-Surface potential in case of common grounding, t=15.873μs; c-Surface potential in case of separate grounding with multiple points, t=200.855μs; d-Surface potential in case of common grounding, t=200.855μs).

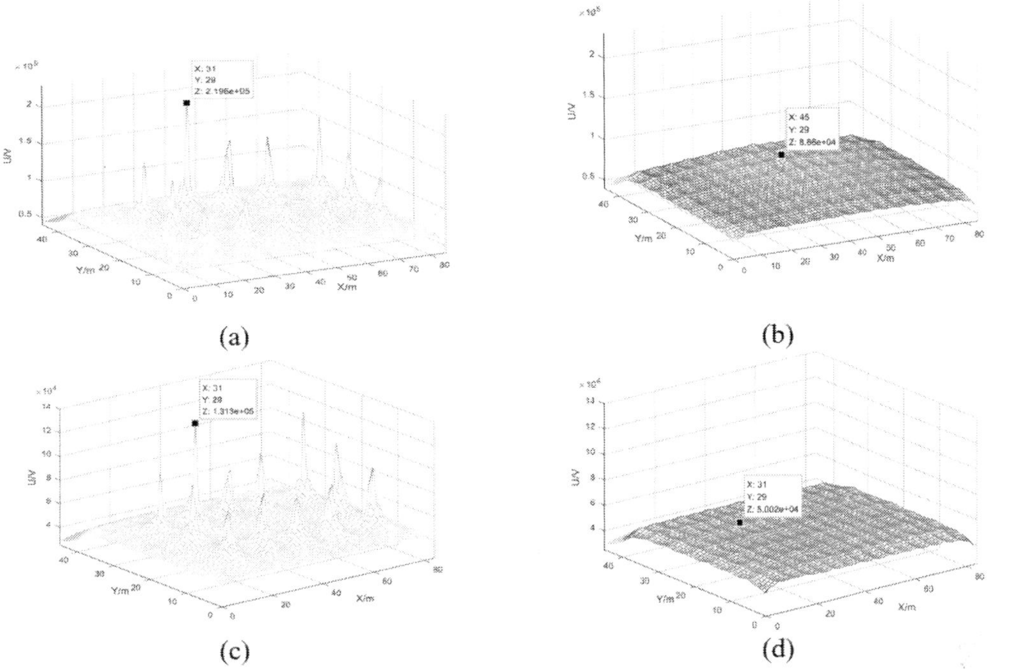

Figure 5. Surface potential at the point(x=0-85m, y=0-40m) at different time in seperate grounding system and common grounding system.

We can safely come to the conclusion as follows by contrast of the grounding potential.

- The peak of the grounding potential in separate grounding system reaches as 230kV, higher than that in common grounding system. The peak comes at t=7.937μs, later than in the latter system.
- The grounding potential has a peak value at all grounding points in single grounding system and the surface potential distribution is extremely uneven. In case of common grounding system, the grounding potential fluctuates moderately without multiple peaks.
- Separate grounding has poor discharge capacity and the potential peak lasts for a longer period. The lightning current could be discharged much more quickly in common grounding system.

3.2.2. Separate Grounding with a Single Earthing Point. In this case, we assume that there is only one downlead connected to the grounding grid. Hypothetical conditions are the same with 3.2.1. The distribution characteristics of surface potential are shown in figure 6. (a-t=15.875μs; b-t=1000μs).

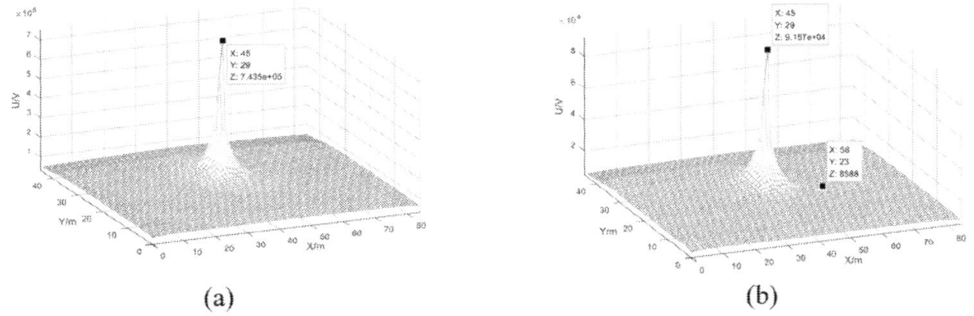

Figure 6. Surface potential at the point(x=0-85m, y=0-40m) at different time in separate grounding with a single earthing point.

As is shown in figure 6, the surface potential has a peak at the grounding electrode. 15.875μs after the lightning strike the peak is as high as 743.5kV. It is 91.57 kV 1000μs later, which is also dangerous to human and device safety. So we can see a worst discharge capacity in this system.

4. Impact Time of Lightning Current

In order to study how long lightning current would impact on the grounding potential, we take nearby grounding with multiple earthing points as an example to simulate the grounding potential distribution. The lightning-struck point on the metal roof was assumed as B, which was the same as discussion in 3.2. As is seen from Fig.8, 3500μs after the lightning strike on the roof the grounding potential all over the grid is around 36V, which is safe to humans and equipment inside. So we can say that in common grounding system, including both nearby grounding and external grounding, the lightning impact on the substation would be within several microseconds due to discharge via the grounding grid.

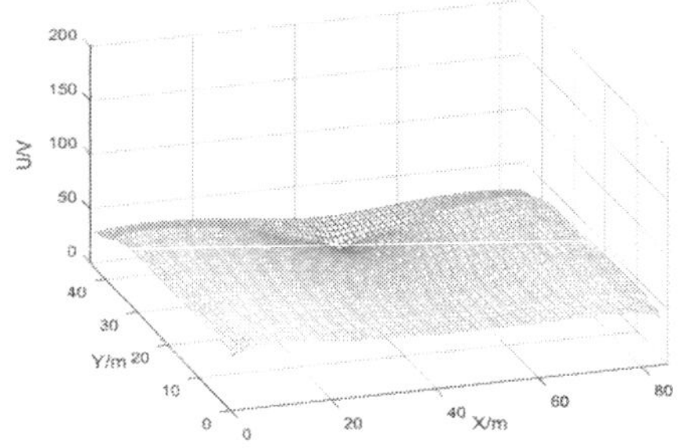

Figure 7. Surface potential of the common grounding system, t=3500μs.

5. Conclusions

We took a metal structure substation in Nanjing as an example and analyzed its ground potential in case of different grounding system by simulation and calculation. Nearby grounding, external grounding, separate grounding and common grounding are taken into consideration. We came to conclusions as follows.

- When the lightning downlead is insulated from steel structure and grounding grid of the substation, the grounding potential could reach as high as 743.5kV and 230kV with a single earthing electrode and multiple electrodes respectively. 1000μs after a lightning strike on the roof the ground potential is 91.57 kV, which is still a significant threat to humans and equipment inside. So discharging capacity of this grounding system is much worse than common grounding. It is not a sound choice of grounding system for steel structure substations.
- Nearby grounding and external grounding are both common grounding system. In the former system, the downlead is directly conducts lightning current to the nearest grounding grid conductor. In the latter one, the downlead is wrapped with insulating material and connected to the outer edge of the grounding grid. Except for the grounding point, the downlead is insulated with the substation. The peak of ground potential is 101.4kV for external grounding and 109kV for nearby grounding. Waveform and peak time of the two kinds of system are similar to each other. So they have similar discharging capacity. From the perspective of cost and environmental protection, nearby grounding is a more reasonable scheme.
- 3500μs after the lightning strike on the roof the grounding potential all over the grid is around 36V, which is safe to humans and equipment inside. So we can say that in common grounding system, the lightning impact on the substation would be within several microseconds due to discharge via the grounding grid.
- The point of the downlead connecting to the grounding grid tends to have the highest potential when the roof of the metal substation is struck by lightning current. So it is recommended to keep the equipment inside away from this earthing electrode as far as possible.

Acknowledgments
This paper is co-sponsored by Natural Science Foundation of Jiangsu Province (BK20191094) and Scientific Research Programme of State Grid Nanjing Power Supply Company (SGTYHT/18-JS-202).

References
[1] He J L, Zeng R. 2007 Ground technology in power system [M] Beijing: Science Press 80-161(in Chinese).
[2] Zhang B, He Jinliang, Zeng Rong. 2015 State of art and prospect of grounding technology in power system [J] *High Voltage Engineering* **41** (8) 2569-2582(in Chinese).
[3] Liu Y, Cui X, Zhao Z B. 2009 Method of Corrosion Diagnosis of Substations' Grounding Grids Based on Electromagnetic Induction Theory [J] *Proceedings of the CSEE* **29** (4) 97-103.
[4] Li Q. 2013 Condition evaluation of large grounding grid of Power Plant and Substation [M] Beijing: China Electric Power Press 133-185(in Chinese).
[5] Li Q, Zhang B. 2018 Multi-aspect Assessment and Parameter Calculation Method of Large Grounding Grid [J] *Proceedings of the CSEE* 38 129-136.
[6] Yuan Tao, Li T X, Sima W X. 2018 Analysis of Soil Discharge Image Characteristics under Impulse Currents [J] *Proceedings of the CSEE* **38** (17) 5266-5274.
[7] Zhang Z H, Shi W D, Kang P. 2020 Research on Measurement of Impact Characteristics of UHV Converter Station Grounding Grid [J] *Proceedings of the CSEE* 40 140-148.
[8] Peng X Y, Wen X S, Wang R. 2019 Experimental Study on Lightning Triggering Ability of Lightning Towers With Different Materials[J] *Power System Technology Apr* **43** (4) 1322-1328.
[9] Yang W, Zhu T Y, Tian Y. 2020 Analysis for Transient Electric Field Computation of GIS Basin Insulator Under Lightning Impulse Voltage [J] *High Voltage Engineering* **46** (3) 807-814.
[10] Sima W X, Zhang H J, Yang M. 2021 Statistical Analysis of the Amplitude and Waveform Characteristics of Intrusive Lightning Over-voltages Measured From the 10 kV System of a Substation [J] *High Voltage Engineering* **47** (1) 214-222.

Study on Temperature Control of Gravity Anchorage without Cooling Water

Hanzheng Xu, Yazhou Yin*, Kaihan Xu and Chongjin Li

School of Highway, Chang'an University, Xi'an, Shaanxi 710064, China
*Corresponding Author, Email:15847263426@163.com

Abstract. This paper uses Midas Fea simulation software to analyze the hydration heat of a suspension bridge anchorage mass concrete construction without cooling water. According to specific boundary conditions and convection coefficients, the concrete heating process and cooling process are simulated. Analyze the influence of surface air convection coefficient on the surface tensile stress of the cast layer, and the influence of the pouring interval on the interlayer stress of the anchor block, and the temperature difference between the inside and outside of the concrete when the anchor block is layered. It is found that reducing the surface convection coefficient of the pouring layer can effectively improve the stress condition, and the pouring interval has little effect on the stress.
Keywords. Anchorage hydration heat without cooling water air convection coefficient pouring interval.

1. Introduction

During the curing period of the integrally poured mass concrete, the temperature shrinkage and deformation caused by the cooling is relatively large. Temperature shrinkage deformation will generate tensile stress in the structure and its surface under constrained conditions. When the tensile stress exceeds the tensile strength of the concrete at the corresponding age, or the confined shrinkage deformation of the concrete exceeds the ultimate tensile strain capacity of the concrete, cracks will occur. [1] Therefore, in the process of mass concrete construction, to avoid excessive temperature stress, prevent the generation of temperature cracks or control the cracks within a certain limit, temperature control must be carried out. The main goal of temperature control is to make the temperature field change inside the mass concrete develop according to the expected goal, which can be divided (1) reduce the maximum adiabatic temperature rise and the maximum temperature peak; (2) control the temperature gradient of the surface concrete to reduce The temperature difference between inside and outside prevents the cold shock effect; (3) Control the cooling rate of concrete; (4) Control the temperature difference between upper and lower layers to prevent possible interlayer cracks and reduce interlayer restraint. [2] The construction temperature control of large-volume concrete structures should be carried out from the design stage. The main contents are: (1) perfect structural design and improve constraint conditions; (2) optimize design mix ratio and rational raw material selection; (3) Optimize construction technology and strengthen construction management; (4) Adopt heat reduction and insulation measures; (5) Scientific prediction and accurate monitoring. [3] The general principle of temperature control can be summarized as "internal reduction and external protection". "Internal drop" refers to reducing the amount of cement by optimizing the mix ratio and reducing the maximum adiabatic temperature rise of concrete. Arrange the cooling water pipes before pouring, keep the water pipes with cold water during the pouring process, to achieve the purpose of reducing the temperature peak and delay the time of the temperature peak, and in the cooling stage after the completion of all the concrete pouring, the temperature of the cooling water pipe is not low. The cooling rate is controlled by circulating water with a maximum temperature of 10 degrees Celsius inside the concrete. "External protection" refers to covering the surface of the concrete with plastic film and geotextile for heat preservation and moisturizing after the final set of the concrete, to avoid the excessive temperature difference between the inside and the outside and the cracking of the concrete surface caused by the evaporation of water. The use of non-cooling water pipes for mass concrete construction can only reduce the maximum adiabatic temperature rise of concrete from the aspect of optimizing the mix ratio. For the construction that can meet the requirements of the specification, not using the cooling water pipe can effectively save materials and manpower, and at the same time, it can also avoid the rust of the water pipe caused by the late grouting of the cooling water pipe and the lack of compactness.

2. Project Overview

Gravity rock-socketed anchorage is used on both sides of a ground anchor suspension bridge, and an expanded foundation is used for anchorage foundation. Gravity anchorage is composed of anchor block, anchor chamber, loose cable saddle foundation, saddle, and anchorage system. C30 concrete is used for the rest of the anchorage except the support pier and anchor chamber, and C30 micro-expansion concrete is used for the post-pouring section. The anchor block is a C30 mass concrete structure of 50×24×28m (length × width × height of the maximum section). The mass uncooled water pipe casting process is adopted for the casting of anchorage concrete, and the construction process of "left and right blocks and vertical layers" is adopted. The vertical casting thickness is 2m, and each casting is 28m for left and right sides, which is finished 14 times.

3. Calculation of Heat of Hydration and Simulation of Construction

3.1. Material and Mix Ratio

Fly ash: F class I grade ash
Admixture: water reducing agent
Crushed stone: two-stage graded gravel
Sand: river sand
Water: Groundwater

Table 1. A slightly more complex table with a narrow caption.

Material consumption (kg/m^3)						Compressive strength (MPa)			Setting time (h)	
Cement	Fly ash	Sand	Gavel	Water	Admixture	7d	28d	60d	Initial set	Permanent set
280	160	813	1077	140	4.4	30.9	35	42	7	10

3.2. Definition of Model Parameters

C30 concrete has an elastic modulus of 30MPa, a density of 240kg/m^3, a Poisson's ratio of 0.2, a thermal conductivity of 10.6kJ/(m•h•°C)and specific heat of 0.96kJ/(kg•°C). When defining the shrinkage and creep of concrete, according to the CEB-FIP specification, the 28d compressive strength of C30 concrete is 30000kN/m^2, the annual average relative humidity is 65%, and the concrete material is set to 3d when the material starts to shrink.

3.3. FEA Model

According to the actual construction situation, choose the right anchor block, establish the anchor foundation, and four-layer anchor block model for simulation calculation. The model sets a fixed boundary at the bottom of the foundation, takes the average atmospheric temperature of 20°C, and applies it to the bottom of the foundation according to the forced temperature. A convection boundary is set for the contact between the anchor block and the template and the contact surface with the surrounding rock mass. The convection coefficient is 14W/ (m²•°C). Considering that the top surface of the cast layer is in contact with air, the convection boundary is also set, and the air convection coefficient is 4W / (m²•°C), the model in figure 1.

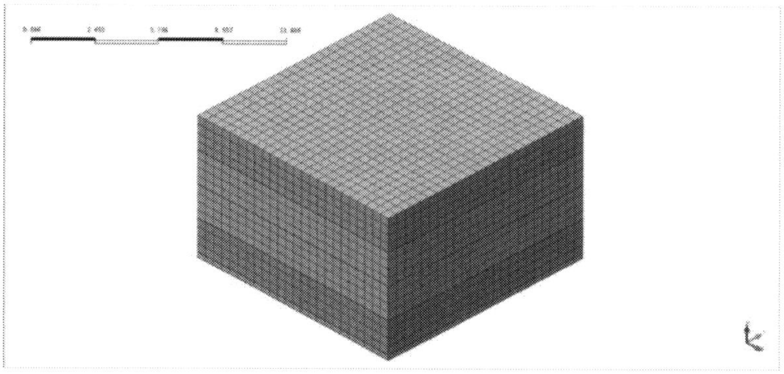

Figure 1. FEA Model.

Working condition 1, the temperature of fly ash cement concrete is 20°C, and the maximum adiabatic temperature rise of C30 concrete is 36.8°C and the temperature conductivity coefficient is 0.529 according to the Japanese JSCE2002 standard.

Working condition 2, changing the temperature of fly ash cement concrete to 10°C under other conditions unchanged, its adiabatic temperature rise is 33°C, and the temperature conductivity coefficient is 0.309.

Analyze the calculation results of working conditions one and two. When other conditions remain unchanged, the temperature of fly ash concrete changes from 20°C to 10°C. During the pouring of each layer of concrete, the peak temperature and the temperature difference between inside and outside will be reduced. The stress value has been reduced.

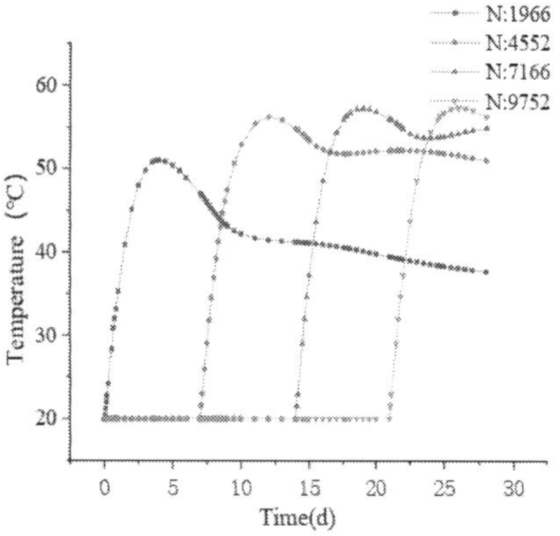

Figure 2. Temperature change curve of key points in the centre of each layer under working condition 1.

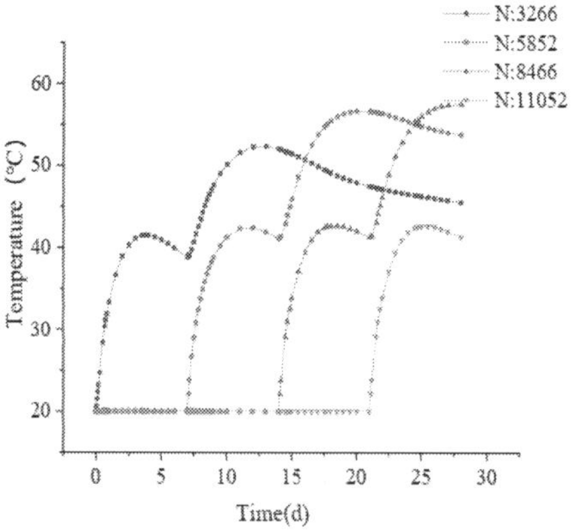

Figure 3. Temperature change curve of key points on the surface of each layer under working condition 1.

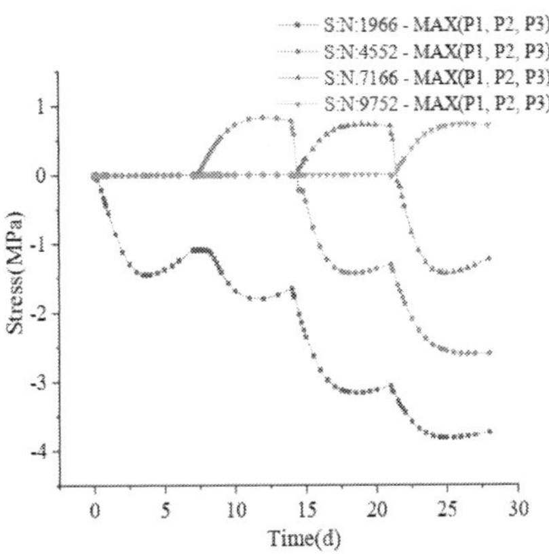

Figure 4. Stress development curve of key points in the centre of each layer ender working condition 1.

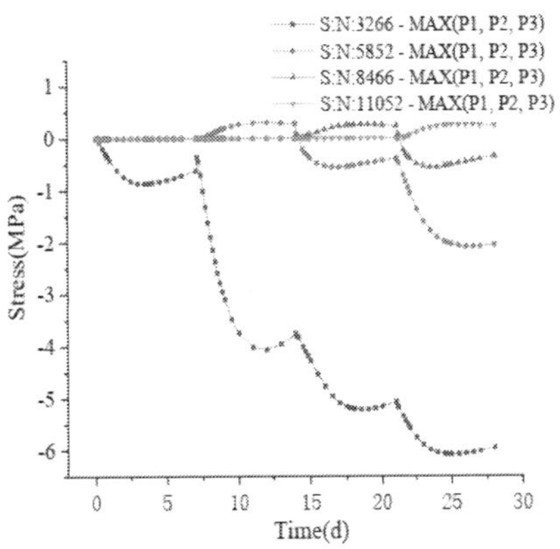

Figure 5. Stress development curve of key points on the surface under working condition 1.

Analyze the calculation result of working condition 2, under the condition that the temperature difference between the inside and outside of each layer of concrete is not more than 25°C, the stress of the key points of the concrete 4d-6d after the completion of the pouring exceeds the allowable tensile stress, indicating that the external environment temperature is 25°C. When the cement dosage is 240kg/m3, the adiabatic temperature rise of fly ash cement concrete at 20°C is 36.8°C, and the temperature conductivity coefficient is 0.529. If the cooling pipe is not used, the stress will exceed the allowable stress.

The calculation results show that the temperature of the first layer of anchor blocks reaches a peak of 44°C 5.5 days after the completion of pouring, and then enters a slow cooling stage. The second layer of anchor blocks reached a peak temperature of 49.6°C 6.5 days after pouring, and then slowly cooled to 48°C, and stabilized. The temperature of the third layer of anchor blocks reached a peak of 50.7 °C. 7 days after the completion of pouring, and then entered a slow cooling stage, and then stabilized at 50°C. With the heating process of the fourth layer of anchor blocks, the central temperature of the third layer increased to 51.6 °C. The temperature of the fourth layer of anchor blocks reached a peak of 50.8°C 7 days after the completion of pouring. The temperature change curves of the key points of each layer of concrete are shown in figure 6 and figure 7.

Considering the influence of the concrete's weight, the compressive stress of the first layer will increase with the application of subsequent pouring layers. The maximum stress of the first layer does not exceed the allowable stress. The pouring of the second layer anchor block is completed, and the maximum stress of the key point exceeds the allowable stress at this point. When the third layer is completed, the stress of the key point is less than the allowable stress. The pouring of the third layer of anchor blocks is completed, and the maximum stress at the key point exceeds the allowable stress at this point. When the third layer is poured, the stress at the key point is less than the allowable stress. The stress development curves of key points of each layer of concrete are shown in figure 8 and figure 9.

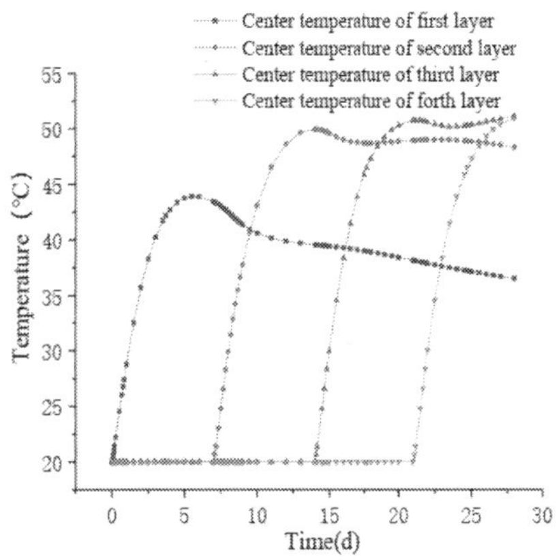

Figure 6. Temperature change curve of key points in the centre of each layer under working condition 2.

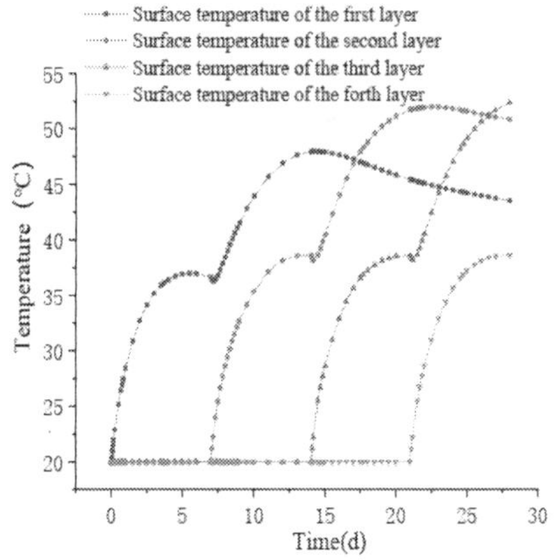

Figure 7. Temperature change curve of key points on the surface of each layer under working condition 2.

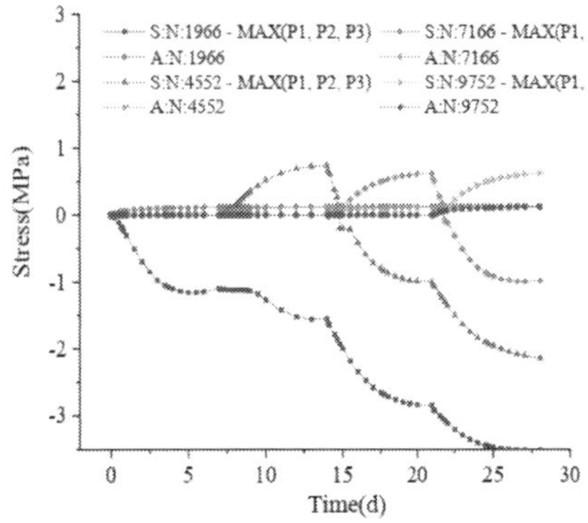

Figure 8. Stress development curve of key points in the centre of each layer in working condition 2.

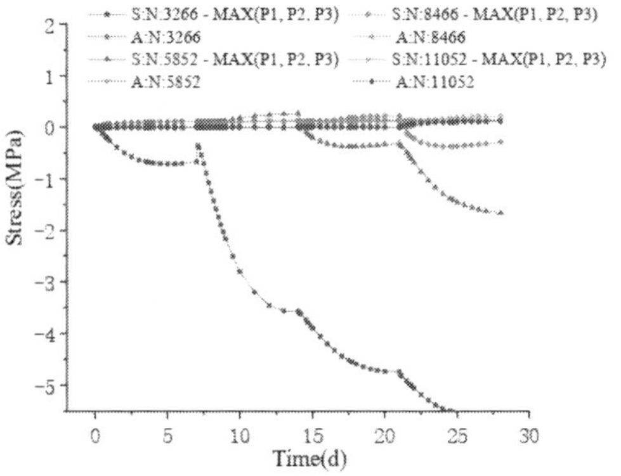

Figure 9. Stress development curve of key points on the surface working condition 2.

4. The Influence of Air Convection Coefficient on the Surface of the Pouring Layer

In the actual construction process, the thermal insulation material used on the surface of the cast layer has its specific thermal conductivity. By keeping other parameters constant in the model and changing the size of the air convection coefficient, the influence of the thermal insulation measures on the surface of the cast layer on the stress and temperature of the cast concrete was analysed.

4.1. The Influence of Air Convection Coefficient on the Stress of Each Layer of Concrete

The calculation results show that the tensile stress at the central point of the cast layer surface increases with the increase of the convection coefficient, the weakening of the thermal insulation effect of the concrete surface will increase the risk of concrete surface cracking. The stress development curves of each layer of concrete under different air convection coefficients are shown in figures 11-14. Analyzing the relationship between the stress of the second layer surface point and the convection coefficient, it can be seen that when the air convection coefficient is greater than equal to 6, the stress value of the surface central point is greater than the allowable stress at this point, as shown in figure 10.

 2148 (2022) 012050 doi:10.1088/1742-6596/2148/1/012050

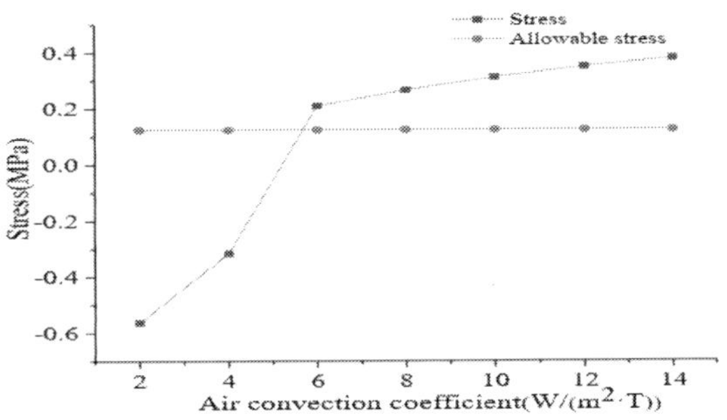

Figure 10. The relationship between the maximum stress on the second layer surface and different air convection coefficients.

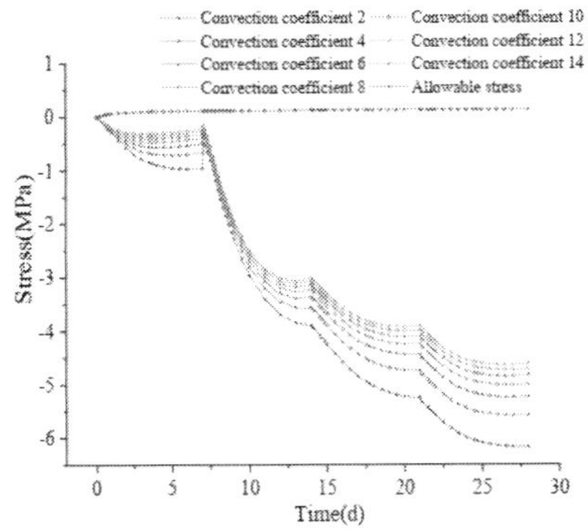

Figure 11. Stress development curve of the central point on the first layer surface under different convection coefficients.

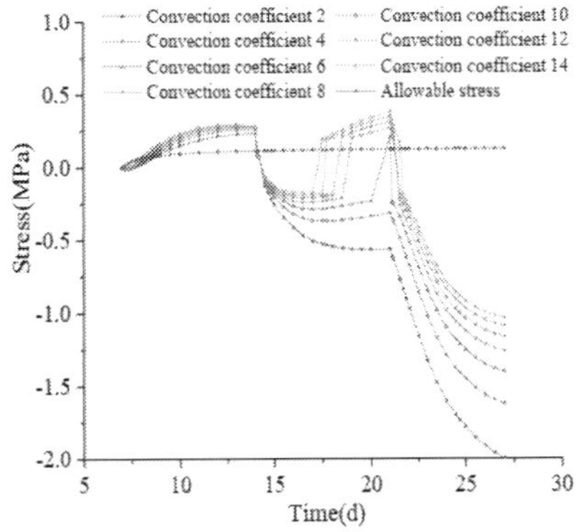

Figure 12. Stress development curve of the central point on the second layer surface under different convection coefficients.

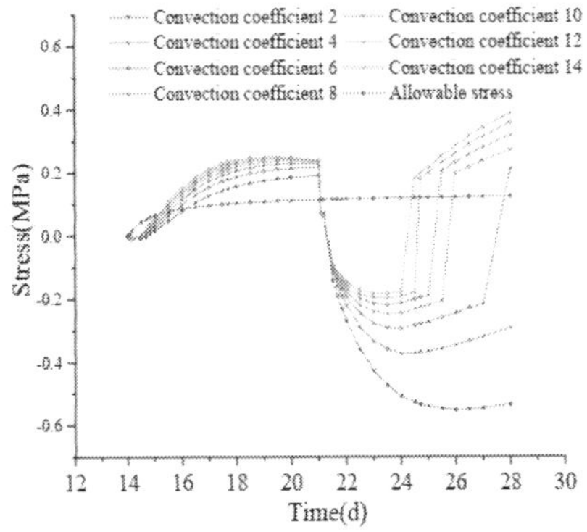

Figure 13. Stress development curve of the third layer surface central point under different convection coefficients.

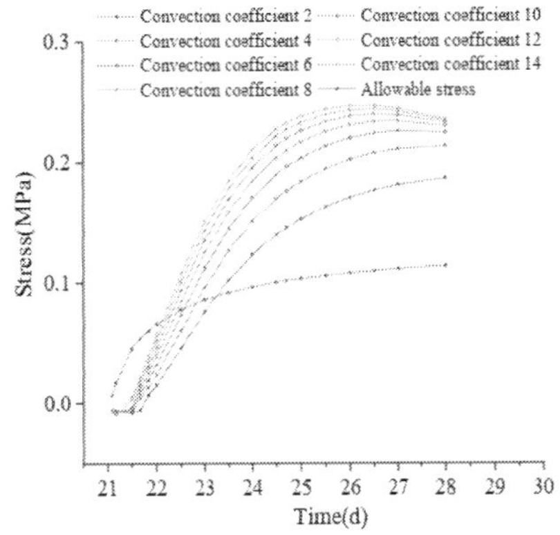

Figure 14. Stress development curve of the central point of the fourth layer surface under different convection coefficients.

4.2. The Influence of Air Convection Coefficient on the Temperature of Each Layer of Concrete

When the surface air convection coefficient of the last layer is 2, the highest temperature in the center of the fourth layer is 53.1°C, and when the surface air convection coefficient is 14, the highest temperature in the center is 47°C, as shown in figure 15. The peak temperature of the center of the cast layer decreases with the increase of the air convection coefficient on the surface of the cast layer. The temperature difference between the inside and outside of the concrete layer decreases with the increase of the air convection coefficient, as shown in figure 16.

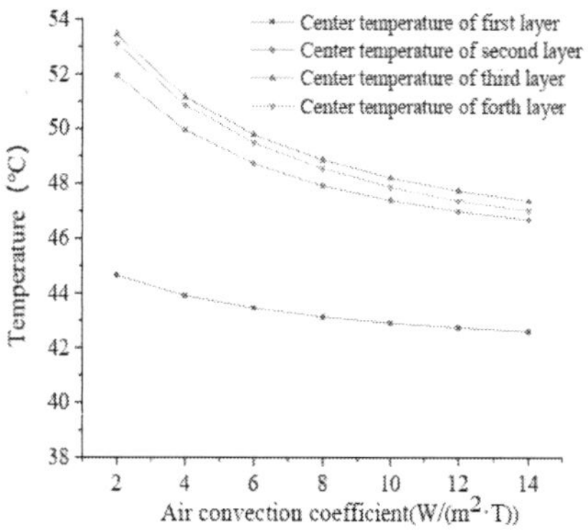

Figure 15. These two figures have been placed side-by-side to save space. Justify the caption.

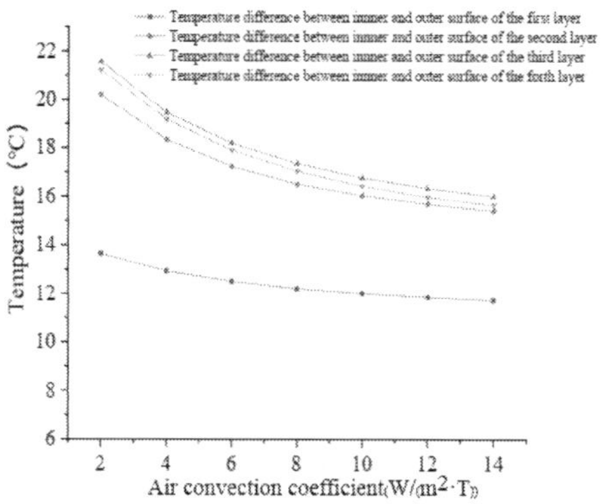

Figure 16. These two figures have been placed side-by-side to save space. Justify the caption.

5. The Influence of Pouring Time Interval

The mass concrete specification requires that the time interval for layered pouring should not exceed seven days, but in actual construction, due to the influence of the construction environment (such as the rainy season), it is difficult to ensure that the pouring interval is within the range of 7 days. Through model and analysis of the impact of different interlayer pouring intervals, it is found that after the fourth layer pouring is completed, the surface central temperature of the anchor blocks of the first and second layers will decrease with the increase of the interval time, which is roughly in a linear relationship, see Figure 18. The maximum temperature difference between the highest temperature at the center and the surface temperature of the layering decreases with the extension of the pouring interval, as shown in Figure 19. The influence of different pouring time intervals on the stress of each layered interface is quite small. Take the stress at the central point of the top surface of the second layer as an example, as shown in Figure 17. The maximum tensile stress is 0.247 MPa when the pouring interval is five days, and the maximum tensile stress is 0.25 MPa when the interval is ten days.

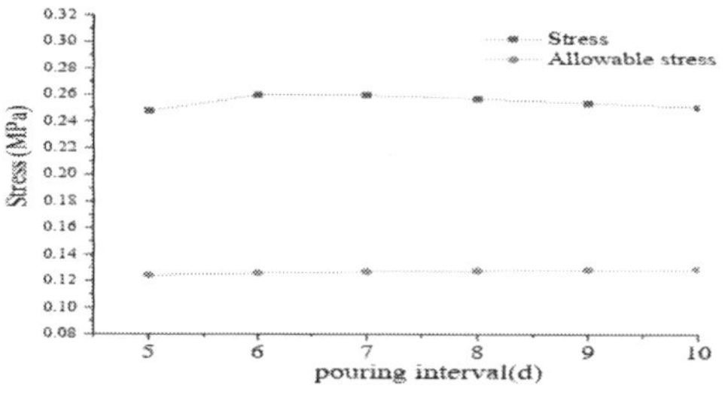

Figure 17. In this case simply justify the caption so that it is as the same width as the graphic.

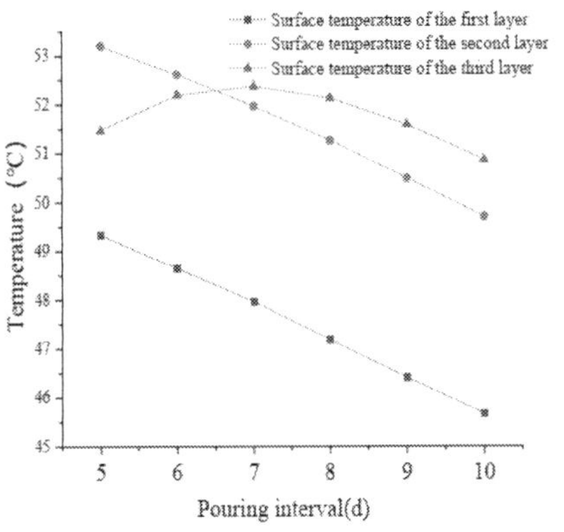

Figure 18. These two figures have been placed side-by-side to save space. Justify the caption.

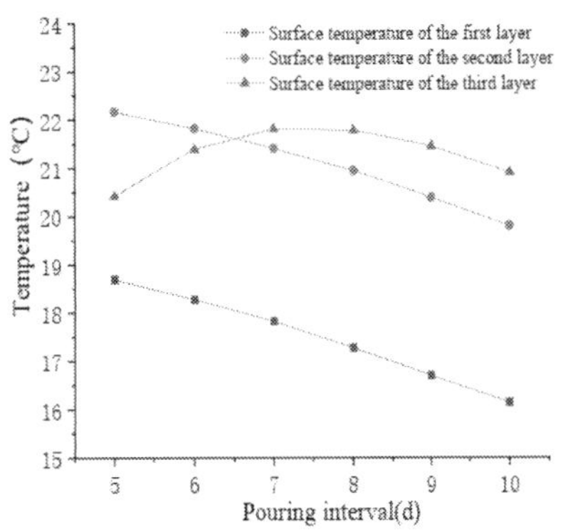

Figure 19. These two figures have been placed side-by-side to save space. Justify the caption.

6. Conclusion

This paper uses Midas FEA software to analyze the effects of three factors: materials, surface insulation measures, and interlayer pouring interval on the construction of suspension bridge's anchorage mass concrete without cooling water, the following conclusions can be drawn.

(1) The reduction of the temperature of fly ash cement can reduce the maximum adiabatic temperature rise and thermal conductivity of concrete, reduce the temperature difference between the inside and outside of the concrete, and improve the surface tensile stress during layered pouring.

(2) The surface tensile stress of the pouring layer is proportional to the air convection coefficient. Reducing the air convection coefficient and strengthening the thermal insulation effect can improve the surface tensile stress of the concrete.

(3) For large-volume concrete constructed by layered pouring without cold water pipes, the interlayer pouring interval is in the range of 5 to 10 days, which has no effect on the maximum tensile stress on the surface. The maximum temperature of the center of the interlayer and the surface temperature will vary with the interval time increases and decreases.

When mass concrete is constructed without cooling water, strictly control the raw materials, reduce the hydration heat and thermal conductivity of the concrete itself, strengthen the thermal insulation effect and control the temperature difference between inside and outside, which can effectively reduce the tensile stress on the concrete surface and reduce the risk of cracking.

For large-volume concrete constructed without cooling water, it is necessary to determine the raw materials, ambient temperature, concrete molding temperature, insulation measures, boundary constraints, and layer thickness according to the actual construction environment to formulate a complete construction plan.

References

[1] Zhu B F. 1999 Mass concrete temperature stress and temperature control [M] Beijing: China Electric Power Press 1 5-15, 272-279, 320-331.

[2] Yu Y N, Zhang W, Shen Y G. 2010 Early surface cracking control measures for large-volume concrete caps [J] *Journal of Zhejiang University* (Engineering Science Edition) **44** (08) 1621–1628.

[3] Chen Z X, Tang L. 2001 Temperature control of mass concrete in large bridges [J] *Bridge Construction* (01) 14-16+20.

Analysis of Situ Test Results of Earth Pressure and Pile Internal Force under Pile-Anchor Support in a Deep Foundation Pit in Xi'an

Ruibin Yang[1,*], Xinsheng Li[1], Dongzhou Xie[2] and Hongte Meng[3]

[1] College of Geological Engineering and Geomatics, Chang'an University, Xi'an 710054, Shaanxi, China
[2] Institute of Rock and Soil Mechanics, Chinese Academy of Sciences, Wuhan 430071, Hubei, China
[3] China Water Resources Beifang Investigation, Design and Research Co. Ltd. (BIDR), Tianjin 300222, Tianjin, China
Email: yangrb729@126.com

Abstract. At present, in deep foundation pit engineering, on the one hand, practice is ahead of theory, and on the other hand, theory can not correctly reflect the actual construction process and environmental effects. In order to further study the distribution and change law of earth pressure and internal force of pile body in deep foundation pit pile-anchor supporting system, field monitoring test of earth pressure and pile body reinforcement stress was carried out. The monitoring results show that before excavation, the distribution of earth pressure has a great relationship with the layering of the soil, and it is distributed in sections along the depth. Compared with the theoretical static earth pressure, the measured data of the upper depth is relatively small; after excavation, the overall earth pressure is distributed along the depth in a "z" shape under the non-limiting state. As the excavation progresses, the magnitude of the reduction of the earth pressure varies from place to place, and the magnitude of the decrease of the soil with better properties is not large; after the excavation, the stress and earth pressure of the pile reinforcement correspond to each other, and the distribution is also nonlinear. The existence of anchor tension has an obvious effect on improving the internal force of the pile. The selected earth pressure calculation methods have some discrepancies in the calculation of the earth pressure value of the project, and they need to be further improved. The research in this paper can provide reference and reference for the calculation of earth pressure and support design of pile-anchor supported foundation pit.
Keywords. Foundation pit; pile-anchor support; earth pressure; pile internal force; earth pressure calculation.

1. Introduction

At present, the classical Coulomb and Rankine soil pressure theory is still used in the design of foundation pit support in our country, Which is clear, easy to use, combined with a lot of engineering experience, and construction monitoring, and it is widely used. However, it is not uncommon for engineering accidents to be caused by unclear understanding of earth pressure, and the development of the national economy requires more accurate and reasonable design theories. Therefore, the development of earth pressure calculation theory needs to be carried out urgently to understand the non-limit state, nonlinear distribution, and distribution of earth pressure. The influence of non-single factors needs to be further deepened.

The research and exploration of earth pressure by domestic and foreign experts and scholars has mainly focused on theoretical derivation, model testing, numerical simulation and other aspects for several years, and certain results have been achieved. In the calculation of non-limiting earth pressure considering displacement, the relationship curve of earth pressure changing with displacement was mostly based on geotechnical tests earlier, and then its physical mechanism and unique relationship were summarized [1-6]; There are many researches on the calculation of non-limiting earth pressure of non-cohesive soils in rigid retaining walls, and they are in-depth and rich [7-12]. Afterwards, using the theoretical concept of shear strength exerted value, many researchers considered the influence of soil displacement changes on the calculation of earth pressure, and put forward and deduced some calculation formulas of non-limiting earth pressure theory. Xu Riqing et al. [13] deduced and proposed the calculation method of non-limiting active earth pressure of cohesive soil, and compared with the model test, the results are quite consistent. Li Chaoyang et al. [14] deduced the non-limit active earth pressure calculation formula of flexible pile cohesive soil by summarizing and considering various earth pressure influence factors, and compared it with the classical Rankine calculation theory. In addition, In recent years, scholars have made efforts to study the calculation of the earth pressure in the pile anchor support, the unloading stress path of the excavation of the foundation pit, the additional earth pressure of the wall displacement, and the accurate design of the retaining structure of the foundation pit [15-18]. In terms of the calculation and monitoring of the internal force of the pile-anchor supporting pile, many studies focused on speculating and deducing the change and development of the internal force based on the horizontal displacement of the pile, the change of nearby ground settlement, and the analysis of finite element data.

Based on the field monitoring test of soil pressure and pile internal force under deep foundation pit pile-anchor support, combined with observation data of foundation pit dewatering and pile top displacement, this paper analyzes and discusses the soil pressure and pile internal force of layered soil under pile-anchor support, force changes and distribution laws. Several existing non-limiting soil pressure calculation methods are selected, calculated and compared with the actual measured values, and the design calculation methods that are in line with the actual working conditions of the foundation pit engineering are explored.

2. Earth Pressure Field Monitoring Test

2.1. Experimental Ideas

In the non-limiting soil pressure theory research problem, the field measurement of the soil pressure of the foundation pit under the cohesive soil flexible support system is closer to the actual situation than the model test, and the data of the structure in the actual working state can be obtained. Analogous rock-soil bearing capacity test, the bearing capacity value obtained from the in-situ test is always greater than that of the remodeled disturbed soil sample. The reason is inseparable from the strength of the original structure of the soil. Distribution is particularly important.

Due to the strict limitation of the displacement of the sidewall of the foundation pit at the actual construction site, the flexible supporting structure and the supported soil are coordinated, and the soil pressure of the foundation pit is in a non-limiting state [15], as shown in Figure 1. To ascertain the specific value, time and space, the reasonable arrangement and smooth burying of the earth pressure cell is very important. In addition, on-site earth pressure measurement has always been subject to material and technical difficulties, large investment, long period, and measurement accuracy is greatly affected by environmental factors.

This test relies on a large-scale deep foundation pit excavation project in Xi'an. Compared with the model test with a single test condition, this test relies on the actual in-situ test of the project, which is unmatched by the model test in terms of scale, time, and site effects. The actual engineering conditions are more complicated. On this basis, the results obtained may be caused by the coupling of multiple factors. Discussing the law of its distribution changes is of great benefit to promoting the development of non-limiting earth pressure calculation theory.

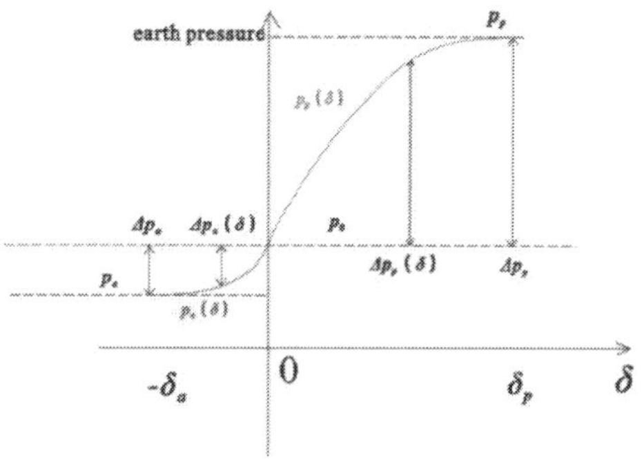

Figure 1. Earth pressure-displacement curve.

Figure 2. Sketch map of test site.

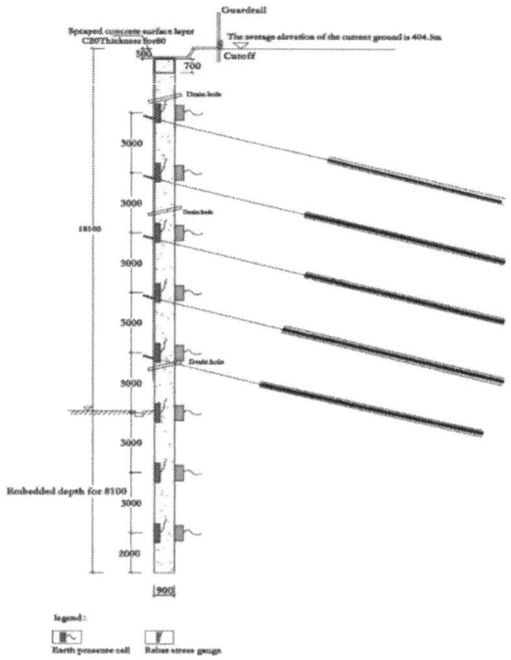

Figure 3. Layout of monitoring test sensors.

2.2. Project Overview and Test Plan

The Xi'an area is located in the middle of the Weihe Basin in the faulted depression. The foundation pit in this test is located on the first-level alluvial terraces on the east side of the Fenghe River. The excavation depth is about 17.1~19.8m, and the circumference is about 530m. The supporting structure is a composite flexible pile supporting mode with anchor rods, and the foundation pit adopts the open excavation method. This test refers to the foundation pit support design drawing and the corresponding geotechnical engineering survey report of the site. The specific physical and mechanical parameters of the soil layer beside the pile are shown in table 1. The site contains aquifers and requires precipitation during excavation. The water level at the location is about 16.9m. The foundation pit is next to the building load (the first phase of the project), as shown in figure 2. The height of the first phase of the project is about 24m. Its structure is a steel structure, and the foundation is a pile foundation.

This test was carried out on the supporting piles (section A-B, small box position in figure 2) of the foundation pit on the east side of the first phase of the project. The excavation depth of this section of the foundation pit is 18.1m, the supporting structure is pile-anchor support, the supporting pile is 25.0m long and the pile diameter is 0.9m. There are 5 anchor cables.

In this test, the WY-E06B vibrating wire earth pressure box is used for earth pressure monitoring. The earth pressure box is shown in figure 3(a). The vibrating wire sensor has good stability and high sensitivity, and is used in construction, railway, transportation, water and electricity, Dams, tunnels and other engineering fields. The working principle of the vibrating wire earth pressure cell is to determine the earth pressure by the natural frequency of the steel wire. The WY-CXX type steel bar stress meter is used to measure the internal force of the pile body. The steel bar stress meter is shown in figure 3(b). The working principle and working process of the vibrating wire steel bar stress meter are basically the same as those of the vibrating wire earth pressure cell. The difference is that the earth pressure box can only measure pressure, while the steel bar strain gauge can measure both pressure and tension. The main parameters are shown in table 2.

As shown in figure 4, it is a schematic diagram of the installation and burial of the soil pressure box and the reinforcing steel stress gauge in this experiment. In the figure, the soil pressure box is buried from top to bottom on the outside of the pit to measure the earth pressure in the active area on the unexcavated side outside the pit. And the steel bar stress gauge near the inner side of the pit is welded to the pile body to measure the internal force of the pile body. Number 16 earth pressure boxes and steel bar stress gauges with different line lengths in sequence, according to the buried depth of the earth pressure box according to T1, T2···T7, T8, and according to the buried depth of the steel bar stress gauges in turn according to G1, G2···G7, G8 are numbered, and the numbers are respectively attached to the surface of the earth pressure box, the steel bar strain gauge and the corresponding wires.

Table 1. Mechanical parameters of soil layer.

Number	Geotechnical name	γ(KN/m^3)	C(kPa)	φ(°)
(1)	Fill soil	15.0	5	10
(2)	Loess like soil	17.6	26	21
(3)	Silty clay	18.2	28	22
(3)-1	Medium sand	19.0	2	30
(4)	Silty clay	19.2	28	22
(4)-1	Medium sand	19.0	2	30

Table 2. Main parameter of sensors.

Sensor	Type	Range (MPa)	Deviation (kPa)	Installation method
Earth pressure cell	WY-E06B	0.6	1	Hanging cloth method
Rebar stress gauge	WY-CXX	±260	1	Welding method

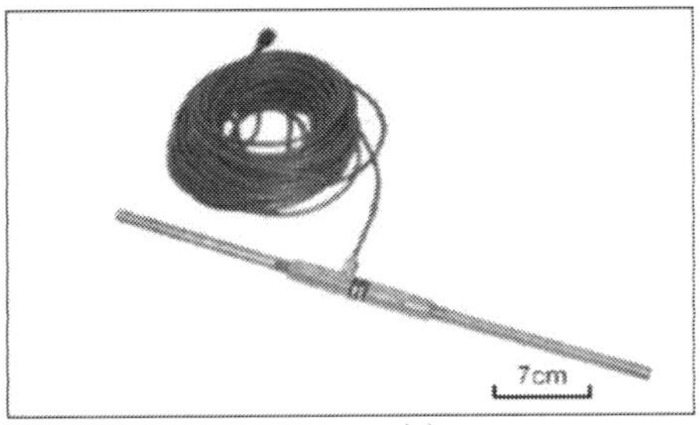

(a)

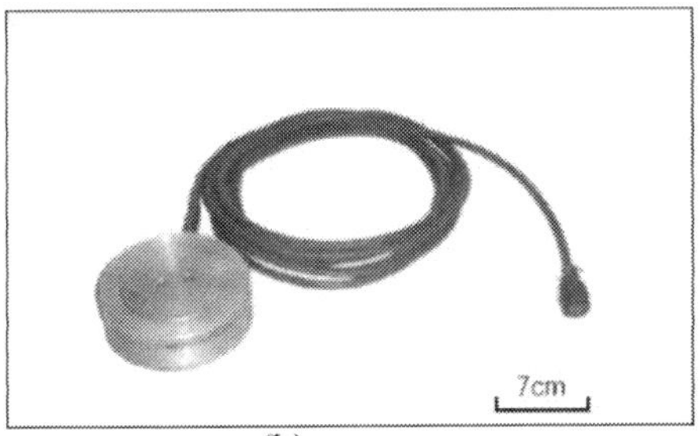

(b)

Figure 4. Photos of sensors.

2.3. Implementation Process

(1)Send all test equipment to the construction site, and communicate and coordinate with the construction unit. The steel bar strain gauge should be installed first, and then the earth pressure box should be installed.

(2)The steel bar stress gauge is installed by welding and the following steps, as shown in figure 5(a). Transport the reinforcement cage to a flat area, arrange the points at equal intervals from the pile end according to the layout drawing, and mark them with a marker. Install the rebar at both ends of the reinforcement strain gauge, and place it on the reinforcement cage according to the number. The strain gauge is bound to the main reinforcement of the steel cage with a fine loop of wire. The strain gauge and test lead should be bound on the bottom side of the structural steel bar to avoid damage to the strain gauge and lead during vibrating. Weld the rebar on both sides of the strain gauge with electric welding. On the main bar, use a wet cloth to wrap the stress gauge during welding to prevent excessive temperature from damaging the stress gauge. 10cm should be reserved for the wire at the meter.

(3) The earth pressure box adopts the hanging method and installs according to the following steps, as shown in figure 5(b). Turn the steel cage 90°, and install the earth pressure box on the other side of the welded steel strain gauge. Spread the nylon cloth smoothly on the construction site, and cut it to 27m×1m. The supporting pile length is 25m, the pile diameter is 0.9m, and 1m nylon cloth should be reserved at the pile end. Cut the nylon cloth into 8 small square nylon cloths of 0.25m×0.25m, and sew 8 small pockets on the center line of the paved nylon cloth at equal intervals, mark the corresponding earth pressure number on the mouth of the bag, pay attention to the seam When making, start from the end of the pile and sew one every 3m. Put the earth pressure box into the corresponding pocket in turn with the stressed side facing out. After placing it, fill the pocket with fine sand. If there is no fine sand on site Sand, it is not necessary to fill, and finally sew the pockets of the earth pressure box. Paste the wires of each earth pressure box to the nylon cloth with transparent large tape, and paste them to the top of the pile in turn. The wires at the earth pressure box are not It should be too tight, and the wire should be reserved for 20cm to prevent the side pressure from breaking the connection between the earth pressure box and the wire when the concrete is poured. Spread the nylon cloth and the earth pressure box together on the steel cage. Nylon cloth should reserve 1m of surplus at the end of the pile so as to wrap the end of the pile, and fix the two long sides of the nylon cloth on the adjacent steel bars with ties or plastic buckles, but should not be tied too tightly, and should be reserved enough The space for grouting can squeeze the earth pressure box into the soil.

(a) (b)

Figure 5. Photos of installation steps.

(4) After all the sensors are installed in the steel cage, put the steel cage into the drilled hole, and then pour the concrete. After the concrete is poured, bundle all the wires together and put them in the wire box. Use the reader to read the sensor data and check whether each sensor is installed successfully.

After the arrangement of various monitoring equipment is completed, data collection starts, and the collection frequency is determined by the excavation process of the foundation pit. Data collection should be adjusted or optimized in conjunction with the construction process. When the test pile side soil is excavated or the anchor cable is constructed, data collection should be carried out within 1 day after construction. When data is collected each time, in addition to collecting and using a reader to record stress data, the time of each data collection, excavation depth, construction progress and other information should also be recorded. The change law of earth pressure should be analyzed in time. If there is a sudden change in earth pressure that does not conform to the normal law, vigilance should be increased, the monitoring frequency should be intensive, and the monitoring data of the horizontal displacement of the pile top and the ground settlement of the monitoring unit should be combined.

3. Test Results

This study collected and summarized the measurement data of soil pressure and pile internal force for a series of working conditions of foundation pit excavation. The origin software was used to draw the relationship between the two kinds of stress, depth, and time. Can have a complete and clear study and understanding of the earth pressure under the pile-anchor supporting system of deep foundation pits and the internal force of the pile body.

3.1. Data Processing and Analysis

In this field test, a total of 8 soil pressure monitoring points and 8 pile internal force monitoring points were set up on the selected supporting piles of the foundation pit, for details, see the monitoring point layout figure 4. The monitoring data of the important working condition change points are selected for analysis. And the change law of the soil pressure and the stress of the pile body reinforcement during the excavation of the foundation pit is summarized.

(1) The determination of earth pressure is the prerequisite for the design of deep foundation pit supporting structure, and its accuracy directly affects the rationality of the design. The earth pressures measured in this test at different excavation depths are of great significance to support system design, support system deformation prediction, support structure safety, and economic rationality of foundation pit support. In the earth pressure test in this paper, 8 measuring points are arranged, all of which are distributed on the non-airside side (the side of the foundation pit wall), with depths of 3.1m, 6.1m, 9.1m, 12.1m, 15.1m, 18.1m, and 21.1m. , 24.1m, see figure 4 for details. Based on the actual measurement data of each measurement point before the excavation of the foundation pit as an example, the static earth pressure change is analyzed in terms of time, stratum distribution, and water level.

It can be seen from figure 6(a)(b). (1) During the construction of the bored pile, the measured static earth pressure is much larger than the earth pressure of the other groups (except T8). After the concrete is solidified, all The measured earth pressure value is closer to the actual situation, and the measured earth pressure value decreases slowly with time. (2)After the earth pressure is stabilized, the last static earth pressure is selected for analysis, and the distribution of the soil layer is plotted. It is found that the magnitude of the earth pressure is obviously affected by the stratum, and the change of the distribution form is consistent with the stratum change, which is a segmented distribution and not a continuous change. From the perspective of depth as a whole, it is roughly a zigzag change. (3)Compared with the theoretical earth pressure, it is found that at positions above 2/3 of the depth of the foundation pit, the theoretically calculated earth pressure is slightly larger than the actual measured value, while the results in the following parts are opposite.

Analyze the reasons for the above phenomenon: (1) When the pile is poured into concrete, a series of chemical reactions between the concrete and water are still going on. The contact relationship between the construction pile and the surrounding soil is relatively complicated. Most of the pressure measured by the earth pressure box is applied by the concrete to the soil. This phenomenon detailed explanations are subject to further analysis and research. Regarding the abnormality of the T8 earth pressure box, the analysis may be due to its deep location, which is the bottom position of the cast concrete, and the concrete is not in close contact with the surrounding soil, so the earth pressure can be measured well. After that, the measured value of earth pressure gradually decreases, which may be related to the solidification of the pile body and the pressure dissipation caused by the previous chemical reaction; (2)

Due to the difference in physical and mechanical properties of the layered soil on the site, the earth pressure of each layer of soil is also different are not the same. If the values of cohesive forces and friction angle of the lower layer soil are significantly greater than the upper layer, the earth pressure will be lower than the upper layer [7]; (3) Because the physical and mechanical parameters and the coefficient of static earth pressure used in the calculation are comprehensively selected from the experimental and empirical values of the soil in the site. They cannot be consistent with the actual values. In addition, the actual distribution of earth pressure does not increase linearly with depth. Inferring from this law, it is likely to conform to the characteristics of the static earth pressure theory within a certain depth range, but the increase in the lower earth pressure may decrease; the presence of groundwater at a depth below 15m increases the total earth pressure.

(2) After the excavation of the foundation pit, the soil is displaced in the direction of the pit, and the earth pressure changes from the static earth pressure to the active earth pressure, as shown in Figure 1. At this time, the earth pressure is in a non-limiting state, which is the external load received by the supporting structure when it is working, and is an important reason for the internal force and deformation of the supporting structure. Therefore, the monitoring and analysis of earth pressure during this period is particularly important for the design and construction of supporting structures. Taking the measured data of each measuring point after excavation as an example, the earth pressure changes will be analyzed from the aspects of stratum distribution, excavation of foundation pit, application of anchor cables, and theoretical comparison.

It can be seen from figure 6(c) (d): After the excavation of the foundation pit, the distribution and change of the measured earth pressure. (1)On the whole, the earth pressure gradually decreases with the increase of the excavation depth, and the shape of the change along the depth still presents a zigzag shape. Excavation to the deepest part of the foundation pit, the reduction of soil pressure in each layer is different, the reduction of soil pressure of loess-like soil, medium sand is smaller, and the reduction of soil pressure of silty clay is larger; The earth pressure below the bottom of the foundation pit decreases first and then increases with the excavation of the foundation pit; (2)With the increase of the excavation depth, the earth pressure value decreases as a whole, and changes toward the active earth pressure value, which is in a state of non-limiting earth pressure. Compared with the active earth pressure of Rankine, the measured earth pressure at the end of excavation is not very consistent. The measured earth pressure of the upper soil body is obviously higher than the theoretical value, and the lower soil pressure is slightly lower than the theoretical value;

In figure 6(e), the earth pressure is connected, and then the different excavation depths and anchor cable positions are drawn, and the laws can be compared and analyzed. After the excavation depth is about 4m, precipitation is carried out. When the water level drops to a depth of 26m, the earth pressure will decrease to a certain extent. And at the place where the anchor cable is applied, the earth pressure decreases greatly, but there is a certain time lag.

Analyze the reasons for the above phenomenon: (1) from the perspective of mechanism, the decrease in earth pressure is mainly caused by the displacement of the soil body into the pit due to the excavation of the foundation pit. As the excavation depth of the foundation pit continues to increase, the soil outside the pit is displaced to the inside of the pit due to additional earth pressure [14]. However, due to the existence of the supporting system, the soil is coordinated with its deformation, the displacement is controlled, and the earth pressure is in a non-limiting state. The reduction of soil pressure with different physical and mechanical parameters is quite different. According to this test, for the Xi'an area, the range of variation is silty clay>medium sand>loess-like soil; for the part below the bottom of the foundation pit, the earth pressure first decreases and then increases mainly because the active earth pressure state changes to the passive earth pressure state. (2) Rankine earth pressure theory does not consider wall friction, so its value is too small [10]. The measured value may indicate the true state of the development of non-limiting earth pressure. (3)The decrease of water level causes the water pressure to dissipate and the total earth pressure decreases. The anchor tension provided by the anchor cable is the direct cause of the reduction of the local earth pressure.

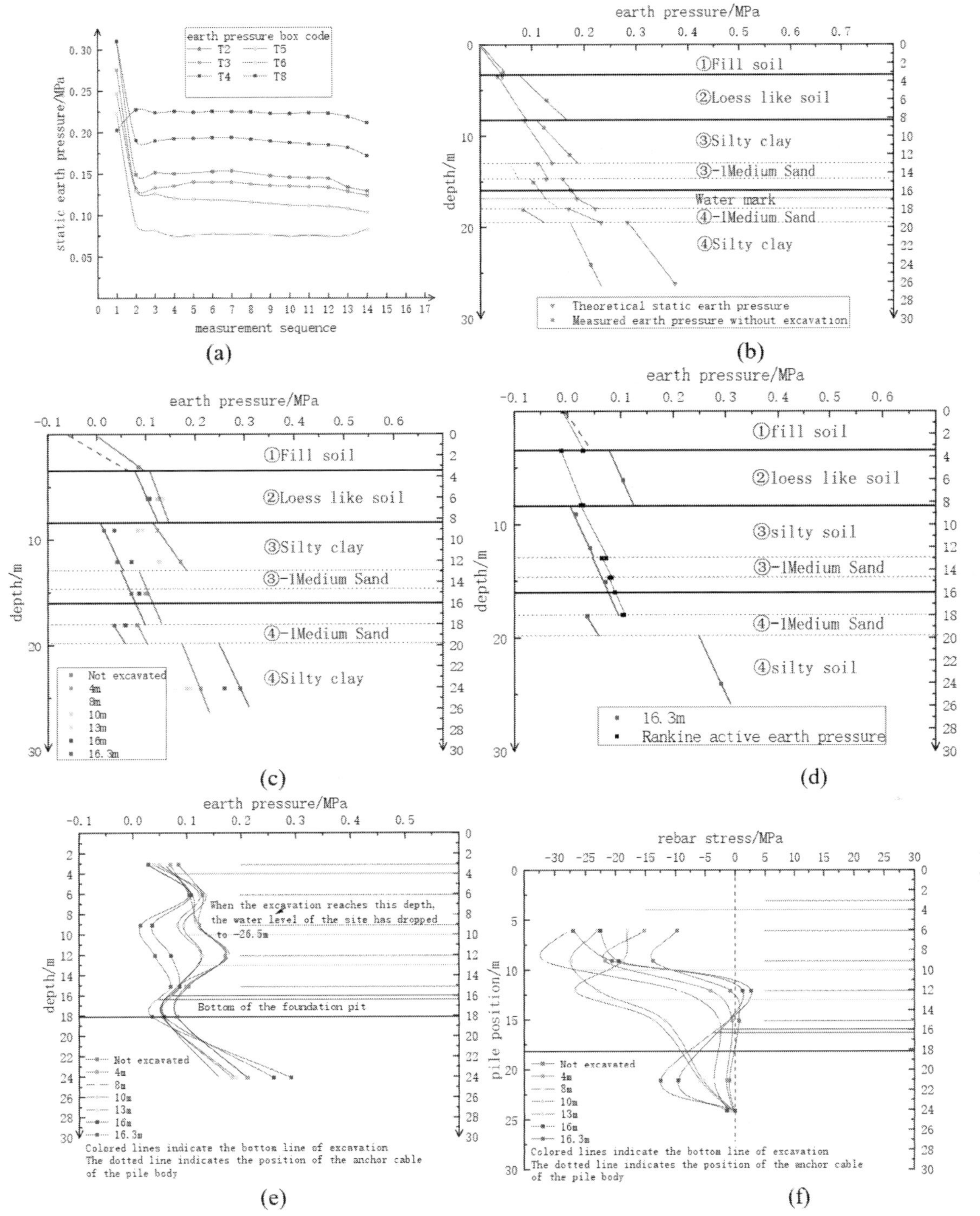

Figure 6. Tidying and analysis for test data.

(3) The internal force of the pile body is an important indicator reflecting the force status and safety status of the supporting structure of the foundation pit. A reasonable analysis of the measured results is of great significance for studying the force characteristics of the supporting structure and ensuring the smooth implementation of the project. In the pile internal force test in this paper, a total of 8 measuring

points are arranged, all of which are arranged on the side of the empty space (the side of the foundation pit), with depths of 3.1m, 6.1m, 9.1m, 12.1m, 15.1m, 18.1m, 21.1m, 24.1m, see figure 4 for details. During the actual measurement, it was found that the steel bar stress gauges G1 and G6 failed due to unknown reasons. Taking the measured data of each measuring point as an example, the analysis is carried out from three aspects: the distribution of the stress of the steel bars along the pile at each stage of the excavation of the foundation pit, the relationship with the excavation of the foundation pit, and the application of anchor rods.

It can be seen from figure 6(f) that as the excavation depth of the foundation pit increases, the displacement of the pile top gradually increases. The distribution and change of the stress of the pile body steel bar with depth and time: (1)In the initial stage of excavation, with the increase of the excavation depth, on the whole, the compressive stress is at the depth of about 5m to about 12m, showing an increasing trend. The law of change with depth first increases and then decreases; the overall change law of the stress of the steel bar at the depth of 12m to the bottom of the pile is slowly increasing from 0, with little change, and no change along the depth; (2)Mid-term excavation , the excavation depth is about 10-13m, and the compressive stress at the depth of about 5m to about 12m decreases as a whole. Along the depth, it first increases and then decreases and gradually increases. From the depth of 12m to the bottom of the pile, the compressive stress of the steel bars increases as a whole, and the distribution law along the depth is gradually decreasing to 0 from top to bottom; (3)In the later stage of excavation, the excavation depth is about 13-18m. From about 5m to about 12m, the overall compressive stress gradually increases again, and gradually decreases from top to bottom along the depth to near 0. From about 12m to the bottom of the pile, the overall compressive stress of steel reinforcement suddenly increases, and the distribution is regular along the depth. In order to increase from top to bottom, it first increases and then decreases to 0; (4) There are also obvious changes in the stress of the steel bar caused by the application of the anchor cable. At the place where the anchor cable is applied, the stress will be reduced considerably.

There is no real-time monitoring data for the depth of the pile from 0 to 5m. However, Related studies [21] show that if there are anchor cables and other components that limit the displacement of the pile top, tensile stress may occur in the stress of the steel bar on the top of the pile. Analyze the reasons for the above phenomenon: (1) At the beginning of excavation, the load on the pile body was mainly imposed by earth pressure. At this time, two anchor cables were applied, and the effect was not obvious. With the excavation, the soil continued to move into the pit. The upper pile support is stressed, and its internal force gradually increases. The lower part of the pile at the anchoring section has not yet received force. (2)In the middle of excavation, the soil on the upper part of the pile has displacement, which causes the earth pressure to decrease, and it is in harmony with the deformation of the pile. At this time, 4 anchor cables have been applied. The cable effect is generated. The external load on the upper pile body decreases, so the internal force of the pile body decreases. Due to overall tilt of the lower pile body, the passive earth pressure on the inside of the pit decreases from top to bottom along the depth to zero at the bottom of the pile; (3) In the later stage of excavation, the soil displacement reaches the maximum, the pile tilt reaches the limit, all anchor cables have been applied, and the effect has been achieved. Due to the coordinated deformation of the upper pile body, the internal force experienced slightly increased. In the middle part, the internal force of the pile body is greatly reduced due to the existence of anchor tension of the anchor cable. However, due to the inclination of the lower part, the passive earth pressure on the pile further increases, and the internal force of the pile body further rises to zero at the bottom of the pile. (4) The existence of anchor tension of anchor cable restricts the displacement of the soil, thereby reducing the soil pressure, which has a great effect on the improvement of internal force of pile.

3.2. Analogy of Existing Non-Limiting Earth Pressure Calculation Methods

There are still no complete answers about the calculation method of the earth pressure on the pile-anchor supporting system. Under flexible support, the study of the development of earth pressure considering displacement is essentially the search for the state of non-limiting earth pressure and its accurate calculation. Therefore, this paper selects two theories from many non-limiting earth pressure theories, and verifies and compares them through calculation and test data, in order to achieve the research purpose of discussing engineering application.

(1) In the study (Xu Riqing, Liao Bin, Wu Jian, et al.) of the active earth pressure calculation method, starting from the stress Mohr circle, the formula of the relationship between the internal friction angle of the cohesive soil and the wall displacement in the intermediate state is derived. At the same time, the external friction angle and cohesion of the wall-soil contact surface are considered. According to the geometric relationship of the cohesive soil stress Mohr circle, the relationship between the soil cohesion and the wall displacement is obtained. Finally, the horizontal layering method is used to obtain the active earth pressure calculation formula of the cohesive soil in the non-limit state [12]:

$$p=(\gamma z+q_0)\frac{\tan\theta}{\tan(\theta+\varphi_m)}-c_m\frac{1+\tan\theta\tan(\theta+\varphi_m)}{\tan(\theta+\varphi_m)} \quad (1)$$

Equation (1) γ: Soil density, z: depth, q0: upper loads, θ: Assuming the angle between the fracture surface and the vertical direction, cm, φm: Soil strength considering wall displacement.

Figure 7 is the comparison between the soil pressure calculated in Theory 1 and the measured value. From the figure, it can be seen that the shape of the calculated value curve is quite consistent with the measured value, and the law and trend are very similar. This reflects the calculated value curve expressing the soil pressure. The characteristics of body stratification; The calculated value is generally slightly larger than the measured value. The preliminary consideration may be that the anchoring force provided by the anchor cable is not considered in the calculation theory.

Therefore, in practical applications, the theory has a good consistency in reflecting the earth pressure under the excavation of multi-layer soil foundation pits. However, the effect of anchor tension under the pile-anchor support system is not considered. So when the measured value is compared, there are still discrepancies in local locations. As for how to add anchor tension to its calculation theory to make it more realistic, relevant derivation studies may need to be further carried out.

(2) In the study (Li Chaoyang, Xie Qiang, Kang Jingwen, etc.) of the non-limiting active earth pressure of the flexible pile cohesive soil, consider the soil arching effect behind the pile, the pile-soil internal friction angle and the cohesive force development value under the non-limit state, and the pile After the friction angle and cohesive force play value in the soil body, starting from the cohesive soil stress Mohr circle, the static balance is established by the micro-layer analysis method, and the potential sliding surface of the soil body behind the pile is searched. The calculation formula of non-limiting active earth pressure of flexible pile cohesive soil is derived [13]:

$$E_{ai}=\sqrt{\left(p_{yi}K_i-c_{mi}\cot\varphi_{mi}\right)^2+\left(\left(p_{yi}K_i-c_{mi}\cot\varphi_{mi}\right)\tan\delta_{qmi}+c_{qmi}\right)^2} \quad (2)$$

Equation (2) pyi: Vertical stress on the soil surface, Ki: Positive pressure coefficient of pile-soil interface, cmi, φmi: Soil strength considering wall displacement, cqmi, δqmi: Cohesion and friction angle between pile and soil considering wall displacement.

Figure 8 shows the comparison of the soil pressure calculated by theory 2 with the measured value. It can be seen from the figure that the calculated value curve shape is at the upper part, which is quite different from the measured value. The theoretical value is smaller than the measured value; below about 6m, the curve is extremely Close to the measured value, the calculated value is close to the actual measured value.

This theory is seriously distorted when reflecting the soil pressure of the upper soil under the pile-anchor support system. It is speculated that it may be caused by the greater cohesion of the soil in the process of deriving the formula. However, the theoretical value of the lower layer is compared with the actual measurement. It is quite consistent. In practical applications, this type of theory is more in line with the actual value of the earth pressure of the pile support, but the anchor tension stress is still not well reflected.

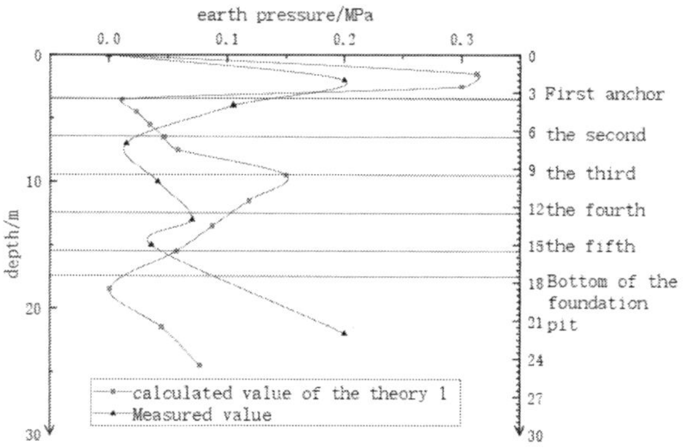

Figure 7. Comparison curve between theory 1 and measured data.

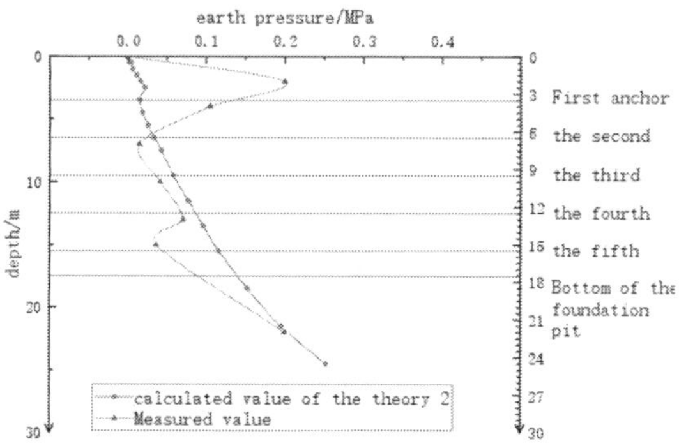

Figure 8. Comparison curve between theory 2 and measured data.

4. Concluding Remarks

Through the statistical analysis of on-site monitoring data of deep foundation pit soil pressure and pile internal force in Xi'an area, and the comparison of two soil pressure calculation theories, the following conclusions are obtained:

(1) For the excavation of cohesive soil foundation pits, the soil pressures of layered soils with different properties are distributed in sections along the depth, and the overall distribution is in the shape of a "Zig". The magnitude and change of the value are closely related to the strength parameters of the soil, the change of the water level, and the excavation stage of the foundation pit. The soil pressure under the pile-anchor supporting system is obviously affected by the anchor tension along the depth curve.

(2) In the deep foundation pit supporting system, the temporal and spatial variation and distribution of the internal force of the pile body are in a one-to-one correspondence with the earth pressure, and they develop simultaneously. Regarding the internal force of the pile body, its changes are more complicated, not one direction or trend. It changes with time and depth. The increase and decrease of internal force at the same position of the pile body may alternately occur, so special attention should be paid to this part in the design.

(3) Select the existing non-limiting earth pressure calculation methods, apply calculations and compare them, and the results show that the theory can roughly reflect its changing laws. Among them, Xu Riqing et al. [12]'s calculation method reflects the layered changes of soil close to reality; Li Chaoyang et al. [13] in the study of non-limit active earth pressure of flexible pile cohesive soil, the calculated value is very reflective at a certain depth in fact, it has certain value and significance for construction reference. However, because the two calculation methods have not considered the existence of anchor tension in the derivation process, they are slightly inconsistent with the stress conditions of the engineering conditions in this experiment, so the earth pressure calculation curves are not completely consistent.

References

[1] Liu Y F. 2010 Field Test and Numerical Simulation Research on Single Support Pile-anchor Support for Deep Foundation Pit [D] *Qin Huangdao Yan Shan University.*

[2] Lu K L and Yang Y.2009 Approximate calculation method of active earth pressure considering displacement [J] *Rock and Soil Mechanics* **30** (2) 553-557.

[3] Lu K L and Yang Y. 2010 Preliminary study of active earth pressure under nonlimit state [J] *Rock and Soil Mechanics* **31** (2) 615-619.

[4] Mei G X and Zai J M. 2001 Earth pressure calculating method considering displacement [J] *Rock and Soil Mechanics* **22** (4) 83-85.

[5] Xu R Q. 2000 Calculation method of earth pressure considering displacement and time [J] *Journal of Zhejiang University (Engineering Science)* **34** (4) 370-375.
[6] Yang B, Hu L Q. 2000 Test study on Relationship between Lateral Earth Pressure Acting on Retaining Structures and Horizontal Displacement [J] *Building Science* **16** (2) 14-20.
[7] Wang Y Z, Li X G and Chen N N. 2005 Active earth pressure on a retaining wall and lateral coefficient of earth pressure [J] *Rock and Soil Mechanics* **26** (7) 1019-1022.
[8] Xu R Q, Gong C, Wei G, Wang J C. 2005 Theory of earth pressure against rigid retaining walls considering translational movement effect [J] *Journal of Zhejiang University(Engineering Science)* **39** (1) 119-122.
[9] Chen Y K, Wang Y M, Xu R Q, etc. 2004 Numerical analyses of active earth pressure on rigid retaining wall [J] *Chinese Journal of Rock Mechanics and Engineering* **23** (6) 989-995.
[10] Xu R Q, Chen Y K, Yang Z X, etc. 2002 Experimental research on the passive earth pressure acting on a rigid wall [J] *Chinese Journal of Geotechnical Engineering* **24** (5) 569-575.
[11] Wang W Z and Wu Y Z. 1992 Study on earth pressure and its distribution behind gravity retaining wall [J] *Highway Engineering* 60 22-29.
[12] Xu R Q, Liao B, Wu J, et al. 2013 Computational method for active earth pressure of cohesive soil under nonlimit state[J] *Rock and Soil Mechanics* **34** (1) 148-154.
[13] Li C Y, Xie Q, Tang J W, et al. 2018 Active earth pressure of cohesive soil against flexible pile under nonlimit state [J] *Journal of Civil and Environmental Engineering* **40** (5) 64-70.
[14] Wang C H and Wang X X. 2019 Experimental study on physical model of additional earth pressure theory [J] *Journal of Nanchang University(Engineering & Technology)* **41** (1) 40-50.
[15] Wu L L. 2015 An experimental study and numerical analyses of the relationship between additional eath pressure and displacement for retaioning structures [D] *Tianjin: Tianjin University*.
[16] Sheng Z Q and Cui T. 2018 Engineering characteristics of foundation soil under the action of stress path of foundation pit excavation [J] *Building Science* **34** (7) 1-11.
[17] Chen J, Zhou T H. 2015 Theoretical Study and Field Monitoring Analysis on Lateral Soil Pressure Acting on Pile—Anchor Excavation Retaining Structure [J]. *Building Science* **31** (9) 108-114.
[18] Sheng Z Q, Teng Y J, Li P. 2021 Discussion on several problems in design of retaining structures of deep excavation [J] *Chinese Journal of Geotechnical Engineering* **43** (1) 94-101.

Journal of Physics: Conference Series **2148** (2022) 012052 doi:10.1088/1742-6596/2148/1/012052

Bearing Capacity Analysis of Transmission Tower Structure Considering Corrosion Damage

Ruizong Lin, Benzhao Fu, Xinmin Yu, Hui Ma*, Zeyan Wu and Chenqing Yang

State GRID Fujian Economic Research Institute, Fuzhou 350000, China
Email: machengbing1123@163.com

Abstract. As an important part of power grid equipment, transmission towers are in direct contact with the external environment for a long time. As a coastal city in China, Fujian has been affected by marine environment and industrial pollution for a long time, which directly affects the safety of transmission towers in long-term service. In order to explore the changes of the ultimate bearing capacity of the tower structure after corrosion, this paper uses finite element software to analyse the mechanical properties of the tower structure during long-term service, and finds that the 45° wind direction is the control condition, and the overall stiffness of the tower decreases with the growth of corrosion time, and the increment of tower top displacement reaches 7% at 12 years of corrosion. The corrosion-sensitive members of the tower were clearly identified, and their stress ratios increased from 0.78, 0.79, and 0.83 to 0.97, 0.98, and 0.99, respectively, at 12 years of corrosion.

1. Introduction

As an important part of the power grid equipment, transmission towers are usually made of lightweight and high-strength steel structure, whose safety and reliability are the basis for ensuring the safe and stable operation of the power grid, and is also a key part of the "three types of two networks" construction goal. However, with the rapid development of China's power system construction and service time growth at the same time, the operation of transmission towers has some safety hazards.

Chen Weiming et al. studied the load carrying capacity of transmission pylons underground deformation in different directions [1], Yu Cong et al. studied the effect of transmission pylon systems on ultimate load carrying capacity under different wind angles [2], Dai Kaoshan et al. studied the form of failure of transmission pylon systems under combined earthquake and wind loads [3], Wang Jia et al. studied the ultimate load carrying capacity of pylons with local defects [4], Zhang Yihu et al. studied the form of transmission tower damage when considering nonlinearity [5], and Jian Zhang et al. studied the form of failure of transmission towers under strong wind loads [6]. All the above studies are based on external load effects.

Fujian Province is located in the southeast coast with hot and humid climate, and its transmission towers have been affected by the marine environment for a long time, coupled with serious industrial pollution and frequent acid rain in recent years, its corrosion environment is quite different from other regions. Secondly, the current research on towers is limited to the material level, and the relationship between the material properties after corrosion and the performance deterioration of the tower structure has not been established. In this paper, by establishing a finite element model of a typical tower structure and introducing a typical corroded material degradation ontology, the mechanical property degradation law of the tower structure under corrosive environment is systematically analysed, and suggestions are made for the operation and maintenance of the tower.

2. Project Overview

In this paper, we choose the transmission level of 220kV general tension-resistant tower, model JC2-30, transporting double circuit, alternating current, recorded as N1 tower (hereinafter referred to as N1). N1 tower's front gear distance 700m, rear gear distance is 300m, the total height of the tower 47.7m, and the N1 tower as the center of the establishment of the line tower system model, as shown in figure 1. The nodes are connected by high strength bolts, and the foundation is inserted is the foundation. The ground wire in the line tower system is TS_JLB20A-150, with an average running tension ratio of 0.18 and a cross section of 148[mm]^2, the average height of the ground wire is 15m, and the conductor is JL/LB20A-630/45 with an average running tension ratio of 0.25 and a cross section of 666.55[mm]^2 . The average height of the conductor is 15m, and the specific dimensional information of the tower is shown in figure 2.

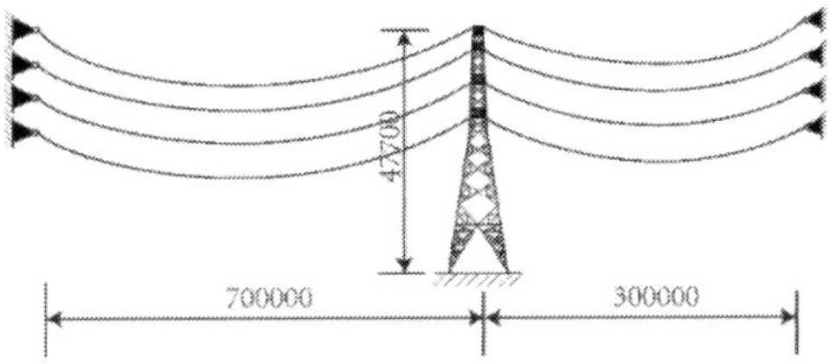

Figure 1. Schematic diagram of line tower system (mm).

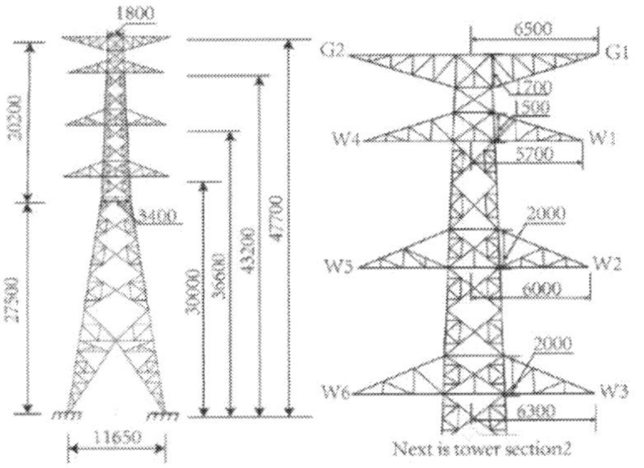

(a) Tower composition (b) Cross arm composition

Figure 2. Overall schematic diagram of the tower (mm).

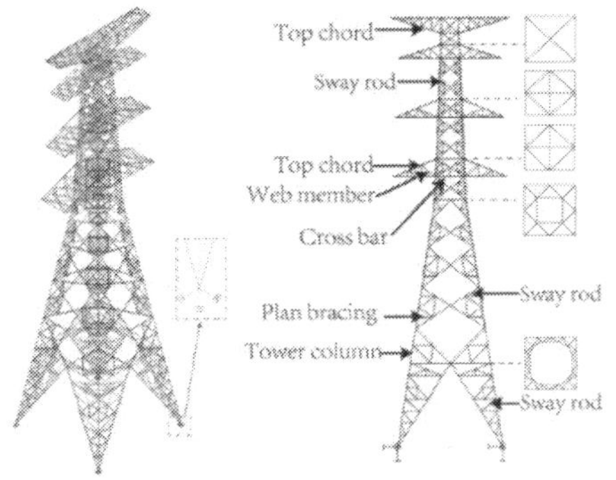

(a) 3D3S Model (b) Name of Tower Member

Figure 3. Tower model.

3. Finite Element Model

3.1. Model Establishment

The tower model is established based on the tower design module selected in 3D3S software. The tower is a system of one tower and two lines (as shown in table 1). In the software, the tower model library provided by the software is used to establish the model, and the key members and material information of the tower are added. The types of tower members are tower columns, sway rods, secondary sway rods, cross bars, web members, chords and other main components. The cross sections of members are equilateral angle steels, as shown in table 1 and figure 3.

Table 1. Tower members.

Member name	sections	materials
Tower column	L200×20, L180×16, L140×12	Q420B Q345B
Sway rod	L140×10, L125×10, L110×10, L100×10 L100×8, L100×7, L90×7, L90×6	Q345B
Secondary swat rod	L63×5, L40×4, L40×3	Q235B
Cross bar	L63×5, L40×3	Q235B
Web member	L90×7, L63×5, L56×5, L45×4, L40×3	Q345B Q235B
Other	L20×3, L40×3, L40×4, L45×4, L56×6, L63×5 L75×5, L75×6, L80×6, L90×6, L75×10	Q235B

3.2. Boundary Condition

Inputing the basic wind pressure in the loading module of the software. According to the "Load code for building structure" GB50009-2012, the basic wind pressure should use the basic wind pressure with a return period of 50 years. When the transmission tower is calculated in 3D3S structural design software, the ground roughness is class B, the wind load calculation code is high-rise code, the building structure type is high-rise structure, the house type is non maintenance steel structure, and the terrain is flat. Connection between tower and foundation is rigid connection. According to the relevant standards and requirements of code for design of 110kV ~ 750kV overhead transmission lines (GB 505445-2010), the basic wind pressure is calculated according to formula (1).

$$W_0 = \frac{V^2}{1600} \tag{1}$$

W0 is the standard value of reference wind pressure and V is the wind speed with height of 10m.

According to the control standard for basic wind speed conditions of tension type tower load in GB 505445-2010 and considering the influence of wind speed, tower height and tower type of transmission tower, this paper only checks and calculates the four working conditions of wind load of 0 °, 45 °, 60 ° and 90 ° under strong wind, and the schematic diagram is shown in figure 4. Since the structure is a non-maintenance steel structure, and the member type of the tower tip is a truss element, and the member is only subjected to axial force, the load acts directly on the node, as shown in figure 4 (b). The load bearing direction and range of wind load under four working conditions of tower 0 °, 45 °, 60 ° and 90 ° are shown in figure 5.

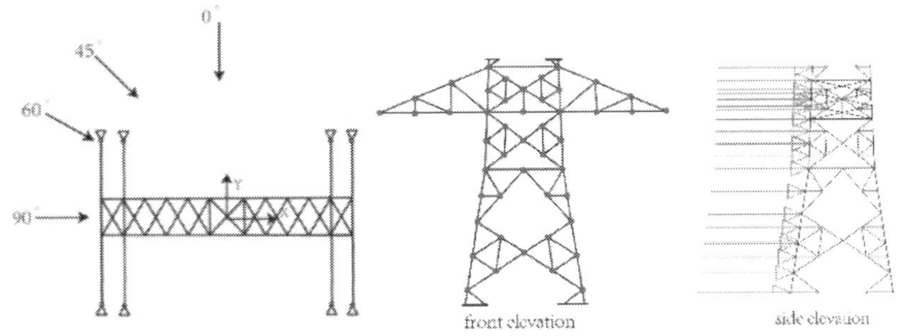

(a) Schematic diagram of wind (b) Location of wind load

Figure 4. Diagram of wind load application

(a)0°wind load (b)45°wind load (c)60°wind load (d)90°win load

Figure 5. Structural wind load action condition.

The parabolic model in the literature [3] is used for the linear tower system model, as shown in figure6. Taking the left hanging line point A as the origin, the horizontal line passing through point A as the x-axis, and the left vertical downward line as the y-axis, then the relationship is shown in formula (2). The horizontal tension, approximate values of conductor and ground wire length and approximate values of conductor and ground wire tension can be calculated by the same formula(3)(4). Table 2 for conductor tension.

$$y = \frac{4fx(l-x)}{l^2} + \frac{h}{l}x \tag{2}$$

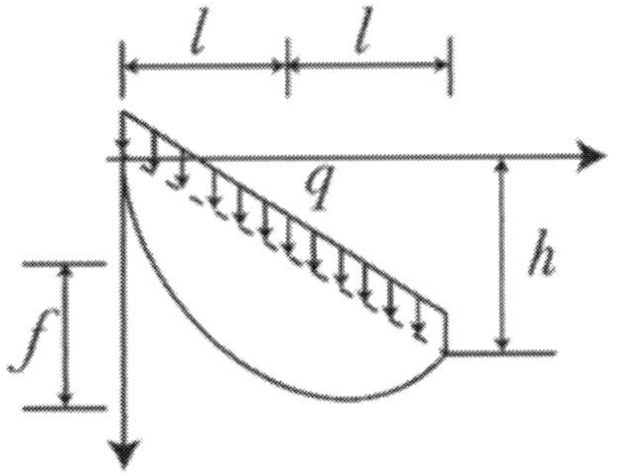

Figure 6. Parabolic.

Relationship between mid-span sag and horizontal tension:

$$H = \frac{ql^2}{8f\cos\theta} \tag{3}$$

Approximate cable length:

$$S = L\left[1 + \frac{1}{2}\left(\frac{h}{l}\right)^2 + \frac{8}{3}\left(\frac{f}{l}\right)^2\right] \tag{4}$$

Approximate maximum cable tension:

$$T = H\sqrt{1 + 16\frac{f^2}{l^2} + \frac{h^2}{l^2} + \frac{8fh}{l^2}} \tag{5}$$

Model of tower system q is the unit load along the conductor and ground wire, h is the height difference of suspension point, and f is the mid span sag.

The standard values of wind load of conductor and ground wire and wind load of insulator string are calculated according to relevant standards of "code for design of 110kV ~ 750kV overhead transmission lines" (GB 50545-2010). The standard value of horizontal wind load of conductor and ground wire shall be calculated according to formula (6), and the standard value of wind load of insulator string shall be calculated according to formula (11) without considering icing.

$$W_{\mathrm{x}} = \alpha_L \mu_z \mu_{sc} W_0 \beta_c d L_{\mathrm{p}} \sin^2(\theta) \tag{6}$$

$$\beta_{\mathrm{c}} = \gamma_{\mathrm{C}}(1 + 2g \cdot I_{\mathrm{Z}}) \tag{7}$$

$$I_Z = I_{10} \cdot \left(\frac{z}{10}\right)^{-\alpha} \tag{8}$$

$$\alpha_{\mathrm{L}} = \frac{1+2g\cdot\varepsilon_{\mathrm{c}}\cdot I_Z\cdot\delta_{\mathrm{L}}}{1+5I_z} \tag{9}$$

$$\delta_{\mathrm{L}} = \frac{\sqrt{12L_{\mathrm{X}}L_{\mathrm{P}}^3+54L_{\mathrm{x}}^4-36L_{\mathrm{X}}^3L_{\mathrm{P}}-72L_{\mathrm{x}}^4e^{\frac{L_{\mathrm{p}}}{L_{\mathrm{x}}}}+18L_{\mathrm{x}}^4e^{-\frac{2l_{\mathrm{p}}}{L_{\mathrm{x}}}}}}{3L_{\mathrm{P}}^2} \tag{10}$$

$$W_I = \mu_Z w_0 A_I \tag{11}$$

W is the sum of the standard values of wind load of conductor and insulator string. $W_{\mathrm{x}}$is the standard value of horizontal wind load perpendicular to the direction of conductor and ground wire.α_L is the span reduction coefficient. $\beta_{\mathrm{c}}$is the gust coefficient of conductor and ground wire, where 1.0 is taken. μ_z is the wind pressure height variation coefficient. μ_{sc} is the shape coefficient of conductor and ground wire. d is the outer diameter of conductor and ground wire. $L_{\mathrm{p}}$is the horizontal span of tower. θ is the included angle between wind direction and conductor and ground wire. W_0 is the standard value of reference wind pressure. V is the basic wind speed with a reference height of 10m. W_I is the standard value of wind load of insulator string. A_I is the calculated value of wind pressure bearing area of insulator string.

The vertical load, horizontal load and tension load on the front and rear sides of the conductor and ground wire act on the tower and act on the hanging point according to the node load. The vertical load is the load generated by the icing along the length of the conductor and the dead weight of the insulator string on the conductor and ground wire, and the direction is vertical and downward. Horizontal load is the load of conductor, ground wire and insulator string under wind load, and the tension is perpendicular to the tower along the conductor direction, as shown in figure7. The calculated load is shown in table 2.

Table 2. Load at hanging point.

name	f(m)	q(N/m)	Tension load(N)	horizontal load(N)	Vertical load(N)
wire 1 / 2 front	14.01	16.6676	41884	8192	8183
wire 1 / 2 rear	2.33	16.6676	38073	3851	2156
Wire 1 / 4 front	24.20	32.7282	110359	28664	31867
Wire 1 / 4 rear	4.45	32.7282	95951	13618	12380
Wire 2 / 5 front	24.20	32.7282	110359	27127	31867
Wire 2 / 5 rear	4.45	32.7282	95951	12975	12380

Wire 3 / 6 front	24.20	32.7282	110359	25313	31867
Wire 3 / 6 rear	4.45	32.7282	95951	12240	12380

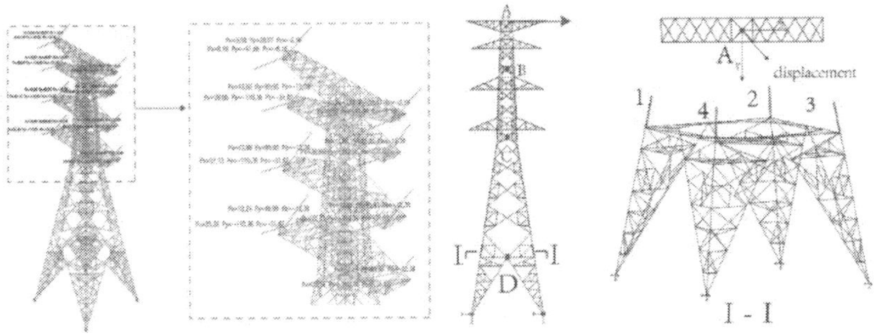

(a) X direction (b) Y direction

Figure 7. Relationship between lateral displacement of tower top and corrosion time.

4. Material Corrosion Damage Constitutive Model

After the model is established, the constitutive models of the main members and other members of the tower are modified in the material attribute module of the software and input according to the ideal elastic constitutive model. Based on the three-stage corrosion model of low alloy steel in [11], this paper analyses the relationship between mass loss and corrosion time of Q420B and Q345B in atmospheric environment. The first stage of the three-stage model is the rapid corrosion stage, the second and third stage is the corrosion mitigation stage. There are obvious differences in the corrosion models of the three stages. Kuang Jinxin studied the corrosion behavior of Q420 under atmospheric corrosion and accelerated corrosion. Based on the atmospheric corrosion data of low alloy steel in Jiangjin area of Chongqing, a three-stage mathematical model between mass loss and corrosion time is obtained as shown in formula (12): the corrosion time in the first stage is linear with the mass loss rate, the second stage is parabolic, and the third stage is also linear. Compared with the first stage, the corrosion rate becomes smaller.

$$\rho_m = \begin{cases} 2.09t & t \le 1 \\ -8.5^{-2}t^2 + 1.66t + 0.52 & 1 \le t \le 8 \\ 0.52t + 4.15 & t \ge 8 \end{cases} \tag{12}$$

w is the mass loss, t is the corrosion time, m is the mass before corrosion [11], ρ_m is the mass loss rate.[12]

By studying the corrosion behavior of Q235B in atmospheric environment and doing static tensile test on the corroded material, the changes of mechanical properties of the material are obtained. It is found that the elastic modulus, yield strength, tensile strength and mass loss rate of Q235B after corrosion have the following linear relationship, and the constitutive damage model of Q235B in atmospheric corrosion environment is established.

$$\frac{f_{ny}}{f_y} = 1.011 - \beta_{nfy} \cdot \rho_{\mathrm{m}}(\%) \quad ,\beta_{nfy} = 0.022 \tag{13}$$

$$\frac{f_{nu}}{f_u} = 1.007 - \beta_{nf_u} \cdot \rho_{\mathrm{m}}(\%) \quad ,\beta_{nfu} = 0.015 \tag{14}$$

$$\frac{E_c}{E} = 0.980 - \beta_E \cdot \rho_{\mathrm{m}}(\%) \qquad \beta_E = 0.004 \tag{15}$$

f_{ny} is the yield strength, f_{nu} is the ultimate tensile strength, E_c is the elastic modulus after corrosion, β_{nfy} is the reduction factor of nominal yield strength, β_{nf_u} is the reduction factor of the nominal ultimate tensile strength.

Therefore, the above three-stage model and constitutive damage model are used to determine the damage constitutive models of Q420B, Q345B low alloy steel and Q235B after corrosion for 2, 4, 6, 8, 10 and 12 years, as shown in table 3.

Table 3. Material Constitutive.

t	ρ_m	Q235B			Q345			Q420B		
		f_y(MPa)	f_u(MPa)	E_c(GPa)	f_y(MPa)	f_u(MPa)	E_c(GPa)	f_y(MPa)	f_u(MPa)	E_c(GPa)
0	0	235.0	370.0	206	345.0	470.0	206	420.0	520.0	206
2	2.09	226.8	360.9	200.1	332.9	458.5	200.1	405.3	507.3	200.1
4	3.5	219.5	353.1	199.0	322.2	448.6	199.0	392.2	496.3	199.0
6	5.8	207.6	340.4	197.1	304.7	432.4	197.1	371.0	478.4	197.1
8	7.42	199.2	331.4	195.7	292.4	420.9	195.7	356.0	465.7	195.7
10	8.36	194.3	326.1	194.9	285.3	414.3	194.9	347.3	458.4	194.9
12	9.35	189.2	320.7	194.1	277.8	407.3	194.1	338.2	450.7	194.1

5. Results and Discussion

According to the finite element calculation results we can see that the direction of wind load 45 degrees controls the structure, the displacement of the top of the tower the internal force of structure internal bar is biggest at this time, therefore, the displacement of key points A, B, C and D in the structure in the wind load direction of 45 degrees and the change of stress ratio of key members 1, 2, 3 and 4 are selected. The positions of key points and key members in the tower are shown in figure 8.

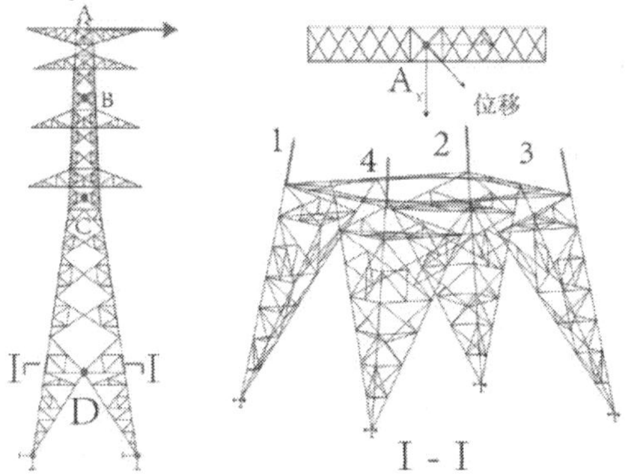

Figure 8. Key point, key member position

The displacement of the top of the tower in X and Y directions is shown in Figure 9.Displacement in both directions increases with the increase of service time of tower, change trend can be divided into three stages, the stage I is a rapid growth stage (0~2 years), the stage II is faster growth stage (2~10 years), the stage III is slow growth phase (10 years~), among them, the displacement of the top of the tower in X direction reached 182mm after 12 years of service., the direction of the displacement Y direction displacement reached 33 mm, it can be inferred that the X direction plays a controlling role.

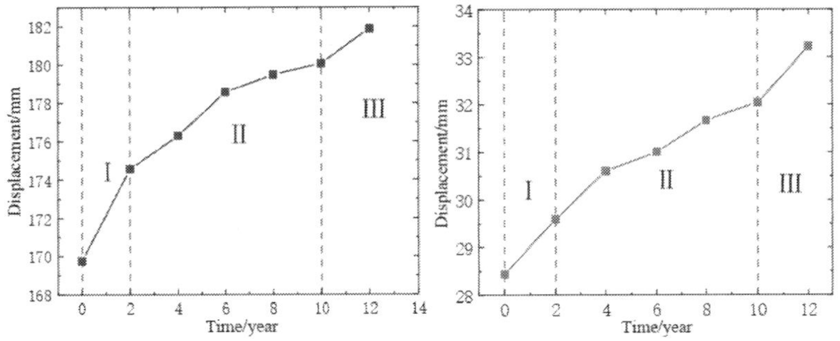

(a) X direction (b) Y direction

Figure 9. Relationship between lateral displacement of tower top and corrosion time.

Figure 10 shows the structure's displacement changes in the X direction. It can be seen that with the longer service time of the tower, the overall displacement of the tower shows an increasing trend, compared with the displacement of the initial structure, when the structure is in service for 12 years, the displacement of point A increases by about 7%, point B by about 6.5% and point C by about 6%. The curve can be divided into two stages with 2 years as the cut-off point. The stageIis 0-2 years, and its displacement growth rate is the fastest, the displacement of the structure has increased by about 4% after only two years of service, and the growth rate has gradually slowed down in 2-10 years, reaching 9% in 10 years, and then the growth rate has further decreased. This is because in the first two years of service, the corrosion degree of the main material of the tower is the largest, the change of the elastic modulus of the main material is the fastest, and the stiffness of the overall structure of the tower is the most reduced. As the corrosion time increases, protective rust layer is formed on the surface of the main material, and the corrosion rate decreases, it caused a decrease in the displacement change rate of the structure.

In figure 10(a) and (b), the relative positions of points A, B and C are opposite in the above two figures. During the same period of service, the overall displacement in the X direction of the same position is less than that of the overall structure in Y, which has a controlling effect on the tower. The figure shows that the overall Y-direction displacement of the structure gradually increases with the increase of service time. Its growth trend can be divided into two stages: the first stage is rapid growth, the second stage is slow growth. Take the second year as the demarcation point, two years before belongs to the rapid growth stage, two years after belongs to the slow growth stage. This is because the structure of advocate material corrosion rate at first, the fastest advocate material elastic modulus decreases faster, the overall stiffness decrease rate, along with the increase in time tower advocate material formed on the surface of protective rust layer, advocate material corrosion rate is reduced, the elastic modulus change rate is reduced, the structure of the whole stiffness change rate is reduced, displacement rate would decrease gradually. It can be seen from the figure that the change rate of the structure's displacement in the Y direction is the largest at point D, followed by two points B and C, and the smallest at point A. After 12 years of service, the displacement of the structure at Point A increases by about 7%, that at point B and C by about 8% and that at point D by about 11%.This is because in the initial state, the displacement of the whole structure in the Y direction is compared, but the displacement growth is the same.

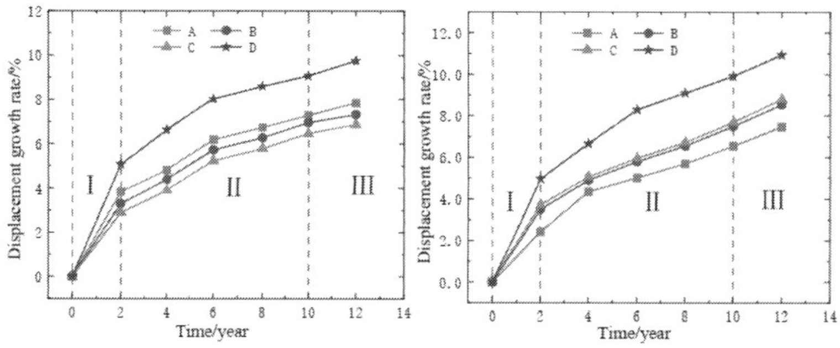

(a) X direction (b) Y direction

Figure 10. Relationship between lateral displacement of typical parts and corrosion time.

Figure 11 shows the changes of the stress ratio of the rods. It can be seen that the stress of all the rods presents three growth stages: I, II and III, the stress before 6 years belongs to the rapid growth stage, and after 6 years belongs to the slow growth stage. This is because the corrosion rate in the initial stage of the tower main material is fast, the rate of reduction of the cross-sectional area of the tower components increases, and the rate of increase of stress ratio increases. With the increase of service time, a protective rust layer is formed on the surface of the tower main material, which reduces the corrosion rate of the material. The increasing rate of area loss rate decreases, for this reason the increasing rate of stress ratio decreases. In addition, in figure 3 and 4, curve of stress than similar trends, stress ratio were greater than 1 and 2, this is because in the structural design to ensure the bearing capacity of the lower structure is bigger than the upper structure, the material resistance of the lower member is larger than that of the upper member, and the residual value of the material resistance of the lower member is larger than that of the upper member under load. However, after 10 years, the stress ratio of No.2 is greater than that of No.3 and No.4, this is because when the stress ratio of No.3 and No.4 members reaches the limit load, other members share the load generated by the superstructure load. The stress ratio of No.2 bar is greater than that of No.1 bar, because No.1 bar bears tension and both bars bear pressure, and the stress on the section of No.2 bar is greater than that on No.1 bar. It is observed that after 12 years, the stress ratio of all members reaches about 1.

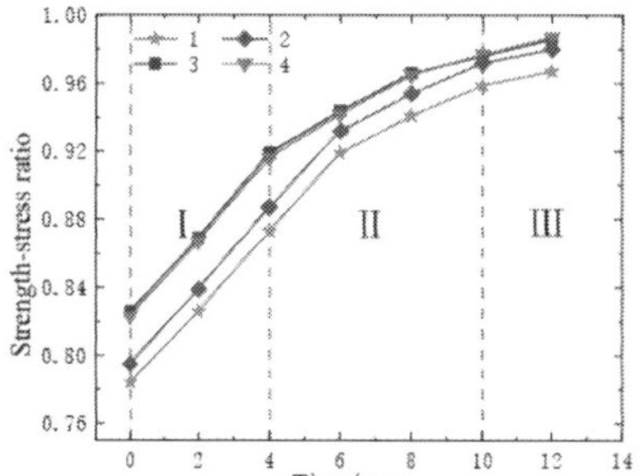

Figure 11. Relationship between stress ratio of key members and corrosion time.

6. Conclusion

The tower is in direct contact with the external environment for a long time during service. However, there is a large amount of acidic ions in the external environment, and the tower under the action of acidic ions the main material undergoes serious corrosive effects. Using 3D3S structural analysis software to analyze the displacement changes in the tower shut-in position and the stress ratio changes in the key bars of the transmission tower after corrosion, the following conclusions were drawn:

1. The overall trend of the growth rate of the lateral movement of the structure and the amount of lateral movement as the tower becomes longer in service can be divided into three phases, the

first phase is a rapid growth phase, the second phase is a faster growth phase, and the third phase is a slow change growth phase.

2. The displacement of the tower in the X-direction is greater than the displacement in the Y-direction when the wind load is 45 degrees wind load direction, and the X-direction plays a controlling role. As the tower becomes longer in service, the lateral displacement is gradually increasing, after 12 years of service the lateral displacement increases by 12 mm compared to uncorroded, with a growth rate of 7%, that is, the tower stiffness is also gradually decreasing.
3. As the time in service of the towers increased, all the bars reached stress ratios that gradually increased, from 0.78 to 0.97 for tower 1, from 0.79 to 0.98 for tower 2, and from 0.825 to 0.985 for towers 3 and 4 after 12 years in service.

Acknowledgments
This work is supported by research project of State Grid Fujian Economic Research Institute (No. B3130N210014).

References

[1] Chen W M,Chen W G,Wang X. 2015 Research on the influence law of ground deformation on transmission tower system in mining area [J] *Steel Structure* **30** (8) 38-42.
[2] Yu C, Li Z L, Wang Z S. 2018 Analysis of damage process of transmission tower line system under equivalent static wind load [J] *Special Structures* **35** (4) 38-45.
[3] Dai K S, Zhao Z, Meng J Y. 2021 Evaluation of failure probability of wind power towers under earthquake and wind loads [J] *Engineering Science and Technology* **53** (02) 38-44.
[4] Zhang Y H, Dai Y, Wang Y Q. 2019 Analysis of damage forms of power transmission towers in mining areas considering material nonlinearity [J] *Jiangsu Construction* (03) 24-28.
[5] Liu H Y, Li Z L, Huang Z L. Study on the compressive stability load bearing capacity of angular steel members of power transmission towers [J/OL] *Advances in building steel structures* 1-11 [2021-09-25].
[6] Qu K M, Li Z L, Huang Z L. 2021 Bearing capacity tests and finite element analysis of cross-slanting materials of transmission towers [J] *Special Structures* 38 (04) 73-84.
[7] Liu S, Huang Z, Yang Y, Zhang C, Gao Q S. 2020 Analysis of ultimate bearing capacity of transmission tower system under equivalent static wind load of typhoon [J] *Journal of Disaster Prevention and Mitigation Engineering* **40** (05) 757-763.
[8] Kuang J X. 2020 Study on impact resistance of Q420 steel pipe considering corrosion damage [D] Mianyang:Southwest University of Science and Technology.
[9] Wu H Y. 2020 Experimental and theoretical study of static and fatigue performance of structural steel based on atmospheric and accelerated corrosion correlation [D] Taiyuan University of Technology.
[10] Jafari M, Sarkar P P. 2021 Buffeting and Self-Excited Load Measurements to Evaluate Ice and Dry Galloping of Yawed Power Transmission Lines [J] *Journal of Structural Engineering* 147 (11).
[11] Jia J J, Liu C K, Jie L B, Shu Q J, Zhong C S, Yuan G L. Study on the deformation resistance of transmission tower line system in mining area under horizontal deformation of ground surface in different directions [J/OL] *Industrial Building* 1-8[2021-09-25].
[12] Fu C Y. 2021 Uncertainty response and collapse analysis of transmission tower-line system under earthquake [D] Shandong University.
[13] Zhong W J, Xu H W, Lou W J, Ma Y P, Guo G P. 2020 Wind vibration analysis of offshore high-voltage transmission structures under typhoon wind fields [J] Spatial Structure **26** (04) 91-96.
[14] Du W L. 2021 Theoretical, empirical and simulation studies on wind vibration coefficients of transmission lines under the action of typhoon [D] Dalian University of Technology.

An Analysis of Slope Stability Based on Finite Element Method and Distinct Element Method

Bowen Xu[1*], Shijie Liu[1] and Jing Wang[2]

[1] College of Geological Engineering and Geomatics, Chang'an University, Xi'an 710054, Shaanxi, China
[2] Housing and Urban-rural Construction Bureau of Shaanxi Province Xi'an New City Peace, Xi'an 710005, Shaanxi, China
Email: 843075301@qq.com

Abstract. The analysis of slope stability involves complex geological and topographical boundary conditions, nonlinear behavior of material stress-strain, coupling analysis of initial in-situ stress, water pressure and seismic load, etc., and in most cases, analytical solutions cannot be obtained. Under the background of the continuous development of computer and calculation method, the numerical analysis method represented by finite element has been gradually popularized and applied in geotechnical engineering in 1970s, and has developed into a powerful calculation and analysis tool. Among them, the finite element strength reduction method and the discrete element method are the two most widely used slope numerical analysis methods. In this paper, two typical cases, Ankang reservoir landslide and Wenma Highway slope, are simulated by the two methods.

Taking Ankang reservoir landslide as the research object, this paper would use MIDAS / GTS finite element analysis software, and two-dimensional finite element numerical simulation would be carried out to study the influence of reservoir water level periodic fluctuation on the reinforcement effect of anti-slide pile. Under the condition of water saturation and water loss cycle, main material of landslide body and landslide belt, namely the strong weathered phyllite, displays obvious deterioration phenomenon, showing the trend of rapid decline first and then slow decline; after the anti-slide pile is set in the middle and front of the slope, the stability of it has been greatly improved, but with the increasement of the number of water level changes, the reinforcement effect of the anti-slide pile continues to weaken, and the weakening speed is fast at first, and then slows down.

Taking the bedding slope of Wenma Highway as the research object, this pater would adopt UDEC discrete element software to simulate the deformation and failure process of the slope after excavated, and analysize the failure mechanism at the same time. The failure process of bedding slope can be divided into four stages: the formation of tension cracks caused by excavation, the expansion of cracks and the formation of deformation body, the sliding of deformation body and the accumulation of damaged rock mass at the foot of slope. Tensile failure is the main failure mode, and shear failure occurs locally. The failure of bedding slope starts from the foot of slope, which is traction sliding.
Keywords. finite element method, discrete element method, rock slope, stability analysis.

1. Introduction

The analysis of slope stability is the prerequisite of slope design, which determines whether the slope is unstable and how much thrust exists when the slope is unstable, so as to provide scientific basis for support structure design. However, this problem has not been properly solved, because in order to solve this problem, we must first find out the geological conditions and strength parameters of the slope, and at the same time, have a scientific and reasonable analysis method [1]. For homogeneous soil slope, the traditional methods mainly include: limit equilibrium method, limit analysis method, slip line field method, etc.; for the current engineering application, the main method is limit equilibrium method, but the position and shape of sliding surface need to be known in advance. For homogeneous soil slopes, various optimization methods can be used to search for dangerous sliding surfaces. However, for rock slopes, due to the fact that there are a large number of discontinuities with different structures, occurrences and characteristics, the traditional limit equilibrium method is not qualified enough to figure out the dangerous sliding surfaces and the corresponding stability safety factors.

In 1975, the British scientist Zienkiewicz proposed to use the method of increasing load or reducing geotechnical strength to calculate the ultimate load and safety factor of geotechnical engineering [1]. In the 1980s and 1990s, it was used for slope and foundation stability analysis [2]. However, due to the lack of strict, reliable and powerful large-scale finite element program at that time, as well as the selection of strength criteria and the lack of specific operation technology, the calculation accuracy was insufficient, which was not widely adopted by geotechnical engineering.

At the end of the 20th century, many papers have been published in the world to study the solution of the stability safety factor F of homogeneous soil slope by the strength reduction method of finite element method [3, 4]. Because the results of some calculation examples are close to the results of traditional method, they are gradually recognized by the academic community. Some foreign scholars think that the finite element strength reduction method has brought the analysis of slope stability into a new era. Especially in 1999, D.V. Griffith and others of Colorado Institute of mining and technology used their own finite element program to analyze the stability of homogeneous soil slope [5, 6], which is different from other programs in that it can simulate the influence of water level and pore water pressure, and can also analyze the slope stability under the condition of reservoir water falling.

Distinct Element Method (DEM) was proposed by Cundall in the early 1970s [7]. At first, it mainly studied the mechanical behavior of rock and other discontinuous media. Cundall proposed the first practical DEM and used it to simulate the progressive movement of rock blocks [8]. Later, Cundall and Strack put forward a two-dimensional program to simulate particles, and the results were in accordance with the dynamic phofootlastic experiments [9]. So far, through the joint efforts of many scholars, the DEM has been greatly developed. The early DEM can only deal with the distinct stiffness block system, but it has been extended to simulate the deformed block. In 1980, Itasca Consulting Group developed UDEC and put it into the market [10]. Lorig and Brady have developed a distinct element boundary element coupled calculation program [11]. Cundall and other scientists developed a 3D distinct element program (3DEC) to simulate jointed rock mass [12, 13].

Finite element strength reduction method (FEM) and distinct element method (DEM) are widely used in slope numerical analysis. Therefore, in this paper, two typical cases, Ankang reservoir landslide and Wenma Highway slope, are used for numerical simulation.

2. Method

2.1. Finite Element Reduction Method

With the development of computer technology, the finite element strength reduction method has been concerned at home and abroad recently. When the slope reaches the failure state through the finite element strength reduction, the displacement on the sliding surface will change suddenly and produce large and unlimited plastic flow. The finite element program cannot find one from the finite element equations which can satisfy the static balance and the stress-strain relationship. Therefore, it is reasonable to use the convergence criterion of force and displacement as the criterion of slope failure. The calculation accuracy and influential factors of the finite element strength reduction method are analyzed in details, including yield criterion, flow rule, finite element model itself and calculation parameters. The specific measures to improve the calculation accuracy are given: the error between the stability safety factor obtained by Mohr Coulomb equal area circle yield criterion and the traditional

Spencer method is about 5 %, which proves the feasibility of this method in engineering; in the case of plane strain, the mole matching DP criterion can be used; and the finite element method can also be applied to the stability analysis of rock slope. Therefore, in this paper, the strength reduction program of MIDAS/GTS finite element software is used to simulate the Ankang reservoir landslide, and the sliding surface and safety factor of the landslide are obtained.

2.2. Distinct Element Method

Distinct element belongs to the category of discontinuous mechanics, which describes the continuous elements and discontinuous elements in the research medium respectively. For example, the two basic components of rock mass, rock block and structural surface are described by the law of continuous mechanics and contact law respectively. The contact (structural plane) is the boundary of continuum (rock block), and a single continuum can be treated in the process of mechanical solution. It forms an independent object and interacts with other continuum through contact. For rock slope, the block often deforms and destroys along the structural plane, while UDEC (Universal Distinct Element Code) distinct element software (a kind of analysis program based on distinct element theory) has great advantages in dealing with discontinuous medium, which is suitable for dealing with the dynamic response of solid medium under external action, and can better simulate the failure process of discrete medium along the structural plane, such as sliding or tensile crack. Therefore, this paper uses UDEC distinct element software to simulate the bedding slope of Wenma Highway to reflect the specific process of slope deformation and failure and explore the mechanism of slope failure.

3. Engineering Cases

3.1. Ankang Reservoir Landslide

Ankang reservoir landslide is located in the slope section of "V" valley slope on the East Bank of Ankang reservoir, which is a typical bedrock landslide. The rear edge of the landslide is clear, which is a 2 m high landslide scarp. The shear outlet is located at the front edge of the river scarp, and the main sliding direction is about 276°. The plane shape of the landslide (figure 1) is approximately long fan-shaped, with a length of 50~80 m, a width of 300~320 m, and a slope of about 25° to 35°. The landslide mass is mainly strongly weathered phyllite (O_d), overlying silty clay with gravel. The average thickness is about 13 m, and the total volume is about 27.94×10^4 m^3. According to the landslide volume classification, it belongs to medium-sized landslide; the sliding zone is located at the contact surface of strongly weathered phyllite and moderately weathered phyllite, which is nearly arc-shaped; the sliding bed is moderately weathered Ordovician phyllite (O_d). It is easy to form a sliding surface between the weathering boundary of strong and medium phyllite. The treatment measures are to set up a row of anti-slide piles in front of the landslide mass, and arrange drainage works on the slope.

3.1.1. Model Building. Madis numerical simulation is carried out for the typical section of Ankang reservoir landslide. The negative direction of X axis is the main sliding direction (horizontal), and the negative direction of Y axis is the gravity direction (vertical). The origin is the left corner of the model. The length of the model in X direction is 148.40 m, and the height in Y direction is 81.40 m. The geological conditions in the model are mainly based on field survey data, and are mainly divided into sliding mass, sliding zone, strongly weathered phyllite and moderately weathered phyllite. The anti-slide pile is set in the front of the slope, which is also the area with large displacement. The post grid model has a total of 1325 elements and 1478 nodes.

3.1.2. Boundary Condition. Displacement boundary: the slope is a free surface, and fixed displacement constraints are set on the right side and bottom surface. Water head boundary: referring to the reasonable water level of Ankang reservoir in each month, it is proposed that the water level changes at a constant speed between 300 m and 330 m. Therefore, taking the lower left corner of the model as the origin, the relative height of water level change is 20~50 m. In addition to the self-weight load, houses are built in the middle and upper part of the slope, and the house loads are set as 60 kPa, 60 kPa, and 40 kPa respectively.

Table 1. The physical and mechanical parameters of rock-soil mass.

Materials	cycles	Modulus of Elasticity (MPa)	Poisson's ratio	Volume weight (kN/m³)	Cohesion (kPa)	Friction (°)	Permeability coefficient (m/d)	Constitutive model
Sliding belt	0	25	0.35	22.71	24.73	17.49	6.50e^{-2}	Mohr-Coulomb
	1			21.47	24.50	17.48		
	5			21.45	22.37	17.24		
	10			21.42	20.05	16.91		
	20			21.39	17.99	15.59		
	40			21.36	15.63	14.58		
Sliding body	/	50	0.35	21.50	25.50	18.00	6.00e^{-2}	
Strong weathered phyllite	/	50	0.32	25.70	45.20	28.00	4.50e^{-3}	
Moderately weathered phyllite	/	5000	0.20	26.50	4500	36.20	4.32e^{-8}	

3.1.3. Parameter Selection. The physical and mechanical parameters of each layer in the model are comprehensively determined by the indoor physical and mechanical test results and the previous research results (table 1). The anti-slide pile adopts the beam element model, and its physical and mechanical parameters are shown in table 2.

Table 2. The physical and mechanical parameters of slide pile.

Materials	Properties	Model type	Modulus of Elasticity (MPa)	Poisson's ratio	Volume weight (kN/m³)
Anti-slide pile	Isotropy	Elasticity	30000	0.3	28
Materials	**Properties**	**Model type**	**Final shear stress**	**Shear stiffness (kN/m³)**	**Normal stiffness (kN/m³)**
Implant-abutment interface	Linear	Elasto plasticity	500	50000	500000

3.2. Wenma Highway Slope

The slope of K28+466~K28+758 section is about 232 m high and 324 m wide, which is located at the foot of left bank of Zagunao River. The slope is a bedding steep slope with a slope of about 44° and the lithology is phyllite with thin layer. The schistosity plane is extremely developed and the occurrence is 23°∠48°. There are two groups of joints in the slope, the occurrence of which are 237°∠72° and 274°∠50° respectively. At the same time, there are many weathering fractures and unloading fractures on the slope.

3.2.1. Model Building. According to the actual location and scale of the slope, combined with the goal of simulation, the scope of simulation is finally established. The details are as follows: the horizontal distance from the west boundary to the west of the excavation line is 46 m, and the horizontal distance from the east boundary to the east of the excavation line is 257 m. The horizontal direction (x direction) of the whole model is 337 m, and the vertical direction (Y direction) is 232 m. The geological interface in the model includes original slope line, schistosity surface and joint contact surface. The influence of joint cutting and slope excavation on the overall stability is mainly reflected in the sliding of the surface rock layer of the slope. In order to facilitate the calculation during the modeling, the rock strata and joints on the surface of the slope are densified, and the interior of the slope is simplified.

3.2.2. Boundary Condition. In order to balance the system of the model and meet the final requirements of the simulation, the boundary conditions of the slope model need to be limited:

(1) There is no external load on the slope surface, and it is free surface;

(2) Only the influence of gravity field is considered in the slope, and the influence of temperature field, tectonic stress field and seepage field is ignored;

(3) The bottom surface and both sides of the slope are set as fixed, and the bottom surface (X direction) is set to make the velocity in Y direction is 0; the constraint on both sides (Y direction) is set to make the velocity in X direction is 0.

Table 3. Physical and mechanical parameters of slope rock mass.

Materials	Density	Cohesion	Friction	Bulk modulus	Shear modulus
	ρ(kg/m^3)	(MPa)	φ(°)	(GPa)	(GPa)
Phyllite	2660	1.7	42	12.4	5.7

3.2.3. Parameter Selection. Before the calculation and simulation analysis, it is necessary to define the contact surface. In order to make the established slope calculation model consistent with the actual slope layer structure, the equivalent treatment is made when the physical and mechanical parameters of rock block and structural plane are given, and the final parameters are shown in tables 3 and 4.

Table 4. Parameters of structural surface in slope.

Materials	Normal stiffness	Shear stiffness	Cohesion	Friction
	Kn(GPa)	Gt(GPa)	C(MPa)	φ(°)
Joint	5.8	6.5	0.6	35

4. Discussion

4.1. Ankang Reservoir Landslide

The calculation conditions are divided into 5 cycles: 1, 5, 10, 20 and 40. The stress-strain characteristics of slope stability and stress-strain relationship of anti-slide pile under different water saturation loss cycles are analyzed.

4.1.1. Stability Coefficient. The stability coefficient of the slope with anti-slide pile begins to decrease with the increase of the number of cycles (figure 1). Between 1 and 20 times, the stability coefficient decreases approximately linearly; when 20-40 times, the downward trend slows down, and the stability coefficient is 1.06, and the slope is still in a basically stable state, but the reinforcement effect of the anti-slide pile cannot meet the design requirements.

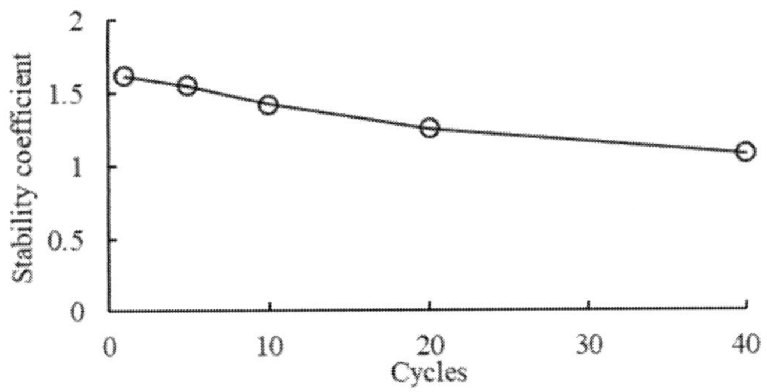

Figure 1. The tendency of stability coefficient in different cyclical times.

4.1.2. Displacement. After 5 cycles (figure 2 (b)), the maximum displacement range of the slope is larger than that of the first cycle, and it is concentrated in front of the pile. After 10 cycles (figure 2 (c)), the displacement in front of the pile extends to the back of the pile. After 20 cycles (Figure 2 (d)), the maximum displacement range in front of the pile does not further expand. After 40 cycles (figure 2 (e)), the displacement of the slope changes obviously. The displacement range of the pile before and after the pile intersects, and the range of the displacement behind the pile increases, but the maximum displacement increases slowly, which is only 0.004 m higher than the maximum displacement of 20 cycles. The maximum displacement at the back of pile increases gradually at 1 to 20 cycles, and tends to be horizontal after 20 to 40 cycles (figure 3).

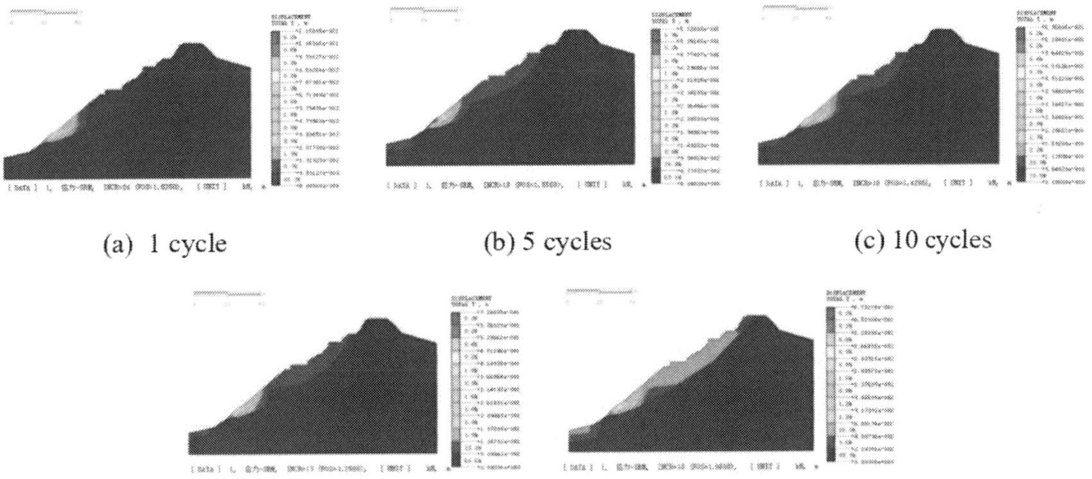

(a) 1 cycle (b) 5 cycles (c) 10 cycles

(d) 20 cycles (e) 40 cycles

Figure 2. The displacement nephogram of landslide in different cyclical times.

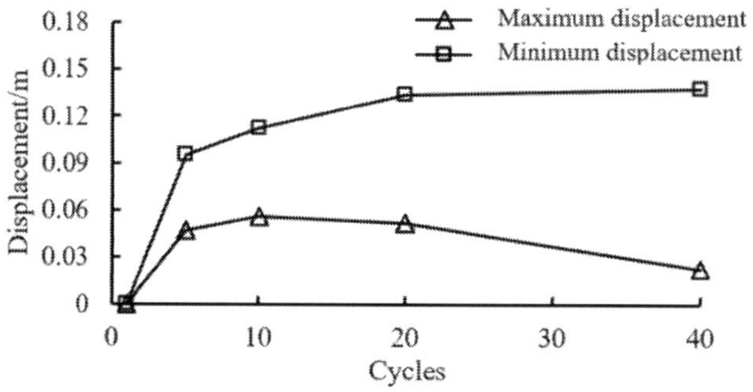

Figure 3. The tendency of displacement in different cyclical times.

4.1.3. Shear Strain. After one cycle (figure 4 (a)), the maximum shear strain of slope body is mainly distributed at the boundary between strongly weathered and moderately weathered phyllite at the foot of slope. At 5 cycles (figure 4 (b)), the maximum shear strain extends upward. After 10 cycles (figure 4 (c)), the shear strain is distributed at the weathering boundary of phyllite in the middle of the slope, which is not connected. After 20 cycles (figure 4 (d)), the shear strain band behind the pile is further extended, but it is still not through. After 40 cycles (figure 4 (e)), a through shear strain band is formed behind the pile, that is, the soil behind the pile is destroyed and the slope body slides along the position. The value of the post pile shear strain band is analyzed and counted. With the increase of the number of cycles (figure 5), the post pile shear strain increases continuously. When the number of cycles is 1~20, the growth rate is fast, but when the cycle number is 20~40, the growth rate is slow.

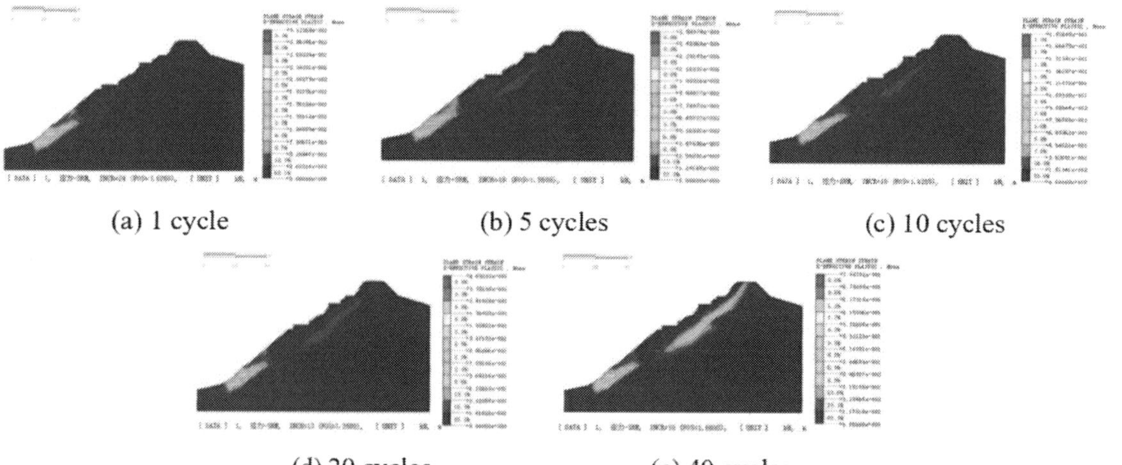

(a) 1 cycle (b) 5 cycles (c) 10 cycles

(d) 20 cycles (e) 40 cycles

Figure 4. The shear strain nephogram of landslide in different cyclical times.

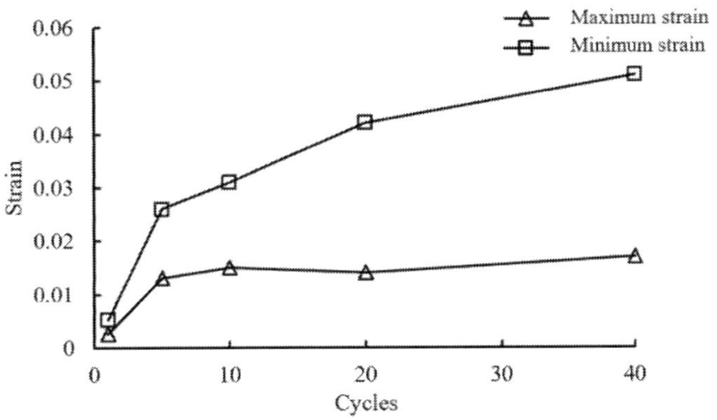

Figure 5. The tendency of shear strain in different cyclical times.

4.2. Wenma Highway Slope

4.2.1. Stress Analysis. The maximum principal stress nephogram formed after slope excavation is shown in Figure 6, which is similar to the maximum principal stress nephogram under the initial state. It is distributed nearly parallel to the slope surface, and gradually increases from the slope surface to the interior of the slope body. However, the maximum principal stress at the foot of the slope increases from -1 MPa to -2 MPa in this state, and the range is also expanded under the influence of excavation.

It is easy to increase the sliding force of rock mass and reduce the stability of slope. The tensile stress value in the minimum principal stress nephogram did not increase, but the tensile stress range at the slope began to expand to the interior of the slope (figure 7), indicating that the scope of rock mass with tensile failure trend increased, while the minimum principal stress inside the slope did not change significantly. However, the interaction between the compressive stress in the maximum principal stress and the tensile stress in the minimum principal stress on the slope surface easily leads to new tension cracks and sliding failure.

Figure 8 shows the shear stress nephogram formed after excavation. Because of the influence of excavation, the upper rock mass loses support, the shear stress at the foot of slope increases to -1 MPa, and the shear stress concentration zone extends along the schistosity towards the upper part, connecting with the joint surface and tensile fracture. Due to the free surface formed at the foot of the excavation slope, the locking capacity of the rock mass is lost, and the sliding failure is easy to occur.

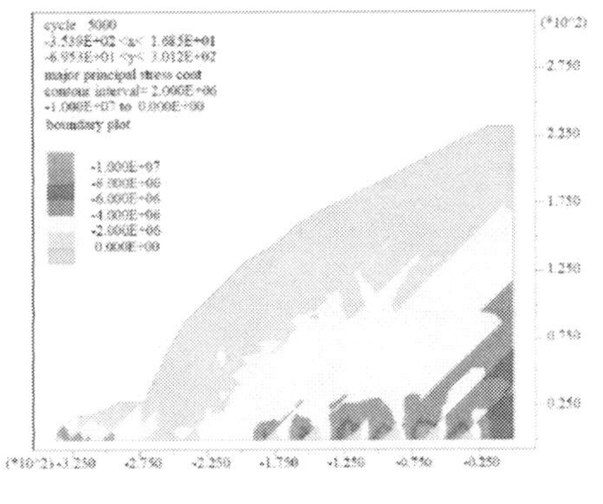

Figure 6. Maximum principal stress after slope excavation.

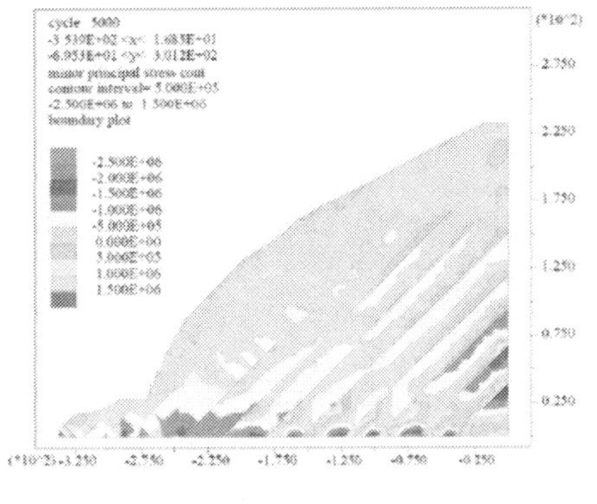

Figure 7. Minimum principal stress after slope excavation.

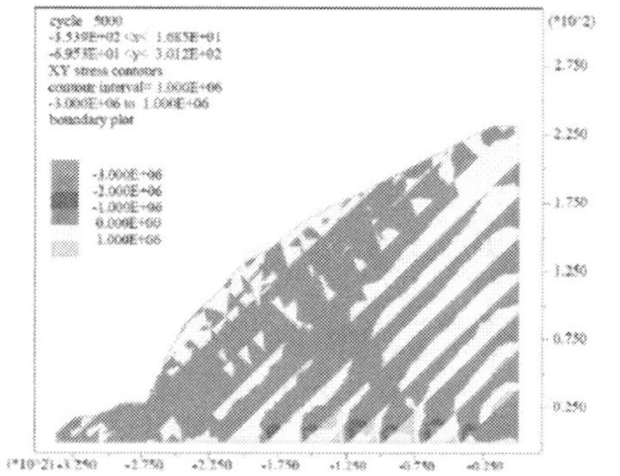

Figure 8. Contour map of shear stress after slope excavation.

4.2.2. Displacement Analysis. For the bedding slope, to excavate the foot of the slope is easy to cause the rock mass to slide along the layer, and the slope is also destroyed rapidly after excavation. It can be seen from the displacement vector diagram (Figure 9) that the surface and near surface rock layers after excavation tend to slide along the structural plane. Among them, the displacement vector at the foot of the slope is the largest, and the initial displacement is 0.052 m. As time goes on, the displacement change will further increase, indicating that the sliding effect brought by excavation of slope foot is very obvious.

The horizontal displacement nephogram of the slope after excavation (Figure 10) reveals the process from the beginning of slope failure to the end of it, and shows the actual sliding mode, sliding range and sliding distance. When step=5e^4, the rock strata near the foot of the slope slide, and the horizontal displacement of each part of the sliding is different. The top broken rock block at the foot of the slope first displaces and the displacement is the largest, which indicates that the failure occurs at the foot of the slope, and then the upper rock layer begins to slide. Therefore, the failure mode belongs to traction sliding failure; In the stage when step=1e^6~5.5e^6, the rock layer continues to slide, and the scope of the sliding rock layer does not expand. The rock layer sliding occurs only in the empty part of the slope foot caused by excavation, and the whole lower slope body does not have deformation phenomenon; when step=7e^6, the maximum horizontal displacement is 48 m and the minimum horizontal displacement is 8 m. Finally, part of the sliding mass is accumulated at the foot of the slope, and part of it remains in the upper part of the slope.

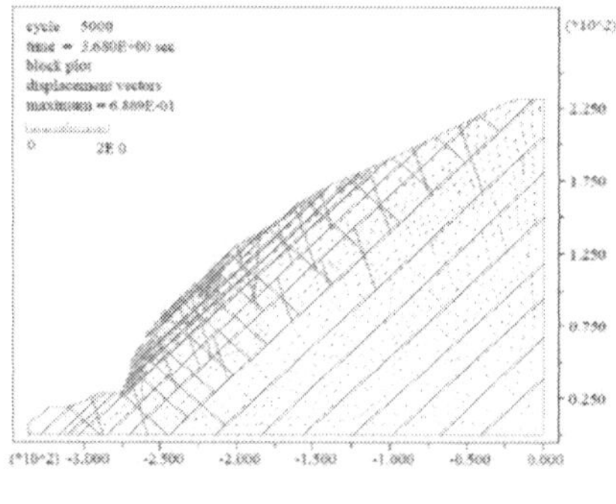

Figure 9. Vector diagram of displacement after slope excavation

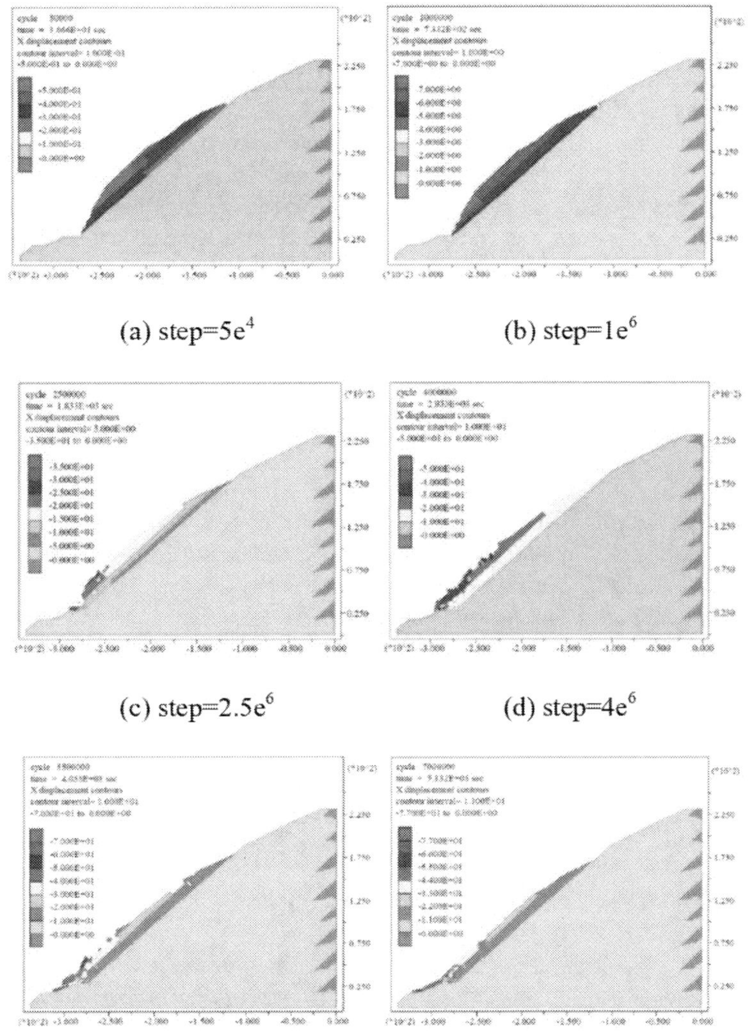

(a) step=5e[4] (b) step=1e[6]

(c) step=2.5e[6] (d) step=4e[6]

(e) step=5.5e[6] (f) step=7e[6]

Figure 10. Contour map of horizontal displacement after slope excavation.

The effect of excavation at foot on the bedding slope is very obvious. The sliding process can be divided into four stages: the formation of tension cracks caused by excavation, the expansion of tension cracks and the penetration of sliding surface, the sliding of rock mass cut by structural plane, and the accumulation of materials.

5. Conclusion

In this paper, the finite element strength reduction method and distinct element method are employed to simulate the Ankang reservoir landslide and Wenma Highway slope respectively, and the conclusions are as follow:

(1) Using the finite element method to simulate the reservoir water level landslide, the following conclusions are obtained: after the reinforcement of anti-slide pile, the slope is relatively stable in the initial stage, the displacement and shear strain are mainly concentrated in the front of the slope, and the reinforcement effect is obvious. However, with the increase of the number of water level cycles, the displacement and plastic strain band gradually move back to the back edge of the landslide, and the

maximum values are concentrated in front of the pile, and the force on the pile body at the sliding zone is the largest and its value increases slowly. The reinforcement effect of anti-slide pile has experienced two stages, that is, before the number of cycles is 20, the reinforcement effect is weakened quickly; when the number of cycles is 20 to 40, the weakening effect is slow. The reinforcement effect of anti-slide pile is weakening. When the water level of the reservoir changes for many years, the slope still has the possibility of damage, so there should be enough safety reserve when designing it.

(2) The distinct element method is used to simulate the bedding rock slope. According to the above analysis, the following conclusions can be obtained: the failure process of bedding slope is mainly caused by tension, and shear failure occurs at the foot of slope, resulting in the loss of locking capacity and sliding of rock mass; the interaction between the maximum principal stress and the minimum principal stress leads to the formation of many tension cracks at the slope surface, which enhances the possibility of rock mass sliding; the rock mass has the characteristics of high strength, high stability and high stability The sliding range is heavily controlled by the excavation, only the free part of the slope foot slides and does not extend into the slope.

References

[1] Zienkiewicz O C, Humpheson C, Lewis R W. 1975 Associated and non-associated visco-plasticity and plasticity in soil mechanics [J] *Geo technique* **25** (4) 671~ 689.

[2] Matsui T, San K-C. 1992 Finite element slope stability analysis by shear strength reduction technique [J] *Soils and Foundations* **32** (1) 59~ 70.

[3] Griffiths D V, Lane P A. 1999 Slope stability analysis by finite elements [J] *Geo technique* **49** (3) 387~ 403.

[4] Lane P A, Griffiths D V. 2000 Assessment of stability of slopes under drawdown condition [J] *Journal of Geotechnical & Geo environmental Engineering* **126** (5) 443~ 450.

[5] Smith I M, Griffiths D V. 1982 Programming the Finite Element Method. John Wiley & Sons, Hoboken.

[6] Dawson E M, Roth W H, Drescher A. 1999 Slope stability analysis by strength reduction [J] *Geo technique* **49** (6) 835~ 840.

[7] Z_SOIL.PC 2003 User manual [R] *Zace Services Ltd. Report 1985-2003*. Lausanne: Switzerland.

[8] Cundall P A. 1971 A computer model for simulating progressive large-scale movements in blocky rock systems. *Proceedings of the Symposium International Society of Rock Mechanics* **1** 132~ 150.

[9] Cundall P A, Strack O. 1979 A discrete numerical model for granular assembles [J] *Geo technique* **29** (1) 47~ 65.

[10] Cundall P A. 1980 UDEC— a generalized distinct element program for modelling jointed rock. *Report PCAR-1-80* Peter Cundall Associates, European Research Office, US Army Corps of Engineers.

[11] Lorig L J, Brady B, Cundall P A. 1986 Hybrid distinct element-boundary element analysis of jointed rock [J] *International Journal of Rock Mechanics & Mining Science & Geomechanics Abstracts* **23** (4) 303~ 312.

[12] Cundall P A. 1988 Formulation of a three-dimensional distinct element model - Part I: a scheme to detect and represent contacts in a system composed of many polyhedral blocks [J] *International Journal of Rock Mechanics & Mining Science & Geomechanics Abstracts* **25** (3) 107~ 116.

[13] Hart R, Cundall P A, Lemos J. 1988 Formulation of a three-dimensional discrete element model - Part II: mechanical calculation formation and interaction of a system composed of many polyhedral blocks [J] *International Journal of Rock Mechanics & Mining Science & Geomechanics Abstracts* **25** (3) 117~ 125.

Evaluation of Concrete Durability Based on Cloud Model and D-S Evidence Theory

Sai Yang [1], Jinlong Huang [2], Yawei Qin [1, 3] and Xianguo Wu [1]

[1] School of Civil And Hydraulic Engineering, Huazhong University of Science and Technology, Wuhan, Hubei, 430074, China
[2] China Construction Third Engineering Bureau Group Co., Ltd, Wuhan, Hubei, 430064, China
[3] Wuhan Huazhong University of science and Technology Testing Technology Co., Ltd, Wuhan, Hubei, 430074, China
yangsai0522@163.com

Abstract. The environment in alpine region is complex and harsh, and the durability of concrete is seriously affected by freeze-thaw and salt invasion. In this paper, the relative dynamic modulus of elasticity, chloride permeability coefficient, mass loss rate and carbonation depth are selected as the evaluation indexes of concrete durability in Northeast China. Based on prior knowledge and expert group decision-making, the durability grade is divided and the evaluation standard is established; The evaluation model of concrete durability based on cloud model and D-S evidence theory is established. According to the engineering experimental data, the membership degree of concrete durability evaluation index in different grades is obtained through the correlation measurement of cloud model. The normalized evidence is formed and fused by D-S evidence theory. The results show that the durability grade of concrete is grade I, which is consistent with the actual project. It shows that using cloud model and evidence theory evaluation model to evaluate the durability of concrete is a new and effective method.
Keywords. Concrete; Freeze thaw cycle test; Evaluation criteria; Cloud model; D-S evidence theory; Durability evaluation.

1. Introduction

As an excellent building material, concrete has been widely used in civil engineering industry, and its durability has a direct impact on the safe use of engineering structure [1]. In practical engineering, because of the durability of concrete, a lot of structural deterioration and huge economic losses are caused. Therefore, how to evaluate the durability of concrete reasonably and improve the durability of concrete has become an urgent need for engineering construction and concrete development [2].

At present, many scholars have carried out a series of researches on the durability of concrete.。 Aït-Mokhtar et al.[3] conducted an experimental study on the change of concrete durability. Wang et al.[4] studied the influence and coupling effect of carbonation, dry wet cycle and freeze-thaw cycle on the durability of three kinds of recycled polymer concrete. Zhu et al.[5] evaluated the durability of recycled polymer concrete in complex environment based on freeze-thaw cycles and chloride coupling experiments. Choi et al.[6] Put forward a method to determine the early frost damage of concrete by measuring the permeability change of concrete, and studied the influence of early frost damage on the durability of concrete. Alireza Habibi et al. [7] evaluated the mechanical properties and durability of recycled polymer concrete via RSM method. Considering the current research, most of the research is mainly based on experimental and theoretical analysis methods. For concrete durability evaluation, a set of practical, reasonable and complete evaluation methods has not been established yet. How to effectively mine the collected information, so as to carry out scientific risk identification, early warning and prevention and control of concrete durability has become an urgent problem.

Cloud model can transform qualitative concepts and quantitative values, and deal with fuzzy and random problems effectively [8,9]. D-S evidence theory is a multi information fusion method which can deal with uncertain decision and reasoning. It is widely accepted and applied in the field of information fusion [10,11].In this paper, a concrete durability evaluation method based on cloud model and D-S evidence theory is proposed. Through the cloud model, the index evidence of concrete durability is constructed, and the improved D-S evidence theory is used to fuse the index evidence for many times to realize the evaluation of concrete durability, which provides the basis for the prevention and control management decision of concrete durability.

2. Preliminaries

2.1. Cloud Model

Cloud model is a kind of uncertainty transformation model between qualitative concept and accurate value, which can deal with both fuzzy and random events. Let $U=\{x\}$ be a quantitative universe and C be a qualitative concept over U. Element x is a random implementation of qualitative concept C.The membership degree $\mu(x)$ is a random number with a value of $0\sim 1$. Cloud is actually a mapping from the universe U to interval [0, 1], that is

$$\mu : U \to [0,1] \quad \forall x \in U \quad x \to \mu(x) \tag{1}$$

Then, (x,μ) is cloud drop. The generation process of cloud droplets realizes the mapping between qualitative concept and quantitative value. Expectation Ex, Entropy En and Hyper entropy He are the three characteristic values of cloud model.

2.2. Dempster-Shafer Evidence Theory

D-S evidence theory can solve some important problems such as "uncertainty" and "prophet". Through uncertainty reasoning, the fusion idea is adopted to synthesize the information represented by multi-source data as the basis of evaluation, so as to obtain the most likely result from imprecise and incomplete information, and then get more reliable and effective conclusions. It is a simple method of information fusion and decision-making.

Suppose Θ is an identification framework, for $\forall A \subseteq \Theta$, $m_i\,(i=1,2,\cdots,n)$ is the basic probability assignment (BPA) of $A_i\,(i=1,2,\cdots,n)$,thus, the combination rule of multiple evidences is as follows:

$$m(A)=\frac{\sum\limits_{A_1\cap A_2\cap\cdots\cap A_n=A} m_1(A_1)\cdot m_2(A_2)\cdots m_n(A_n)}{1-k} \tag{2}$$

$$k=\sum_{A_1\cap A_2\cap\cdots\cap A_n=0} m_1(A_1)\cdot m_2(A_2)\cdots m_n(A_n) \tag{3}$$

k is the conflict coefficient between n evidences.

3. Modeling

3.1. Evaluation Framework of Concrete Durability

In this paper, the relative dynamic modulus of elasticity, chloride ion permeability coefficient, mass loss rate and carbonation depth are selected as the indexes to evaluate the durability of concrete under freeze-thaw cycle test in northeast cold region, which can well reflect the change of internal microstructure and damage degree of concrete after freeze-thaw cycle test[12].The durability of concrete is divided into three grades, I (very good), II (good) and III (fairly good), which respectively indicate that the effective service life of concrete can reach 200-250 years, 150-200 years and 100-150 years. On this basis, the prior knowledge and expert group decision are used to identify and divide the attribute interval of concrete durability index and determine the evaluation standard.

Based on the obtained experimental data rules, assuming that n experts participate in decision-making, the i-th expert believes that$[c_i(L), c_i(R)]$ $i = 1,2,\dots,n$is the interval of an evaluation index belonging to a certain level. The reverse cloud algorithm is used to estimate the judgment of N experts as a whole. In order to promote the convergence of decision-making results, n experts are arranged for multiple rounds of inquiry. Each expert needs to carry out the next round of feedback on the basis of the previous round of inquiry cloud, so as to gradually eliminate the differences of expert opinions and obtain more reasonable attribute interval division results.

To determine the comprehensive durability grade of concrete specimens to be tested, the variable weight theory is used to determine the comprehensive weight of each index. Based on the above steps, the results of durability evaluation grade standard grade and index weight are obtained as shown in table 1.

Table 1. Evaluation index grade and criterion of concrete durability under freeze-thaw cycles.

Evaluating index	Weight	Classification Criterion					
		I		II		III	
		Value	Expected service life /year	Value	Expected service life /year	Value	Expected service life /year
Relative dynamic modulus of elasticity /(%)	0.35	[95,100]	200-250	[90,95]	150-200	[85,90]	100-150
Chloride ion permeability coefficient / ($10^{-12}m^2/s$)	0.35	[0,1]		[1,2.5]		[2.5,4]	
Mass loss rate /(%)	0.2	[0,2]		[2,4]		[4,5]	
Carbonation depth /(mm)	0.1	[0,10]		[10,20]		[20,30]	

3.2. Evidence Generation Via Cloud Model

In the process of concrete durability evaluation, the important premise to determine the basic probability assignment is to get the three digital features of cloud model. The parameters of cloud model $(Ex_{ij}, En_{ij}, He_{ij})$ of evaluation index X_i in response interval$[C_{ij}(\min), C_{ij}(\max)]$ of concrete durability grade T_j $(j = \text{I,II,III})$ are obtained by formula (4).

$$\begin{cases} Ex_{ij} = \left(C_{ij}(\min) + C_{ij}(\max)\right)/2 \\ En_{ij} = \left(C_{ij}(\max) - C_{ij}(\min)\right)/2 \\ He = s \end{cases} \tag{4}$$

s is a constant, which is determined by engineering experience and the fuzzy threshold of the index, indicating the uncertainty of the membership degree. In this paper, s = 0.002.

The membership degree of the observation value x_i to the j-th durability grade is obtained by formulas (5) and (6). Considering the uncertainty of concrete durability evaluation caused by measurement error and external interference, the membership degree of global uncertainty φ_i is calculated by equation (7).

$$\mu_i\left(T_j\right)=\exp\left(\frac{-\left(x_i-Ex_{ij}\right)^2}{2\left(En'_{ij}\right)^2}\right) \tag{5}$$

$$En'_{ij}\sim N\left(En_{ij},He\right) \tag{6}$$

$$\varphi_i=1-\max\left(\mu_i\left(T_j\right)\right), j=\mathrm{I,II,III} \tag{7}$$

The membership degree is normalized to be transformed into the basic credibility allocation function in D-S evidence theory to receive the evidence $\{m_i(T_{\mathrm{I}}),m_i(T_{\mathrm{II}}),m_i(T_{\mathrm{III}}),m_i(\Theta)\}$.

3.3. Information Fusion Via D-S Theory

The conflict coefficient k between multiple evidences is calculated by formula (3). When $k\le T_1$, it's regarded as credible evidence and kept the original evidence information unchanged; when $k\ge T_1$, it's considered as high conflict evidence, which needs to be revised and reused. Based on many experiments and literature review, the threshold T1 = 0.85 was preliminarily determined.

For the evidence with high conflict, firstly calculate the distance d and similarity Sim between each pair of evidences. Then by summing the similarity between one piece of evidence and other evidences, the support degree $Sup(m_i)$ of the evidence m_i is obtained. All $Sup(m_i)$ are normalized to get the weight $W(m_i)$ of the evidence m_i .The weighted evidence is used to replace the conflict evidence, and the conflict detection is carried out again. If there is no conflict, the evidence fusion is carried out. If there is conflict, the evidence replacement is continued until there is no conflict.

After dealing with high conflict evidence, Dempster rule by formula (2) is used to realize multiple evidences fusion. The fusion results of a certain durability index evidence is separately acquired, then the weighted fusion of the above four index evidence is realized using the weights of the evaluation indexes in Table 2 to perceive the durability grade status of the batch of concrete specimens.

4. Case Study

4.1. Data Collection

The data samples obtained in this study are from seven sections of a highway project in the northeast cold region.25 groups of prismatic concrete specimens with dimensions of 100mm×100mm×400mm were randomly selected for freeze-thaw cycle test. The durability evaluation index was measured in the same cycle. The test results are shown in table 2.

Table 2. Test results of durability index of concrete specimen.

Evaluation Index / Specimen number	Relative dynamic modulus of elasticity / (%)	Chloride ion permeability coefficient /($10^{-12}m^2/s$)	Mass loss rate / (%)	Carbonation depth /(mm)
1	97.2	1.3	1.5	6.7
2	93.4	1.8	3.2	16.7
3	96.2	0.9	1.7	8.9
...	...	...	...	...
24	88.2	3.8	4.6	24.3
25	93.8	2.1	3.1	15.3

4.2. Results Analysis

A normal cloud model is constructed for each durability grade interval of the evaluation index, and the characteristic diagram of the grade cloud model of the four indexes is shown in figure 1.

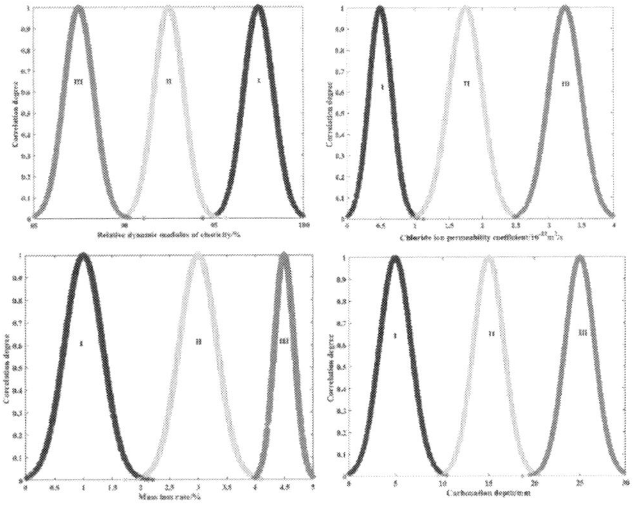

Figure 1. Cloud chart of durability index grade.

Engineering data cannot effectively characterize the concrete durability grade status. Combining the actual values in table 2 and the fuzzy set constructed in Figure 1, it is transformed into the degree of membership of the index X_i to the durability grade T_j by formulas (4)-(6). The membership degree is normalized to construct BPAS. Taking the relative dynamic modulus of elasticity as an example, the basic reliability distribution is shown in table 3.

Table 3. Basic reliability distribution of relative dynamic modulus of elasticity.

Grade / Number	I	II	III	Θ
1	0.9373	0.0000	0.0000	0.0627
2	0.0000	0.5579	0.0000	0.4421
3	0.2961	0.0000	0.0000	0.7039
...	...	...	...	...
24	0.0000	0.0000	0.7026	0.2974
25	0.0000	0.2972	0.0000	0.7028

Conflict detection is performed on the 25 pieces of evidence in table 3, and the k between these evidences is calculated to be 0.927>0.85. It is considered that this group of evidence is high conflict evidence and should be corrected and reused.

For high conflict evidence, the Euclidean distance of pairwise evidence is calculated to get the distance matrix, and further the evidence support degree is calculated and normalized to obtain the weight of each evidence. The support degree and weight of 25 evidences are shown in figure 2; Finally, the conflict evidence is weighted.

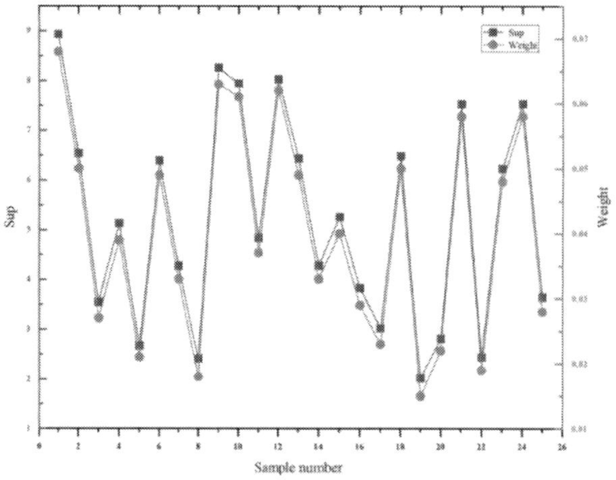

Figure 2. Support degree and weight of 25 evidences for relative dynamic modulus of elasticity.

Conflict detection is performed on the revised evidence again. At this time, the conflict coefficient of the 25 pieces of evidence is 0.429 <0.85, therefore, there is no conflict between the modified evidences. The modified evidences are fused according to the combination rule of equation (2) .For the remaining three indexes like chloride ion permeability coefficient, mass loss rate and carbonation depth, the corresponding evidences of 25 groups of concrete specimens are obtained and fused with reference to the above steps to obtain the evaluation results of each index of concrete durability. The evaluation results of the four indexes are weighted and fused again to obtain the durability grade state of the batch of concrete specimens, as shown in figure 3.

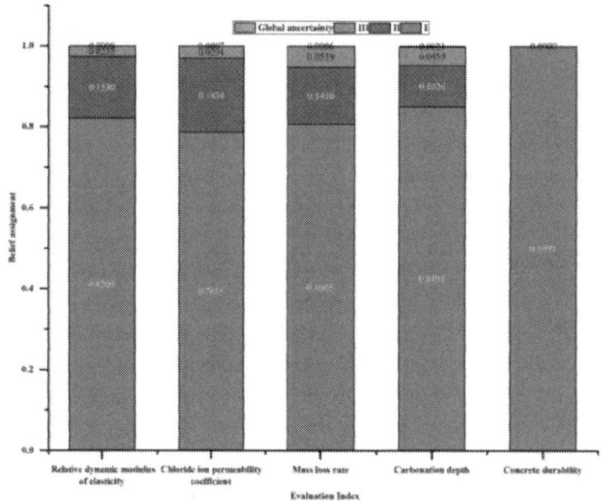

Figure 3. Evaluation results of concrete durability under freeze-thaw cycles.

Figure 3 shows that the durability of this batch of concrete specimens is good. The evaluation results of relative dynamic elastic modulus, chloride ion permeability coefficient, mass loss rate and carbonization depth index all have the largest reliability distribution at level I, which are 0.8209, 0.7865, 0.8065 and 0.8494 respectively. According to the principle of maximum membership degree, the final evaluation result of concrete durability is determined by the level of the maximum reliability distribution value. It can be seen from figure 3 that the final fusion result of the concrete specimens under the freeze-thaw cycle experiment has the largest reliability distribution at level I, which is 0.9991. Therefore, the durability level of the concrete in the service life of the project is considered to be level I. The results are consistent with the actual inspection results of the project and the expected service life is 200-250 years .It indicates that the model selected in this study has a good effect on the evaluation of concrete durability.

5. Conclusion

1. In this paper, the relative dynamic elastic modulus, chloride ion permeability coefficient, mass loss rate and carbonization depth are selected as the concrete durability evaluation indicators in the severe cold area of Northeast China. The inverse cloud model is used to realize the exchange and integration of the knowledge decision of the expert group. The durability level is divided as level I (very good), II (good), and III (fairly good), and the evaluation criteria and the corresponding concrete effective service life is given.

2. This paper constructs a concrete durability evaluation model based on cloud model and D-S evidence theory. The cloud model is established one by one for the durability grade interval in response to each evaluation index and the membership degree of the measured data to each grade cloud is obtained. Conflict detection and conflict evidence replacement are carried out, and information fusion is realized by D-S evidence theory.

3. The concrete specimens of Song-Tong expressway project are selected for durability evaluation. The results show that the reliability distribution of each index in grade I is the largest under the freeze-thaw cycle test, which indicates that the concrete durability is excellent. The evaluation results are consistent with the actual test results, and the expected service life is 200-250 years. This proves the rationality and accuracy of the model proposed in this paper for concrete durability evaluation.

References

[1] Li K, Zhang D, Li Q, et al. 2019 Durability for concrete structures in marine environments of HZM project: Design, assessment and beyond[J] *Cement and Concrete Research* 115 545-558.
[2] Guo H, Dong Y, Gu X. 2020 Durability assessment of reinforced concrete structures considering global warming: A performance-based engineering and experimental approach [J] *Construction and Building Materials* 233 117251.
[3] Aït-Mokhtar A, Belarbi R, Benboudjema F, et al. 2013 Experimental investigation of the variability of concrete durability properties[J] *Cement and Concrete Research* 45 21-36.
[4] Wang J G, Zhang J X, Cao D D. 2021 Pore characteristics of recycled aggregate concrete and its relationship with durability under complex environmental factors [J] *Construction and Building Materials* 272 12.
[5] Zhu P, Hao Y, Liu H, et al. 2020 Durability evaluation of recycled aggregate concrete in a complex environment [J] *Journal of Cleaner Production* 273 122569.
[6] Choi H, Zhang W, Hama Y. 2017 Method for determining early-age frost damage of concrete by using air-permeability index and influence of early-age frost damage on concrete durability [J] *Construction and Building Materials* 153 630-639.
[7] Habibi A, Ramezanianpour A M, Mahdikhani M, et al. 2021 RSM-based evaluation of mechanical and durability properties of recycled aggregate concrete containing GGBFS and silica fume[J] *Construction and Building Materials* 270 121431.
[8] Wu H-W, Zhen J, Zhang J. 2020 Urban rail transit operation safety evaluation based on an improved CRITIC method and cloud model [J] *Journal of Rail Transport Planning & Management* 16 100206.
[9] Wang J, Wang Y, Zhao G, et al. 2020 The active learning multi-task allocation method in mobile crowd sensing based on normal cloud model [J] *Pervasive and Mobile Computing* 67 101181.
[10] Xiao F. 2020 A new divergence measure for belief functions in D–S evidence theory for multisensor data fusion [J] *Information Sciences* 514 462-483.
[11] Li P, Wei C. 2019 An emergency decision-making method based on D-S evidence theory for probabilistic linguistic term sets [J] *International Journal of Disaster Risk Reduction* 37 101178.

[12] Dong F, Wang H, Yu J, et al. 2021 Effect of freeze–thaw cycling on mechanical properties of polyethylene fiber and steel fiber reinforced concrete[J] *Construction and Building Materials* 295 123427.

Experimental Investigation on Influence of Molding Methods on Properties of Pervious Concrete

Qiuxiang Li [1], Jie Zheng [1] and Mingfu Fu [2]

[1] Jiangxi Vocational Technical College of Industry & Trade, Nanchang, Jiangxi, 330038, China.
[2] School of Civil Engineering and Architecture, Nanchang Institute of Technology, Nanchang, Jiangxi, 330099, China.
Email: 253793943@qq.com

Abstract. Despite the common use of pervious concrete (PC), there is no standard way of producing the test specimens, which undergo testing to infer the behaviour of PC in the field. Vibrating table is the most common method but greatly reduced in vibration time compare with normal concrete in the laboratory. Marshall compaction and superpave gyratory compactor (SGC) are recommended standard molding methods for porous asphalt mixtures manufactured in the laboratory environment. Three kinds of pervious concrete samples with three target porosities were prepared by the above three methods, and the effects of the molding method on the physical properties, mechanical properties and durability of the samples were investigated in the study. Experimental results showed, with different molding methods adopting, pervious concrete with the same mixture design exhibits slightly different physical and mechanical properties. After analysis and comparison, SGC is the best choice to obtain concrete with high permeability, good freeze-thaw resistance and high strength, followed by Marshall compaction molding, and vibration molding is the last one. As a result, a win-win situation of the hydraulic characteristics and mechanical properties of pervious concrete can be achieved due to both optimized mix-design and appropriate molding method.

1. Introduction

Perious concrete, a kind of mixture with continuous voids, in which coarse aggregates with a specific particle size gradation are cemented into concrete by cementitious material, with little or no fine aggregate added. Compared with conventional concrete, this material is light and porous, and requires less energy to put into production [1], so it is a leader in emerging environmentally friendly products. Its characteristics make it as good sound-absorbing and heat-insulating material, a bed material for vegetation and sewage purification. Of course, the most widely and longest-lasting material applied as permeable pavement [2-10]. It is preferred by engineers to use it in road engineering with the following benefits: precipitation enters the soil through multi-porous while filtering for contaminants in water [11], allow more water to evaporate from the soil and natural recharge of groundwater [12, 13], effectively effectively suppress the loss of surface water [14], reduce the urban heat island effect [15], limit the budget cost of building road drainage systems [12], and at the same time, porous materials can effectively absorb noise [16]. Just as everything has two sides, compared with ordinary concrete, the durability and strength of pervious materials are not high. It is limited to areas with light traffic or low levels of congestion such as local roads, streets, parking lots, shoulders of heavy pavements etc. [17, 18]. Overcoming the performance bottlenecks of pervious concrete, especially its low strength, which is an inspiring but challenging task. Fortunately, since the advent of permeable concrete, many scholars have done a lot of work to improve its mechanical properties on the basis of firmly retaining the hydraulic properties of permeable concrete. Such as varying aggregate particle size gradation curve [17, 19], the type and amount of fine sand added to the mixture [20-22], the content of cement and water, admixtures, especially for some certain types of fibers [20, 21, 23], various polymer modified water-permeable concrete materials [24-28] and so on. In addition to the above studies of the materials that make up concrete, construction techniques adopted is another key factor affecting the performance of pervious concrete [29-31]. Therefore, both researching proper molding methods in the laboratory and full scale applications is still a topic of constant discussion. As mentioned above, PC is a non-skeleton structure formed by a special gradation aggregate cemented with reduced cementitious material, which has poor fluidity and large frictional resistance between aggregates. Therefore, the choice of molding process has a significant impact on the mechanical properties and destructiveness of the PC. At present, vibration molding is a molding method commonly used in laboratories. Unfortunately, the low fluidity and high viscosity of fresh pervious concrete and the formation of high-quality voids during the forming process make it unsuitable for long-term vibration consolidation like ordinary concrete mixtures. On the other hand, vibration molding also creates local large cavities in some areas of the material, while low voids appear in some areas around it, making the distribution of voids very uneven. Fortunately, the strength of the fresh infiltration mixture is very low when initially placed and low compaction energy can be immediately applied. Accordingly, Marshall compaction and superpave gyratory compactor (SGC) commonly used in permeable asphalt can be adopted for pervious concrete. Some scholars have been committed to the research and achieved certain results. Sengün et al. [32] compared and analyzed the effects of four molding methods including SGC on the properties of roller compacted concrete (RCC) and revealed the correlation of mechanical properties with compaction ratio of RCC, cementitious materials and w/c. It is concluded that the permeable void is affected by the compaction method but hardly affected by the content of the cementing material, but its mechanism needs to be further studied. Lang et al. [33] studied the influence of three kinds of molding methods on the permeability of magnesium phosphate cement steel slag permeable concrete (MSPC): vibration molding (VM), tamping molding (TM) and hydrostatic molding (HM). The comparative analysis shows that the mechanical properties of MSPC first increase and then decrease with the porosity, and the compressive strength of the sample formed by vibration molding is the largest, but the reason is unknown. Bonicelli et al. [31] used Marshall compactor with different blows to simulate compaction tools with different impact forces in the field, and tested and evaluated the performance of the samples. The results show how three factors including compaction energy affect the indirect tensile strength and porosity of the mixture without hydrological properties of PC. Kevern et al. [34] evaluated and analyzed the performance of Marshall-formed pervious concrete under different compaction energy by Marshall with its mixture design with no or small addition of fine sand. Prediction formulas related to the correlation between porosity (Vc), permeability coefficient (k), stiffness (E), strength (ITS), particle loss (CL) and bulk density (ρ), but their application needs to be studied. There are differences in the impact of molding methods on the performance of permeable concrete, and the molding methods are uncertain and diverse. In summary, the existing research on pervious concrete molding method and its effect on its properties is not comprehensive enough. At present and to the author's best knowledge, superpave gyratory compactor (SGC) only used for RCC but scarcely for pervious concrete. This article uses different molding methods to make samples, compares the effects of different methods on the performance of permeable concrete, analyzes the causes of the effects, and tries to find out the best molding method and proportion to achieve the best balance of mechanical and hydraulic performance. That has practical guiding significance for on-site construction.

This paper reports a laboratory study conducted to balance the mechanics and hydraulic performance of pervious concrete as a comprehensive result of mixture design and molding method. The experiment program considered three target porosity levels for each of the three molding methods: 12%, 15% and 20%. Then evaluate the properties of pervious concrete produced by three molding methods: vibrating table, Marshall compaction and superpave gyratory compactor (SGC). The properties of PC involving effective porosity, water permeability cofficient, dynamic elastic modulus, compressive strength and freeze-thaw resistance. At last, comparative analysis the influence of molding method adopted in several pervious concrete mixtures and find the optimization of molding method balance the hydraulic and mechanics properties of pervious concrete.

2. Materials and Methods

2.1. Materials and Mixture Design

A common pavement limestone produced in Jiangxi, with a grain-size distribution of the aggregate is 13.2-9.5 mm, 9.5-4.7 mm, and 4.75-2.23 mm account for 10%, 80% and 10%, respectively. They were preliminary characterized according to ASTM C131-06, EN 933-3, and EN 933-4 resulting in a Shape Index (S.I.) from 4% to 9.8%,Los Angeles Index (L.A.) between 18% and 23% and Flakiness Index (F.I.) between 9.5% and 14.5%. The apparent density and bulk density of the aggregate are 2600kg/m^3 and 1600kg/m^3, respectively.

Cono brand P.O 42.5 ordinary portland cement was used as cementitious material, with chemical composition and physical properties complying with the requirements of ASTM C150-07, as shown in table 1. SR-5 type ecological concrete reinforcing agent produced by Nanjing Jiuhe Run Technology Co.Ltd., which can participate in cement hydration reaction to form polymer cement cement hydrate. Its quality is 4% of the cement. Samples were produced with three different target porosity with volume method, as shown in table 2. Inspired by the literature [21, 32, 34], the water-cement ratio for all the mixtures remained unchanged at 0.28.

Table 1. Chemical composition and physical properties of cement.

Chemical composition (wt%)		Physical properties	
CaO	62.1	Specific gravity	3.10
SiO_2	20.3	Specific surface blain (m^2/kg)	3.45
Al_2O_3	5.1	Initial setting time (min)	155
Fe_2O_3	2.9	Final setting time (min)	245
MgO	1.6	3 days compressive strength (MPa)	25.9
SO_3	2.0	28 days compressive strength (MPa)	50.1
K_2O	0.4	3 days flexural strength (MPa)	5.1
Na_2O	0.7	28 days flexural strength (MPa)	7.9
Loss on ignition	2.36		

Table 2. Mix proportions of pervious concrete with three target porposity.

Mix	Target porosity (%)	Water-cement ratio	Cement (kg/m^3)	Water (kg/m^3)	Coarse aggregate (kg/m^3)	Concrete reinforcing agent (kg/m^3)
PC-12	12	0.28	422	118	1600	17
PC-15	15	0.28	374	105	1600	15
PC-20	20	0.28	294	83	1600	12

2.2. Specimen Preparation

With comprehensive consideration of the factors involved in the study and its post-experimental requirements, 108 PC specimens (3 molding methods×3 target porosity×12 replicates) need to be prepared. All pervious concrete specimens were cast to the same dimension of Φ 100×120mm. The sequence for fabricating the mixture was follows: the weighted aggregates and 1/4 water were mixed for 30 seconds and cement, concrete reinforcing agent and remaining water were added to the mixture with additional 2.5 min.

As aforementioned, the effects of different molding methods on porosity and compressive strength of PC were investigated to obtain the target performance. A steel rod with 14mm diameter, vibrating table, standard Marshall electric compactor, superpave gyratory compactor (SGC) were used to achieve the desired height. Different molding methods were shown as figure 1.

Vibration molding method (VM): 2 layer_ rodding 10 times each layer, then vibrating 10 seconds. Marshall compactor molding method (MCM): 2 layer_ rodding 10 times each layer, then compacting one side of the specimen to the desired height with standard Marshall compactor (determine the number of hits with the desired height by trial experiments. When the sample is compacted to a height of 120mm, the corresponding number of hits were 23, 18, 12 with three porosity levels). Superpave gyratory compactor molding method (SGCM): 2 layer_ rodding 10 times each layer, then compacting to the desired height with superpave gyratory compactor (SGC).

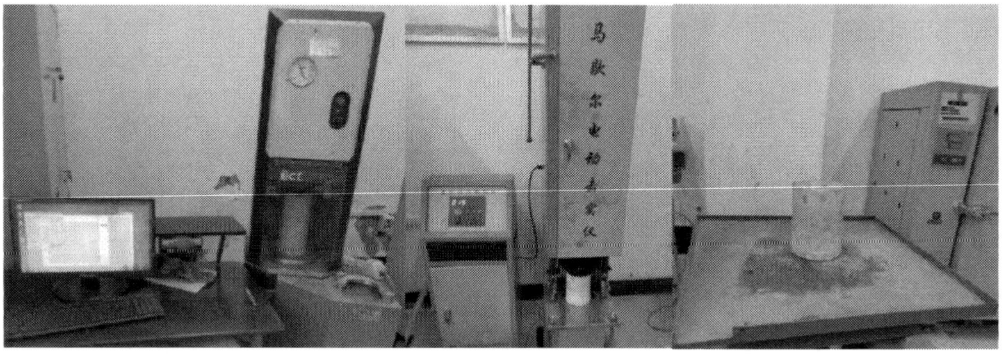

Figure 1. Three different molding methods.

All pervious concrete specimens were immediately moved to the curing room, whose temperature and humidity were maintained at 20°C ± 2°C and 95% ± 5%, respectively. After one day, molds were removed and specimens were cured in water until the days for testing.

2.3. Test Methods

2.3.1. Effective Porosity and Unit Weight Test. Closed pores, semi-connected pores and connected pores were constituted the pore structure of pervious concrete. According to Chinese standard CJJ / T 253–2016, the effective porosity were measured by volumetric drainage method including the latter two.

For each group, after 28 days of curing, effective porosity tests of specimens was measured according to the following procedure. First, the specimen volume V was obtained. After 24 hour of submerging in water, the weight of the sample was recorded as m_1 as figure 2. Then removed the specimen from the water and wiped with a damping cloth immediately. Then specimen was dried in blast drying oven with temperature maintained at 50 ± 5 °Cuntil the weight was constant. Then constant weight was recorded as m_2. After that, the effective porosity and the unit weight PC were calculated according to equation (1) and equation (2), respectively.

$$P = (1 - \frac{m_2 - m_1}{V.\rho_w}) \times 100\% \tag{1}$$

$$\rho = \frac{m_2}{V} \tag{2}$$

where P is effective porosity (%), m_1 is weight submerged in water (g), m_2 is dry weight (g), V is sample volume (cm^3), ρ_w is water density (1g/cm^3), ρ is the unit weight.

2.3.2. Permeability Test. According to Chinese Standard CJJ/T 135–2009, constant head method was used for *determination* of permeability coefficient in figure 3. All samples were used in effective porosity tests. In the permeability test, The circumferential direction of the cylindrical specimen is tightly sealed with the plastic sleeve to ensure that the water can only pass from the upper and lower surfaces of the specimen. The permeability coefficient k was calculated using equation (3) by Darcy's law:

$$\mathrm{k} = QL / AHt \tag{3}$$

where the unit of k is mm/s, Q is the volume of water that flowed out (mm^3), L is the height of sample (mm), A is the cross-sectional area of sample (mm^2), H is the height of constant water head (mm), t is the time (s).

Figure 2. Weighing of pervious concrete samples in water.

Figure 3. Measuring the permeability coefficient of pervious concrete samples.

2.3.3. Compressive Test. Specimens after 28 days curing were selected for compressive strength test referring to the Chinese Standard GB/T 50081-2016. The test was loaded at a constant rate of 0.5 KN/s, until the sample was *destroyed* and the maximum load was recorded. The compressive strength value of each group was obtained from the results processing of the four samples.

3. Results and Discussion

3.1. Comparision of Porosity ,Unit Weight and Permeability Values of Specimens

Table 3. Unit weight and permeability coefficent of pervious concrete specimens with different molding methods.

	VM			MCM			SGCM		
Target porosity (%)	Unit weight (g/cm³)	Effective porosity (%)	Permeability coefficient (mm/s)	Unit weight (g/cm³)	Effective porosity (%)	Permeability coefficient (mm/s)	Unit weigt (g/cm³)	Effective porosity (%)	Permeability coefficient (mm/s)
12	2.16	10.69	0.99	2.06	12.80	2.89	2.15	10.89	1.8
	2.14	11.11	1.54	2.05	13.62	3.10	2.15	11.06	2.36
	2.16	11.05	1.54	2.09	13.50	2.80	2.17	11.91	4.00
	2.11	13.64	2.35	2.12	11.87	2.30	2.15	12.04	2.30
15	2.06	15.35	3.57	2.03	16.70	3.58	2.04	16.15	5.00
	2.04	16.97	4.03	2.01	17.10	6.00	2.07	16.90	5.45
	2.03	17.02	3.17	2.06	15.96	3.50	2.05	16.83	4.85
	2.07	15.97	3.24	2.05	16.83	4.50	2.09	15.82	4.45
20	1.97	19.86	5.71	1.98	20.50	6.50	1.97	21.45	9.05
	2.09	18.36	5.86	1.93	23.50	8.50	1.94	20.89	8.24
	2.00	17.18	4.36	1.96	22.00	8.00	1.91	21.13	8.56
	1.97	19.53	5.48	1.92	23.12	9.00	1.93	22.05	8.50

The correlation between the unit weight and effective porosity of specimens with different molding methods are plotted in figure 4, figure 5 and figure 6, respectively.

These three plots show that, the unit weight decreases linearly with the effective porosity,as represented by three regression (R^2) equations as shown in figures. As the test result in table 3, the relationship between unit weight and effective porosity of pervious concrete formed by three molding methods can be obtained. From figure 4 to 6, some important observations can be identfied as follows:

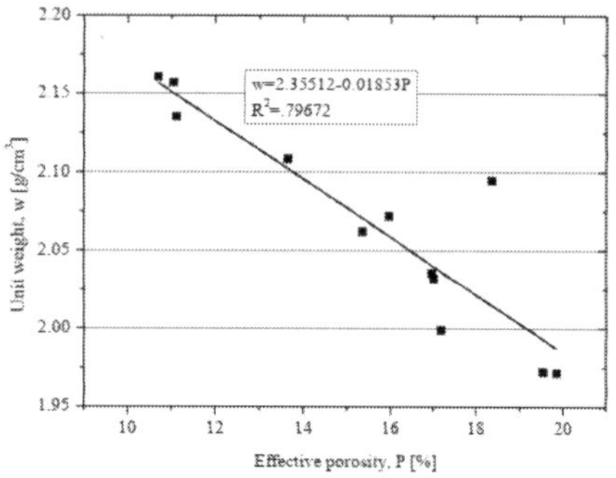

Figure 4. Correlation between unit weight and effective porosity of pervious concrete by VM.

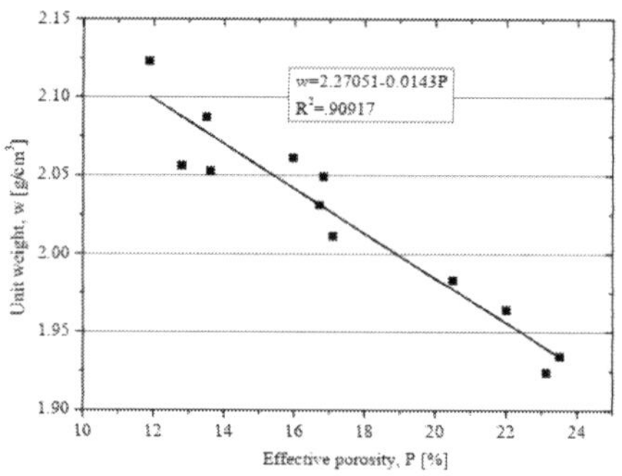

Figure 5. Correlation between unit weight and effective porosity of pervious concrete by MCM.

The difference in the relationship between unit mass and effective porosity due to the different porosity generated in different molding methods, the experimental data show that pervious concrete formed by SGCM has the closest relationship, the vibration forming method is the weakest, and Marshall compaction is in the middle.

The difference in unit weight of pervious concrete formed by three different molding methods is not obvious. When the effective porosity is 12%, pervious concrete formed by SGCM was found to have little higher unit weight than the one formed by the other two molding methods, and the difference became small as the porosity level increased. As table 4 shows, the difference decreased from 0.05 g/cm^3 at 12% porosity to 0.008 g/cm^3 at 20% porosity.

The rate of decrease of unit weight of pervious concrete formed by SGCM was higher than the two others. table 4 and Figure 4-6 showed that, within the porosity range of this study, the rate of decrease of unit weight with porosity is 0.02114 g/cm^3 per 1% rise in porosity for pervious concrete fromed by SGCM. Corresponding the values of the rates of decrease for MCM and VM were 0.01853 and 0.0143, respectively. That is, for every 1% increase in effective porosity, pervious concrete formed by SGCM would suffer a higher loss in unit weight than the other two.

Table 4. Rate of increase of unit weight of PC formed by different methods.

	VM		MCM		SGCM			
Porosity level,P(%)	Unit weight w(g/cm^3)	Rate of increase of w per % rise in P	Unit weight w(g/cm^3)	Rate of increase of w per % rise in P	Unit weight w(g/cm^3)	Rate of increase of w per % rise in P	Different between w of SGCM and VM	Different between w of SGCM and VM
12	2.133	0.0185	2.099	0.0143	2.146	0.0211	0.013	0.047
15	2.077	0.0185	2.056	0.0143	2.082	0.0211	0.005	0.026
20	1.985	na	1.985.	na	1.976	na	-0.008	-0.008

The relationship between the permeability coeffcient and effective porosity of PC with different molding methods are plotted in figure 7, figure 8 and figure 9, respectively.

The above three plots show that, the permeability coefficient value increases exponentially with the effective porosity level, as represented by three regression (R^2) equations as shown in figures.

From the test result on unit permeability coefficient and effective porosity obtained for VM, MCM and SGCM molding method, the following conclusions can be drawn.

Within the porosity range of this study, pervious concrete formed by SGCM had the highest permeability cofficient among the three molding methods at any given porosity. At the three porosity levels tested, the permeability cofficient of pervious concrete formed by SGCM was the largest, MCM was the second one and VM was minimal, and the difference became larger as the porosity value increased. As result shows, the difference between SGCM and VM increased from 0.982 mm/s at 12% porosity to 1.864 mm/s at 20% porosity, and the corresponding value between SGCM and MCM increased from 0.367 mm/s at 12% porosity to 1.153 mm/s at 20% porosity.

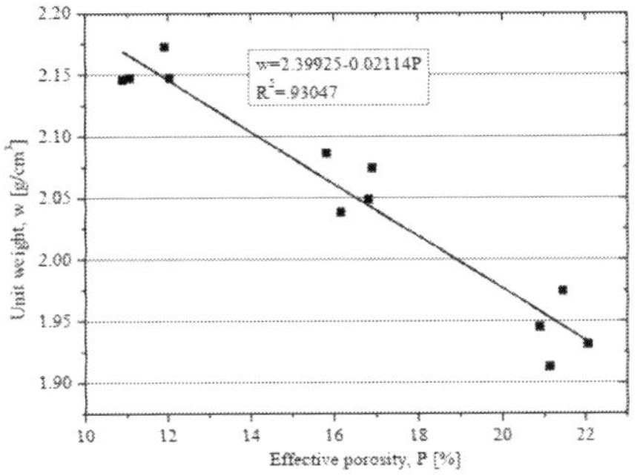

Figure 6. Correlation between unit weight and effective porosity of pervious concrete by SGCM.

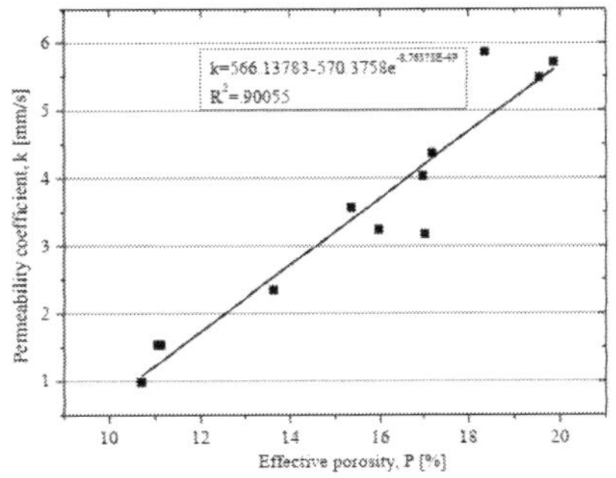

Figure 7. Relationship between effective porosity and permeability coefficient of PC by VM.

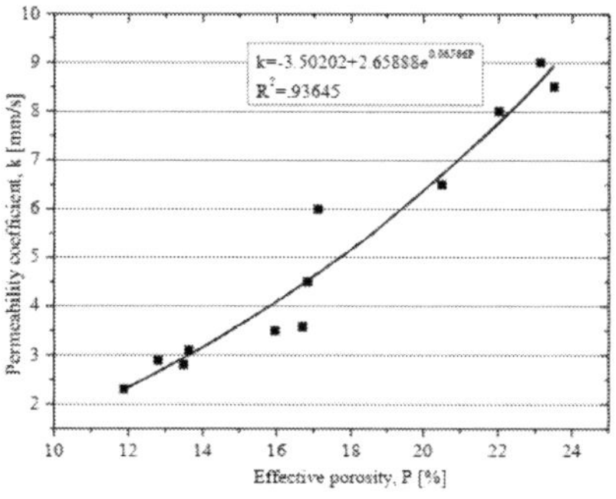

Figure 8. Relationship between effective porosity and permeability coefficient of PC by MCM.

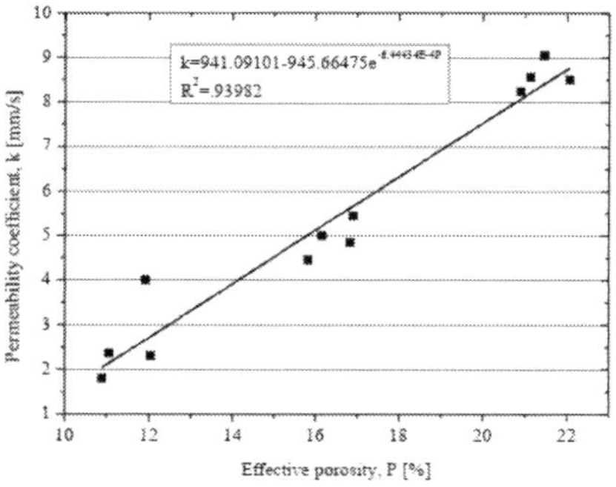

Figure 9. Relationship between effective porosity and permeability coefficient of PC by SGCM.

Pervious concrete formed by SGCM had the highest rate of increase of the permeability cofficient with porosity. Table 5 indicates that, Within the porosity range of the study, the rate of increase of permeability coefficient with porosity was 0.604 mm/s per 1% rise in porosity made by SGCM. Corresponding thc values of the rates of increase for MCM and VM were from 0.424 to 0.553 mm/s and 0.494, respectively. That is, pervious concrete formed by SGCM achieved hightest gains in permeability coefficient from increasing effective porosity in the three motheds. Conversely, for every 1% drop in effective porosity, it would suffer a higher loss in permeability cofficient than the other two.

Table 5. Rate of increase of permeability with porosity of PC formed by different methods.

Porosity level p(%)	VM		MCM		SGCM		Different between k of SGCM and VM	Different between k of SGCM and MCM
	Permeability coefficient k(mm/s)	Rate of increase of k per % rise in p	permeability coefficient k(mm/s)	Rate of increase Of k per % rise in p	Permeability coefficient k(mm/s)	Rate of increase of k per % rise in p		
12	1.729	0.494	2.344	0.424	2.711	0.604	0.982	0.367
15	3.211	0.492	3.617	0.553	4.523	0.603	1.313	0.906
20	5.672	na	6.384	na	7.536	na	1.864	1.153

3.2. Comparative Analysis of Compressive Strength And Permeability Coefficient

Based on many studies [35-37], there is a direct conection between porosity and compressive strength. The result of compressive strength test of PC formed by different methods are shown in figure 10. For the range of porosity values studied, the range of compressive strength of PC mixtures made by SGCM were from 12.85Mpa to 32.14 MPa, the ones made by MCM were 17.46 MPa to 27.85 MPa, and the corresponding values of pervious concrete formed by VM were 9.79 MPa to 19.94 MPa. From the test result, the compressive strength decreases as the porosity increases, and the rate of decrease of concrete formed by SGCM was the hightest. According to the above analysis, the permeability coefficient and porosity of PC was positively related. With the porosity, the compressive strength and permeability coefficient of PC can be studied.

Table 6. Compressive strength and permeability coefficient of PC with different molding methods.

	VM		MCM		SGCM	
Porosity p(%)	Compressive strength c(MPa)	Permeability coefficient k(mm/s)	Compressive strength c(MPa)	Permeability coefficient k(mm/s)	Compressive strength c(MPa)	Permeability coefficient k(mm/s)
12	19.94	1.729	27.85	2.344	32.14	2.711
15	16.55	3.211	25.84	3.617	25	4.523
20	9.79	5.672	17.46	6.384	12.85	7.536

As table 6 shows, pervious concrete made by SGCM has good water permeability and high strength, but VM is the worst choice, and MCM is centered. For the same material with the same original ratio, it is necessary to carry out in-depth analysis why different molding methods get such results.

The aggregates with the same original ratio were respectively compacted and rotated by MCM or SGCM for the same number of times as the concrete molding, and then the aggregate was sieved to obtain the new particle size gradation as shown in figure 11.

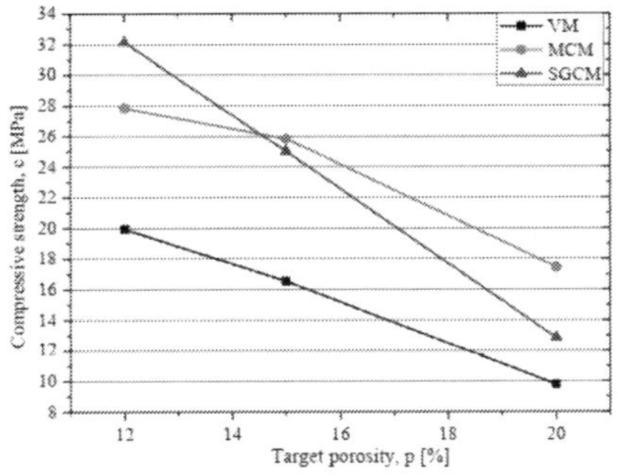

Figure 10. Compressive strength of PC formed by three different molding methods.

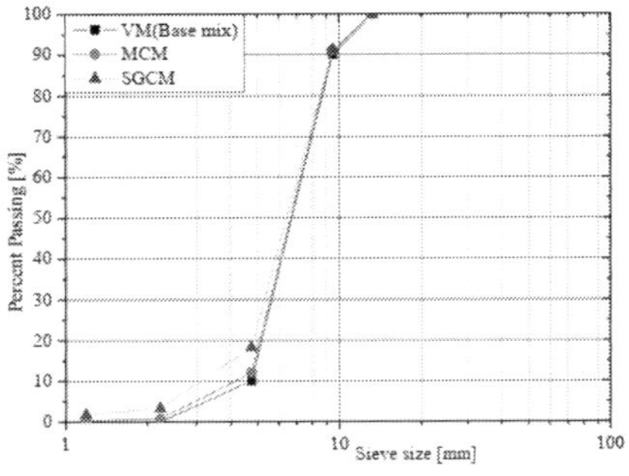

Figure 11. Sieve distribution curve of aggregates after MCM and SGCM treatment.

As shown in figure 11, when pervious concrete manufactured by MCM or SGCM, the particle size gradation of the aggregate changes significantly. Although the particle size distribution is still concentrated in the range of 9.5-2.23 mm, the aggregate of less than 2.23 mm after MCM accounts for 1%, and the corresponding value of SCGM is more than 5%. As revealed in literature [21, 38], the addition of sand into PC would improve the concrete's strength. It can be speculated that the change in particle size is one of the reasons why the strength and permeability coefficient of pervious concretes formed by MCM and SGCM are higher than that of VM. Second, the materials that make up the concrete move in different forms during different molding processes. Under the action of gravity and vibration, the displacement of aggregate and cement material is mostly concentrated in the vertical direction. There are also vertical and horizontal differences in the gap formed by the accumulation of aggregate and the filling of cement material. In VM process, the distribution of the slurry is also uneven, resulting in low strength and low permeability of PC. In MCM process, aggregate and cement in mixture compacted mutually in vertical and horizontal direction, and vertical was dominant. For this reason, the strength of the pervious concrete formed by this method is significantly higher than that of VM. In process of MCM, the distribution of the cemented material is relatively uniform, and the permeability coefficient is also large. In process of SGCM, each substance in the mixture is centrifuged under vertical pressure. As analyzed in literature [39], at the first 25 gyrations approximately, around 70% vertical displacement and 40% horizontal displacement were produced. Such a process facilitates the mutual compaction of the aggregate and the uniform distribution of the cementitious material. The concrete formed by this method has a compressive strength and a permeability coefficient which are significantly larger than those of the other two methods at a lower target porosity. Third, during SGCM process, the particle size changes and the surface area of aggregate increases, and the required cementing material should be increased. However, in the mix design, the cement cementing material decreases with the increase of the target porosity, as a result, the strength of PC formed by SGCM is significantly lower than that of MCM.

4. Summary and Conclusions

The properties of pervious concrete formed by different molding methods have been reported and studied in this paper. It compared the unit weight, effective porosity, permeability, compressive strength and freeze-thaw durability of pervious concrete manufactured by three different methods at three levels of target porosity: 12%, 15% and 20%. The three molding methods were vibration molding method, Marshall compactor molding method, superpave gyratory compactor molding method. The causes of different mechanical and hydraulic characteristics of pervious concrete in different molding methods were also detailed analyzed. As a result, the following conclusions can be drawn:

For pervious concrete formed by the three different molding methods studied in this paper, the decrease of unit weight (w) with effective porosity (P) can be described by a linear relationship of the form $\mathrm{w} = a + bP$, where a and b are positive and negative regression coefficients, respectively. Within the range of porosities studied, the unit weight of pervious concrete formed by SGCM is slightly larger than that of the other two methods, but the gap decreases with the increase of porosity. Unit weight of pervious concrete of the three molding methods tends to be consistent as the porosity approaches 20%.

For the range of porosity values studied, the increase of permeability coefficient (k) with effective porosity (P) can be described by an exponential ralationship of the form $k = k_0 + me^{nP}$, where k_0 , m and n are regression coefficients. Among the three molding methods, permeability coefficient of pervious concrete formed by SGCM is significantly higher than the other two at the same porosity level, and this gap increases with the increase of porosity, and the gap between SGCM and VM is greater than SGCM and MCM.

In the range of pores studied, the mix ratio design of the same target porosity, SGCM is the best way to obtain concrete with good permeability and high strength, MCM is the second preferred method, and VM is the least one. This advantage of SGCM is even more pronounced at low target porosity. Comparative analysis of the material movement in the three molding methods, the compaction effect of SGCM and MCM is conducive to the improvement of concrete strength and uniform distribution of cementing materials, and the particle size gradation distribution occurs in the process. When the porosity is 20%, the increase in small particle size aggregate during SGCM process and the decrease in cementing materials result in a compressive strength of 12.85 MPa, which is 0.74 for pervious concrete manufactured by MCM.

References

[1] Chandrappa A and Biligiri K. 2016 Pervious concrete as a sustainable pavement material-research findings and future prospects: A state-of-the-art review *Construction and Building Materials* **111** 262-274.

[2] Mullaney J and Lucke T. 2014 Practical review of pervious pavement designs *Clean Soil Air Water* **42** 111-124.

[3] Jang J, Ahn Y, Souri H and Lee H. 2015 A novel eco-friendly porous concrete fabricated with coal ash and geopolymeric binder: heavy metal leaching characteristics and compressive strength *Construction and Building Materials* **79** 173-181.

[4] Kim G, Jang J, Naeem F and Lee H. 2015 Heavy metal leaching, CO_2 uptake and mechanical characteristics of carbonated porous concrete with alkali-activated slag and bottom ash *International Journal of Concrete Structures and Materials* **9** 283-294.

[5] Barnhouse P and Srubar W. 2016 Material characterization and hydraulic conductivity modeling of macroporous recycled-aggregate pervious concrete *Construction and Building Materials* **110** 89-97.
[6] Kim G, Jang J, Khalid H and Lee H. 2017 Water purification characteristics of pervious concrete fabricated with csa cement and bottom ash aggregates *Construction and Building Materials* **136** 1-8.
[7] Nasser I, Khalil W and Abbas W. 2020 Strength and thermal conductivity of geopolymer pervious concrete made from artificial lightweight aggregate *IOP Conference Series Materials Science and Engineering* **737** 012074.
[8] Alshareedah O, Haider M and Nassiri S. 2020 Correlating laboratory and field compaction levels to achieve optimum in situ mechanical properties for pervious concrete pavements *Journal of Materials in Civil Engineering* **32** 04020278.
[9] Rahman M and Kamal N. 2021 Effect of the pervious concrete and scoria rock permeability to micro-hydropower generation from stormwater *IOP Conference Series: Earth and Environmental Science* **690** 012040.
[10] Rao Y and Yang T. 2021 Three volumetric methods for measuring the vertical porosity distribution of pervious concrete pavement *International Journal of Pavement Engineering* **6** 1-12.
[11] Luck J, Workman S, Coyne M and Higgins S. 2008 Solid material retention and nutrient reduction properties of pervious concrete mixtures *Biosystems Engineering* **100** 401-408
[12] Andrew 2007 United states environmental protection agency *Proceedings of the Water Environment Federation* **36** 726-737.
[13] Nemirovsky E, Welker A and Traver R. 2011 Evaporation from a pervious concrete stormwater control measure: parameterization, quantifying, and evaluation *World Environmental and Water Resources Congress* May 22-26 3610-3619 (American Society of Civil Engineers World Environmental and Water Resources Congress 2011 - Palm Springs, California, United States).
[14] Wang K, Schaefer V and Kevern J. 2006 Development of mix proportion for functional and durable pervious concrete, NRMCA Concrete Technology Forum: Focus on Pervious Concrete *Nashville.*
[15] Haselbach L, Boyer M, Kevern J and Schaefer V. 2011 Cyclic heat island impacts on traditional versus pervious concrete pavement systems *Transportation Research Record Journal of the Transportation Research Board* **2240** 107-115.
[16] Neithalath N, Weiss J and Olek J. 2006 Characterizing enhanced porosity concrete using electrical impedance to predict acoustic and hydraulic performance *Cement and Concrete Research* **36** 2074-2085.
[17] Yang J and Jiang G L. 2003 Experimental study on properties of pervious concrete pavement materials *Cement and Concrete Research* **33** 381-386.
[18] Shabalala A. 2020 Efficacies of pervious concrete and zero-valent iron as reactive media for treating acid mine drainage *14th International Mine Water Association Congress- Mine Water Solutions.*
[19] Hung V, Seo S, Kim H and Lee G. 2021 Permeability and strength of pervious concrete according to aggregate size and blocking material *Sustainability* **13** 426.
[20] Bonicelli A, Giustozzi F, Crispino M and Borsa M. 2015 Evaluating the effect of reinforcing fibres on pervious concrete volumetric and mechanical properties according to different compaction energies *European Journal of Environmental and Civil Engineering* **19** 184-198.
[21] Bonicelli A, Giustozzi F and Crispino M. 2015 Experimental study on the effects of fine sand addition on differentially compacted pervious concrete *Construction and Building Materials* **91** 102-110.
[22] Kevern J, Wang K, Suleiman M and Schaefer V. 2008 Pervious concrete mixture proportions for improved freeze-thaw durability *Journal of Astm International* **5** 101320-31.

[23] Made A and Rogge S. 2013 Development of high quality pervious concrete specifications for maryland conditions *Maryland State Highway Administration* **MD-13** SP009B4F.
[24] Huang B, Wu H, Shu X and Burdette E. 2010 Laboratory evaluation of permeability and strength of polymer-modified pervious concrete *Construction and Building Materials* **24** 818-823.
[25] Shu X, Huang B, Wu H, Dong Q and Burdette E. 2011 Performance comparison of laboratory and field produced pervious concrete mixtures *Construction and Building Materials* **25** 3187-3192.
[26] Pindado M, Aguado A and Josa A. 1999 Fatigue behavior of polymer-modifified porous concrete *Cement and Concrete Research* **29** 1077-1083.
[27] Beeldens A, Gemert D, Winne E, Caestecker C and Messern M. 1997 Development of porous polymer cement concrete for highway pavements in Belgium *Proceedings of the 2nd East Asia Symposium on Polymers in Concrete EASPIC* (Koriyama) 121-129.
[28] Giustozzi F. 2016 Polymer-modified pervious concrete for durable and sustainable transportation infrastructures *Construction and Building Materials* **111** 502-512.
[29] Chopra M, Wanielista M, Ballock C and Spence J. 2007 Construction and maintenance assessment of pervious concrete pavements *Cross* **201** 8105-11.
[30] Kevern J, Wang K, Suleiman M and Schaefer V. 2006 Mix design development for pervious concrete in cold weather climates *Binder Content*.
[31] Bonicelli A, Crispino M, Giustozzi F and Shink M. 2013 Laboratory analysis for investigating the impact of compaction on the properties of pervious concrete mixtures for road pavements *Advanced Materials Research* **723** 409-19.
[32] Sengün E, Alam B, Shabani R and Yaman I. 2019 The effects of compaction methods and mix parameters on the properties of roller compacted concrete mixtures *Construction and Building Materials* **228** 11680701-12.
[33] Lang L, Duan H and Chen B. 2019 Properties of pervious concrete made from steel slag and magnesium phosphate cement *Construction and Building Materials* **209** 95-104.
[34] Kevern J, Schaefer V and Wang K. 2011 Mixture proportion development and performance evaluation of pervious concrete for overlay applications *ACI Materials Journal* **108** 439-48.
[35] Schaefer V, Wang K, Suleiman M and Kevern J. 2006 Mix design development for pervious concrete in cold weather climates *Center Transp, Res. Educ. Iowa State Univ* **2.**
[36] Meininger R. 2004 No-fines pervious concrete for paving *Concr. Int.* **10** 20-27.
[37] Tennis P, Leming M and Akers D. 2004 Pervious concrete pavements, portland cement association skokie illinois and national ready mixed concrete association *Silver Spring* (Maryland).
[38] Khankhaje E, Salim M, Mirza J, Hussin M and Rafieizonooz M. 2016 Properties of sustainable lightweight pervious concrete containing oil palm kernel shell as coarse aggregate *Construction and Building Materials* **126** 1054-65.
[39] Gong F Y, Zhou X D, You Z P, Liu Y and Chen S Y. 2018 Using discrete element models to track movement of coarse aggregates during compaction of asphalt mixture *Construction and Building Materials* **189** 338-51.

Journal of Physics: Conference Series **2148** (2022) 012056 doi:10.1088/1742-6596/2148/1/012056

Stress Analysis of Vertical Bar Additional Connector at the Joint of Fabricated Shear Wall

Haowen Jiang, Yueguo zhang, Handuo Fan, Qikai Liu and Linghang Du

School of Engineering, Shenyang Jianzhu University, Shenyang 110168, China
Email: ceygzhang@sjzu.edu.cn

Abstract. Although there are many mainstream reinforcement connection methods, the realization of force transfer between reinforcement is the starting point of all connections and the most important inspection standard. Aiming at the vertical connection of the fabricated shear wall, this paper proposes a new type of additional connector connection, and two fabricated shear walls connected by new types of additional connectors and a cast-in-situ shear wall were established. The dimensions and material parameters of the fabricated shear wall connected by the new additional connector were the same as those of the cast-in-situ shear wall. The finite element software ABAQUS is used to perform static finite element analysis on fabricated shear walls and cast-in-situ shear walls connected by new additional connectors. By comparing the static finite element simulation results, the results show that the new additional connector can ensure the effective force transfer of reinforcement.

1. Introduction

Based on the advantages of high mechanization, convenient construction, energy conservation, and environmental protection, prefabricated buildings have been widely used in the continuous development of urbanization. Prefabricated building is a building construction technology with sustainable development generally recognized by society. Compared with cast-in-situ structures, there are joints between components in the prefabricated structure. To a certain extent, joint connection technology determines the quality of prefabricated building structures, so the joint connection technology is the key to researching prefabricated building structures. At present, scholars at home and abroad have made many attempts and studies on the connection technology of prefabricated shear wall structures. Among them, Yu Zhiwu and others had studied the anchorage performance of u-hoop reinforcement connections in the prefabricated shear wall structure [1]. The results show that the buried depth of U-shaped reinforcement directly affects the failure mode of the specimen. When the anchorage depth of U-shaped reinforcement is large, the specimen has good ductility and reliable performance. Chen Le et al. made the whole process static elastic-plastic analysis of the shear wall dominated by bending deformation and explained the feasibility of the calculation method [2]. Liu Liping et al. verified the effectiveness of the combined connection of longitudinal reinforcement through the quasi-static test of three shear wall specimens [3]. Brian J Smith et al. conducted an experimental study on the feasibility of vertical connection of precast concrete shear wall through unbonded prestressed reinforcement and gave the calculation formula of bearing capacity and deformation according to the test results [4]. James F. et al. studied the dynamic performance of bolted connections of fabricated structures [5]. Ertas et al. conducted experimental research on several ductile connection forms of prefabricated frame structures, such as welding, bolt connection, composite chain connection, etc [6]. Although the existing joint connection methods of the fabricated shear wall have their characteristics, there are also various deficiencies. For example, sleeve grouting connection can obtain a good force transfer effect, but the construction accuracy is high, and the construction quality is not easy to detect. The mechanical performance of the connection at the horizontal joint of the fabricated shear wall mainly depends on the reliability of the vertical reinforcement connection. Reliable force transmission, easy construction, and controllable quality are the key points of assembled shear wall joint connection. Based on the existing connection, a reinforced U-shaped connection with additional connectors is studied in this paper. The new joint connection is numerically simulated by using ABAQUS finite element software to study its force transmission reliability.

2. Connection of Additional Connectors For Horizontal Joint And Vertical Reinforcement of The Fabricated Shear Wall

The U-shaped reinforcement connection form of the additional connector is shown in Figure 1. After the U-shaped reinforcement is overlapped, the H-shaped connector is supported on the horizontal section of the U-shaped reinforcement. When the reinforcement is in tension, the reinforcement tension at the butt joint is transmitted through the H-shaped connector. When the reinforcement is compressed, the force is transmitted through the concrete anchorage in the overlapping range. To understand the feasibility of this connection mode, the solid modeling analysis of the additional connector with 50mm U-lap reinforcement was carried out. The results show that the stress of the H-shaped connector is 47.2MPa when the reinforcement yields in tension, while the stress of the H-shaped connector is 25.01MPa when the reinforcement yields in compression, both of which are in an elastic working state. It can be seen that this connection method can ensure the tensile and compressive yield of reinforcement with a small lap length, and the stress of the connector is small. The local analysis results of the additional connector reinforcement connection are shown in figure 1.

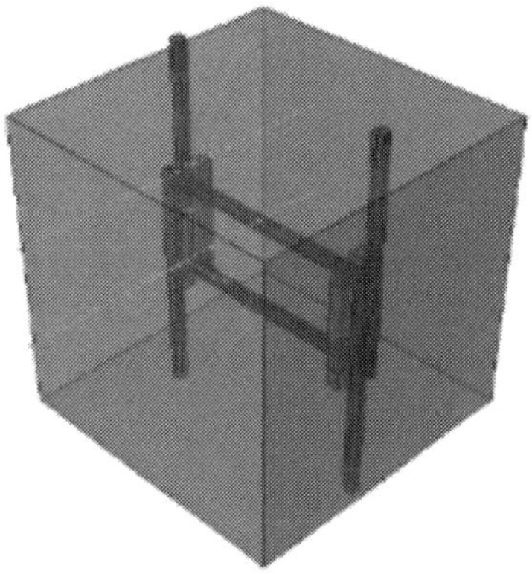

(a) Solid model.

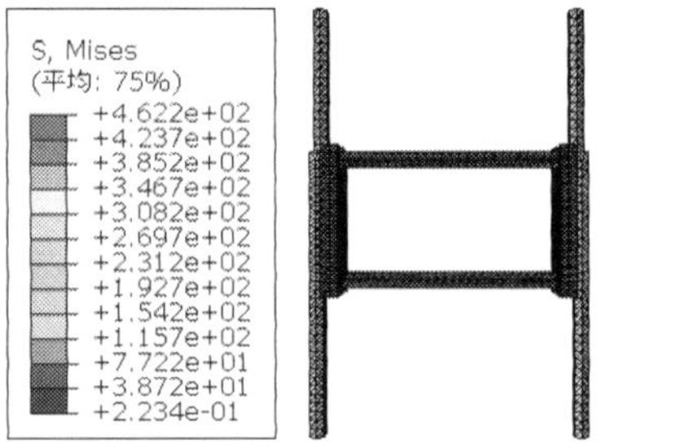

(b) Stress program of reinforcement and connector under compression.

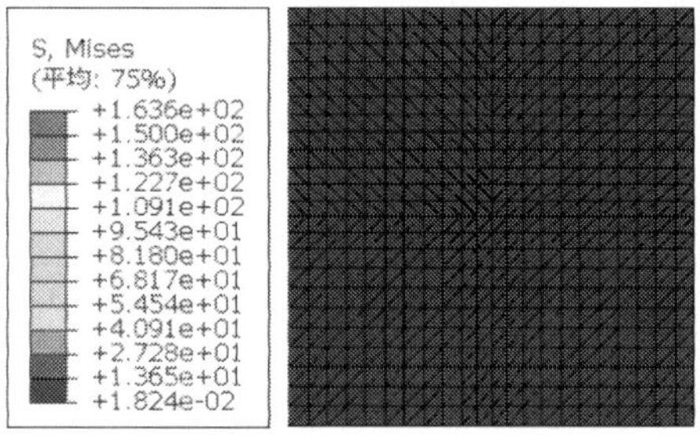

(c) Stress program of concrete under compression.

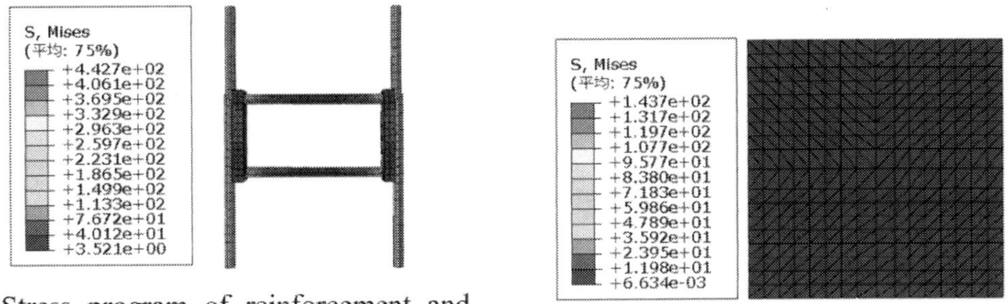

(d) Stress program of reinforcement and connector under tension. (e) Stress program of concrete in tension.

Figure 1. Reinforcement connection of additional connectors.

3. Integral Model of Fabricated Shear Wall

3.1. Basic Parameters of The Model

To study the mechanical performance of the reinforcement connection form at the joint of the shear wall proposed in this paper, three analysis models of the reinforced concrete shear wall were designed, including one integral shear wall XW-1 and two assembled shear walls ZW-1 and ZW-2. The change of joint width was considered in the fabricated shear wall, and the reinforcement amount of the integral shear wall was the same as the fabricated shear wall. The specific model parameters are shown in table 1 and figure 2.

Table 1. Model design parameters.

Test piece	Material Science			Connection mode	Joint position	Joint height / mm
	Concrete	Rebar	Connector			
XW-1	C30	HRB400	—	Cast in situ member	—	—
ZW-1			Q235B	H-type connector	bottom	50
ZW-2				H-type connector	bottom	100

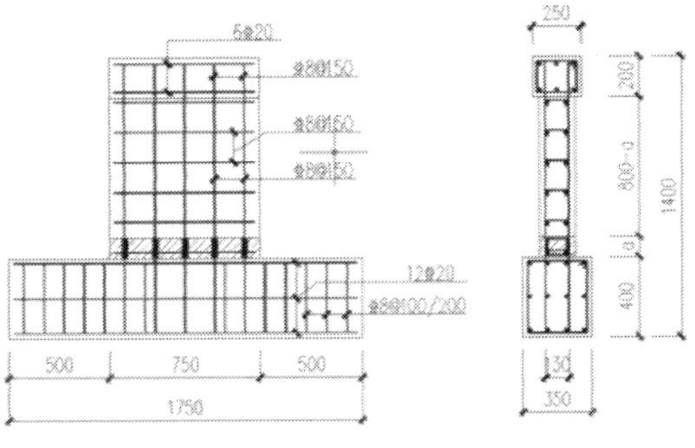

Figure 2. Dimension and reinforcement diagram of the fabricated model (unit: mm).

3.2. Element Types and Boundary Conditions of Finite Element Model

The element type of finite element model was as follows: three-dimensional solid reduced integral element (C3D8R) was selected for concrete, and truss element (T3D2) was selected for steel bar element. According to the force characteristics of the connector, the H-shaped connector is simplified as a steel bar element for the convenience of modeling and analysis.

The boundary conditions of the model are: the end face of the bottom beam of the model is hinged to restrict the out-of-plane displacement and torsion of the wall. A general contact pair is set at the joint between the new and old concrete in the fabricated shear wall, the tangential friction coefficient is taken as 0.7, and the normal direction is hard contact.

The load conditions of the shear wall model are as follows: according to the axial compression ratio of the wall, the vertical compression load is applied to the top-loading beam, and the horizontal load is applied at the coupling point of the end face of the top-loading beam. The analysis is realized by applying horizontal displacement.

3.3. Mechanical Properties of Materials

The compressive constitutive relationship of concrete adopted the stress-strain relationship given by Saenz [7], and the tensile constitutive relationship adopted the stress-strain relationship given in code for design of concrete structures (GB50010-2010) [8]. The stress-strain relationship of reinforcement and H-shaped connector adopted the bilinear constitutive model, and the curve slope of the strengthened section was 0.01 times that of the elastic section, i.e $E_S' = 0.01E_S$. See table 2 for specific values of material mechanical properties.

Table 2. Mechanical properties of materials.

Material Science	Elastic modulus / MPa	Poisson's ratio	Yield strength / MPa
C30	3×10^4	0.2	20.1
HRB400	1.98×10^5	0.3	452
Q235B	2.06×10^5	0.3	380

4. Analysis Results and Analysis of Prefabricated Shear Wall

4.1. Reinforcement Stress Analysis

Figure 3 shows the stress program of reinforcement when XW-1, ZW-1, and ZW-2 reach yield load and ultimate load. The bearing capacity and deformation results of each model are shown in table 3. Ultimate load state refers to the bearing capacity of concrete when it reaches the ultimate compressive strain.

Table 3.bearing capacity and deformation of model.

Model	Yield load / kN	Yield displacement / mm	Ultimate load / kN	Ultimate displacement / mm
XW-1	206.60	1.65	268.04	3.98
ZW-1	199.40	1.78	264.30	4.15
ZW-2	193.45	1.68	253.70	4.24

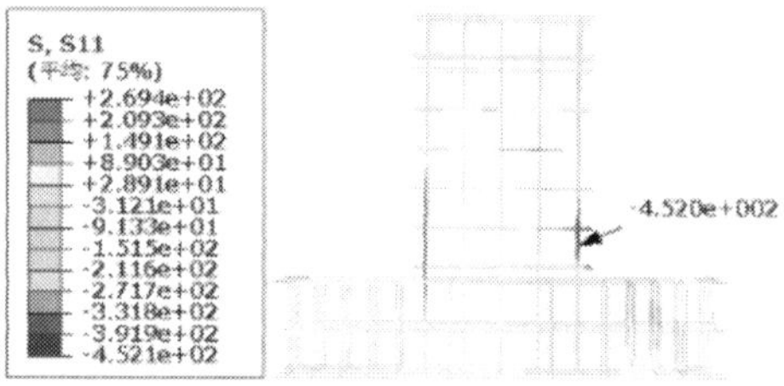

I. Yield state

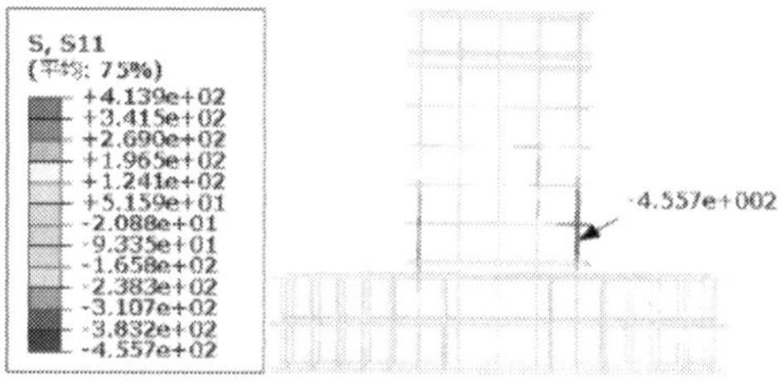

II. Limit state

(a) Reinforcement stress program of XW-1 model.

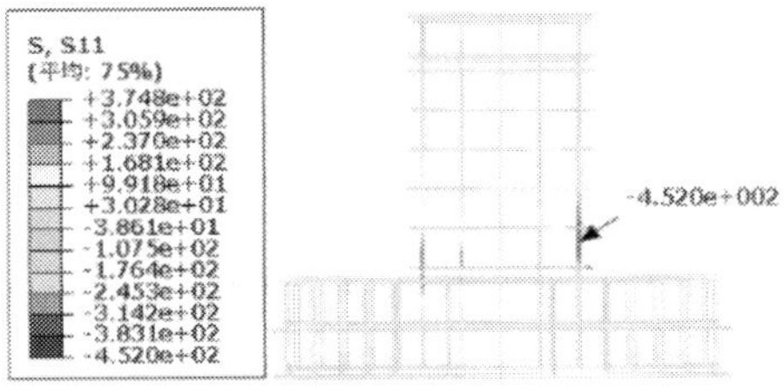

I. Yield state

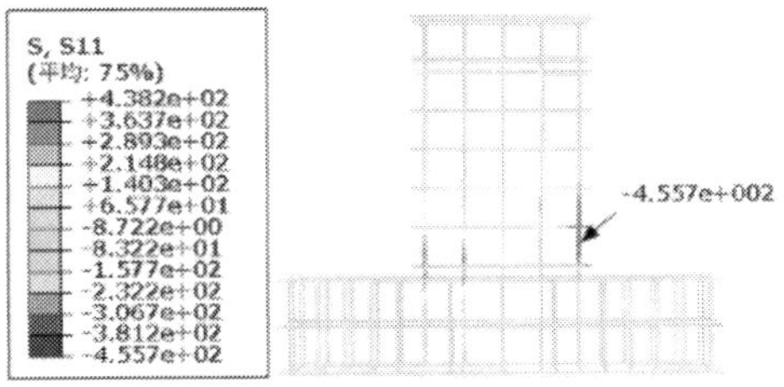

II. Limit state

(b) Reinforcement stress program of ZW-1 model.

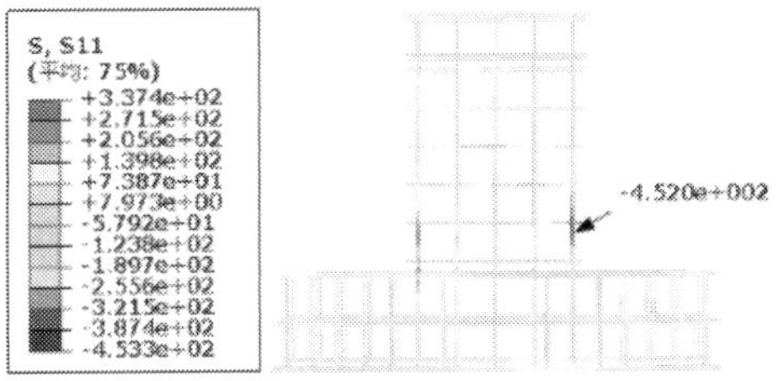

I. Yield state

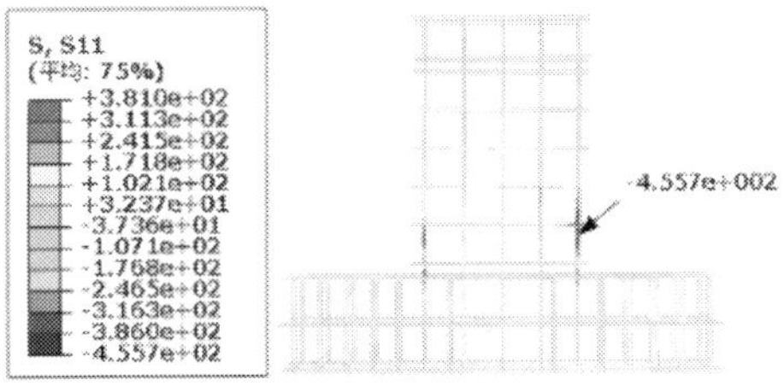

II. Limit state

(c) Reinforcement stress program of ZW-2 model.

Figure 3. Reinforcement stress program of model.

It can be seen from figure 3 that the reinforcement stress development state of the model of the integral shear wall and the fabricated shear wall is the same. Under the combined action of vertical load and horizontal load, the vertical reinforcement on the side with large compression of the wall first yields. When the concrete on this side reaches the ultimate compressive strain, the concrete is crushed and damaged. Table 3 shows that the bearing capacity and deformation performance of the fabricated shear wall with an additional H-shaped connector U-lap joint for vertical reinforcement is equivalent to that of the overall shear wall.

4.2. Concrete Stress Analysis

Figure 4 shows the program of normal stress of concrete under yield and limit states of model reinforcement. It can be seen from the model concrete stress program in Figure 4 that the compressive stress of the overall shear wall and the fabricated shear wall is mainly concentrated at the bottom of the compression side of the wall, and the oblique compressive stress decreases gradually along the wall. The maximum tensile stress is at the bottom of the tensile side of the wall, and the oblique tensile stress decreases gradually along the wall. The main compressive and tensile stress traces of concrete of integral and fabricated shear walls are X-shaped, showing the same mechanical performance of concrete.

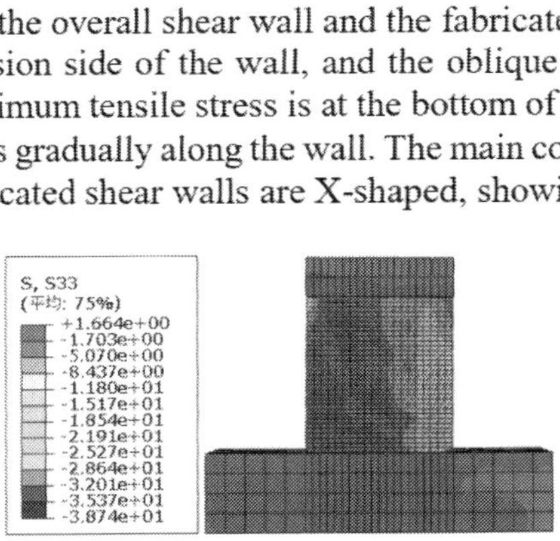

I. Yield state

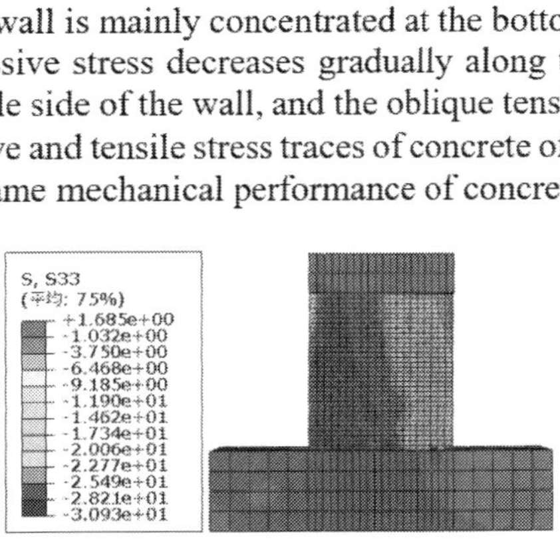

II. Limit state

g(a)Normal stress program of concrete in XW-1 model.

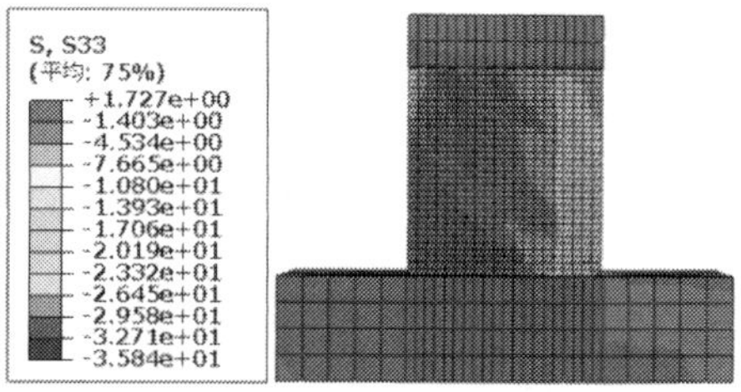

I. Yield state

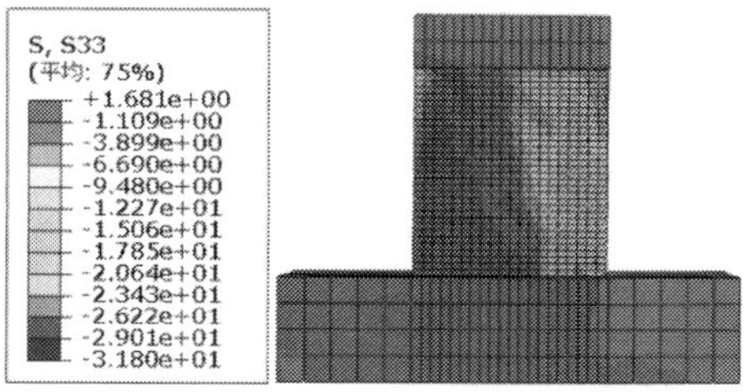

II. Limit state

(b)Normal stress program of concrete in ZW-1 model.

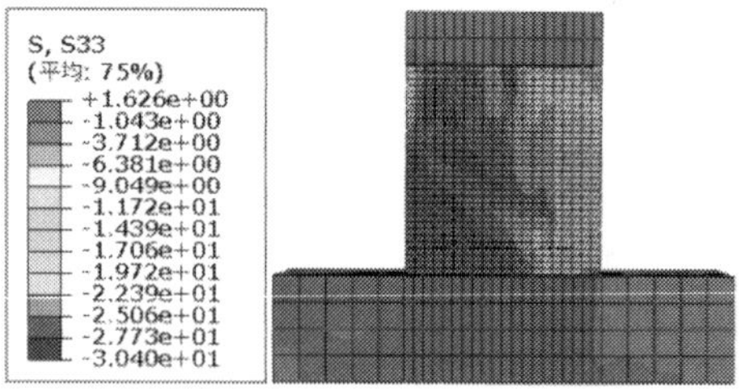

I. Yield state

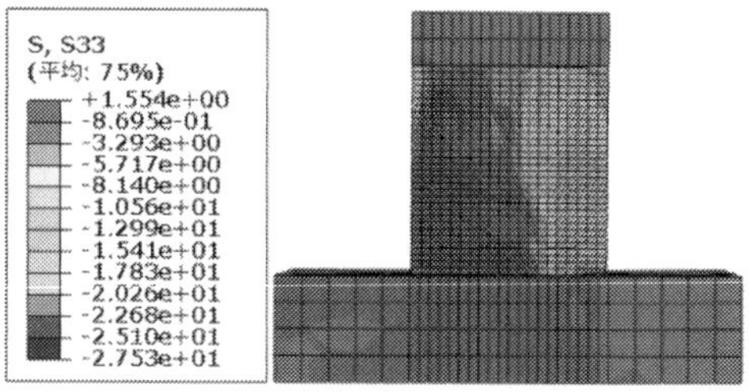

II. Limit state

(c)Normal stress program of concrete in ZW-2 model.

Figure 4. Cloud diagram of concrete stress of model.

5. Load Displacement Curve

From the horizontal load-displacement curve of the shear wall model in figure 5, it can be seen that the horizontal displacement of the integral shear wall and the fabricated shear wall are consistent with the increase of the horizontal load. Below the yield load, the horizontal load-displacement curve of the model changes linearly and maintains the elastic working state. The difference is that the stiffness of the elastic working range of the fabricated shear wall is lower than that of the integral shear wall due to the existence of the horizontal joint of the fabricated shear wall, which shows that the slope of the horizontal load-displacement curve of the fabricated shear wall is lower than that of the integral shear wall. After the reinforcement yield, the slope of the horizontal load-displacement curve of the model decreases rapidly, and the slope of the curve of the fabricated shear wall model decreases faster than that of the integral shear wall. The peak load of the integral shear wall is greater than that of the fabricated shear wall, but the horizontal displacement corresponding to the peak load is not different, which is about 4mm. After all shear wall models exceed the peak load, the load-displacement curve is a horizontal straight line, showing a plastic stress state.

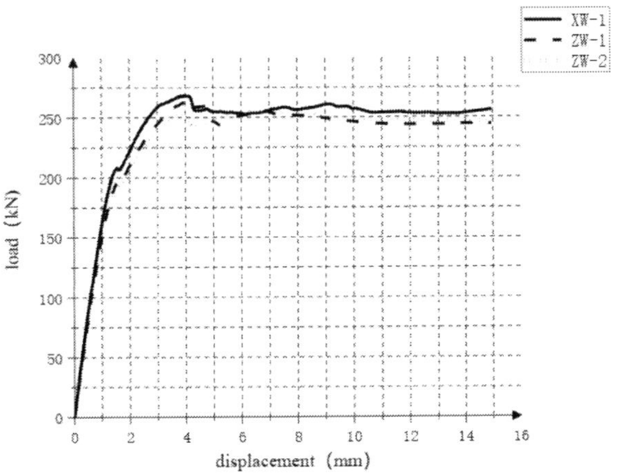

Figure 5. Load-displacement curve of the model.

6. Conclusion

In this paper, the static load analysis of the steel bar lap connection model and the integral shear wall model with H-shaped connectors in the horizontal joint of the fabricated shear wall is carried out, and the analysis results are compared. The following conclusions can be drawn:

(1) The horizontal joint of the fabricated shear wall adopts the U-shaped overlapping mode of reinforcement with H-shaped connectors, which can reliably realize the force transfer of steel bar between the components at the joint.

(2) The mechanical performance of the fabricated shear wall with additional H-shaped connectors at the butt joint is consistent with that of the cast-in-situ shear wall. Due to the existence of joints, the bearing capacity and shear stiffness of fabricated shear walls are lower than those of cast-in-situ shear walls.

(3) From the beginning of loading to concrete crushing, the additional H-shaped connector is always in an elastic working state, the connector is less stressed, and the section size can meet the equal strength of reinforcement.

Acknowledgments

This work was supported by Mentoring Program Project of Natural Science Foundation of Liaoning Province (2019-ZD-0674).

Reference

[1] Yu Z W, Peng X D, Guo W, Peng M P. 2015 Study on anchorage performance of u-hoop reinforcement connection in fabricated shear wall structure [J] *Journal of Railway Science and Engineering* **12** (04) 879-886.

[2] Chen Q, Qian J R, Li G Q. 2004 Static elastic-plastic analysis of shear wall with macro model [J] *Journal of civil engineering* (03) 35-43.

[3] Liu L P, Yu J, Liao D F, Liu J J, Deng X H, Li Y M. 2019 In-plane mechanical behavior of fabricated reinforced concrete shear wall with combined longitudinal reinforcement [J] *Journal of Civil and Environmental Engineering* (Chinese and English) **41** (06) 135-142.

[4] Smith B J, Kurama Y C. 2010 Analytical Model Validation of a Hybrid Precast Concrete Wall for Seismic Regions[C] *Proceedings of the 2010 ASCE Structures Congress* 2914-2924.

[5] Ozturan T, Ozden S, Ertas O. 2006 Ductile Connections in Precast Concrete Moment Resisting Frames[J] *PCI Journal Prestressed Concrete Institute Jou* 51 66-76.

[6] Wilson J F, Callis E G. 2004 The dynamics of loosely jointed structures [J] *International Journal of Non-Linear Mechanics* **39** (3) 503-514.

[7] Saenz L P. 1964 Discussion of 'Equation for the Stress Strain Curve of Concrete' by Desayi and Krishnan [J] *Journal of the American Concrete Institute* **61** (9) 1229-1235.
[8] GB 50010-2010, code for design of concrete structures [S].

Experimental Investigation on Dynamic Performance of Micro-Pile with Predrilled Oversize Hole on Shaking Table Test

Laixiu Cheng and Gumai Chen

Department of Gannan University of Science and Technology, Ganzhou Jiangxi 341000, China
Email: 9599253@qq.com

Abstract. For the situation of lacking research on micro-pile with predrilled oversize hole, the key part of semi-integral abutment bridge, the micro-pile-soil interaction shaking table test is carried out by considering the reaming pore diameter, depth, packing and other parameters in the end of the micro-pile to obtain the acceleration, pile moment, displacement and pile-system response frequency and other basic dynamic response and dynamic interaction law. Results show that: 1) the change of predrilled-hole parameters has litter effect on the dynamic properties of soil outside oversize hole; 2) The change of predrilled-hole parameters can cause the change of structural frequency, so led to the change of inertia force of pile head; 3) Inertial interaction has an important influence on the response of the upside part of pile and little influence in the downside part (lower than 15D). These conclusions will provide reference for dynamic response of interaction between pile with predrilled oversize hole and soil and make contribution to the practical application and designing of micro-pile with predrilled oversize hole.
Keywords. Bridge engineering, micro-pile with predrilled hole, parameters analysis, shaking table test, dynamic response

1. Introduction

Micro pile can effectively improve the seismic performance of seamless bridge with new semi integral abutment [1]. Under the seismic action, part of the seismic action of the bridge is transmitted to the micro pile by the connecting plate. Through the pile-soil interaction, part of the seismic energy is absorbed and dissipated by the soil around the pile, so as to reduce the damage of the seismic action to the bridge. At the same time, enlarging the hole at the top of the micro pile can improve the ability of the bridge to absorb longitudinal deformation in the normal service period.

Scholars at home and abroad have carried out many studies on the dynamic response characteristics of pile-soil. Xu Yanfei[2] simulated the shaking table test of pile group soil structure dynamic interaction by using the finite element software ANSYS, studied the dynamic responses of pile group soil system such as displacement time history, acceleration time history and pile internal force under earthquake, and considered that increasing the pile length is helpful to earthquake resistance. Sato [3] and Yang [4] carried out shaking table test of single pile in dry sand, focusing on the analysis of pile-soil dynamic interaction from the perspective of input seismic wave frequency and pile-soil system frequency. Wei Chunli [5] conducted five pile-soil-structure shaking table tests, processed the shaking table data with MATLAB, and analyzed the influence of inertial interaction and motion interaction on pile-soil dynamic interaction by using the method of frequency decomposition. Wang Jianhua [6] and others studied the dynamic response of micro aluminum pile in saturated sand through shaking table test, and mainly analyzed the pile bending moment and pore pressure at different depths during vibration. Zhang Zhongle[7] studied the shaking table test of micro pile group and single pile in dry sand, and mainly analyzed the acceleration time history, displacement time history and p-y hysteretic curve. Li Yurun[8] completed a large number of shaking table tests of micro aluminum piles in dry sand, and compared and analyzed the dynamic responses of pile shaft acceleration, pile-soil interaction force and pile shaft bending moment in pile-soil interaction. Fan

In conclusion, there are few studies on the dynamic response of micro pile-soil dynamic interaction under reaming. Therefore, this paper will further discuss its dynamic response under earthquake. The influence of reaming parameters on the dynamic response characteristics of micro pile foundation is revealed, which provides experimental and theoretical support for the reaming design technology of integral abutment bridge.

2. Shake Table Test

Test materials and equipment mainly include model soil, model pile, model box, reaming filler, measuring equipment and test equipment. The model soil is dry sand. The whole test adopts a single layer of soil. During the test, the dry sand is backfilled in layers (with a thickness of 20cm) into the model box, and the sand is compacted after each layer is paved. Physical and mechanical parameters of model soil are shown in table 1. At the same time, in order to avoid the influence of model pile body material and size effect on the test, the model pile is made as the actual pile, and its parameters are shown in table 2.

Table 1. Physical and mechanical parameters of model soil.

Compaction density ρ(g/cm^3)	Relative compaction D_r(%)	Angle of internal friction φ(°)	Void ratio e	Coefficient of uniformity C_u
1.601	43	31	0.663	3.17

Table 2. Parameters of model pile.

cross section form	diameter(mm)	wall thickness(mm)	pile length(mm)	slenderness ratio
hollow section	60	4	2300	38
density(kg/m^3)	material	flexural rigidity(kN·m^2)	poisson ratio	
7850	Q235	58.20	0.3	

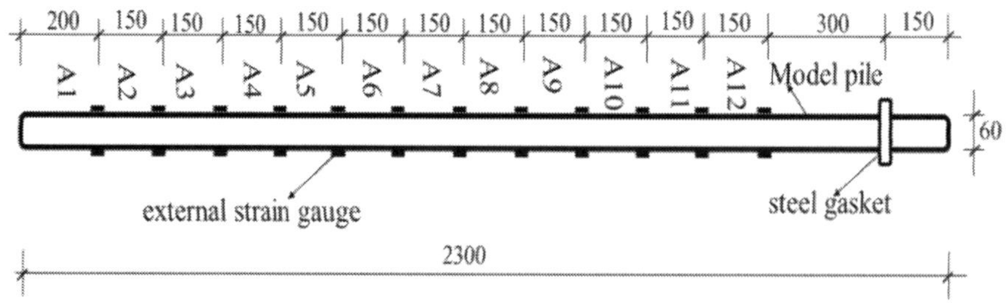

Figure 1. Arrangement of micro-pile strain gauge unit (mm).

The size of model steel box is 2m (length) × 2m (width) × 2.1m (high). Channel steel is welded at the bottom of the box to reduce the relative sliding between the model soil and the bottom plate of the steel box. In order to reduce the influence of reflected waves, reduce the restraint on the model soil and increase the softness of the side wall of the box, a layer of 100mm thick polystyrene foam board is lined on the inside surface of the steel box perpendicular to the ground motion. At the same time, in order to reduce the friction between the inner surface of the side wall parallel to the seismic direction of the steel box and the model soil, the inner surface of the side wall is pasted with polyvinyl chloride film and coated with lubricating oil. The test and measurement equipment mainly includes strain gauges and accelerometers. 24 reinforcement strain gauges, numbered A1 ~ A12 and B1 ~ B12, are symmetrically pasted along the depth direction of the pile at the edge of the outer wall of the steel pipe pile, as shown in Figure 1. At the same time, five accelerometers are arranged on the table and soil respectively, as shown in Figure 2(a). A medium is set at the top of the pile within a certain depth and width for reaming, and the soil around the pile is divided into internal and external parts. The compactness of the filler in the reaming is controlled to be less than that of the soil outside the reaming, so as to increase the deformation capacity and energy consumption capacity of the pile under lateral load. See Fig. 2 (b) for the local photos of reaming. In order to eliminate the boundary effect, when the ratio of soil lateral boundary value D to pile diameter D is greater than or equal to 6, the error caused by lateral boundary is very small and tends to be stable [9]; Previous studies have shown that [10], when the pile group spacing is greater than or equal to 10 times the pile diameter, it has little impact on the pile group effect. Therefore, the distance between the pile and the boundary is greater than 6D, and the pile spacing is greater than 10d. The test photos are shown in Figure 2C. In order to simulate the influence of pile top quality of micro pile in actual situation, a mass block is set at the pile top of model pile. Referring to reference [11], according to the weight of the actual approach slab, the weight of the pile top counterweight of a single micro pile is 56.4kg.

In addition, 2mm thick Q_{235} steel sleeve is selected as the reaming material, and the reaming filler is divided into loose sand and rubber particles. For loose sand, since the volume of reaming iron sleeve is known, its density is determined by controlling the quality of loose sand (its density is ρ= 1.48 g / cm^3), so as to obtain the relative compactness of sand, and then obtain the sand compactness and mechanical parameters (internal friction angle) through the k-value curve [12] (figure. 3) φ= 29°, initial stiffness of sand k = 5000 kN / mm^3). The recycled rubber particles produced by Hubei huayitong Rubber Co., Ltd. are extracted from pure radial steel tire, with particle size of 1.2 ~ 2.3 mm, shore a hardness of 68, elastic modulus of 10.6 MPa and density of 0.65 g / cm^3.

In the test, 0.4g 5Hz sine wave was used to simulate 9-degree earthquake. Due to the limited test time and funds, Q235 steel pipe pile with 0.06m pile diameter and 4mm wall thickness was selected as the research object, the model soil was dry medium dense sand, and the loading waveform was representative load condition (0.4g sine wave). Considering the reaming hole diameter Shaking table test of pile-soil dynamic interaction of reaming depth and reaming backfill material. The reaming parameters are mainly changed during loading, and the loading scheme is shown in table 3.

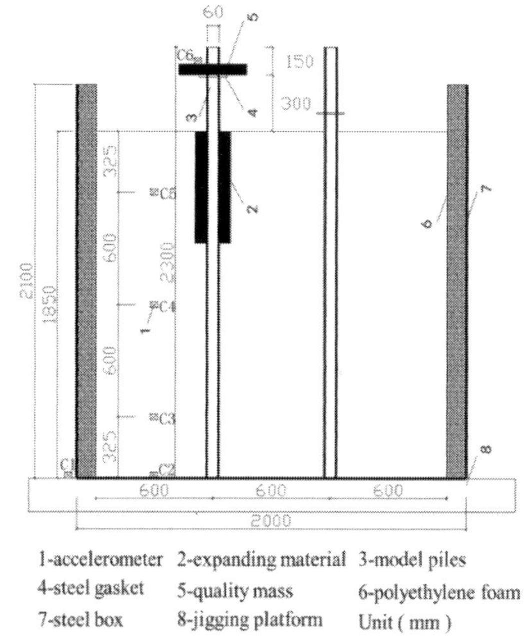

(a) Test device for vibration table of Micro-Pile with enlarged hole.

(b) Detail structure of Micro-Pile with enlarged hole (c) Solid Diagram of Micro-Pile Test

Figure 2. Micro-pile with predrilled oversize hole on shaking.

Table 3. Loading case (The quality of top of pile is 56.4kg).

working condition	seismic wave	peak acceleration(g)	Expanding material	Expansion width(mm)	Hole Enlarging Depth(mm)
—	white noise	—	loose sand	180	600
1	5Hz sine wave	0.4g	loose sand	180	600
—	white noise	—	rubber	180	600
2	5Hz sine wave	0.4g	rubber	180	600
—	white noise	—	loose sand	180	450
3	5Hz sine wave	0.4g	loose sand	180	450
—	white noise	—	loose sand	180	450

4	5Hz sine wave	0.4g	rubber	180	450
—	white noise	—	loose sand	180	300
5	5Hz sine wave	0.4g	loose sand	180	300

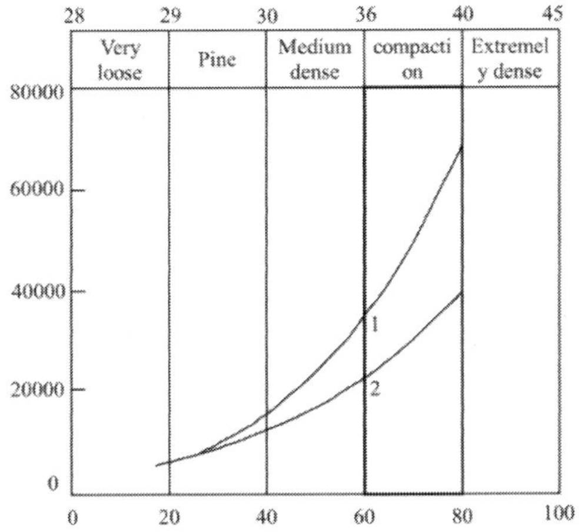

Figure 3. Curve of *K*-value (1 - water, 2 – Under water).

3. Calculation Method of Relative Displacement of Pile Side

In the p-y curve, y refers to the displacement of the pile relative to the sand. In the analysis process, the displacement y_{pile} of the pile is calculated separately according to the strain of the pile body, and then the displacement y_{sand} of the sand is obtained according to the acceleration of the soil. Finally, the displacement y of the pile relative to the sand is obtained by subtracting the displacement of the sand from the displacement of the pile.

3.1. Displacement of Pile

The curvature moment method is used to solve the pile side displacement [13]. According to the basic theory of material mechanics, the following equation can be obtained:

$$EI\frac{d^2y}{dx^2} = -M \tag{1}$$

$$EI\frac{1}{\rho} \approx EI\frac{d^2y}{d^2x} = -M(x) \tag{2}$$

$$EI\theta \approx EI\frac{d^2y}{d^2x} = -\int_0^x M(x)dx + C_1 \tag{3}$$

$$EIy = -\int_0^x dx\int_0^x M(x)dx + C_1x + C_2 \tag{4}$$

Where: EI—bending stiffness of beam; ρ—radius of curvature of beam; θ—Inclination of beam; Bending moment of beam at $M(x)-x$.

When the beam is subjected to external load, the following expression can be obtained:

$$\frac{d\theta}{dx} = \frac{d^2 y}{d^2 x} = -\frac{M(x)}{EI} \tag{5}$$

$$\int_{\theta_C}^{\theta_D} d\theta = -\int_{x_C}^{x_D} \frac{M(x)}{EI} dx \tag{6}$$

$$\theta_D - \theta_C = \int_{x_C}^{x_D} \frac{M(x)}{EI} dx \tag{7}$$

Thus, the moment area theory is obtained, that is $\theta_{D/C}$ is equal to the area between C and D in the relationship between M / EI of beam and pile length X

According to the tangent lines of point P and P' on the curve, take a length of dt line segments on the vertical line of point C, as shown in figure. 4, and calculate the relative horizontal displacement $t_{C/D}$ of point d relative to point C in the vertical plane.

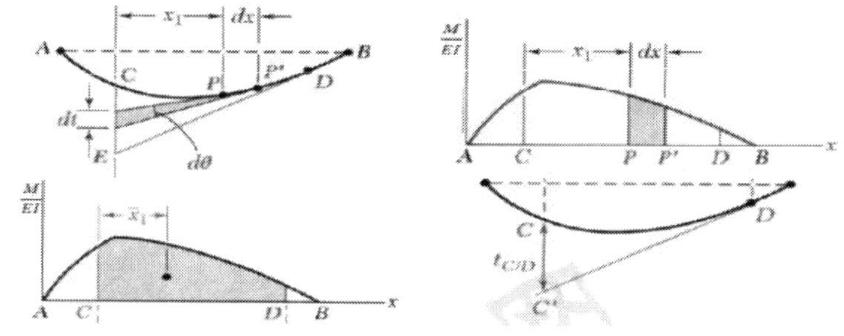

Figure 4. The derivation formula of curvature-area theories.

$$dt = x_1 d\theta = -x_1 \frac{M}{EI} dx \tag{8}$$

$$t_{C/D} = -\int_{x_C}^{x_D} x_1 \frac{M}{EI} dx \tag{9}$$

$$y = t_{C/D} = \int_{x_C}^{x_D} x_1 k(x) dx \tag{10}$$

$$k(x) = \frac{1}{\rho}(x) \tag{11}$$

The curvature area theory is obtained. $t_{C/D}$ is the tangent offset of point d relative to point C, and x_1 is the distance between point D and the integration unit. Assuming that the relative displacement between the pile bottom and the steel box is zero and has a certain initial inclination angle, the lateral displacement of the pile body is equal to the first-order moment of inertia of the area between points C and D on the curvature diagram relative to the calculation point. In the calculation of curvature area, linear interpolation method is used between nodes. The essence of this method is based on piecewise double integration and depends on the boundary conditions of the pile body. During the test, the pile bottom is consolidated with the steel box to meet the basic conditions of calculation requirements.

3.2. Displacement of Sand

In order to obtain the relative displacement y of pile-soil, the displacement response of far-field soil layer in the test needs to be calculated y_{sand}. Therefore, the acceleration time history recorded in the far-field soil layer is used to calculate the displacement time history of the excavated layer. Finally, the pile-soil relative displacement time history is obtained according to the pile displacement time history obtained by the curvature area method and the sand displacement time history.

4. Dynamic Response of Reamed Micro Pile Soil System

4.1. Fundamental Frequency Response of Reamed Micro Pile Soil System

Before the start of each working condition, input white noise with peak acceleration of 0.1g to measure the dynamic characteristics of pile-soil system. Based on the spectrum analysis of the acceleration data in the soil, the soil fundamental frequency can be obtained. It is found that the soil fundamental frequency is basically maintained at about 22 Hz, and changing the reaming parameters has little effect on the soil frequency. This is because the fundamental frequency of soil is mainly affected by compactness, and the compactness of sand is not greatly affected when loading and changing reaming parameters. This again shows that the dynamic performance of soil is relatively stable and the repeatability of the test is good.

4.2. Acceleration and Displacement Time History Curve of Mesa And Soil

During the test, 0.4g sine wave is selected to analyze the law of dynamic interaction between reamed micro pile and soil. Figure 5 shows the measured table acceleration and displacement time history curve under condition 5. It can be seen from figure 5 (a) that the table acceleration curve rapidly increases to the peak value of 0.4214g (basically close to 0.4g) at the initial stage of loading, remains constant during the loading process, and rapidly decreases to 0 at the later stage of loading. Similarly, it is obvious from figure 5 (b) that the table displacement curve increases rapidly to the peak value of 0.004 m at the initial stage of loading, and there is no sudden change in the curve during the loading process, indicating that the loading is relatively stable. Under other working conditions, the peak acceleration of the table is basically about 0.4 g, meeting the input requirements.

The law of soil acceleration and displacement under other working conditions is similar to that under working condition 5 (figure 6). Due to space limitation, this paper only gives the peak value and amplification factor of soil acceleration under various working conditions, as shown in table 5. It can be seen from table 5 that the soil acceleration from the pile bottom to the pile top shows an increasing trend, and the soil acceleration at 925mm and 1525mm is greater than the table acceleration, which indicates that the seismic response of the middle and upper layer of soil is large. According to the soil acceleration under different working conditions, the change of reaming parameters has little effect on the peak value of soil acceleration at the same height.

4.3. Time History Curve of Pile Top Acceleration and Displacement

Figure 7 shows the time history curve of pile top acceleration and displacement under condition 5. Limited to space, only the peak acceleration, peak displacement, amplification factor and the maximum inertial force obtained from the peak acceleration under various working conditions are given, as shown in Table 4. It can be seen from figure 7 that the variation law of pile top acceleration and displacement time history curve is also similar to that of table top acceleration and displacement time history curve, with peak acceleration of 1.0119g and peak displacement of 0.0079m. It can be seen from table 4 that the peak acceleration and peak displacement of the pile top are significantly greater than the peak acceleration and peak displacement of the table top. The amplification factor of the pile top is the largest compared with the acceleration amplification factor at other heights of the pile body, which is caused by the inertial force effect.

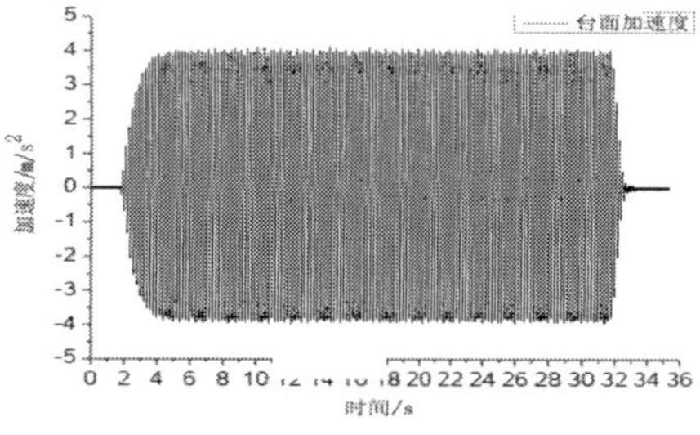

(a) Time history curve of table acceleration.

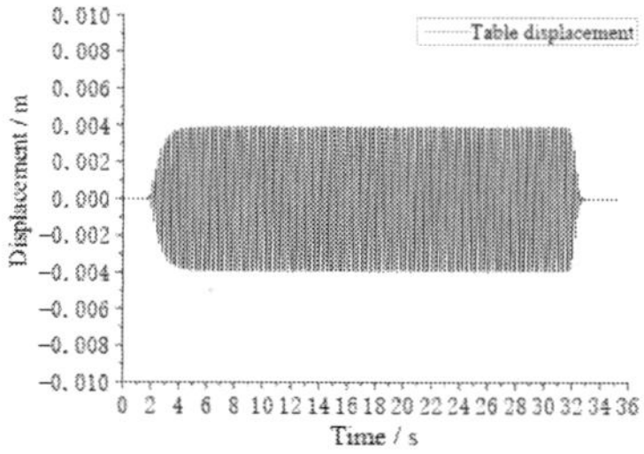

(b) Table displacement time-history curve.

Figure 5. Time - history curves of acceleration and displacement for working condition 6.

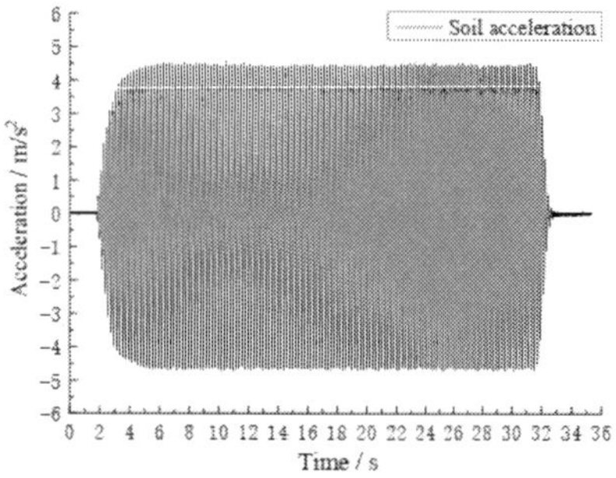

(a) Time-history curve of soil acceleration at 925 mm from the bottom of soil

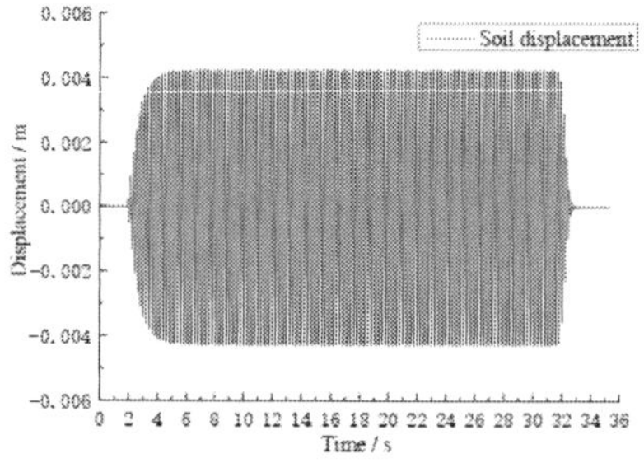

(b) Time history curve of soil displacement at 925 mm from bottom of soil

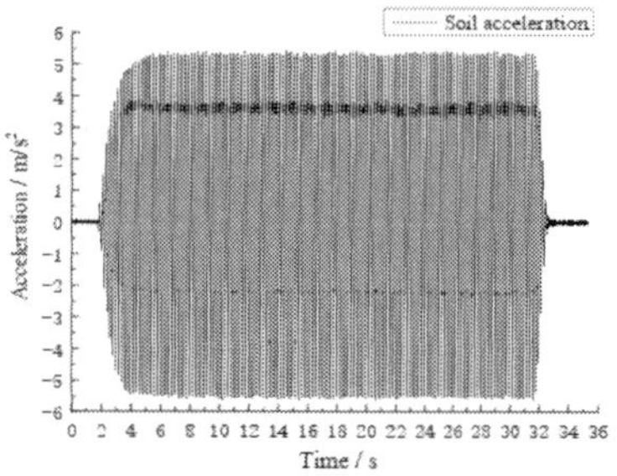

(c) Time-history curve of soil acceleration at 1525mm from the bottom of soil

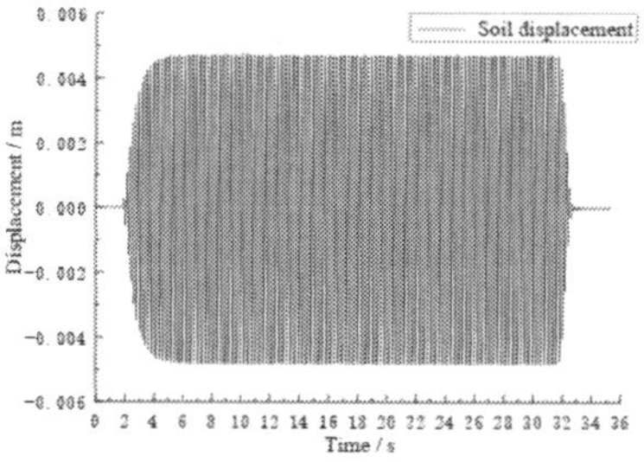

(d) Time history curve of soil displacement at 1525mm from the bottom of soil

Figure 6. Time-history curves of soil acceleration and displacement under condition 5.

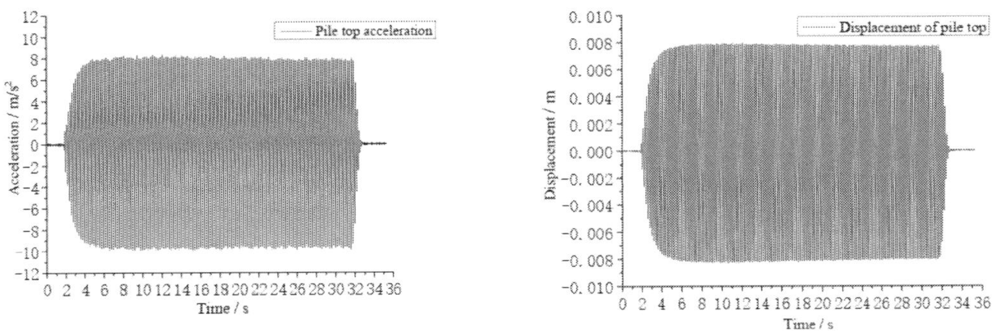

(a) Time-history curve of pile top acceleration (b) Time-history curve of pile top displacement

Figure 7. The time history curve of acceleration and displacement of pile.

Table 4. Peak value of acceleration, amplification factor, inertia force and peak displacement of pile under all cases

working condition	peak acceleration(m/s^2)	amplification coefficient	inertia force(N)	peak displacement(m)
1	1.1597g	2.7945	65.4044	0.0085
2	1.6109g	3.8363	90.8518	0.0143
3	1.0311g	2.4620	58.1530	0.0085
4	1.5074g	3.5814	85.0184	0.0134
5	1.0120g	2.4014	57.0734	0.0079
6	1.3082g	3.1953	73.7798	0.0120
7	0.9918g	2.3893	55.9373	0.0075
8	1.2367g	2.9829	69.7524	0.0113
9	0.9324g	2.2323	52.5890	0.0081
10	1.0107g	2.3950	57.0031	0.0104
11	0.8535g	2.0444	48.1386	0.0072

In addition, the inertia force exerted by the pile top after reaming is greater than that without reaming, indicating that reaming plays an amplifying role in the inertia force. When the reaming depth and hole diameter are unchanged, the inertia force is obviously less than that of rubber when the reaming material is loose sand; When the reaming material and hole diameter remain unchanged and the reaming depth increases, the inertial force increases; When the reaming material and depth remain unchanged and the reaming diameter increases from $2.5D$ to $4D$, the inertial force increases first and then decreases. Because the loading conditions and pile top mass are the same, and the soil dynamic performance is also very stable according to the soil response frequency and acceleration response, the change of inertia force is the reaming parameter. The change of reaming parameters leads to the change of pile-soil system frequency, which leads to the change of inertial force.

References

[1] Allen C, Jesus G, Donald B.et al. Mircopiles: Recent Advances and Future Trends [A] Deep Foundation. ASCE

[2] XU Y F. 2010 Dynamic Response Analysis of Group Piles in Earthquake Based on ANSYS [D] Hefei: Hefei University of Technology.

[3] Sato H, Kouda M, Yamashita T. Study on Nonlinear Dynamic Analysis Method of Pile Subjected to Ground Motion-Part2: Comparison Between Theory and Experiment[C] Proc. 1lth World Conf. Earthquake Eng, Acapuleo, No.1289.
[4] Yang J, Mcmanus K.J, and Benrill J. B.Kinematic Soil. Micropilc Interaction[C] Proc.12th World conf.Earthquake Eng, New Zealand No.1501
[5] Wei C L. 2008 Shaking Table Test of Seismic Pile-Soil-Structure Interaction [D] Chongqing: Chongqing Jiaotong University.
[6] Wang J H, Feng S L. 2004 The shake table test on soil-pile interaction [J] *Chinese Journal of Geotechnical Engineering* **26** (50) 616-618.
[7] Zhang Z L. 2013 Experimental Research on Lateral Dynamic Response Characteristics and P-Y curves of pile Groups [D] Tianjin: Hebei University of Technology.
[8] LI Y R. 2006 Study on lateral dynamic response of pile foundation in liquefiable soil [D] Harbin: Institute of Engineering Mechanics, China Earthquake Administration.
[9] Chen Q J, Zhao Y F, Wang H D. 2002 Numerical Simulation for Soil Domain in Shaking Table Model Test [J] CHINESE QUARTERLY OF MECHANICS **23** 407-410.
[10] Min-Taek Y, In-Woo J, Jin-Tae H. 2011 Myoung-Mo Kim. Pile Installation Effect on Dynamic p-y Curves in Dry Sand[C] Proceedings of the Twenty-first (2011) International Offshore and Polar Engineering Conference.
[11] Chen Y. 2014 Shaking table testing research on micro-pile soil dynamic interaction [D] Fuzhou: Fuzhou University.
[12] Hang L A. 2004 Calculation of horizontally loaded piles [M] Changsha, Central South University Press.
[13] Tang L. 2010 P-Y model of dynamic pile-soil interaction in liquefying ground [D] Harbin: Harbin Institute of Technology.

CFD Study on the Air Temperature Field inside Fire-Rated Duct of Pressurisation System

Hang Yin and Longfei Tan

Sichuan Fire Research Institute of Ministry of Emergency Management, No. 69 Jinke South Road, Jinniu District, Chengdu, China.
Email: bank2002yinhang@126.com

Abstract. Avoiding overheating of the supply air inside the fire-rated duct of pressurisation system is crucial for the escape and rescue of the building occupants during the development of a fire. The simulation scenarios are established concerning three cross-sectional dimensions of ducts and four inner air velocities. Based on the performance criteria of insulation specified in the international standard of ISO 834-1, the temperature fields of supply air inside different fire-rated ducts and corresponding safety are analysed based on the assumption that the exterior of fire-rated ducts is in a consistent state of being subjected to fire.

1. Introduction

The pressurisation system is a special form of smoke control system that should be maintaining the air supply state for restricting the penetration of smoke into certain critical areas, such as stairways, lobbies and corridors. Generally, the ductwork of pressurisation system shall be capable of resisting the anticipated fire temperatures generated during the development of a fire. Meanwhile, it is noteworthy that the overheating of the supply air inside the fire-rated duct should be avoided as well, because the supply air is directly exported to the escape and rescue area of the building occupants. However, previous studies mainly focused on the technology and strategy of smoke control on the systems level [1-5], the research on the flow field inside the ductwork of smoke containment system are less comparatively. In this study, the inner air temperature fields of fire-rated duct of pressurisation system are numerically simulated using the CFD software of FLUENT.

2. Geometrical Model and Computational Domain

A common height of standard floor of buildings, five meters, is set as the length of study objects. As shown in figure 1, three types of fire-rated ducts of initial section are built in the cross-sectional dimensions of 0.5×0.5 m, 1.0×1.0 m and 1.5×1.5 m, respectively.

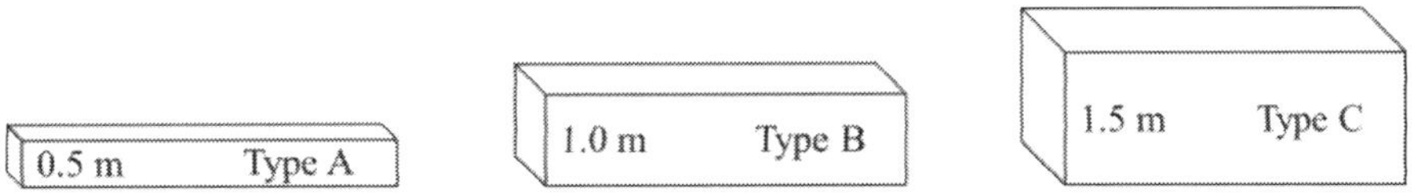

Figure 1. Three types of ducts with different cross-sectional dimensions.

The cross section of ducts is vertical to the gravity direction. Additionally, two kinds of computational domains are constructed based on different inner velocity fields of ducts. For the ducts

Journal of Physics: Conference Series **2148** (2022) 012058 doi:10.1088/1742-6596/2148/1/012058

in air supply state, the computational domain only contains the volume of geometrical model. For the ducts in off-duty state, the computational domain is extended appropriately for obtaining a better calculation accuracy, as illustrated in table 1.

Table 1. Dimensions of computational domain for the ducts in off-duty state.

Cross-sectional dimensions of ducts	Length	Width	Height
0.5×0.5 m	4.5 m	4.5 m	15 m
1.0×1.0 m	5.0 m	5.0 m	15 m
1.5×1.5 m	5.5 m	5.5 m	15 m

3. Simulation Scenarios and Boundary Conditions

As shown in table 2, 12 simulation scenarios are designed concerning three cross-sectional dimensions of ducts and four inner air velocities. The direction of inner air velocities is opposite to the gravity direction. According to the provisions specified in the international standard of ISO 834-1, the criterion to confirm the specimen has lost its insulation performance is any of the following: (1) The average temperature increase above the initial average temperature of the surface unexposed to fire is more than 140°C; (2) The temperature increase above the initial temperature at any location of the surface unexposed to fire is more than 180°C. To analyse the most unfavourable condition, four external surfaces of each fire-rated duct mentioned above are assumed to be subjected to fire simultaneously. Therefore, the average temperatures of the corresponding four internal surfaces are set as the critical temperature, 160°C, by assuming the initial average surface temperatures are 20°C. The detailed simulation scenarios and boundary conditions of geometrical model are illustrated in table 2 and table 3, respectively.

Table 2. Simulation scenarios with different cross-sectional dimensions and inner air velocities.

Inner air velocities	0 m/s	5 m/s	10 m/s	20 m/s
Cross-sectional dimensions: 0.5×0.5 m	Scenario 1	Scenario 2	Scenario 3	Scenario 4
Cross-sectional dimensions: 1.0×1.0 m	Scenario 5	Scenario 6	Scenario 7	Scenario 8
Cross-sectional dimensions: 1.5×1.5 m	Scenario 9	Scenario 10	Scenario 11	Scenario 12

Table 3. Boundary conditions of geometrical model.

	Inner air velocities: 0 m/s	Inner air velocities: 5, 10, 20 m/s
Inlet of ducts	Pressure-in	Velocity-in
Outlet of ducts	Pressure-out	Outflow
Internal surface of ducts	Wall: 160°C, 871 J/(kg·K), 202.4 W/(m·k)	

4. Sections of Temperature Analysis and Safety Criterion

As shown in figure 2, the temperature fields of middle longitudinal section A-B-C-D and middle transverse section E-F-G-H are the objects of investigation. According to the research findings described in SFPE (Society of Fire Protection Engineers) Handbook of Fire Protection Engineering, the tolerance time for exposure to hot air of 60°C and 100% relative humidity is only 30 min, and will be shorten to 12 min when the air temperature is 100°C and relative humidity is less than 10%. Therefore, in this study, the criterion to confirm the safety of the supply air inside the fire-rated duct is defined strictly that the average temperatures of above-mentioned sections are no more than 60°C.

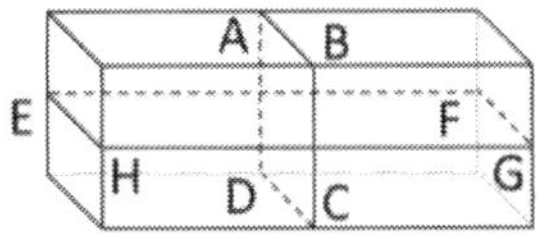

Figure 2. Two sections used for temperature analysis and safety determination.

5. Changing Process of Temperature Field of Target Sections

Taking Scenario 1 and 4 for example, the changing process of the temperature fields of section A-B-C-D and E-F-G-H are shown in figure 3 and 4.

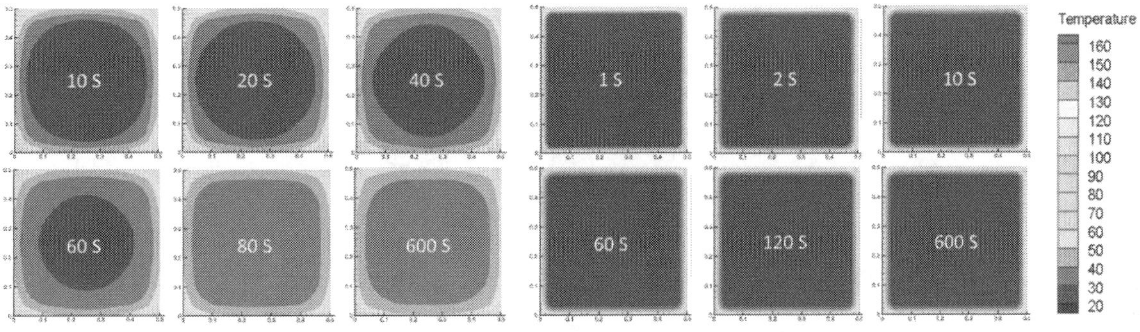

(a) Scenario 1 (b) Scenario 4

Figure 3. Temperature fields of section A-B-C-D for Scenario 1 and 4.

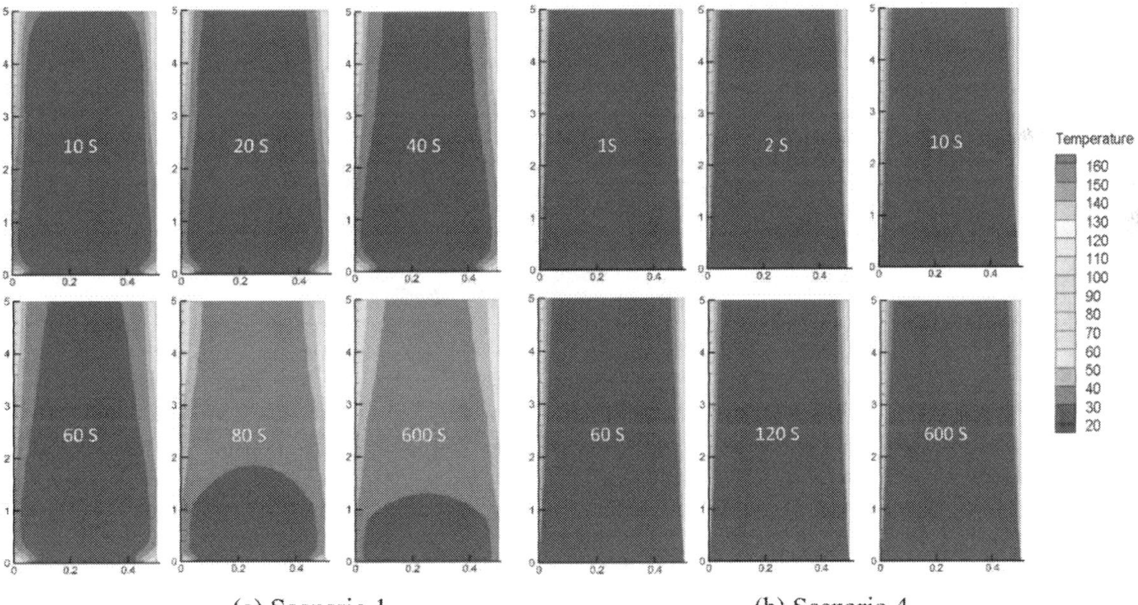

(a) Scenario 1 (b) Scenario 4

Figure 4. Temperature fields of section E-F-G-H for Scenario 1 and 4.

It can be seen that, for Scenario 1, the static air under normal temperature inside the duct is heated by the high-temperature duct wall and rises under the action of buoyancy. The annular thermal boundary layer thickens gradually along the flow direction of heated air. When the temperature field of inner air reaches steady state, the temperatures of main zone for the entrance region and fully developed region are 20-30°C and 30-40°C, respectively. For Scenario 4, since the inner air velocity is 20m/s, the heat exported from the duct wall is taken out rapidly along with the air. Therefore, only a thin thermal boundary layer appears near the duct wall. The temperature of main zone always maintains 20-30°C.

6. Temperature Field of Target Sections under Steady State

The temperature fields of section E-F-G-H under steady state are shown in figure 5. For the scenarios without initial air velocity, as the cross-sectional dimension of duct increases, the length of entrance region flow increases and the thickness of thermal boundary layer decreases. For the scenarios with initial air velocity, the thickness of thermal boundary layer decreases with the increase of cross-sectional dimension of duct and inner air velocity.

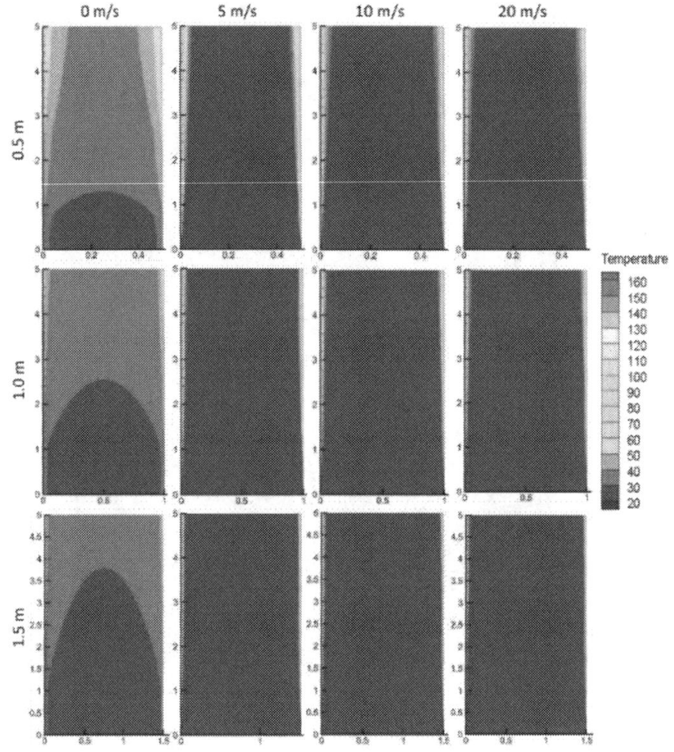

Figure 5. Temperature fields of section E-F-G-H under steady state.

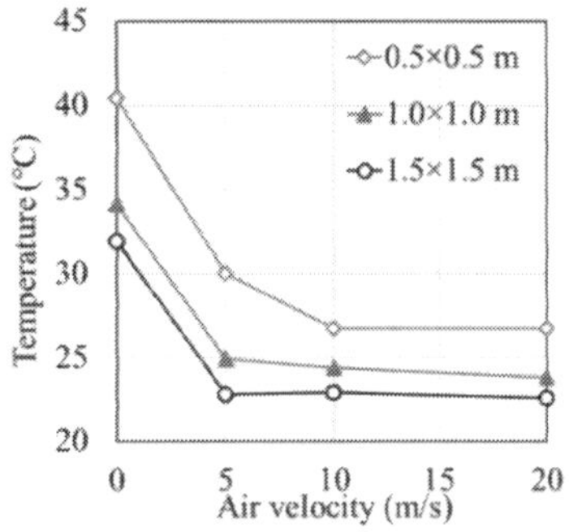

Figure 6. Average temperature of section A-B-C-D under steady state.

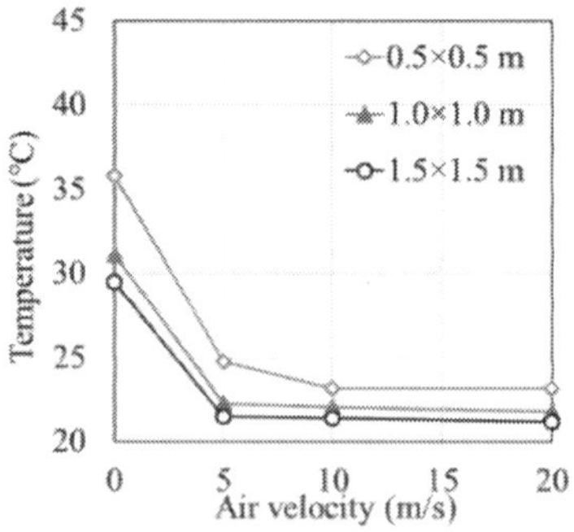

Figure 7. Average temperature of section E-F-G-H under steady state.

The average temperatures of section A-B-C-D and E-F-G-H under steady state are shown in figure 6 and figure 7, respectively. It can be seen that the average temperatures of section A-B-C-D and E-F-G-H both decrease with the increase of cross-sectional dimension of duct and inner air velocity. Moreover, since the heat convection rate in the state of natural convection is smaller than that in the state of forced convection, a large amount of heat still exists in the duct without initial air velocity. Therefore, the average temperatures of the two sections for the scenarios in off-duty state are obviously higher than those for the scenarios in air supply state. However, the maximum average temperatures of section A-B-C-D and E-F-G-H are only 40.4°C and 35.8°C, respectively.

7. Conclusion

The temperature fields of supply air inside different fire-rated ducts of pressurisation system, concerning three cross-sectional dimensions of ducts and four inner air velocities, are simulated in this study. The temperatures inside the ducts in off-duty state are obviously higher than those in air supply state. It can be seen from the simulation results that the average temperature of the middle longitudinal section and middle transverse section of the ducts are less than 40.4°C and 35.8°C, respectively, when the fire-rated ducts are in a state of air supply and being subjected to fire. Based on the performance criteria of insulation specified in existing standard, the supply air is safe enough to be exported from the pressurisation system during the period in which the fire-rated duct maintains its insulation performance.

Acknowledgements

This research was financially supported by Research Program of Fire and Rescue Department of Ministry of Emergency Management (No.2018XFGG18) and Central Public-interest Scientific Institution Basal Research Fund (NO.20218805Z).

References

[1] Lu S, Wang Y H, Zhang R F and Zhang H P. 2011 *Procedia Eng* **11** 369–378.
[2] Merci B, Shipp M. 2013 *Fire Safety J* **57** 3–10.
[3] Węgrzyński W. 2017 *Int. J. Heat Mass Tran* **114** 483–500.
[4] Byrne E, Georgieva K and Carvel R. 2018 *Tunn. Undergr. Sp. Tech* **81** 306–314.
[5] Li J, Prétrel H, Suard S, Beji T and Merci B. 2021 *Fire Safety J* **125** 103426.

Review of Research on Physical and Mechanical Properties and Engineering Application of Metal Tailings

Shengfu Wang, Lechen Yan, Kaixi Xue*, Liang Lv, Dongjie Zhang and Dongwei Li

Department of Civil & Architecture Engineering, East China University of Technology, Nanchang, Jiangxi Province, 330013, China
* Corresponding author, E-mail: xkx1257@foxmail.com

Abstract. Processing and storage requirements for metal residues are becoming stricter to achieve the carbon neutralization target. The physical and mechanical properties of tailings affect the stability of tailing dams. Metal tailings can be used as secondary resources, and it is easy to pollute the environment under poorly managed conditions. Therefore, it is necessary and urgent to reuse these deposits such as iron tailings, copper tailings, zinc tailings et al. This article discusses the current research on the mechanical properties of metal tailings and its engineering application. Based on previous research, it is pointed out that there still needs more attention on the mechanical properties of metal tailing sands, especially under different conditions like dry-wet, freeze-thaw, dynamic loads and large-scale application. In the future, research on the filling of metal tailings as roadbed and new building materials will be one of the directions to solve the problem of tailing pond accumulation.
Keywords. Tailings, Geotechnical engineering, physical properties, mechanical properties, comprehensive utilization.

1. Introduction

Metal tailings are wastes that discharged after the ore is crushed, ground, sorted through multiple processes and finally piled up in the form of tailing ponds [1-2]. These wastes contain secondary resource components such as re-selectable metal ore and non-metal ore. After mine reclamation process these tailings can also be used in other engineering fields such as road construction and civil engineering. Pilled tailings not only take lots of land, wastes a big sum of money, but also severely pollutes the surrounding environments. Tailings dam break accidents have occurred sometimes during past decades, causing enormous and irreparable loss of life and property. At the end of 2020, the cumulative tailings in China have exceeded 20 billion tons, but the total annual comprehensive utilization of tailings is only 335 million tons [3-4]. The annual comprehensive utilization rate is less than 30% [5]. In recent years, the prevention and control of tailing pond pollution have been listed in the national major development strategy in China. The Ministry of Emergency Management of the People's Republic of China clearly stated that the number of tailings pond must decline from 2020, it also demands that the comprehensive utilization rate of tailings must be increased. Therefore, based on the deep and comprehensive understanding of the mechanical properties of metal tailings. It is of great significance to continuously improve the utilization rate to solve the risk of tailing leakage and meet the requirements of carbon neutralization.

For now, papers concern the characteristics of metal tailings are becoming more and more, however, most study of tailings focus on one material. External factors under different conditions also introduced as another way to broad their research field. The integrity study for different tail ore is relatively less, most of literature focused on iron ore and copper ore [6]. Based on the previous research, engineering characteristics and applications were summarized. This article also proposes further research direction in order to enrich the content of the mechanical properties of tailings sand and engineering applications.

2. Composition and Particle Size of Metal Tailings Sands

Different types of metal deposits are affected by geological conditions resulting in the diversity of material composition. Therefore, the mineral composition and chemical composition of metal deposits are relatively complex. The main mineral components of iron tailings sand are plagioclase, quartz, and chlorite, calcite, flaky biotite [7]. Common components mainly include TFe, SiO_2, Fe_2O_3, CaO, Al_2O_3, MgO, K_2O, MnO, Na_2O, et al [8]. Among which the content of SiO_2 between copper tailings fluctuates from 31.88% to 80.10%. The composition of SiO_2 lists at the top.

Due to the difference in mine types and mining processes, the distribution of particle size after beneficiation varies greatly. Studies have shown that the particle size of metal tailings sand usually varies from 0.075mm to 5mm. The particle size of metal tailings sand has a great influence on its engineering characteristics. Existing research mainly focus on the classification of tailings sand and particle size. The grouping of metal tailings has not yet formed a unified standard. According to the classification of rock and soil particle groups in classical soil mechanics, Wu [9] defined the tailings particle size less than 75μm as fine sand. Cao et al [10] studied the selection of physical parameters of tailings and found 12 parameter indicators that can represent the physical properties of tailings. These 12 indicators were divided into 3 categories and the particle size can be used as a basic indicator. 795 sets of data were statistically analysed and discussed separately [11]. By using Scanning Electron Microscopy and image processing technology, Yang et al [12] quantified the flatness, roundness, convexity, and roughness of tailings sand. As the particle size increases, the flatness of the tailings decreases, indicating that the shape of the tailings particles changes from needle-like to spherical.

Under different particle gradation conditions, the strength of tailings is different [13]. Saturated samples were subjected to isotropic consolidation undrained shear tests under confining pressures of 100kPa, 200kPa, and 300kPa respectively. The stress-strain curve of the tailings sand sample is softened, and as the confining pressure gradually increases, the softening degree gradually weakens. Under the same confining pressure, the stress-strain curve of the same density and different gradation samples have different softening degrees. The triaxial shear test shows that the fine particles of the sample are more likely to lose their shear strength under the action of continuous pore water pressure. Therefore, the finer particles in the sample with the same density, the lower the effective stress under the same confining pressure. The faster it is, the easier it is to soften. The sedimentation characteristics of tailings sand under different particle size gradation conditions are different. With the increase of fine particles of tailings sand, the maximum underflow concentration of tailings sand after sedimentation shows a downward trend [14]. Yang et al [15] studied the effect of the particle size of tailings sand on the shear strength of tailings. The results showed that as the median diameter increases from 83.23μm to 149.9μm, the unsaturated tailings cohesive force decline and the friction angle increases. Ma et al [16] investigated the influence of ore phase transition after roasting at different temperatures.

Particle characteristics, especially the particle size distribution of tailings sand have an important influence on tailings sand. The graphic visualization of tailings sand particles can reflect the microscopic morphology of particles. However, the microcosmic study of tailings sand particles is still in the state of two-dimensional interface, and the three-dimensional microcosmic simulation is important in the next stage. The properties of tailings sand particles change under the condition of high temperature roasting, so it is necessary to further study the mineral phase changes of tailings sand particles under different gradation conditions at high temperature. Most of the current research focus on iron ore but the research on copper ore and other minerals is also very important. The classification of tailing sand particle size has been discussed in the above literatures, but there are still relatively few studies on viscous particles with different particle sizes. It is necessary to further study the distribution of viscous particles with different particle sizes.

3. Strength Characteristics of Tailings Sand

3.1. Strength Characteristics

The high pressure has an important influence on the strength of tailing sand [17]. Yang et al [18] conducted research on the strength characteristics of tailings, and through consolidated drained test and compression test, respectively, explored the strength characteristics at different degrees of compaction. They pointed out that as the porosity ratio of tailings increases, tailings ore compressibility increased, but the compressive modulus is negatively correlated. The change of soil sample plasticity index affects the dynamic strength of tailing sand particles [19]. The tailing sand was tested with a dynamic triaxial instrument and it was pointed out that the existence of initial static shear stress reduced the strength of tailing sand particles. The residual strength is lower than the initial strength and its decay rate decreases with the increase in the plasticity index. The change characteristics of the particle strength of tailings sand in the saturated state are different from tailing sand under unsaturated state [20].

Different vibration frequencies affect the strength of tailings sand [21]. At the same time, the shear rate under different stress also influences the strength characteristics of tailings sand [22]. When the shear rate increases, the shear strength of tailings sand particles will decrease. Changes in the low confining pressure state are less pronounced than the high confining pressure state. An et al [23] conducted an experimental to study the shear strength of tailings sand under dry-wet cycles. During the dry-wet cycle, the internal particle structure of tailings changed due to the repeated action of dry-wet. After several cycles, the effective cohesion c' and the internal friction angle φ' have a certain regularity.

3.2. Compression Characteristics

Cao et al [24] studied the influence of the compression and consolidation coefficient. The results showed that the consolidation coefficient increases with the change of the consolidation pressureand decreases gradually after reaching the peak, finally tends to be stable. The consolidation coefficient decreases with the particle size. The change of compaction degree and water content also have an important effect on the compression characteristics of tailing sand particles [25]. The standard consolidation test found that both compaction and moisture content have an impact on the compressibility and consolidation coefficient. With the increase of the degree of compaction, the compressibility and consolidation coefficient of tailings sand particles decrease simultaneously, but the change of the compression coefficient and consolidation coefficient caused by water content is relatively complicated. Wu et al [26] conducted a correlation study on the particle properties of tailings sands on compression and consolidation characteristics. The results showed that tailing sands with different particle compositions have different characteristics, compression characteristics of tailings sand are greatly different from each other in composition, particle size and shape of tailing particles.

Different types of tailing sand have great differences in mechanical properties, and the mechanical properties of tailings sand particles are a key factor to meet the specific purpose especially in highway construction and civil engineering. The current research on the mechanical properties of tailings sand particles is only focused on certain factors. Comprehensive consideration of various factors is of great value for the compression characteristics of tailings sand.

4. Study on the Stability of Tailings Dams

The research on tailings stability mainly focuses on the tailings dam under external forces especially nature force. The stability analysis of the Dashi river tailings dam under the action of the earthquake shows that the earthquake can cause cracks in the tailings pond and water spraying, sanding leakage [27]. Psarropoulos et al [28] found that the stability of tailings dams gradually decreased under static and dynamic cycling conditions. Uniaxial compression test and triaxial shear laboratory are used to study the stability of tailings dams. The stability of tailings dams is also affected by the following factors such as permeability, capillary water, dry and wet cycles [29]. Unify these parameters into a model to describe the stability of tailings dams is one of the directions. The external force such as earthquake acting on the internal factors of the tailings remains to be studied. The coupling of a single factor or double factors is the current research hotspot. The research on the correlation between internal factors and external factors is a key point.

4.1. The Influence of Earthquake on the Stability of Tailings Dams

The liquefaction of tailing sand often occurs under earthquake which has a negative effect on the stability of the dam. The discriminant method of seismic liquefaction is obtained by considering several different working conditions and analysing the results of two-dimensional seismic response analysis of tailing dam [30]. Based on the relationship between the anti-liquefaction degree and the pore pressure ratio, a simplified calculation formula for the excess pore pressure on the tailings dam is proposed [31], the calculation indicates that the area near the water's edge of the upstream slope is the easiest area to liquefy, followed by the initial dam. The length of the dry beach has a great influence on the seismic liquefaction stability of the tailings dam. Through sensitivity analysis and reliability calculation analysis [32]. It is found that the internal friction angle and severity need to be included in the stability analysis of the tailings pond for further improvement. Hu et al [33] pointed out that the main indicator of anti-sliding stability is the shear strength and used separate calculations to evaluate the stability of its tailings. The calculation results show that the infiltration line affects its stability. Comparative studies have found that tailings with different particle sizes have different stability performance under seismic action [34]. Under high intensity test, large area deformation occurs at the top of the tailings dam slope, resulting in large amounts of sand blasting and dam liquefaction caused by cracks [35]. The numerical simulation software also analysed the stability of tailing dams. Despite the overall stability of the dam body behaves well, there is still a risk of liquefaction and dam failure under the action of earthquake [36]. Conventional treatment methods for tailings dam reinforcement usually adopt gravel pile or sand pile [37].

4.2. The Influence of Freeze-Thaw on the Stability of Tailings Dams

According to the geographical distribution of tailings ponds 91.4% of tailings ponds suffer different degrees of freezing in winter. Field surveys revealed [38] that in winter, except surface freezing, there are deep frozen layers of tailings dam in alpine areas. These solid frozen layers have a great influence on the stability of tailings dams. They will affect the seepage field of the tailings dam and reduce the stability of the dam body. At worst, they will lead to deformation and failure of tailings dam when the temperature rises. However, the research on the mechanical properties and factors of frozen tailings is rarely reported. The research results showed that there are three types of uniaxial compression failure of frozen tailings. The uniaxial stress-strain curve of frozen tailings can be divided into 4 stages: initial strain softening stage, linear strain hardening stage, nonlinear strain hardening stage and nonlinear strain softening stage. Among the 4 influencing factors, the uniaxial compressive strength of frozen tailings has logarithmic relationship with average particle size, exponential relationship with dry density, linear increasing relationship with water content, and quadratic parabolic growth relationship with loading rate. However, the deformation modulus of frozen tailings has a natural logarithmic relationship with average particle size, an exponential relationship with dry density, a quadratic parabolic relationship with water content, and an exponential growth relationship with loading rate [39]. In addition to the conventional influencing factors, the stability of the mine should also consider the hazards caused by seasonal freezing and thawing. This freeze-thaw cycle will significantly change the connection structure among the tailings sand particles and directly affect the changes in physical and mechanical properties [40-41].

The construction of tailings projects in China often adopts the upstream method to build dams, but the upstream method is prone to geological disasters and dam break accidents. The study of the stability of tailings dam shows that many factors have an important impact on the tailings dam, the maintenance and management of the tailings pond also bring the increase of cost to the mining enterprise. It is one of the effective way to change the disposal way of tailings and increase the exploitation of tailings resources to eliminate the hidden trouble that may happen in the operation of tailings ponds.

5. Comprehensive Utilization of Secondary Resources of Tailings

Li et al [42] proposed to strengthen the utilization and protection of mineral resources and provided

Journal of Physics: Conference Series **2148** (2022) 012059 doi:10.1088/1742-6596/2148/1/012059

relevant ideas for the development and utilization of non-metallic minerals. Gu [43] proposed the possibility of comprehensive utilization of tailings and proposed some methods for the use of building materials. Li [44] discussed the development and utilization of metal mines in China and briefly introduced the current situation and characteristics of metal mine tailings resources. Ding et al introduced more specific examples of the utilization of different types of tailings such as iron ore tailings and copper ore tailings [45]. Direct application of metal ore tailings and non-metallic ore tailings also carried. Xu et al [46] used asbestos tailings to conduct research on glass-ceramic decorative panels and the parameters of asbestos tailings as the main raw material can be as high as 60%. The physical and chemical properties of the product are better than natural marble and granite. Its simple production method and reasonable parameters is a new way for the development and utilization of asbestos tailings. Ye et al [47] briefly introduced the utilization status of kaolin from pyrite tailings in southern Sichuan and commented on its comprehensive utilization methods. They pointed out that kaolin from pyrite tailings in southern Sichuan can also be used as a preparation method. Carbon black, activated clay, synthesis of zeolite, production of refractory materials etc. Yang et al [48] pointed out that the valuable mineral resources in the beneficiation plant are poor, fine, and hybrid. Limited by conventional beneficiation technology, resulting in low concentrate grades, recovery rates, and poor economic benefits.

There are few discussions on copper tailings and most of them focus on the utilization in the mine engineering fields [49]. The discussion of lead-zinc tailings is mostly focused on material, not on it is application in other engineering. At present, the engineering construction of China is in and will be in the period of large-scale construction for a long time. Therefore, it is necessary to combine the comprehensive utilization of tailings with civil engineering construction and highway construction. Through large-scale application of engineering and rapid iteration of products, the cost of development and application will constantly reduce, and finally a new effective way of comprehensive utilization of tailings is explored.

6. Engineering Application Tailings Sand

At present, the method of comprehensive utilization of tailings in China is mainly used as filling of underground mining, followed by the construction of roads, pavement materials, anti-skid materials, coastal land reclamation, and construction materials. Little tailing materials are used to re-select [50]. In 2018, the total quantity of tailings used for building materials and value components in China was approximately 184.43 million tonnes.

6.1. Technology for Manufacturing Cement Clinker Using Tailings Sand.

Li et al [51] observed the thermal reaction characteristics of fine tailings under different high temperature conditions through high-temperature experiments. The higher the temperature, the greater the characteristic kurtosis intensity. The test after tailings and CaO blended combustion showed that: The main phase composition after calcining at 1400 °C is basically the same as that of ordinary Portland cement clinker. Barati et al [52] used phosphorous tailings to configure a sintering aid suitable for new dry process production and appropriately adjusted cement clinker. Tests showed that the clinker has high strength. The solubilizing effect improves the strength of the clinker. Huang et al [53] discussed the possibility of using lead-zinc tailings as cement ingredients based on the opinion that the composition of the oxides in lead-zinc tailings are similar to that of cement raw materials. The test results show that the lead-zinc tailings under different proportions have different characteristics and some lead-zinc tailings cannot be used as ingredients for cement production because they contain elements that are not conducive to the combustion of clinker. With better lead-zinc tailings, it can improve the flammability of the raw cement material, and increase the strength of the fired cement, while meeting the national environmental protection requirements. Zhao et al [54] conducted experimental research on copper tailings calcined cement and pointed out that it is feasible for copper tailings to replace iron powder to calcinate cement clinker and determine the appropriate conditions for copper tailings to burn Portland cement are 1350°C, the holding time is 30 minutes.

Wei et al [55] used low-silicon iron tailings to study the preparation of cement. The results showed that the strength index of bellite cement for 3 and 28 days reached the requirement of PI42.R cement by calcining iron tailings with a mass fraction of 30.90% and limestone with a mass fraction of 69.10% in 1350°C for 1h and adding activator gypsum equal to 0.60% of clinker mass. Liu et al [56] studied the preparation of magnesium portland cement from magnesite tailings. The test showed that the composition point of the mixture fell within the line of $2CaO \cdot SiO_2$, $3CaO \cdot SiO_2$ and MgO, and the M/S ratio in the raw material was 2/1 which is feasible to make cement tailing sand within 2.1~2.56.

6.2. Used As Subgrade Filling

Subgrade filling has certain requirements for the selection of filling material [57]. The selection of tailing sand should consider both the influence of radioactivity and the influence on the water environment. Considering the difference between the actual situation of subgrade compaction and the laboratory test, it is necessary to carry out the soil-mixed test on tailings. In the pilot project, the mixing ratio is lime: fly ash: tailing sand =8:22:70. The unconfined compressive strength of 7d field sampling test is 0.56 MPa, the compaction degree is 97%, and the measured bending and subsidence value of the top of the base is 0.91mm, which meets the design requirements and can be used for the road bottom base [58]. In the same type of project, the stability of the clay-bound tailings sand subgrade was analysed, and the analysis results showed that the stability of the clay-bound tailings sand subgrade can meet the requirements of specification [59]. The construction of highway engineering has higher requirements for roadbed materials. Under the premise of meeting environmental and construction technical requirements, through quality control during the construction process and monitoring of key points, stability of roadbed filled with tailings sand turned out to be good, which has obvious economic and social benefits [60].

7. Summary and Conclusions

To sum up, the current research on tailings mainly focuses on the tailings dam in different regions, and its strength characteristics stability characteristics, pore water migration law, freeze-thaw law and other studies have been relatively rich. Much has been achieved in the use of tailings secondary resources, but relevant research on the scale of application has yet to be completed. Based on current research, it is suggested that further investigation of tailings engineering characteristics should include:

1) In order to provide design basis for tailings construction in freeze-thaw area, it is necessary to expand the range of negative temperature to conduct strength and thawing test of tailing sand and study the influence of pore water on strength characteristics and deformation characteristics before and after the secondary phase transformation in tailing sand

2) At present, the research on the stability of tailings is mainly focus on the macro scale and most of the research is only considering the influence of geological forces, mainly represented by earthquakes. There is a lack of research on the key mechanisms of the permeability, void content, water content, etc in the tailings sand. Therefore, it is necessary to study the correlation between internal factors and external factors of tailing sand, investigate the influence of external forces represented by earthquake on various internal elements, as well as the correlation degree between different elements, so as to build the connection of mechanical properties in macro and micro states.

3) The stacking and disposal of tailings in mining enterprises is only a temporary measure. limited by the development of technology, the secondary extraction technology resources from tailings cannot be qualitatively changed in a short time. Consequently, it is necessary to extend the mind of using mine residues both on the supply side and on the demand side. There is a strong demand for materials in road construction. It is meaningful to further explore the use of tailing sand as roadbed filling in order to achieve economic and social benefits while solving the problem of unhealthy and danger tailing reservoirs.

Acknowledgments

The work was supported by the National Natural Science Foundation of China (Nos.51869001,

41977236, 42061011), Natural Science Foundation of Jiangxi Province of China (No. 20202BABL214044), Foundation of Jiangxi Educational Committee (No. GJJ200701).

References

[1] Blight G E. 1997 Destructive mudflows as a consequence of tailings dyke failures [J] *Proceedings of the Institution of Civil Engineers-Geotechnical Engineering* 125 9-18.
[2] Kossoff D, Dubbin W E, Alfredsson M, et al. 2014 Mine tailings dams: Characteristics, failure, environmental impacts, and remediation [J] *Applied Geochemistry* 51 229-245.
[3] Pastor M, Quecedo M, Merodo J A F, et al. 2002 Modelling tailings dams and mine waste dumps failures [J] *Geo-technique* 52 579-591.
[4] Zhang Q, Yin G, Wei Z, et al. 2015 An experimental study of the mechanical features of layered structures in dam tailings from macroscopic and microscopic points of view [J] *Engineering Geology* 195 142-154.
[5] Wang H. 2019 *National Mineral Resources Conservation and Comprehensive Utilization* chapter 1 pp 12-17.
[6] He C. 2018 Seismic safety evaluation of tailing pond engineering site [J] *Metal mine* 8 153-158.
[7] Lin X, Wang C, Zhang K, et al. 2019 Study on the preparation of autoclaved lime-sand bricks from iron tailings and waste rocks [J] *Mining Research and Development* 39 109-114.
[8] Mao K, Cai L, Wu X, et al. 2019 Research on performance and hydration mechanism of aerated concrete prepared from several typical iron tailings [J] *Bulletin of the Chinese Ceramic Society* 38 3719-3725.
[9] Wu S. 2017 *Study on the effect of particle size on the physical and mechanical properties of tailings and the stability of the dam* (China: Chongqing University Press) p 85.
[10] Cao Z, Zhang M, Zhu J, et al. 2014 Research on Representative Parameters of Physical Properties of Tailings [J] *Metal mine* 10 152-156.
[11] Zhou Q, Sun J, Li Y H, et al 2009 Comparison of physical and mechanical properties of several tailings [J] *Geotechnical Engineering World* 12 42-45.
[12] Yang Y, Wei Z, Chen Y, et al. 2017 Study on the morphology of tailings particles based on image processing technology. [J] *Chinese Journal of Rock Mechanics and Engineering* 36 3689-3695.
[13] Pan J, Liu X, Wang Y. 2014 Experimental Study on Steady State Strength Characteristics of Tailings with Different Grain Sizes [J] *Journal of Safety Science and Technology* 10 39-44.
[14] Zhu J, Liu F, Xiong Y. 2020 Experimental Research on Settlement Characteristics of Tailings with Different Gradations [J] *Mining Research and Development* 40 41-44.
[15] Yang Y, Xu J, Yang Y. 2013 Study on the effect of particle size on the shear strength of unsaturated tailings [J] *Chinese journal of underground space and engineering* 9 861-864.
[16] Ma D, Zhang M, Guo M. 2014 Effect of roasting on phase transformation, weight loss rate, average particle size and specific surface area of bauxite tailings [J] *Bulletin of the Chinese Ceramic Society* 33 2153-2157.
[17] Yin J, Lai A, Jiang P. 1980 Strength and deformation characteristics of tailings sand under high pressure [J] *Journal of Geotechnical Engineering* 2 1-10.
[18] Yang K, Lv S, Zhang Y. 2014 Experimental Study on the Strength Characteristics of Tailings in Tailings Dam [J] *Metal Mine* 2 166-170.
[19] Xin H, Wang Y. 1995 Dynamic Deformation and Strength Characteristics of Dashihe Tailings Cohesive Soil [J] *Journal of Hydraulic Engineering* 11 56-62.
[20] Chen J, Jing S, Zhang J, Sun X. 2006 Test Results and Analysis of Dynamic Strength Characteristics of Saturated Tailings Sand [J] *Journal of Hydraulic Engineering* 5 603-607.
[21] Sun C, Shao L, Guo X. 2015 Research on Dynamic Triaxial Test of Tailings Sand with Different Frequency [J] *Mining Research and Development* 35 69-73.
[22] Zhang C, Wang W, Cao G. 2016 Study on the influence of shear rate on the strength characteristics of tailings sand under different stresses [J] *Chinese journal of underground space and engineering* 12 463-469.

[23] An P, Dou M, Cao B, et al. 2019 Experimental Research on Shear Strength of Tailings Sand under the Action of Dry and Wet Cycle [J] *Mining Research and Development* 39 82-86.
[24] Cao G, Wang W, Chen P, et al. 2016 Experimental Study on Compression and Consolidation Coefficient of Copper Tailings [J] *Chinese Journal of Rare Metals Nonferrous metals* 6 63-67.
[25] Liang B, Wang X, Jiang L, et al. 2018 Experimental study of tailings compression and consolidation characteristics and e-P curve model analysis [J] *Journal of Disaster Prevention and Mitigation Engineering* 38 109-117.
[26] Wu S, Yang C, Hu X, et al. 2017 Research on the Correlation between the Properties of Tailings Particles and Compression and Consolidation [J] *Journal of Huazhong University of Science and Technology (Natural Science Edition)* 45 121-126.
[27] Xin H B et al. 1996 Analysis of seismic response of Dashihe tailing dam in 1976 [J] *Journal of rock mechanics and geotechnical engineering* 4 51-59.
[28] Psarropoulos P, Tsompanakis Y. 2008 Stability of tailings dams under static and seismic loading [J] *Canadian Geotechnical Journal* 45 663-675.
[29] Jin J, Cui H, Liang L, et al. 2018 Variation of Pore Water Pressure in Tailing Sand under Dynamic Loading [J] *Shock and Vibration*.
[30] Wan S, Wang K, Kong S, et al. 2017 Study on the Influence of Elevation of Tailing Dam on Stability [J] *Earth and Environmental Science* 100 012058.
[31] Jin J, Cui H, Liang L et al. 2018. Variation of pore water pressure in tailing sand under dynamic loading [J] *Shock and Vibration*
[32] Zhang C, Yang Ce, Xu W. 2004 Reliability analysis of the stability of tailings dam [J] *Rock and soil mechanics* 11 26-31.
[33] Hu M, Guo A, Chen S, 2004 Some thoughts on the anti-sliding stability analysis of an upstream tailing *dam* [J] *Rock and soil mechanics* 5 97-101.
[34] Wei Z, Wan L. 2006 Seismic stability analysis of fine-grained tailings accumulation dam [J] *Transactions of Nonferrous Metals Society of China* 1 79-81.
[35] Wang G, Duan X, Yang Y, et al. 2008 Research on the Influence of High Intensity Earthquakes on the Stability of Longdu Tailing Dam [J] *Journal of Kunming University of Science and Technology (Science and Technology)* 4 1-6.
[36] Wang B, Zhang L, Zhao Y, et al. 2010 Seismic stability analysis of tailing pond based on FLAC~(3D) [J] *Metal mine* 4 159-162.
[37] Liu Y, Liu E. 2020 Study on cyclically dynamic behaviour of tailing soil exposed to freeze-thaw cycles [J] *Cold Regions Science and Technology* 171 102984.
[38] Liu Y, Huang R, Liu E, et al. 2008 Mechanical behaviour and constitutive model of tailing soils subjected to freeze-thaw cycles [J] *European Journal of Environmental and Civil Engineering*.
[39] Ke X, Chen J, Shan Y. 2019 A new failure criterion for determining the cyclic resistance of low-plasticity fine-grained tailings [J] *Engineering Geology* 261 105273.
[40] Wei Z, Yang Y, Xu J, et al. 2016 Experimental study on the mechanical properties of frozen tailings by uniaxial compression tests [J] *Journal of North-eastern University (Natural Science)* 37 123.
[41] Wang Q, Qi J, Wang S, et al. 2020 Effect of freeze-thaw on freezing point of a saline loess [J] *Cold Regions Science and Technology* 170 102922.
[42] Li Z. 1984 Several understandings and suggestions on the utilization of mineral resources [J] *Geology and Exploration* 10 29-30.
[43] Gu S.1989 Comprehensive Utilization of Tailings in Anshan-type Iron Mine [J] *Metal Mine* 7 48-51.
[44] Li Z.1992 Development and Utilization of Tailings in my country's Metal Mines [J] *Geology and Exploration* 7 25-30.
[45] Ding Q, Yang Q, Wang J. 1994 Some achievements and directions of tailings utilization in my country [J] *Multipurpose Utilization of Mineral Resources* 6 28-34.

[46] Xu W, Wen X, Wei J, et al. 2018 Feasibility of kaolin tailing sand to be as an environmentally friendly alternative to river sand in construction applications [J] *Journal of Cleaner Production* 205 1114-1126.
[47] Ye Q, Liu J, Zhang Q. 2002 Comprehensive Utilization of Kaolin from Pyrite Tailings in South Sichuan Coal *Measures* [J] *Multipurpose Utilization of Mineral Resources* 1 41-45.
[48] Yang J. 2002 Comprehensive recovery and utilization of valuable mineral resources in tailings [J] *The Chinese Journal of Nonferrous Metals*.
[49] Wang Y, Luo S, Jiang M, et al. 2019 Process conditions for preparing geo-polymers from iron tailings [J] *Multipurpose Utilization of Mineral Resources* 5 121-126.
[50] Yang J, Mao Y, Chen X, et al. 2019 Research on the disposal and comprehensive utilization of a certain tailings resource [J] *Multipurpose Utilization of Mineral Resources* 6 117-122.
[51] Li N, Lv S, Wang W, et al. 2020 Experimental investigations on the mechanical behaviour of iron tailings powder with compound admixture of cement and nano-clay [J] *Construction and Building Materials* 254: 119259.
[52] Barati S, Tabatabaie S, Samani N, et al. 2020 Stabilization of iron ore tailings with cement and bentonite: a case study on Golgohar mine [J] *Bulletin of Engineering Geology and the Environment*.
[53] Huang Q, Liu T, Zhang J et al. 2020 Properties and pore-forming mechanism of silica sand tailing-steel slag-coal gangue based permeable ceramics [J] *Construction and Building Materials* 253 118870.
[54] Zhao J, Zheng X, Peng Y. 1996 Experimental study on copper tailings calcined cement [J] *Journal of Harbin University of Civil Engineering and Architecture* 6 75-78.
[55] Shi W, Zhang Y, Chen T, et al. 2012 Preparation of Belite Cement from Low Silicon Iron Tailings [J] *Metal Mine* 7 165-168.
[56] Liu Y, Sun J, Meng Q. 2013 Study on the Preparation of Magnesium Portland Cement from Magnesite Tailings [J] *Bulletin of the Chinese Ceramic Society* 6 1126-1130.
[57] Hosney M S, Rowe R K. 2017 Performance of polymer-enhanced bentonite–sand mixture for covering *arsenic*-rich gold mine tailings for up to 4 years [J] *Canadian Geotechnical Journal* 4 588-599.
[58] Luo H, Yan H. 2005 Application of Phosphate Mine Tailings in Road Base (Cushion) [J] *Highway* 5 146-148.
[59] Chou Y, Rowe R, Brachman R W I. 2018 Erosion of silty sand tailings through a geomembrane defect under filter incompatible conditions [J] *Canadian Geotechnical Journal* 11 1564-1576.
[60] Bing L, Zhong Z, Biao T, et al. 2018 Comprehensive utilization of iron tailings in China [J] *Earth and Environmental Science* 4 042055.

Water-glass Module on Mechanical Properties of Geopolymer Recycled Aggregate Concrete

Zhaoyang Ding[1], Qun Su[1], Hongguan Bian[2], Qing Wang[2] and Jinghai Zhou[1,*]

[1] Shenyang Jianzhu University, School of Civil Engineering, Shenyang, 110168, China

[2] Shenyang Jianzhu University, School of Materials Science and Engineering, Sh enyang, 110168, China

Email: zhoujinghai@sjzu.edu.cn

Abstract. Geopolymer recycled aggregate concrete (GRAC) was prepared by replacing cement with geopolymer and natural aggregate with wast concrete. The effect of water-glass modules on mechanical properties of GRAC was studied. It was found that there are tow kind of binding structures in geopolymer hydration product: C-A-S-H and N-A-S-H, they both contribute to the strength of GRAC. The value of size conversion coefficient of current national standard is inapplicable for GRAC, the calculation method of which is given in this paper. Elasticity modulus and peak stress of GRAC is proportional to water-glass modulus, and peak strain is inversely proportional and its constitutive equation was established.

Keywords. Geopolymer recycled aggregate concrete; water-glass modules; size conversion coefficient; stress-strain curve; constitutive equation.

1. Introduction

The concept of geopolymer was proposed by French materials scientist Joseph Davidovits in 1978 [1], raw materials of which are industrial residue, such as slag and fly ash. Van Deventer [2], John Provis [3], and Caijun Shi [4] make great contribution to the development of geopolymer, and the compressive strength of which can be proved to 90 MPa at present. GRAC basically does not consume natural resources, and also reuses industrial wastes. Researches have been conducted that water-glass module is the main factor affecting the mechanical properties of geopolymer [5, 6]. In this paper the effect of water-glass module on mechanical properties of GRAC, such as compressive strength, size effect, and stress-strain curve, is studied.

2. Experimental Program

2.1. Materials

Slag and fly ash were used to make geopolymer samples, the chemical composition is given in table 1. Water-glass was selected from Shandong Yousuo Chemical Technology Co., Ltd. The original modulus is 3.3, and the chemical composition of which is given in table 2. Coarse aggregate was waste concrete, and sand was natural medium-fine river sand.

Table 1. Chemical composition of slag and fly ash /wt. %.

	CaO	SiO_2	Al_2O_3	Fe_2O_3	MgO	Na_2O	K_2O	Ignition loss
Slag	43.10	32.26	14.69	2.06	6.19	-	-	0.97
Fly ash	5.51	48.54	28.35	6.37	2.42	3.01	3.90	0.96

Table 2. Chemical composition and property of water-glass /wt. %.

Chemical composition/w%			$n(SiO_2)/n(Na_2O)$	pH	$\rho/g\cdot cm^{-3}$
Na_2O	SiO_2	H_2O			
7.96	26.1	66	3.3	13.1	1.47

2.2. Mix Proportion Preparation Method

This experiment mainly studied the influence of water-glass module on the mechanical properties of GRAC. Water-glass module is 0.6~2.4, the mix proportion is shown in table 3. After the mixing, GRAC was cured in standard maintenance system for 28 days. Cube concrete specimens of $100mm^3$, $150mm^3$, and $200mm^3$ were prepared to study its size effect. The stress-strain curves of GRAC was test by sample size is 150mm×150mm×300mm, and the loading rate is 0.01mm/min. SEM and XRD are also used to test its morphology and crystal structure.

Table 3. Test grouping and mixing proportions /$kg\cdot m^{-3}$.

No.	Water-glass module	Slag	Fly-ash	Water-glass	Water	Sand	Recycled aggregate
WG06	0.6	315	135	54	126	681	1075
WG09	0.9	315	135	54	126	681	1075
WG12	1.2	315	135	54	126	681	1075
WG15	1.5	315	135	54	126	681	1075
WG18	1.8	315	135	54	126	681	1075
WG21	2.1	315	135	54	126	681	1075
WG24	2.4	315	135	54	126	681	1075

3. Results and Discussion

3.1. Structure of Water-Glass with Different Module

The internal structure of water-galss is shown in figure 1[7]. It can be seen that there is an electric double-layer, where zNa^+ is adsorbed in the dense layer, (y-z) OH^- is distributed in the diffusion layer. Different water-glass module is different thickness of its double-layer[7] According to the "alkali activation" theory[8] and "depolymerization condensation" theory[1], Ca, Si and Al ions in mineral raw materials are activated by OH^- and Na^+ in the solution, among which geopolymer monomers, such as $[SiO_4]$ and $[AlO_4]$, form by Si ions and Al ions. As the concentration is saturation, these monomers converge on the precursor and undergo polycondensation reaction. Ca ions will also react with free Si ions and free Al ions. Then, geopolymer gel is formed. The hydration product of the geopolymer contains two kinds of gels, N-A-S-H and C-A-S-H.[9] N-A-S-H mainly composed of silicon and aluminum with a three-dimensional network structure, C-A-S-H is a layered calcium silicate hydrate structure mainly formed by calcium and silicon[3], which is shown in figure 2.

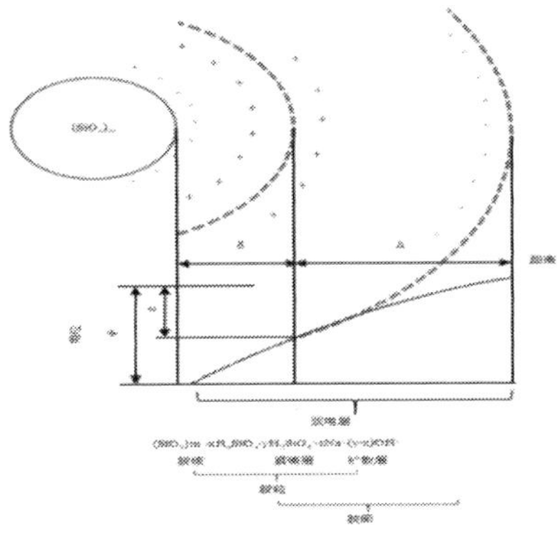

Figure 1. Double electron layer structure of water-glass [7].

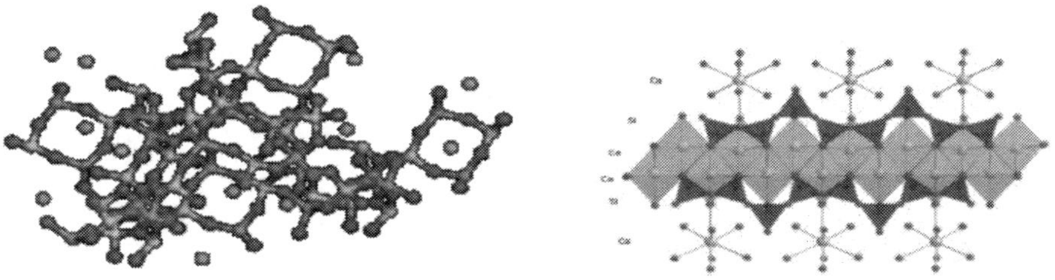

(a) N-A-S-H structure (b) C-A-S-H structure

Figure 2. Two main hydration products of geopolymer [3].

3.2. Water-glass Module on Compressive Strength of GRAC

The influence of water-glass module on the compressive strength of GRAC is shown in figure 3. As water-glass module is too high (n=2.4), the electric double-layer structure of water-glass is very thin, and there is not enough Na^{+} and OH^{-} to activate the mineral raw materials, so it does not condense and cannot form strength. However, aswater-glass module is 0.6, the polycondensation reaction occurs in advance, flashing occurs, the strength is extremely low or there is no strength. Moreover, as water-glass module is gradually reduced from 2.1 to 0.9, the compressive strength of the GRAC gradually increases. It is because that as water-glass module decrease, the concentration of hydroxide and sodium ions in the system gradually increases, so that more Ca, Si, and Al ions can be activated in the system, generating more zeolite and hydrate calcium silicate which lead to a tighter and complete structure of hydration products. Thus, the strength of GRAC gradually increases.

Figure 4 presents the XRD patterns of geopolymer cementing materials under different water-glass module conditions. The main hydration products of geopolymer are calcium silicate hydrate (Tobermorite 9A, PDF card number: 89-6458), a type containing calcium zeolite (Zeolite, PDF card number: 38-0232) and mulite (Mulite, PDF card number: 74-2419). It can be seen that the hydration products of geopolymer are layered structure (C-A-S-H) and net structure (N-A-S-H). As water-glass module changes, mullite and zeolite in the geopolymer change significantly, while the calcium silicate hydrate basically does not change, indicating that the strength of the geopolymer depends on the N-A-S-H gel content. And the N-A-S-H gel content increases with decreasing modules. The higher the N-A-S-H gel content, the more complete the geopolymer network structure and the higher the overall strength of the structure.

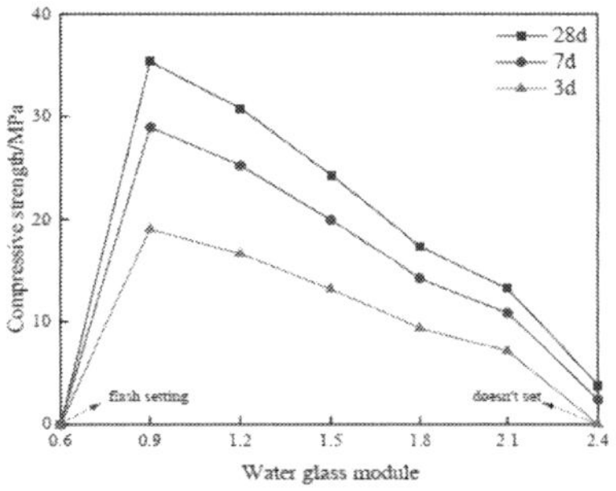

Figure 3. Compressive strength of GRAC on water-glass module.

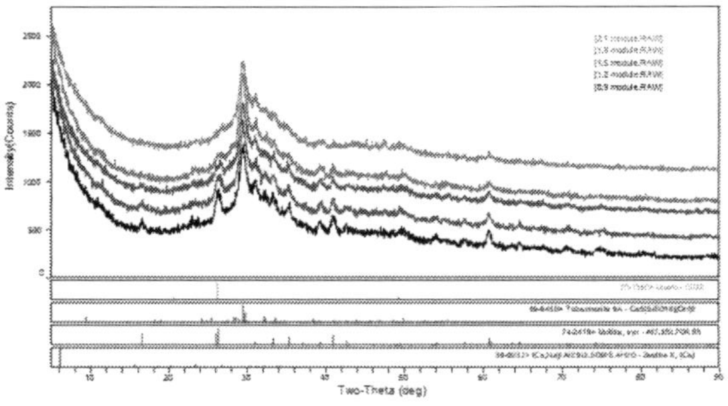

Figure 4. X-ray diffraction pattern of geopolymer on different water-glass module.

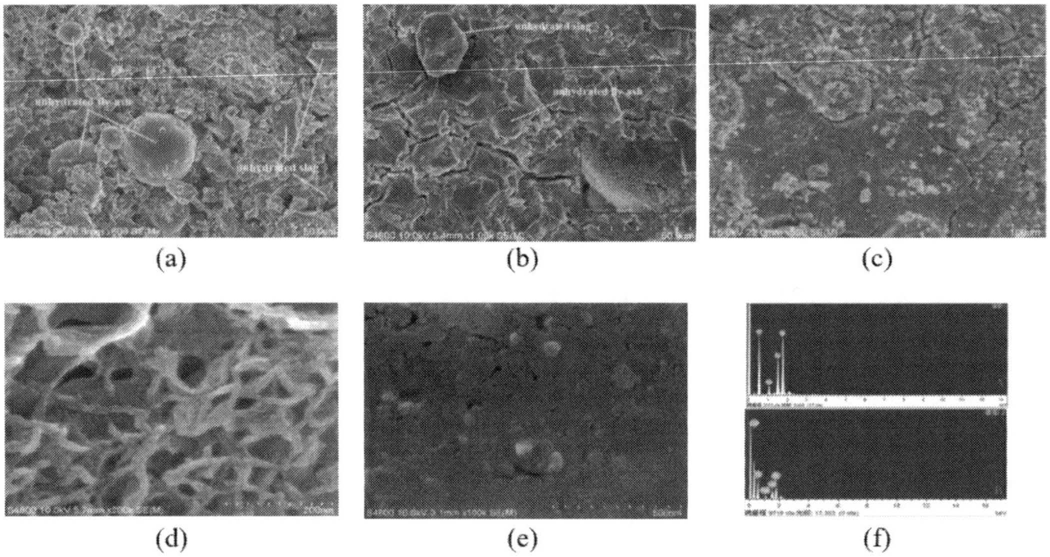

Figure 5. SEM micrographs of different water-glass modules of GRAC.

Figure 5 shows the micro-structure images of GRAC as water-glass module is 2.1, 1.5, and 0.9 respectively. It can be seen from figure 5(a), as a result of small diffusion layer of water-glass under high module (2.1), there are not enough free Na^+ and OH^- to activate the raw material, and the hydration product of geopolymer is very loose. As water-glass module is 1.5, shown in figure 5 (b), the structure is more denser, this is because there are more free Na^+ and OH^- caused by the thick electron layer of water-glass. As water-glass module is 0.9, shown in figure 5 (c), the structure becomes very dense. Point A and B in figure 5 (c) is two different gel structure of the geopolymer hydration product, which is shown in figure 5 (d) and figure 5 (e), and figure 5 (f) is the EDS of these two gel. It shows that point A and B is different both in morphology and elementary composition, which evidences the XRD and strength analysis.

3.3. Water-glass Module on the Size Conversion Coefficient

The compressive strength of GRAC with different side length cube specimens (100mm^3, 150mm^3, 200mm^3) is shown in figure 6. According to the current national standard "Standard for Test Methods for Mechanical Properties of Ordinary Concrete" GB/T50081, the standard specimen refers to a cube specimen with a side length of 150mm, and the size conversion coefficient (α) of other non-standard specimens is as shown in equation (1) and equation (2):

$$\alpha_{100} = f_{cu,150}/f_{cu,100} \tag{1}$$

$$\alpha_{200} = f_{cu,150}/f_{cu,200} \tag{2}$$

The influence of water-glass module on the size conversion coefficient of GRAC is given in figure 7. The national standard stipulates that α_{200} is 1.05, α_{100} is 0.95 (the horizontal dotted line). However, it should be noted that none of the data in figure 7 is between 0.95-1.05, indicating that the standard of ordinary cement concrete does not apply to the size conversion coefficient of GRAC. This may be attributed to the fact that geopolymer is different from cement, and the properties of recycled aggregates are also different from natural aggregates. It can be seen directly that when the water-glass module reach 0.9-1.5, the size conversion coefficient of the two sizes of specimens change relatively smoothly; while the water-glass module reach 1.5-2.1, the size conversion coefficient changes drastically. Therefore, this article uses a segmented method to calculate the size conversion coefficient of GRAC. The segment point selects the water-glass module (ε) =1.5. And linear fit the size conversion factor, the result is: $0.6 \leq \varepsilon < 1.5$, $\alpha_{200}=0.79063+0.23305\times\varepsilon$, $\alpha_{100}=1.0293+0.0737\times\varepsilon$; $1.5 \leq \varepsilon < 2.4$, $\alpha_{200}=0.845-0.0833\times\varepsilon$, $\alpha_{100}=1.22951-0.19296\times\varepsilon$.

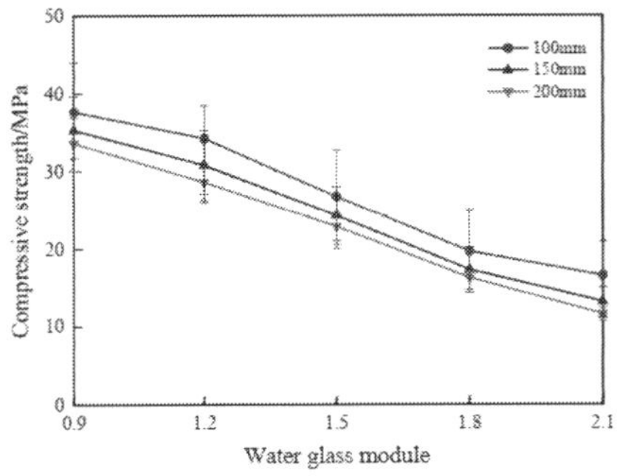

Figure 6. Compressive strength of GRAC on different sizes.

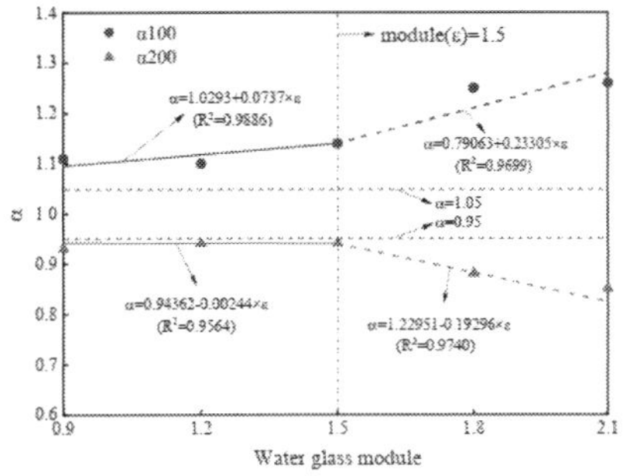

Figure 7. Size conversion coefficient on water-glass modules.

3.4. Stress-Strain Curves of GRAC

Figure 8 presents stress-strain curves of GRAC on different water-glass module. It can be observed that the stress-strain curve of GRAC has the same pattern to the ordinary cement concrete's, which can be divided into up-portion and down-portion. Figure 9 shows the influence of water-glass module on the elastic modulus, peak stress and corresponding peak strain of GRAC. It can be observed that the elastic modulus increases with the increasing of water-glass module. As water-glass module were reduced from 2.1 to 0.9, the elastic modulus increased. The reason for this observation is that the increase in the water-glass module will reduce the internal porosity of the geopolymer cementing material, forming a denser structure. The decrease in the water-glass module will also significantly increase the peak stress, and the trend is directly proportional to the water-glass module. As water-glass modules reduced to 0.9, the peak stress is increased to 1.38 times compared with the water-glass module is 2.1. The analysis of the compressive strength of GRAC shows that the water-glass module decrease, and the content of Na^+ and OH^- in the system increases, which activates freer $[SiO_4]$ and $[AlO_4]$. Hence, the formed skeleton structure is more complete, the higher the compressive strength, the higher the peak stress and the lower the peak strain.

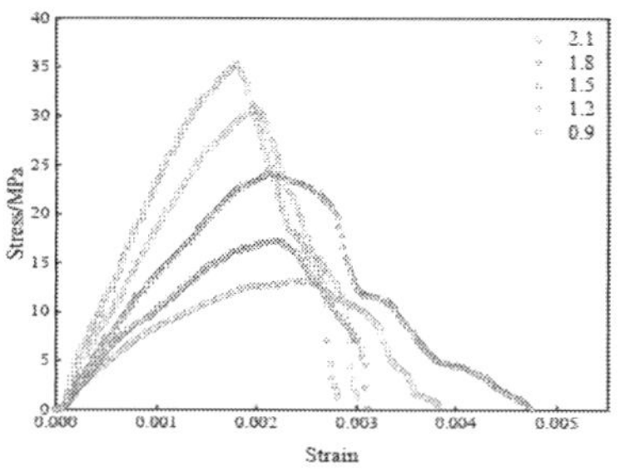

Figure 8. Stress-strain curves of GRAC on different water-glass module.

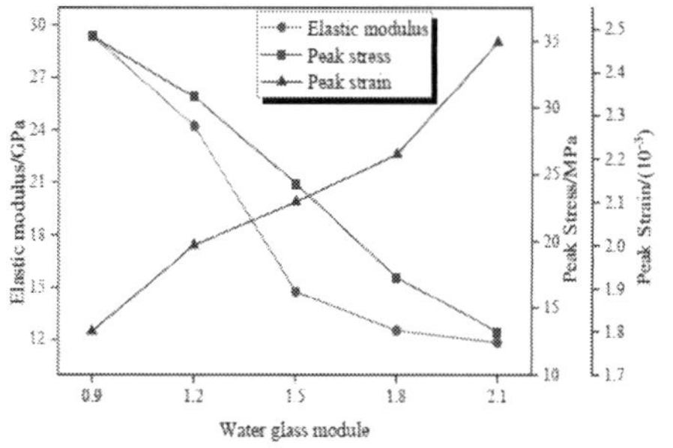

Figure 9. Elastic modulus, peak stress and peak strain of GRAC on different water-glass module.

3.5. Establishment of Constitutive Equation

The stress-strain curve of GRAC under uniaxial compression has similar geometric characteristics to the ordinary portland cement concrete (in Figure 8.). Thus, the stress-strain behavior model of Guo Zhenhai concrete uniaxial compression is adopted in this paper, as shown in equation (3).

$$y=\begin{cases} ax+(3-2a)x^2+(a-2)x^3 & 0\le x<1 \\ \dfrac{x}{b(x-1)^2+x} & x\ge 1 \end{cases} \tag{3}$$

where a is up-portion parameter,b is down-portion parameter.The parameter a and b are related to the water-glass module, calculated using the equations (4) and (5):

$$a=2.2026-0.7466m+0.1340m^2 \tag{4}$$

$$b=10.3277m-5.5359 \tag{5}$$

Combining equations (4), (5) and (6), and then introducing $x=\varepsilon/\varepsilon_{pr}$, $y=\sigma/f_{pr}$, constitutive equation of compressive strength of GRAC with water glass module as variable was obtained, as shown in equation (6):

$$\sigma=\begin{cases} f_{pr}\times[(2.02026-0.7466m+0.1340m^2)\times(\dfrac{\varepsilon}{\varepsilon_{pr}}-2\dfrac{\varepsilon^2}{\varepsilon_{pr}^2}+\dfrac{\varepsilon^3}{\varepsilon_{pr}^3})+3\dfrac{\varepsilon^2}{\varepsilon_{pr}^2}-2\dfrac{\varepsilon^3}{\varepsilon_{pr}^3}] & (\varepsilon\le\varepsilon_{pr}) \\ f_{pr}\times\dfrac{\varepsilon/\varepsilon_{pr}}{(10.3277m-5.5359)\times(\varepsilon/\varepsilon_{pr}-1)^2+\varepsilon/\varepsilon_{pr}} & (\varepsilon\ge\varepsilon_{pr}) \end{cases} \tag{6}$$

4. Conclusions

Water-glass has an electric double-layer structure, high module results in a thin layer and low module results in a thick layer, the thickness of the electric double-layer in water-glass dominates the alkali activate effect and the mechanical strength of GRAC. There are two different gel structure in geopolymer hydration product, C-A-S-H and N-A-S-H, they both contribute to the compressive strength of GRAC.

Size conversion coefficient of GRAC cannot adopt the value specified in the current national standard "Standard for Test Methods for Mechanical Properties of Ordinary Concrete" GB/T50081.The calculation method of size conversion coefficient of GRAC is given in this paper.

GRAC exhibits similar stress–strain behavior to ordinary portland cement concrete, the elastic modulus and peak stress of GRAC are inversely proportional to the water-glass module, the peak strain is proportional to the water-glass module. Moreover, the constitutive equation of the compressive strength of GRAC based on the water-glass module is given.

Acknowledgments
This research was funded by the National Natural Science Foundation of China, grant number 51678374; the Science and Technology Program of the Ministry of Housing and Urban-Rural Development of China, grant number 2019-K-48; State Key Laboratory of Silicate Materials for Architectures (Wuhan University of Technology), grant number SYSJJ2021-13.

References
[1] Davidovits J. 1991 Geopolymers:inorganic polymeric new materials [J] *Journal of Thermal Analysis* and *Calorimetry* **37** (8) 1633-56.
[2] Jsj V D, Alireza K, Rackel S N, Greg G. Q. 2014 Modelling the yield stress of ternary cement-slag-fly ash pastes based on particle size distribution [J] *Powder Technology* **266** (7) 203-9.
[3] Provis J L. 2014 Geopolymers and other alkali activated materials: why, how, and what? [J] *Materials and Structures* **47** (1-2) 11-25.
[4] Shi C, He F, Fernandez J A, et al. 2012 Classification and characteristics of alkali-activated cements [J] *Journal of The* Chinese *Ceramic Society* **040** (001) 69-75.
[5] Liu Y, Zhang Z, Shi C, et al. 2020 Influence of silica fume on performance of high-strength geopolymer [J] *Journal of The Chinese Ceramic Society* **48** (11) 1689-99.
[6] Shi C, Krivenko P V, Roy D M. 2006 Alkali-activated cements and concretes *M. Taylor & Francis, Abingdon*, UK.
[7] Zhu C, Lu C. 2001 The progress to recognize the hardening mechanism of waterglass [J] *Inorganic Chemicals* Industry **33** (1) 22-25.
[8] Purdon A O the action of alkalis on bast-furnace slag [J] *Soc Chem Ind*, **59** (1940) 191~202.
[9] Ding Z, Zhou J, Su Q, et al. 2021 Aggregate content on size effect of geopolymer recycled aggregate concrete [J] *Journal of Shenyang Jianzhu University (Natural Science)* **37** (04) 709-15.
[10] Zhang W, Ye J, Wang Y, et al. 2010 Study on low alkali content and high strength geopolymer [J] *Materials China* **29** (09) 13-18+33.
[11] Ding Q, Wu J, Lv L, et al. 2007 Study on new type geopolym er-based lightw eighthigh-tem perature concrete [J] *Concrete* (03) 1-3.
[12] Luo X, Xu J, Su H, et al. 2014 Strength of properties highly fluidized geopolymer concrete under impact loading [J] *Journal of Building Materials* **17** (01) 72-77.
[13] Li C, Wang L, Zhang T, et al. 2016 Preparation of geopolymeric materials by pisha sandston [J] *Journal of Building Materials* **19** (02) 373-8.
[14] Ding Z, Zhou J, Su Q, et al. 2021 Mechanicalproperties of geopolymer recycled aggregate concrete [J] *Journal of Shenyang Jianzhu University (Natural Science)* **37** (01) 138-146.
[15] Zhu C. 1998 Process of understanding the hardening mechanism of CO_2 sodium silicate sand [J] *Hot Working Technology* 000 (006) 46-47.

Soil Pressure and Deformation Analysis of Small Size Deep Foundation Pit

Zhao Long[1,2], Yilei Shi[1,2], Weili Li[1,2] and Shuaihua Ye[3]

[1] Gansu CSCEC Municipal Engineering Investigation and Design Institute Co. Ltd., Gansu Lanzhou 730000, China
[2] CSCEC AECOM CONSULTANTS CO., LTD, Gansu Lanzhou 730000, China;
[3] Lanzhou University of Technology, Gansu Lanzhou 730000, China.
Email: 1321781334@qq.com

Abstract. In this paper, the influence of space effect on soil pressure and deformation of deep foundation pit was considered, and the finite soil pressure calculation model was established. The soil pressure of deep foundation pit was calculated by assuming the slip surface and using the finite soil limit equilibrium theory. Then, PLAXIS 3D finite element software was used to establish finite element models of different plane sizes and depths. The distribution regulation of side wall soil pressure and deformation of deep foundation pit was calculated. Finally, the results of finite soil pressure calculation was compared with finite element method. The results shown that: The soil pressure of small deep foundation pit was affected by space effect, and the soil pressure and deformation decrease significantly along the foundation pit depth. Shear fracture Angle was related to the ratio of width to depth of foundation pit, and it was no longer a constant value of $45°+\varphi/2$. Therefore, the spatial effect should be considered in the calculation of soil pressure of small deep foundation pit. The research results can provide some guidance for the design and calculation of similar small size deep foundation pit.

1. Introduction

With the continuous acceleration of urbanization in China, a large number of small size deep foundation pit projects, such as pipe jacking shaft and subway shaft, have encountered in municipal engineering construction. The main characteristics of this kind of deep foundation pit had small excavation section size, large depth, complex surrounding environment. All these factors bring some difficulties to the excavation and support of foundation pit. In the design of foundation pit support, the distribution of earth pressure along the depth of foundation pit should analysis. Since the small size deep foundation pit was obviously affected by size effect, the calculation of earth pressure by traditional method was inevitably different from the actual earth pressure. Therefore, it had great engineering and practical significance to study the soil pressure calculation and deformation distribution regulation of small deep foundation pit.

At present, some scholars have studied the earth pressure distribution of caisson working well, and the calculation formula of earth pressure and deformation distribution regulation under different conditions had be proposed [1-2]. Through indoor model test and FLAC 3D numerical simulation, the spatial distribution of earth pressure behind pit wall and the influence of earth pressure Angle effect on internal force of retaining structure were studied [3]. The pit Angle effect and its main influencing factors of retaining structure deformation were studied by using FLAC-3D software, and a calculation method of retaining structure deformation considering pit Angle effect was proposed [4]. The influence of different width-depth ratio on foundation pit deformation and the control measures based on scale effect were analysed [5]. With the development of earth pressure research, different earth pressure calculation methods were developed according to actual engineering conditions [6-8]. Considering the friction force between the wall back and soil, the formula for calculating Rankine passive earth pressure of vertical retaining wall and inverting retaining wall was derived by using limit equilibrium theory [9]. The relation between friction Angle and relative displacement of rigid retaining wall in non-limit state was deduced, and the calculation formula of active earth pressure in non-limit state was obtained [10]. By establishing the relationship between the internal friction Angle, cohesion and displacement of finite viscous soil, the force analysis was carried out by horizontal stratification method [11]. Based on the theory of limit equilibrium and the assumption of plane sliding, the formula for calculating the active earth pressure of finite soil was obtained considering the cohesion of soil [12]. The active earth pressure of retaining wall adjacent to bedrock was studied by PLAXIS finite element software. The earth pressure of retaining wall under different dip angles of bedrock was calculated and analysed [13]. Based on the assumption that a circular arc earth arch was formed in the soil, a formula for calculating the active earth pressure strength of finite non-viscous fill was obtained by using the equilibrium condition of horizontal micro-element force [14]. The formula for calculating soil pressure in finite range was derived by limit analysis method. It was concluded that there was a certain difference between the finite earth pressure obtained by traditional Rankine earth pressure theory and finite analysis method [15].

In this paper, based on the influence of finite soil on sliding crack surface, a calculation model of soil pressure for small deep foundation pit was established. Based on the limit equilibrium theory and assuming that the sliding surface was a plane, the formula for calculating active earth pressure of small size deep foundation pit was derived. Then PLAXIS 3D finite element software was used to establish calculation models with different plane sizes and depths. Finite element calculation of soil pressure and deformation was carried out. The distribution of soil pressure along the depth and width was analyzed by comparing the results of finite element calculation with Rankine active earth pressure theory. The research results can provide a basis for the calculation of soil pressure and the optimization of retaining structure design of small size deep foundation pit.

2. Finite Soil Pressure Calculation

At present, the earth pressure theory of Rankine and Coulomb was widely used in geotechnical engineering. The application condition of the theory was that the soil behind the wall was assumed to be a semi-infinite space. Therefore, the above theory has a large error in the calculation of soil pressure of deep foundation pit with a small width to depth ratio. When there was an existing building basement or a small deep foundation pit near the excavation (as shown in figure 1), the shear fracture Angle θ of the sliding surface of the side wall of the excavation will no longer be $45°+\varphi/2$. In this case, it was obviously problematic to apply the classical Rankine or Coulomb earth pressure theory to calculate the soil pressure of the side wall of foundation pit.

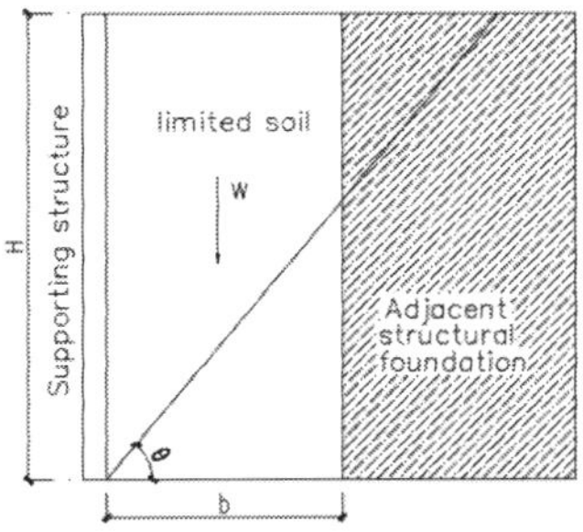

(a) Affected by existing buildings

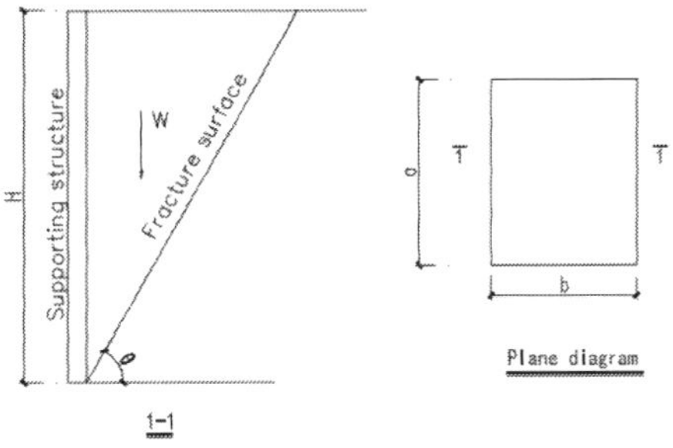

(b) Affected by spatial effects

Figure1. Diagram of slippage surface with limited soils.

2.1. Calculation and Analysis of Finite Soil Pressure in Foundation Pit

Foundation pit belongs to spatial stress system, and its stress and deformation have obvious spatial effect. The plane size, excavation depth, soil distribution and geometric shape of excavation have direct influence on the calculation and distribution of soil pressure on the side wall. According to the engineering experience, when the traditional method was used for the support design and calculate of pipe jacking Wells, the earth pressure value was too large, which leaded to the large section and reinforcement of the support structure. The monitoring data often deviate from the calculation. The results shown that the soil pressure and deformation regulation of foundation pit with different plane size and depth were obviously different due to the size effect of foundation pit. Therefore, this paper mainly consider the deep foundation pit with excavation depth greater than 1.0 times the excavation plane size, and calculates the magnitude and distribution of earth pressure along the depth direction of the foundation pit. The three-dimensional finite element calculation of soil pressure along the depth and width of foundation pit was carried out by using finite element mode.

2.2. The Basic Assumptions

In order to simplify the calculation, the following assumptions were made:
(1)Shear failure surface was plane;
(2)The influence of groundwater was not considered;
(3)Soil was considered as an ideal elastoplastic material and only a single layer of soil was considered. The friction between soil and support structure was not considered;
(4)The top load around the foundation pit was not considered;
(5)The soil behind the wall was horizontal and the back of the wall was vertical and smooth.

2.3. Establishment of Computational Model

For the small-size deep foundation pit as shown in Fig1 (b), the shear slip surface failure model was established as shown in figure 2. The slip Angle θ was assumed to be a variable affected by the plane size and excavation depth of the foundation pit. Then the sliding soil ABC was in the ultimate equilibrium state under the action of the support structure resistance E, the sliding weight force W and the force R of the lower soil on the sliding body. The relationship between the resistance of supporting structure and the earth pressure exerted by sliding body on the supporting structure was acting force and reaction force. R was related to the internal friction Angle φ of the soil. The Angle between the direction of the acting force R and the normal line of the slip surface was φ, and the magnitude was unknown.

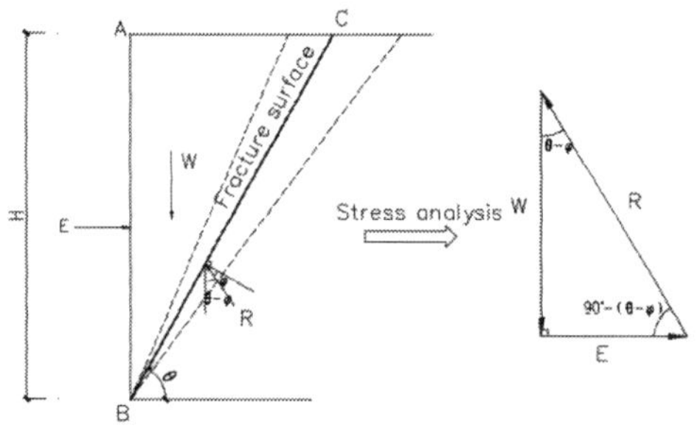

Figure 2. Diagram of soil pressure calculation.

2.4. Soil Pressure Calculation of Small Size Deep Foundation Pit

As shown in figure 2, the following formula can be obtained according to the limit equilibrium principle and the triangle sine theorem:

$$\frac{E}{\sin(\theta-\varphi)}=\frac{W}{\sin[90^\circ-(\theta-\varphi)]} \tag{1}$$

The calculation expression of earth pressure can be obtained from the above formula as follows:

$$E = \frac{W\sin(\theta-\varphi)}{\cos(\theta-\varphi)} = W\tan(\theta-\varphi) \tag{2}$$

Gravity W of the triangular sliding body ABC per meter of soil can be expressed as:

$$W = \gamma V = \gamma \cdot \frac{1}{2} H \cdot \frac{H}{\tan\theta} = \frac{\gamma H^2}{2\tan\theta} \tag{3}$$

Substitute (3) into Equation (2) to obtain:

$$\begin{aligned} E_a &= f(\theta,\varphi,\gamma,H) = \gamma V\tan(\theta-\varphi) \\ &= \frac{1}{2}\gamma H^2 \frac{\tan(\theta-\varphi)}{\tan\theta} \end{aligned} \tag{4}$$

In the above types, γ was the weight of soil, and φ was the internal friction Angle of soil. H was the excavation depth of the foundation pit, and θ was the Angle between the sliding crack surface and the horizontal surface.

According to Equation (4), active earth pressure *Ea* was related to shear fracture Angle θ, foundation pit depth and soil shear strength parameters. To calculate earth pressure using limit equilibrium theory, the rupture Angle should be determined first. In this paper, the influence of space effect on soil pressure of foundation pit was considered, and the failure Angle was assumed to be a variable affected by the ratio of width to depth of foundation pit, rather than the traditional 45°+φ/2. Therefore, the earth pressure calculation of small size deep foundation pit affected by space effect can be optimized according to the following finite element calculation results and formula (4) to make the calculation results more reasonable.

When the pit depth H was constant and the soil parameters was known, the active earth pressure E was only the function of the fracture Angle θ. According to the limit equilibrium principle, when dE/dθ=0, the extreme value of the active earth pressure E can be obtained. The calculated result was the active earth pressure under this condition. By substituting it into the above equation, the corresponding shear fracture Angle θ value can also be obtained. The derivative of Equation (4) can be obtained as follows:

$$\frac{df}{d\theta} = \gamma H^2 \frac{\sec^2(\theta-\varphi)\tan\theta - \tan(\theta-\varphi)\sec^2\theta}{2\tan^2\theta} \tag{5}$$

If dE/dθ=0, the expression of θ can be obtained as follows:

$$\theta' = f(\varphi,\gamma,\mathrm{H}) \tag{6}$$

In order to explain the rationality of the above calculation process, a special case was considered. When the plane size of foundation pit was large and the size effect was not taken into account. Rankine or Coulomb active earth-moving pressure theory was satisfied according to fracture Angle θ. When θ=45°+φ/2, it can be obtained by substituting it into Equation (4):

$$E_a = \frac{1}{2}\gamma H^2 \frac{\tan(45°+\varphi/2-\varphi)}{\tan(45°+\varphi/2)} = \frac{1}{2}\gamma H^2 \tan(45°-\varphi/2)\cdot\cot(45°+\varphi/2) = \frac{1}{2}\gamma H^2 \tan^2(45°-\varphi/2)$$

Put the active earth pressure coefficient *Ka* into the equation above, and get: $E_a = \frac{1}{2}\gamma H^2 K_a$

The above formula was the calculation formula of active earth pressure resultant force obtained by Rankine earth pressure theory. Therefore, it was reasonable to use the above formula to calculate earth pressure of small size deep foundation pit affected by size effect.

The above calculation process was two dimensional plane calculation with fracture Angle θ as the variable. The theoretical calculation results cannot reflect the influence of foundation pit width on earth pressure. Therefore, it was necessary to combine three-dimensional calculation or model test to calculate the soil pressure of small deep foundation pit. The model test was time-consuming and labor-intensive, so the model test should be carried out on the basis of certain research. Therefore, this paper mainly calculated the earth pressure and deformation of small size deep foundation pit by establishing a three-dimensional finite element model, and the change distribution regulation was obtained.

3. Finite Element Analysis Model

Combined with engineering experience and existing studies, the spatial effect had a significant impact on the soil pressure of foundation pit. In order to correspond to the above calculation in this paper, firstly, considering the sand layer of non-cohesive soil, foundation pit models with different width-depth ratios was established by using PLAXIS 3D finite element software. The lateral earth pressure and lateral movement of foundation pit were calculated. According to the calculated results, the relationship between the soil pressure and the ratio of width to depth of foundation pit was analyzed. By comparing the finite element calculation results with the classical Rankine earth pressure calculation results, the influence of foundation pit size effect on the size and distribution of soil pressure of small deep foundation pit was obtained.

3.1. Soil Pressure Calculation of Small Size Deep Foundation Pit

The constitutive model of soil had great influence on the calculation of soil pressure. This paper mainly carried out the simulation calculation of the excavation problem of foundation pit engineering, so the soil hardening model was adopted in the finite element calculation of soil constitutive relation. The model can consider the change of soil stiffness during initial loading and unloading and reloading, so it was suitable for the simulation calculation of foundation pit excavation. The main soil parameters were shown in table 1.

Table 1. Table of soil parameters.

soil	γ(kN/m^3)	E_a(MPa)	c(kPa)	φ(°)
sand	16.8	40	2	35

3.2. Soil Pressure Calculation of Small Size Deep Foundation Pit

Foundation pit engineering was a temporary project, so the finite element stress calculation was set as the condition of no drainage. In order to eliminate the influence of model size on calculation results, the model plane size and model height were 3 times of the excavation depth. Finite element model of small size deep foundation pit was established as shown in figure 3.

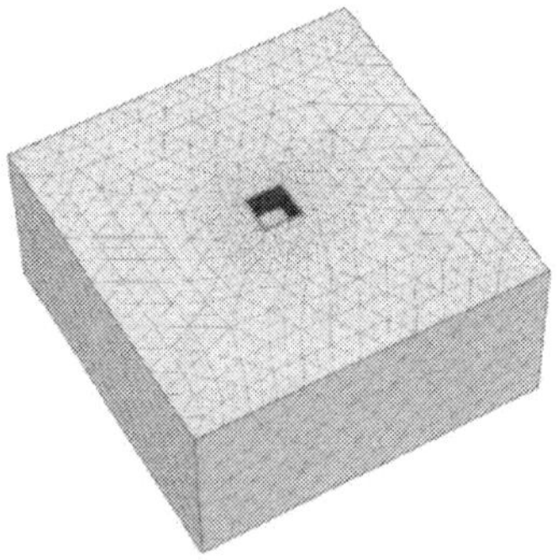

Figure 3. Finite element calculation model of foundation pit.

In order to analysis and explain the influence of size effect on soil pressure and lateral movement of deep foundation pit, six three-dimensional finite element models with different plane sizes and depths were established in this paper, and the calculated results were compared and analyzed. The finite element models size were as follows: (1) Plane dimension: 8m×8m, 12m×12m, 50m×50m; (2) Depth of foundation pit: 8m, 12m.

3.3. Parameter Setting of Support Structure

In the calculation models, the supporting structure was simulated by 6-node plate element converted according to the equivalent stiffness. The soil-structure interaction was simulated by 12-node interface element. The interface strength was connected with the soil strength by setting the interface reduction coefficient. Based on previous experience, the interface reduction coefficient was set as 0.70. By setting reduction coefficient, both strength and stiffness were reduced by default in finite element calculation. Then, the interface cohesion and internal friction Angle were calculated according to Equation (7).

$$c_i = 0.70c_{soil}; \quad \tan\varphi_i = 0.70\tan\varphi_{soil} \tag{7}$$

Parameter setting of plate element in the model was calculated as follows:

$$D = 0.838d\sqrt{1/(1+t/d)} = 0.838\times0.8\sqrt{1/1+1.2/0.8} = 0.424\text{m}$$

$$EA = E_c D = 1.5\times10^7\times0.424 = 6.36\times10^6\text{kN/m}$$

$$EI = \frac{1}{12}E_c D^3 = 9.53\times10^4\text{kN}\cdot\text{m}^2/\text{m}$$

4. Analysis of Finite Element Calculation Results

4.1. Spatial Distribution of Horizontal Lateral Movement of Foundation Pit

The three-dimensional finite element models of different sizes were respectively calculated. The horizontal lateral distribution of the calculation results with the plane size of 12m×12m and depth of 12m at the same depth was shown in figure 4. The shadow width in the horizontal lateral movement cloud image only represented the relative lateral movement at the same depth. The distribution rule of horizontal lateral movement along the depth of foundation pit with different plane dimensions and depth of 12m shown in figure 5.

As shown in figure 4, the horizontal displacement of the side wall along the width direction of the foundation pit at different excavation depths roughly presented a broken line distribution. According to the calculation results, the maximum displacements at sections of different pit depths were as follows: 31 mm, 43 mm, 49 mm, 66 mm, 69 mm, 62 mm. In the same foundation pit depth plane, the maximum displacement roughly appeared in the middle of each edge, and the minimum displacement appeared in the pit Angle. It shown that the horizontal lateral movement of foundation pit was obviously influenced by the size effect of foundation pit and adjacent supporting structure. According to the comparison results of horizontal lateral displacement of foundation pits of different plane sizes in this paper, it was also found that the larger the width to depth ratio of foundation pits, the more obvious the spatial effect.

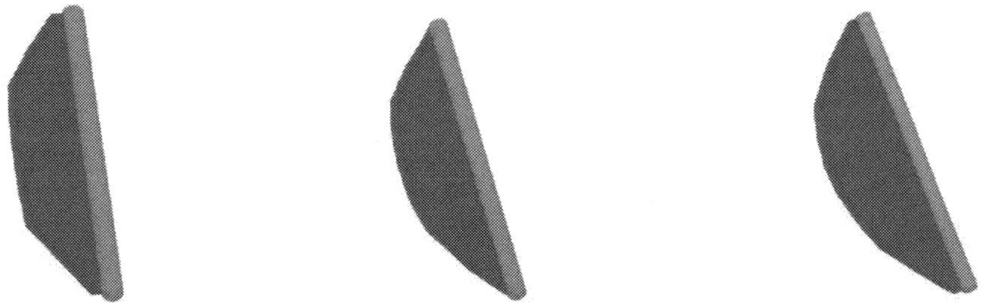

(a)z=-12 (m) (b)z=-10 (m) (c)z=-8 (m)

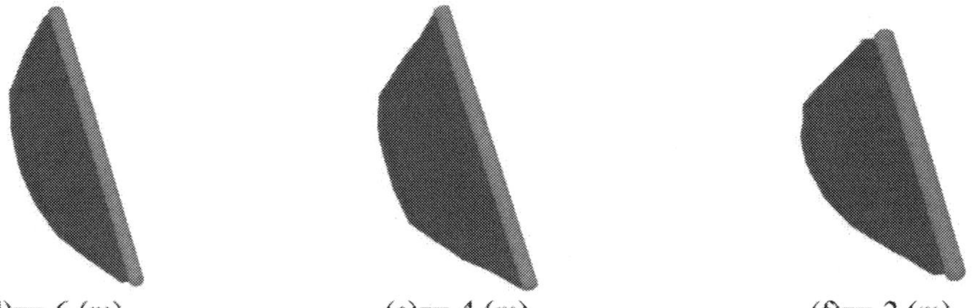

(d)z=-6 (m) (e)z=-4 (m) (f)z=-2 (m)

Figure 4. Horizontal lateral diagrams of sections at different depths.

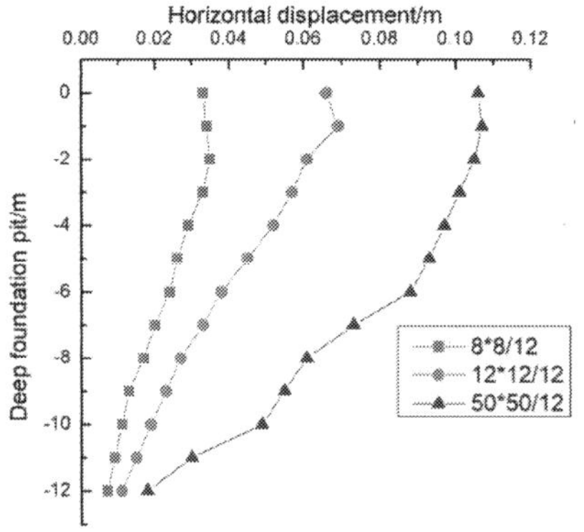

Figure 5. Horizontal displacement diagram of foundation pit with different plane dimensions.

As shown in figure 5, along the direction of foundation pit depth, the horizontal lateral movement along the foundation pit depth presents a gradually decreasing distribution regulation. There was obvious difference in displacement of foundation pit with different plane dimensions. The maximum displacements all appeared at approximately the top of the pit, and the variation regulation was consistent with the deformation law of the cantilever rigid support structure, which also indicated the rationality of the calculated results. The maximum displacements of the three models in figure 5 were 0.035m, 0.069m and 0.107m, respectively. The comparison results also shown that the displacement of foundation pit with the same depth decreased obviously with the decrease of plane size. The main reason was that under the influence of foundation pit size effect, the deformation of adjacent side wall of small deep foundation pit had a certain mutual restraint. The distribution of earth pressure was obviously different from traditional semi-infinite space calculation. Finally, the horizontal displacement of foundation pit decreased obviously. It shown that the size effect of foundation pit had a significant effect on deformation

When the plane size of the foundation pit was smaller than the pit depth, although the plane size decreased slightly, the horizontal lateral movement had an obvious downward trend. This shown that the size effect was quite obvious, which was beneficial to the stability of the foundation pit. In the support design of municipal small size deep foundation pit, the displacement can be appropriately reduced when the traditional method was used to calculate.

4.2. Spatial Distribution of Horizontal Lateral Movement of Foundation Pit

Neither the classical earth pressure theory nor the two-dimensional finite element calculation can consider the spatial effect of foundation pit on the foundation pit deformation and earth pressure distribution. In order to analysis and explain the size effect of small size deep foundation pit deformation and earth pressure distribution, PLAXIS 3D, PLAXIS 2D and classic Rankine active

earth pressure theory were respectively used to calculate the earth pressure on the side wall of foundation pit. The calculation results were shown in table 2.
Rankine active earth pressure can be calculated as follows:

$$e_a = \gamma H K_a = \gamma H \tan^2(45 - \varphi / 2) \tag{8}$$

Table 2. Calculation results of normal effective stress of base.

Plane dimension /(m)	Depth /(m)	PLAXIS 3D /(kN/m²)	PLAXIS 2D /(kN/m²)	Rankine Theory /(kN/m²)
8×8	8	49.42	61.69	58.88
12×12	8	50.90	61.69	58.88
50×50	8	60.78	61.69	58.88
8×8	12	72.40	90.36	88.30
12×12	12	75.50	90.36	88.30
50×50	12	97.20	90.36	88.30

As seen from the calculation results in table 2, when the plane size of foundation pit was close to or smaller than the excavation depth, the finite element calculation results of the normal effective stress of the foundation were significantly smaller than the theoretical calculation value of Rankine earth pressure. When the plane size was much larger than the excavation depth, the finite element calculation results were close to those of Rankine earth pressure. The results indicated that due to the size effect, the surrounding soil deformation caused by the small size deep excavation was small and the influence range was reduced, which lead to the change of the potential failure angle and the reduction of side wall earth pressure. It was also proved that the calculation of soil pressure of small deep foundation pit was considered as finite soil.

4.3. Spatial Distribution of Horizontal Lateral Movement of Foundation Pit

In order to analysis the influence of size effect of foundation pit with different plane size on the stress of supporting structure, different foundation pit models were calculated respectively. According to the calculation results, the cloud diagram of the variation regulation of the normal force of the supporting structure along the horizontal section at a depth of 8m in different models was shown in figure 6.

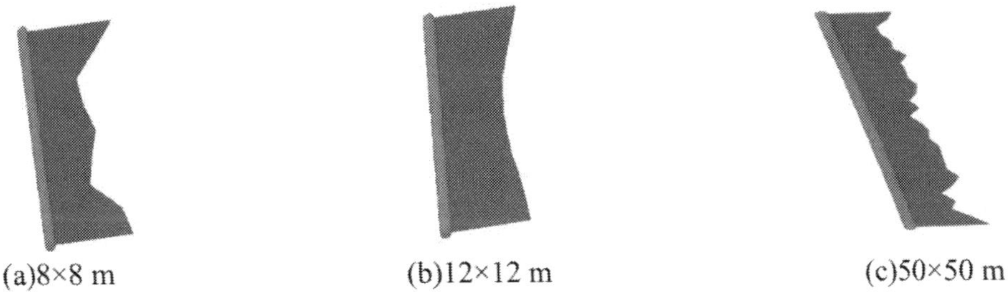

(a)8×8 m (b)12×12 m (c)50×50 m

Figure 6. Normal force diagram of support structure.

As shown in figure 6, the spatial effect of foundation pit had an obvious influence on the stress situation and distribution of supporting structure. The maximum normal force of the three model support structures at 8m pit depth was 100.9kN /m, 89.42kN /m and 81.78kN /m, respectively. The distribution regulation of normal stress on supporting structure was large pit Angle and small in the middle. The maximum normal stress of supporting structure appeared in the corner of foundation pit, and the minimum stress appeared in the middle of foundation pit. The main reason was that: affected by the spatial effect, the pit conner position was like adding a slanting brace inside the supporting structure, which inhibited the deformation of the supporting structure. In the middle position, the restraining effect weaken, the displacement of supporting structure increased, and the active earth pressure decreases. This phenomenon was also consistent with the above horizontal lateral displacement results, indicating the rationality of the calculation results.

5. Conclusions

In this paper, the earth pressure of small deep foundation pit was solved by considering the influence of space effect on the slip Angle of foundation pit. Three finite element models of foundation pit with different plane dimensions were established to calculate the soil pressure and deformation. The influence of space effect on soil pressure and deformation of foundation pit was compared and analyzed. The main conclusions were as follows.

(1) The soil pressure of small deep foundation pit was affected by the space effect, and the shear fracture angle cannot be calculated as $45°+\varphi/2$. It was affected by the depth and plane size of the foundation pit.

Along the horizontal direction, the maximum horizontal lateral movement occurred in the middle of each side, and the minimum horizontal lateral movement occurred in the pit Angle. It shown that the larger the ratio of depth to width, the more obvious of spatial effect.

(3) Along the direction of foundation pit depth, the horizontal lateral movement presented a gradually decreasing distribution regulation. The calculation results of different models were quite different. It shown that the deformation of adjacent side wall of small deep foundation pit was mutually restricted by the spatial effect of foundation pit. The spatial effect of foundation pit had a direct and significant influence on the earth pressure and deformation.

(4) The size and distribution of soil pressure on the side wall of foundation pit were closely related to the plane size of foundation pit. The actual earth pressure of small size deep foundation pit was less than that of Rankine active earth pressure. The two dimensional finite element calculation of foundation pit earth pressure was close to that of Rankine theory. It shown that the small size deep foundation pit can be calculated according to the theory of finite soil pressure.

References

[1] Ying H W, Liang W P. 2018 Spatial Active Earth Pressure Against Translating Retaining Walls in Square Shaft Excavations [J] *Journal of Shanghai Jiao Tong university* **11** (52) 1459-1566.
[2] Wei G, Xu R Q, Gong C. 2004 Research on earth counterforce calculation for circular caisson working well during pipe jacking project [J] *Journal of Zhejiang University (Engineering Science)* **11** (38) 1474-1478.
[3] Jia M C, Yang X H, Ye J Z. 2016 Corner effect of active earth pressure for small-sized excavation [J] *Journal of harbin institute of technology* **11** (48) 95-102.
[4] Wang X W, Tong H W. 2011 Deformation Calculation of Supporting Structure Considering Deep Excavation Pit Corner [J] *Chinese Journal of Underground Space and Engineering* **3** (7) 479-484.
[5] Wang H, Jia M C, Yang X H. 2016 Influence of size effect on deformation of foundation pit and control measures [J] *Sichuan Building Science* **1** (42) 69-74.
[6] Xie M X, Zheng J J, Cao W Z. 2019 Study of active earth pressure against embankment retaining wall of limited backfill [J] *J Huazhong Univ of Sci & Tech (Natural Science Edition)* **2** (47) 1-6.
[7] Xie T, Luo Q, Zhang L. 2018 Relation of Wall Displacement earth Pressure from Active to Passive State [J] *China Journal of Highway and Transport* **2** (31) 181-190.
[8] Wang D J, Tang H M, Wu Q. 2016 Method for Calculating Active Earth Pressure of Unsaturated Soil Considering Arching Effect [J] *Journal of Yangtze River Scientific Research Institute* **8** (33) 69-74.
[9] Zhu Y P, Wei P Y, Ma X R. 2019 Exploration of Rankine Passive Earth Pressure Based on Sliding-cracking Surface of the Soil Behind the Wall [J] *Science Technology and Engineering* **8** (19) 225-230.

[10] Huang B, Yang H, He X M. 2007 Research on Active Earth Pressure under Non-limit State [J] *Journal of Yangtze River Scientific Research Institute* **4** 46-49.
[11] Zhao X, Ma S Z, Tang W M. 2017 A Computation Method for Earth Pressure of Limited Soils under Non-limit State [J] *Journal of Yangtze River Scientific Research Institute* **12** (34) 89-93.
[12] Ma P, Qin S Q, Qian H T 2008 Calculation of active earth pressure for limited soils [J] *Chinese Journal of Rock Mechanics and Engineering* **27**(S1) 3070-3074.
[13] Fan C C, Fang Y S 2010 Numerical solution of active earth pressures on rigid retaining walls built near rock faces [J] *Computer and Geotechnics* **6** (37) 1023-1029.
[14] Liu Z Y 2018 Active Earth Pressure Calculation of Rigid Retaining Walls with Limited Granular Backfill Space [J] *China journal of Highway and transport* **2** (31) 154-164.
[15] Gao Y L 2001 Calculation of finite earth pressure by limit analysis [J] *Building Structures* **8** (31) 66-68.

Experimental Study on Mechanical Properties of Repair Materials for Underwater Rapid Construction

Shangchuan Zhao, Longlong Liu*, Xindai Zuo and Shaopeng Wang

Bridge Technology Research Center, Research Institute of Highway Ministry of Transport, Beijing 100088, China
Email: liulong4517@126.com

Abstract. Several strength grades of sulphoaluminate cement concrete were designed, and the mechanical properties with different mix ratios were studied. This paper mainly analyzes the mechanical properties of sulphoaluminate cement concrete with the setting time test, compressive strength test and flexural test. The test results show that the setting time of cement concrete can be controlled by mixing different admixtures. One hand, the initial setting time increases from 34 min to 340 min, and the final setting time increases from 57 min to 580 min when the incorporated borax content changes from 0 to 1.0 %. Other hand, the initial setting time decreased from 34 min to 11 min, and the final setting time increased from 57 min to 18 min, when the incorporated borax content changed from 0 to 0.5 %. Furthermore, the compressive strength can reach 40 MPa, and the flexural strength can reach more than 2.5MPa after 6 hours of curing. The experiment results illustrate that the setting time and the mechanical properties satisfy the needs of the rapid construction requirements under normal temperature conditions.

1. Introduction

The number and scale of hydraulic structures in China have caught up the world level. What's more, the Three Gorges Dam, Baihetan Hydropower Station and the South-to-North Water Diversion Project under construction have become the largest water conveyance and diversion projects in the world. However, natural disasters occur frequently in our country. Landslides and breaches caused by flood disasters, and long-term use often destroy the underwater parts of deep-water buildings, most of them begin to leak. The leakage of underwater concrete structure is mainly divided into point leakage, linear leakage and flow leakage. Point leakage is divided into one or more points, most of which are multi-point leakage. Most leakage points are dispersed and the range is relatively large. The flow leakages usually appear in the contact area between the water conveyance building and the adjacent building. Due to the long-term washing and erosion, leakage occurs in the contact area, and leakage channels are formed along the edge of the water conveyance building. Linear leakage can be divided into leakage of disease cracks, contact and expansion joints. The crack-leakage is mainly caused by penetrating cracks inside the concrete structure, while the contact joint and expansion joint leakage is mainly caused by the failure of waterproof structure used in the construction process of water conveyance buildings. The conventional plugging methods include direct plugging method, lower pipe plugging method, wood wedge plugging method and grouting plugging method. Different plugging methods are used according to different construction environments and conditions. The direct plugging method is mainly aimed at the leakage point with water pressure less than 1 m. The lower pipe plugging method is mainly aimed at the leakage channel corresponding to the water head of 1 m - 4 m, when the wood wedge plugging method is mainly aimed at the leakage channel with depth greater than 4 meters. The grouting method can be used to seal holes with large water pressure, large holes and large water leakage. The leakage passages of underwater concrete structures in deep-water buildings are often located in the deep underwater position, especially for the leakage passages at 300 m water depth. Under the conditions of low temperature and high pressure, the grouting materials should meet the requirements of rapid hardening at low temperature, rapid development of strength and micro expansion.

Therefore, it is of great significance to develop an early-strength, high-strength and micro-expansion material for the repairment of underwater leakage channel at low temperature conditions. To overcome technical difficulties for the micro-expansion grouting materials, which can rapidly harden under low temperature conditions of underwater engineering, the project team mainly studies the theory and practice of grouting materials. The setting time, early strength, high strength and micro-expansion of grouting materials are studied, because these indexes are important to repair leakage points under low temperature conditions [1]. Usually, different types of leakage channel grouting materials have different characteristics. Ordinary Portland cement-based grouting materials are suitable for general ground engineering due to their poor water resistance and chemical corrosion resistance. Thanks to the short setting time, Sulphoaluminate cement grouting material are often used to repair the emergency construction, shotcrete anchor support, slurry anchor node, cementing plugging, winter construction application. What's more, Sulphoaluminate cement grouting material can meet the requirements of fast hardening and early strength under the condition of low temperature [2]. Compared with the ordinary Portland cement-based grouting materials, sulphoaluminate cement-based grouting materials can be rapidly constructed at low temperature. However, the setting time is so fast that the sulphoaluminate cement concrete have been hardened before the construction is finished. Due to this reason, the sulphoaluminate cement can't be directly used to construct. Otherwise, the only sulphoaluminate cement can't achieve the effect of rapid construction. Instead, it blocks equipment and repair pipelines, affecting the construction period. In order to control the setting time of sulphoaluminate cement, the heating of sulphoaluminate cement should be reduced [3]. Meanwhile, the indexes of early strength, high strength, no shrinkage, and micro expansion should meet the requirements. Most researchers mixed silicate cement and sulphoaluminate cement to prepare the cement-based grouting material of silicate cement and sulphoaluminate cement. Similar preparation methods also include the cement-based grouting material composed of silicate cement and aluminate cement, but the performance is not stable when two or more types of cement are mixed, and the production processes are complex and the cost is high, which is not the best choice [4].

Firstly, the setting time of the sulphoaluminate cement should satisfy the needs of construction. Simultaneously, the subsequent strength of grouting material can increase stably. Furthermore, the construction technology of grouting material should be improved for leakage passage of underwater concrete structure. In this paper, early strength and high strength of the sulphoaluminate cement grouting material is developed. The setting time of cement clinker is reasonably controlled by the retarder and accelerator, and the preparation technology of early strength grouting material at normal temperature is developed.

2. Materials and Test Methods

2.1. Materials

Sulphoaluminate cement is produced by Beijing Arctic Bear New Building Materials Co., Ltd, and its strength grade is 42.5. The additives used in the experiment are sodium silicate nanahydrate (molecular formula: $Na_2SiO_3·9H_2O$), aluminium sulfate ($Al_2(SO_4)_3·18H_2O$), Sodium borate ($NaB_4O_7·10H_2O$), gluconic acid sodium salt ($C_6H_{11}NaO_7$), see tables from 1 to 4 for details.

$Na_2SiO_3·9H_2O$ is colorless orthogonal bipyramidal crystal or white to gray-white block or powder. At 100 °C, 6 molecules of crystal water were lost, as shown in table 1. The content of sodium oxide (Na_2O) is 19.3 ~ 22.8 %, and the ratio of sodium oxide and silica content are 1.03 and 0.03. It should be sealed in a cool place to save.

Table 1. Technical conditions for $Na_2SiO_3 \cdot 9H_2O$ (Maximum impurity content, %).

Chloride (cl)	Sulfate (SO_4)	Carbonate	Aluminium (Al)	Iron (Fe)	Heavy metal (Pb)
0.01	0.01	eligible	0.05	0.05	0.01

The aluminium sulfate (Al2 $(SO_4)_3 \cdot 18H_2O$), the maximum impurity content isn't more than 1 %, and the result of pH is 2.7(≥2.5, 25g/L, 25°C), see table 2.

Table 2. Technical conditions for $Al_2(SO_4)_3 \cdot 18H_2O$ (Maximum impurity content, %).

Water insoluble	Chloride	Ammonium (NH_4)	Na	Mg	K	Ca	Fe	Heavy metal
0.02	0.002	0.005	0.02	0.002	0.005	0.01	0.003	0.001

Sodium borate ($NaB_4O_7 \cdot 10H_2O$), also known as sodium tetraborate, its properties are colorless transparent crystalline powder and it can be weathering in dry air. See table 3.

Table 3. Technical conditions for $NaB_4O_7 \cdot 10H_2O$ (Maximum impurity content, %).

HCL insoluble	Chloride	SO_4	PO_4	Ca	Fe	Cu	Pb
0.005	0.002	0.01	0.002	0.005	0.0003	0.001	0.001

Gluconic acid sodium salt ($C_6H_{11}NaO_7$), the maximum impurity content can't be more than 8 %, see table 4.

Table 4. Technical conditions for $C_6H_{11}NaO_7$ (Maximum impurity content, %).

Chloride	Sulfate (SO_4)	Heavy metal (PO_4)
<0.005	<0.02	<0.002

2.2. Test Methods

2.2.1. Setting Time. Admixtures used in the test include borax, aluminum sulfate, sodium silicate and lithium carbonate. The performance of sulfate aluminate cement was investigated when the content of admixtures changes from 0% to 0.8%. According to GB/T 1346-2001, the sand and binder ratio is 1:3, and the water consumption is determined by JC 9333-2003. Based on the water-cement ratio of 0.47, the fluidity of cement-based cement is regulated to the range of 165-175mm, and the residual test methods is carried out according to GB/T 17671-1999.

2.2.2. Compressive Test. In this experiment, a total of five groups of new rapid hardening retarding concrete were prepared, and the effects of water-cement ratio and water reducer content on rapid hardening concrete were studied.

2.2.3. Bending Test. Dimensions of 100mm×100mm×400mm concrete specimen were used in the bending test. The concrete specimens should be cured in the standard curing room, and the dimensions of these concrete specimens should be checked before the bending test. The test device is shown in figure 1. During the bending test, if the fracture position of the lower edge of the specimen is between two concentrated load action lines, the flexural strength (MPa) of the specimen should be calculated as follows:

$$f_f = \frac{Fl}{bh^2} \tag{1}$$

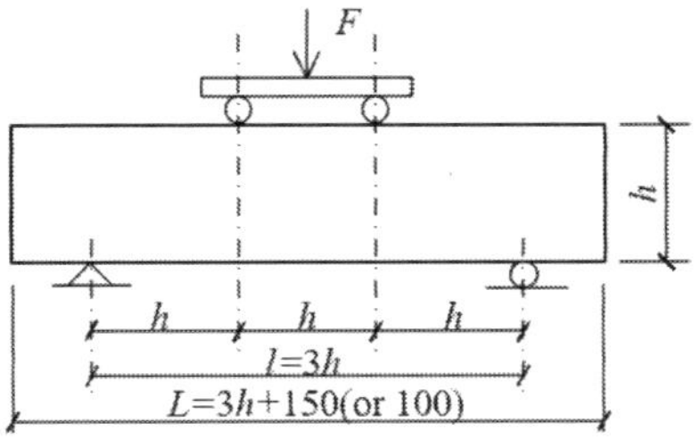

Figure 1. Bending test device.

3. Test Results

3.1. Setting Time

It can be seen from figure 2 and table 5 that the initial setting time and final setting time of sulphoaluminate cement without borax are 34 min and 57 min, respectively. With the increase of borax content, when the borax content changes from 0 to 1.0 %, the initial setting time increases from 34 min to 340 min, and the final setting time increases from 57 min to 580 min.

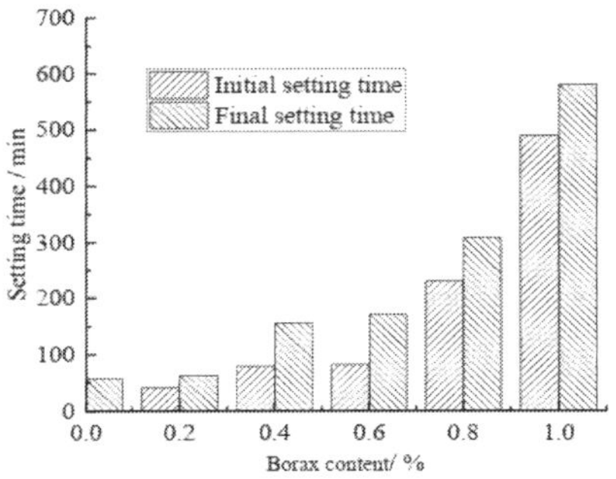

Figure 2. Setting time of sulphoaluminate cement with borax.

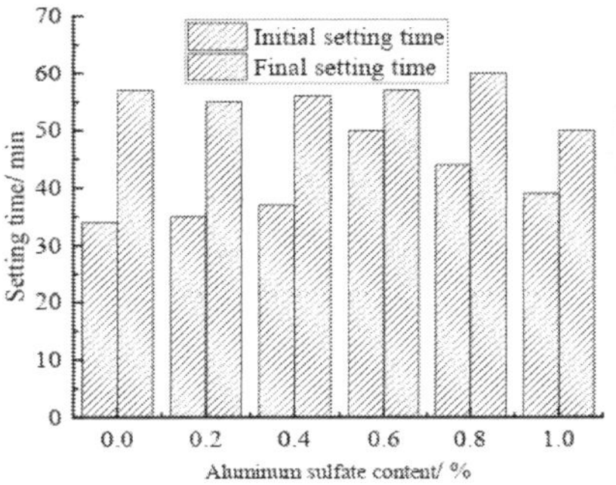

Figure 3. Setting time of sulphoaluminate cement containing aluminum sulfate.

As shown in figure 3 and table 6, there is no obvious change in the setting time of sulphoaluminate cement, the test results indicate that the aluminum sulfate has little effect on the setting time of sulphoaluminate cement.

Table 5. Setting time of sulphoaluminate cement with borax (experiment temperature 20°C).

Content of borax (%)	0	0.2	0.4	0.6	0.8	1.0
Initial setting time(min)	34	41	79	82	230	490
Final setting time(min)	57	63	156	171	308	580

Table 6. Setting time of sulphoaluminate cement with aluminum sulfate (experiment temperature 20°C).

Aluminum sulfate (%)	0	0.2	0.4	0.6	0.8	1.0
Initial setting time(min)	34	35	37	50	44	39
Final setting time(min)	57	55	56	57	60	50

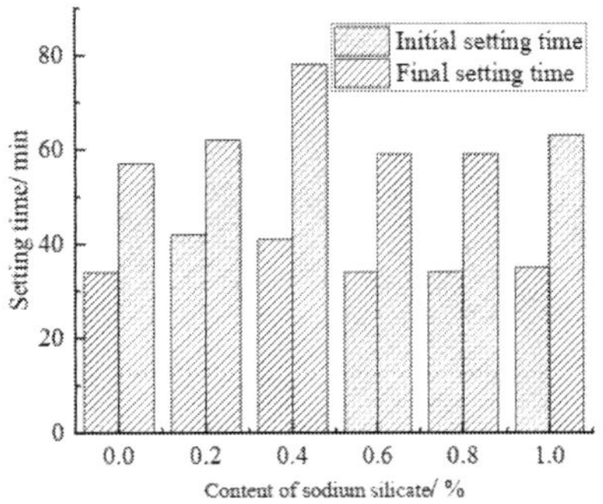

Figure 4. Setting time of sulphoaluminate cement with sodium silicate.

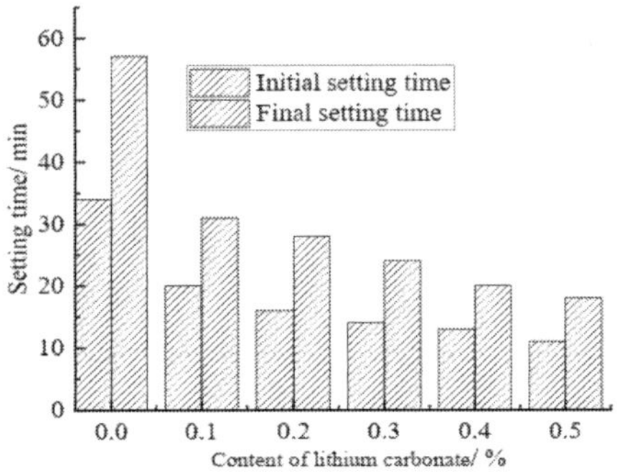

Figure 5. Setting time of sulphoaluminate cement with lithium carbonate.

From figure 4 and table 7, it can be seen that the initial setting time and final setting time increase first and then decrease when the sodium silicate content changes from 0 to 1.0 %. When the sodium silicate content is 0.2 %, the initial setting time is 42 min. Compared with the setting time without sodium silicate, the change was not significant. The test results indicate that the sodium silicate was not stable for the setting time of sulphoaluminate cement.

Table 7. Setting time of sulphoaluminate cement with sodium silicate (experiment temperature 20°C).

Content of sodium silicate (%)	0	0.2	0.4	0.6	0.8	1.0
Initial setting time(min)	34	42	41	34	34	35
Final setting time(min)	57	62	78	59	59	63

Table 8. Setting time of sulphoaluminate cement with lithium carbonate (experiment temperature 20°C).

Content of lithium carbonate (%)	0	0.1	0.2	0.3	0.35	0.4	0.5
Initial setting time(min)	34	20	16	14	15	13	11
Final setting time(min)	57	31	28	24	20	20	18

As shown in figure 5 and table 8, it can be seen that the initial setting time reduces from 34 min to 11 min, and the final setting time reduces from 57 min to 18 min when the borax content changes from 0 to 0.5 %. During the setting time test, it is found that the sulphoaluminate cement concrete specimen is prone to cracks in the later curing process, if the lithium carbonate in the amount of admixture increases so much. Therefore, it is necessary to reasonably control the content of lithium carbonate when the lithium carbonate is considered as the accelerator of sulphoaluminate cement.

3.2. Compressive Test

Dimensions of 100 mm×100 mm×100 mm cube specimens were used in the compressive strength test. The compressive strength at 6 h, 1 d, 3 d, 7 d and 28 d were tested by the universal testing machine, as shown in figure 6. According to GB / T 50081-2019, if the difference between the maximum and minimum values and the intermediate values exceeds 15 % of the intermediate value, the test results of this group of specimens should be eliminated.

(a) Compressive failure specimens at 1h (b) Compressive failure specimens at 1h

Figure 6. Compressive failure specimens.

Table 9. Results of compressive test (MPa).

Specimen number	Water-cement ratio	Loading time				
		1h	3h	6h	1d	3d
1	0.36	8.9	28.3	39.5	51.8	56.0
2	0.4	7.8	22.3	31.9	44.5	48.2
3	0.45	5.9	20.4	29.0	41.5	44.8
4	0.48	-	14.8	24.4	37.5	41.2
5	0.60	-	5.6	17.8	31.4	36.5

3.3. Bending Test

The experimental results of the influence of different water-cement ratios on the flexural strength of low sulfur aluminate cement concrete are shown in figure 7 and table 10.

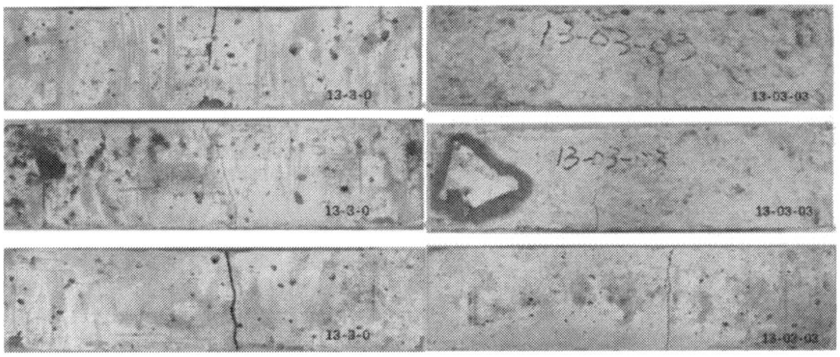

Figure 7. Damage specimen of bending test.

As can be seen from figure 7, the concrete specimens remain integrated after the bending test. The test results indicates that the sulphoaluminate cement concrete has good toughness. As can be seen from table 10, with the increase of water-cement ratio, the flexural strength shows an increasing trend with the curing age.

Table 10. Test results of bending test (MPa).

Specimen number	Water-cement ratio	Loading time			
		6h	1d	3d	7d
1	0.36	4.5	5.3	6.2	6.5
2	0.4	3.8	4.5	4.9	5.4
3	0.45	3.6	4.6	5.1	5.3
6	0.48	3.4	4.3	4.8	5.2
7	0.60	2.5	3.3	3.7	4.1

4. Conclusions

In this paper, the effects of four additives on the setting time of sulphoaluminate cement concrete were studied, and the compressive strength and flexural strength of sulphoaluminate cement concrete under different mixing ratios were studied. The test results and conclusions are as follows:

(1) Through the setting time test, the influence of four admixtures on the setting time of sulphoaluminate cement concrete was investigated. The test results show that the sodium borate for the effect of controlling setting time was better than the other admixtures. When the content of borax changed from 0 to 1.0 %, the initial setting time can increase from 34 min to 340 min, and the final setting time can increase from 57 min to 580 min. While the borax content changed from 0 to 0.5 %, the initial setting time can reduce from 34 min to 11 min, and the final setting time can reduce from 57 min to 18 min.

(2) At the normal temperature, the compressive strength of the repair material reaches 40MPa in 6h when the flexural strength reaches 2.5 MPa at least. As a repair material, the compressive strength and flexural strength satisfy the needs of rapid hardening.

(3) The mix design of different cement-based grouting materials and related physical and mechanical properties was investigated at normal temperature; however, the corresponding rapid construction technology at low temperature also need further research in the next step.

Acknowledgements

This study is supposed by the National Key Research and Development Program of China (Grants Nos. 2020YFC1511902).

References

[1] Gui Y, Liao Y, Pengfei X. 2020 Effect of admixtures on hydration of sulfoaluminate cement based materials [J] *Concrete* (02) 111-113+117.

[2] Wang Y, Shi C, Ma Y, et al. 2021 Accelerators for shotcrete-Chemical composition and their effects on hydration, microstructure and properties of cement-based materials [J] *Constr. Build. Mater* 281.

[3] Tao Y, Rahul A V, Lesage K, et al. 2021 Stiffening control of cement-based materials using accelerators in inline mixing processes: Possibilities and challenges[J] *Cem. Concr. Compos* 119 103972.

[4] Paglia C. 2003 The influence of alkali-free and alkaline shotcrete accelerators within cement systems: I. Characterization of the setting behavior [J] *Cem. Concr. Res* **33** (3) 387-395.

Overview of Calculation Methods of Structural Time-Dependent Reliability

Lisheng Luo, Xinran Xie*, Yongqiang Zhang and Wenyuan He

Department of Civil Engineering, College of Civil Engineering and Architecture, Hainan University, Haikou, 570228, China
*Email: xxr_912@163.com

Abstract. Under the action of natural erosion, the strength, durability and other safety performance of structures and elements gradually decrease with time, which has a great impact. To solve the above problem, a series of time-dependent reliability analysis methods were proposed. Based on different structural performance functions, this paper analyzes and discusses different time-dependent reliability theories, including outcrossing-based reliability method, Monte Carlo simulation method, extremum method and other new methods proposed in recent years, which provides reference for later research.
Keywords. Structural reliability; Time-dependent reliability; Calculation method; Performance function of structure.

1. Introduction

The traditional reliability problems are mainly load-based stationary binomial stochastic process model [1] and the specific calculation methods include Monte Carlo method, first-order second-moment theory [2], high-order moment method and so on. Traditional methods usually assume that the structural resistance and loading effect are fixed, and they are random variables at a certain time. However, because of an oversized variety of uncertainties within the actual project, the reliability of the structure is constantly changing during the service period, resulting in that the traditional theories cannot meet the requirements of the particular project. Thus, factors such as time and resistance degradation should be considered in the design [3]. To solve the influence of uncertain factors on the safety performance of structures or elements, researchers have proposed a series of time-dependent reliability analysis methods. However, there has not been an accurate and unified theoretical method, nor a lot of practice.

Starting from different structural performance functions, this paper analyzes and discusses different time-dependent reliability theories, including outcrossing-based reliability method, Monte Carlo simulation method, extremum method and other new methods proposed in recent years, so as to provide reference for later research.

2. Performance Function of Structure

As shown in figure 1, the performance functions of structure in service include monotonic and non-monotonic variations, which are applicable to different specific situations.

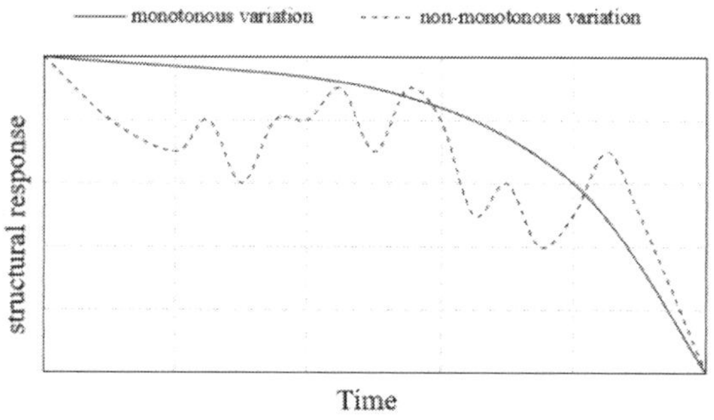

Figure 1. Structural response-time curves.

(1) Monotonic variation is principally applicable to the case of structural aging and cumulative damage, which is generally described by random variable model and random process model [4]. The structural performance function can be expressed as a random process g(x, t), where *x* is the vector of structural input random variables, and *t* is time. The instantaneous failure probability of structures is defined as:

$$P_{f,i}(t) = P_r\{g(x,t) < 0\} \tag{1}$$

When *t* is assumed to be a fixed value, the formula (1) is expressed as the failure probability of time-invariant reliability. Therefore, the traditional time-invariant reliability analysis method can be used to replace the time-dependent reliability at the endpoint of the time interval, such as Monte Carlo simulation method [5] and first-order second-moment method [6-7].

(2) Non-monotonic variation is mainly applicable to wind load and seismic load. Compared with monotonic variation mode, the performance function is more changeable and the calculation is more difficult. The mature theories are outcrossing-based reliability method [8], extremum method [9] at present and some novel methods, which are shown in the detailed analysis below.

3. Theories of Time-Dependent Reliability

There are two main ways to analyze the time-dependent reliability at present: (1) The service time of the structure is divided into several discrete time periods, and the problem is transformed into time-invariant reliability problem respectively [10]. (2) Based on the stochastic process theory, the resistance and load of the structure are treated as stochastic processes [11]. Researchers have applied and further analyzed the above theories in different fields for different research objects.

3.1. Outcrossing-Based Reliability Method

Outcrossing-based reliability method [8] is characterized by whether the dynamic response (e.g., displacement and velocity) exceeds the safety limit or not. The structure fails when the boundary is first exceeded, which is generally applicable to special stochastic processes such as Gaussian processes. The commonly used analysis methods are Poisson model method, modified Poisson model method, Volterra series method [12], stochastic averaging method [13] and integral equation method.

On the basis of this method, Andrieu-Renaud et al. [14-15] proposed and improved the PHI2 method for establishing the reliability numerical equation in the discrete time period, which reduced the influence of time step on the results. Zheng et al. [16] considered the memory of the first crossing time and proposed a calculation method for brittle structures based on integral equation method. Li et al. [17-19] gave the expressions of structural crossing probability and failure probability under different loads, and proposed a modified method for calculating the time-dependent reliability of structures according to the correlation of crossing occurrence. Considering the randomness of resistance degradation process, Zhang et al. [20] gave the calculation formula of time-varying reliability of structure by combining point estimation method. Considering discrete load process based on the boundary method of penetration threshold, an outcrossing-based reliability assessment method was recommended by Wang et al. [21-22] in order that the load model was not restricted to Poisson process.

However, the outcrossing-based reliability method is only applicable to some specific stochastic processes such as Gaussian processes, which requires complex multidimensional numerical integration and requires more efficient and wider application methods.

3.2. Monte Carlo Simulation Method

Monte Carlo simulation method is one of the most effective methods to evaluate time- dependent reliability, especially for high-dimensional problems [5] [23]. This method obtains the performance function or structural response value of the structure through random sampling test, and calculates the time-dependent reliability of the structure according to the number of tests for structural failure and the total number of sampling tests.

The Monte Carlo simulation method does not need to consider the dimension of random variables and omits the complex mathematical deduction process. However, the sampling calculation is huge, and sufficient precision sampling and tests are required to support its accuracy. Researchers have developed different selection and variance reduction techniques for these shortcomings, and combined them with other methods to improve the efficiency of Monte Carlo method [24-28]. Luo et al. [29] simplified the calculation process by Monte Carlo sampling method combined with series system model, and gave the time-dependent reliability calculation method of corroded elements. An improved response surface method combined with Monte Carlo sampling method was put forward by Zhang [30], which can effectively improve calculation accuracy of the reliability and reduce the design sample points. Considering the statistical uncertainty, Zhang et al. [31] theorized to combine Bootstrap method with weighted analysis of probability density to calculate the reliability of foundation pit deformation by Monte Carlo method. Ji et al. [32] improved the WUS (Weighted Uniform Simulation) method by combining Monte Carlo method to calculate the failure probability of slope system more efficiently.

3.3. Extremum Method

The extremum method [9] focuses on the global maximum and minimum of structural response. When the extreme value exceeds the safety range, it is the structural failure, which is a typical method to transform the time-dependent problem into the time-invariant problem. In recent years, researchers have proposed different methods: Gong et al. [33-34] put forward the concept of time synthesis of load to calculate cumulative failure probability. Li et al. [35] developed the concept of equivalent extreme events and analyzed the system reliability of dynamic and static problems. NERS (Nested Extreme Response Surface) was proposed by Wang et al. [36] based on Kriging model and nested extreme response surface was established. Considering the time-dependent limit state equation involving stochastic processes, Hu et al. [37] estimated the corresponding distribution of the extreme values of stochastic processes by sampling and substituted the extreme values for stochastic processes. Huang [4] theorized that the extremum method combined with AK-SS method was used to analyze the time-dependent reliability of the structure, and the Kriging model of the time-varying function was established by the mixed EGO algorithm to improve the accuracy of the extreme response surface and calculate the failure probability.

3.4. Reliability Theory Based on Probability Density Evolution

In order to solve the problem that the outcrossing-based reliability method is only applicable to certain random processes, researchers have proposed the reliability theory based on the evolution of probability density, which has become one of the most popular theories in recent years. This method constructs a virtual random process, and takes the extreme value of the required solution as the cross-sectional random variable of the random process, so the obtained structural dynamic reliability has high accuracy [38-39].

Based on Lagrange description of dynamic system, Li et al. [40-42] incorporated deterministic system and stochastic system into a unified framework, gave a generalized density evolution equation with universal significance and developed different point selection methods [43-44]. Fan [45] introduced the Dirac sequence solution of the generalized probability density evolution equation without the first-order derivative of the target quantity with respect to time, which simplified the calculation process. Zhou [1] put forward a hybrid method using the Dirac sequence to. Wang et al. [46] used covariance matrix to obtain joint probability density function based on KDE (Kernel Density Estimation) method to calculate reliability and prove its accuracy. The reliability theory based on probability density evolution has been widely concerned because of its wide application and high precision.

However, there are still some limitations, which are mainly reflected in the strong experience and subjectivity in tangential radius, the number of selected points and the selection of the median value of the Dirac sequence solution. and needs further improvement. Further improvement is a problem that researchers need to pay attention to in the future.

3.5. New Theories of Time-Dependent Reliability

In addition to the above classical time-dependent reliability theories, researchers have also proposed many new reliability theories in recent years, introduced different models or theories, and applied them to different fields: Mourelatos et al. [47] theorized the limit state equation of explicit time-dependent function including random variables and random processes, and calculated the time-dependent conditional probability by combining FORM and CLS methods, which was suitable for solving low dimensional failure probability problems. Wang et al. [48] introduced a non-probabilistic time-dependent reliability estimation model and a sequential multidisciplinary optimization, combined with SMO-NTRA (sequential multidisciplinary optimization and non-probabilistic time-dependent reliability assessment) method. Gong et al. [49] improved the Poisson outcrossing rate method of the first-order reliability, to analyze the time-dependent reliability more easily. TSM method (two-step method) was developed by Shi et al. [50], which simplified analysis of time-dependent reliability and some deterministic optimization.

4. Conclusion

In this paper, monotonic and non-monotonic structural functions are introduced, and different time-dependent reliability theories are analyzed and discussed, including several classical methods such as outcrossing-based reliability method, Monte Carlo simulation method, extremum method, as well as novel methods proposed in recent years such as reliability theory based on probability density evolution. Basic problems in engineering can be solved using those several theories at present, but the limitations are also obvious, such as the low efficiency of Monte Carlo simulation method, the subjectivity of probability density evolution method, and the special limitation of outcrossing-based reliability method. Thus, it is necessary to apply a wider and more convenient theoretical method as a support, and it is also a problem that researchers need to consider in the future.

Acknowledgements

The authors gratefully acknowledge the support from National Natural Science Foundation of China (Grant No.51808176) and High-level Natural Science Foundation of Hainan Province of China (Grant No. 2019RC055).

References

[1] Li G Q. 2005 *Load and Reliability Design Principle of Engineering Structure* (Beijing: CN. Architecture Publishing &Media Co., Ltd.).

[2] Standard Press of China. 2001 *Unified Standard for Reliability Design of Build.ing* Structures: *GB50068-2001* (Beijing: CN. Architecture Publishing &Media Co., Ltd.)

[3] Zhou Q Y. 2017 Study on Time-variant Reliability Method of Deteriorating Structures and its Application ph.D. Chongqing: Chongqing Univ.

[4] Huang X X. Study on the Structural Time-Dependent ReliabilityAnalysis and the Optimization Design ph.D. Wuhan: Huazhong Univ. of Sci. and Technol.

[5] Fishman G S. 1996 Monte Carlo: concepts, algorithms, and applications *Technometrics* **39** (3) 338-338.
[6] Hasofer A M and Lind N C. 1974 Exact and invariant second-moment code format *J. Eng. Mech.* **100**(1) 111-121.
[7] Rackwitz R and Flessler B. 1978 Structural reliability under combined random load sequences *Comput. Struct* **9** (5) 489-494.
[8] Veneziano D, Cornell C A and Grigoriu M 1977 Vector-process models for system reliability *J. Eng. Mech.* **103** (3) 441-460.
[9] Wang Z Q, Chen W. 2016 Time-variant reliability assessment through equivalent stochastic process transformation *Reliab. Eng. Syst. Safe.* 152 166-175.
[10] Thoft-Christensen P. 1983 *Reliability Theory and Its Application in Structural and Soil Mechanics* (Kluwer Boston: M. Nijhoff).
[11] Melchers R E. 2017 *Structural Reliability Analysis and Prediction* (Hoboken: John Wiley & Sons Ltd).
[12] Naess A. 2001 Crossing rate statistics quadratic transformations of Gaussian processes *Prob. Engng. Mech.* **16** 209-217.
[13] Zhu W Q, Deng M L and Huang Z L. 2002 First-passage failure of quasi-integrable hamiltonian systems *J.* Appl. *Mech.* **69** 274-282.
[14] Andrieu-Renaud C, Sudret B and Lemaire M. 2004 The PHI2 method: a way to compute time-variant reliability. *Reliab. Eng. Syst.* Safe **84** (1) 75-86.
[15] Sudret B. 2008 Analytical derivation of the out-crossing rate in time-variant reliability problems *Struct.* Infrastruct. *Eng* **4** (5) 353-362.
[16] Zheng C L, Chen J, He J and Kou X J. 2006 A new method for first-passage time probability density function of structures *Earthq. Eng. Eng. Vibrat* **26** (4) 52-55.
[17] Li C Q, Firouzi A and Yang W. 2016 Closed-form solution to first passage probability for nonstationary lognormal processes *J. Eng. Mech.* 04016103.
[18] Wang C, Zhang H and Beer M Structural. 2019 Structural time-dependent reliability assessment with new power spectral density function *J. Struct. Engi* **145** (12) 04019163.
[19] Wang C 2020 Stochastic process-based structural reliability considering correlation between upcrossings ASCE-ASME *J. Risk Uncertainty Eng. Syst., Part A: Civ. Eng.* **6**(4) 06020002
[20] Zhang Y T and Li Q W. 2021 Time-dependent reliability analysis method based on outcrossing rate *Proc. of the 30th National Conf. on Structural Engineering No.II*(Beijing: Engineering Mechanics) pp 142-147.
[21] Wang C, Li Q, Zhang H and Ellingwood B R. 2017 Modeling the temporal correlation in hurricane frequency for damage assessment of residential structures subjected to climate change *J. Struct. Eng.* 143(5) 04016224.
[22] Wang C and Zhang H. 2018 Roles of load temporal correlation and deterioration-load dependency in structural time-dependent reliability *Comput. Struc* **194** (Jan) 48-59.
[23] Wang C, Beer M and Ayyub B M. 2021 Time-Dependent Reliability of Aging Structures: Overview of Assessment Methods *J. Risk Uncertainty Eng. Syst., Part A: Civ. Eng* **7** (4) 03121003.
[24] Mahadevan S and Raghothamachar P. 2000 Adaptive simulation for system reliability analysis of large structures. *Comput. Struct* **77** 725-34.
[25] Yang K and Younis H. 2005 A semi-analytical Monte Carlo simulation method for system's reliability with load sharing and damage accumulation *Reliab. Eng. Syst. Safe.* **87** 191-200.
[26] Mori Y and Ellingwood B R 1993 Time-dependent system reliability analysis by adaptive importance sampling *Struct. Safe* **12** (1) 59-73.
[27] Singh A, Mourelatos Z and Nikolaidis E. 2011 Time-dependent reliability of random dynamic systems using time-series modeling and importance sampling *Int. J. Mater. Manuf* **4** (1) 929-946.

[28] Wang J, Cao R and Sun Z. 2021. Importance sampling for time-variant reliability analysis *IEEE Access9* 20933-20941.
[29] Luo L S and Chen Z H. 2019 Time-dependent reliability analysis of corroded steel members based on the important sampling method *Ind. Const.* **49** (8) 1-5+107.
[30] Zhang X G. 2018 Structural reliability analysis based on an improvement of the response surface method *J.* Mech. *Strength* **40** (6) 1382-1388.
[31] Zhang L S, Li D Q, Cao Z J and Tang X S. 2019 Efficient reliability analysis of excavation deformation considering statistical uncertainty using Monte Carlo simulation *E. J. Wuhan Univ.* **52** (3) 207-215.
[32] Ji J, Wang L P, Liao W W, Zhang W J, Zhu D S and Gao Y F. 2021 System reliability analysis of slopes based on weighted uniform simulation method *CN. J. Geotech. Eng* **43** (8) 1492-1501.
[33] Gong J X and Zhao G F. 1998 Reliability analysis for deteriorating structures *J. Build. Struct.* **19**(5) 43-51.
[34] Singh A, Mourelatos Z P and Li J. 2010 Design for lifecycle cost using time-dependent reliability. *J. Mech. Design Trans.* **132**(9) 0910081-09100811.
[35] Li J, Chen J B and Fan W L. 2007 The equivalent extreme-value event and evaluation of the structural system reliability *Struct. Safe* **29** (2) 112-131.
[36] Wang Z and Wang P A. 2013 New approach for reliability analysis with time-variant performance characteristics *Reliab. Eng. Syst. Safe* **115** (7) 70-81.
[37] Hu Z and Du X A. 2013 sampling approach to extreme value distribution for time-dependent reliability analysis *J. Mech. Design* **135** (7) 071003.
[38] Chen J B and Li J. 2004 The extreme value probability density function based method for dynamic reliability assessment of stochastic structures *Earthq. Eng. Eng. Vibrat* **24** (6) 29-44.
[39] Guo H Y, Gu X L, Zhou B B and Zhang W P. 2019 Time-dependent reliability analysis for corroded RC beams based on probability density evolution theory *J. Build. Struct.* **40** (1) 67-73.
[40] Li J and Chen J B. 2002 Probability density evolution of nonlinear response of concrete structures *J. Southeast Univ* **31** 295-298.
[41] Li J and Chen J B. 2006 Generalized density evolution equation in stochastic dynamic system *Prog. Nat. Sci* **16** (6) 712-719.
[42] Li J and Chen J B. 2008 The principle of preservation of probability and the generalized density evolution equation *Struct. Safe* **30** (1) 65-77.
[43] Chen J B and Li J. 2008 Strategy for selecting representative points via tangent spheres in the probability density evolution method *Int. J. Numer. Meth. Eng* **74** (13) 1988-2014.
[44] Li J and Chen J B. 2007 The number theoretical method in response analysis of nonlinear stochastic structures *Comput. Mech.* **33** (6) 693-708.
[45] Fan W L, Chen J B and Li J. 2009 Solution of generalized density evolution equation via a family of δ sequences. *Comput. Mech.* **43** (6) 781-796.
[46] Wang Z H , Wang Z L, Yu S and Zhang K W. 2019 Time-dependent mechanism reliability analysis based on envelope function and vine-copula function *Mech. Mach. Theory* **134** 667-684.
[47] Mourelatos Z P, Majcher M, Pandey V and Baseski I. 2015 Time-dependent reliability analysis using the total probability theorem *J. Mech. Design* **137** (3) 031405.
[48] Wang L and Xiong C 2019 A novel methodology of sequential optimization and non-probabilistic time-dependent reliability analysis for multidisciplinary systems *Aerosp. Sci. Technol.* 94 105389.
[49] Gong C Q and Frangopol D M. 2019 An efficient time-dependent reliability method *Struct. Safe.* **81** 101864.
[50] Shi Y, Lu Z Z, Xu L Y and Zhou Y C. 2020 Novel decoupling method for time-dependent reliability-based design optimization *Struc. Multidiscip. O.* **61** 507-524.

Calculation and Analysis of Through Concrete Filled Steel Tubular Tied Arch Bridge

Zhanyi Zhang and Peiheng Long*

Civil and hydraulic engineering (bridge and Tunnel Engineering) of Beijing University of civil engineering and architecture, Beijing University of Civil Engineering and Architecture, Beijing, 102616, China
Email: 763312476@qq.com

Abstract. In order to understand the stress of small through tied arch bridge. In this paper, the finite element simulation analysis of Lu Shanqu bridge is carried out in the completion stage by using MADIS / civil software, and the tie bars, arch ribs and suspenders of the superstructure are monitored. The results show that the axial force of the arch rib of this bridge is reduced by the balance of the tie rod, and the bending moment of the tie rod is greatly reduced by the action of the suspender. The stress characteristics of the bridge type are internal statically indeterminate and external statically indeterminate structure.
Keywords. Through concrete filled steel tube tied arch bridge; Tie Bar; Arch Rib; structure

1. Introduction

Through tied arch bridge is a kind of composite bridge combining arch and beam. The interior of the bridge is statically indeterminate and the exterior is statically indeterminate. Its remarkable feature is that the interaction between arch and beam through arch foot greatly reduces the bending moment borne by arch and the axial force borne by beam. The mechanical characteristics of arch compression and beam bending are brought into full play.

The bridge adopts the construction method of beam before arch. After the completion of the bridge, the focus of this paper is whether the alignment of the arch ring and the bending moment of the beam meet the design specifications. Taking lushanqu bridge as an example, this paper introduces the stress state of arch and beam in the completion stage.

2. Project Summary

The structural form of the bridge is a through concrete-filled steel tubular simply supported tied arch bridge with a calculated span of 80m. The rise span ratio is 1 / 5 and the calculated rise height is 16m. In order to improve the overall stiffness of the structure, the bridge deck adopts longitudinal and transverse beams and integral bridge deck system. The arch rib is fixedly connected with the stiffening longitudinal beam. The two arch ribs are provided with concrete-filled steel tubular wind bracing. The cross section of the wind bracing is circular, and the wind bracing steel pipe is not filled with concrete. One suspender is set every 5m. The suspender adopts the form of single end tension, with a total of 14 suspenders. The tensioning end is set at the bottom of the longitudinal beam and the fixed end is set at the top of the arch rib.

The substructure of the bridge adopts a zigzag abutment, the thickness of the abutment body is 1m, the foundation is bored pile foundation, the thickness of the abutment cushion cap is preliminarily proposed to be 3M, and the pile foundation is designed with friction pile. Basin type rubber bearing

Journal of Physics: Conference Series **2148** (2022) 012064 doi:10.1088/1742-6596/2148/1/012064

with high bearing capacity and good turning performance is adopted for the bridge bearing. The expansion joint adopts D80 expansion joint.

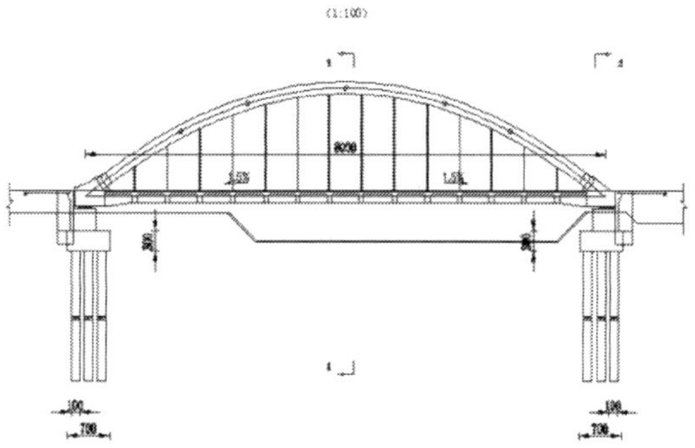

Figure 1. Elevation.

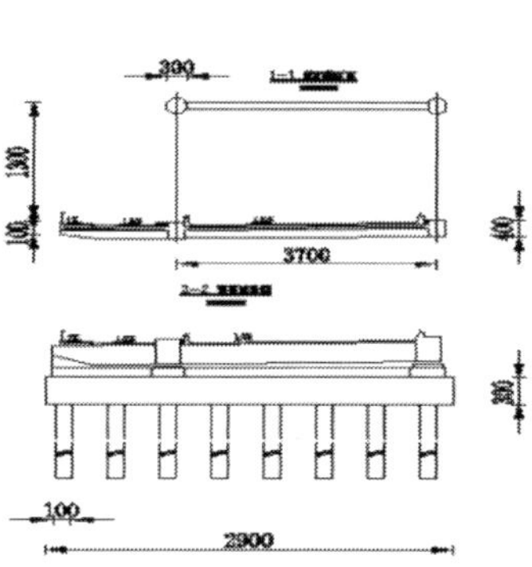

Figure 2. Cross section.

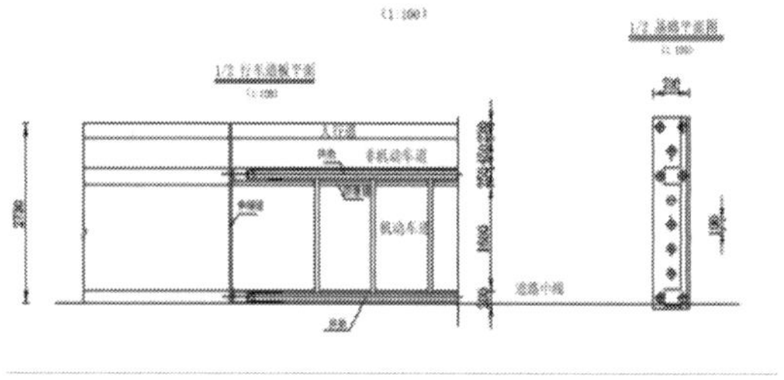

Figure 3. Plan.

3. Finite Element Analysis Model

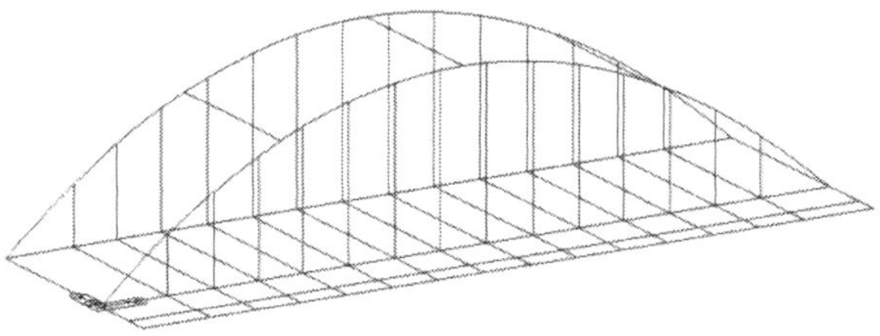

Figure 4. Finite element analysis model.

4. Bridge Completion Status Detection Content

The bridge detects the stress of arch rib, tie rod and suspender in the completed state to ensure that the structure in the completed state is under absolute safety control. According to the actual state of the structure, the most unfavorable situation of bending moment and shear force of each member in the completion stage is obtained, and whether each member is in the stress state after the completion of the bridge is checked.

4.1. Internal Force Detection

When the bridge is completed, the bending moment and axial force of the arch rib, the bending moment and axial force of the tie rod and the axial force of the suspender are tested.

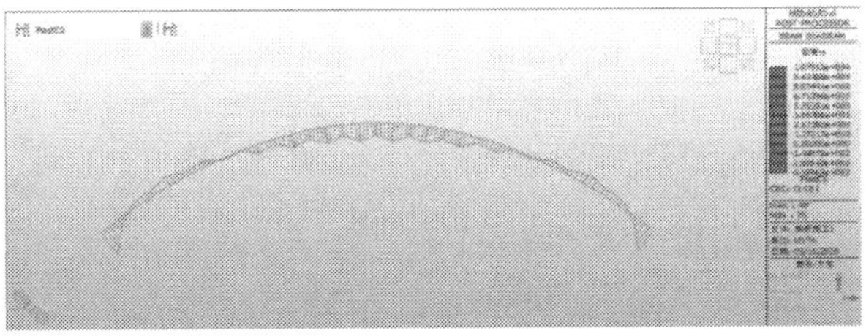

Figure 5. Arch bending moment diagram.

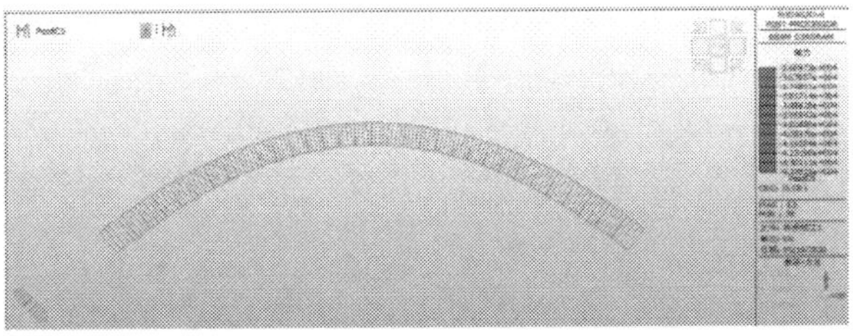

Figure 6. Arch axis diagram.

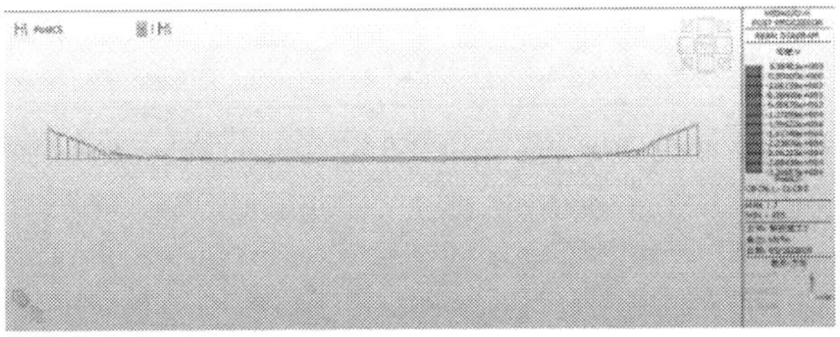

Figure 7. Tie bar bending moment diagram.

Figure 8. Tie bar axial force.

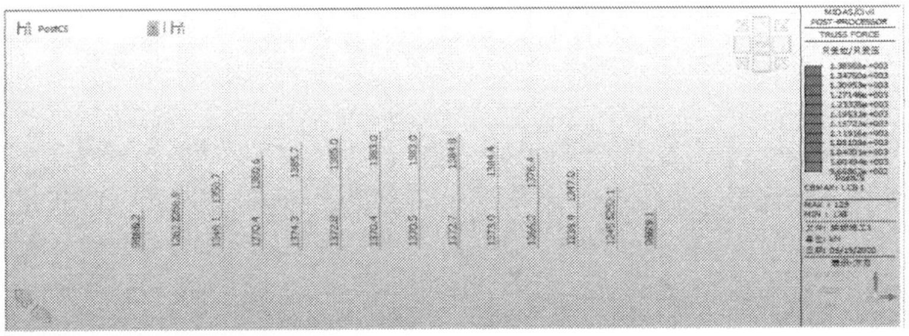

Figure 9. Hanger shaft diagram.

4.2. Deflection Detection

The deflection of arch rib and tie bar is detected, and whether it meets the requirements of design specifications is calculated through simulation analysis.

5. Test Results of Bridge Completion Status

5.1. Internal Force Test Results

The section of concrete-filled steel tubular arch rib is equivalent to beam column for checking calculation. The effective length of the equivalent beam column shall be 0.36s without hinged arch, and the force at both ends of the equivalent beam column shall be the bending moment and axial force at the L / 4 (or 3L / 4) section of the arch. Due to the symmetry of the structure, the most unfavorable internal force of L / 4 section is equal to that of 3L / 4 section. Therefore, only the most unfavorable internal force of L / 4 section needs to be checked during the overall checking calculation of the structure.

Axial compression section resistance:

$$N_D = \sum (N_0^i + N_f^i) = 2 \times 29301.91 + 2 \times 2462 = 63527.82kN$$

For maximum combination of axial forces:

$$\phi \cdot \phi_e \cdot k_c \cdot k_p N_D = 0.831 \times 0.825 \times 0.8 \times 0.742 \times 63527.82$$
$$= 25853.11kN > \gamma_O N_s = 1.1 \times 18091.51 = 20791.66kN$$

Therefore, the stable bearing capacity meets the requirements.

While the tie bar longitudinal beam balances the horizontal thrust, it is also subject to the bending moment generated by the beam and suspender. Therefore, the tensile bearing capacity of the normal section, flexural bearing capacity of the normal section and shear bearing capacity of the inclined section of the tie bar are checked.

Table 1. Calculation and checking results of shear bearing capacity of inclined section.

Section position	Max / min	Bearing capacity limit state combination	Structural resistance	Checking results
End	Max	-8090.3295	25516.0185	Check calculation qualified
	min	-11113.1848	25516.0185	Check calculation qualified
L/8	Max	1623.7117	15740.9467	Check calculation qualified
	min	-2032.4187	20198.6573	Check calculation qualified
L/4	Max	736.3899	15975.0600	Check calculation qualified
	min	-2494.1450	20368.3365	Check calculation qualified
3L/8	Max	3827.3776	16573.1727	Check calculation qualified
	min	-2.3326	20819.8592	Check calculation qualified
L/2	Max	3691.5474	16741.7523	Check calculation qualified
	min	-124.0382	20953.1046	Check calculation qualified

Table 2. Calculation and checking results of flexural bearing capacity of normal section.

Section position	Max / min	Bearing capacity limit state combination	Structural resistance	Checking results
End	Max	-2365.5539	149051.9011	Check calculation qualified
	min	-3477.8345	149051.9011	Check calculation qualified
L/8	Max	-782.7658	89749.6430	Check calculation qualified
	min	-1588.4749	98349.4167	Check calculation qualified
L/4	Max	-5.3722	89749.6430	Check calculation qualified
	min	-708.3785	89749.6430	Check calculation qualified
3L/8	Max	251.4853	89749.6430	Check calculation qualified
	min	-468.6064	89749.6430	Check calculation qualified
L/2	Max	459.0979	89749.6430	Check calculation qualified
	min	-279.5187	89749.6430	Check calculation qualified

The checking calculation of finished cable suspender of concrete-filled steel tubular tied arch bridge shall meet the following requirements:$\sigma \ll [\sigma] = \frac{f_{pk}}{\alpha}, f_{pk} = \frac{1670}{3} = 556.67MP$

Table 3. Checking calculation of suspender bearing capacity.

Boom No	Boom unit no	Stress calculation value	Stress limit
1(15)	79(125)	24.10	556.67
2(16)	80(126)	30.24	556.67
3(17)	81(127)	31.97	556.67
4(18)	82(128)	32.41	556.67
5(19)	83(129)	32.49	556.67
6(20)	84(130)	32.47	556.67
7(21)	85(131)	32.42	556.67
8(22)	86(132)	32.42	556.67
9(23)	87(133)	32.47	556.67
10(24)	88(134)	32.49	556.67
11(25)	89(135)	32.39	556.67
12(26)	90(136)	31.92	556.67
13(27)	91(137)	31.92	556.67
14(28)	92(138)	24.01	556.67

The stress under the most unfavorable combination of suspender endurance condition and bearing capacity limit state meets the requirements.

5.2. Deflection Calculation Results

Through calculation, the deflection of arch rib and tie bar of arch bridge meet the design requirements.

6. Conclusion

Taking Lushanqu Bridge as an example, this paper discusses the stress of the arch bridge in the completed state, and the conclusions are as follows:

(1) Through concrete tied arch bridge is a bridge with reasonable structure and giving full play to the mechanical characteristics of each component. Among them, it gives full play to the good compression characteristics of concrete and the bending characteristics of reinforcement. During the design, pay special attention to the rise span ratio of the bridge. If the span is too large, the dead weight load of the bridge will increase, so as to increase the stress of the arch rib and tie rod. If the span is too small, the mechanical characteristics of the combination of arch rib and tie rod will not be brought into full playIn the construction process of the bridge, the stress of adjacent points of arch rib and beam tension is close to the theoretical value as a whole, and the average cross section of key points on the arch rib and beam is close to the theoretical value as a whole.

(2) In the completion stage, the arch rib mainly bears compressive stress. In the design, the reasonable arch axis of the arch ring should be obtained through simulation calculation, so as to reduce the bending moment borne by the arch ring as much as possible and give full play to the compressive characteristics of the arch bridge. The tie bar mainly bears the bending moment, and the joint action of suspender and arch rib reduces the bending moment of tie bar. In the completion stage, the suspender can be regarded as the support of the continuous beam bridge and act on the tie rod longitudinal beam, so as to reduce the bending moment of the tie rod.

References

[1] Fan L J. 2012 Bridge engineering. Beijing: People's Communications Press.
[2] Yao L S. 2008 Bridge engineering. Beijing: People's Communications Press.
[3] Ye J S. 2005 Structural design principle. Beijing: People's Communications Press.
[4] Ministry of communications of the people's Republic of China. Technical standard for Highway Engineering (jtgb01-2014). Beijing: People's Communications Press, 2014.
[5] Ministry of communications of the people's Republic of China. General code for design of highway bridges and culverts (JTG d60-2015). Beijing: People's Communications Press, 2015.
[6] Ministry of communications of the people's Republic of China. Code for design of foundation and foundation of highway bridges and culverts (JTG d63-2007). Beijing: People's Communications Press, 2004.
[7] Wang R X, Li Z L. 2011 Examples of bridge engineering design guidance. Wuhan: Huazhong University of science and Technology Press.
[8] Xu Y. 2000 Concrete filled steel tubular tied arch bridge. Beijing: People's Communications Press.
[9] Wang X M. 2003 Basic engineering. Beijing: People's Communications Press.
[10] Ling Z P. 1996 Basic engineering. Beijing: People's Communications Press.
[11] Zhao M H. 2004 Bridge foundation and foundation. Beijing: People's Communications Press.
[12] Shao X D. 2007 Bridge design and calculation. Beijing: People's Communications Press.

Analysis of Longitudinal Force of Beam and Rail of Continuously Welded Rails (CWR) on Railway Bridge

Zhiping Zeng[1, 2], Ji Hu[1], Qiang Zeng [1], Zhibin Huang[1, 3*], Huatuo Yin[4] and Hui Huang[4]

[1] School of Civil Engineering, Central South University, Changsha, Hunan 410075, China
[2] MOE Key Laboratory of Engineering Structure of Heavy Haul Railway, Central South University, Changsha, Hunan 410075, China
[3] Southeast Coastal Railway (Fujian) Co., Ltd, Fuzhou, Fujian, 350001, China
[4] Guangzhou Metro Design & Research Institute Co., Ltd., Guangzhou, Guangdong 510010, China
Email: 1146401456@qq.com

Abstract. To study the longitudinal force of CWR on viaduct, a track-bridge-pier finite element model is established. Taking a multi-span simply supported beam with a maximum span of 32.7m of an elevated CWR as an example, the additional expansion and contraction forces, displacement between rail and beam and the force of pier are calculated, and whether the rail stress meets the requirements when setting constant resistance fasteners is checked. The results show that: (1) For the left and right lines, the maximum additional expansion forces of single strand rail are both 211.13kN, and the maximum relative displacements between beam and rail are both 6.572mm. (2) The maximum value of the additional expansion and contraction forces and the relative displacement between beam and rail of the same line occur at the same position. The left line is at ZFZ29 pier and the right line is at ZFS31 pier. (3) The maximum force of pier in this section is 500.80kN, and the pier numbers are ZFZ27 and ZFS29. (4) The rail stress is less than the allowable stress of 352MPa, and the rail strength meets the requirements.

1. Introduction

Under the influence of train load, temperature change and other factors, the rail and beam will have longitudinal interaction, produce relative displacement and additional internal force [1-3].

The researches on Continuously Welded Rails (CWR) on bridge mainly study the longitudinal interaction mechanism between track and bridge and the structural design of track and bridge to ensure that the track and bridge structure meet the requirements of strength and stability under temperature and train load [4-5].

In this paper, taking the CWR on a viaduct as an example, the spatial structure model is established by using the finite element software ANSYS [6-8], and the longitudinal mechanical properties are calculated and analyzed, which could provide a reference basis for the design of laying CWR on the bridge.

2. Track-Bridge-Pier Finite Element Calculation Model

2.1. Model Assumptions and Treatment

The finite element model of beam-rail interaction under temperature change is established. There is no regulator in the design scheme, and the longitudinal force of CWR is calculated according to the concrete simply supported beam. In order to eliminate the influence of boundary conditions, 180 m subgrade sections are set at both ends. The rail and bridge are simulated by beam element respectively, the vertical characteristic of the fastener is simulated by linear spring element, and the longitudinal characteristic of the fastener is simulated by nonlinear spring. The spring action direction is to prevent the displacement of rail relative to sleeper.

2.2. Model Parameters

2.2.1. Bridge Parameters. The finite element model of track-bridge structure is established based on ANSYS. The ballastless track structure is adopted. The bridge is a simply supported beam bridge, and the bearings are arranged in the form of 'fixed-movable', the basic parameters of the bridge are shown in table 1.

Table 1. Bridge parameters.

line	Pier number	Span(m)	Stiffness(kN/cm)	Bearing
	ZFZ9	24.7	26149	Fixed- Movable
	ZFZ10	24.7	2178	Fixed- Movable
	ZFZ11	24.7	1918	Fixed- Movable
	ZFZ12	24.7	2056	Fixed- Movable
Left line	ZFZ13	24.7	2178	Fixed- Movable
	ZFZ14	/	28149	/
	ZFZ27	32.7	28149	Fixed- Movable
	ZFZ28	32.7	1667	Fixed- Movable
	ZFZ29	/	28149	/
	ZFS9	24.7	28149	Fixed- Movable
	ZFS10	24.7	2878	Fixed- Movable
	ZFS11	24.7	2358	Fixed- Movable
	ZFS12	24.7	2167	Fixed- Movable
Right line	ZFS13	24.7	5649	Fixed- Movable
	ZFS14	/	28149	/
	ZFS29	32.7	28149	Fixed- Movable
	ZFS30	32.7	1667	Fixed- Movable
	ZFS31	/	28149	/

2.2.2. Track Structure Parameters

(1) Standard 60kg/m rail is adopted, and the elastic modulus of rail is 2.06×10^5Mpa, the cross-sectional area of a single rail is 77.45cm^2; The linear expansion coefficient is taken as 1.18×10^{-5}/°C.

(2) SFC type constant resistance fastener is adopted, the fastener spacing is 0.6m, the longitudinal resistance of fastener is taken as 9kN for each group, and the slip value is 2mm.

(3) The elastic modulus of concrete bridge is 35.5GPa, and the linear expansion coefficient of concrete beam is 1×10^{-5} /°C calculation.

(4)According to *Code for Design of Railway Continuous Welded Rail* (TB 10015-2012) [9], the maximum rail temperature in this area is 61.7 °C and the minimum rail temperature is - 1.7 °C. The daily temperature difference of concrete beam is calculated as 30 °C, the locking rail temperature is selected as 30 ± 5 °C, and the maximum temperature change range is 36.7 °C.

2.2.3. Model Establishment. The overall beam and rail interaction models are shown in figure 2.

(a) Left line ZFZ9~ZFZ14

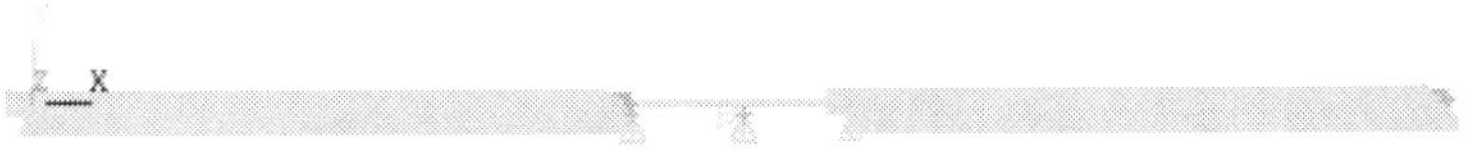

(b) Left lineZFZ27~ZFZ29

(c) Right lineZFS9~ZFS14

(d) Right lineZFS29~ZFS31

Figure 2. Finite element model of expansion force and temperature force of beam rail interaction.

3. Calculation and Analysis

3.1. Calculation of Additional Expansion and Contraction Force of Rail, Relative Displacement Between Beam and Rail And The Force of Pier

When the concrete bridge temperature rises 30 °C, the additional expansion and contraction forces of the rails and relative displacements between the beam and rail are shown in figure 3 to 10. The force of each pier is shown in table 2.

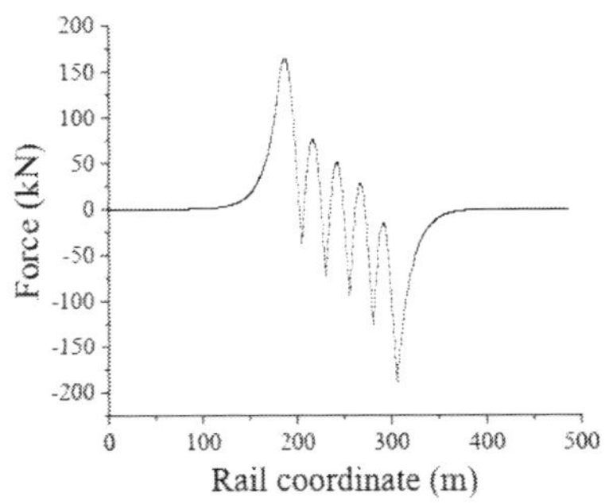

Figure 3. Additional expansion and contraction force of left rail (ZFZ9~ZFZ14).

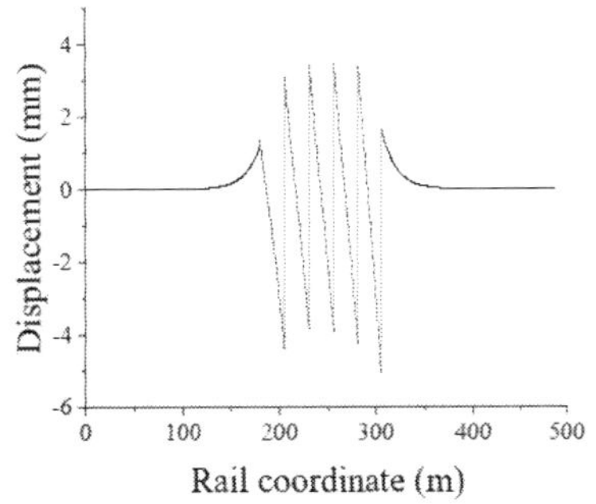

Figure 4. Relative displacement of left track between the beam and rail (ZFZ9~ZFZ14).

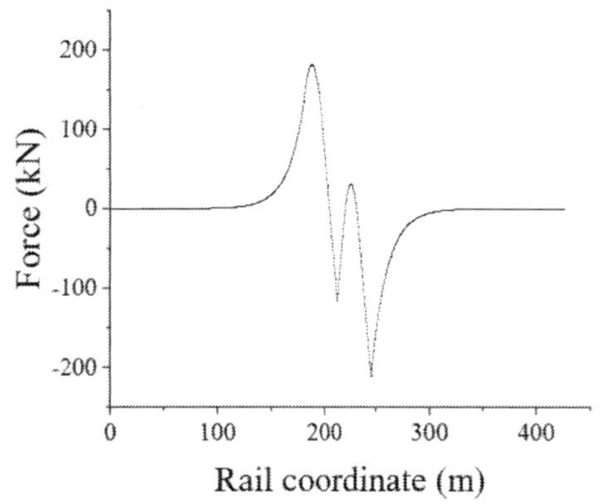

Figure 5. Additional expansion and contraction force of left rail (ZFZ27~ZFZ29).

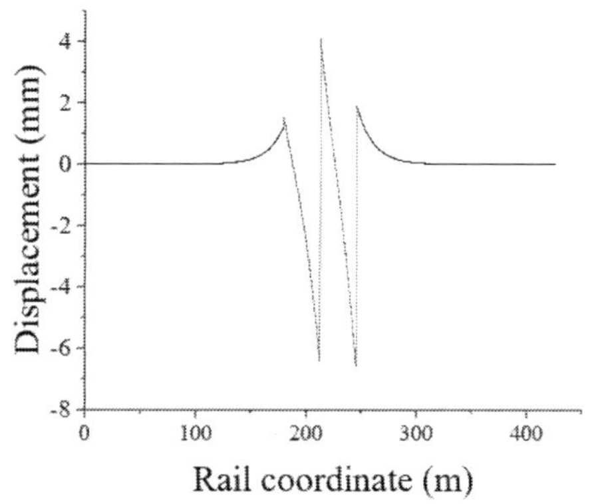

Figure 6. Relative displacement of left track between the beam and rail (ZFZ9~ZFZ14).

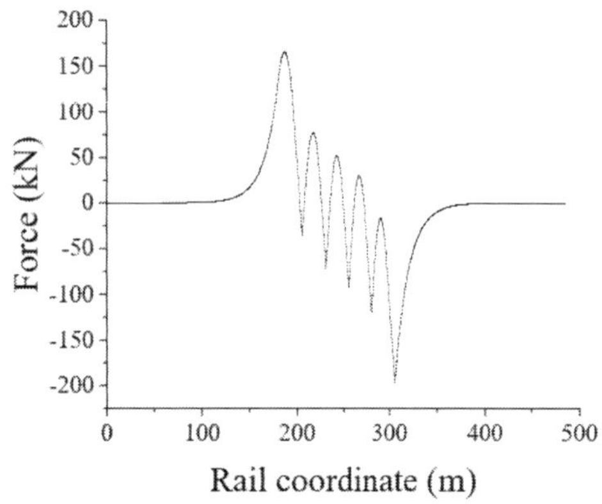

Figure 7. Additional expansion and contraction force of left rail (ZFS9~ZFS14).

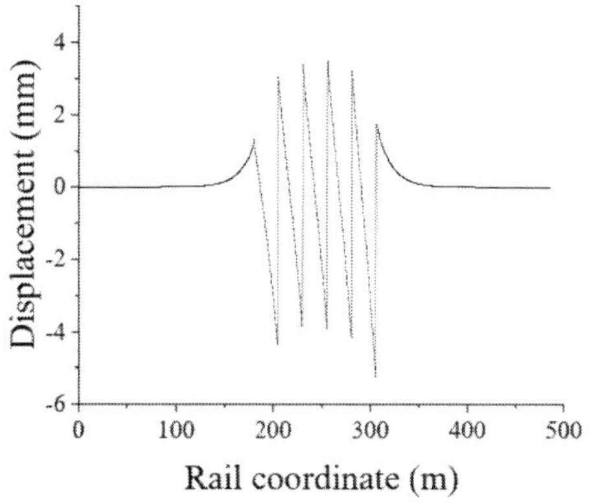

Figure 8. Relative displacement of left track between the beam and rail (ZFS9~ZFS14).

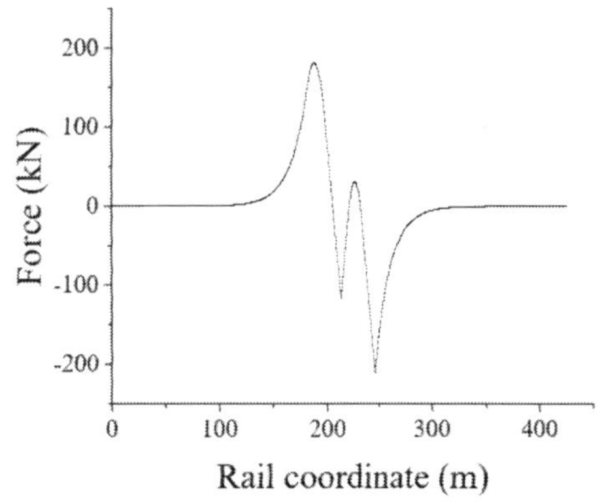

Figure 9. Additional expansion and contraction force of left rail (ZFS29~ZFS31).

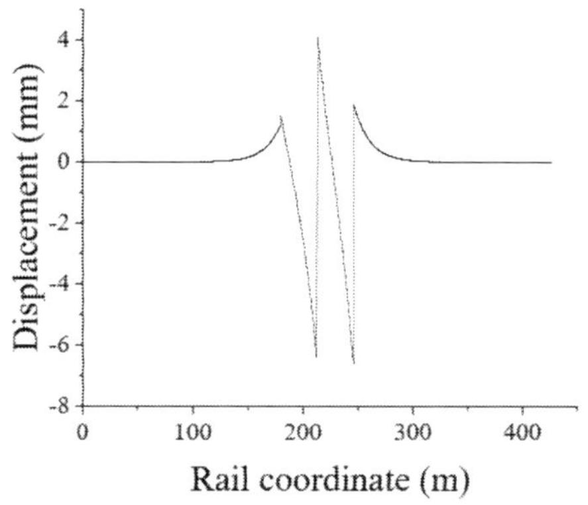

Figure 10. Relative displacement of left track between the beam and rail (ZFS29~ZFS31).

According to figure 2 to 9, for the left line, the maximum additional expansion force on the single strand rail is 211.13kN, which occurs at the pier numbered ZFZ29; The maximum relative displacement between beam and rail also occurs at this position, which is 6.572mm. For the right line, the maximum additional expansion force on the single strand rail is 211.13kN, which occurs at the pier numbered ZFS31; The maximum relative displacement between beam and rail also occurs at this position, which is 6.572mm. Therefore, for the left and right lines, the values of the maximum additional expansion force of single strand rail and the values of the maximum relative displacement between beam and rail are the same. The maximum value of the additional expansion and contraction forces and the relative displacement between beam and rail for each line occur at the same position.

As is shown in table 2, the maximum force of pier in the section is 500.80kN, and the pier numbers are ZFZ27 and ZFS29.

Table 2. Longitudinal force of pier (kN / rail).

Pier number	Force	Pier number	Force
ZFZ9	165.08	ZFS9	164.12
ZFZ10	34.50	ZFS10	36.14
ZFZ11	22.10	ZFS11	21.00
ZFZ12	30.76	ZFS12	26.86
ZFZ13	63.52	ZFS13	78.10
ZFZ14	0	ZFS14	0
ZFZ27	250.40	ZFS29	250.40
ZFZ28	95.10	ZFS30	95.10
ZFZ29	0	ZFS31	0

Therefore, in the design, construction and maintenance of CWR on the bridge, more attention should be paid to the bridge-track structure at these piers, ZFZ29, ZFS31, ZFZ27 and ZFS29.

3.2. Checking and Calculation of Rail Strength

Each rail shall have sufficient strength to ensure that it will not fail under the joint action of dynamic bending stress, temperature stress, expansion additional stress, train starting / braking stress and other additional stresses. At this time, the sum of various stresses borne by the rail shall not exceed the specified allowable value [σ], and the calculation formula is shown in formula (1):

$$\sigma_d+\sigma_t+\sigma_z+\sigma_f\leq[\sigma] \tag{1}$$

Where:

σ_d—— Maximum dynamic bending tensile stress of rail (MPa);

σ_t—— Temperature stress (MPa);

σ_f—— Maximum additional stress on rail (MPa);

σ_z—— Traction (braking) stress on rail (MPa);

[σ] —— the allowable stress of rail (MPa), according to *Code for Design of Railway Continuous Welded Rail* (TB 10015-2012) [9]. The yield strength of U71Mn rail is 457MPa, and the safety factor K=1.3, so [σ]=352MPa.

The checking calculation of rail strength is shown in table 3.

Table 3. Statistics of rail stress (MPa).

indexes	Dynamic bending stress	Temperature stress	Additional expansion stress	Additional braking stress	Total stress	Allowable stress
Tension	119.99	91.02	27.26	11.80	250.07	352
compression	154.63	91.02	27.26	11.80	284.71	352

It can be seen from the table that the sum of tensile stress and compressive stress borne by the rail is 250.07MPa and 284.71MPa, both less than the allowable stress 352MPa, and the rail strength meets the requirements.

4. Conclusion
Taking the CWR on the multi-span simply supported beam bridge as an example, the following conclusions can be drawn after analyzing the longitudinal mechanical properties of the beam-rail interaction model:

(1) For the left line, the maximum additional expansion force on the single strand rail is 211.13kN, which occurs at the pier numbered ZFZ29; The maximum relative displacement between beam and rail also occurs at this position, which is 6.572mm. For the right line, the maximum additional expansion force on the single strand rail is 211.13kN, which occurs at the pier numbered ZFS31; The maximum relative displacement between beam and rail also occurs at this position, which is 6.572mm.

(2) For the left and right lines, the values of the maximum additional expansion force of single strand rail and the values of the maximum relative displacement between beam and rail are the same.

(3) The maximum value of the additional expansion and contraction forces and the relative displacement between beam and rail for each line occur at the same position, and both are located at the pier.

(4) The maximum stress of pier in the section is 500.80kN, and the pier numbers are ZFZ27 and ZFS29.

(5) The total tensile stress borne by the rail is 250.07MPa, and the total compressive stress borne by the rail is 284.71MPa, both less than the allowable stress 352MPa, so the rail strength meets the requirements.

Acknowledgments
The work described here has been supported by the Project of Science and Technology Research and Development Plan of China Railway Nanchang Bureau Group Co., Ltd (Grant 202007). To smoothly complete this paper, I would like to sincerely thank my teacher for his guidance, my senior fellow disciples for answering some questions and solving some doubts for me, and my friends for their help and encouragement when I felt depressed. I really feel very lucky to meet them on my growth path!

References
[1] Lu Y R. 2004 *Research and Application of CWR* [M] Beijing: China Railway Press.
[2] Wei F, Gao L, Zhao L, et al. 2014 Research about mechanical characteristics of position restriction structure on beam end of CRTSI slab ballastless track on long-span bridge[C] *ICRE, Beijing* 191-200.
[3] Wang P. 2001 Behavior of FRP Jacketed Concrete Columns Under Eccentric Loading [J] *Journal of Composts for Construction* **5** (3) 146-152.
[4] Bae Y, Hamad M. 2019 Quantification of Ballasted Track-bridge Interaction Behavior Due to the Temperature Variation through Field Measurements [J] *NDT and E International* **103** (12) 84-97.
[5] Anna M R, Andrzej N. 2018 Live Load Spectra for Railway Bridges in USA [J] *Konferencja Naukowa* **15** (18) 3-7.
[6] Xuan N. 2014 A Finite Element Model of Vehicle-Cable Stayed Bridge Interaction Considering Braking and Acceleration: The 2014 World Congress on Advances in Civil, Environmental, and Materials Research, Busan Korea[C].
[7] Xie K Z, Zhao W G, Cai X P, Wang P, Zhao J, Giovanni L. 2020 Interaction between Track and Long-Span Cable-Stayed Bridge: Recommendations for Calculation [J] *Mathematical Problems in Engineering* 2020.

[8] Dai G L, Ge H, Liu W S, Chen Y F. 2017 Interaction analysis of Continuous Slab Track (CST) onlong-span continuous high-speed rail bridges [J] Structural Engineering and Mechanics **63** (6).

[9] Ministry of Railways of the People's Republic of China. Code for Design of Railway Continuous Welded Rail: TB 10015-2012 [S] Beijing China Railway Publishing House 2013. (in Chinese)

Physical and Mechanical Properties of Cemented Soil Fill At Liquid and Hardened States

Cheng Shi, Jinghu Ding, Junjie Chen and Fangzhou Chu

Economic Research Institute, State Grid Jiangsu Electric Power Co.,LTD. , No. 251, Zhongshan Road, Gulou District, Nanjing, Jiangsu Province, China
Email: 278783442@qq.com

Abstract. Cemented soil fill is a new backfilling technology developed for the problems of narrow foundation trenches and uncompacted backfilling. It has good fluidity before solidification and higher strength and stiffness after solidification. This type of fill materials makes full use of the waste soils. The proportioning test was carried out on excavated soil on a construction site. Liquid property tests and unconfined compressive strength tests was carried out. The results show that the cemented soil fill can meet the requirement of foundation trenches backfilling, which has great prospect for future applications.

1. Introduction

In recent years, the development of urban areas have continued to rise and a large number of projects have been built. Many open-cut pits in the city are generally narrow in space, and it is difficult to achieve high compaction requirements for backfill construction. This kind of backfilling project generally has the problem that the backfilled soil settles and deforms more later, and the buried pipelines and upper roads or structures are damaged due to subsidence. It affects the function of the facilities, and the repeated maintenance later causes high repair costs. Cement mixing soil can only be used for in-situ soil improvement [1, 2]. Cemented soil fill is a new type of filling materials. It consists of excavated waste soils, Portland cement, water, and possibly other waste materials and cementing agents [3, 4]. Cemented soil fill has high strength and stiffness after curing to meet the requirements of bearing capacity and deformation resistance of construction projects[5].

Although there have been some successful field trials of using cemented soil fill, its number of practical applications is still limited. In order to achieve a better understanding of the physical and mechanical properties of the material, experimental studies were carried out on the fluidity and strength of the material with varying soil, cement, and water contents.

2. Test Methods

Since the cemented soil fill is filled without rolling or compacting, it must have certain fluidity before initial setting in order to ensure that the material can fully fill various narrow spaces and irregularly-shaped spaces. If the fluidity is too small, the self-compacting effect of the material cannot be guaranteed, resulting in the holes and voids in the filled volumes. While the fluidity is too large, it will easily lead to the problems such as low strength and bleeding during or after curing[6]. This test uses the slump test to evaluate the fluidity of the cement soil, and unconfined compressive test to obtain the strength of the material.

2.1. Materials
Portland cement and excavated soil on a construction site in Suzhou were used for sample preparation. The soil is dried before the experiment, and the mix proportion is determined by cement/soil ratio and water/soil ratio. The excavated soil used in the test was silty clay, and had natural moisture content of around 28%.

2.2. Methods
The slump cylinder is used to measure the fluidity, and 70.7mm cubic samples is prepared after the fluidity measurement, which is completed within 30min to prevent hydration of cement. The cubic samples were used for the compression tests.

3. Results

3.1. Slump Results
The results show that the fluidity of the cemented soil fill increases with the increase of water consumption and decreases with the increase of cement content, as shown in table 1 and table 2. Based on our field trial, it has been found that the material at the fluid state with a slump of 120mm can be used for filling without laminating.

Table 1. Self-compacted cemented soil proportioning table with 8 % cement admixture.

Sample Number	Water consumption per kilogram of soil (kg)	Measured slump (mm)	Curing Way
A-1	220	72	Standard
A-2	270	123	Standard
A-3	310	142	Standard
A-4	380	195	Standard

Table 2. Self-compacted cemented soil proportioning table with 10 % cement admixture.

Sample Number	Water consumption per kilogram of soil (kg)	Measured slump (mm)	Curing Way
B-1	370	77	Standard
B-2	410	112	Standard
B-3	460	160	Standard
B-4	500	203	Standard
W-1	420	123	Water
W-2	450	148	Water

3.2. Strength Results
The samples were placed in standard curing conditions and demoulded after 3d. Unconfined compressive strength tests were carried out on 7, 14, 28 and 90 days. The results are shown in the following figure 1 and figure 2.

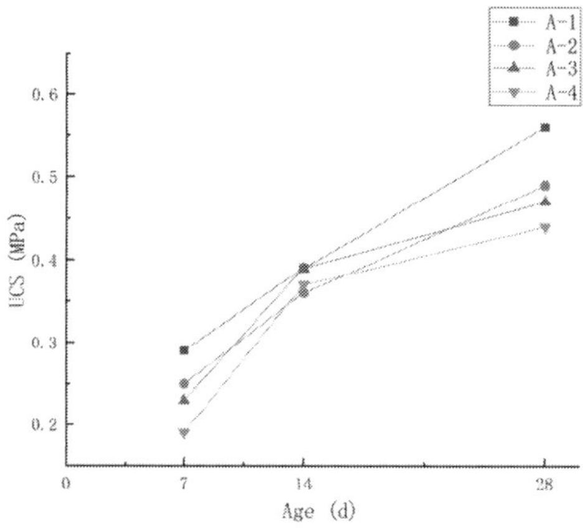

Figure 1. UCS results with 8% content cement.

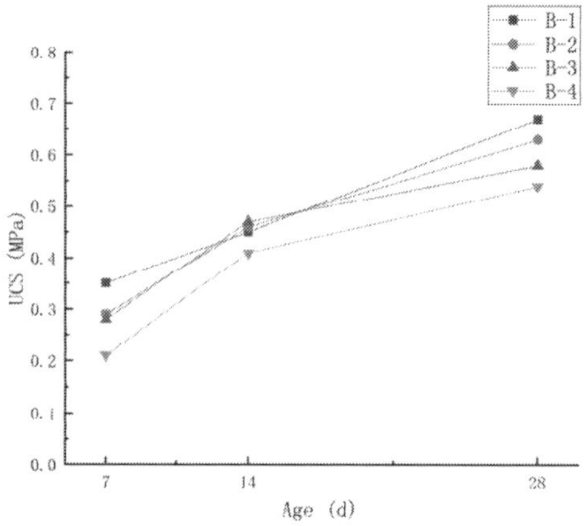

Figure 2. UCS results with 10% content cement.

In order to verify the effect of the cemented soil fill curing under groundwater level on strength, two groups of samples with 10% content cement are made and the results are shown in figure 3. These samples were cured at the submerged condition under around 25°C.

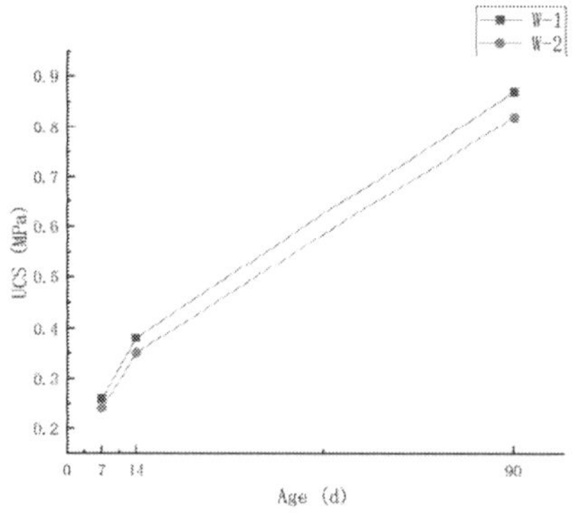

Figure 3. UCS results curing under groundwater.

The laboratory test results show that the unconfined compressive strength of the test block with silty clay as raw material soil can reach about 0.20 ~ 0.30 MPa at 7 days, 0.37 ~ 0.46 MPa at 14 days and 0.44 ~ 0.56 MPa at 28 days; With 10% cement content, the unconfined compressive strength can reach about 0.21 ~ 0.35 MPa at 7 days, 0.41 ~ 0.54 MPa at 14 days and 0.54 ~ 0.67 MPa at 28 days. The underwater curing method has no obvious effect on the strength.

3.3. Results of Site-Collected Samples

Two groups of test samples were collected for each site filling of about 200 m^3 during construction and the slump value is estimated to be 10mm. The experimental results of unconfined compressive strength at different ages are shown in the figure 4 and figure 5.

Table 3. Self-compacted cemented soil proportioning table with 10 % cement admixture.

Sample Number	Cement content	Measured slump (mm)	Curing Way
S-1	8%	90	Standard
S-2	8%	120	Standard
S-3	10%	80	Standard
S-4	10%	110	Standard

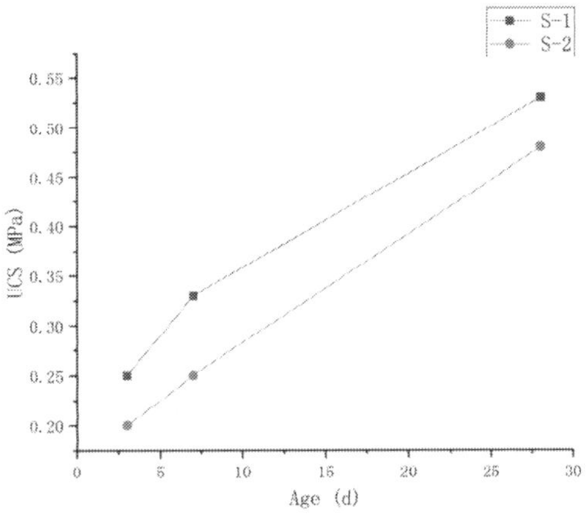

Figure 4. UCS results with 8% content cement on site.

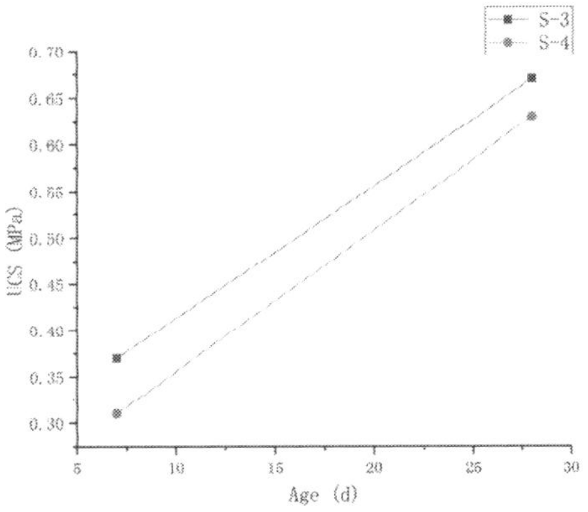

Figure 5. UCS results with 10% content cement on site.

The results showed that the unconfined compressive strength of site samples could reach about 0.35 MPa at 7 d and 0.5 ~ 0.67 MPa at 28 d under the field conditions with 8 ~ 10% cement admixture. Its higher strength indicates a proper on-site mixing condition, where soil and cement can be mixed homogeneously.

4. Conclusions

Cemented soil fill can be used to fill narrow trenches and irregularly shaped voids by utilizing local excavated soils and other waste materials. According to the laboratory test results, the UCS should reach 0.30 MPa in 7 days and not less than 0.50 MPa in 28 days. Furthermore, the results on the field collected samples show relatively high strength, indicating a the good mixing condition at the field construction.

Acknowledgments

Financial support for this study was primarily provided by Research on the Application of Cemented Soil Fill in the Construction of Power Transmission and Transformation Projects of STATE GRID JIANGSU ELECTRIC POWER CO., LTD. ECONOMIC RESEARCH INSTITUTE.

References

[1] Zhou J ,Yu J, Gong X, El Nagger M H and Zhang R. 2020 *Acta Geotech.* **15** 3271.
[2] Ni P, Yi Y and Liu S. 2021 *Soils and Founda* **61** 416.
[3] Ling T C, Kaliyavaradhan S K and Poon C S. 2018 *Construct. and Building Materials* **158** 535.
[4] Santos Nascimento E S, Souza P C, Oliveira H A, Melo Júnior C M, Oliveira Almeida V G and Melo F M C. 2021 *J. Cleaner Product.* **321** 129002.
[5] Liu X, Zhang X, Kong L, Li X and Wang G. 2021 *Soils and Founda.* **61** 520.
[6] Pham T A, Koseki J and Dias D. 2021 *Transport. Geotech.* **28** 100544.

AUTHOR INDEX